Fish biology in Japan: an anthology in honour of Hiroya Kawanabe

Developments in environmental biology of fishes 18

Series Editor
EUGENE K. BALON

Fish biology in Japan: an anthology in honour of Hiroya Kawanabe

Guest Editors:

Masahide Yuma, Izumi Nakamura & Kurt D. Fausch

Reprinted from *Environmental biology of fishes*, Volume 52 (1–3), 1998
with addition of species and subject index

SPRINGER-SCIENCE+BUSINESS MEDIA, B.V.

A C.I.P. Catalogue record for this book is available from the Library of Congress

ISBN 978-90-481-5048-9 ISBN 978-94-015-9016-7 (eBook)
DOI 10.1007/978-94-015-9016-7

Cover design by Yasuko Segawa
inspired by the tail end of ayu, *Plecoglossus altivelis*

The logos
are parts of Kawanabe's family kamon taken from the embroideries on his kimono

The caligraphy
on p. 10 is the honoured person's signature

Printed on acid-free paper

Originally published by Kluwer Academic Publishers in 1998
Softcover reprint of the hardcover 1st edition 1998

CONTENTS

Environmental Biology of Fishes **52**: 7–8, 1998.

Prelude to the anthology in honour of Hiroya Kawanabe

Few English-speaking biologists are likely to know the history and scope of work on fish biology in Japan. This is not surprising given the cultural barriers that have hampered communication, especially via written language. This volume is a collection of papers assembled to honour Hiroya Kawanabe, an eminent Japanese ecologist who studied fishes. Through it we hope to further Kawanabe's goal of fostering communication among ecologists and fish biologists throughout the world.

Kawanabe retired from his position of Professor at Kyoto University in March 1996. During his 40-year career his interests grew from the unique ecology of algae-scraping ayu in rivers of central Japan, to the complexity of biotic factors that structure fish communities in Lake Tanganyika, to conservation of global biodiversity. In Japan, he is perhaps best known for the hundreds of essays he wrote for newspapers during his long career, which helped develop popular support for preserving the natural environment. For his lectures and essays, Hiroya often draws from his enormous knowledge of various other fields, including literature, Japanese history, Oriental philosophy, fine art, calligraphy, drama, music, folk-lore, and anthropology, to support or illustrate a point. This has resulted in ecological works with unique titles that beg to be read, such as 'Ecology with a bias' (Kawanabe[1]) and 'Complexity in biological communities, or the importance of being equivocal' (Kawanabe[2]). When hearing or reading some of his lectures and essays, however, one can sometimes become bewildered by the labyrinth of arcane metaphors that he weaves through his material. His point can be lost if one does not happen to be acquainted with the sources he cites. Much later, though, the meaning of his citations may suddenly become evident, giving one a sense of great delight. It is a fitting tribute to his unusual personality that while at the university he was nominated as one of the three most curious professors for often wearing Japanese traditional dress (kimono), as shown in Figures 13 and 16 of his biography (p. 23, 25) in this volume, and for reading a book while walking down the street.

Despite a career that has so far spanned from 1952 to the present, during which he wrote more than 780 essays, papers, and books, Hiroya Kawanabe appears young both externally and internally. He seems to derive his youthfulness from a wide variety of interest in many fields, and frequent trips to local study sites, such as the rivers U-kawa, Okinawa, and Lake Naka-umi, and overseas study areas, such as lakes Tanganyika, Malawi, and Baikal, as well as to Asia, Australia, Europe, and North America. In fact, one of his hobbies is to use time tables to find the best connections for trains and international flights, and thereby find extra hours to enjoy operas and orchestras, hot springs, good meals, and fine wines.

The idea for this volume began in 1994, when Eugene Balon stayed during his sabbatical in Japan as an invited professor of the Japan Society for the Promotion of Science (JSPS). In that same year, Balon and his colleagues pointed out in the prelude to another volume (Balon et al.[3]) that '...we know there is a history of accomplishments and contributions by colleagues in Asia, eastern Europe, South America, Africa and elsewhere'. They exhorted others to '...accept the challenge and bring forth the more comprehensive historical and conceptual treatment of our subject'. We hope this volume provides some of this foundation for Japan.

The groundwork of international communication among Japanese and North American scientists that resulted in this volume was actually laid by Hiroya Kawanabe himself, but began much earlier than 1994. One of Kawanabe's most brilliant and lasting contributions was to organize and motivate Japanese ecologists to

[1] Kawanabe, H. 1987. Henken no seitaigaku (Ecology with a bias). Nosangyson Bunka-kyokai, Tokyo, 250 pp. (in Japanese).

[2] Kawanabe, H. 1990. Complexity in biological communities, or the importance of being equivocal. pp. 186–194. *In:* H. Shimizu (ed.) Biological Complexity and Information, World Scientific, Singapore.

[3] Balon, E.K., M.N. Bruton & D.L.G. Noakes. 1994. Prelude to the anthology in honour of women ichthyologists. Env. Biol. Fish. 41: 7–8.

sponsor several international conferences on ecology and fish biology, starting with the International Symposium on Charrs and Masu Salmon in Sapporo in 1988 (Kawanabe et al.[4], Noakes[5]). This conference marked the first visit to Japan for Eugene Balon, David Noakes, and Kurt Fausch, all of whom developed lasting friendships and professional ties with Hiroya Kawanabe and many of the authors in this volume. For example, after conducting collaborative research with Kawanabe and his colleagues during 1990–1992, Fausch also visited Japan during June and July 1994 as a JSPS Visiting Research Scholar. After learning of Balon's plans for this special volume from a Japanese newsletter, Fausch decided to honor his hosts Kawanabe and by collaborating with Shigeru Nakano on a review of research on fish ecology in Japan, which appears herein.

In the beginning, Masahide Yuma and Izumi Nakamura of Kyoto University, and Tamotsu Iwai of Kinki University began planning the volume, but unfortunately Tamotsu fell ill thereafter for a few months. Kurt Fausch was invited to help complete the work, and conducted the final editing of all manuscripts and prepared the index. We also thank many other colleagues who helped with the project, especially Yasuko Segawa who drew the caudal half of the body of the ayu for the cover page of the hardcover volume. Our sincerest thanks are due to Eugene Balon and David Noakes who proposed the volume and continued tenaciously to encourage us throughout its long gestation.

Otsu and Fort Collins, 1 March 1998

Masahide Yuma
Izumi Nakamura
Kurt D. Fausch

[4] Kawanabe, H., F. Yamazaki & D.L.G. Noakes (ed.). 1989. Biology of charrs and masu salmon. Physiology and Ecology Japan, Special Volume 1, Kyoto University, Kyoto, 711 pp.
[5] Noakes, D.L.G. (with photographs by E.K. Balon). 1989. Symposium to be remembered. Env. Biol. Fish. 24: 313–317.

Part 1: Biography, bibliography, and interview

川那部浩哉

Environmental Biology of Fishes **52**: 11–35, 1998.

The life and work of Hiroya Kawanabe: the priest ecologist

Masahide Yuma[1] & Eiji Harada[2]
[1]*Center for Ecological Research, Kyoto University, Otsu 520-0105, Japan*
(e-mail: yuhma@ecology.kyoto-u.ac.jp)
[2]*Sakae, Shirahama, Wakayama 649-2332, Japan*

Received 31.7.1997 Accepted 27.11.1997

Key words: biography, Japan, scholar, example, ichthyology

Synopsis

Hiroya Kawanabe was born the son of a Buddhist priest and teacher of Japanese literature, who died when Kawanabe was very young. Kawanabe also studied Buddhism by himself, and passed the examination to be a priest of his sect while still in high school. He studied zoology and ecology at Kyoto University and earned his doctorate under the guidance of Denzaburo Miyadi, a well-known Japanese ecologist, in 1960. During his academic career at Kyoto University, Kawanabe advanced to hold the chair of Animal Ecology as Professor in the Department of Zoology. Kawanabe's doctoral research concerned the social behavior and population ecology of the ayu, *Plecoglossus altivelis*, an amphidromous fish that lives in streams as adults and grazes algae. His research lead to the discovery that social structure changed from territoriality to schooling as population density increased, and also varied with changes in food and habitat. During this work, he pioneered the use of underwater observation to study ecology of freshwater fishes in streams. Kawanabe also observed ayu social structure from the northern to southern limits of their range, and advanced the theory that the more stable territoriality in the Lake Biwa population was a relic social structure to guarantee food supply during earlier glacial periods when productivity was lower. Additional work on stream fishes in central Japan and Okinawa Island led Kawanabe to propose that interactions among individuals affect interspecific relationships, and thereby, community structure. Discussions with Charles Elton, the famous British ecologist of Oxford University, strengthened Kawanabe's view that communities could be best understood as the 'whole of interrelationships' among organisms. Kawanabe advanced these ideas during a joint study he led with a host of Japanese and Zairean scientists on the fishes of Lake Tanganyika, beginning in 1979. This work, as well as additional research on Lake Biwa in Japan, led to a deeper understanding of the complexity of biotic interactions (including competition, predation, mutualism, commensalism, and indirect effects) that promote the high species diversity in these ecosystems. In addition to basic research, Kawanabe was part of research teams organized during the 1960s by D. Miyadi to study the effects of public works projects on natural environments and biota in Lake Naka-umi and Lake Biwa. During the late 1980s he expanded his network to an international venue, both by organizing and hosting important international ecological meetings in Japan, such as the Fifth International Ecological Congress, and by increasing his international activities to promote global biodiversity. In 1991, Kawanabe founded the Center for Ecological Research at Kyoto University to study the interrelationships among organisms and their environments. Recently retired from the University, he became Director General of the new Lake Biwa Museum in 1996, and continues to promote conservation of biodiversity worldwide through an international network of scientists and organizations.

Introduction

Hiroya Kawanabe (HK) was born in Kyoto on 10 May 1932. His father, Osamu Kawanabe, was a teacher of Japanese literature and also a Buddhist priest, who died in 1934 at an early age. His mother, Tsuya Fujii-Kawanabe, was a teacher of the tea ceremony and of flower arrangement, and had given birth to him at age 38. He attended elementary school from 1938 to 1945, junior high school from 1945 to 1948, and Kyoto City's Ohki High School, from 1948 to 1951. In March 1945, military requisition of the Kawanabe property, which was situated in downtown Kyoto, and the subsequent demolition of the family home necessitated the tearful and hasty removal of only a fraction of the family's belongings in a cart (HK 490; numbers refer to the subsequent 'Lifetime list of publications by Hiroya Kawanabe'). He and his mother were able to save only a portion of his father's books; his passion for reading may be due to the influence of his father. Interestingly enough, during his junior high school and high school years, Kawanabe tried to improve the method used to catalogue the books in his school libraries, though he subsequently learned that the NDC (Nihon Decimal Classification) was the most convenient cataloguing method in common use.

As Kawanabe was born into a priestly family, he studied the Buddhism of the Jôdo Sinshû sect by himself, and passed the examination for the priest's license at Kyoto's Higashi-Hongan-ji temple while he was still in high school. From 1951 he studied at the Faculty of Science, Kyoto University, receiving respectively B. Sc. (1955), M. Sc. (1957), and D. Sc. (1960) degrees in Zoology. In 1955 he began scientific works on the social structure and bio-economy of the ayu-fish, *Plecoglossus altivelis* (Plecoglossidae), and the interspecific relationships (especially the interactive food segregation) among river fishes.

He was employed in the Department of Zoology, Faculty of Science, Kyoto University, as instructor from 1960 to 1961, as lecturer from 1961 to 1967, as associate professor from 1967 to 1977, and as professor from 1977 to 1993 in the chair of Animal Ecology at the Department of Zoology. He devoted his last years at the university to being Director of the Center for Ecological Research, Kyoto University, from 1991 to 1996. After his retirement, he was employed as the Director General of the Lake Biwa Museum, Shiga Prefecture, from 1996 until the present. He is undoubtedly one of world's most eminent ecologists, and his far-ranging contributions encompass various aspects of ecology, e.g., population ecology of the ayu, interspecific relationships among freshwater fishes in Japanese rivers and Lake Tanganyika, freshwater conservation biology, biodiversity and biological complexity in ecosystems, and also literature and music. He received the Honourary Degree of Doctor of Science, University of Guelph, Canada in 1995 (Figure 1), the Professor Emeritus of Kyoto University and the Japan Academy Duke of Edinburgh Prize in 1996, and the Foreign Honorary Member of the American Academy of Arts and Science in 1997.

Figure 1. Hiroya Kawanabe during his convocation address on the occassion of receiving Doctor of Science honoris causa at the University of Guelph, 5 October 1995. Photograph by M. Schwalbe.

Figure 2. Ayu, *Plecoglossus altivelis altivelis*, repelling an invader. Photograph by T. Taguchi.

Figure 3. Unique ayu-fishing — tomo-duri — along the River Karino, Shizuoka Prefecture. Photograph by T. Taguchi.

A field biologist floating down the river

Hiroya Kawanabe devoted his early work to the social and population ecology of the ayu (Figure 2). The ayu is one of the most abundant fishes in the rivers of the Japanese and Ryukyu archipelagos, and is also found on the Korean Peninsula, mainland China, and Taiwan Island. It is an annual fish as well as being amphidromous. Eggs are laid in late autumn, and fish grow, spawn, and die the following

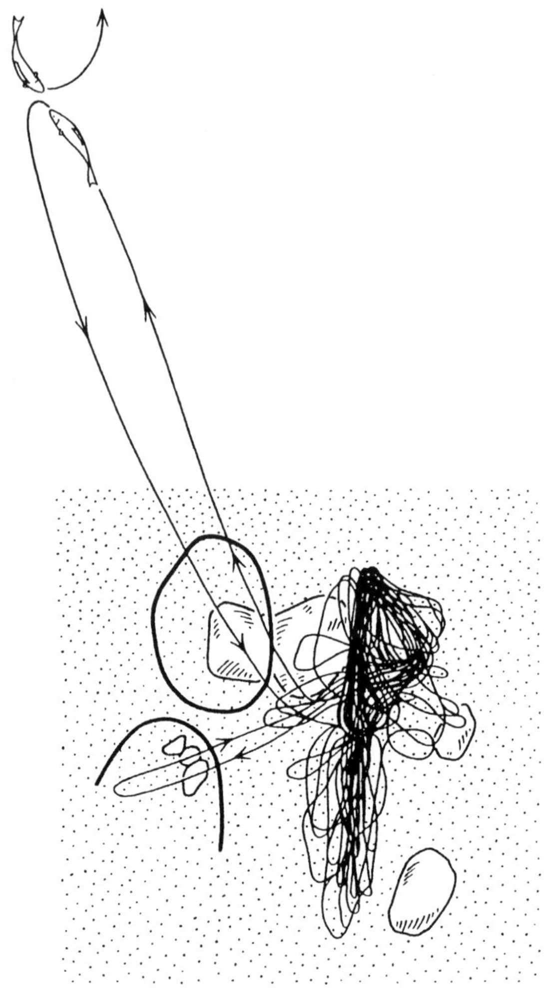

Figure 4. Swimming track of an individual ayu in three minutes, during which the fish chased invaders twice, observed along the River Kurama, Kyoto City, in 1951 (after Miyadi 1994 and HK 88).

autumn. Spawning occurs in the lower reaches of rivers, and hatched young descend to the ocean, live along the coast during winter, ascend rivers in the spring, and grow in the middle reaches of rivers by grazing on diatoms and blue-green algae growing on rocks and stones. The ayu supports the most important freshwater fisheries in Japanese rivers, and is also popular among anglers who capture them using decoy fish with hooks which are attacked by the territorial ayu (Figure 3; also see Yuma et al. 1998 this volume).

In 1951, prior to Kawanabe's research, a project began in the rivers of Kyoto Prefecture to determine a standard stocking density for young ayu in natural rivers, at the request of the Kyoto Prefectural Government Fisheries Department. The project was organized by Denzaburo Miyadi (1901–1988) at the Department of Zoology, Faculty of Science, Kyoto University, who was a professor of Kawanabe's and who was working on various aspects of freshwater ecology and animal sociology (see Harada 1998 this volume). D. Miyadi also introduced SCUBA equipment for ecological underwater observation.

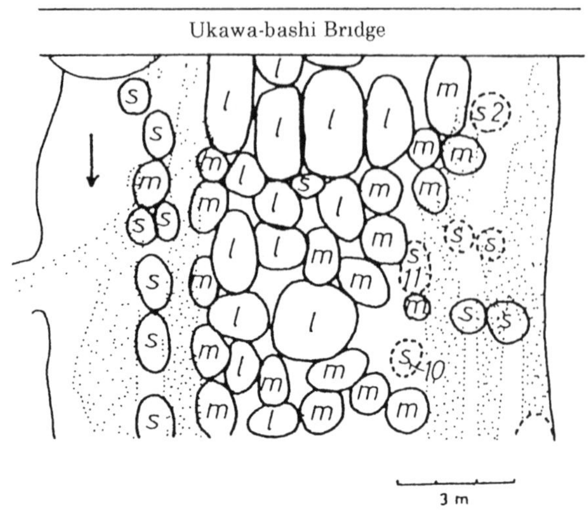

Figure 5. Distribution pattern of the territories of ayu observed from a bridge on the U-kawa in June 1956. Solid and broken lines show territory boundaries and areas used by non-territorial fish, respectively. Symbols *l*, *m* and *s* indicate fishes of large, intermediate and small sizes (after Miyadi et al. 1953 and HK 88).

Because the river water was so transparent, it was easy to observe and trace the behavior of individual ayu from bridge and riverbank (Figures 4, 5). Denzaburo Miyadi, Masakazu Kawabata, Shoichi Yagi, Hiroki Mizuhara and others observed the territorial behavior of ayu in riffles and river-pools, and estimated their average territory size to be ca. 1 m^2 (Miyadi et al. 1952). This territory area was used as the basis for estimating a standard density of territorial ayu on a riffle, by which fishery cooperative societies determine the number of ayu to stock in rivers even today.

Miyadi and his colleagues clarified the social structure of the ayu, and found that territorial ayu occupied riffles and non-territorial fish lived in peripheral parts of riffles or river pools (Figure 5).

Figure 6. Research team at the U-kawa on 16 July 1956. From left to right: J. Itani, N. Nishimura, Z. Nozaki, K. Onodera, D. Miyadi, S. Yagi, N. Mizuno, H. Ohta, Y. Furuya, and H.Kawanabe.

They also conducted field experiments to examine the effect of density on social structure of the ayu by manipulating fish density in the river (Miyadi et al. 1953). They found that, when more ayu were released, non-territorial fish gathered in pools. Moreover, when territorial fish in riffles were excluded, a proportional number of ayu immigrated from pools to establish territories. These results suggested that pools were subordinate habitats, supporting individuals that supplemented the social structure of

Figure 7. Hiroya Kawanabe taking underwater photographs in the River Inukai, Kyoto Prefecture, on 30 July 1958. Photograph by E. Harada.

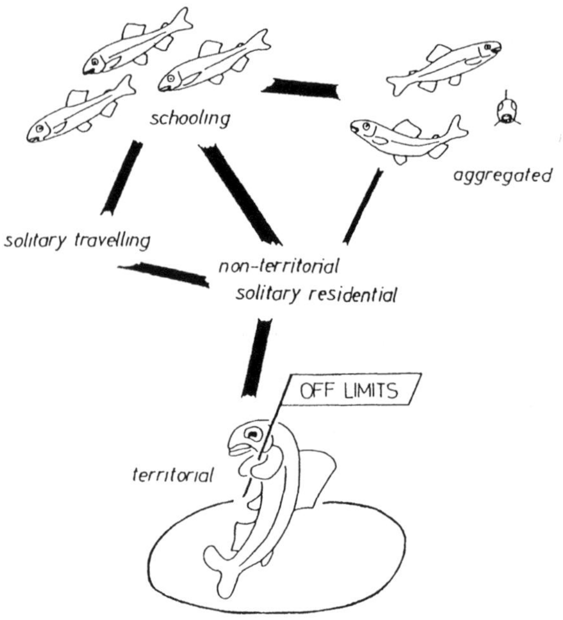

Figure 8. Types of social behavior of ayu and their shifts between these types. Width of lines between the types shown in the figure indicates degree of shifting (after HK 7 and 88).

the ayu. However, non-territorial ayu also formed schools in pools and often grew large independent from territorial ayu in riffles; meanwhile, ayu in riffles used pools as a resting site at night. Thus pools seemed to be not subordinate habitats but important ones for ayu (Mizuhara et al. 1955). Based on these preliminary studies, Miyadi requested Hiroya Kawanabe and Eiji Harada to continue the study of ayu, especially the importance of pools. However, instead of the requested, Kawanabe proposed a much broader research principle to guide the study, 'the whole river system, the whole life cycle of the ayu, and the whole biological community' (Miyadi 1994, HK 89).

Kawanabe and Harada visited many systems to find a typical medium-sized river where juvenile ayu naturally ascended from the sea, and finally selected the U-kawa, a small river (ca. 22 km in length) on the Sea of Japan coast of Kyoto Prefecture, as a main study site (HK 626). Kawanabe began his study of ayu in 1955 (Figures 6, 7). He identified individual ayu by marking each with a piece of nylon thread sewn through the dorsal muscle, and observed their feeding behavior such as feeding time, feeding frequency, and feeding site. He found that territorial ayu grazed algae growing on rocks and stones from 5:00 to 19:00 h at a relatively constant frequency of 20 times per minute, whereas non-territorial ayu grazed intermittently at a higher frequency of more than 30 times per minute, but that total food intake differed little between the two types (HK 8). The population density of ayu in the river was fairly high that year and Kawanabe also found that territorial ayu gave up trying to repel intruders when large schools of non-territorial ayu entered their territory, and that they often abandoned the territory and joined the school (HK 8, 10). From these results the first year, Kawanabe concluded that territorial behavior of ayu was unstable, with frequent exchange of individuals between territorial and schooling groups, and presumed a similar growth between territorial and non-territorial ayu (Figure 8). He therefore rejected the previous theory that the territory was the principal social structure of ayu and non-territorial fish were subordinates in the population (HK 6, 11).

In 1956, the second year of Kawanabe's study, he again visited the U-kawa. On the way from bus stop to the lodge, he observed the river from the bridge, and was then astonished to find that most ayu lived in territories, and that the riffle was filled by their territories (Figure 5, HK 10). At that moment, he could not understand the behavior of the ayu compared to the previous year, and puzzled, stood on the bridge for a long time. In fact, the population density of 5.4 indiv. m^{-2} observed in 1955 was the highest ever recorded to the present, and the second and third highest densities in the U-kawa were 3.2 indiv. m^{-2} in 1968 and 0.9 indiv. m^{-2} in 1956, respectively (HK 97, 541). Then, Kawanabe considered that the social structure of ayu was influenced by their population density: When the population density is low, territoriality is a stable life style and territorial individuals grow better than non-territorial ones. On the contrary, when the population is very dense, schooling is a better style than territoriality because of the high cost of repelling intruders. Moreover, the growth of ayu is less variable when the schooling individuals are dominant, and the production of algae is large. These suppositions were confirmed by subsequent field observations and experiments through 1968 (HK 9, 16, 26, 88, 89).

Thus, Kawanabe examined the social structure of ayu and established the importance of variable social structure, including both schooling and territoriality, the latter being considered the single most important social structure at that time. He also showed the interacting influences of habitat conditions, including the amount of food, and population densities on social structure. To clarify the territorial and feeding behavior of ayu, Kawanabe observed them underwater, a method adopted from D. Miyadi, M. Kawabata, and H. Mizuhara. He pioneered the use of underwater observations in the ecological studies of freshwater fishes (cf. Fausch & Nakano 1998 this volume). He often floated down the U-kawa from its upper reaches to its mouth, counting the number of ayu while wearing a mask and snorkel. This direct underwater observation led Kawanabe into the world of interrelationships among fishes in rivers.

Dreaming about the behavior of ayu during past glacial periods

Ayu are generally territorial when the overall population density is lower than 3 indiv. m^{-2} (HK 20, 88). If food-linked territoriality is self-regulating, aggressive behavior and spatial requirements should be quite rigid at higher population densities. However, in the case of ayu it was the reverse, and Kawanabe concluded that territoriality in ayu was not currently self-regulating (HK 97, 145).

Kawanabe also compared the consumption rate of ayu and the average production rate of algae in summer and concluded that the algae were usually sufficient to feed 9–12 ayu per m^2 (HK 20, 88). Nevertheless, why do individual ayu establish territories as large as ca. 1 m^2? To examine this, Kawanabe tried to observe the territorial behavior of ayu near the southern and northern limits of its distribution.

From 1968 to 1971, Kawanabe visited Okinawa Island three times to observe the territorial behavior of ayu at the southern end of its distribution (unfortunately, the ayu in Taiwan Island had been extinct since the early 1960s). There, he found that the territoriality of ayu was unstable, and so proposed that territoriality of ayu might have played an effective part in population self-regulation during previous glacial ages but might be a 'relic' type of social structure in the inter- and post-glacial ages (HK 137).

The ayu was known to be distributed from the Korean Peninsula to the Liaoning, Shantung, Chekiang, and Fukien districts in China. In 1972, Kawanabe visited the Korean Peninsula to observe the territorial behavior of ayu near the northern limit of its distribution. He also surveyed the behavior of ayu in Hokkaido, and confirmed that ayu in these northern areas had a similar social style as the populations in Kyoto (HK 317).

In 1979, Kawanabe visited the Institute of Hydrobiology, Chinese Academy of Sciences, at Wuhan, China, to examine specimens of ayu collected in the River Peilong near the border of Vietnam in 1978; these specimens expanded the previous record of distribution to the south more than 1000 km (HK 351). Although Kawanabe again visited China in 1980 and 1983, he was unable to observe the ayu in Chinese rivers.

To compare the behavior of related species, Kawanabe tried to observe *Prototroctes* spp. in Australia, Tasmania, and New Zealand from December 1978 to January 1979, but unfortunately the fish in Australia was an omnivorous species and that in New Zealand probably became extinct in the 1930s (HK 235, 262).

He thereafter expanded his hypothesis to include the land-locked population of ayu in Lake Biwa which has territorial behavior much stronger and more stable than that of the marine type. Kawanabe considered that the case of ayu in Lake Biwa was applicable to his hypothesis, because fish that evolved in the lake could not escape glaciation by dispersing southward into warmer waters, but instead were forced to develop stronger territoriality to gain sufficient food resources. In contrast, the marine type could migrate into southern areas with warmer waters (HK 207). He also proposed that the unique photoperiodism and morphology of eggs in ayu of Lake Biwa were relic characters from previous glacial ages.

Later, M. Nishida (1985, 1986) reported a large difference in allozyme patterns between three populations, located in the Ryukyus, from Yakushima

Island to Hokkaido including the southern Korean Peninsula, and in Lake Biwa.

Toward interactive relationships among species

Besides the study of ayu, Kawanabe and his colleagues observed the behavior of other fishes in the rivers to analyze species interactions. Previously, there had been little research on species interactions in natural freshwater fishes except the study of habitat segregation between Japanese charr, *Salvelinus leucomaenis* subsp., and masu salmon, *Oncorhynchus masou* subsp., by K. Imanishi (1951), and none on feeding segregation.

Kawanabe's research was based on the belief that interactions among individuals should affect interspecific relationships and in turn, community structure in a river system, similar to the case for ayu. He found in the late 1950's that the shift and segregation in food habits was caused by competitive social relationships among riverine fishes (HK 21). Free-swimming cyprinids depended chiefly on terrestrial insects when territorial ayu were abundant, whereas they tended to feed mainly on bottom algae when ayu were in more dense populations and less aggressive. This indicated that the cyprinids changed their diet based on the aggressiveness of ayu grazing on algae. Such changes in diet were, in part, related to habitat shifts by these fishes. For example, the pale chub, *Zacco platypus*, inhabited the central parts of riffles and fed mainly on bottom algae when ayu were absent, but when ayu ascended the river and established territories in riffles the pale chub was driven away to peripheral areas. Subsequently, the pale chub drove out dark chub, *Z. temmincki*, which shifted to the central parts of riffles where ayu were grazing on bottom algae but where the dark chub fed upon drifting invertebrates. The two cyprinids returned to their former habitats when ayu descended the river (Miyadi et al. 1952, HK 21). Kawanabe also found that the presence or absence of bottom-dwelling gobionids influenced the diet of gobies, loaches and crucian carps (HK 21). Thus, Kawanabe found that the diets of coexisting species could be explained based on feeding segregation that resulted from interactive relations among individual fishes (Figure 9).

Prior to these works, W.E. Frost (1946) reported for eels and T. Miura (1959) suggested for cyprinids that the foods eaten might be correlated with the abundance of possible prey. T. Miura also suggested that the realized diet might possibly change in accordance with the fishes' situation in a community. During the same years that Kawanabe was conducting his research competitive feeding segregation in fishes was being studied independently in experimental ponds in Russia by V.S. Ivlev (1955) and in Swedish lakes by N.-A. Nilsson (1955, 1957).

Another opportunity to find the interactive shift in food habits among freshwater fishes came when Kawanabe visited the Benoki-gawa, a river in Okinawa Island. In summer 1971, he wrote that: 'When I had a lunch on the river bank, I dropped a piece of boiled rice onto a rock. I took it and threw it into the river. As soon as the boiled rice reached the water surface, a fish came up from the river bottom and ate it. It was a common phenomenon. I continued eating my lunch, and took a piece of roast pork. Then suddenly, I became surprised. Was the fish that ate a piece of boiled rice at the water surface *Rhinogobius*? There could be no doubt. Only two gobies, *Sicyopterus japonicus* and *Rhinogobius* sp., lived in this area. In the evening, I also observed that *Rhinogobius* jumped up from the water and took adult chironomids flying above the river surface' (HK 160, 382).

In other locations with more fish species, *Rhinogobius* usually preyed on attached algae and aquatic insects on the river bottom, whereas the other species foraged in the mid-water zone. However, in the upper reaches of the river above a large waterfall where there were no mid-water fishes, *Rhinogobius* ate a substantial amount of terrestrial insects flying or floating on the water surface in addition to attached algae and benthic insects (HK 160, 230, 382, 417).

Water resource development plans and research groups at Lake Naka-umi and Lake Biwa

In 1958, an official of the Shimane Prefectural Gov-

Figure 9. Hiroya Kawanabe presenting his paper on species interactions among river fishes in the symposium on river ecology at the conference of The Ecological Society of Japan held in Gifu City (9 April 1958). Photograph by T. Miura.

ernment requested D. Miyadi to investigate the biological communities in Lake Naka-umi and its adjacent areas. Lake Naka-umi is a shallow brackish water lake between Miho Bay on the Sea of Japan and the slightly salty freshwater lake, Shinji-ko. At that time, the Japan Ministry of Agriculture, Forestry, and Fisheries planned to construct a dam at the outlet of Lake Naka-umi to stop the inflow of marine water and make the lake into a freshwater reservoir, and also planned to reclaim part of the lake for agricultural purposes. The request was to assess the effects of these plans on fisheries in the lake.

Miyadi organized a scientific study team for this project, with the collaboration of 19 scientists from 5 universities from 1958 to 1962 (Figure 10). They greatly advanced the knowledge of the biology of the lake, and D. Miyadi, S. Mori, E. Harada, and H. Kawanabe were in charge of studying the fishes and the biological community of the lake.

Many fishes lived in, or migrated between, two or three of Shinji-ko, Lake Naka-umi and Miho Bay. These fishes fed on a few kinds of foods in the fresh water of Shinji-ko and in the saline water of Miho Bay, but on various foods in the brackish water of Lake Naka-umi. Although the abundance of food organisms in brackish water was similar to those in fresh- and saline waters on average, it fluctuated markedly from season to season. Fishes in brackish water fed mainly on foods that were seasonally abundant, and thus only generalist with a broad diet had an advantage there. During the same years, R.M. Darnell (1958) found that most fishes in the estuary of the Mississippi River had a broad diet. Moreover, most micro- and benthic organisms in Lake Naka-umi were eaten by fishes, even though C.G.J. Petersen (1914, 1915, 1918) pointed out, in the study of a *Zostera* bed, the existence of 'useless' organisms that were not eaten by any other organisms. Thus the efficiency of biological production through food chains was much higher in brackish water, which resulted in a high production of fishes in Lake Naka-umi (HK 31–44, 57, 75, 89).

Changing such a productive body of brackish water into freshwater, especially one as wide as Lake Naka-umi, might cause the collapse of the biological community and a great decline in fish production. Thereafter, the government constructed a dam to freshen Lake Naka-umi, but the dam has not yet been operated.

A similar ecological study program was carried out in Lake Biwa during the early 1960s. The Japan Ministry of Construction developed a plan to con-

Figure 10. Colleagues of the Lake Naka-umi research project, gathering at a meeting held in Matsue City on 22 July 1961. First row from left to right, M. Azuma, S. Fuse, T. Sunaga, T. Iwai, T. Habe, S. Mori, and Y. Tezuka (later Saito); back row from left to right, E. Harada, I. Maki, H.Kawanabe, D. Miyadi, Y. Saijo, T. Kikuchi, K. Negoro, H. Asano, S. Takamatsu, I. Yamazi, and M. Sakamoto.

struct a dam between the northern and southern basins of the lake. The purpose of the dam was to maintain a stable supply of lake water for more than 10 million people in the downstream area by regulating the lake water level, or in drought, by pumping lake water from the northern basin into southern basin. To assess the effects of the plan on fisheries, an intensive ecological survey was conducted from 1961–1965 by the Research Group for Biotic Resources of Lake Biwa (BST: Biwako Seibutsusigen Tyousadan, in Japanese, or Lake Biwa Survey Team) a group of 60 scientists under the direction of D. Miyadi. The group covered taxonomy, ecology, and physiology of phytoplankton, macrophytes, zooplankton, benthic invertebrates, molluscs, and fishes, and also meteorology, chemical and physical sciences, fisheries, and aquaculture. Kawanabe belonged to the subgroups investigating the biology of fishes and molluscs, and he and Harada held the post of secretary of the BST.

Many young researchers and graduate students engaged in studying fish biology in Lake Biwa matured into professional fish biologists as a result of their participation in BST: Yoshiro Tomoda on ontogenic development of crucian carps, *Carassius* spp.; Taizo Miura on population dynamics of the ayu, in relation to its competitor, the Biwa goby, *Chaenogobius isaza*; Hiroya Kawanabe on defecation rate of the round crucian carp, *Carassius auratus grandoculis*, and population ecology and production of a freshwater mussel, *Corbicula sandai* (Corbiculidae); Susumu Tanaka on food habits and growth of the piscivorous chub, *Opsariichthys uncirostris uncirostris*; Tetsuo Sunaga on seasonal changes in food habits of juvenile *O. u. uncirostris* and 16 other fish species; Iwao Maki on the relationship between fishes and their foods in a bay of the lake and population dynamics of the Biwa gudgeon, *Gnathopogon caerulescens*; Makoto Nagoshi on population dynamics of the deepbodied crucian carp, *Carassius cuvier*, and the Biwa goby; *C. isaza*; Ken-ichi Hirai on the habitat use by juvenile cyprinids and the reproductive ecology of bitterlings, *Acheilognathus* spp.; Mikio Azuma on variation in the life history of ayu; Norio Suzuki on metabolic rates of fishes of the lake.

These researchers stressed the importance of shallow areas in the attached lakes and bay of Lake Biwa as production and nursery areas of fishes. Although they did not actually evaluate the effects of the government plan (see E. Harada 1998 this volume), fortunately the plan to construct a dam in the lake was abandoned. However, land development in the years that followed largely altered the habitats in attached lakes and along the shoreline, and the fish community in the lake changed drastically. Recently, there has again been keen interest in the importance of such habitats.

In addition to these projects, D. Miyadi, H. Kawanabe, and N. Mizuno compiled a book of colored illustrations of the freshwater fishes of Japan (1976, HK 52), which provided useful and convenient information on taxonomy and biology of freshwater fishes and stimulated many amateur scientists. Later, Kawanabe and Mizuno edited another colored guidebook, 'Freshwater Fishes of Japan' (1989, HK 459). A total of 69 fish biologists and 10 photo-ecologists participated in production of the book which provides more precise behavioral and ecological information on freshwater fishes in Japan.

Social problems and a scientist — from public nuisance to natural conservation

During the 1960s, Japan developed greatly both industrially and economically. With such developments came many environmental problems, ranging from public nuisances to the destruction of natural environments and extinction of species. Minamata disease is one case of serious public endangerment caused by poisoning via organic methyl mercury discharged from chemical factories and through biomagnification of chemicals in the food chain. Kawanabe served as an expert witness for the plaintiff in the case of Minamata disease in the river Agano-gawa in 1968, and explained the ecological mechanism concerning the disease. He also had another opportunity to stand in the witness box in a suit against the government by a huge number of local residents over the dam at the mouth of the river Nagara-gawa in 1978, and explained the ecological influence of dam construction.

After the 1970s, nature conservation boomed in Japan. Kawanabe has been a leader in this social movement, serving as member of many committees concerned with nature conservation, such as the Japanese Committee of the World Wildlife Fund (WWF), the Monument Conservation Council and the Monument Conservation Planning Special Committee of the Agency for Cultural Affairs, the Cultural Properties Preservation Councils of Kyoto Prefecture and Kyoto City, and the Natural Environment Conservation Council of Kyoto City. He was concerned with major government projects, such as Lake Naka-umi mentioned earlier, and a dam at the mouth of the Nagara-gawa built by the Ministry of Construction as a barrier against sea water, a project still in dispute (e.g. HK 198, 480, 539). He also addressed many local problems, such as conservation of the Taiwan masu salmon, *Oncorhynchus masou formosanus*, and the Ryukyu ayu, *Plecoglossus altivelis ryukyuensis*, and preservation of a moor, Mizoro-ga-ike, in Kyoto City. In the course of these activities, Kawanabe frequently stated that 'Nature is the culture of human beings. That is, culture has developed in connection with Nature. A part of Nature has been preserved through human utilization and management. The conservation of nature inevitably requires conserving the entirety of relations between environments and organisms, including humans' (HK 261, 531, 627). His opinions on nature and culture have recently been incorporated in the new Lake Biwa Museum, which opened in 1996, and have been further developed there.

Ecological complexity, or the whole of interrelationships

Hiroya Kawanabe first met Charles S. Elton (1900–1991) at the Bureau of Animal Populations, Department of Zoology and Comparative Anatomy, Oxford University, in September 1963 (Figure 11). During his days at Oxford, Kawanabe discussed a wide range of topics in ecology with Elton, and remembers his words: 'Researchers in Europe and the U.S.A. do their best to find general scientific theories. To do so, they assume a simple system instead of a complicated natural system, and pay attention to the common phenomena irrespective of species and places. Researchers in the UK, at least me, intend to study in different way by trying to un-

Figure 11. Charles S. Elton (right), E.J. Scovell (Mrs. Elton, left), and Hiroya Kawanabe near Wytham Woods in Oxford on 31 August 1984. Photograph by A. Kawanabe.

derstand the complicated system itself. And furthermore, we want to ask: Of what importance is the complexity in a natural system? Why does the biological community differ depending on species composition and environmental changes? The most important rule might be the "local" one which depends on particular species and places. A "general" rule is merely based on such local rules. I am less concerned about these general phenomena which can be found in any species and any place in the world. Although I don't like to evaluate the applicability of problems, the phenomena which humans can manage and control are not those relating to "general" rules and commonly found at any place and time, but those depending on "local" rules.' To this Elton added a question: 'How do most Japanese researchers think about this matter ?' (HK 118). The encounter with Elton and his writings deeply influenced Kawanabe's ideas and logic on scientific matters. Thereafter, Kawanabe and his colleagues translated Elton's books into Japanese; 'The ecology of invasions by animals and plants' (1958) in 1971 (HK 117), 'The ecology of animals' (1933) with 3 articles in 1978 (HK 241), and 'The pattern of animal communities' (1966) in 1990 (HK 488).

Kawanabe first stated in 1960 that 'I wish to understand a biological community as the whole set of relations' (HK 567). His belief was probably drawn from his study of relations among fishes in the U-kawa, and seemed to largely coincide with Elton's ideas. Calling himself an Eltonian, he wrote in the postscript of the translation of 'The ecology of animals' that he was fascinated with Elton's essay, and that the struggle to be free from simply following Elton's views might help him develop his own ideas (HK 242). Kawanabe still persists in the theory above from 1960, and has expanded his views by incorporating many new scientific ideas.

Kawanabe had the good fortune to first be employed as lecturer at the Otsu Hydrobiological Station and associate professor in the Department of Zoology, both at Kyoto University. Later, when the Laboratory of Animal Ecology was formed, he was promoted to the chair post (Figure 12). There he was able to interact with many colleagues and graduate students studying a wide variety of animals (Figure 13). Through such discussions he developed the idea that there were many differences in modes

Figure 12. Long-haired Hiroya Kawanabe in his office. His hair had been cut by his wife, Aya, and he is seen wearing a sweater hand-made by her (13 February 1975). Photograph by H. Hasegawa.

of life that could be observed not only among populations of the same species, but also among individuals in a population, and in the same individuals in different situations (HK 470). Doubtless these enjoyable discussions led Kawanabe to confirm his theory about the whole of relations in a biological community. His ideas on ecological complexity also advanced following his work at Lake Tanganyika

Figure 13. Under the cherry blossoms at the Joushoukouji temple, Kyoto Prefecture, in April 1982. From left to right, K. Hori, M. Yuma, and H.Kawanabe. Photograph by M. Hinoue.

Figure 14. An evening party at C.R.H./Uvira, Zaire, on 21 July 1988. From left to right, N. Abe, I. Kinoshita, H.Kawanabe, M.W. Tshittende, W. Amissi, and M. Yuma.

(after 1979), and more recently the work done through the Scientific Research on Priority Areas, 'Symbiotic biosphere: an ecological interaction network promoting the coexistence of many species' (1991–1995).

Touching Lake Tanganyika — the start of a joint study

Hiroya Kawanabe first visited Lake Tanganyika in 1963 as a fellow of the primatologist, Jun'ichiro Itani. Itani intended to establish a national park ranging from the mountains to the lake at Mahale, Tanzania, and requested that Kawanabe survey the underwater habitats of Lake Tanganyika. Preliminary work was conducted by Masakazu Kawabata and Teruo Doi in 1967–1968 and Kawabata in 1977–1978. Kawanabe visited the lake for the second time in 1977 as a member of a study team organized by Kawabata, but was forced to spend almost all the time at the shore of Lake Kivu because of political problems in Zaire, and could stay only three days at Uvira along the shore of Lake Tanganyika. Following this, he led the survey of Lake Tanganyika in 1979. It was started as a joint study with scientists at the Uvira Station of the Scientific Research Institute of Central Africa (Institut de Recherche Scientifique en Afrique Centrale; now Centre de Recherche en Hydrobiologie) in Zaire (Figure 14). The members of the study team in 1979 were Michio Hori, Kousaku Yamaoka, Kenji Takamura, Yuji Ankei, Masakazu Kawabata and Hiroya Kawanabe from Japan, and Nseu B. Mbomba, Menga K. Kwetuenda and Ngabo y.K. Mihigo from Zaire. The joint study with Zairian researchers continues today with the aid of the Japan Ministry of Education, Science, Sport, and Culture; the Japan Society for the Promotion of Science; and the Japan International Cooperation Agency.

The research in Lake Tanganyika was also conducted at Mahale (Tanzania) in 1981 and 1983. At Mahale, Kawanabe's members were supported by Japanese primatologists who set up a base camp to study chimpanzees in 1966. Since 1983 an intermittent survey has also been carried out along the Burundian coast by Japanese and Zairian researchers, with scientists from the National University of Burundi.

Since 1988, another joint study was started

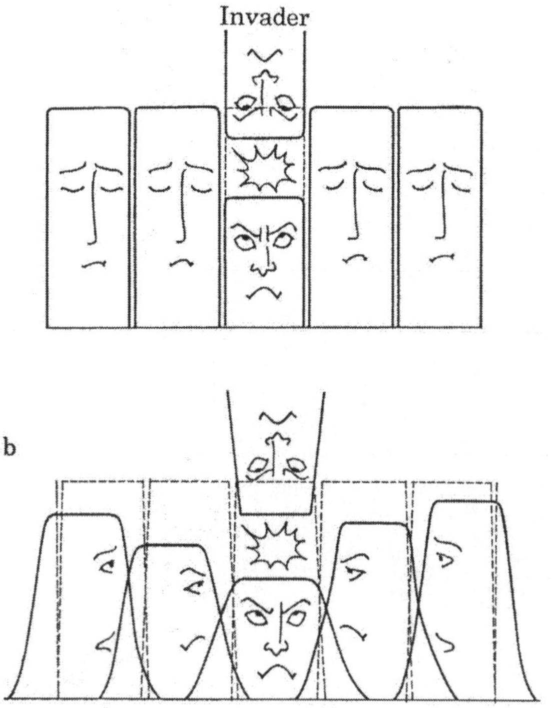

Figure 15. Typical responses of biological communities against an invading species: a–Fixed community comprising species with separate niches, in which an invader affects only one competitive species. b–Adjustable community comprising species with overlapping niches, in which all species are influenced by an invading species and thus may stand against the invader in cooperation (after HK 266).

Figure 16. Hiroya Kawanabe in kimono is presenting his opening lecture at the 'International Symposium on Charrs and Masu Salmon' in Sapporo, 4 October 1988. Photograph by E.K. Balon.

around Mpulungu, Zambia, with scientists at the Lake Tanganyika Research Unit, Fisheries Research Institute (now Department of Fisheries) in Zambia. One benefit of these joint research programs was that African scientists from different countries around the lake freely visited one another.

For the joint study, Kawanabe organized the 'Maneno team' ('Maneno' means 'word' in Swahili and its plural term means many, or too many words and discussion, but actually it was the name of the 8-year old daughter of a family at a diving site in Luhanga, Zaire). It now consists of more than 50 Japanese scientists, and more than 20 Zairian, Zambian, and European scientists, all dealing with fishes, aquatic insects, molluscs, shrimp, zooplankton, primary production, physico-chemical features in and around the lake, fisheries, marketing, and anthropology. However, Kawanabe stressed that research in Lake Tanganyika to promote ecologically sound development should be based on a joint study with African scientists. He stated: 'What is the best course for development? This should be decided by the resident people. To develop a sound plan, the resident scientists have to stand on their own feet, understand the importance of environmental problems, and perceive the mistakes by so called developed countries. Our role is to assist these procedures. However, although the general public praises the transfer of new scientific techniques, they miss the importance of transfering the basic principles of the scientific method that underpins research' (HK 498).

Figure 17. An excursion at the Hakusan Nature Conservation Center, Ishikawa Prefecture, Japan, on 14 October 1988, after the 'International Symposium on Charrs and Masu Salmon'. First row from left to right, A. Mizuno, T. Furukawa-Tanaka, T. Kato, E. Nozaki, E. Yamaguchi, and R. Gydemo; back row from left to right, A. Klemetsen, G. Power, H. Nordeng, D.L.G. Noakes, M.-A. Svenning, Y. Machino, P.-A. Amundsen, E.K. Balon, J.D. Reist, H.Kawanabe, L. Johnson, J. Hammar, and T. Maruyama.

Beyond interactive relationships — cooperative relationships in fish communities

Lake Tanganyika is very famous as one of the oldest lakes in the world and has an extraordinarily rich fauna. When the study of the lake began, Kawanabe focused his research around two questions. First, 'how do the many species segregate in their intrinsic habits?', and second, 'are there any interspecific relations between fishes with overlapping habits?' (HK 266, 711). Kawanabe wanted to compare his findings to what he observed in Japanese freshwater fish communities (Figure 15).

Along the lakeshore, Kawanabe observed the territorial behavior of an algal browser, *Tropheus moorii* (Cichlidae). Because he could not dive into the lake, he lay on rocks on the shoreline and looked into the water using a mask and snorkel. Using the projection lines of rocks and stones as boundaries, he was able to map *T. moorii* feeding territories, and discussed its territoriality compared to that of ayu (HK 275, 280).

Thereafter, Kawanabe's colleagues continued intensive studies with the aid of SCUBA on interspecific relations, which in many cases suggested that facultative mutualism was important in structuring the fish communities (see HK 757). Although previous scientists believed that it is a disadvantage for organisms to share resources, and that 'competitive exclusion' should usually occur instead, the complex biotic interactions observed in Lake Tanganyika indicated that such an understanding is one-sided (HK 567, 711). Based on these results, Kawanabe became convinced that the existence of a third species changed the relationship between two interacting species.

Kawanabe's ideas on biological communities were investigated further by a group research pro-

⟶

Figure 18. Some participants of the 'charr symposium' before the visit of the Hakusan (13 October 1988) spend the night at a nearby resort famous for its hot springs (see Harada 1998): a–During an evening rain, from left to right, E. Balon, R. Gydemo, J. Hammar, and H. Kawanabe walk to the hot springs area where b–Kawanabe was the first to submerge. Photographs by J. Hammar (a) and E. Balon (b).

a

b

ject titled 'Scientific research on priority areas, symbiotic biosphere: an ecological interaction network promoting the coexistence of many species' (1991–1995). In this project, Kawanabe's colleagues revealed that many relationships between two species were mediated by a third species, by means of direct and indirect or chemical interactions between individuals (see HK 582). Kawanabe proposed that the analysis of such indirect effects should be a next step for further development of community ecology in terms of multi-species interactions (HK 665, 708, 711).

Toward an international network in ecology

In 1963 Hiroya Kawanabe started his activities abroad, visiting the United States, the United Kingdom, Germany, and Lake Tanganyika. He says, however, that the most important opportunity to expand his international activities was in 1968 when he was invited by Thomas G. Northcote to the University of British Columbia, Canada, where he participated in the 'Symposium on salmon and trout in streams'. At the symposium he delivered a paper on the significance of social structure to the production of ayu (HK 88), and met many North American and European fish biologists.

Kawanabe's international activities increased abruptly after the late 1980s when he became President of the Ecological Society of Japan (1988–1991). It was, unexpectedly, also in response to a plan for a new 'National Institute of Ecology', a part of which was later realized when the 'Center for Ecological Research' was established by Kyoto University in 1991.

Another international link was forged in November 1986, when Kawanabe invited G. Evelyn Hutchinson (1903–1991) to receive the Kyoto Prize for his contributions to biology and ecology, sponsored by the Inamori Foundation. Related to Hutchinson's contributions, he and E. Harada published an issue on the 'Ecological play on the Earth: ecological niche and its related problems' (HK 429).

In 1988 Kawanabe organized the 'International Symposium on Charrs and Masu Salmon' at Sapporo, Japan, which took place on 3–9 October (Figure 16). This symposium was suggested by Lionel Johnson whom Kawanabe met in 1968 and 1981 in Canada, and was carried out in collaboration with Eishige Komiyama, Akira Goto, Koji Maekawa, and Fumio Yamazaki who invited more than 90 sponsors and supporters (Figure 17). More than 50 scientists from 12 countries, and more than 80 Japanese scientists participated in the symposium and the three excursions (Figure 18) offered by the local hosts after the meeting (see Noakes 1989, HK 463). In the same year, Kawanabe was invited to a meeting of theoretical ecology, the U.S.-Japan Seminar in the Environmental Sciences, organized by Ei Teramoto and held at the East-West Center, Hawaii, on 25–30 April, where he met many ecologists.

In 1989 Kawanabe served as chairperson for an international symposium 'Ecology for Tomorrow', held at Otsu, at the southern end of Lake Biwa. This was a pre-symposium of the 5th International Congress of Ecology (INTECOL '90) at which 13 ecologists from abroad and 4 Japanese ecologists gave papers (HK 475).

Kawanabe was the convener and chair of the organizing committee for INTECOL '90, held at Yokohama, Japan, on 23–30 August 1990. This congress was proposed by Kawanabe at the 4th INTECOL in United States in 1986, when he was the Secretary General of the Ecological Society of Japan. More than 2300 scientists from 82 countries participated in the 5th INTECOL, which was the first time such a large group of ecologists had met in Asia.

Just before INTECOL '90, Kawanabe also organized the Oji International Seminar at Gifu, held on 18–22 August 1990, where 23 scientists from abroad and 33 Japanese scientists discussed mutualism, cooperation and organization in natural communities (see HK 578). Kawanabe collaborated with Takuya Abe, Keiji Iwasaki, and other colleagues in his laboratory to carry out the seminar, which was sponsored by the Fujiwara Foundation of Science and the Japan Society for the Promotion of Science.

Through these meetings Kawanabe enlarged his network of international scientists. Many foreign scientists sent him letters encouraging the establishment of an institution to develop ecological research in Japan. Then, in 1991, the Center for Ecological Research (CER) was established at Kyoto

Figure 19. Some invited speakers at the Canadian Society of Zoologists symposium 'Great Lakes of the World' held at the University of Guelph in May 1993. From left to right, I. Bailey, R.H. Lowe-McConnell, J. Magnuson, G. Fryer, H.Kawanabe, and A. Rossiter.

University to study the interrelationships among organisms and environments, and to contribute the International Geosphere-Biosphere Programme (IGBP). Kawanabe was the founder of the CER, and as the first Director further increased his international activities to develop the international joint study around the CER (Figures 19, 20).

In 1991, the same year that CER began, an international programme of biodiversity science, DIVERSITAS, was established with the partnership of the International Union of Biological Sciences (IUBS), the Scientific Committee on Problems of the Environment (SCOPE), and the United Nations Educational, Scientific and Cultural Organization (UNESCO). Otto T. Solbrig published a book relating to the programme titled 'From genes to ecosystems: a research agenda for biodiversity' in 1992. Kawanabe countered with his opinion 'from ecosystems to genes' which led him to organize the programme of the 'SymBiosphere'. The term 'SymBiosphere' was produced for the project title by Hiroya Kawanabe, Takuya Abe, Tamiji Inoue, Masahiko Higashi and Takayuki Ohgushi in 1989.

In June 1992 Kawanabe visited the Earth Summit, the 'United Nations Conference on Environment and Development (UNCED)' at Rio de Janeiro, and attended the 'Symposium on Biodiversity' organized by the Scientific Meeting of Brazil and UNESCO. In October 1992, Kawanabe attended the meeting of DIVERSITAS in Costa Rica and proposed an international joint study program on the 'SymBiosphere: ecological complexity for promoting biodiversity' which was planned in collaboration with Takayuki Ohgushi, M. Higashi, and other colleagues mainly at the CER (HK 582). In the proposal, Kawanabe stressed that: 'Biodiversity should be the basis of the physical and intellectual life of human beings, and should support all culture and arts. Although scientists continue to look for common phenomena on which to base a principal theory, we should hereafter try to clarify why each ecosystem is so different and to develop general

Figure 20. David Noakes, Hiroya Kawanabe, and Ro Lowe-McConnell at the hot tub in Balon's residence. Photograph by E.K. Balon.

rules to answer this question. This is my belief, as a scientist' (HK 573). Fortunately, the proposed program was adopted by the IUBS, UNESCO and other organizations in October 1992, with the help of Talal Younès, the Executive Director of IUBS, and Tokindo S. Okada, the Vice President of IUBS. Kawanabe then organized an international workshop on 'SymBiosphere: ecological complexity for promoting biodiversity' at Kyoto in December 1992 (HK 582).

In December 1993, Kawanabe and his colleagues at the CER organized the international symposium on 'Ecological perspective of biodiversity' at Kyoto with papers by 35 scientists from Japan and abroad, to celebrate the Japan International Biological Prize of 1993 which honored Edward O. Wilson. Thereafter, Japanese scientists took the initiative for a part of the DIVERSITAS program and developed the international network for DIVERSITAS in the Western Pacific and Asia (DIWPA), with the secretariat being located at the CER since 1994 (Figure 21).

In October 1992 Kawanabe visited the Organization for Tropical Sciences (OTS) at La Selse in Costa Rica, an excellent institution that provides lectures and exercises on tropical sciences for select students from universities throughout the Americas. Based on this experience, Kawanabe led the CER to start the 'International summer seminar on the global environment and ecology', later the 'International field biology course in western Pacific Asia', for graduate students, beginning in 1992. He stated: 'Our purpose in the international seminar is not only to educate young scientists in Asia. When we continue such seminars for 20 years, can you imagine the result? We should see the appearance of a wonderful network among leading scientists. That is our real purpose' (HK 592). By saying this, Kawanabe expressed his intention to build inter- and intra-national networks of scientists, based on the premise that international cooperation among the scientists is necessary primarily to solve problems of the environment and biodiversity on both a global and local scale (Figure 22).

From biodiversity to cultural diversity as whole of relations

Hiroya Kawanabe's activities were supported by many people; e.g., his major professor and colleagues mentioned earlier, secretaries and graduate students in his laboratories, and friends abroad. During his years at the Laboratory of Animal Ecology at Kyoto University, he had more than 50 graduate students dealing with various animals. Compared to his huge number of other contributions, he produced relatively few papers with his students. He treated his students as fully-fledged researchers, and encouraged them to produce papers independently rather than publishing papers with him. He insisted that they refer to him not as 'Professor Kawanabe' or 'Kawanabe-sensei', but as 'Kawanabe-san', the more familiar way of addressing a person of equal professional stature in Japanese. This was

Figure 21. On the top terrace, 35 m above ground of a tree-tower aside a dipterocarp, *Dryobalanops lanceolata*, in the Lambir Hills National Park, Sarawak, Malaysia, on 25 December 1994. This site is one of the core field stations of the DIVERSITAS in the Western Pacific and Asia (DIWPA) programme. First row from left to right, T. Inoue, E. Wada, S. Sakai; back row from left to right, R. Harrison, T. Hiromi, H. Nagamasu, M. Kawai, H.Kawanabe, and K. Watanabe. Tamiji Inoue, a professor at CER and an excellent tropical ecologist, had a fatal accident in an airplane in Sarawak on 6 September 1997.

Figure 22. At the last dinner after Balon's sabbatical in Japan, given by Kawanabe. From left to right, Hiroya Kawanabe, Masahide Yuma, Yosikazu Nagata, and Christine Flegler-Balon, 7 June 1994. Photograph by E.K. Balon.

Figure 23. After the visit at the private Kyouto Seika University with its President Sibatani, dinner was served at a small local restaurant near the Kino station: left to right, Eugene Balon, Hiroya Kawanabe, Atuhiro Sibatani, and Masahide Yuma (27 April 1994).

actually the style of D. Miyadi, who told Kawanabe-san that 'because you came here to become a fully-fledged researcher, you are not my student but rather a researcher equal to me. So you have to call me Miyadi-san!' (HK 445). Kawanabe enjoyed discussions with young researchers, giving criticisms and accepting interesting results, but puzzling them with his unique and paradoxical words. The latter should be considered to be Kawanabe's style of training, based on his belief that scientists must have plural viewpoints on natural phenomena (Figure 23).

Hiroya Kawanabe also enjoys discussions beyond science, especially about literature and music. He has produced many reviews and essays on books and music. He has a passion for reading, has a personal library of more than 10 000 books, and often reads books while walking down the street. Kawanabe also frequently attends operas and classical music concerts, and is an avid collector of music recordings. He wrote: 'Opening a music book, I consider what kind of structure in music gives us the "complex" feeling. Complexity in communities of organisms may be likened to chamber music, which contains complex melodies. And, the basic structure of Western music seems to draw on rigid relations, to be constructed in a regular manner, while that of Japanese music is loose in structure and seems to not have any regular rules. Hypotheses or methods for understanding the complexity in communities of organisms might be found in this analogy comparing Eastern and Western music' (HK 567).

Above all, Hiroya Kawanabe has been supported by his family (Figures 24, 25). His wife, Aya, often knits and selects his fascinating clothes using her unique sense as a professional modern artist (Figure 12). He is also much obliged to his secretaries, Kiyoko Hori at the Laboratory of Animal Ecology, and Yuko Kosuga, Yumiko Yamamoto, Naoko Tanaka, and Makiko Kawasaki at the CER and Lake Biwa Museum, depending on them to regulate his

Figure 24. Hiroya Kawanabe and his daughter Nagi in the Imperial Palace Park in Kyoto on 15 September 1967.

Figure 25. Aya, Nagi and Hiroya Kawanabe on Yonaguni Island, 11 August 1973.

schedules. However, he does cause trouble for his office mates and even in his home by keeping rooms too cold because he is intolerant of heat.

During his field work, Kawanabe was obliged to many local people; especially Ryuhei Okada at the Kami U-kawa Fisheries Cooperative, and Kenji Watanabe at the Kami-Katura-gawa Fisheries Cooperative in Keihoku Town, Kyoto Prefecture. He often tells young researchers that 'if you are going to survey organisms, you have first to ask the local residents about the condition of the local biota'. Recently, he has tried to foster further ties with

Figure 26. Lake Biwa Museum at Oroshino opened in 1996 after 10 years of planning and construction. After retirement from Kyoto University Hiroya Kawanabe became the founding Director General of this new museum. Photographs by D.L.G. Noakes.

local residents while expanding the international ties at the Lake Biwa Museum (Figure 26). Based on the theme of the museum — Coexistence of the Lake and Human Life and Culture — he organized the International Conference on Ancient Lakes (ICAL) in June 1997, where many scientists and resident audiences discussed matters ranging from biological processes to cultural aspects in and around the lakes.

Acknowledgements

We are grateful to Aya and Hiroya Kawanabe for

their kindness in allowing us to interview them, and to permit us to see their well-organized and catalogued photographs. We are also grateful to many colleagues for their information relating to Kawanabe's biography. We thank Kurt Fausch, Alfred Eberle, Izumi Nakamura, and Eugene Balon for their kind revision of language and style.

References cited

(for the publications by HK, see the 'Lifetime list of publications by Hiroya Kawanabe')

Darnell, R.M. 1958. Food habits of fishes and larger invertebrates of Lake Pontchartrain, Louisiana, an estuarine community. Publ. Inst. Mar. Sci., Univ. Texas 5: 353–416.

Fausch, K.D. & S. Nakano. 1998. Research on fish ecology in Japan: a brief history and selected review. Env. Biol. Fish. 52: 75–95 (this volume).

Frost, W.E. 1946. On the food relationships of fish in Windermere. 13th Biol. Jaarb. Dodonaea: 216–231.

Harada, E. 1998. An interview with Hiroya Kawanabe. Env. Biol. Fish. 52: 67-72 (this volume).

Imanishi, K. 1951. Japanese chars and masu-salmon. Nihon Ringyo Fukyuu-kyoukai, Tokyo. 86 pp. (in Japanese).

Ivlev, V.S. 1955. Experimental ecology of the feeding of fishes. Yale University Press, New Haven. 302 pp.

Miura, T. 1959. Some ecological studies on fish populations in Lake Sagami, an impoundment in Kanagawa Prefecture, Japan. Bull. Freshw. Fish. Res. Lab. 1959: 23–39.

Miyadi, D. 1994. Story of ayu fish. Iwanami Shoten, Tokyo. 208 pp. (in Japanese).

Miyadi, D., M. Kawabata & K. Ueda. 1952. Standard density of the ayu, *Plecoglossus altivelis*, on the basis of its territorial behavior and grazing unit area. Contr. Physiol. Ecol. Kyoto Univ. 75: 1–23 (in Japanese).

Miyadi, D., S. Yagi & S. Tani. 1953. Field experimental studies of the standard density of the ayu, *Plecoglossus altivelis*. Rep. Fish. Dep. Kyoto Pref. Govern. 1952: 1–50 (in Japanese).

Mizuhara, H., D. Miyadi, S. Mori & S. Kawai. 1955. Ecology of the ayu, *Plecoglossus altivelis*, in river pools. Rep. Fish. Dep. Kyoto Pref. Govern. 1954: 1–34 (in Japanese).

Nilsson, N.-A. 1955. Studies on the feeding habits of trout and char in north Swedish lakes. Rep. Inst. Freshw. Res. Drottningholm 36: 163–225.

Nilsson, N.-A. 1957. On the feeding habits of trout in a stream of northern Sweden. Rep. Inst. Freshw. Res. Drottningholm 38: 154–166.

Nishida, M. 1985. Substantial genetic differentiation in ayu *Plecoglossus altivelis* of the Japan and Ryukyu islands. Nippon Suisan Gakkaishi 51: 1269–1274.

Nishida, M. 1986. Geographic variation in the molecular, morphological and reproductive characters of the ayu *Plecoglossus altivelis* (Plecoglossidae) in the Japan-Ryukyu Archipelago. Japan. J. Ichthyol. 33: 232–248.

Noakes, D.L.G. (with photographs by E.K. Balon) 1989. Symposium to be remembered. Env. Biol. Fish. 24: 313–317.

Petersen, C.G.J. 1914, 1915, 1918. The animal associations of the sea-bottom in the North Atlantic. Kobenhavn Ber. Biol. Sta. 22: 89–98; 23: 1–28; 26: 1–62.

Solbrig, O.T. 1992. From genes to ecosystems: a research agenda for biodiversity. IUBS, SCOPE & UNESCO, Paris. 124 pp.

Yuma, M., K. Hosoya & Y. Nagata. 1998. Distribution of the freshwater fishes of Japan: an historical overview. Env. Biol. Fish. 52: 97-124 (this volume).

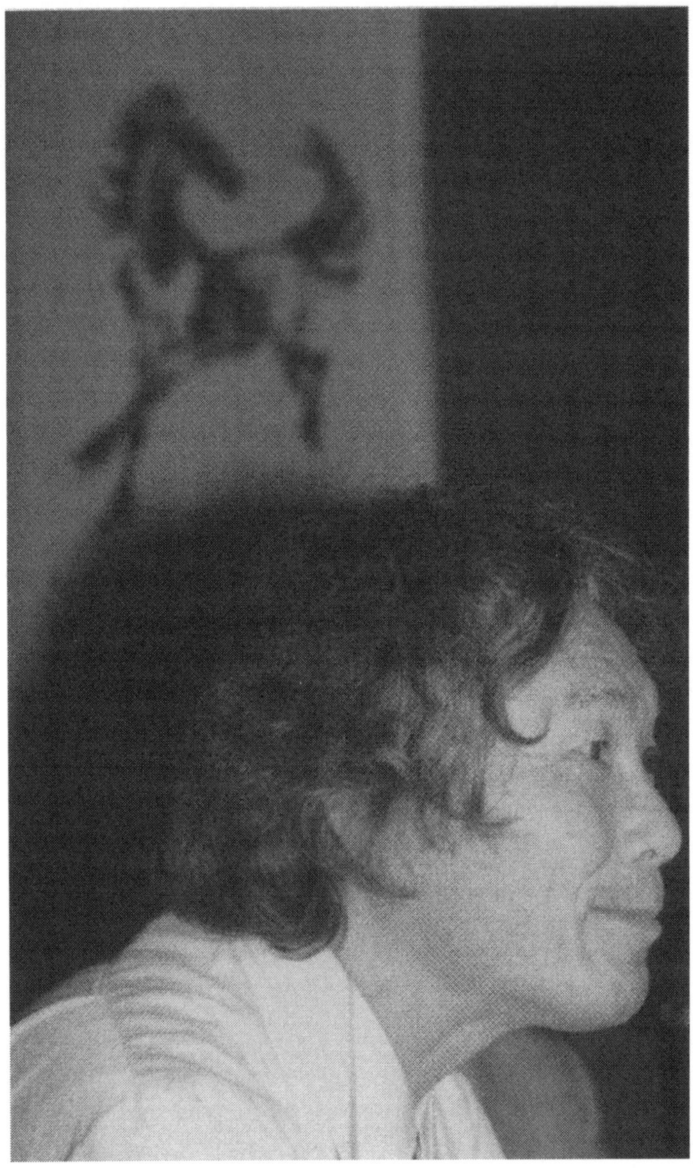

Environmental Biology of Fishes **52**: 37–66, 1998.

Lifetime list of publications by Hiroya Kawanabe

Masahide Yuma
Center for Ecological Research, Kyoto University, Otsu 520-01, Japan (e-mail: yuhma@ecology.kyoto-u.ac.jp)

Hiroya Kawanabe produced more than 780 scientific papers, popular articles, governmental reports, chapters in books, edited or co-edited books, encyclopedia entries, and Japanese translations of books from 1952 until the compilation of this bibliography (Fall 1997). He was the sole or first author of 88% of these publications and shared authorship with 286 collaborators. Eighty-nine percent of his publications were in Japanese, others were in English, German, Chinese, Korean and Italian. His publications were devoted mainly to four topics: (1) ecology of freshwater fishes, especially the territorial behavior of ayu, *Plecoglossus altivelis,* and interspecific relationships and food segregation among fishes from 1956; (2) fish community ecology in Lake Tanganyika since 1979; (3) political articles on ecological research after 1991; and (4) museum activities after 1996. He also produced a number of newspaper articles (over 25% of total publications) addressing topics not only of science but literature, culture and philosophy. Kawanabe's articles in newspapers were produced mainly when he was the Director of the Center for Ecological Research of Kyoto University, and most of these articles concentrated on promoting the development of ecological research in Japan. The publications are arranged chronologically by year. The decision to add Japanese titles was based on the fact that their English translations, which follow in parentheses, are often too loose and imprecise. Unfortunately, our printers could not insert these titles in the Japanese characters so Roman transliterations had to be substituted.

1. Kawanabe, H. 1952. Patinko no suikeigaku (Stochastics of 'Patinko', the pinball game). Shingaku 1952 (10): 22–27 (in Japanese).
2. Kawanabe, H. 1952. Seibutu-toukeiho ni yosete (For biological statistics). Bulletin of the Kyoto Branch for Biology, Democratic Scientists' Society 8: 2–3 (in Japanese).
3. Harada, E. & H. Kawanabe. 1955. Kometuki-gani no koudou to sougo-sayou (The behavior of the sand-crab, *Scoipimera globosa* de Haan, with special reference to the problem of co-action between individuals). Japan. J. Ecol. 4: 162–165 (in Japanese).
4. Kawanabe, H., D. Miyadi, S. Mori, E. Harada, H. Mizuhara & R. Ohgushi. 1956. Sozyou ayu no seitai: tokuni huti ni okeru ayu no seikatu-yousiki ni tuite (Ecology of natural stock of ayu, *Plecoglossus altivelis*, with special reference to their life in river-pools). Contribution to Physiology and Ecology, Kyoto University 79: 1–37 (in Japanese).
5. Ohgushi, R., H. Kawanabe & E. Harada. 1956. Huti no soko no kontyuu-gunsyuu (Benthic insect communities in river pools). Physiol. Ecol. 7: 61–71 (in Japanese).
6. Mori, S., S. Fuse, E. Harada, H. Kawanabe, J. Kikkawa, H. Mizuhara & T. Miura (translators) 1956. Seitaigaku no kiso (Japanese translation of 'Fundamentals of ecology' by E.P. Odum). Asakura-shoten, Tokyo. 432 pp. (in Japanese).
7. Kawanabe, H. 1957. Ayu no nawabari (Territory of the ayu). Sizen, 1957(8): 3–11 (in Japanese) (also in 240).
8. Kawanabe, H., N. Mizuno, D. Miyadi, S. Mori, R. Ohgushi & N. Nishimura. 1957. Sozyou ayu no seitai (Ecology of natural stock of salmon-like fish, ayu). II. Tokuni seisoku-mitudo to seikatu-yousiki ni tuite (The mode of life in relation to its population density). Physiol. Ecol. 7: 145–167 (in Japanese).
9. Kawanabe, H., S. Mori & N. Mizuno. 1957. Sozyou ayu no seisoku-mitudo to huti no riyou no sikata (Modes of utilizing the river pools by a salmon-like fish, *Plecoglossus altivelis* or ayu, in relation to its population density). Japan. J. Ecol. 7: 22–26 (in Japanese).
10. Mizuno, N. & H. Kawanabe. 1957. Nawabari no missyuu sita tiiki ni okeru ayu no koudou (Behaviour of salmon-like fish 'ayu' in an area with closely established territories). Japan. J. Ecol. 7: 26–30 (in Japanese).
11. Kawanabe, H. 1957. Ayu no syakai-kouzou to seisan: seisoku-mitudo to kanren dukete (Social behaviour and production of a salmon-like fish, *Plecoglossus altivelis*, or ayu, with reference to its population density). Japan. J. Ecol. 7: 131–137 (in Japanese).
12. Kawanabe, H., N. Mizuno & N. Nishimura. 1957. Ayu wa kasyou-gata wo ikani riyou suruka: ayu no mitudo to taityou-bunpu (Population density of a salmon-like fish, *Ple-*

coglossus altivelis, or ayu, in different river beds). Bull. Japan. Soc. Sci. Fish. 23: 430–434 (in Japanese).

13. Kawanabe, H. 1957. Saisyusya no tame no tansuigyo kensaku-hyou (Keys to freshwater fishes in Japan for collectors). Published by the author. 34 pp. (in Japanese).
14. Kawanabe, H. 1957. Nihon-zaru: sono migotona syakaikouzou (Book review: 'The Japanese macaque: its wonderful social structure' by H. Mizuhara). Nihon Dokusho Shinbun (precise information not available) (in Japanese).
15. Harada, E. & H. Kawanabe. 1958. Kurama-yama no syokusei to doubutu-sou (The fauna found in accordance with the successional stages of plants at Mt. Kurama). Physiol. Ecol. 8: 67–79 (in Japanese).
16. Kawanabe, H. 1958. On the significance of the social structure for the mode of density effect in a salmon-like fish, 'ayu', *Plecoglossus altivelis* Temminck et Schlegel. Mem. Coll. Sci., Univ. Kyoto, Ser. B 25: 171–180.
17. Mizuno, N., H. Kawanabe, D. Miyadi, S. Mori, H. Kodama, R. Ohgushi, A. Kusakabe & Y.Y. Furuya. 1958. Kawa no sakana no seikatu (Studies on the life of river fish) I. Koi-ka 4 syu no seikatusi wo tyuusin ni site (Life history of some stream fishes with special reference to four cyprinid species). Contribution to Physiology and Ecology, Kyoto University 81: 1–48 (in Japanese).
18. Kawanabe, H. 1958. Sakana no seikatu (The life of fishes). Kagaku Yomiuri (Yomiuri Science) 10(8): 25–28 (in Japanese).
19. Kawanabe, H. & K. Osaki. 1958. Seibutu-kagaku sinpoziumu wo hurikaette (Retrospective for symposium of biological sciences). Bulletin of the Kyoto Branch for Biology, Democratic Scientists' Society 22: 1–3 (in Japanese).
20. Kawanabe, H., S. Mori & N. Mizuno. 1959. Ayu no seityou to sourui-ryou (On the food economy of 'ayu' fish with relation to the production of algae). Physiol. Ecol. 8: 118–123 (in Japanese).
21. Kawanabe, H. 1959. Food competition among fishes in some rivers of Kyoto Prefecture, Japan. Mem. Coll. Sci., Univ. Kyoto, Ser. B 26: 253–268.
22. Kawanabe, H. 1959. Sakana no syakai-seikatu (Social life of fishes). pp. 49–61. *In:* K. Imanishi (ed.) Doubutu no syakai to kotai (Society and individuals in animals). Iwanami-shoten, Tokyo (in Japanese).
23. Kawanabe, H. 1959. Seitaigaku-zissyu no kiroku wo yonde: koukou no seitaigaku-zissyu muyou-ron (Critics for ecological practice in high schools). Bulletin of the Kyoto Branch for Biology, Democratic Scientists' Society 25: 1–3 (in Japanese).
24. Kawanabe, H. 1959. Nihon gyomei no kenkyuu (Book review: 'Names of Japanese fishes' by K. Shibusawa). The Tosho (precise information not available) (in Japanese).
25. Kawanabe, H. 1960. Kawa no doubutu gunsyuu wo dou toraeruka (An opinion on the study of the lotic animal community). Physiol. Ecol. 9: 1–10 (in Japanese) (also in 382).
26. Miyadi, D., S. Mori, H. Kawanabe, N. Mizuno, H.N. Kodama, Y. Tezuka, K. Onodera, G. Shiro & T. Tomita. 1960. Kawa no sakana no seikatu (Studies on the life of river fish). II. Gyorui 5 syu no seikatusi to gunsyuu-kouzou (Life histories of five species of river fish and community structure). Fisheries Department of Kyoto Prefectural Government, Kyoto. 20 pp. (in Japanese).
27. Miyadi, D., S. Mori, H. Kawanabe, N. Mizuno, H.N. Kodama, S. Tomoda, R. Ohgushi & Y.Y. Furuya. 1961. Gyorui no genzyou (Present status of fishes). pp. 14–20. *In:* D. Miyadi & M. Tsuda (ed.) Amagase-damu kensetu ni tomonau gyogyou-hosyou no tame no seibutu tyousa houkoku (Biological Report for Compensation of Fisheries Under the Amagase-dam Construction), Amagase-dam Construction Office, Kinki Construction Bureau, Kyoto (in Japanese).
28. Kawanabe, H. 1961. Sinka-gaku (Book review: 'Evolution study' by T. Shikama). Sizen 16(10): 56 (in Japanese).
29. Nagoshi, M., H. Kawanabe, N. Mizuno, D. Miyadi, S. Mori, Y. Sugiyama, I. Maki & Y.T. Saito. 1962. Kawa no sakana no seikatu (Studies on the life of river fish). III. Oikawa no seikatusi wo tyuusin ni site (The life history of pale chub, 'oikawa' *Zacco platypus*). Contribution to Physiology and Ecology, Kyoto University 82: 1–19 (in Japanese).
30. Kawanabe, H. 1962. Tora yon-dai (Four tales of the tiger). Heian 1962(1): 6–7 (in Japanese).
31. Miyadi, D. & H. Kawanabe. 1962. Hazime ni (Introduction). pp. 1–12. *In:* D. Miyadi (ed.) Naka-umi kantaku tansui-ka zigyou ni tomonau gyozoku seitai tyousa houkoku (Ecological Report of Fishes and Other Organisms in Relation to Freshening and Land Construction in Naka-umi and its Adjacent Area), Shimane Prefectural Government, Matsue (in Japanese).
32. Kawanabe, H. & Y. Tezuka. 1962. Gyogyou-nissi kirokutyousa to kikitori-tyousa ni yoru gyozoku seitai no gaiyou (General feature of fish ecology by fisheries statistics and interviews with fishermen). pp. 106–112. *In:* D. Miyadi (ed.) Ibidem (ibidem), Shimane Prefectural Government, Matsue (in Japanese).
33. Kawanabe, H. 1962. Gyorui no genzonryou ni tuite (Standing stock of fishes). pp. 113–122. *In:* D. Miyadi (ed.) Ibidem (ibidem), Shimane Prefectural Government, Matsue (in Japanese).
34. Kawanabe, H. 1962. Gyorui to esa-seibutu tono kankei: hazime ni (Introduction for relationship of fishes to their food items). pp. 123. *In:* D. Miyadi (ed.) Ibidem (ibidem), Shimane Prefectural Government, Matsue (in Japanese).
35. Sakamoto, M., E. Harada & H. Kawanabe. 1962. Seisan kankei (Production relationships). pp. 135–147. *In:* D. Miyadi (ed.) Ibidem (ibidem), Shimane Prefectural Government, Matsue (in Japanese).
36. Kawanabe, H. 1962. Bora (Mullet, *Mugil cephalus*). pp. 160–162. *In:* D. Miyadi (ed.) Ibidem (ibidem), Shimane Prefectural Government, Matsue (in Japanese).
37. Kawanabe, H. & H. Asano. 1962. Sonota no gyosyu (Other fishes). pp. 170–179. *In:* D. Miyadi (ed.) Ibidem (ibidem), Shimane Prefectural Government, Matsue (in Japanese).
38. Kawanabe, H. & D. Miyadi. 1962. Kantaku no eikyo: hazime ni (Introduction for effects of land construction). pp.

203. *In:* D. Miyadi (ed.) Ibidem (ibidem), Shimane Prefectural Government, Matsue (in Japanese).
39. Mori, S. & H. Kawanabe. 1962. Suisitu henka to seisan (Change of water quality and production). pp. 203–204. *In:* D. Miyadi (ed.) Ibidem (ibidem), Shimane Prefectural Government, Matsue (in Japanese).
40. Kawanabe, H. 1962. Seisan kankei no henka (Change of production relationship). pp. 209. *In:* D. Miyadi (ed.) Ibidem (ibidem), Shimane Prefectural Government, Matsue (in Japanese).
41. Kawanabe, H. 1962. Gyorui no henka: syubetu (Change of each species of fish). pp. 210- 215. *In:* D. Miyadi (ed.) Ibidem (ibidem), Shimane Prefectural Government, Matsue (in Japanese).
42. Kawanabe, H. 1962. Gyorui to koukakurui no suiiki-betu higai soukatu (Summary of damage of fishes and crustacians by certain bodies of water). pp. 215–217. *In:* D. Miyadi (ed.) Ibidem (ibidem), Shimane Prefectural Government, Matsue (in Japanese).
43. Kawanabe, H. 1962. Gyokakuryou no henka (Change in fishery yields). pp. 217. *In:* D. Miyadi (ed.) Ibidem (ibidem), Shimane Prefectural Government, Matsue (in Japanese).
44. Kawanabe, H. 1962. Seisanryoku zoukyou e no suusyu no kokoromi (Several trials for increasing biological production). pp. 217–218. *In:* D. Miyadi (ed.) Ibidem (ibidem), Shimane Prefectural Government, Matsue (in Japanese).
45. Azuma, M., E. Harada, T. Ishiguro, K. Izawa, H. Kawanabe, I. Maki, M. Nagoshi, T. Sunaga, S. Takamatsu, R. Takeda, T. Tomoda, S. Yanagishima, K. Mizuto, O. Murakami, O. Nakamura, J. Nishigaki, Y. Sato & II. Tsuji. 1962. Doubutu seitaigaku (Book review: 'Animal ecology' by D. Miyadi, M. Kato, S. Mori, M. Morisita, K. Shibuya & Y. Kitazawa). Japan. J. Ecol. 12: 157–160 (in Japanese).
46. Kawanabe, H. 1962. Kawa no seibutu-gunsyuu (Biological community in rivers). Kagaku no Zikken 13: 441–446 (in Japanese).
47. Kawanabe, H. 1962. Ayu no seikatu (Life of the ayu). Kagaku-no-zikken 13: 471–476 (in Japanese) (also in 240 and 319).
48. Mori, S., H. Kawanabe & N. Mizuno. 1962. Damu-ko oyobi sore ni turanaru kasen ni okeru gyorui no seitaigakuteki kenkyuu (Ecological study of fishes in a dammed lake and its related rivers) 1. pp. 5–7. *In:* M. Tsuda (ed.) Damu-ko no rikusui-seibutugakuteki kenkyuu (Biolimnological Study of Dammed Lakes) (in Japanese).
49. Mizuno, N., S. Mori & H. Kawanabe. 1963. Idem (idem) 2. pp. 8–10. *In:* M. Tsuda (ed.) Ibidem (ibidem) (in Japanese).
50. Kawanabe, H. 1963. Syokurinti ni hi wo ireru: yamayakiya tati no ben (Fire ecology). The Asahi 1.9.1963 (in Japanese).
51. Kawanabe, H. 1963. Kawa no seibutu seisanryou to seisankankei (On the relation between the rate of production and related phenomena in rivers). Japan. J. Limnol. 24: 1–15 (in Japanese).
52. Miyadi, D., H. Kawanabe & N. Mizuno. 1963. Gensyoku Nihon tansui-gyorui zukan (Color illustrations of the freshwater fishes of Japan). Hoikusha, Osaka. 275 pp. (in Japanese).
53. Kawanabe, H. 1964. Ayu no nawabari, tuki, Kyoto Kitayama no tansuigyo (Territory of ayu and freshwater fishes in the northern hills of Kyoto). pp. 106–108. *In:* Kyoto Nature Study Group (ed.) Kyoto no sizen (Nature in Kyoto), Rokugatsusha, Osaka (in Japanese).
54. Kawanabe, H. 1964. Biwako to Yodo-gawa no tansuigyo (Fishes in Lake Biwa and the Yodo River). pp. 156–157. *In:* Kyoto Nature Study Group (ed.) Ibidem (ibidem), Rokugatsusha, Osaka (in Japanese).
55. Kawanabe, H. & E. Harada. 1964. Seibutugakuteki seisan ni kansuru syo-gainen no kentou (The use and abuse of the production concepts and terms). Physiol. Ecol. 12: 300–316 (in Japanese).
56. Mizuno, N., S. Mori & H. Kawanabe. 1964. Damu-ko oyobi sore ni turanaru kasen ni okeru gyorui no seitaigakuteki kenkyuu (Ecological study of fishes in a dammed lake and its related rivers). Tansui-seibutu (Freshwater Organisms) 9: 4–8 (in Japanese).
57. Miyadi, D., H. Kawanabe & E. Harada. 1964. Ecology and biological production of Lake Naka-umi and adjacent regions. 1. General introduction. Spec. Publ., Seto Mar. Biol. Lab. Ser. 2 1: 1–7.
58. Kawanabe, H. 1964. Biwako no genzai no sugata: sono seitaigakuteki sokumen (Present features of Lake Biwa, its ecological aspects). Kagaku no Zikken 15: 684–691 (in Japanese) (also in 240).
59. Kawanabe, H. 1964. Doubutu seitaigaku (Ecology of animals) 1. Nihon Dokusho Shinbun (precise information not available) (in Japanese).
60. Kawanabe, H. 1964. Doubutu seitaigaku (Ecology of animals) 2. Nihon Dokusho Shinbun (precise information not available) (in Japanese).
61. Kawanabe, H. 1964. Doubutu seitaigaku (Ecology of animals) 3. Nihon Dokusho Shinbun (precise information not available) (in Japanese).
62. Kawanabe, H. 1964. Tabenai turi (Angling without feeding). The Mainichi (evening) 28.5.1964 (in Japanese).
63. Kawanabe, H. 1964. 1970 nendai no doubutu-seitaigaku (Animal ecology in the 1970s). pp. 34–41. *In:* Division of Biological Sciences, Japan Science Council (ed.) Seibutukagaku no hatten no tame no genzyou to syourai ni tuite kataru sinpoziumu (Symposium Report on the Present and Future of Biological Sciences) (in Japanese).
64. Kawanabe H. 1965. Seitaigaku no tatiba (On the standpoint of ecology). Nature (Tokyo) 35: 232–238 (in Japanese).
65. Kawanabe, H. 1965. Kasen (Rivers). Kagaku no Zikken 16: 40–43 (in Japanese).
66. Isemura, J., G. Ota, T. Kato, A. Kawakita, H. Kawanabe, H. Kikkawa, K. Kihara, Y. Kondo, H. Sato, K. Dan, T. Tokizane & I. Watanabe (ed.) 1966. Iwanami koza, gendai no seibutugaku (The Iwanami series on modern biology), 9 volumes. Iwanami Shoten, Tokyo (in Japanese).

67. Kawanabe, H. 1966. Seitaigaku no rekisi to tenbou (History and perspectives in ecology). pp. 1–18. *In:* J. Isemura, G. Ota, T. Kato, A. Kawakita, H. Kawanabe, H. Kikkawa, K. Kihara, Y. Kondo, H. Sato, K. Dan, T. Tokizane & I. Watanabe. (ed.) Iwanami koza, gendai no seibutugaku (The Iwanami Series on Modern Biology) vol. 9. Iwanami Shoten, Tokyo (in Japanese) (also in 382).
68. Kawanabe, H. 1966. Seibutu-gunsyuu ni okeru syu no seisan (Species production in biological communities). pp. 123–152. *In:* J. Isemura, G. Ota, T. Kato, A. Kawakita, H. Kawanabe, H. Kikkawa, K. Kihara, Y. Kondo, H. Sato, K. Dan, T. Tokizane & I. Watanabe (ed.) Ibidem (ibidem) vol. 9. Iwanami Shoten, Tokyo (in Japanese).
69. Kawanabe, H. 1966. Gunsyuu ni okeru seibutu-seisan (Biological production in communities). pp. 5–12. *In:* T. Hanaoka & S. Mori (ed.) JPM-JPF Kyoudou Sinpoziumu 'Gunsyuu wo ikani toraeruka' (Organizing Committee of the Joint Symposium of Production in Freshwater Ecosystems and Production of Marine Ecosystems in Japan International Biological Programme at the Department of Fisheries), Tohoku University, Sendai (in Japanese).
70. Kawanabe, H. 1966. Ayu no seitai kenkyuu syousi (A short history of ecological study on the ayu). Japan. J. Michurin Biol. 2: 235–255 (in Japanese).
71. Kawanabe, H. 1967. Defecation ratio of a young crucian carp, *Carassius carassius*, in relation to its ingestion rate. Mem. Coll. Sci., Univ. Kyoto, Ser. B 33: 157–162.
72. Mori, S. & H. Kawanabe. 1967. Biwako no sizen to sono hogo (Lake Biwa and its natural conservation). pp. 15–16. *In:* M. Tsuda (ed.) Rikusui-seibutu narabini rikusui-seibutu-gunsyuu no hogo no houhou ni kansuru kenkyuu (Study on Conservation Methods for Freshwater Organisms and Freshwater Communities), vol. 1, Department of Biology, Nara Women's University, Nara (in Japanese).
73. De La Cruz, A.A. & H. Kawanabe. 1967. The population and food habits of fish in a small estuarine pond. Nat. Appl. Sci. Bull. 20: 473–477.
74. Kawanabe, H. 1967. Huna-rui yougyo no bussitu-taisya sokutei-hou (Method of measurement of metabolism in crucian carp juveniles). pp. 135–140. *In:* Discussion Group on Ecological Practice (ed.) Seitaigaku zissyuusyo (Ecological Practice), Asakura-shoten, Tokyo (in Japanese).
75. Kawanabe, H., Y.T. Saito, T. Sunaga, I. Maki & M. Azuma. 1968. Ecology and biological production of Lake Naka-umi and adjacent regions. 4. Distribution of fishes and their foods. Spec. Publ., Seto Mar. Biol. Lab. Ser. 22: 44–73.
76. Kawanabe, H. 1968. Kisui-iki ni okeru seibutu-seisan: Naka-umi suikei ni okeru seibutu-kankei wo tyuusin ni site (Biological productivity and production relationships in estuarine waters with reference to the relation between ecology and paleoecology). Geol. Bull. Japan 3: 15–22 (in Japanese).
77. Kawanabe, H. 1968. Aga-no-gawa ni okeru 'Minamata-byo' ni tuite ('Minamata disease' in the Aga River). pp. 95–99. *In:* M. Tsuda (ed.) Rikusui-seibutu narabini rikusui-seibutu-gunsyuu no hogo no houhou ni kansuru kenkyuu (Study on Conservation Methods for Freshwater Organisms and Freshwater Communities), vol. 2, Department of Biology, Nara Women's University, Nara (in Japanese).
78. Kawanabe, H. 1968. Kogai-mondai ni iti seitaigakusya tosite omou (An ecologist's opinion of pollution). Science (Tokyo) 38: 629 (in Japanese) (also in 417).
79. Kawanabe, H. 1968. Nihon Doubutu Gakkai dai 38 kai taikai kizi (On the 38th annual meeting of the Japanese Society of Zoology). Zool. Mag. 76: 351–352 (in Japanese).
80. Kawanabe, H. 1969. Seisan ni kansuru yougo to kigou (Terms and symbols of biological production). pp. 1–7. *In:* S. Mori, M. Sakamoto, M.Tsuda, K. Onodera, H. Kadota, H. Kawanabe & M. Morikawa (ed.) Rikusui-seibutu seisan kenkyuu-hou (Research Methods of Biological Production in Freshwaters), Kodansha, Tokyo (in Japanese).
81. Kawanabe, H. & K. Hirai. 1969. Sisui-iki no kogata hutyaku doubutu no sokutei-hou (Measurement for small benthic animals in lentic habitats). pp. 115–119. *In:* S. Mori, M. Sakamoto, M.Tsuda, K. Onodera, H. Kadota, H. Kawanabe & M. Morikawa (ed.) Ibidem (ibidem), Kodansha, Tokyo (in Japanese).
82. Morisita, M. & H. Kawanabe. 1969. Seisan ni kansuru syoryou to syoryou-kan no kankei (Items in biological production and their relationship). pp. 345–351. *In:* S. Mori, M. Sakamoto, M.Tsuda, K. Onodera, H. Kadota, H. Kawanabe & M. Morikawa (ed.) Ibidem (ibidem), Kodansha, Tokyo (in Japanese).
83. Kawanabe, H. & K. Onodera. 1969. Bussitu-zyunkan to energi-ruten (Material cycles and energy flow). pp. 352–357. *In:* S. Mori, M. Sakamoto, M.Tsuda, K. Onodera, H. Kadota, H. Kawanabe & M. Morikawa (ed.) Ibidem (ibidem), Kodansha, Tokyo (in Japanese).
84. Kawanabe, H. 1969. Gunsyuu zentai no bussitu to enerugi no syuusi (Balance of material and energy in the whole community). pp. 367–374. *In:* S. Mori, M. Sakamoto, M.Tsuda, K. Onodera, H. Kadota, H. Kawanabe & M. Morikawa (ed.) Ibidem (ibidem), Kodansha, Tokyo (in Japanese).
85. Kawanabe, H. 1969. Seisan-kankei tyousa to sono mondaiten: gairon (Research on production relationships and its problems: general features). pp. 374–375. *In:* S. Mori, M. Sakamoto, M.Tsuda, K. Onodera, H. Kadota, H. Kawanabe & M. Morikawa (ed.) Ibidem (ibidem), Kodansha, Tokyo (in Japanese).
86. Kawanabe, H. & Y. Nagata. 1969. Detoritasu to sono zyuuyousei: gairon (Detritus and its importance: general aspects). pp. 388–389. *In:* S. Mori, M. Sakamoto, M.Tsuda, K. Onodera, H. Kadota, H. Kawanabe & M. Morikawa (ed.) Ibidem (ibidem), Kodansha, Tokyo (in Japanese).
87. Kawanabe, H. 1969. 'Gunsyuu no seisansei' kara 'gunsyuu ni okeru seibutu-seisan' e (From 'community productivity' to 'biological production in communities'). pp. 440–447. *In:* S. Mori, M. Sakamoto, M.Tsuda, K. Onodera, H. Kadota, H. Kawanabe & M. Morikawa (ed.) Ibidem (ibidem), Kodansha, Tokyo (in Japanese).
88. Kawanabe, H. 1969. The significance of social structure in

production of the 'ayu', *Plecoglossus altivelis*. pp. 243–251. *In:* T.G. Northcote (ed.) Symposium on Salmon and Trout in Streams, H.R. MacMillan Lectures in Fisheries, University of British Columbia, Vancouver.

89. Kawanabe, H. 1969. Kawa to mizuumi no uo-tati (Fishes in rivers and lakes). Chuoukouronsha, Tokyo. 196 pp. (in Japanese).
90. Kawanabe, H. 1969. Tyosya tono 1 zikan 'Kawa to mizuumi no uo-tati' (One hour with the author of 'Fishes in rivers and lakes'). The Asahi 15.4.1969 (in Japanese).
91. Kawanabe, H. 1969. Tyosyo to tyosya: 'Kawa to mizuumi no uo-tati' (A book and its author: 'Fishes in rivers and lakes'). The Chunichi-Tokyo 18.4.1969 (in Japanese).
92. Kawanabe, H. 1969. Tyosya to no taiwa: 'Kawa to mizuumi no uo-tati' (A conversation with the author of 'Fishes in rivers and lakes'). The Kyodo 20.4.1969 (in Japanese).
93. Kawanabe, H. 1969. Ayu no nawabari: sono sikumi to esa to no kankei (The territory of the ayu: its design and relation to food). Fishing 1969(5): 40–41 (in Japanese).
94. Kawanabe, H., S. Kitamura, M. Kondo, M. Tsuda, T. Nishihara & I. Morishita. 1969. Kawa to ningensei syakai: tosikasen no kako, genzai, mirai (Rivers and human society: past, present, and future of rivers in urban areas). Kougai to Taisaku (J. Pollution Control) 5: 495–506 (in Japanese).
95. Kawanabe, H. 1969. Oti-ayu (The downstream ayu). Izin Yakuzin (Medical and Pharmaceutical Persons) 18(11): 22–23 (in Japanese) (also in 319).
96. Kawanabe, H. 1970. Bunpu nangen ni ayu no seitai wo miru: Taiwan to Okinawa 23-niti kan no tabi kara (Observation of ayu at the southern end of its distribution: 23 days' travel in Taiwan and Okinawa). Fishing 6: 100–103 (in Japanese).
97. Kawanabe, H. 1970. Ayu no syakai-kouzou to seisan II: 15 nenkan no henka wo mite (Social behaviour and production of ayu in the U-kawa River between 1955 and 1969, with reference to the stability of its territoriality). Japan. J. Ecol. 20: 144–151 (in Japanese).
98. Kawanabe, H. (ed.) 1970. Gogawa suikei no seibutu seitai (Ecology of the Gogawa River System) 1. Shimane Prefectural Government, Matsue. 121 pp. (in Japanese).
99. Kawanabe, H. & K. Hori. 1970. Hazime ni (Introduction). pp. 2–4. *In:* H. Kawanabe (ed.) Gogawa suikei no seibutu seitai (The Ecology of the Gogawa River System) 1, Shimane Prefectural Government, Matsue (in Japanese).
100. Kawanabe, H. & N. Mizuno. 1970. Gogawa honryuu kasenkeitai no genkyou to kako no zyoutai no hukugen no kokoromi (The present situation of river morphology in the main course of the Gogawa river, and its recovery the past conditions). pp. 5–20. *In:* H. Kawanabe (ed.) Ibidem (ibidem), 1, Shimane Prefectural Government, Matsue (in Japanese).
101. Kawanabe, H. & N. Mizuno. 1970. Ayu no kizyun-mitudo to seisoku-ryou (Standard and actual density of ayu). pp. 111–114. *In:* H. Kawanabe (ed.) Ibidem (ibidem), 1, Shimane Prefectural Government, Matsue (in Japanese).
102. Azuma, M. & H. Kawanabe. 1970. Huka-sigyo ryuukaryou no nissyuu-henka to kakou hukin ni okeru sigyo no bunpu (Diel change of downstream migration of fry of ayu and its distribution around the river mouth). pp. 115–121. *In:* H. Kawanabe (ed.) Ibidem (ibidem), 1, Shimane Prefectural Government, Matsue (in Japanese).
103. Kawanabe, H. 1970. Sugata naki minami-tomiyo (An extinct freshwater fish, *Pungitius sinensis kaibarae*). The Asahi 5.4.1970 (in Japanese).
104. Kawanabe, H. 1970. Kyakkou abiru seitaigaku no kage kara: aru kenkyusya no nikki kara (Spotlighting the shadow of ecology: from a researcher's diary). Asahi Journal 12 (30): 31–35 (in Japanese) (also in 240).
105. Kawanabe, H. 1970. Ningen-teki sizen wo hukkatu saseuruka?: zokuryuu 'kogai-ron' e no iti hihan (Can we recover 'humanistic nature'?: a critical essay on vulgar opinion of 'environmental pollution'). Ushio (Special Issue) 1970(Autumn): 74–80 (in Japanese) (also in 240).
106. Kawanabe, H. 1970. Kawa no seibutu no seisansei (Productivity of stream organisms). Kagaku to Seibutu (Chemistry and Biology) 8: 103–108 (in Japanese) (also in 240).
107. Kawanabe, H. 1970. Kono sizen wa dare no mono ka? (Who owns nature?). Tape Lecture Co. Ltd., Kyoto (in Japanese).
108. Kawanabe, H. & M. Yoshida. 1970. Kankyou-kanri zidai no ningen isiki (Human consciousness in the age of environmental control). Tape Lecture Co. Ltd., Kyoto (in Japanese).
109. Kawanabe, H. 1970. Kasen-suiiki no seitaigakuteki kousatu (A heterodoxical essay on the ecology of the stream community), I. Hurui tyosaku kara (From an old paper of minc). Kougai to Taisaku (J. Pollution Control) 6: 589–594 (in Japanese).
110. Kawanabe, H. 1970. Idem (idem), II. Kawa wa nagareru (River runs). Ibidem (ibidem) 6: 723–726 (in Japanese).
111. Kawanabe, H. 1970. Idem (idem), III. Ayu no seikatu no tayousei (Diversity of life of the ayu). Ibidem (ibidem) 6: 877–882 (in Japanese).
112. Kawanabe, H. 1971. Idem (idem), IV. Tansuigyo no tiriteki bunpu (Geographical distribution of freshwater fishes). Ibidem (ibidem) 7: 51–55 (in Japanese).
113. Kawanabe, H. 1971. Idem (idem), V. Aganogawa no 'Haika-rakuzitsu' (On Minamata disease of the Agano River). 1. Ibidem (ibidem) 7: 333–339 (in Japanese).
114. Kawanabe, H. 1971. Idem (idem), VI. Aganogawa no 'Haika-rakuzitu' (On Minamata disease of the Agano River) 2. Ibidem (ibidem) 7: 489–496 (in Japanese).
115. Kawanabe, H. 1971. Idem (idem), VII. Damu no kensetu to ayu (Dam construction and the ayu) 1. Ibidem (ibidem) 7: 613–623 (in Japanese).
116. Kawanabe, H. 1971. Idem (idem), VIII. Damu no kensetu to ayu (Dam construction and the ayu) 2. Ibidem (ibidem) 7: 843–849 (in Japanese).
117. Kawanabe, H., M. Osawa & T. Abe. (translators) 1971. Sinryaku no seitaigaku (English into Japanese translation of 'The ecology of invasions by animals and plants' by C.S. Elton). Shisakusha, Tokyo. 223 pp. (in Japanese).

118. Kawanabe, H. 1971. Yakusya no yaya nagame no atogaki (Rather long postscripts by the senior translator). pp. 215–223. *In:* Sinryaku no seitaigaku ('The ecology of invasions by animals and plants' by C.S. Elton). Shisakusha, Tokyo (in Japanese) (also in 417).
119. Kawanabe, H., K. Tanaka, I. Dan, H. Furukawa, Y. Yamashina & H. Obara. 1971. Ningen mo ikiru tameni (Human life as a part of the ecosystem). Asahi Doubutu Hyakka (Asahi Encyclopedia of Animals) 2: 2–4 (in Japanese).
120. Kawanabe, H. & H. Obara. 1971. Nani ga sizen no gaku ka?: 'seitaigaku'-teki sizen-zou wo tou (What is the study of nature?: questions on natural status in 'ecology'). Space Design 6: 39–52 (in Japanese) (also in 417).
121. Kawanabe, H. & J. Ui. 1971. Kougai-genron: kougai to seitaigaku (Principles of pollution: pollution and ecology). Group of Instructors, Fac. Engineering, Univ. Tokyo, Tokyo. 42 pp. (in Japanese).
122. Kawanabe, H. (ed.) 1970. Gogawa suikei no seibutu seitai (The ecology of the Gogawa River System) 2. Shimane Prefectural Government, Matsue. 57 pp. (in Japanese).
123. Kawanabe, H. & K. Hori. 1971. Hazime ni (Introduction). pp. 2–5. *In:* H. Kawanabe (ed.) Gogawa suikei no seibutu seitai (The ecology of the Gogawa River System) 2, Shimane Prefectural Government, Matsue (in Japanese).
124. Maruyama, T., H. Kawanabe & M. Goto. 1971. Gogawa seibu ni okeru gogi no bunpu (Distribution of a special charr, 'gogi', in the western part of the Gogawa River). pp. 33–37. *In:* H. Kawanabe (ed.) Ibidem (ibidem) 2, Shimane Prefectural Government, Matsue (in Japanese).
125. Kawanabe, H. & M. Azuma. 1971. Ayu no kizyun-mitudo to seisoku-ryou (Standard and actual density of the ayu) 2. pp. 39–42. *In:* H. Kawanabe (ed.) Ibidem (ibidem) 2, Shimane Prefectural Government, Matsue (in Japanese).
126. Azuma, M. & H. Kawanabe. 1971. Gogawa ni okeru hukasigyo no ryuuka-ryou to kakou hukin ni okeru sigyo no bunpu (Amount of downstream migration of juveniles of ayu and its distribution around the river mouth). pp. 43–50. *In:* H. Kawanabe (ed.) Ibidem (ibidem) 2, Shimane Prefectural Government, Matsue (in Japanese).
127. Kawanabe, H. 1971. Gogawa ni okeru iwayuru 'sougoukaihatu' ni tuite (So-called 'synthesized development' in the Gogawa River System). pp. 51–57. *In:* H. Kawanabe (ed.) Ibidem (ibidem) 2, Shimane Prefectural Government, Matsue (in Japanese).
128. Kawanabe, H. 1971. Syohyou: 'Seibutu-syakai no ronri' (Book review: 'Logic of biological societies' by K. Imanishi). pp. 14–18 *In:* Shisakusha syohyo-syuu ('Book reviews' published by Shisakusha) (in Japanese) (also in 240).
129. Kawanabe, H. 1972. Syohyou: 'Kagaku to zinrui no seizon' (Book review: 'Science and survival' by Barry Commoner). Sizen 1972(2): 106 (in Japanese).
130. Kawanabe, H. 1972. Syohyou: 'Saru no kuni no rekisi' (Book review: 'History of a monkey nation' by H. Mizuhara). Monkey (precise information not available) (in Japanese).
131. Kawanabe, H. 1972. Oikawa (A fish, *Zacco platypus*). The Chunichi Sports 4.2.1972 (in Japanese).
132. Kawanabe, H. 1972. Oikawa, zoku (A fish, *Zacco platypus*, the continuation). The Chunichi Sports 11.2.1972 (in Japanese).
133. Kawanabe, H. 1972. Biwako no ko-ayu (Small-sized ayu in Lake Biwa). The Chunichi Sports 18.2.1972 (in Japanese).
134. Kawanabe, H. 1972. Syourai-keikaku-iinkai 'Seitaigaku-kenkyu-syo' wo kangaeru guruupu tyuukan houkoku (Preliminary report of the group for a 'National Institute of Ecology' in the Future Project Committee of the Ecological Society of Japan). Japan. J. Ecol. 21: 83–88 (in Japanese).
135. Kawanabe, H. 1972. Sizen to hito: ayu no nawabari (Nature and human beings: territory of the ayu). Denshin Denwa 24(5): 22–23 (in Japanese).
136. Kawanabe, H. 1972. Osakana to tori to musi to: seitai hikaku no omosirosa (Fishes, birds and insects: interests of comparative ecology). The Yomiuri 8.6.1972 (in Japanese) (also in 417).
137. Kawanabe, H. 1972. Ayu no syakai-kouzou no sinkasiteki igi ni tuite: hu, bunpu nangen ni okeru syakai-kouzou (An evolutionary aspect of the territoriality of the ayu, *Plecoglossus altivelis*, with the social structure at the southern end of its distribution). Japan. J. Ecol. 22: 141–149 (in Japanese).
138. Mori, S. & H. Kawanabe. 1972. U-kawa (The U-kawa River). pp. 44–48. *In:* Nihon-seitaigakkai rikusui sizen-hogo tiiki an (Plan of Lakes and Rivers for Nature Conservation), Ecological Society of Japan, Sendai (in Japanese).
139. Kawanabe, H., T. Maruyama & K. Tanida. 1972. Kasen seibutu no tyousa (Research report of river organisms). Hakusan Sigen-tyousa Houkoku (Research Report of Mt. Hakusan Area) 5: 24–49, 3 special tables. Hakusan Sigen Tyousa-dan, Kanazawa (in Japanese).
140. Kawanabe, H. 1972. Kisetu no sisya ayu (The ayu as a season's messenger). The Kyoto 1.6.1972 (in Japanese).
141. Kawanabe, H. 1972. Sakana no siniyuku hi (The day fishes are going to die). Tenbo 1972(8): 38–55 (in Japanese) (also in 240 and in 'Ecology' published by Kawade Shobo, Tokyo, 1980).
142. Kawanabe, H. 1972. Kurasi no nakade: 'Nihon-rettou kaizou-ron' wo hihan suru (In my life: a critic for 'Re-organization of the Japanese Archipelago'). The Kyoto 9.10.1972 (in Japanese).
143. Kawanabe, H. 1972. Nawabari wa hyouki no nokosimono? (Territory of the ayu as an endemism of the glacial age?). The Asahi 21.10.1972 (in Japanese).
144. Kawanabe, H. 1972. Kougai to seitaigaku (Pollution and ecology) II. pp. 269–326. *In:* J. Ui (ed.) Gendai-kagaku to kougai (Modern Science and Pollution), Keiso Shobo, Tokyo (in Japanese).
145. Kawanabe, H. 1973. Ayu no 'nawabari' towa nanika?: gunsyuu no ronri ni kansuru hitotu no kokoromi (What is the 'territory' of the ayu?: a logical opinion on biological com-

munities). Science (Tokyo) 43: 74–83 (in Japanese) (also in 382).

146. Kawanabe, H. 1973. 'Nawabari' to doubutu-syakaigaku ('Territory' and animal sociology). Miyadi Denzaburo Doubutu-ki, Geppou ('History of Animals' by D. Miyadi, vol. 2, attached pamphlet) 4–7. Chikuma Shobo, Tokyo (in Japanese) (also in 240).
147. Kawanabe, H. & Y. Takahashi. 1973. 'Rettou-kaizou' no mizu-mondai sindan (A diagnosis of water problems in 'Re-organization in the Japanese Archipelago'). Kougai to Taisaku (J. Pollution Control) 9: 29–40 (in Japanese).
148. Kawanabe, H. 1973. Syokumotu rensa (Food chain). pp. 2292–2294. *In:* Alpha Great World Encyclopedia, Alpha Enterprise, Tokyo (in Japanese).
149. Kawanabe, H. 1973. Nawabari (Territory). pp. 4314–4315. *In:* Ibidem, Alpha Enterprise, Tokyo (in Japanese).
150. Kawanabe, H. 1973. Kyouzon to kyousou (Coexistence and competition). University Press 2(4): 18–20 (in Japanese) (also in 240).
151. Kawanabe, H. 1973. Minamata-byou hanketu no yokuyo-kuzitu ni omou (Thoughts on the second day following the judicial decision on the case of Minamata disease). The Kyoto University News 22.3.1973 (in Japanese) (also in 417).
152. Kawanabe, H. 1973. Kasen no seitaigaku suudai (Tales of river ecology). Hanadokei (Kobe City) 8: 15–16 (in Japanese).
153. Kawanabe, H. 1973. Syohyou: 'Seitaigaku e no syoutai' (Critical review of current books on ecology: 'Introduction to ecology' vol. 1–6). J. Michurin Biol. 9: 53–56 (in Japanese).
154. Kawanabe, H. 1973. Eruton no kotaigun to seisoku-basyo no sougoteki-sanzai wo megutte (An introduction to the population and habitat interspersion by C. S. Elton). J. Michurin Biol. 9: 100–109 (in Japanese) (also in 417).
155. Kawanabe, H. 1973. Furuta Mankichi 'Ayu no hanasi' wo yonde (Thoughts after reading 'Tales of the ayu' by M. Furuta). The Sankei (evening) 20.8.1973 (in Japanese).
156. Kawanabe, H. 1973. Okinawa no kawa (Rivers in the Ryukyus) 1. The Yomiuri 1.10.1973 (in Japanese).
157. Kawanabe, H. 1973. Idem (idem) 2. The Yomiuri 2.10.1973 (in Japanese).
158. Kawanabe, H. 1973. 'Tetugaku' to siteno seitaigaku (Ecology as 'philosophy'). Riso 482: 32–41 (in Japanese) (also in 382).
159. Miyadi, D., H. Kawanabe & T. Iwai. 1973. Gyorui (Pisces). pp. 637–656. *In:* M. Ueno (ed.) Kawamura Nihon tansui seibutugaku (The Late Tamiji Kawamura's Freshwater Biology of Japan). Hokuryukan Publ., Tokyo (in Japanese).
160. Kawanabe, H. 1973. Okinawa-tou ni okeru yosinobori no tokuina sessyoku-koudou ni tuite (Food and feeding habits of a freshwater goby, *Rhinogobius brunneus*, in Okinawa Island, the Ryukyus, with some discussion on the interactive food segregation). Biol. Mag. Okinawa 10: 1–6 (in Japanese) (also in 382).
161. Kawanabe, H. 1973. Sin doubutu-si (New animal history) 33. Ayu. The Asahi 12.11.1973 (in Japanese).
162. Kawanabe, H. 1973. Idem (idem) 37. Gori (Gobies). Ibidem 19.11.1973 (in Japanese).
163. Kawanabe, H. 1973. Idem (idem) 43. Itasenpara (A bitterling, *Acheilognathus longipinnis*). Ibidem 29.11.1973 (in Japanese).
164. Kawanabe, H. 1973. Idem (idem) 45. Ta-unagi (Mud eel, *Monopterus albus*). Ibidem 3.12.1973 (in Japanese).
165. Kawanabe, H. 1973. Idem (idem) 48. Iwana to yamame (Charr and masu-salmon). Ibidem 11.12.1973 (in Japanese).
166. Obara, H., A. Sugimoto, S. Kouhara, M. Ohmori, M. Numata, M. Horikoshi, H. Kawanabe, J. Nemoto, I. Mototani, I. Hiura & T. Han'ya (ed.) 1973. Nihon rettou (Japanese Archipelago). Time Life Books, Tokyo. 192 pp. (in Japanese).
167. Obara, H., A. Sugimoto, S. Kouhara, M. Ohmori, M. Numata, M. Horikoshi, H. Kawanabe, J. Nemoto, I. Mototani, I. Hiura & T. Han'ya (ed.) 1973. Nihon no seibutu (Organisms in Japan). Time Life Books, Tokyo. 192 pp. (in Japanese).
168. Obara, H., A. Sugimoto, S. Kouhara, M. Ohmori, M. Numata, M. Horikoshi, H. Kawanabe, J. Nemoto, I. Mototani, I. Hiura & T. Han'ya. (ed.) 1973. Nihon-zin to sizen (The Japanese and nature). Time Life Books, Tokyo. 192 pp. (in Japanese).
169. Kawanabe, H. 1973. Seitai no mihon-iti, Biwako (Lake Biwa, an exposition place for ecology). pp. 45–55. *In:* H. Obara, A. Sugimoto, S. Kouhara, M. Ohmori, M. Numata, M. Horikoshi, H. Kawanabe, J. Nemoto, I. Mototani, I. Hiura & T. Han'ya (ed.) Nihon no seibutu (Organisms in Japan), Time Life Books, Tokyo (in Japanese).
170. Kawanabe, H., M. Azuma, K. Ishigaki, N. Mizuno & M. Uramoto. 1973. Mizu to sora no ikimono (Organisms living in water and air). pp. 119–154. *In:* H. Obara, A. Sugimoto, S. Kouhara, M. Ohmori, M. Numata, M. Horikoshi, H. Kawanabe, J. Nemoto, I. Mototani, I. Hiura & T. Han'ya (ed.) Ibidem (ibidem), Time Life Books, Tokyo (in Japanese).
171. Kawanabe, H. 1974. Seikatu no hen'i wo torae-naosou (Let's study life's variation). Anima 1974(3): 2–3 (in Japanese) (also in 417).
172. Abe, H., T. Iwai, M. Uramoto, R. Okuno, H. Kawanabe, M. Konishi & T. Hidaka. 1974. Sin doubutu-si (New animal history). Asahi Shinbunsha, Tokyo. 207 pp. (in Japanese).
173. Kawanabe, H., T. Maruyama & K. Tanida. 1974. Takasegawa suikei to sono gyorui ni tuite (The Takase River System and its fishes). pp. 223–232. *In:* Takasegawa ryuuiki sizen sougou tyousa houkokusyo (Research Report of Nature in the Takase River Basin), Comprehensive Research Group of Nature in the Takase River Basin, Nagano (in Japanese).
174. Kawanabe, H. 1974. Kotaigun (no) seitaigaku ni tuite omou koto (An opinion on population ecology and the ecology of populations). pp. 164–170. *In:* T. Miwa, H. Oka, I. Watanabe & S. Itimura (ed.) 'Seibutu II' Koukou-kyousi-yo sidousyo (Guidance for High School Teachers on Textbook 'Biology II'), Sanseido, Tokyo (in Japanese) (also in 382).

175. Kawanabe, H., T. Maruyama, K. Tanida & T. Tomita. 1974. Kyoto-hu-ka no tansui-gyorui (Freshwater fishes in Kyoto Prefecture). pp. 79–96. *In:* M. Morisita (ed.) Kyoto-hu-ka no yasei doubutu (Wild Animals in Kyoto Prefecture), Bureau for Environmental Pollution, Kyoto Prefectural Government, Kyoto (in Japanese).

176. Kawanabe, H. 1974. Syokumotu rensa (Food chain). Ban'yu-hyakka daiziten (Shogakkan's Encyclopedia), Shogakkan, Tokyo 20: 338–339 (in Japanese).

177. Kawanabe, H. 1974. Sumiwake (Habitat segregation). Ibidem (ibidem), Shogakkan, Tokyo 20: 359 (in Japanese).

178. Kawanabe, H. 1974. Nawabari (Territory). Ibidem (ibidem), Shogakkan, Tokyo 20: 466 (in Japanese).

179. Nishijima, S., S. Shokita & H. Kawanabe. 1974. Ryukyu rettou no kasen no gaikyou (A general description of the streams of the Ryukyu Islands). pp. 113–143. *In:* S. Ikehara (ed.) Ryukyu rettou no sizen to sono hogo ni kansuru kisoteki kenkyuu (An Ecological Study of Nature Conservation in the Ryukyu Islands), vol. 1, Department of Biology, University of Ryukyus, Naha (in Japanese).

180. Kawanabe, H., T. Itai, S. Nishijima & S. Shokita. 1974. Syokumotu-rensa kara mita Ryukyu rettou kasen seibutu-gunsyuu no tokutyou: yohou (On the structure of the stream community in the Ryukyus with reference to food-chain relationships: preliminary report). pp. 87–193. *In:* S. Ikehara (ed.) Ibidem (ibidem), vol. 1, Department of Biology, University of Ryukyus, Naha (in Japanese).

181. Kawanabe, H., S. Nishijima & S. Shokita. 1974. Ryukyu rettou rikusui-seibutu bunken-mokuroku (Publications on the freshwater organisms of the Ryukyu Archipelago: a tentative list). pp. 255–275. *In:* S. Ikehara (ed.) Ibidem (ibidem), vol. 1, Department of Biology, University of Ryukyus, Naha (in Japanese).

182. Kawanabe, H. 1974. Hon-mono to nise-mono (Truths and falsehoods). The Mainichi (evening) 21.6.1974 (in Japanese).

183. Kawanabe, H. 1974. Syohyou: 'Biwako I' (Book review: 'Lake Biwa I' edited by Lake Laboratory, Shiga University). Kagaku-Asahi (precise information not available) (in Japanese).

184. Kawanabe, H. 1974. Ayu. The Asahi Family 12.7.1974 (in Japanese).

185. Kawanabe, H. 1974. Yodo-gawa no itasenpara (A bitterling, *Acheilognathus longipinnis*, in the Yodo River). The Yomiuri (evening) 12.10.1974 (in Japanese) (also in 417).

186. Kawanabe, H., T. Inoue & M. Matsui. 1974. Hatyuurui to ryouseirui (Reptiles and amphibians). pp. 114–137. *In:* Y. Tsutsui (ed.) Yodo-gawa no kasenziki ni okeru seitai tyousa houkokusyo (Ecological Report for the Yodo River), Yodo-gawa Construction Office, Ministry of Construction, Osaka (in Japanese).

187. Kawanabe, H. 1974. Ryukyu rettou no kasen seibutu (Stream organisms in the Ryukyus). pp. 69–71. *In:* M. Tsuda (ed.) Rikusui hueiyouka no kisoteki kenkyuu (Fundamental Research of Eutrophication in Freshwater), Department of Biology, Nara Womens University, Nara (in Japanese).

188. Kawanabe, H. (ed.) 'Gogawa suikei no seibutu seitai' ni kansuru tuika houkokusyo (Appendicised report for 'The ecology of the Gogawa River system'). Gogawa Fisheries Cooperative Association, Kawamoto, Shimane. 52 pp. (in Japanese).

189. Kawanabe, H. 1974. Gogawa 'sougou-kaihatu-keikaku' no gyogyou tou ni ataeru eikyou (Effects of 'comprehensive reorganization' of the Gogawa River on its fisheries and others). pp. 1–9. *In:* H. Kawanabe (ed.) 'Gogawa suikei no seibutu seitai' ni kansuru tuika houkokusyo (Appendicised Report for 'The ecology of the Gogawa River System'), Gogawa Fisheries Cooperative Association, Kawamoto, Shimane (in Japanese).

190. Kawanabe, H. 1975. Nagara-gawa no seibutu sindan (Biological diagnosis of the Nagara River). The Chunichi 25.2.1975 (in Japanese).

191. Kawanabe, H. 1975. Nagara-gawa wa yandemasuna (The Nagara River is sick!). Kawaboe 13: 1–4 (in Japanese).

192. Kawanabe, H. 1975. Doubutu kara mita Yodo-gawa no sizen-hogo (Nature conservation of the Yodo River from the viewpoint of animals). Sizensi Kenkyuu (Occasional papers from the Osaka Museum of Natural History) 1: 74–75 (in Japanese).

193. Kawanabe, H. 1975. Ota-gawa Kake-cho hukin ni okeru kasen-seibutu (River organisms in Kake Town along the Ota River). pp. 1–8. *In:* Y. Monma (ed.) Sanson-sinkou konsarutanto ikensyo, Hiroshima-ken Yamagata-gun, Kake-cho (Report on the Promotion of a Mountain Village, in the Case of Kake Town, Hiroshima Prefecture), Sanson-shinko Chosakai, Tokyo (in Japanese).

194. Kawanabe, H. 1975. Kawa no 'katati' no hakai (Destruction of a river 'morphology'). Monkey 138: 4–5 (in Japanese).

195. Kawanabe, H., S. Kimura, T. Maruyama & K. Tanida. 1975. Damu kensetu ni tomonai yosou sareru syo-eikyou to sore e no taiou (The estimated effects of dam construction on local environments, and reaction). pp. 39–48. *In:* Y. Takahashi (ed.) Shishimuda Damu eikyou-tyousa houkokusyo (Research Report on Environmental Effects of the Construction of Shishimuda Dam), Kokonoe Town Government, Kokonoe, Ooita (in Japanese).

196. Kawanabe, H., S. Kimura, T. Maruyama & K. Tanida. 1975. Kusugawa seibutu tyousa houkoku memo (A memorandum for biological research of the Kusugawa River). pp. 76–85. *In:* Y. Takahashi (ed.) Ibidem (ibidem), Kokonoe Town Government, Kokonoe, Ooita (in Japanese).

197. Kawanabe, H. 1975. Sougou-touron no tame ni: 'kotaigun' to 'koudou' (For general discussion: 'population' 'and' 'behavior'). Bull. Popul. Ecol. 26/27: 37–39 (in Japanese).

198. Kawanabe, H. & T. Tanaka. 1975. Nagara-gawa ni kansuru seitaigakuteki yobi kousatu (Ecological observations concerning the Nagara River: the effects of construction of a dam at the mouth of the river). Research on Environmen-

tal Disruption toward Interdisciplinary Cooperation 5(2): 1–14 (in Japanese).

199. Kawanabe, H. 1975. Nihon no tansuigyo, Biwako no tansuigyo (Freshwater fishes in Japan and in Lake Biwa). Tansuigyo (Freshwater Fishes) 1: 4–6 (in Japanese) (also in 240).

200. Tanida, K. & H. Kawanabe. 1975. Yodo-gawa no tisui-kouzi to tosi no naka no sizen-hogo (Flood prevention works in the Yodo River and nature conservation in a city area). Tansuigyo (Freshwater Fishes) 1: 31–33 (in Japanese).

201. Kawanabe, H. 1975. Nishijima Shinsho-si hakken ni kakaru Ryukyu-rettou sin-bunpu no tansui haze-rui (Four species of goby in the Ryukyus, newly discovered by Dr. S. Nishijima). Tansuigyo (Freshwater Fishes) 1: 79 (in Japanese).

202. Kawanabe, H. 1975. On the origin of ayu-fish (Pisces, Osmeridae) in Lake Biwa. pp. 317–320. *In:* S. Horie (ed.) Paleolimnology of Lake Biwa and the Japanese Pleistocene 3, Kyoto University, Kyoto.

203. Kawanabe, H. 1975. Biwako, Yodo-gawa no sizen to ningen (The nature of Lake Biwa, the Yodo River, and human beings). Karin (precise information not available) (in Japanese).

204. Ito, S., R. Ishikawa, H. Kawanabe, Y. Takahashi, K. Hanayama, I. Kitagawa, K. Sahara & I. Hatate. 1975. Komento to touron (Comments and discussion). pp. 53–130. *In:* Y. Takahashi & Y. Monma (ed.) Mizu wo meguru ziti to zyuumin (Autonomous People and Their Control of Rivers), Zenkoku Mizumondai Kyogikai, Tokyo (in Japanese).

205. Kawanabe, H. 1976. Chikugo-gawa zyouryuu Nakatsue, Shimotsue ryou-son ni okeru kasen-seibutu (Biology of the Chikugo River in the villages of Nakatsue and Shimotsue). pp. 32–36. *In:* Y. Takahashi (ed.) Sanson-sinkou konsarutanto ikensyo: Ooita-ken Hita-gun Nakatsue, Shimotsue Mura (A Suggestion made to the Villages of Nakatsue and Shimotsue, Hita-gun, Ooita Prefecture, from the Promotion Consultant for Mountain Villages). Sanson-sinkou Tyousa-kai, Tokyo (in Japanese).

206. Kawanabe, H. 1976. Nesu-ko no kaibutu wa nanbiki iruka? (How many Nessies in Loch Ness are there?). Anima 1976 (5): 70–71 (in Japanese).

207. Kawanabe, H. 1976. Biwako ayu no nawabari ni tuite, hyouki-izon-syuusei-setu ni yoru iti kousatu (A note on the territoriality of the ayu, *Plecoglossus altivelis* Temminck et Schlegel (Pisces: Osmeridae) in the Lake Biwa stock, based on the 'relic social structure' hypothesis). Physiol. Ecol. Japan 17: 395–399 (in Japanese) (also in 382).

208. Kawanabe, H. 1976. Sakuramasu gun no gakumei ni tuite (Scientific names of *Salmo (Oncorhynchus) masou* complex). Tansuigyo (Freshwater Fishes) 2: 58–62 (in Japanese).

209. Kawanabe, H. & Y. Matsuura. 1976. Nitiyou intabyuu: keisou wo kutta ayu wa umai (Sunday interview: the diatom-fed ayu is delicious!). The Sankei 15.6.1976 (in Japanese).

210. Kawanabe, H. 1976. 'Sizen (= sougo-kankei no soutai) kansatu' no susume (Encouragement of 'nature (= whole of interrelations among organisms) observation'). Anima 1976(7): 98 (in Japanese) (also in 240).

211. Kawanabe, H. 1976. Hakubutu-si wa hatasite hatten-sita no ka? (Has natural history developed?). Anima 1976(8): 98 (in Japanese) (also in 240).

212. Kawanabe, H. 1976. Hakubutu-si no hatten no tame ni (For the development of natural history). Anima 1976(9): 98 (in Japanese) (also in 240).

213. Kawanabe, H. 1976. Dokusou-teki natyurarisuto e no miti (The road to becoming a creative naturalist). Anima 1976 (10): 98 (in Japanese) (also in 240).

214. Kawanabe, H. 1976. Kagaku no houhouron ni tuite (On methodology of science). Anima 1976(11): 98 (in Japanese) (also in 240).

215. Kawanabe, H. 1976. Nihon hakubutu-si no 'dentou' (A 'tradition' of natural history of Japan). Anima 1976(12): 102 (in Japanese) (also in 240).

216. Kawanabe, H. 1976. Sizenkankyou-hozen wa mazu kawa no tyousa kara: Kansa-gawa Chubu Gyogyou-kyoudou-kumiai (The nature conservation of any environment begins with the study of its rivers: The Fisheries Cooperative Association of the Central Kansa River). Mizu to hito to: Zensuikyo News (Water and Human Beings: News from Zenkoku Mizumondai Kyogikai) 4: 1 (in Japanese).

217. Miyadi, D., H. Kawanabe & N. Mizuno. 1976. Gensyoku Nihon tansui gyorui zukan (Color illustrations of the freshwater fishes of Japan), 3rd edition. Hoikusha, Osaka. 462 pp. (in Japanese).

218. Kawanabe, H. 1976. Hyouki no kurikaesi no nakade: ayu no keisei to sinka – nawabari kouzou wo tyuusin ni (During successive ice ages: origin and evolution of the ayu with a special reference to territorial structure). Anima 1976(10): 21–28 (in Japanese) (also in 240 and 319).

219. Iwai, T., M. Azuma, H. Kawanabe, Y. Kuwaba & E. Harada. 1976. Ayu no sekai (The world of the ayu). Anima 1976 (10): 35–44 (in Japanese) (also in 319).

220. Kawanabe, H. 1976. Ayu (The ayu, *Plecoglossus altivelis*). pp. 95. *In:* H. Aragaki (ed.) Nihon no siki (Japan's Four Seasons), Asahi Shinbunsha, Tokyo (Second edition was published as H. Aragaki (ed.) 1988. Siki no hakubutu-si (The natural history of the four seasons). Asahi Shinbunsha, Tokyo) (in Japanese).

221. Kawanabe, H. 1976. Haze (Gobies, *Rhinogobius* spp.). pp. 139. *In:* H. Aragaki (ed.) Ibidem (ibidem), Asahi Shinbunsha, Tokyo (in Japanese) (also in 1988 edition).

222. Kawanabe, H. 1976. Hasu (The piscivorous chub, *Opsariichthys uncirostris*). pp. 146. *In:* H. Aragaki (ed.) Ibidem (ibidem), Asahi Shinbunsha, Tokyo (in Japanese) (also in 1988 edition).

223. Kawanabe, H. 1976. Sake to masu (chum salmon and masu salmon, *Oncorhynchus keta* and *O. masou*). pp. 197. *In:* H. Aragaki (ed.) Ibidem (ibidem), Asahi Shinbunsha, Tokyo (in Japanese) (also in 1988 edition).

224. Kawanabe, H. 1976. Huna (Crucian carps, *Carassius* spp.). pp. 258. *In:* H. Aragaki (ed.) Ibidem (ibidem), Asahi Shinbunsha, Tokyo (in Japanese) (also in 1988 edition).

225. Kawanabe, H. 1976. Gaku to gakkai to gakusya to (Science, scientific society, and scientists). Japan. J. Limnol. 37: 154 (in Japanese) (also in 417).
226. Kawanabe, H. 1977. Kawa-uo: sono 'uo-gokoro' to 'mizu-gokoro' (River fishes: their 'minds' and 'minds of water'). pp. 44–50. *In:* Bunshun De Rakkusu, me de miru 'Tyougyo taizen' (Bunshun De Luxe, 'Complete Angler' with Eyes), Bungeishunju-sha, Tokyo (in Japanese) (also in 240 and 319).
227. Kawanabe, H., E. Harada & S. Iwao (ed.) 1977. Doubutu seitaigaku kankei koumoku (Items of animal ecology). *In:* T. Yamada, F. Maekawa, F. Egami, R. Yasugi, H. Ozeki, M. Furuya & T. Hidaka (ed.) Iwanami seibutugaku ziten (Iwanami's Encyclopedia of Biology), 2nd ed., Iwanami Shoten, Tokyo. 1654 pp. (in Japanese).
228. Ito, S., R. Ishikawa, H. Kawanabe, K. Sahara, Y. Takahashi & I. Hatate. 1977. Situgi (Discussions). pp. 36–69. *In:* Damu no mura wa horobiruka? (Will a Dam Destroy a Village?), Zenkoku Mizumondai Kyogikai, Tokyo (in Japanese).
229. Kawanabe, H. 1977. Relic social structure hypothesis on the territoriality of the ayu, *Plecoglossus altivelis* (Pisces: Osmeridae). Proc. Japan. Acad. 53 (Ser. B): 74–77.
230. Kawanabe, H. 1977. Kawa no 'ue' to 'sita': Ryukyu-rettou no kasen gunsyuu bekken ('Up' and 'down' in rivers: a glimpse of river communities in the Ryukyu Archipelago). Sizen 1977(8): 42–53 (in Japanese) (also in 417).
231. Kawanabe, H. 1977. Yodo-gawa no tennen-ayu ni kansuru go-situmon e no kaitou (Answer to a question about natural stock of the ayu in the Yodo River). Yodo-gawa no Sizen wo Mamoru-kai Kaihou (Bulletin of the Yodo River Conservation Society) 20: 10–13 (in Japanese).
232. Kawanabe, H. 1977. Tansuigyo bunpu-zu no sakusei no tame ni (Mapping the distribution of freshwater fishes). Tansuigyo (Freshwater Fishes) 3: 48–49 (in Japanese).
233. Kurihara, Y., H. Iwaki & H. Kawanabe (ed.) 1977. Seimei no rensa to ningen: seitaigaku (Life chains and human beings: ecology). NHK (Japan Broadcasting Corporation), Tokyo. 128 pp. (in Japanese).
234. Tanaka, T. & H. Kawanabe. 1977. Sizen no renzokusei to seibutu, ningen no seikatu (Continuity of nature and organisms and human beings). pp. 7–11. *In:* Y. Kurihara, H. Iwaki & H. Kawanabe (ed.) Seimei no rensa to ningen: seitaigaku (Life Chains and Human Beings: Ecology). NHK (Japan Broadcasting Corporation), Tokyo (in Japanese).
235. Kawanabe, H. 1977. Seitai-teki tii no seiritu: ayu to minami-ayu wo rei to site (Establishment of an ecological niche: in the case of the ayu, *Plecoglossus altivelis*, and the southern ayu, *Prototroctes* spp.). pp. 47–50. *In:* Y. Kurihara, H. Iwaki & H. Kawanabe (ed.) Ibidem (ibidem), NHK (Japan Broadcasting Corporation), Tokyo (in Japanese).
236. Tanaka, T. & H. Kawanabe. 1977. Seitaigaku to gizyutu (Ecology and engineering). pp. 73–76. *In:* Y. Kurihara, H. Iwaki & H. Kawanabe (ed.) Ibidem (ibidem), NHK (Japan Broadcasting Corporation), Tokyo (in Japanese).
237. Kawanabe, H. (ed.) 1978. Kyoto-si-nai no kasen no seitaigaku-teki kenkyuu (Ecological research on rivers in Kyoto City). Kyoto-si Kougai Taisaku-situ (Bureau of Measures against Environmental Pollution, Kyoto City Government), Kyoto. 189 pp. (in Japanese).
238. Kawanabe, H. 1978. Hazime ni (Introduction). pp. 2–3. *In:* H. Kawanabe (ed.) Kyoto-si-nai no kasen no seitaigaku-teki kenkyuu (Ecological research on rivers in Kyoto City), Kyoto-si Kougai Taisaku-situ (Bureau of Measures against Environmental Pollution, Kyoto City Government), Kyoto (in Japanese).
239. Kawanabe, H. 1978. Syohyou: 'Kani Tokichi zensyuu' (Book review: 'Works of Tokichi Kani'). Sizen 26 (precise information not available) (in Japanese).
240. Kawanabe, H. 1978. Seibutu to kankyou (Organisms and their environment). Jinbun-shoin, Kyoto. 230 pp. (in Japanese).
241. Kawanabe, H., H. Matsui, T. Ihara, M. Hori & K. Tanida. (translators) 1978. Doubutu no seitai ('Ecology of animals' by C. S. Elton). Shisakusha, Tokyo. 293 pp. (in Japanese).
242. Kawanabe, H. 1978. Yakusya no yaya mizikame no atogaki (Rather shorter postscripts of the translator). pp. 290–293. *In:* Doubutu no seitai ('Ecology of animals' by C. S. Elton), Shisakusha, Tokyo (in Japanese) (also in 709).
243. Kawanabe, H. 1978. Tori ni yoseru mousou: saezuri-wake wa nanno tame? (A daydream about birds: what do different bird songs mean?). The Yomiuri (evening) 22.3.1978 (in Japanese).
244. Kawanabe, H. 1978. Masu to ayu no sirabe (Melodies of the trout and the ayu). Tansuigyo (Freshwater Fishes) 4: 104 (in Japanese).
245. Kawanabe, H. 1978. Ongaku wo meguru wa ni natta hanasi (Circular stories around music). Gesangsthema 3: 8–11 (in Japanese).
246. Kawanabe, H. 1978. Some biological problems in deep core of Lake Biwa. Verh. internat. Verein. Limnol. 20: 2674–2677.
247. Furukawa, T., M. Nishida, H. Matsubara & H. Kawanabe. 1978. Some morphological characters of sea-run and lacustrine forms of the red-spot masu-trout, *Salmo (Oncorhynchus) masou macrostomus* (?). Paleolimnology of Lake Biwa and the Japanese Pleistocene 6: 277–280.
248. Kawanabe, H., T. Iwai, J. Shirataki & E. Kimura. 1979. Doubutu bunpu tyousa houkokusyo: tansui-gyorui (Report on animal distribution: freshwater fishes). Kyoto Prefectural Government, Kyoto. 59 pp. (in Japanese).
249. Kawanabe, H. & K. Takamura. 1979. Fish fauna near Katana, southwest coast of Lake Kivu. pp. 7–8. *In:* M. Kawabata (ed.) Preliminary Report on the Aquatic Community with Special Reference to the Fish in Lake Tanganyika and Lake Kivu. Shizuoka Women's College, Shizuoka.
250. Kawanabe, H. 1979. Yon-satu no doubutu syasin-syuu (Four books of animal photographs). Anima 1979(9): 100 (in Japanese).
251. Kawanabe, H. 1979. Minami-ayu no hanasi (Story of the southern ayu, *Prototroctes* spp.). Tansuigyo (Freshwater Fishes) 5: 26–28 (in Japanese) (also in 319).
252. Kawanabe, H. 1979. Kagaku no yougo (The vindication of

science). pp. 6–8. *In:* S. Kato (ed.) Kato Syuiti tyosaku-syuu vol. 14 geppou (Works of Syuiti Kato, vol. 14, Monthly Bulletin), Heibonsha, Tokyo (in Japanese).
253. Kawanabe, H. 1979. Rikusui seibutu no syu-seitaigaku (Species ecology of freshwater organisms). pp. 94–95. *In:* S. Mori (ed.) Seitaigaku no genzyou-bunseki to syourai no sinro (Analysis of the Present State and Future Development of Ecology), Hydrobiological Station, Kyoto University, Otsu (in Japanese).
254. Kawanabe, H. & S. Iwao. 1979. Kaidai (Bibliographical introduction). pp. 555–560. *In:* Morisita Masaaki seitaigaku ronsyuu (Ecological Works of Masaaki Morisita), vol. 1, Shisakusha, Tokyo (in Japanese).
255. Iwao, S. & H. Kawanabe. 1979. Kaidai (Bibliographical introduction). pp. 581–585. *In:* Ibidem (ibidem), vol. 2, Shisakusha, Tokyo (in Japanese).
256. Kawanabe, H. 1979. Shimanto-gawa no seiryu to uo (The Shimanto: a crystal river with fishes). pp. 103–116. *In:* Hokoraka ni katare kono Shimanto wo (Speaking proudly about the Shimanto River), Zenkoku Mizumondai Kyogikai, Tokyo (in Japanese) (also in 417).
257. Asou, T., H. Kawanabe, H. Yasuoka, Y. Takahashi, I. Hatate, I. Kitagawa & N. Mizuno. 1979. Sougou touron (General discussions). pp. 117–160. *In:* Ibidem (ibidem), Zenkoku Mizumondai Kyogikai, Tokyo (in Japanese).
258. Furukawa, T., H. Matsubara, M. Nishida & H. Kawanabe. 1979. Studies on the masu salmon, *Salmo (Oncorhynchus) masou*, complex. II. Paleolimnology of Lake Biwa and the Japanese Pleistocene 7: 358–368.
259. Kawanabe, H. 1980. Recent progress of limnology in Japan: fishes. Recent Progress of Natural Science in Japan 5: 231–236.
260. Kawanabe, H. 1980. Tamenteki seitaigaku no genzyoutyousa to kaiseki (The present status and analysis of the wide field of ecology). pp. 1–8. *In:* S. Mori (ed.) Seitaigaku no sougouka to syourai-keikaku (Synthesis of Ecology and Future Plans), Hydrobiological Station, Kyoto University, Otsu (in Japanese).
261. Kawanabe, H. 1980. Bunkazai-kouen ni naru Kyoto Narabi-ga-oka (Narabi-ga-oka Hill in Kyoto as a cultural asset park). The Asahi (evening) 17.4.1980 (in Japanese) (also in 417).
262. Kawanabe, H. & K. Miura. 1980. Ayu ni miru tansuigyo sinka no nazo (The mystery of the evolution of a freshwater fish: the ayu). Asahi Science 1980(8): 77–82 (in Japanese) (also in 417).
263. Kawanabe, H. 1980. Tennen-kinenbutu (Natural monuments). pp. 24–26. *In:* Kyoto-si Bunka-kankou-sigen Tyousakai houkokusyo (Report of the Committee of Cultural and Sightseeing Resources in Kyoto City), Kyoto City Government, Kyoto (in Japanese).
264. Kawai, T., H. Kawanabe & N. Mizuno (ed.) 1980. Nihon no tansui-seibutu: sinryaku to kakuran no seitaigaku (Freshwater organisms in Japan: their ecology of invasion and disturbance). Tokai University Press, Tokyo. 194 pp. (in Japanese).
265. Kawanabe, H. 1980. Nizi-masu (Rainbow trout, *Oncorhynchus mykiss*). pp. 44–48. *In:* T. Kawai, H. Kawanabe & N. Mizuno (ed.) Nihon no tansui-seibutu: sinryaku to kakuran no seitaigaku (Freshwater Organisms in Japan: Their Ecology of Invasion and Disturbance), Tokai University Press, Tokyo (in Japanese).
266. Kawanabe, H. 1980. Gunsyuu no taiou (Community response). pp. 183–194. *In:* T. Kawai, H. Kawanabe & N. Mizuno (ed.) Ibidem (ibidem), Tokai University Press, Tokyo (in Japanese) (also in 711).
267. Kawanabe, H. 1981. Chou-kou seibutu-gunsyuu no kyoudou-tyousa ni kansuru teian (A plan for a joint study on the Yangtze River communities). Bulletin of Kyoto Society on Japan-China Scientific Exchange 9: 12–13 (in Japanese).
268. Kawanabe, H. 1981. Owari ni (Postscripts). Bulletin of Kyoto Society on Japan-China Scientific Exchange 9: 15 (in Japanese).
269. Kawanabe, H. 1981. Kougai wo yakamasiku iu no wa zyakusya no inaori de aru? (Is strong protest against pollution only the desire of small group?). Kyoto University News 16.1.1981 (in Japanese) (also in 417).
270. Kawanabe, H. 1981. Konogoro (These days). The Mainichi (evening) 7.4.1981 (in Japanese).
271. Takahashi, Y., K. Hanayama, H. Shimura, K. Sahara, I. Kitagawa, H. Kawanabe & N. Abe. 1981. Zyogensya to no koukan (Exchange with advisors). pp. 105–154. *In:* Y. Takahashi & Y. Monma (ed.) Aka-gawa: Shonai wo uruosu mono (The Aka-gawa: a River Beneficial to the Shonai Plain), Zenkoku Mizumondai Kyougikai, Tokyo (in Japanese).
272. Kawanabe, H. 1981. Tanganiika-ko ni syukan-kankei wo tazuneru (Searching for interspecific relationships in Lake Tanganyika). 1. Naze Ahurika nanoka? (Why is Africa a target?). Anima 1981(2): 66–72 (in Japanese) [also in M. Hori (ed.) 1993. Tanganiika-ko no uo-tati (Fishes in Lake Tanganyika). Heibonsha, Tokyo].
273. Kawanabe, H. 1981. Idem (idem). 2. Ko-han de no meisou (Meditation along a lake shore). Anima 1981(3): 72–77 (in Japanese) [also in M. Hori (ed.) 1993. Ibidem].
274. Kawanabe, H. 1981. Idem (idem). 3. Gunsyuu no kabe to ana (Walls and holes in community). Anima 1981(4): 76–82 (in Japanese) [also in M. Hori (ed.) 1993. Ibidem].
275. Kawanabe, H. 1981. Territorial behaviour of *Tropheus moorei* (Osteichthyes: Cichlidae) with a preliminary consideration on the territorial forms in animals. Afr. Stud. Monogr. 1: 101–108.
276. Kawanabe, H. 1981. Tanganiika-ko no sakana-tati (A research note on fishes of Lake Tanganyika). J. Afr. Stud. 20: 66–78 (in Japanese) [also in M. Hori (ed.) 1993. Ibidem].
277. Morimura, S., K. Suei, H. Kawanabe, S. Itasaka & A. Naoe. 1981. Sinpoziumu 'Hakusan no kawa to uo no hogo' (Symposium on 'Rivers and fishes around Mt. Hakusan'). Hakusan 8 (Spec. Issue): 3–15 (in Japanese).
278. Kawanabe, H. (ed.) 1981. Ecological and limnological study on Lake Tanganyika and its adjacent regions 1. Kyoto University, Kyoto. 58 pp.

279. Kawanabe, H. & M.K. Kwetuenda. 1981. Introduction. pp. 4–6. *In:* H.Kawanabe (ed.) Ecological and Limnological Study on Lake Tanganyika and its Adjacent Regions 1, Kyoto University, Kyoto.

280. Kawanabe, H. 1981. Territories of *Tropheus moorei* and *Petrochromis polyodon* with a general discussion on peripheral and central systems territoriality. pp. 14–16. *In:* H. Kawanabe (ed.) Ibidem 1, Kyoto University, Kyoto.

281. Kawanabe, H. 1981. Characteristics of the communal relationship among fishes in Lake Tanganyika in comparison with usual freshwater communities. pp. 23–25. *In:* H. Kawanabe (ed.) Ibidem 1, Kyoto University, Kyoto.

282. Kawanabe, H. 1981. Atogaki ni kaete (As the postscripts). pp. 302–303. *In:* S. Kitamura (ed.) Mizoro-ga-ike no sizen to hito (Nature and Human Beings of Mizoro-ga-ike Pond), Bureau of Culture and Sightseeing, Kyoto City Government, Kyoto (in Japanese) (also in 417).

283. Kawanabe, H. 1981. Gendai no kiseki Mizoro-ga-ike: seibutu subete ga kityou (The moor, Mizoro-ga-ike, a miracle: all the organisms are valuable). The Kyoto 2.6.1981 (in Japanese).

284. Tabata, H., S. Kitamura, G. Murata, H. Kawanabe & A. Endo. 1981. Gendai no kiseki Mizoro-ga-ike: eien no kataribe (A contemporary miracle, Mizoro-ga-ike Pond: an eternal storyteller) 1. The Kyoto 16.6.1981 (in Japanese).

285. Tabata, H., S. Kitamura, G. Murata, H. Kawanabe & A. Endo. 1981. Idem (idem) 2. The Kyoto 17.6.1981 (in Japanese).

286. Kawanabe, H. 1981. Wakamono wa dokuritu-dokkou suru! (Young people striking out on their own!). The Kyoto (evening) 21.7.1981 (in Japanese) (also in 417).

287. Kawanabe, H. 1981. Doubutu (?) samazama (Various animals (?)). The Kyoto (evening) 1.9.1981 (in Japanese) (also in 417).

288. Kawanabe, H. 1981. Hukusuu no kokoro wo motu to (Having plural minds). The Kyoto (evening) 22.10.1981 (in Japanese) (also in 417).

289. Kawanabe, H. 1981. Usso! (Incredible!). The Kyoto (evening) 20.11.1981 (in Japanese) (also in 417).

290. Kawanabe, H. 1981. Syohyou: 'Hon no hanasi: seisyun ni okuru tyouhatu-teki dokusyo-ron' (Book review: 'The story of books: provocative essays for young people' by K. Sirakami). Asahi Journal 4.9.1981 (in Japanese).

291. Kawanabe, H. 1981. Nobita ayu no hokugen (The ayu's northern limit extends). The Hokkaido (evening) 7.9.1981 (in Japanese) (also in 417).

292. Kawanabe, H. 1981. Kansin dekinu keiryuu-gyo no houryuu: suiiki mamori sodateru doryoku ga daizi (Releasing stream fishes for artificial stocking is not recommended: efforts for conservation of rivers are more important). The Asahi (evening) 25.9.1981 (in Japanese) (also in 417).

293. Kawanabe, H. 1981. Mudai (no title). pp. 11–12. *In:* Kirokueiga 'Kusuri ni yamu' Kyoto zyouei panhuretto (A Pamphlet for the Private Showing of the Movie 'Ill-Introduced by Medicine' in Kyoto), Kyoto Kirogu-eiga wo Miru Kai (Kyoto Group for Documentary Films), Kyoto (in Japanese) (also in 417).

294. Kawanabe, H. & T. Nakamura. 1981. Tori-tati no iti-nen wa aki ni hazimaru (The birds' year begins in autumn). Anima 1981(10): 25–29 (in Japanese) (also in 417).

295. Kawanabe, H. 1981. Mizikana hassou de kaku-rongi ni genzitukan wo (Realistic discussion of nuclear problems). Asahi Journal (a supplement issue) 10.12.1981 (in Japanese).

296. Mizuno, N. & H. Kawanabe. 1981. A topographical classification of streams, with an introduction of the system widely used in Japan. I. Reach type, stream zone and stream type. Verh. intern. Verein. Limnol. 21: 913.

297. Kawanabe, H. 1982. Uo-tati ni hanasi-kakeru koto (Saying a few words to fishes). The Kyoto (evening) 7.1.1982 (in Japanese) (also in 417).

298. Kawanabe, H. 1982. Kokugo no nyuusi-mondai ni yosete (On the Japanese language for university entrance examinations). The Kyoto (evening) 10.3.1982 (in Japanese) (also in 417).

299. Kawanabe, H. 1982. Kunsi wa hyouhen suru (?) (A wise man changes his mind, a fool never (?)). The Kyoto (evening) 1.5.1982 (in Japanese) (also in 417).

300. Kawanabe, H. 1982. Ayu-gari ni yosete (On ayu fishing). The Kyoto (evening) 23.6.1982 (in Japanese) (also in 417).

301. Kawanabe, H. 1982. 'Ko no hogo' wa doko made susumu ka? (To what extent has 'infant care' progressed?). The Kyoto (evening) 28.8.1982 (in Japanese) (also in 417).

302. Kawanabe, H. 1982. Nagai rekisi no naka no mizikai zikan (A short time in a long history). The Kyoto (evening) 18.10.1982 (in Japanese) (also in 417).

303. Kawanabe, H. 1982. Siwasu san-dai (Three stories for December: the month of the Running Teacher). The Kyoto (evening) 7.12.1982 (in Japanese) (also in 417).

304. Kawanabe, H. 1982. Tanganiika-ko moude (A pilgrimage to Lake Tanganyika). Kokuritu-daigaku-kyoukai Kaihou (Bulletin of the Union of National Universities) 95: 131 (in Japanese).

305. Kawanabe, H. 1982. Seitaigaku ni kansuru taiwa no susume (Encouragement of dialogues on ecology). Science (Tokyo) 52: 153 (in Japanese).

306. Kawanabe, H. 1982. Hutatu no seitaigaku: keizizyougakuteki oboegaki (Two types of ecology: a metaphysical memorandum). Science (Tokyo) 52: 154–159 (in Japanese) (also in 382).

307. Kawanabe, H. 1982. Saisyoku nawabari to seisyoku nawabari to no syunbetu (Differences between feeding and mating territories). Science (Tokyo) 52: 390–392 (in Japanese) (also in 382).

308. Kawanabe, H. 1982. Toriaezu go-aisatu made (A brief immediate reply to readers). Science (Tokyo) 52: 464–465 (in Japanese).

309. Kawanabe, H. 1982. Syu-bunka wa seikatukan-zyou no tokutei no dankai de okoru (Speciation occurred at a particular stage in life history). Science (Tokyo) 52: 527–529 (in Japanese).

310. Kawanabe, H., T. Sunaga & T. Uematsu. 1982. Medaka-rui no sessyoku koudou (Feeding behavior of medaka, *Oryzias latipes*, and related species). pp. 181–198. *In:* H. Morita & K. Kubota (ed.) Sessyoku koudou no mekanizumu (Mechanisms of Feeding Behavior), Department of Biology, Kyushu University, Fukuoka (in Japanese).
311. Kawanabe, H. 1982. Syokumotu-rensa to eiyou-kan (The food chain and trophic cycle). pp. 109–128. *In:* M. Numata (ed.) Seitaigaku-dokuhon (A Book of Ecology), Toyo Keizai Shinposha, Tokyo (in Japanese) (also in 382).
312. Kawanabe, H., M. Shikadake, M. Onishi & K. Ohta. 1982. Ayu no hanasi (The story of the ayu). 1. Nawabari ni tuite (Regarding its territory). Turi no Tomo 1982(6): 59–64 (in Japanese).
313. Kawanabe, H., M. Shikadake, M. Onishi & K. Ohta. 1982. Idem (idem). 2. Ayu syosou komogomo (Its various phases). Turi no Tomo 1982(7): 68–74 (in Japanese).
314. Kawanabe, H., M. Shikadake, M. Onishi & K. Ohta. 1982. Idem (idem). 3. Kawa no kankyou-mondai ni tuite (On environmental problems in rivers). Turi no Tomo 1982(8): 67–71 (in Japanese).
315. Kawanabe, H. 1982. Ko Iwao Syun'iti-si no seitaigaku (The ecology of the late Professor S. Iwao). pp. 6–11. *In:* E. Kuno (ed.) Iwao Syun'iti Sensei tuito-bun-syuu (Collection of Obituaries for Professor Syun'iti Iwao), Laboratory of Entomology, Faculty of Agriculture, Kyoto University, Kyoto (in Japanese) (also in 709).
316. Kawanabe, H. & N. Mizuno. 1982. Toyo-kawa suikei de no gyorui no seisoku-zyoutai (Living conditions of fishes in the Toyo-kawa River System). pp. 89–148. *In:* Y. Takahashi (ed.) Toyo-kawa suikei ni okeru mizu-sigen kaihatu to Horai-cho (The Development of Water Resources in the Toyo-kawa River System and Horai Town), Horai Town Government, Horai, Aichi (in Japanese).
317. Kawanabe, H. & S. Horie. 1982. 'Kitahankyuu ni okeru daiyon-ki hyouga-sayou' kokusai-syuukai kaisai-houkoku (Report of the International Symposium on 'the Quaternary glacial action in the northern hemisphere'). Japan. J. Ecol. 32: 260 (in Japanese).
318. Kawanabe, H. 1982. Ayu: sono bunpu to syuzoku (The ayu: its distribution and variation). Hon (Kodansha, Tokyo) 1982(8): 30–33 (in Japanese) (also in 319).
319. Kawanabe H. & A. Sakurai. 1982. Ayu no hakubutu-si (The natural history of the ayu). Heibonsha, Tokyo. 142 pp. (in Japanese).
320. Kawanabe, H. 1982. 'Iwana Kokusai Sinpoziumu' wo meguru nan to iu koto mo nai hanasi (Some stories on the 'International Symposium of Arctic Charr'). Tansuigyo (Freshwater Fishes) 8: 91–96 (in Japanese).
321. Kawanabe, H. 1982. Beetooben no koukyou-kyoku wa amehuri no hi ni husawasii (Beethoven's symphonies are good for listening on rainy days). pp. 1–5. *In:* Kyoukyou dai 249 kai teiki-ensokai panhuretto (The Pamphlet for the 249th Regular Concert of Kyoto Symphony Orchestra), Kyoto City Government, Kyoto (in Japanese).
322. Kawanabe, H. 1982. 'Ayu no hakubutu-si' no koto (On my book 'The natural history of the ayu'). The Sankei 15.11.1982 (in Japanese).
323. Kawanabe, H. 1983. Huyu no sizen (Nature in winter). The Kyoto (evening) 3.2.1983 (in Japanese) (also in 417).
324. Kawanabe, H. 1983. Hukurou ni yoseru (Regarding owls). The Kyoto (evening) 18.3.1983 (in Japanese) (also in 417).
325. Kawanabe, H. 1983. 'Husei' nyuugaku to wa nani ka (What is the 'illegal' entrance to universities). The Kyoto (evening) 6.5.1983 (in Japanese) (also in 417).
326. Kawanabe, H. 1983. Tendai Kokuseiji ni mouderu (Visiting the Buddhist temple of Tendai Kokuseiji in China). The Kyoto (evening) 21.6.1983 (in Japanese) (also in 417).
327. Kawanabe, H. 1983. Kyouzon to kyoudou (Coexistence and cooperation). Kaiyo to Seibutu (Aquabiology) 5: 1 (in Japanese) (also in 417).
328. Tsurumi, S., M. Nishibe, H. Kawai & H. Kawanabe. 1983. Hoo-due wo tuite kanrisya wo gyousi suru (Gazing at local authorities with chins resting in hands). Asahi Journal 25 (13): 10–19 (in Japanese).
329. Kawanabe, H. 1983. Syo wa mangan (On a vast number of books). Shiso-no-kagaku 32(5): 20–21 (in Japanese).
330. Kawanabe, H. (ed.) 1983. Ecological and limnological study on Lake Tanganyika and its adjacent regions 2. Kyoto University, Kyoto. 68 pp.
331. Kawanabe, H. & M.K. Kwetuenda. 1983. Introduction. p.6. *In:* H. Kawanabe (ed.) Ecological and Limnological Study on Lake Tanganyika and its Adjacent Regions 2, Kyoto University, Kyoto.
332. Kawanabe, H. 1983. Further observation of territoriality of *Tropheus moorei* and *Petrochromis polyodon* (Osteichthyes: Cichlidae). pp. 10–11. *In:* H. Kawanabe (ed.) Ibidem (ibidem) 2, Kyoto University, Kyoto.
333. Kawanabe, H. & M.K. Kwetuenda. 1983. Introduction. pp. 18–19. *In:* H. Kawanabe (ed.) Ibidem (ibidem) 2, Kyoto University, Kyoto.
334. Kawanabe, H., E. Harada, T. Sunaga, M. Azuma & I. Sakamoto. 1983. Shinji-ko, Naka-umi no gyorui (Fishes in Lake Shinji and Naka-umi). pp. 623–642. *In:* I. Minami (ed.) Shinji-ko, Naka-umi tansuiko-ka ni kanren suru suiri, suisitu oyobi seitai no kyodou ni tuite (On Water Physics, Water Quality and Ecological Changes in Relation to the Freshening of Lake Shinji and Naka-umi), Committee of Water Management and Ecological Change by the Freshening of Lake Shinji and Naka-umi, Japanese Society of Agricultural Engineering, Tokyo (in Japanese).
335. Kawanabe, H. 1983. Kawamura Tamiji seitan 100 nen no kai (The hundredth anniversary of the late Professor T. Kawamura). Sizen 1983(6): 16 (in Japanese).
336. Kawanabe, H. 1984. Ayu no kakeibo (An account book of the ayu). The Kyoto 1.1.1984 (in Japanese).
337. Kawanabe, H. 1984. Yuki 'hiroge' ('Scattering' snow). The Daihorin 1984(5): 36–37 (in Japanese).
338. Kawanabe, H. 1984. Tya-banasi (Tea-talk). The Mainichi (evening) 3.4.1984 (in Japanese) (also in 417).
339. Kawanabe, H. 1984. Humi-dukai (Emissary of the letter).

The Mainichi (evening) 10.4.1984 (in Japanese) (also in 417).
340. Kawanabe, H. 1984. Isana-tori (Catching whales). The Mainichi (evening) 17.4.1984 (in Japanese) (also in 417).
341. Kawanabe, H. 1984. Umare-izuru nayami (The burden of having been born). The Mainichi (evening) 24.4.1984 (in Japanese) (also in 417).
342. Kawanabe, H. 1984. Sakura no mori no mankai no sita (A forest of cherry blossoms in full bloom). The Mainichi (evening) 1.5.1984 (in Japanese) (also in 417).
343. Kawanabe, H. 1984. Syayou (A declining sun). The Mainichi (evening) 4.5.1984 (in Japanese) (also in 417).
344. Kawanabe, H. 1984. Hi no tori (Fire birds). The Mainichi (evening) 15.5.1984 (in Japanese) (also in 417).
345. Kawanabe, H. 1984. Ikiteiru heitai (Living soldiers). The Mainichi (evening) 22.5.1984 (in Japanese) (also in 417).
346. Kawanabe, H. 1984. Yama no oto (The sound of mountains). The Mainichi (evening) 29.5.1984 (in Japanese) (also in 417).
347. Kawanabe, H. 1984. Kikyou (Homecoming). The Mainichi (evening) 5.6.1984 (in Japanese) (also in 417).
348. Kawanabe, H. 1984. Sansyou-uo (A great salamander). The Mainichi (evening) 12.6.1984 (in Japanese) (also in 417).
349. Kawanabe, H. 1984. Kayoi-komati (Man's visiting lover). The Mainichi (evening) 19.6.1984 (in Japanese) (also in 417).
350. Kawanabe, H. 1984. Sore kara (And then). The Mainichi (evening) 26.6.1984 (in Japanese) (also in 417).
351. Kawanabe, H. 1984. Ayu no kenkyuu, sonogo (Research on the ayu: thereafter). Anima 1984(6): 12–13 (in Japanese) (also in 417).
352. Tokui, T. & H. Kawanabe. 1984. Fishes. pp. 339–360. *In:* S. Horie (ed.) Lake Biwa, Dr W. Junk Publishers, Dodrecht.
353. Kawanabe, H. 1984. Idainaru kisyo 'Tisyou no seitaigaku' (A wonderfully odd book, 'Ecology of lakes and swamps' by T. Mizuno). pp. 334–336. *In:* Commemoration Committee for Professor T. Mizuno Retirement (ed.) Mizuno Toshihiko Kyouzyu tankan kinen-si (Memorial Volume for the Retirement of Professor Toshihiko Mizuno), Osaka College of Education, Osaka (in Japanese) (also in 417).
354. Kawanabe, H. 1984. Sinpi na uo tati: Tanganiika-ko no kawasuzume (Mysterious fishes: cichlid fishes in Lake Tanganyika). 1. Nettai no mizuumi, seitai wo saguru (Research on ecology in a tropical lake), The Shinano Mainichi 23.5.1984 (in Japanese).
355. Kawanabe, H. 1984. Idem (idem). 12. Samazamano kyoudou-kankei (Various cooperative relations). The Shinano Mainichi 6.6.1984 (in Japanese).
356. Kawanabe, H. 1984. Aru si ni yoseru (To a poem). Animals and Zoo 1984(7): 3 (in Japanese).
357. Kawanabe, H. 1984. Tanganiika-ko gyorui ni mirareru kyoudou-kankei (Cooperative relations in fishes of Lake Tanganyika). Seibutu no Syakai-kouzou Nyuusu (News for Social Structure in Biological Organisms) 4: 4–5 (in Japanese).
358. Kawanabe, H. 1984. Ayu no me de Tanganiika-ko no uo wo miru (Fishes in Lake Tanganyika from the ayu's point of view). Souzou no Sekai (Creative World) 51: 6–25 (in Japanese).
359. Kawanabe, H., T. Iwai, T. Umehara, A. Endo, M. Kawai & K. Sakuda. 1984. Sinka no husigi (The mystery of evolution). Souzou no Sekai (Creative World) 51: 26–51 (in Japanese).
360. Kawanabe, H. 1984. Ahurika kenkyuu no kaiko to tenbou: doubutu seitaigaku no baai (African studies in Japan: retrospect and prospects in animal ecology). J. Afr. Stud. 25: 28–41 (in Japanese).
361. Kawanabe, H. 1984. Yooroppa kara no tegami: uo no torikata ni tuite (A letter from Europe: on fishing). Kenshuu 1984(10): 15–20 (in Japanese) (also in 417).
362. Kawanabe, H. 1984. Chikugo-gawa no kigyo 'etu' (Japanese tapertail anchovy, *Coilia nasus nasus:* a curious fish living in the Chikugo River). pp. 131–134. *In:* Y. Suyehiro & S. Kato (ed.) Uo no saiziki 2, Syoka no uo (A Glossary of Seasonal Terms of Fishes for Haiku Composers, 2. Fishes in Early Summer), Gakushu-kenkyu-sha, Tokyo (in Japanese).
363. Kawanabe, H. 1984. Ayu no nawabari wo kenkyuu site (Studying the ayu's territory). pp. 102. *In:* M. Kimura, Y. Higuchi, T. Matsumoto, S. Yasugi & R. Yasugi (ed.) Iwanami zyunia kagaku koza 4, Seibutu no sekai wo saguru (The Iwanami Science Lecture Series for Juniors 4, Exploring Biological Worlds), Iwanami Shoten, Tokyo (in Japanese).
364. Kawanabe, H., T. Furukawa-Tanaka & T. Maruyama. 1985. Interrelations among charrs and trout in Japanese streams. Intern. Soc. Arctic Charr Fanatics (ISACF) Information Series 3: 65–71.
365. Kawanabe, H. (ed.) 1985. Ecological and limnological study on Lake Tanganyika and its adjacent regions 3. Kyoto University, Kyoto. 83 pp.
366. Kawanabe, H. & M.K. Kwetuenda. 1985. Introduction. pp. 9–10. *In:* H. Kawanabe (ed.) Ecological and Limnological Study on Lake Tanganyika and its Adjacent Regions 3, Kyoto University, Kyoto.
367. Kondo, T. & H. Kawanabe. 1985. Distribution and abundance of insects and algae in the northwestern part of Lake Tanganyika. pp. 30. *In:* H. Kawanabe (ed.) Ibidem (ibidem) 3, Kyoto University, Kyoto.
368. Kawanabe, H. 1985. Introduction. pp. 45. *In:* H. Kawanabe (ed.) Ibidem (ibidem) 3, Kyoto University, Kyoto.
369. Iino, T., H. Iguti, S. Ueda, H. Esawa, H. Kawanabe, Y. Taki, K. Toyoshima & Y. Hoshino (ed.) 1985. Kagaku no ziten (Encyclopedia of sciences), 3rd ed. Iwanami Shoten, Tokyo. 1432 pp. (in Japanese).
370. Ueda, T., A. Endo & H. Kawanabe. 1985. Koudou to syakai (Behavior and society in animals). pp. 339–349. *In:* T. Iino, H. Iguti, S. Ueda, H. Esawa, H. Kawanabe, Y. Taki, K. Toyoshima & Y. Hoshino (ed.) Kagaku no ziten (Encyclopedia of Sciences), 3rd ed., Iwanami Shoten, Tokyo (in Japanese).
371. Endo, A., T. Ueda & H. Kawanabe. 1985. Doubutu no sei-

tai (Ecology of animals). pp. 930–941. *In:* T. Iino, H. Iguti, S. Ueda, H. Esawa, H. Kawanabe, Y. Taki, K. Toyoshima & Y. Hoshino (ed.) Ibidem (ibidem), 3rd ed., Iwanami Shoten, Tokyo (in Japanese) (also in 711).

372. Kawanabe, H. 1985. Seibutu gunsyuu (Biological community). Encyclopedia Heibonsha 8: 420–423 (in Japanese).

373. Kawanabe, H. 1985. Biwako no seibutu-sou (Fauna and flora in Lake Biwa). Encyclopedia Heibonsha 12: 805–806 (in Japanese).

374. Kawanabe, H. 1985. Nawabari-kouzou no hikaku-seitaigakuteki kenkyuu (Comparative ecology of territorial structure). Kyoto University, Kyoto. 32 pp. (in Japanese).

375. Kawanabe, H. 1985. Syohyou: 'Huukyoo simatu, 'Kyouku kogarasi-no-maki' sinsyaku' (Book review: 'New 'Kyouku' poem, 'volume of cold winter wind', written by Basho Matuo' by T. Ando). The Kyoto University News 16.4.1985 (in Japanese) (also in 417).

376. Kawanabe, H. 1985. Elton, Charles Sutherland. Encyclopedia Nipponica (Shogakukan, Tokyo) 2001, 3: 686 (in Japanese).

377. Kawanabe, H., M. Hori & K. Takamura. 1985. Cooperative feeding among cichlid fishes along rocky shores of Lake Tanganyika. Verh. intern. Verein. Limnol. 22: 2661.

378. Barel, C.D.N., R. Dorit, P.H. Greenwood, G. Fryer, N. Hughes, P.B.N. Jackson, H. Kawanabe, R.H. Lowe-McConnell, M. Nagoshi, A.J. Ribbink, E. Trewavas, F. Witte & K. Yamaoka. 1985. Destruction of fisheries in Africa's lakes. Nature 315: 19–20.

379. Abe, T., A. Endo, M. Kawai, H. Kawanabe & K. Sakuta. 1985. Mori no naka no kagaku senryaku (Chemical strategy in forests). Souzou no Sekai (Creative World) 55: 40–68 (in Japanese).

380. Kawanabe, H. 1985. Izuko mo onazi (?) 'Iwana Kyoukai' no hanasi (A story on the 'International Society of Arctic Charr Fanatics', being everything's the same all over (?)). Tansuigyo (Freshwater Fishes) 11: 83–88 (in Japanese).

381. Morisita, M. & H. Kawanabe (ed.) 1985. Showa 59 nen-do ayumodoki seitai-bunpu tyousa-houkokusyo (Report on the ecological distribution of the kissing loach, *Leptobotia curta*, in 1984). Department of Fisheries, Kyoto Prefectural Government, Kyoto. 43 pp. (in Japanese).

382. Kawanabe, H. 1985. Kawa to mizuumi no seitaigaku (The ecology of rivers and lakes). Kodansha, Tokyo. 253 pp. (in Japanese).

383. Kawanabe, H. 1985. Tansui-gyorui no bunpu, seitai to suion (The distribution and ecology of freshwater fishes in relation to water temperature). pp. 869–877. *In:* Rivers Division, Japan Ministry of Construction (ed.) 1984-nen Nihon kasen suisitu nenkan (Yearbook of River Water Quality in Japan, 1984), Sankaido, Tokyo (in Japanese) (also in 711).

384. Kawanabe, H. 1986. Tanganiika-ko no hatu-hinode (Sunrise on New Year's Day along Lake Tanganyika). The Yomiuri (evening) 8.1.1986 (in Japanese) (also in 417).

385. Kawanabe, H. 1986. Renku to seibutu-gunsyuu (Linked poetry, Renku, and biological community). The Yomiuri (evening) 22.1.1986 (in Japanese) (also in 417).

386. Kawanabe, H. 1986. Myunhen no Bahha (Bach in Munich). The Yomiuri (evening) 29.1.1986 (in Japanese) (also in 417).

387. Kawanabe, H. 1986. Kikoku tyokugo no tyousyoku no seki de (At breakfast upon returning from abroad). The Yomiuri (evening) 5.2.1986 (in Japanese) (also in 417).

388. Kawanabe, H. 1986. Hutuka-yoi to Syokusanzin (A hangover and Syokusanzin, a writer in Edo Era). The Yomiuri (evening) 12.2.1986 (in Japanese) (also in 417).

389. Kawanabe, H. 1986. 'Kari to gyorou' no sinpo de (On the symposium on 'hunting and fishing'). The Yomiuri (evening) 19.2.1986 (in Japanese) (also in 417).

390. Kawanabe, H. 1986. Uo no sei-tenkan (Sex change in fishes). The Yomiuri (evening) 26.2.1986 (in Japanese) (also in 417).

391. Kawanabe, H. 1986. Daigaku nyuusi no hi (Entrance examination day to universities). The Yomiuri (evening) 5.3.1986 (in Japanese) (also in 417).

392. Kawanabe, H. 1986. Bou-tyuu kan ari; kan-tyuu bou ari (Leisurely moments on the busiest day; hectic moments of press on the leisure day). The Yomiuri (evening) 12.3.1986 (in Japanese) (also in 417).

393. Kawanabe, H. 1986. Tori no hansyoku (The breeding of birds). The Yomiuri (evening) 19.3.1986 (in Japanese) (also in 417).

394. Kawanabe, H. 1986. Shinji-ko no haru (Spring at Lake Shinji-ko). The Yomiuri (evening) 26.3.1986 (in Japanese) (also in 417).

395. Kawanabe, H. 1986. Usiro-sugata (A figure from behind). The Yomiuri (evening) 2.4.1986 (in Japanese) (also in 417).

396. Kawanabe, H. 1986. Iwana to yamame no sinpo (Symposium on charrs and masu-salmon). The Yomiuri (evening) 9.4.1986 (in Japanese) (also in 417).

397. Kawanabe, H. 1986. Shibuya-san no koto (On the late ecologist, K. Shibuya). The Yomiuri (evening) 16.4.1986 (in Japanese) (also in 417).

398. Kawanabe, H. 1986. Zaiiru to no kenkyuu-kyouryoku (Research cooperation with Zairian scientists). The Yomiuri (evening) 23.4.1986 (in Japanese) (also in 417).

399. Kawanabe, H. 1986. O-hana-mi (Flower viewing). The Yomiuri (evening) 30.4.1986 (in Japanese) (also in 417).

400. Kawanabe, H. 1986. Miwataseba yama-moto kasumu (Overlooking hills spreading with haze). The Yomiuri (evening) 7.5.1986 (in Japanese) (also in 417).

401. Kawanabe, H. 1986. Kokudaikyou no 'kettei' ni omou (An opinion on the 'decision' of the Union of National Universities). The Yomiuri (evening) 14.5.1986 (in Japanese) (also in 417).

402. Kawanabe, H. 1986. Modan-aato-ten nite (At the exhibition of the Modern Art Society). The Yomiuri (evening) 21.5.1986 (in Japanese) (also in 417).

403. Kawanabe, H. 1986. Kioku no yami (A memory of darkness). The Yomiuri (evening) 28.5.1986 (in Japanese) (also in 417).

404. Kawanabe, H. 1986. Sanzyuppun-kan no ha-migaki (Thirty

minutes for brushing teeth). The Yomiuri (evening) 4.6.1986 (in Japanese) (also in 417).
405. Kawanabe, H. 1986. Atama no razieeita (Radiator in brain engine). The Yomiuri (evening) 11.6.1986 (in Japanese) (also in 417).
406. Kawanabe, H. 1986. 'Naoru' to iu go (On the word 'naoru'). The Yomiuri (evening) 18.6.1986 (in Japanese) (also in 417).
407. Kawanabe, H. 1986. Kikyorai no zi (It's time to go home). The Yomiuri (evening) 25.6.1986 (in Japanese) (also in 417).
408. Kawanabe, H. 1986. Syohyou: 'Kakutyou sareta hyougen-gata, Rityaado Dookinsu' (Book review: 'Extended phenotype' by R. Dawkins). Science Salon 10(1): 2 (in Japanese) (also in 417).
409. Kawanabe, H. 1986. Syohyou: 'Kyoto no doubutu vol. 2' (Book review: 'Animals in Kyoto vol. 2' edited by K. Shibuya). The Kyoto Minpo 16.3.1986 (in Japanese).
410. Kawanabe, H. 1986. Toubi wo kazaru beki kantou-gen (The lead article which should adorn the final issue). Tokutei-kenkyuu 'Seibutu no syakai-kouzou to tekiou-senryaku' Nyuusu (News for the Special Research on 'Social structure and adaptive strategy of living organisms') 10: 1–2 (in Japanese).
411. Kawanabe, H. 1986. 'Nihon Seitai-gakkai-si' no sai-syuppatu ni atatte (On the new start of 'Japanese Journal of Ecology'). Japan. J. Ecol. 36: 1 (in Japanese) (also in 417).
412. Kawanabe, H. 1986. Hakkan ni yosete (Preface). pp. 2–3. *In:* Sin Shinano dou-syoku-butu-ki 1. Kemono, sakana (A New History of Animals and Plants in Shinano Region 1. Mammals and Fishes), Shinano-Mainichi Shinbunsha, Nagano (in Japanese).
413. Kawanabe, H. 1986. 'Makino Yonekiti seibutu-gasyuu' ni yoseru (A few words on the 'Biological illustration painted by Y. Makino'). pp. 1–4. *In:* Makino Yonekiti seibutu-gasyuu (Collection of Y. Makino's Biological Illustrations), Kodansha, Tokyo (in Japanese).
414. Ando, T. & H. Kawanabe. 1986. 'Huukyou-simatu' wo megutte (Around 'Particulars of crazy elegance' by T. Ando). Bungaku-kai (Literal world) 1986(9): 166–183 (in Japanese).
415. Kawanabe, H. 1986. Cooperative study on the ecology of Lake Tanganyika between Japanese and Zairian scientists, with special reference to mutual interactions among fishes. Physiol. Ecol. Japan 23: 119–128.
416. Kawanabe, H. 1987. Sumi-wake (Habitat segregation). Encyclopedia Nipponica 2001 13: 200 (in Japanese).
417. Kawanabe, H. 1987. Henken no seitaigaku (Ecology with a bias). Nosangyoson Bunka-kyokai, Tokyo. 250 pp. (in Japanese).
418. Kawanabe, H. & M. Nagoshi (ed.) 1987. Ecological and limnological study on Lake Tanganyika and its adjacent regions 4. Kyoto University, Kyoto. 150 pp.
419. Kawanabe, H., M. Nagoshi & M.K. Kwetuenda. 1987. Preface. p. 4. *In:* H. Kawanabe & M. Nagoshi (ed.) Ecological and Limnological Study on Lake Tanganyika and its Adjacent Regions 4, Kyoto University, Kyoto.
420. Kawanabe, H., M. Nagoshi & M.K. Kwetuenda. 1987. Introduction. pp. 13–15. *In:* H. Kawanabe & M. Nagoshi (ed.) Ibidem 4, Kyoto University, Kyoto.
421. Kawanabe, H., H. Hayashi, Y. Nagata, A. Goto & S. Nishijima. 1987. Fiirudo zukan tansuigyo (Field book of freshwater fishes in Japan). Tokai University Press, Tokyo. 188 pp. (in Japanese).
422. Kawanabe, H. 1987. Hoppou-ken no uo tati: Hokkaido no tansuigyo (Fishes in northern regions: freshwater fishes in Hokkaido) 1. Hoppou-ken 59: 95–98 (in Japanese).
423. Kawanabe, H. 1987. Bikou-rui (On 'Rhinogradentia' by Harold Stuempke in a blurb around 'Rhinogradentia'). Shisakusha, Tokyo (in Japanese).
424. Morisita, M. & H. Kawanabe (ed.) 1986. Showa 60 nen-do ayumodoki seitai-bunpu tyousa houkokusyo (Report on the ecological distribution of the kissing loach, *Leptobotia curta*, in 1985). Department of Fisheries, Kyoto Prefectural Government, Kyoto. 36 pp. (in Japanese).
425. Kawanabe, H. 1987. Kenkyuu-kyouryoku no kati to zyuuyousei e no rikai wo (Understanding the value and importance of international research cooperation). EXPERT (Japan International Cooperation Agency), Tokyo 1987 (2): 16 (in Japanese).
426. Kawanabe, H. 1987. Keiryuu sanka: tondemonai hito (A paean to mountain streams: an unexpected person). The Mainichi 16.6.1987 (in Japanese).
427. Kawanabe, H. 1987. Suisen no kotoba (A few words of recommendation). pp. 7. *In:* Yamame yo eien ni: waga kokoro no genryuu ('Long Live the Masu Salmon: the Source of My Heart' by T. Iizuka), Kosaido Shuppan, Tokyo (in Japanese).
428. Kawanabe, H. 1987. Ayu no sessyoku-syuusei (Feeding habits of the ayu). pp. 66–67. *In:* K.E. Banister & T. Iwai (ed.) Doubutu dai-hyakka (Encyclopedia of animals), 13, Gyorui (Fishes), Heibonsha, Tokyo (in Japanese).
429. Kawanabe, H. & E. Harada (ed.) 1987. Ecological play on the earth: ecological niche and its related problems. A forum for basic sciences of Kyoto Prizes 1986. Physiol. Ecol. Japan (spec. vol.) 24: 1–87.
430. Kawanabe, H. & E. Harada. 1987. Preface. Physiol. Ecol. Japan 24: 51–52.
431. Kawanabe, H. 1987. Niche problems in mutualism. Physiol. Ecol. Japan 24: 75–80.
432. Kawanabe, H. 1988. Sinzirarenai hanasi (Unbelievable story). Rigakubu tosyo Nyuusu (Science Library News), Kyoto University 1: 6 (in Japanese).
433. Kawanabe, H. 1988. Ahurika ni yosete omou (A consideration about Africa). Bulletin of the Kansai Nature Conservation Organization 15: 1 (in Japanese).
434. Kawanabe, H. 1988. Kasen no arubeki sugata (Rivers in their natural condition). Ibidem 15: 37–39 (in Japanese).
435. Hirota, I., H. Kawanabe, H. Sumitomo, G. Shibaike, T. Yoneyama & N. Sumitomo. 1988. 'Sizen-kagaku I' no kougi wo megutte (On the lecture 'Natural sciences I'). pp. 3–39.

In: College of Liberal Arts (ed.) Kankyou-mondai e no sougou-teki apurooti no houhou to hongaku ni okeru kyouiku e no tekiyou no kenkyuu (Methods of a Synthesized Approach for Environmental Problems and its Application for Education in Our University), Kyoto University, Kyoto (in Japanese).

436. Kawanabe, H., M. Ohnishi, T. Ohta & T. Hirai. 1988. Ayutte nani, nani ga tomo-duri? (What is the ayu; what is its decoy fishing?). pp. 27–33. *In:* Sin uo siriizu (New series on fishes), 6, Sin ayu no subete (All of the Ayu), Shuukan Turi Sandee-sha, Osaka (in Japanese).
437. Kawanabe, H. 1988. 'Kokusai tiiki-kenkyuu' ni yoseru (For the 'International area studies'). International Area Studies (Kyoto University) 1: 39–40 (in Japanese).
438. Kawanabe, H. 1988. Denzaburo Miyadi. Encyclopedia Nipponica 2001, 22: 471 (in Japanese).
439. Kawanabe, H. 1988. 'Shiretoko no doubutu' suisen-bun (A recommendation letter to 'Animals in Shiretoko Peninsula'). A pamphlet from Hokkaido University Press, Sapporo (in Japanese).
440. Kawanabe, H. 1988. Seitaigaku zakkan (Equivocal thoughts on ecology). 1. Mizuumi no hotori nite (From a lakeshore). Kaiyo to Seibutu (Aquabiology) 10: 340–341 (in Japanese).
441. Kawanabe, H. & M.K. Kwetuenda (ed.) 1988. Ecological and limnological study on Lake Tanganyika and its adjacent regions 5. Kyoto University, Kyoto. 109 pp.
442. Kawanabe, H. & M.K. Kwetuenda. 1988. Introduction. pp. 11–13. *In:* H. Kawanabe & M.K. Kwetuenda (ed.) Ecological and Limnological Study on Lake Tanganyika and its Adjacent Regions 5, Kyoto University, Kyoto.
443. Kawanabe, H. 1988. Doubutu no miti (The road of animals). Kyoto University Bulletin 360: 552–553 (in Japanese).
444. Kawanabe, H. 1988. Miyadi Denzaburo Sensei wo itamu (Grieving for Professor Emeritus D. Miyadi). The Yomiuri (evening) 25.10.1988 (in Japanese).
445. Kawanabe, H. 1988. Miyadi Denzaburo san to watakusi (Denzaburo Miyadi and I). The Mainichi (evening) 31.10.1988 (in Japanese).
446. Kawanabe, H. 1988. Nagara-gawa no kawa-masu (satukimasu) to ayu ni tuite (Populations of masu-salmon and ayu in the Nagara River). pp. 6–10. *In:* E. Kimura (ed.) 'Nagaragawa kakou-zeki hantai no tame no kokusai sinpoziumu' kiroku (Report of the 'International Symposium Against Dam Construction at the Mouth of the Nagara River'), Association for Conservation of Freshwater Fishes, Osaka (in Japanese).
447. Kawanabe, H. 1988. Mizu to midori (Water and green). pp. 12. *In:* A. Miyawaki (ed.) Midori no mori kara no messeezi (Message from green forests), Kanagawa Prefectural Government and Yamato City Government, Yokohama (in Japanese).
448. Kawanabe, H. 1988. Tosyo to gakuzyutu zyouhou (Books and scientific information). The Kyoto University Library Bulletin 25(3): 1–3 (in Japanese).
449. Kawanabe, H. 1988. Kankyou-youryou to seitaigaku (Environmental capacity and ecology). pp. 49–58. *In:* S. Ebise (ed.) Proceedings of the 1st Symposium on Aquatic Carrying Capacity and its Application, The National Institute for Environmental Studies, Tsukuba (in Japanese) (also in 711).
450. Kawanabe, H. 1988. Seibutu ni mirareru kyousou to kyoudou: Nihon to Ahurika no tansuigyo wo rei to site (Competition and cooperation among living organisms, with special emphasis on freshwater fishes in Japan and Africa). Ohtani Gakuho (The Journal of Buddhist Studies and Humanities) 68(3): 87–92 (in Japanese).
451. Kawanabe H. 1989. Seitaigaku zakkan (Equivocal thoughts on ecology). 2. Miyadi Denzaburo san no si ni omou (Mourning for Professor Emeritus D. Miyadi, a great ecologist). Kaiyo to Seibutu (Aquabiology) 11: 106–107 (in Japanese) (also in 709).
452. Kawanabe, H. 1989. Idem (idem). 3. Iwana sakura-masu no kokusai-kaigi (On the 'International symposium on char(r(r))s and masu-salmon'). Kaiyo to Seibutu (Aquabiology) 11: 196–197 (in Japanese).
453. Kawanabe, H. 1989. Aimai koso ga kanzin: seibutu no gunsyuu to wa douiu mono ka? (The importance being equivocal: what are biological communities?). Hermes 20:166–173 (in Japanese).
454. Kawanabe, H. & N. Mizuno. 1989. Kawa to mizuumi no uo (Fishes in rivers and lakes) 1. Hoikusha, Osaka. 200 pp. (in Japanese).
455. Kawanabe, H. & E. Harada. 1989. Obituary: Denzaburo Miyadi (1901–1988). Zool. Sci. 6: 619–621.
456. Inoue, T., A. Utida, M. Kinoshita, Y. Kondo, M. Shibata, K. Tutumi, T. Muraoka, K. Moriya, M. Yoshimura, K. Kawakami, H. Kawanabe & Y. Murai (ed.) 1989. Bunka-zai yougo ziten (Encyclopedia of terms for cultural properties). Tankosha, Kyoto. 524 pp. (in Japanese).
457. Kawanabe, H. 1989. U-kawa no ko (A son of the U-kawa River). pp. 3–5. *In:* K. Huzisaki (ed.) Kawa to hito to hurusato to: Ukawa (A River, Human Beings and a Native Village: Ukawa), Kami-ukawa Kominkan, Tango (in Japanese).
458. Kawanabe, H. 1989. Gunsyuu-seitaigaku kara (From a view point of community ecology). pp. 28–42. *In:* H. Shimizu (ed.) Sinpoziumu 'Atarasii seimei-zou wo saguru: zyouhou no kanten kara' (Symposium on 'New Figure of Life: from the Viewpoint of Information'), National Institute for Research Advancement and Toyota Foundation, Tokyo (in Japanese).
459. Kawanabe, H. & N. Mizuno (ed.) 1989. Yamakei karaa meikan: Nihon no tansuigyo (Freshwater fishes of Japan). Yama-to-Keikoku-sha, Tokyo. 720 pp. (in Japanese).
460. Kawanabe, H. 1989. Dakara kono hito kono 5-saku, dokusya no seisin wo undou saseru sakuhin: Kawanabe Hiroya ga susumeru Matusita Ryuiti (A person and 5 books: Ryuiti Matusita's works for the promotion of readers' spirits, as recommended by H. Kawanabe). Asahi Journal 31(29): 99 (in Japanese).

461. Kawanabe, H. 1989. Aquaculture and fisheries management. Japan Branch of Blackwell Scientific Publishers, Tokyo (pamphlet).
462. Kawanabe, H. 1989. Ueno Masuzo Sensei wo sinobu (In memory of the late Professor Emeritus Masuzo Ueno). Gakuto (Maruzen, Tokyo) 86(10): 13–15 (in Japanese) (also in 709).
463. Kawanabe, H., F. Yamazaki & D.L.G. Noakes (ed.) 1989. Biology of charrs and masu salmon. Physiol. Ecol. Japan, Spec. vol. 1. 711 pp.
464. Kawanabe, H. 1989. Japanese char(r(r))s and masu-salmon problems: a review. pp. 3–12. *In:* H. Kawanabe, F. Yamazaki & D.L.G. Noakes (ed.) Biology of Charrs and Masu Salmon, Physiol. Ecol. Japan, Spec. vol. 1.
465. Kawanabe, H. 1989. Seitaigaku zakkan (Equivocal thoughts on ecology). 4. Kokusai Seitai-gakkai no tameni (Regarding the INTECOL meeting at Yokohama in 1990). Kaiyo to Seibutu (Aquabiology) 11: 466–467 (in Japanese).
466. Kawanabe, H. (ed.) 1989. Ecological and limnological study on Lake Tanganyika and its adjacent regions 6. Kyoto University, Kyoto. 110 pp.
467. Kawanabe, H. & M.K. Kwetuenda. 1989. Introduction. pp. 11–14. *In:* H. Kawanabe (ed.) Ecological and Limnological Study on Lake Tanganyika and its Adjacent Regions 6, Kyoto University, Kyoto.
468. Kawanabe, H & M.J. Pearse. 1989. Introduction. pp. 61–62. *In:* H. Kawanabe (ed.) Ibidem 6, Kyoto University, Kyoto.
469. Azuma, M., H. Kawanabe, S. Nishijima & M. Nishida. 1989. Ryukyu-ayu ni tuite no memo (Memorandum of the Ryukyu-ayu, *Plecoglossus altivelis ryukyuensis*). Tansuigyo Hogo (Freshwater Fish Conservation) 2: 98 (in Japanese).
470. Kawanabe, H. 1990. Complexity in biological communities, or the importance of being equivocal. pp. 186–194. *In:* H. Shimizu (ed.) Biological Complexity and Information, World Scientific, Singapore.
471. Kawanabe, H. 1990. The SIL workshop of the African Great Lakes: an international symposium on 'Resource use and conservation of the African Great Lakes'. International Lake Environment Committee Newsletter (ILEC), Otsu 12: 1–2.
472. Kawanabe, H. 1990. Seitaigaku zakkan (Equivocal thoughts on ecology). 5. Nagara-gawa kakou-zeki no koto (The Nagara River and its estuary-dam construction). Kaiyo to Seibutu (Aquabiology) 11: 120–121 (in Japanese) (also in 709).
473. Kawanabe, H. 1990. 'Kyoto minzoku-si', 'Hontyou syokkan' nado (Book review: 'Folklore in Kyoto' by Y. Inoue, 'Encyclopedia of Japanese foods' by H. Hitomi, etc.). Monthly Encyclopedia 1990(2): 36 (in Japanese).
474. Kawanabe, H. & M. Yuma. 1990. Tanganiika-ko gyorui no sigen, seitai, sizen-hogo (Resource, ecology, and conservation of fishes in Lake Tanganyika). Annual Report of African Area Studies, Kyoto University 4: 18–19 (in Japanese).
475. Kawanabe, H., T. Ohgushi & M. Higashi (ed.) 1990. Ecology for tomorrow. Physiol. Ecol. Japan 27 (spec. vol.). 205 pp.
476. Kawanabe, H. 1990. Tansui-ka go no gyo-zoku no seisoku-zyoukyou no yosoku tuika: Tottori, Shimane ryou-ken no 'Zyogensya-kaigi kenkai' ni taisuru iken wo tyuusin ni (Supplemental prediction of lives of fishes and other important fishery-related animals after freshening, with special emphasis to answer 'The opinion of advisers' of Tottori and Shimane Prefectural governments). pp. 82–88. *In:* I. Minami (ed.) Shinji-ko, Naka-umi no gyokai-rui (Fishes and Other Important Fishery-related Animals in Lake Shinji-ko and Naka-umi), Gyorui Syou-iinkai, Teisei-doubutu Syou-iinkai; Shinji-ko, Naka-umi Tansui-ka ni tomonau Mizu-kanri oyobi Seitai-henka ni kansuru Kenkyuu-iinkai (Subcommittees of Fishes and of Benthic Animals; Committee of Water Management and Ecological Change by the Freshening of Lake Shinji and Naka-umi), Japanese Society of Agricultural Engineering, Tokyo (in Japanese).
477. Kawanabe, H. 1990. Tokusyuu ni atatte (For special articles of new ecology). Heredity 44(6): 9–10 (in Japanese).
478. Kawanabe, H. 1990. Hansu Burinkaa (On 'Hans Brinker'). The Tosho 1990(7): 18 (in Japanese).
479. Kawanabe, H. & N. Mizuno. 1990. Kawa to mizuumi no uo (Fishes in rivers and lakes) 2. Hoikusha, Osaka. 216 pp. (in Japanese).
480. Kawanabe, H. (ed.) 1990. Nagara-gawa kakou-zeki zigyou no mondai-ten: tyuukan houkokusyo (Problems of the dam construction at the mouth of the Nagara River). Nature Conservation Society of Japan, Tokyo. 133 pp. (in Japanese).
481. Kawanabe, H. 1990. Hazime ni (Introduction). pp. 5–6. *In:* H. Kawanabe (ed.) Nagara-gawa kakou-zeki zigyou no mondai-ten: tyuukan houkokusyo (Problems of the Dam Construction at the Mouth of the Nagara River), Nature Conservation Society of Japan, Tokyo (in Japanese).
482. Kawanabe, H. 1990. Koidzumi Kiyoaki hen 'Kiso-san-sen kakou sigen tyousa houkoku' ni tuite (On 'Report on resources of the river mouths of the Kiso, Nagara and Ibi rivers' edited by K. Koidzumi). pp. 34–38. *In:* H. Kawanabe (ed.) Ibidem (ibidem), Nature Conservation Society of Japan, Tokyo (in Japanese).
483. Kawanabe, H. 1990. Kasen-gizyutusya no tame no (?) seitaigaku: honmono no kawa wo motomete (Ecology for (?) river engineers: searching for natural rivers). Sui-ko-gaku siriizu (Water engineering series, Japanese Society of Civil Engineering, Section of Water Physics, Tokyo) 90-A–2: 1–7 (in Japanese) (also in 709).
484. Kawanabe, H. & Z. Mohri. 1990. Seitaigaku kara mita kakou-zeki no kiken-sei (Danger of a river mouth dam, from an ecological point of view). BE-PAL 10(9): 83–87 (in Japanese).
485. Kawanabe, H. 1990. Dai–5-kai Kokusai Seitaigaku Kaigi ni tuite (On the 5th International Congress of Ecology). Bulletin of Japan Science Council 31: 111–112 (in Japanese).
486. Kawanabe, H. 1990. Kakou-zeki kouzi no itizi tyuusi wo: seibutu idensi no tayousei mamore (Protesting the dam construction at the mouth of the Nagara River: conserving

the genetic diversity). Newspapers under Kyodo Tsushin 15.12.1990 (in Japanese).

487. Kawanabe, H. 1990. Seitaigaku zakkan (Equivocal thoughts on ecology). 6. Kokusai Seitaigaku Kaigi no tameni (For the INTECOL meeting at Yokohama in 1990) 2. Kaiyo to Seibutu (Aquabiology) 12: 470–471 (in Japanese).
488. Kawanabe, H., A. Endo & Y. Ezaki (translators) 1990. C.S. Eruton, 'Doubutu gunsyuu no yousiki' ('The patterns of animal communities' by C.S. Elton). Shisakusha, Tokyo. 649 pp. (in Japanese).
489. Kawanabe, H. 1990. Kanyakusya no hodohodo no nagasa no, sikasi kono hon e no omoiire to site wa mizikasugiru atogaki (Moderately but inadequately short postscripts of the advisory editor of translation of this book). pp. 645–649. *In:* H. Kawanabe, A. Endo & Y. Ezaki (translators) C.S. Eruton, 'Doubutu gunsyuu no yousiki' ('The patterns of animal communities' by C.S. Elton), Shisakusha, Tokyo (in Japanese) (also in 709).
490. Kawanabe, H. 1991. Tuki no sekai (Il mondo della luna, or World of the moon). The Mainichi (evening) 8.1.1991 (in Japanese) (also in 709).
491. Kawanabe, H. 1991. Rikouna me-gitune no monogatari ('Příhody lišky bystroušky', or 'The cunning little fox'). The Mainichi (evening) 22.1.1991 (in Japanese) (also in 709).
492. Kawanabe, H. 1991. Ningen no koe (La voix humaine, or Human voice). The Mainichi (evening) 29.1.1991 (in Japanese) (also in 709).
493. Kawanabe, H. 1991. Forusutahhu (Falstaff). The Mainichi (evening) 5.2.1991 (in Japanese) (also in 709).
494. Kawanabe, H. 1991. Supeedo no zyoou ('Pikovaya dama', or 'The queen of spades'). The Mainichi (evening) 12.2.1991 (in Japanese) (also in 709).
495. Kawanabe, H. 1991. Tasukeyo, syou-yousei-tati (Help, help, the Globolinks!). The Mainichi (evening) 19.2.1991 (in Japanese) (also in 709).
496. Kawanabe, H. 1991. Mori no koe (Forests' voice). The Mainichi (evening) 26.2.1991 (in Japanese) (also in 709).
497. Kawanabe, H. 1991. Yuki (La neige ou le nouvel Eginarde, or Snow). The Mainichi (evening) 5.3.1991 (in Japanese) (also in 709).
498. Kawanabe, H. 1991. Kozyou no bizin ('La donna del lago' or 'The lady of the lake'). The Mainichi (evening) 12.3.1991 (in Japanese) (also in 709).
499. Kawanabe, H. 1991. Mateki, Tanhoizaa ('Zauberfloete und Tannhaeuser' or 'The magic flute and Tannhaeuser'). The Mainichi (evening) 19.3.1991 (in Japanese) (also in 709).
500. Kawanabe, H. 1991. Kitai ('Erwartung' or 'Expectation'). The Mainichi (evening) 26.3.1991 (in Japanese) (also in 709).
501. Kawanabe, H. & M. Nagoshi (ed.) 1991. Ecological and limnological study on Lake Tanganyika and its adjacent regions 7. Kyoto University, Kyoto. 142 pp.
502. Kawanabe, H. 1991. Opening address of the Oji International Seminars on 'Mutualism, cooperation and organization in natural communities'. pp. 102–103. *In:* H. Kawanabe & M. Nagoshi (ed.) Ecological and Limnological Study on Lake Tanganyika and its Adjacent Regions 7, Kyoto University, Kyoto.
503. Kawanabe, H. 1991. Opening address for the 5th International Congress of Ecology. pp. 106–107. *In:* H. Kawanabe & M. Nagoshi (ed.) Ibidem 7, Kyoto University, Kyoto.
504. Date, Y. & H. Kawanabe (ed.) 1991. Kisui-iki: sono yutakana seitaikei wo motomete (Brackish water zone: searching their rich ecosystem). Shimane Prefectural Government, Matsue. 187 pp. (in Japanese).
505. Kawanabe, H. 1991. Syusaisya aisatu (Welcome address on behalf of the organizing committee). pp. 3–4. *In:* Y. Date & H. Kawanabe (ed.) Kisui-iki: sono yutakana seitaikei wo motomete (Brackish Water Zone: Searching their Rich Ecosystem), Shimane Prefectural Government, Matsue (in Japanese).
506. Kawanabe, H. 1991. Yama kara umi e no keiretu ni okeru kisui-iki (Brackish water zone, the central chain of ecosystem series from mountains to ocean). pp. 173–177. *In:* Y. Date & H. Kawanabe (ed.) Ibidem (ibidem), Shimane Prefectural Government, Matsue (in Japanese).
507. Kawanabe, H., Y. Date, T. Kikuchi, E. Harada, I. Miyata, F. Koike & H. Kunii (ed.) 1991. For richer ecological systems of brackish water zones. Kokuchosha, Matsue. 23 pp.
508. Kawanabe, H. 1991. Opening address 2. pp. 3–4. *In:* H. Kawanabe, Y. Date, T. Kikuchi, E. Harada, I. Miyata, F. Koike & H. Kunii (ed.) For Richer Ecological Systems of Brackish Water Zones, Kokuchosha, Matsue.
509. Kawanabe, H. 1991. Brackish water zone, the central chain of ecosystem series from mountains to ocean, with special reference to Lakes Shinji-ko and Naka-umi System. pp. 223–224. *In:* H. Kawanabe, Y. Date, T. Kikuchi, E. Harada, I. Miyata, F. Koike & H. Kunii (ed.) Ibidem, Kokuchosha, Matsue.
510. Kawanabe, H. 1991. Seitaigaku zakkan (Equivocal thoughts on ecology). 7. Ueda Akinari san ni yosete (The importance of being uncertain). Kaiyo to Seibutu (Aquabiology) 113: 26–27 (in Japanese).
511. Kawanabe, H. 1991. Seibutu to kankyou to no kankei nado (Relationships of living organisms to environment, and so on). The Mainichi (evening) 26.4.1991 (in Japanese).
512. Kawanabe, H. 1991. Kawa no uo-tati to mori (River fishes and forests). Zuiso Shinrin 25: 28–29 (in Japanese).
513. Kawanabe, H. & M.M. Gashagaza. 1991. Ecological diversity on Lake Tanganyika with special reference to fishes. pp. 69. *In:* A.S. Cohen (ed.) Report on the 1st International Conference on the Conservation and Biodiversity of Lake Tanganyika, Department of Geoscience, University of Arizona, Tuscon (also in French).
514. Kawanabe, H. 1991. Seitaigaku zakkan (Equivocal thought on ecology). 8. Noma Hiroshi san ni yosete (An obituary for Mr. H. Noma, a writer and novelist). Kaiyo to Seibutu (Aquabiology) 13: 174–175 (in Japanese) (also in 709).
515. Kawanabe, H. 1991. Syohyou: 'Madagasukaru sizen kikou' (Book review: 'Naturalist's journey of Madagascar Island'

by S. Yamagishi). The Shinano Mainichi, 12.5.1991 (in Japanese).
516. Kawanabe, H. 1991. Kiso-kenkyuu no ba, Seitaigaku Kenkyuu Sentaa (The Center for Ecological Research as the institute for basic studies). Kyoto University News 20.5.1991 (in Japanese).
517. Kawanabe, H. 1991. Kenkyuu-situ o-zyama houmon (A sudden visit to Professor H. Kawanabe's office). Shiga Minpo 26.5.1991 (in Japanese).
518. Kawanabe, H. 1991. Kyoto-daigaku Seitaigaku Kenkyuu Sentaa no hossoku ni atatte (At the opening of the Center for Ecological Research, Kyoto University). News of Center for Ecological Research, Kyoto University 1: 1–2 (in Japanese).
519. Kawanabe, H. 1991. Kawa-uo no me kara mita kasen kankyou (River environments from the perspective of river fishes). pp. 5–8. *In:* Y. Taniguchi (ed.) Korekara no kasen-kankyou no hozen wo kangaeru fooramu (Forum for Future Conservation of River Environments), Kohchi-ken Naisuimen Gyogyou Kyoudou-kumiai Rengoukai (Union of Inland Fisheries Cooperation in Kohchi Prefecture), Kohchi (in Japanese).
520. Kawanabe, H. & C. Hayashi. 1991. We rabu otoko-tati: kii-waado wo toku (We love men: analyze keywords). 1. Ekorozii (Ecology). The Yomiuri (evening) 20.7.1991 (in Japanese).
521. Kawanabe, H. 1991. Preface for the special issue of plenary lectures at INTECOL '90. Ecol. Res. 6: 127–128.
522. Kawanabe, H. 1991. Seibutu tayousei wo tou: Tanganiika-ko no uo (For biological diversity: fishes in Lake Tanganyika). World Wildlife Fund 1991(11): 11 (in Japanese).
523. Kawanabe, H. 1991. Ima no sizen ni kore ga iitai (I would like to discuss the present status of nature). Sizen Hogo (Nature Conservation) 1991(10): 5 (in Japanese).
524. Kawanabe, H. 1991. Kyousei-kei to site no seibutu-gunsyuu (Biological community as a mutualistic system). Bukkyo Bessatsu (Supplements of Buddhist Journal) 6: 21–30 (in Japanese).
525. Kawanabe, H., E. Harada & A. Endo (ed.) 1991. Seitaigaku to doubutugaku kankei yougo (Words of ecology and zoology). Kozien, 4th ed. Iwanami Shoten, Tokyo. 2858 pp. (in Japanese).
526. Kawanabe, H. 1991. Conservation of freshwater fishes in the Japanese Archipelago, with emphasis on relation to its physical environment and ecological diversity. pp. 18–33. *In:* K.-H. Chang & Y.-S. Lin (ed.) Proceedings of the First International Symposium on Wildlife Conservation, ROC, Council of Agriculture, Taipei.
527. Kawanabe, H. 1991. Kyoto Daigaku Seitaigaku Kenkyuu Sentaa kaisyo-siki aisatu (Opening address of the Center for Ecological Research, Kyoto University) 16.10.1991. News of Center for Ecological Research, Kyoto University 4: 1–2 (in Japanese).
528. Kawanabe, H. 1991. 21-seiki no tikyuu-kankyou sinpoziumu: doubutu-seitaigaku no tatiba kara (Symposium on 'Global environment for the 21st century: from the standpoint of animal ecology'). The Kanagawa 18.11.1991 (in Japanese).
529. Inaba, M., H. Kawanabe & K. Fukushima. 1992. Kankyou denaosi gannen e: mamore Ko-koku no seitaikei (Let us start again to environmental years: conserve ecosystems in the lake country). The Kyoto 1.1.1992 (in Japanese).
530. Kawanabe, H. 1992. 1991-nen dokusyo ankeeto (Questionnaires on reading books in 1991). Misuzu 370: 52 (in Japanese).
531. Kawanabe, H. 1992. 'Sizen' no nani wo mamoru no ka? (What of 'nature' should we conserve?). Bungei-Shunjuu 70(2): 206–213 (in Japanese) (also in 709).
532. Kawanabe, H. 1992. Yume wo katati ni (To make a form from dreams). 70. Tikyuu no kankyou-mondai to wa, tayou na seibutu no kyouzon wo kangaeru koto da (Coexistence of diversified organisms is one of the most important issues in global change). The Nihon Keizai 2.2.1992 (in Japanese).
533. Kawanabe, H. 1992. Zyouhou to senmon-syokuin to (Information and librarians). The Kyoto University Library Bulletin 28(4): 1–3 (in Japanese).
534. Kawanabe, H., N. Akiyama, T. Tanaka & T. Hara. 1992. Nihon no kawa ni mirai wa aruka? (Will there be a future in Japanese rivers?). Sizen Hogo (Nature Conservation) 1992 (3): 15–19 (in Japanese).
535. Kawanabe, H. 1992. Kamo-gawa no 'hukugen' ('Re-construction' of the Kamo River). pp. 159–161. *In:* S. Tanaka (ed.) Dam to osyou (Dam and a Priest), Hokuto Shuppan, Tokyo (in Japanese).
536. Kawanabe, H. 1992. Kyouzon no kikou kaimei e: tikyuu zyou no tayouna seibutu-syu (Analysis of coexistence mechanisms: diversified organisms on the earth). Kagaku Shinbun 10.4.1992 (in Japanese).
537. Kawanabe, H. 1992. Tikyuu-kyousei-kei to seitaigaku: Kyoto Daigaku Seitaigaku Kenkyuu-sentaa (Symbiotic biosphere and ecology: the Center for Ecological Research, Kyoto University). Monbu Jiho 1992(3): 40–43 (in Japanese).
538. Kawanabe, H. 1992. Sizen no mokei, moderu (A model of nature). pp. 2. *In:* 'Garasu no naka no dai-sizen: suisou reiauto sakuhin-syuu (The Greatness of Nature in a Glass: Works of Water Plant Layout) by H. Amano, Marin Kikaku, Tokyo (in Japanese).
539. Kawanabe, H. (ed.) 1992. Nagara-gawa kakou-zeki zigyou no mondai-ten: dai–2-zi houkokusyo (Problems of the dam construction at the mouth of the Nagara River: the second report). Nature Conservation Society of Japan, Tokyo. 73 pp. (in Japanese).
540. Kawanabe, H. 1992. Hazime ni (Introduction). pp. 5–6. *In:* H. Kawanabe (ed.) Nagara-gawa kakou-zeki zigyou no mondai-ten: dai–2-zi houkokusyo (Problems of the Dam Construction at the Mouth of the Nagara River: the Second Report), Nature Conservation Society of Japan, Tokyo (in Japanese).
541. Kawanabe, H. 1992. Doubutu no sigen-ryou kara mita gyorou (Reconstruction of ancient fishing on the basis of population and production of fishes). pp. 80–99. *In:* S. Koyama

(ed.) Syuryou to gyorou: Nihon-bunka no genryuu wo saguru (Hunting and Fishing: Investigations of the Origin of Japanese Culture), Yuzankaku, Tokyo (in Japanese) (also in 711).

542. Kawanabe, H. 1992. Mizu to sakana to ningen to (Water, fishes, and human beings). Hitori-Hutari 42: 16–17 (in Japanese).

543. Kawanabe, H. 1992. 'Sizeon' eui mu od ji ki neum wod in ga (What parts of 'nature' should we conserve). Japan Forum 1992: 150–159 (in Korean).

544. Kawanabe, H. 1992. Hazime ni (Preface). News of Center for Ecological Research, Kyoto University 7: 1–2 (in Japanese).

545. Kawanabe, H. 1992. Dai–1-kai IGBP kokusai sinpoziumu 'Kankyou hendou to seibutu-gunsyuu no tayousei – danryokusei' (The first IGBP international symposium on the 'Environmental variability, diversity, and flexibility of biological communities'). Japanese Scientific Monthly 45:94 (in Japanese).

546. Kawanabe, H., K. Tanida, Y. Takemon, K. Hiramatsu, I. Silva, K. Nakata & T. Kumekawa. 1992. Houhu na doubutu-gunsyuu hozen no tame ni hituyouna kasen-kankyou-youso no kenkyuu (Research on environmental factors in rivers conserving richer animal communities). Tikyuu Kankyou Kenkyuu (Research on the Environment of the Earth) 20: 65–123. (in Japanese).

547. Kawanabe, H. 1992. Cautionary notes for the environmental cause. Japan Echo 19(2): 79–84.

548. Kawanabe, H., M.K. Kwetuenda & M.M. Gashagaza. 1992. Ecological and limnological studies of Lake Tanganyika and its adjacent regions between African and Japanese scientists: an introduction. Mitt. internat. Verein. theor. angewand. Limnologie 23: 79–83.

549. Kawanabe, H. 1992. Sustainable biosphere for the future of the earth. The Earth Summit Times 12.6.1992.

550. Kawanabe, H. 1992. Mamoru beki wa tayouna sizen: Tanganiika-ko ni sakana no seibutu-tayousei wo tou (Diversified nature should be conserved: biodiversity of fishes in Lake Tanganyika) 1. Anima 1992(7): 102–107 (in Japanese) (also in M. Hori (ed.) 1993. Tanganiika ko no uo tati ('Fishes of Lake Tanganyika', Heibonsha, Tokyo).

551. Kawanabe, H. 1992. Idem (idem) 2. Anima 1992(8): 81–87 (in Japanese) (also in M. Hori (ed.) 1993. Ibidem).

552. Kawanabe, H. 1992. Kawa to sakana to ningen (Rivers, fishes, and human beings). Hokkaido Economics 1992(8): 1–8 (in Japanese).

553. Kawanabe, H. 1992. Ato-gaki (Concluding remarks). pp. 246–253. *In:* M. Higashi & T. Abe (ed.) Siriizu tikyuu kyousei-kei (Symbiotic Biosphere Series), 1, Tikyuu kyousei-kei towa nanika? (What are Global Mutual Systems?), Heinbonsha, Tokyo (in Japanese).

554. Kawanabe, H. 1992. Notas de alerta sobre el medio ambiente. Cuadernaos de Japon 5(1): 65–69.

555. Yoneyama, T., N. Aoki, K. Ikegami, G. Iwatsubo, Y. Kaida, M. Kakeya, R. Kawakami & H. Kawanabe. 1992. 'Tiiki-kenkyuu to tiiki-kagaku' no sougou no tameni: touron (Discussions for the synthesizing of 'Area studies and regional sciences'). Kokusai Tiiki-kenkyuu 3: 22–49 (in Japanese).

556. Kawanabe, H. & K. Dai. 1992. 'Tikyuu' ga abunai: ningen ga kankyou wo hakai ('Our planet' is in danger: we have destroyed its environment). Islands Trends (Hawaii) 4.9.1992 (in Japanese).

557. Kawanabe, H. 1992. Seibutu to kankyou no tayousei (Diversity of organisms and environment). Annual Report of Interdisciplinary Research Institute of Environmental Sciences 11: 5–15 (in Japanese) (also in 709).

558. Kawanabe, H. 1992. Imanishi Kinji san no seitaiteki ronri no sekai (World of ecological logic of Professor Emeritus K. Imanishi). Anima 92(10): 34–35 (in Japanese).

559. Kawanabe, H. 1992. La nature n'a que faire de nos bonnes intentions. Cahiers du Japon 53: 66–72.

560. Nakano, S., K.D. Fausch, T. Furukuwa-Tanaka, K. Maekawa & H. Kawanabe. 1992. Resource utilization by bull char and cutthroat trout in a mountain stream in Montana, U.S.A. Japan. J. Ichthyol. 39: 211–217.

561. Ohsawa, M. & H. Kawanabe. 1992. Riodezyaneiro ni okeru kankyou to kaihatu ni kansuru Kokuren Kaigi (Report on activities related to ecology, held along with UNCED in Rio de Janeiro, June 1992). Japan. J. Ecol. 42: 275–282 (in Japanese).

562. Oshima, Y. & H. Kawanabe. 1992. Kokuritu Amerika Kankyou Kenkyuusyo-gun seturitu zyunbi kaigi ni syusseki site (Report of a preparatory meeting on the National Institutes for the Environment (U.S.A.)). Japan. J. Ecol. 42: 283–285 (in Japanese).

563. Kawanabe, H., T. Arima, N. Takada, Y. Nagata & H. Kihira. 1992. Zadankai: Yodo-gawa mondai, seitai tyousa, sizen-hogo (A round-table talk on the Yodo River problems: ecological research and nature conservation). Tansuigyo Hogo (Freshwater Fish Conservation) 1992:175–180.

564. Kafuku, T., T. Miura, Y. Taki, T. Uyeno, M. Matsushima & H. Kawanabe. 1992. Tyuusyutu sita gyosyu no keitai, bunpu, seitai to mondaiten (Morphology, distribution, ecology and problematic points of some fish species). pp. 21–139. *In:* Zenkoku Naisuimen Gyogyou Kyodou-kumiai Rengoukai (Union of Inlandwater Fishery Cooperation) (ed.) Inyuu sureba mondai ni nariuru omona gaikoku-san gyosyu ni kansuru bunken-tyousa (Reference Search on Primarily Dangerous Foreign Fishes in the Future), Fisheries Agency, Tokyo (in Japanese).

565. Kawanabe, H. & K. Watanabe. 1993. Kuni-masu (Black kokanee, *Oncorhynchus nerka kawamurai*). pp. 130–132. *In:* S. Ueno (ed.) Horobi-yuku Nihon no doubutu (Japanese Animals on the Verge of Extinction), Tsukiji Shokan, Tokyo (in Japanese).

566. Kawanabe, H. & K. Watanabe. 1993. Minami-tomiyo (Kyoto ninespine stickleback, *Pungitius kaibarae*). pp. 133–135. *In:* S. Ueno (ed.) Ibidem (ibidem), Tsukiji Shokan, Tokyo (in Japanese).

567. Kawanabe, H. 1993. Seimei tayousei: kankei no soutai-ron kara mita seibutu-gunsyuu to sono sinka (Diversity of life:

biological communities and their evolution from a viewpoint of the whole of interrelationships). pp. 55–87. *In:* H. Kawai, H. Shimizu, Y. Tani, Y. Nakamura, K. Kadowaki & T. Nisikawa (ed.) Iwanami koza, syuukyou to kagaku (Iwanami Series of Religions and Sciences) 6, Seimei to kagaku (Life and Sciences), Iwanami Shoten, Tokyo (in Japanese) (also in 711).

568. Kawanabe, H. 1993. 1992-nen dokusyo ankeeto (Questionnaires on reading books in 1992). Misuzu 382: 68 (in Japanese).

569. Kawanabe, H. 1993. Seimei nettowaaku no hatten to houkai – seimei tayousei: kankei no soutai-ron kara mita gunsyuu to sono sinka (The development and fall of life's networks - diversity of life: communities and their evolution from a viewpoint of the whole of interrelationships). Bussei Kenkyuu 60(1): 25–26 (in Japanese).

570. Kawanabe, H. 1993. Hazime ni (Preface). News of Center for Ecological Research, Kyoto University 14: 1 (in Japanese).

571. Kawanabe, H. 1993. Doubutu no seitai: kyousei-ken to site no tikyuu (Ecology of animals: the earth as symbiotic systems). pp. 80–88. *In:* Y. Hiramoto & H. Mohri (ed.) Sizenkagaku: seimei-hen (Natural Sciences: Life), Japan Open University Press, Tokyo (in Japanese).

572. Kawanabe, H. 1993. Mizoro-ga-ike no sizen (Nature of the moor, Mizoro-ga-ike). 58. Honkakuteki na hogo sisaku wo (Making a genuine policy of conservation). The Kyoto 10.5.1993 (in Japanese).

573. Kawanabe, H. 1993. Seitai-tayousei wo sokusin suru koto no zyuuyousa (The importance of promoting ecological diversity). Kankyou to Kougai (J. Environmental Protection) 22(4): 12–17 (in Japanese) (also in 709).

574. Kawanabe, H. 1993. 'Tikyuu kyousei-ken' wo mamorou (Conserving the 'symbiotic biosphere'). Gekkan Hon no Mado 93(6): 22–25 (in Japanese).

575. Kawanabe, H. 1993. Bunka to site no Shinji-ko (Lake Shinji-ko as a culture). Kisui-ko (Brackish Lakes) 5: 2–5 (in Japanese).

576. Kawanabe, H. 1993. Seitaigaku kara mita tikyuu kankyoumondai no honsitu to kokusai kyouryoku (The essence of global environmental problems from the point of view of ecology and international cooperation). Tikyuu-kankyou Siriizu (Global Environment Series) 12: 1–54 (in Japanese).

577. Nagoshi, M., Y. Yanagisawa & H. Kawanabe (ed.) 1993. Ecological and limnological study on Lake Tanganyika and its adjacent regions 8. Nara Women's University, Nara. 136 pp.

578. Kawanabe, H., J.E. Cohen & K. Iwasaki (ed.) 1993. Mutualism and community organization: behavioural, theoretical and food-web approaches. Oxford Science Publications, Oxford. 426 pp.

579. Kawanabe, H & K. Iwasaki. 1993. Introduction: flexibility and synergism of biological relationships in natural communities. pp. 1–10. *In:* H. Kawanabe, J.E. Cohen & K. Iwasaki (ed.) Mutualism and Community Organization: Behavioural, Theoretical and Food-web Approaches, Oxford Science Publications, Oxford.

580. Kawanabe, H. 1993. Kyoto Daigaku Seitaigaku Kenkyuu Sentaa no dai–2-ki-me ni omou (The second phase of Center for Ecological Research, Kyoto University). News of Center for Ecological Research, Kyoto University 16: 1–2 (in Japanese).

581. Hori, M., M.M. Gashagaza, M. Nshombo & H. Kawanabe. 1993. Littoral fish communities in Lake Tanganyika: irreplaceable diversity supported by intricate interactions among species. Conserv. Biol. 7: 657–666.

582. Kawanabe, H., T. Ohgushi & M. Higashi (ed.) 1993. Sym-Biosphere: ecological complexity for promoting biodiversity. Biology International, Special Issue 29: 1–86.

583. Kawanabe, H. & M. Yuma (ed.) 1993. Seitaigaku kara mita antei syakai (Stable societies from an ecological point of view). 1. Tayousei no izi to sokusin: Tanganiika-ko no gyorui gunsyuu kara (Maintenance and promotion of diversity: the case of the fish community in Lake Tanganyika). Kyoto Seminar House, Keihoku. 278 pp. (in Japanese).

584. Kawanabe, H. 1993. Seminaa kaikai ni atatte (Opening address). pp. 9–11. *In:* H. Kawanabe & M. Yuma (ed.) Seitaigaku kara mita antei syakai (Stable Societies from an Ecological Point of View), 1, Tayousei no izi to sokusin: Tanganiika-ko no gyorui gunsyuu kara (Maintenance and Promotion of Diversity: the Case of Fish Community in Lake Tanganyika), Kyoto Seminar House, Keihoku (in Japanese).

585. Kawanabe, H. 1993. Tanganiika-ko no gyorui gunsyuu (The fish community in Lake Tanganyika). pp. 213–219. *In:* H. Kawanabe & M. Yuma (ed.) Ibidem (ibidem) 1, Ibidem (ibidem), Kyoto Seminar House, Keihoku (in Japanese).

586. Kawanabe, H. 1993. Seminaa heikai ni atatte (Concluding remarks). pp. 231–233. *In:* H. Kawanabe & M. Yuma (ed.) Ibidem (ibidem), 1, Ibidem (ibidem), Kyoto Seminar House, Keihoku (in Japanese).

587. Kawanabe, H. 1993. Dai–3 bunka-kai houkoku (Report from the 3rd Section Meeting). pp. 128–133. *In:* Committee of Hiei Conference (ed.) 1992 Hiei-kaigi houkokusyo: dentou to katuryoku (Report of the 1992 Hiei Conference: Tradition and Activity), Japan IBM Co., West Japan Branch, Osaka (in Japanese).

588. Kawanabe, H. 1993. Kawa wo minagara omou (Thoughts upon looking at a river). Kyoto Tomorrow 2(8): 28–29 (in Japanese).

589. Kawanabe, H. 1993. Seibutu-tayousei wo sokusin saseru (Promoting biological diversity). The Nihon Keizai (evening) 6.7.1993 (in Japanese) (also in 709).

590. Kawanabe, H. 1993. Ayu no azi (The taste of ayu). The Nihon Keizai (evening) 13.7.1993 (in Japanese) (also in 709).

591. Kawanabe, H. 1993. 'Kaika' to 'kaikaku' ('Enlightenment' and 'reformation'). The Nihon Keizai (evening) 20.7.1993 (in Japanese) (also in 709).

592. Kawanabe, H. 1993. Azia no wakamono-tati to (Together with young persons from Asia). The Nihon Keizai (evening) 27.7.1993 (in Japanese) (also in 709).

593. Kawanabe, H. 1993. Hukuhuku-suu no siten (Multiple

points of view). The Nihon Keizai (evening) 3.8.1993 (in Japanese) (also in 709).

594. Kawanabe, H. 1993. Natu no Seitaigaku Kenkyuu Sentaa (This summer at the Center for Ecological Research). The Nihon Keizai (evening) 10.8.1993 (in Japanese) (also in 709).

595. Kawanabe, H. 1993. Kigen no yokunai kikai (A machine in a bad mood). The Nihon Keizai (evening) 17.8.1993 (in Japanese) (also in 709).

596. Kawanabe, H. 1993. Kyokutou Rosia no genzyou (Present status of Far Eastern Russia). The Nihon Keizai (evening) 24.8.1993 (in Japanese) (also in 709).

597. Kawanabe, H. 1993. Kokusai Syokubutugaku Kaigi nite (At the International Congress of Botany). The Nihon Keizai (evening) 31.8.1993 (in Japanese) (also in 709).

598. Kawanabe, H. 1993. Monsuun Azia-iki no kankyou kenkyuu (Environmental research in monsoon Asia). The Nihon Keizai (evening) 7.9.1993 (in Japanese) (also in 709).

599. Kawanabe, H. 1993. Nattoku sigatai koto (The things difficult to assent). The Nihon Keizai (evening) 14.9.1993 (in Japanese) (also in 709).

600. Kawanabe, H. 1993. Seibutu, kankyou no monitaringu (The monitoring of living organisms and environmental factors). The Nihon Keizai (evening) 21.9.1993 (in Japanese) (also in 709).

601. Kawanabe, H. 1993. Tikyuu-kankyou to syuukyou (Global environmental problems and religion). The Nihon Keizai (evening) 28.9.1993 (in Japanese) (also in 709).

602. Kawanabe, H. 1993. Kouyou no onsen nite (Hot springs among red autumn leaves). The Nihon Keizai (evening) 5.10.1993 (in Japanese) (also in 709).

603. Kawanabe, H. 1993. Kyoto no yume, Tokyo no yume (Dream of Kyoto, dream of Tokyo). The Nihon Keizai (evening) 12.10.1993 (in Japanese) (also in 709).

604. Kawanabe, H. 1993. 21-nen-buri no Kankoku (Being in Korea after 21 years). The Nihon Keizai (evening) 19.10.1993 (in Japanese) (also in 709).

605. Kawanabe, H. 1993. Ehon to eiga (Illustrated books and movies). The Nihon Keizai (evening) 26.10.1993 (in Japanese) (also in 709).

606. Kawanabe, H. 1993. Mizoro-ga-ike (The moor, Mizoro-ga-ike). The Nihon Keizai (evening) 2.11.1993 (in Japanese) (also in 709).

607. Kawanabe, H. 1993. Tanganiika-ko no uo (Fishes in Lake Tanganyika). The Nihon Keizai (evening) 9.11.1993 (in Japanese) (also in 709).

608. Kawanabe, H. 1993. Yooroppa no nisyuukan (Two weeks in Europe). The Nihon Keizai (evening) 16.11.1993 (in Japanese) (also in 709).

609. Kawanabe, H. 1993. Fonteenuburoo no mori nite (In the woods at Fontainebleau). The Nihon Keizai (evening) 30.11.1993 (in Japanese) (also in 709).

610. Kawanabe, H. 1993. Kokusai Seibutugaku-syou to sinpoziumu (The International Prize of Biology and its symposium). The Nihon Keizai (evening) 7.12.1993 (in Japanese) (also in 709).

611. Kawanabe, H. 1993. Siwasu no nitiyoubi (A Sunday in December). The Nihon Keizai (evening) 20.7.1993 (in Japanese) (also in 709).

612. Kawanabe, H. 1993. Tikyuu kyousei-ken nettowaaku (Network for research on the symbiotic biosphere). The Nihon Keizai (evening) 21.12.1993 (in Japanese) (also in 709).

613. Kawanabe, H. 1993. Yuki to muhyou no touzi (The winter solstice with snow and frost). The Nihon Keizai (evening) 28.12.1993 (in Japanese) (also in 709).

614. Hori, M. & H. Kawanabe (ed.) 1993. Kawa, ike, sitti (Rivers, lakes and swamps). Asahi hyakka, doubutu-tati no tikyuu (Asahi encyclopedia: the earth for animals), Asahi Shinbunsha, Tokyo 10: 257–288 (in Japanese).

615. Kawanabe, H. 1994. Seimei no tayousei to kankei no soutai: kaisou-huu ni (Diversity of life and the whole of interrelations: a reminiscence). Suuri-kagaku (Mathematical Sciences) 368: 16–19 (in Japanese) (also in 709).

616. Kawanabe, H. 1994. 1993-nen dokusyo ankeeto (Questionnaires on reading books in 1993). Misuzu 394: 26–27 (in Japanese).

617. Kawanabe, H. 1994. Uruka to 'ta-sizen'-gata kasen (Salted entrails of ayu and 'multi-natural' rivers). Nature Conservation Monthly 381: 10 (in Japanese).

618. Kawanabe, H. 1994. 'Suwa-ko ni mouketa kakuri-suitai ni okeru kakusyu sousa-zikken' no saisyuu houkoku (Book review: 'Mesocosm: analysis of a lake ecosystem' edited by Y. Saijo & M. Sakamoto). Nikkei Science 94(3): 152–153 (in Japanese).

619. Urano, E. & H. Kawanabe (ed.) 1994. Tikyuu kyousei-ken (Symbiotic biosphere). Asahi hyakka, doubutu-tati no tikyuu (Asahi encyclopedia, the earth for animals) 14: 289–320 (in Japanese).

620. Kawanabe, H. 1994. Sizen to hito (Nature and human beings). The Heredity 48(4): 4–5 (in Japanese).

621. Kawanabe, H., M.M. Gashagaza & M. Hori. 1994. A conservation issue of biodiversity of Lake Tanganyika, with special reference to inshore-fishes. Verh. intern. Verein. Limnol. 25: 2182.

622. Kawanabe, H., T. Maruyama & Y. Niimura. 1994. Sea-run form of a masu-salmon *Salmo (Oncorhynchus) masou ishikawai:* a conservation problem. Verh. intern. Verein. Limnol. 25: 2191.

623. Kawanabe, H. 1994. Hazime ni (Preface). News of Center for Ecological Research, Kyoto University 21: 1–2 (in Japanese).

624. Kawanabe, H. 1994. Higasi-Azia ni 'Tikyuu kyousei-ken' nettowaaku no koutiku wo (Establishing the network of 'Symbiotic biosphere' in eastern Asia). Syokugyo-noryoku Kaihatu Journal 36(5): 8–9 (in Japanese).

625. Kawanabe, H. 1994. Tikyuu kyousei-kei (Symbiotic biosphere). pp. 44–38. *In:* Hurusato sousei bunka kouen siriizu (Lecture Series on Reconstruction of Hometowns), 9, Beppu-si Hurusato Tyarenzi-zigyou Suisin-kyougikai, Beppu (in Japanese) (also in 709).

626. Kawanabe, H. 1994. Ayu to 'Ayu no hanasi' to watakusi (The ayu, 'The story of the ayu', and I). pp. 293–308. *In:*

Miyadi Denzaburo, ayu no hanasi (D. Miyadi, the Story of the Ayu), Iwanami Shoten, Tokyo (in Japanese).

627. Kawanabe, H. 1994. Sizen wa bunka: Mizoro-ga-ike wo seibutu-tayousei hogo no hitotu no moderu ni (Nature is culture: for the moor Mizoro-ga-ike as a model of conservation of biodiversity). pp. 170–171. *In:* N. Fujita & A. Endo (ed.) Kyoto Mizoro-ga-ike: hyou-ki kara no sizen (The Moor Mizoro-ga-ike in Kyoto: Nature Since Glacial Ages), Kyoto Shinbunsha, Kyoto (in Japanese).

628. Kawanabe, H. 1994. Hazime ni (Preface). pp. 3–4. *In:* H. Morino & N. Miyazaki (ed.) Baikaru-ko: kodai-ko no fiirudo-saiensu (Lake Baikal: Field Science of an Ancient Lake), Tokyo University Press, Tokyo (in Japanese).

629. Kawanabe, H. & A. Hirai. 1994. Kankyou e no teigen (A suggestion for environmental problems). 7. Seitai-kei: fiirudo wa ayu kara 'Kyousei-seibutu-ken' made (Ecosystems: from the ayu to 'SymBiosphere' as my fields). AERA Mook (Asahi-shinbun-sya, Tokyo) 4: 24–25 (in Japanese).

630. Nakai, K., H. Kawanabe & M.M. Gashagaza. 1994. Ecological studies on the littoral cichlid communities of Lake Tanganyika: the coexistence of many endemic species. Arch. Hydrobiol. (Beih. Erg. Limnol.) 44: 375–391.

631. Kyogoku, J., E.R.G. Heath, S.-S. Chen, H. Ueki, H. Kawanabe, H. Wakita, T. Nakanishi & T. Yokoyama. 1994. Sinpoziumu 4. 100-nen go no seizi (Symposium 4: politics 100 years later). pp. 113–135. *In:* T.S. Okada (ed.) Heian-kaigi 'Ningen, bunmei, sosite tikyuu' kaigi-kiroku (Record of the International Forum on the Wisdom of Humanity), Heiankyo 1200th Anniversary Memorial Foundation, Kyoto (in Japanese).

632. Wada, E. & H. Kawanabe. 1994. Kokusai Seibutu-kagaku-rengou dai–25-kai soukai ni syusseki site: tokuni 'Seibututayousei fooramu' to DIWPA ni tuite (Report of the 25th general assembly of International Union of Biological Sciences, with special reference to the 'Forum of biodiversity' and 'International network for DIVERSITAS in the western Pacific and Asia'). News of Center for Ecological Research, Kyoto University 25: 4–5 (in Japanese).

633. Kawanabe, H. 1994. Tikyuu kyousei-ken: seibutu-kankei no tayousei wo saguru (Symbiotic biospheres: searching for diversity of relationships among organisms). Omron Kyoto Forum 39, NHK Kyoto Cultural Center and Omron Co., Kyoto. 40 pp. (in Japanese).

634. Strong, M., T. Umehara, K. Inamori, H. Kawanabe, T. Tachibana, E. Yamada & Z. Fujiyoshi. 1994. Tikyuu samitto: syourai-sedai kara tikyuu no genzai wo tou (Earth Summit: questions from the future generations about the present status of the earth). pp. 39–67. *In*: Kyoto Forum (ed.) Syourai-sedai tono rentai wo motomete (Asking for Solidarity with Future Generations), NHK (Japan Broadcasting Corporation) Enterprise, Tokyo (in Japanese).

635. Kawanabe, H. 1994. Kawanabe Hiroya ga yomu: 'Mori wa yomigaeru' (Reading by H. Kawanabe: 'Forests are coming back' by K. Ishigaki). The Sankei 6.12.1994 (in Japanese).

636. Kawanabe, H. 1994. Kawanabe Hiroya ga yomu: 'Kuraisu-to zensyuu 2, gigkyoku' (Reading by H. Kawanabe: 'Works of H. Kleist, vol. 2, drama'). The Sankei 20.12.1994 (in Japanese).

637. Kawanabe, H., G. Iwatsubo, E. Wada. N. Tachimoto & Y. Oshima. 1994. Azia-Taiheiyou tiiki ni okeru kokusai kankyou kyouryoku no arikata: sinrin to seibutu-tayousei wo kii-waado ni site (Methods for international cooperation on environmental problems in Asia and the Pacific Region: forests and biodiversity as the key words). Kokusai kankyou kyouryoku ni kansuru kenkyu-seika houkoku-kaigi giziroku (The record of the conference on international cooperation to address environmental problems). International Cooperation Sub-committee, Kansai Forum on Global Environment, Osaka. 33 pp. (in Japanese).

638. Kawanabe, H., G. Iwatsubo, E. Wada, N. Tachimoto & Y. Oshima. 1994. Azia-Taiheiyou tiiki ni okeru kokusai kankyou kyouryoku no arikata: houkokusyo-syuu (Methods for international cooperation on environmental problems in Asia and Pacific Region: collected reports). Ibidem. 76 pp. (in Japanese).

639. Kawanabe, H., G. Iwatsubo, Y. Oshima, N. Tachimoto & E. Wada (ed.) 1994. Best ways for Kansai to support the solution of global environmental problems in Asia and the Pacific Area. Ibidem. 74 pp.

640. Kawanabe, H. 1994. Preface: a model of nature. pp. 2. *In:* 'Nature Aquarium World' by T. Amano, T.F.H. Publications, Neptune City (also in 1996 edition and in German edition 'Die Natur als Vorbild'. pp.2. *In:* 'Pflanzenparadiese unter Wasser: Japanisches Gärten im Aquarium' bei T. Amano, Weltbild Verlag, Augsburg).

641. Kawanabe, H. 1995. Tikyuu kyousei-kei to wa nanika? (What is 'Symbiotic biosphere'?). Souzou no Sekai (Creative World) 93: 7–11 (in Japanese).

642. Kawai, M., H. Kawanabe, J. Takabayashi, M. Higashi, E. Wada & K. Nakane. 1995. Seibutu-tayousei no nazo wo saguru (Investigating the mystery of biological diversity). Souzou no Sekai (Creative World) 93: 50–72 (in Japanese).

643. Kawanabe, H. 1995. Fuji-san mo Shinji-ko mo bunka-zai (Mt. Fuji and Lake Shinji-ko are cultural properties). BUILCARE (Hitachi Building System Service Co. Ltd., Tokyo) 109: 6 (in Japanese).

644. Yanagisawa, Y. & H. Kawanabe (ed.) 1995. Ecological and limnological study on Lake Tanganyika and its adjacent regions 9. Ehime University, Matsuyama. 149 pp.

645. Kawanabe, H. 1995. 1994-nen dokusyo ankeeto (Questionnaires on books read in 1994). Misuzu 406: 55 (in Japanese).

646. Kawanabe, H. 1995. 'Tikyuu kyousei-kei' to wa nanika? (What is the 'Symbiotic biosphere'?). University of Tokyo News 5.2.1995, Kyoto University News 20.2.1995 (in Japanese).

647. Kawanabe, H. 1995. Mazu wa teusuna tiiki no tyousa no sokusin kara (Promoting research in local regions with little information). News of Japan Center for Area Studies 1:6 (in Japanese).

648. Yanagisawa, Y., M. Yuma & H. Kawanabe. 1995. Tanganiika-ko no gyorui wo tyuusin to suru seibutu-gunsyuu no

tayousei (Diversity of the biological community among the fishes of Lake Tanganyika). Japanese Scientific Monthly 48:47–53 (in Japanese).

649. Kawanabe, H. 1995. Usinawareru 'inoti' no nigiwai: 'sougo-sayou' de tayou-ka (Bustle of declining 'life': diversified by 'interrelationship'). The Shinano Mainichi 30.3.1995 (in Japanese).
650. Kawanabe, H. 1995. Hokkaidou-tou no ayu ni yosete (On the ayu of Hokkaido Island). Yuurappu 7: 2–3 (in Japanese).
651. Kawanabe, H. & N. Ishii. 1995. Seibutu tayousei wo sokusin suru koto kara, tikyuu kyousei-kei wa sarani sinka suru (Promoting biological diversity: global mutual systems are developing more and more). Big Smile 10: 6–7 (in Japanese).
652. Kawanabe, H. & M. Yuma (ed.) 1995. Seitaigaku kara mita antei syakai (Stable societies from an ecological point of view). 2. Antei to kakuran: tansui-iki no seibutu gunsyuu kara (Stability and disturbance: the case of biological communities in freshwater). Kyoto Seminar House, Keihoku. 207 pp. (in Japanese).
653. Kawanabe, H. 1995. Seminaa kaikai ni atatte (Preface). pp. 7–8. *In:* H. Kawanabe & M. Yuma (ed.) Seitaigaku kara mita antei syakai (Stable Societies from an Ecological Point of View), 2, Antei to kakuran: tansui-iki no seibutu gunsyuu kara (Stability and Disturbance: the Case of Biological Communities in Freshwater), Kyoto Seminar House, Keihoku (in Japanese).
654. Kawanabe, H., T. Akimichi, K. Watanabe, M. Yuma & T. Yokoyama. 1995. Kawa, soko ni sumu ikimono to kurasu hito (Rivers: living organisms and human beings are living together). pp. 114–179. *In:* H. Kawanabe & M. Yuma (ed.) Ibidem (ibidem), Kyoto Seminar House, Keihoku (in Japanese).
655. Kawanabe, H. 1995. Zyuuten-ryouiki-kenkyuu 'Tikyuu kyousei-kei' wo oeru ni atatte (Epilogue for the priority area research titled 'Symbiotic biosphere'). 1. Kore wa tannaru kansoubun desu (It is merely description of my impression). Communication Tikyuu Seitaikei 28:1–2 (in Japanese).
656. Kawai, M., H. Kawanabe, N. Takahata, I. Tayasu, K. Minoura & E. Wada. 1995. Seibutu-sinka no mekanizumu wo saguru (Investigating mechanisms of biological evolution). Souzou no Sekai (Creative World) 94: 40–60 (in Japanese).
657. Kawanabe, H. 1995. Nihon Seibutu-kagakusya Kyoukai-si 'Seibutu-kagaku' no Nobunkyou kara no kankou ni atatte (Upon the publication of the journal 'Biological Sciences', edited by the Japan Society of Biological Sciences published from Nobunkyo Co. Ltd.). Publishing Digest 556: 7 (in Japanese).
658. Kawanabe, H. 1995. Seibutu-tayousei no hozen wo megutte (Conservation of biological diversity). Bulletin of Wildlife Conservation, Philosophical Society of Japan 3: 1–4 (in Japanese).
659. Kawanabe, H. 1995. Seibutu-tayousei ni miru atarasii kagaku-gainen (A new scientific idea of biological diversity). pp. 131–144. *In:* J.-L. Xu (ed.) Asia as the Epicenter of Culture, Center of Japanology, National University of Taiwan, Taipei (in Japanese and Chinese).
660. Kawanabe, H. 1995. Hazime ni (Preface). News of Center for Ecological Research, Kyoto University 29: 1 (in Japanese).
661. Kawanabe, H. 1995. Dai-3-ki-me wo mukaeru ni atatte (The third phase of the Center for Ecological Research, Kyoto University). News of Center for Ecological Research, Kyoto University 30: 1–2 (in Japanese).
662. Kawanabe, H. & T. Kondo. 1995. Seitaigaku (Ecology). Britanica International Encyclopedia 10: 351–361 (in Japanese).
663. Wada, E. & H. Kawanabe. 1995. Dai–25-kai Kokusai Seibutu-kagaku Rengou soukai oyobi seibutu-tayousei fooramu (The 25th general assembly of the International Union of Biological Sciences and Forum on Biodiversity). Japanese Scientific Monthly 48: 888–889 (in Japanese).
664. Kawanabe, H. & E. Wada. 1995. DIVERSITAS in western Pacific and Asia (DIWPA) — an international network. Biology International 31: 30–32.
665. Kawanabe, H. 1995. Tikyuu kyousei-kei to wa nani ka? (What is 'Symbiotic biosphere'?). pp. 8–16. *In:* Organizing Committee of the 9th Open Symposium on 'Universities and Sciences' (ed.) Tikyuu kyousei-kei: tayou na seibutu no kyouzon suru sikumi (Symbiotic Biosphere: an Ecological Interaction Network Promoting the Coexistence of Many Species), Kuba Project, Tokyo (in Japanese) (also in 711).
666. Endo, A., H. Kawanabe, H. Ishikawa, T. Inoue, Z. Kawabata, K. Minoura & J. Takabayashi. 1995. Paneru disukassyon: hutatabi 'Tikyuu kyousei-kei' to wa? (Panel discussion: again, what is 'Symbiotic biosphere'?). pp. 162–174. *In:* Organizing Committee of the 9th Open Symposium on 'Universities and Sciences' (ed.) Ibidem (ibidem), Kuba Project, Tokyo (in Japanese).
667. Kawanabe, H. 1995. Hon kouen no matome to tenbou (Conclusion and perspective). pp. 187–190. *In:* Organizing Committee of the 9th Open Symposium on 'Universities and Sciences' (ed.) Ibidem (ibidem), Kuba Project, Tokyo (in Japanese) (also in 711).
668. Kawanabe, H. 1995. 'Mono' wa 'koto' ga tukuriageru ('Things' created and maintained by 'relations'). pp. 217–225. *In:* K. Hisayama (ed.) Mori no kyousitu: ikimono sanka (Classroom in Forests: Homage to Living Things), Tankosha, Kyoto (in Japanese).
669. Kawanabe, H. 1995. Syohyou: 'Tuusi Nihon no kagaku gizyutu' (Book review: 'History of science and technology in Japan' edited by S. Nakayama). The Mainichi 3.11.1995 (in Japanese).
670. Loreau, M., R. Barboult, H. Kawanabe, M. Higashi, E. Alvaaarez-Buylla & F. Renaud. 1995. Dynamics of biodiversity at the community and ecosystem level. pp. 245–274. *In:* V.H. Heywood & R.T. Watson (ed.) Global Biodiversity Assessment, Cambridge University Press, Cambridge.

671. Kawanabe, H. 1995. Ikimono-dousi no kankei wo dou mite-kita ka? (How have I considered and studied interrelationships among living organisms?). pp. 149–152. *In:* K. Ohba, K. Fujita & K. Chinzei (ed.) Nihon no sizen (Nature in Japan), 6, Kinki District, Iwanami-shoten, Tokyo (in Japanese) (also in 711).
672. Kawanabe, H. 1995. Kawanabe Hiroya ga yomu: 'Sin Nihon koten-bungaku taikei 71. Genroku haikai-syuu' (Reading by Hiroya Kawanabe: 'New collection of classics of Japan 71. Haikai in Genroku Era'). The Sankei 10.1.1995 (in Japanese).
673. Kawanabe, H. 1995. Idem: 'Ainumosiri ni tudou' (idem: 'Gathering in Ainumoshir, the land of Ainu: a message from the indigenous peoples in the world' edited by Nibutani Forum Organizing Committee). Ibidem 24.1.1995 (in Japanese).
674. Kawanabe, H. 1995. Idem: 'Daauinizumu ronsyuu' (idem: 'Collected papers of Darwinism' compiled and translated by R. Yasugi). ibidem 7.2.1995 (in Japanese).
675. Kawanabe, H. 1995. Idem: 'Arihureta oi' (idem: 'A commonplace aging' by R. Matusita). Ibidem 21.2.1995 (in Japanese).
676. Kawanabe, H. 1995. Idem: 'Shincho Nihon bungaku arubamu, Ishikawa Jun' (idem: The Shincho's photo album of Jun Ishikawa). Ibidem 7.3.1995 (in Japanese).
677. Kawanabe, H. 1995. Idem: 'Tetudou geezi sensou' (idem: 'Competition of railway gauge' by S. Koike, illustrated by S. Suzuki). Ibidem 21.3.1995 (in Japanese).
678. Kawanabe, H. 1995. Idem: 'Syousetu Ryukyu syobun' (idem: 'The disposition of the Ryukyus: a novel' by R. Ohshiro). ibidem 4.4.1995 (in Japanese).
679. Kawanabe, H. 1995. Idem: 'Saigyo ka-den' (idem: 'Saigyo: a priest-poet' by K. Tsuji). Ibidem 18.4.1995 (in Japanese).
680. Kawanabe, H. 1995. Idem: 'Bunsi-sinka-gaku e no syoutai' (idem: 'An invitation to the study of molecular evolution' by T. Miyata). Ibidem 2.5.1995 (in Japanese).
681. Kawanabe, H. 1995. Idem: 'Nihon no tennen-kinenbutu' (idem: 'Natural monuments in Japan' edited by M. Kato, M. Numata, K. Watanabe & M. Hata). Ibidem 16.5.1995 (in Japanese).
682. Kawanabe, H. 1995. Idem: 'Barukan ekisupuresu' (idem: 'The Balkan Express: a fragment from the other side of war' by S. Drakulic, translated by K. Mitani). Ibidem 30.5.1995 (in Japanese).
683. Kawanabe, H. 1995. Idem: 'Sensou no kioku' (idem: 'Wages of guilt: memoirs of war in Germany and Japan' by I. Buruma and translated by S. Ishii). Ibidem 13.6.1995 (in Japanese).
684. Kawanabe, H. 1995. Idem: 'Syaarokian satuzin ziken' (idem: 'The case of the Baker Street irregulars' by A. Boucher, translated by K. Jinka). Ibidem 27.6.1995 (in Japanese).
685. Kawanabe, H. 1995. Idem: 'Yuusin-hu' (idem: 'Mind games' by I. Miyazaki). Ibidem 11.7.1995 (in Japanese).
686. Kawanabe, H. 1995. Idem: 'Ninzyou-bon-syuu' (idem: 'Collected love stories of the Edo Era' revised by M. Muto). Ibidem 25.7.1995 (in Japanese).
687. Kawanabe, H. 1995. Idem: 'Seimei to syutai' (idem: 'Gestalt und Zeit, und Anonyma' by V.v. Weitzsaecker, translated by B. Kimura). Ibidem 8.8.1995 (in Japanese).
688. Kawanabe, H. 1995. Idem: 'Sakkyoku-ka no seikatu' (idem: 'Life of a composer of music' by H. Miyake). Ibidem 22.8.1995 (in Japanese).
689. Kawanabe, H. 1995. Idem: 'Seibutu tayousei' (idem: 'Biodiversity' by A. Dohmoto). Ibidem 5.9.1995 (in Japanese).
690. Kawanabe, H. 1995. Idem: 'Unagi to yama-no-imo' (idem: 'Eels and mountain yams' by S. Maruya). Ibidem 19.9.1995 (in Japanese).
691. Kawanabe. H. 1995. Idem: 'Sengo wo yomu: 50 satu no non-fikusyon' (idem: 'Read from post-war Japan: 50 works of non-fiction' edited by S. Sadaka). Ibidem 3.10.1995 (in Japanese).
692. Kawanabe, H. 1995. Idem: 'Watasi-tati no namida de yukidaruma ga toketa: kodomo-tati no Cherunobuiri' (idem: 'Melting snowmen with our tears: Chernobyl for children' by the Supporting Movement for Chernobyl in Kyushu). Ibidem 17.10.1995 (in Japanese).
693. Kawanabe, H. 1995. Idem: 'Bikou-rui' (idem: 'Rhinogradentia' by H. Stuempke, translated by T. Hidaka & S. Haneda). Ibidem 31.10.1995 (in Japanese).
694. Kawanabe, H. 1995. Idem: 'Syounin-tati no kyowa-koku' (idem: 'A republic of/for merchants' by M. Kuroda). Ibidem 14.11.1995 (in Japanese).
695. Kawanabe, H. 1995. Idem: 'Sumibasyo no seitaigaku' (idem: 'Ecology of habitats' by Y. Takemon, K. Tanida, A. Tamaki, H. Mukai & Z. Kawabata). Ibidem 28.11.1995 (in Japanese).
696. Kawanabe, H. 1995. Idem: 'Yami-iti, maborosi no gaidobukku' (idem: 'A guidebook to the black market in Japan during the late 1940s' by M. Matudaira). Ibidem 12.12.1995 (in Japanese).
697. Kawanabe, H. 1995. Idem: 'Edo no yo ni asobu: Ikuta Kosaku syozou syo-ga-ten' (idem: 'Entertainment in the Edo Era: an exhibition of illustrations and books owned by Professor K. Ikuta' edited by T. Sakai). Ibidem 26.12.1995 (in Japanese).
698. Kawanabe, H., T. Nakamura & W. Fujimoto. 1995. Kan-Nippon-kai akademikku (Scientific research in the area around the Japan Sea). pp. 1–33. *In:* Report of the Founding Meeting of the Academic Forum around Japan Sea, Academic Forum around the Japan Sea, Kyoto (in Japanese).
699. Kawanabe, H. 1995. Ayu no seitai to hukugen e no kihonteki siten (Fundamental points of view for the re-establishment of the ayu in Okinawa Island). pp. 9–14. *In:* Committee for Ryukyu-ayu Forum (ed.) 1991, Dam Construction Office in North Okinawa Island, Nago (in Japanese).
700. Kawanabe, H. & K. Ishihara. 1996. Biwako, sekai-isan ni atai suru 'Bunka-zai': honmono no sizen kaihuku-ryoku wo (Lake Biwa, a 'cultural property' being worth World Heritage: we support nature's ability to recover). The Asahi 1.1.1996 (in Japanese).

701. Kawanabe, H., M. Inaba & M. Ishikawa. 1996. Kankyou sin-zidai no maku-ake wo: Biwako no uo-tati mo matte-masu (Opening the curtain for a new environmental age: the fishes of Lake Biwa are waiting). The Shiga Hochi 1.1.1996 (in Japanese).
702. Kawanabe, H. 1996. Bodaiji sanroku no ooiwa (A huge rock on Mt. Bodaiji). Economy and Society in Shiga 78: 2–3 (in Japanese).
703. Kawanabe, H. 1996. 1995-nen dokusyo ankeeto (Questionnaires on books read in 1995). Misuzu 418: 61 (in Japanese).
704. Kawanabe, H. 1996. Seitaigaku Kenkyuu Sentaa no 5-nen-kan wo hurikaette (Looking back upon the first 5 years of the Center for Ecological Research, Kyoto University). News of the Center for Ecological Research, Kyoto University 34: 1–2 (in Japanese).
705. Abe, T. & H. Kawanabe. 1996. Seibutu-tayousei: sono seitaigakuteki kenkyuu no genzyou to tenbou (The ecological aspect of biological diversity). Kankyou Kenkyuu (Environmental Research Quarterly) 100: 133–138 (in Japanese).
706. Kawanabe, H. (ed.) 1996. Tikyuu-kyousei-kei: seibutu no tasyu kyouzon wo sokusin suru sougo-sayou kikou kenkyuu-seika houkokusyo (Progress report on the symbiotic biosphere: an ecological interaction for promoting coexistence of organisms). Center for Ecological Research, Kyoto University, Otsu. 137 pp.
707. Kawanabe, H. (ed.) 1996. Seibutu-tayousei izi, riyou no tame no hozen-tiiki riron kenkyuu-seika houkokusyo (Progress report on the theoretical consideration of a conservation area for the sustaining and use of biodiversity). Ibidem. 124 pp.
708. Kawanabe, H. 1996. Linkage between ecological complexity and biodiversity. pp. 149–152. *In:* F. di Castri & T. Younès (ed.) Biodiversity, Science and Development, CAB International, Wallingford.
709. Kawanabe, H. 1996. Aimai no seitaigaku (Equivocal ecology). Nou-San-Gyo-Son Bunka-kyokai, Tokyo. 230 pp. (in Japanese).
710. Kawanabe, H. 1996. 'Mizu-kankyou' to seitaikei ('Water environment' and aquatic ecosystems). Kankyou (Environment) 21(3): 7–9 (in Japanese).
711. Kawanabe, H. 1996. Seibutu-kai ni okeru kyousei to tayousei (Symbiosis and diversity in a living world). Jinbunshoin, Kyoto. 206 pp. (in Japanese).
712. Kawanabe, H. & H. Tabata (ed.) 1996. Seitaigaku kara mita antei syakai (Stable societies from an ecological point of view). 3. Satoyama to sono sizen no zizoku-teki riyou (Village forests and their sustainable use). Kyoto Seminar House, Keihoku. 192 pp. (in Japanese).
713. Kawanabe, H. 1996. Kaikai ni atatte (Preface). pp. 3–4. *In:* H. Kawanabe & H. Tabata (ed.) Seitaigaku kara mita antei syakai (Stable Societies from an Ecological Point of View), 3, Satoyama to sono sizen no zizoku-teki riyou (Village Forests and their Sustainable Use), Kyoto Seminar House, Keihoku (in Japanese).
714. Kawanabe, H. 1996. Biwako Hakubutukan (Lake Biwa Museum). LAKE 26:3 (in Japanese).
715. Kawanabe, H. 1996. Suihen no seitaigaku (Ecology of the waterfront; an ecotone). Koryuu (Bulletin of Chubu Electric Power Co.) 41: 10–13 (in Japanese).
716. Kawanabe, H. 1996. Syu no hozon: tayousei ga taisetu-nandesu (Species conservation: diversity is important). The Yomiuri (evening) 24.5.1996 (in Japanese).
717. Kawanabe, H. 1996. Syuukan Sincho keizi-ban (Syuukan Sincho notice board). Syuukan Sincho (Sincho Weekly) 30.5.1996 (in Japanese).
718. Kawanabe, H. & S. Tahara. 1996. Kagaku wa dokoe iku? (Where is science going?). 8. Kawanabe Hiroya: Tango no ayu ni osierareta 'kyouzon no tie' (Hiroya Kawanabe: 'the wisdom of coexistence' learned from the ayu in Tango). Chuo Kouron 111(9): 154–167 (in Japanese).
719. Kawanabe, H. 1996. Ikimono no tasaina kurasi: kawa to mizuumi kara (Various organisms in rivers and lakes). pp. 63–83. *In:* K. Inamori, T. Umesao, T.S. Okada, H. Kawai, R. Kotani, J. Murata & E. Yuasa (ed.) 1995 Hiei-kaigi houkokusyo: tikyuu no naka no inoti (1995 Report of Hiei Conference: Global Life), IBM Japan, Osaka (in Japanese).
720. Kawanabe, H. & F. Matsumura. 1996. Hontou no hakubutukan wa hitobito no seikatu sonomono: Biwako Hakubutukan wa sono 'iriguti' nandesu (A true museum reflects people's real lives: The Lake Biwa Museum is an entrance towards it). Realty Time 176: 2–3 (in Japanese).
721. Kawanabe, H. 1996. Gunsyuu seitaigaku: doubutu wo zairyou ni site (Community ecology). Genetics, Spec. vol. 8: 126 (in Japanese).
722. Kawanabe, H. 1996. Importance of the community relationships in biodiversity. pp. 17–22. *In:* R.R. Colwell, U. Shimizu & K. Ohwada (ed.) Microbial Diversity in Time and Space, Plenum Press, New York.
723. Yokoyama, T., H. Kawanabe, J. Fujii & M. Yuma (ed.) 1996. Antei syakai no sougou kenkyuu: mono wo tukuru, tukau (A synthetic study on stable societies: the making and use of materials). Kyoto Seminar House, Keihoku. 205 pp.(in Japanese).
724. Kawanabe, H. 1996. Antei to wa nani ka? (What is stability?). pp. 12–22. *In:* T. Yokoyama, H. Kawanabe, J. Fujii & M. Yuma (ed.) Antei syakai no sougou kenkyuu: mono wo tukuru, tukau (A Synthetic Study on Stable Societies: the Making and Use of Materials), Kyoto Seminar House, Keihoku (in Japanese).
725. Kawanabe, H. 1996. Kaisetu niwa naranai kaisetu (Unexplanatory explanation for Masao Kawai's earliest works). pp. 451–463. *In:* Kawai Masao tyosaku-syuu (Works of Masao Kawai), Shogakukan, Tokyo (in Japanese).
726. Kawanabe, H. 1996. Rekisiteki seimeitai to site no Biwako (Lake Biwa as a historical body of life). pp. 5. *In:* Lake Biwa Museum (ed.) Umindo: Biwako to kurasi no monogatari (Human Beings on/around Lake Biwa: Stories of Lake Biwa and Human Lives), Houhousha, Kyoto (in Japanese).
727. Hidaka, T., T. Yoneyama, H. Kawanabe & Y. Kada. 1996. Biwako kara kangaeru 21 seiki (Lake Biwa: the twenty-

first century and beyond). pp. 64–69. *In:* Lake Biwa Museum (ed.) Ibidem (ibidem), Houhousha, Kyoto (in Japanese).

728. Kawanabe, H. 1996. Yosa Hantou no watakusi (Myself on the Yosa Peninsula). Gakuto 9(10): 28–31 (in Japanese).
729. Kawanabe, H. 1996. Kagaku ni totte tikyuu kankyou mondai towa nanika? (What are global environmental problems in science?). Souzou no Sekai (Creative World) 100: 95–103 (in Japanese).
730. Ishi, H., T. Umehara, H. Kawanabe, T. Sawa, M. Higashi, K. Fukui, I. Furusawa & T. Inoue. 1996. Nizyuu-iti-seiki e ikinokoru tame no senryaku (Strategies for surviving to the 21st Century). Souzou no Sekai (Creative World) 100: 134–137 (in Japanese).
731. Kawanabe, H. 1996. Fiirudo e no sasoi (Introduction to field studies). pp. 1. *In:* Lake Biwa Museum (ed.) Biwako Hakubutukan tenzi gaido (Guide to the Exhibitions of the Lake Biwa Museum), Lake Biwa Museum, Kusatsu (in Japanese).
732. Kawanabe, H. 1996. Biwako Hakubutukan: iriguti to site tanosii mono wo (Lake Biwa Museum: an introduction to the environmental issue). The Kyoto 20.10.1996 (in Japanese).
733. Kawanabe, H. 1996. Asian great lakes, especially Lake Biwa. Env. Biol. Fish. 47: 219–234.
734. Kawanabe, H. 1996. Biwako no uo, sekai no uo (Fishes of Lake Biwa and in the world). pp. 2. *In:* Lake Biwa Museum (ed.) Water and Life 1. Lake Biwa, Fish and People – an East Asian prospect, Lake Biwa Museum, Kusatsu (in Japanese).
735. Kawanabe, H. 1996. Satoyama to watakusitati no kurasi (Village mountains and our lives). pp. 34. *In:* Lake Biwa Museum (ed.) Satoyama, seimei no syou-utyuu: Imamori Mitsuhiko syasinten (Village Mountains, Microcosms of Life: a Photographic Exhibition by M. Imamori), Lake Biwa Museum, Kusatsu (in Japanese).
736. Kawanabe, H. 1996. Go-aisatu (Preface). pp. 3. *In:* Lake Biwa Museum (ed.) The Users' Guide for Lake Biwa Museum, Lake Biwa Museum, Kusatsu (in Japanese).
737. Kawanabe, H. 1996. Hakubutukan wa mijikana sizen ya kurasi no naka ni (Nature and our lives as a museum). Morning Kusatsu 17(19): 2 (in Japanese).
738. Kawanabe, H. 1996. Minamata-byou jiken siryoushuu 1926–1968 (Book review: 'Collection of cases of Minamata Disease' edited by the Minamata Disease Task Force). The Mainichi 3.11.1996 (in Japanese).
739. Kawanabe, H. 1996. 'Mizuumi to ningen', yukkuri kangaete ('Lake and human beings', considering both carefully). The Nihon Keizai (evening) 5.11.1996 (in Japanese).
740. Hidaka, T. & H. Kawanabe. 1996. 21-seiki wo mezasu atarasii Oumi bunka no souzou (Creation of a new Oumi culture towards the 21st century). The Japan Technology 20.11.1996 (in Japanese).
741. Kashiwagi, K., L. Lavoire, E. Wada, H. Kawanabe, M. Utsumi, F. Okuno, T. Morita & R. Kawamura. 1996. Sikkari sei, kimono fooramu (The cheer up kimono forum). pp. 34–35. *In:* Office of Kyoto Kimono Summit (ed.) Nihon wo mituketa kimono samitto 1996 Kyoto houkokusho (Report of the Kyoto Kimono Summit 1996), Kyoto Chamber of Commerce, Kyoto (in Japanese).
742. Kawanabe, H. 1996. Seibutu no tasyu kyouzon: 'seibutu tayousei kenkyuu' suisin no genzyou (Coexistence of species: the present status of 'biodiversity science'). Japanese Science Monthly 49(12): 31–36 (in Japanese).
743. Kawanabe, H. & K. Kishida. 1996. Syousetu to opera to eiga no hanasi (Stories on novels, operas and a movie). Duet 51: 1–8 (in Japanese).
744. Kira, T. & H. Kawanabe. 1996. Biwako no hotori de (On the shore of Lake Biwa). Bulletin of Kansai Organization for Nature Conservation 23: 3–10 (in Japanese).
745. Kawanabe, H. 1996. International Scientific Programme on Biodiversity and DIWPA as its regional network. pp. 25–31. *In:* B.-H. Lee, T.-H. Kim & B.-Y. Sun (ed.) Biodiversity Research and its Perspective in East Asia, Korean Institute for Biodiversity Research (KIBIO), Chonbuk National University, Chonbuk.
746. Kawanabe, H. 1996. Kankei no rensa no soutai wo kangaeru tameno hakubutukan wo mezasite (Toward the museum dealing with chained relations among organisms). pp. 133–140. *In:* Korekara no sizen-si hakubutukan: Seimei no Hosi – Tikyuu Hakubutukan isshuunen kinen ron-shuu (Future of the Natural History Museum: Collected Papers for the First Anniversary of the Museum of Planet Earth), All Planner, Odawara (in Japanese).
747. Kawanabe, H. 1997. 1997 nen watasi no kii-waado: Karei naru tinsen (My keyword for 1997: lavish engrossment in thought). The Kyoto 1.1.1997 (in Japanese).
748. Kawanabe, H. & I. Kondo. 1997. Mizuumi to hito ga tomo ni ikiru yori yoi kankei wo mezasite (Toward better relationships between lake and human beings: coexistence). Mizuumi (Lake) 120: 11–23 (in Japanese).
749. Kawanabe, H. 1997. 'Ningen to mizuumi no tukiai kata' wo kangaeru ba ni nareba (The museum as a guide to 'better thinking of association between human beings and lake'). Shiga-Minpou 5.1.1997 (in Japanese).
750. Kawanabe, H. 1997. Biwako no kankyou saisei (Restoration of the Lake Biwa environment). The Chunichi 9.1.1997 (in Japanese).
751. Kawanabe, H. 1997. 1996 nen dokusyo ankeeto (Questionnaires on books read in 1996). Misuzu 430: 65–66 (in Japanese).
752. Kawanabe, H. & E. Kurimoto. 1997. Rekisi-teki seimeitai: Biwako e no izanai (Invitation to Lake Biwa: the history of a living lake). Gekkan Minpaku (Monthly Bulletin of National Museum for Ethnology) 1997(2): 2–7 (in Japanese).
753. Kawanabe, H. 1997. Biwako wa sizen-isan to douzi ni bunka-isan da to omoimasu ne (I think, Lake Biwa is not merely a natural heritage but also a cultural one). SUN 45: 1 (in Japanese).
754. Kawanabe, H. 1997. Goaisatu (Preface). pp. 1. *In:* Lake Biwa Museum (ed.) Hakubutukan ga dekiru made (Until the

Museum Opened), Lake Biwa Museum, Kusatsu (in Japanese).
755. Kawanabe, H. 1997. Tararu Yunesu Kokusai Seibutu Kagaku Rengou Jumukyoku-chou no hyoukei houmon (Reports of guests from abroad: Dr. Talal Younès, Executive Director of the International Union of Biological Sciences). Gakujutu no Doukou (Trends in the Science, JSC News) 1997(3): 88–89 (in Japanese).
756. Kawanabe, H. 1997. Kongetu no kotoba: toi no sekai ni wa ziyuu ga aru, kotae no sekai ni wa ziyuu ga nai (The words of this month: there's freedom in questions but not in answers). Syougakkou Zihou (Elementary Schools' Bulletin) 1997(547): 2 (in Japanese).
757. Kawanabe, H., M. Hori & M. Nagoshi (ed.) 1997. Fish communities in Lake Tanganyika. Kyoto University Press, Kyoto. 198 pp.
758. Kawanabe, H. 1997. Preface. pp. 5–7. *In:* H. Kawanabe, M. Hori & M. Nagoshi (ed.) Fish Communities in Lake Tanganyika, Kyoto University Press, Kyoto.
759. Kawanabe, H. 1997. Seibutu-tayousei no kagaku (Biodiversity sciences). Gakusi-kai Kaihou (Bulletin of Society for Academicians) 815: 121–125 (in Japanese).
760. Younès, T. & H. Kawanabe. 1997. Seibutu-tayousei wa inoti no nigiwai sonomono desu (Biodiversity as the prosperity of life). Umindo (Bulletin of Lake Biwa Museum) 2: 2–3 (in Japanese).
761. Yuma, M. & H. Kawanabe (ed.) 1997. Ecological factors promoting biodiversity in lakes Biwa, Baikal, Tanganyika and Malawi. Center for Ecological Research, Kyoto University, Otsu. 119 pp.
762. Kawanabe, H. 1997. 'Iro-megane' wo kakete 'ta-syoku' wo miyou (Looking at facts through a different pair of colored glasses). Security 85: 1 (in Japanese).
763. Kawanabe, H. 1997. Biwako kara no hassin (Listening to Lake Biwa). Magazines Today/ Ecology Y24, Y28–29 (in Japanese).
764. Kawanabe, H. 1997. Hazime ni (Preface). pp.5. *In:* M. Yuma, Y. Kada & Y. Fujioka (ed.) Mizube no asobi ni miru seibutu-sou no jidai hensen to isiki henka: juumin sanka ni yoru san-sedai chousa houkokusyo (Historical Changes in People's Attitudes and Aquatic-fauna Related to Childhood-play in Familiar Freshwaters: Survey Report of the Research over Three Generations of People's Participatory Methods), Research Report of the Lake Biwa Museum 9 (in Japanese).
765. Kawanabe, H. 1997. Husigi na hito (A marvelous person). pp. 219–221. *In:* J. Kawakita, K. Kawai, A. Nishimura, N. Takashima, J. Tanaka, A. Kawai & H. Ito (ed.) Tuitou Dogura Kyuuzou (Memorial book of Mr. Kyuuzou Dogura), Dogura Publishing, Kyoto.
766. Kawanabe, H. 1997. Goaisatu (Preface). p. 1 *In:* Lake Biwa Museum (ed.) Kodai-ko no sekai (The World of Ancient Lakes), Lake Biwa Museum, Kusatsu.
767. Kawanabe, H., Y. Takahashi, T. Miyamura, M. Morita, S. Okuda & S. Kitamura. 1997. Kasen-zigyou to seibutu (River impoundment and living organisms) 1. Nihon no kawa (Rivers in Japan) 77: 4–30.
768. Kawanabe, H., Y. Takahashi, T. Miyamura, M. Morita, S. Okuda & S. Kitamura. 1997. Idem (idem) 2. Ibidem (ibidem) 78: 4–30.
769. Kawanabe, H. 1997. Uzi-bunkou no koro (When I was in Uji Junior College). pp. 71–75. *In:* Editorial Committee of the 100th Anniversary of Kyoto University (ed.) Kyoto-Daigaku no seiki (A Century of Kyoto University): 1897–1997, Sisuikai-syuppan, Kyoto.
770. Kawanabe, H. 1997. Environmental conservation in the Asian region. Japanese Scientific Monthly 50: 80–81.
771. Kawanabe, H. 1997. Iti doubutu-seitaigaku-sya no mita Yoda Kyouzi san (Memorial address for Professor Kyoji Yoda as seen by an animal ecologist). pp. 6–9. *In:* Committee for the Mourning Ceremony of Professor Kyoji Yoda, Faculty of Environmental Sciences, Shiga Prefectural University (ed.) Yoda sensei wo sinonde (Obituaries for the Late Professor Kyoji Yoda), Faculty of Science, Shiga Prefectural University, Hikone.
772. Ogawa, S, H. Kawanabe & Y. Kada. 1997. Okisima no gyogyou no hensen nado (Change of fisheries and other things in Oki-sima Island). Umindo: Bulletin of Lake Biwa Museum 3: 2–3.
773. Kawanabe, H. 1997. Why my height has been shortened? Journal of the Institute of Electrical Engineers of Japan 117: 587.
774. Kawanabe, H. & C. Lévêque. 1997. Indicators of ecosystem health. Biodiversity Bulletin 3: 11.
775. Kawanabe, H. 1997. Preface. pp. 1–2. *In:* Biodiversity Issues and Global Environmental Problems: Environmental City Kansai in the 21st Century, Global Environment Forum – Kansai, Osaka.
776. Kawanabe, H. & Y. Fujishima. 1997. 'Aimai' no susume: uwo no seitai ni mita kyousou to kyouzon no genri ('Equivocal' is important: a viewpoint from competition and coexistence of fishes). Kouken 35(9): 42–57.
777. Yokoyama, T., K. Miura, M. Yuma, J. Fujii & H. Kawanabe (ed.) 1997. Antei-syakai no sougou-kenkyuu: koto ga okoru, tuduku, nakadati wo megutte (A synthetic study on stable societies: something happens, continues and interface between them). Kyoto Seminar House, Kyouhoku. 181 pp.
778. Kawanabe, H. 1997. 'Mono' wa 'koto' ga tukuri-ageru (What is stability?). pp.12–20. *In:* T. Yokoyama, K. Miura, M. Yuma, J. Fujii & H. Kwanabe (ed.) Antei-syakai no Sougou-Kenkyuu: Koto ga okoru, tuduku, nakadati wo megutte (A Synthetic Study on Stable Societies: Something Happens, Continues and Interface between Them), Kyoto Seminar House, Kyouhoku.
779. Kawanabe, H. 1997. Nenpou no hakkan ni atatte (Preface). Annual Report of Lake Biwa Museum 1(1996): 1–2.
780. Kawanabe, H. & Y. Kada. 1997. Kantyou ni kikku: Sekai Kodai-ko Kaiagi wo owatte (Concluding remarks at the International Conference on Ancient Lakes). Umindo: Bulletin of the Lake Biwa Museum 4: 2–3.
781. Kawanabe, H. 1997. Biwako hakubutukan de no iti-nen

(One year in Lake Biwa Museum). Human and Environmental Forum (Kyoto University) 3: 1.
782. Kawanabe, H. 1997. 'Ouyou Seitai-Kougaku Zyosetu – Seitaigaku to Doboku-kougaku no yuugou wo mezasite' ni tuite (Book review on 'Introduction to Applied Ecological Engineering'). Kasen (Rivers) 1997(9): 60–61.

Environmental Biology of Fishes **52**: 67–72, 1998.

An interview with Hiroya Kawanabe

Eiji Harada
Shirahama, Wakayama 649-23, Japan

Key words: ecology, interrelationship, community, environmental problem, literature, ayu *Plecogossus altivelis*, Naka-umi, Biwa-ko, Lake Tanganyika, U-kawa

When Eugene K. Balon first suggested publishing the present special volume, Tamotsu Iwai and Masahide Yuma asked me to inverview Hiroya Kawanabe. This was presumably because Kawanabe and I were associated in many ways for a long time, as students attending the same lectures and exercises, as young researchers involved in surveys of the ayu fish, lakes Naka-umi, and Biwa-ko, as co-authors of several papers, as joint revisers of some dictionaries, as commensals in pubs, and so on. Kawanabe loves classical music, literature, and hot-spring baths.

Actually, this is not the record of an interview by me of Kawanabe. During our frequent chats in coffee shops and 'nomiya' pubs in Kyoto and Otsu, I picked out several topics and sent questions incorporating them to Kawanabe. I received written answers from him, rearranged them together with questions, added some comments and further questions, and rendered these for his consideration. The narration in the form of an interview was thus prepared. M.J. Grygier of the Lake Biwa Museum and G.C. Kearn of the University of East Anglia took the job of revising the English, for which we are grateful. I should add one thing. Kawanabe always addresses people as 'san' in Japanese, not 'sensei' – a very common way to refer to professors and teachers – and he himself likes to be called 'san'. So, I adopted '-san' as a suffix for the names of people throughout his narration.

EH = Eiji Harada **HK** = Hiroya Kawanabe

EH: Since I first encountered you in the second-year undergraduate class at Kyoto University in 1952, I have witnessed and shared with you uncountable memorable moments of a kind that are seldom mentioned in typical accounts of 'life and work'. In particular, motivations and circumstances, which are inseparably connected with each other and are of pervasive significance in one's achievements, are hardly described at all. So, I think the interview should emphasize these aspects.

HK: Yes. We were the two of our year class in the postgraduate course to study animal ecology, and we enrolled at and remained in the same university until retirement. Thus, we were associated for more than two thirds of our lives – it has been extraordinary in a sense to maintain such a long-term, close interaction. I will answer straight out any enquiry you put to me; I am like 'a carp on a cooking board', meaning I will not try to jump ship.

EH: To my great surprise, you already had a good knowledge of ecology as an undergraduate. I first came to know the science of ecology, and even the term itself, at the university. I do not think I was taught about ecology at my high school during 1948–1950. Ecology was an utterly new discipline of biological science for me. What stimulated your interest in ecology so early?

HK: That's not correct. I, too, did not know that subject well. If I did so earlier than you, it stems from attending the class 'Natural Science II' in my first year. It was intended for students of literature and social sciences, and the lectures were given in turn by professors of the Zoology and the Botany departments of the Faculty of Science. Among them was Miyadi-san, who later supervised us both; he talked of predator-prey relationships. My astonishment that such common affairs of animal life could be the subject of natural science provoked my interest in ecology, I suppose. I remember I read Imanishi-san's book, 'The logic of organic society', in 1952, and also Kitazawa-san's article, 'Bio-economics', in an ecology textbook, and the latter particularly intrigued me.

It was my understanding that you entered the university with biology in your mind, but in my case, I was considering taking up psychiatry or sociology, and was therefore undecided in my choice between the Faculty of Science and the Faculty of Literature. I had only the slightest interest in biology as a whole, and, during the latter half of the undergraduate course when we specialized in zoology, I asked you occasionally to loan me your notebooks for writing term papers and to work with me in practical classes.

EH: I took my notes word for word in a conventional way, but your notes, in contrast, were instantaneous, diagrammatic summaries of lectures. As for ecology, I recall reading the lectures of Huxley (1863) when we were undergraduate students. He wrote forcefully of the meaning and importance of 'the conditions of existence' and 'the wanderings of matter' in connection with Darwin's 'Origin'. I must confess that I did not understand it well, partly because of my indifference to these aspects of biology.

HK: Ecology deals with the phenomena of the daily life of organisms. They are not occult phenomena that only trained and specially equipped scientists can observe. Ishikawa-san, who first demonstrated that the small-sized ayu, *Plecoglossus altivelis*, of Biwa-ko can be raised by artificial culture in a pond or by transferring them into rivers and who opened the way for the current commercial transplantation of the ayu fingerlings of Biwa-ko, wrote in his textbook of zoology and botany in 1934 that a student watching frogs croak one evening was accused of strange behaviour and eventually arrested by a policeman. Even among our contemporary students of field ecology, many have experienced similar incidents. If they explained what they were doing, they would be greeted with a comment like 'Oh! You have a nice pastime'. This often incited them to think about and debate the use of ecology, the aim of ecology, and the aim of science. The situation has changed a lot, and 'ecology' is now widely known among the general public.

EH: You started your postgraduate work with the ecological study of ayu. Did fish have a special appeal for you?

HK: As you know well, the incentive was a suggestion from Miyadi-san. He asked us to join a survey of ayu, and since none of us were rich, to be able to carry out field work with financial support for travelling, hotel accommodation, and food was quite attractive. I might have once said: 'If monkeys are taken up as the subject, the social relationships tend to be a major concern and interspecific interrelationships and relations with environmental conditions are subordinated. If insects are taken up, the reverse may be the case. Fishes, and particularly river fishes, are easy to observe and both aspects can be pursued'. I might also have said: 'To observe ayu along the whole length of a river, throughout its whole life cycle, in the communities as a whole to which it belongs'; this notion was later printed in my small book. These were views I adopted afterwards, not at the very inception; they are all right, but mentally, I dare say, nothing more than a subsequent rationalization.

The case was similar when I was involved in a survey of Lake Tanganyika in 1977. Eventually, the following central theme developed: 'In the temperate regions, habitat segregation as well as food segregation through interactions are common among freshwater fishes. In a water body with a long, stable history in the tropical region, what would the interspecific interrelationships be like? Are their habits much more firmly established, so that their interactions are greatly reduced, producing, so to speak, a calm world?' The first, simple step was to compare them.

As for the biodiversity problem, I was forced by

Figure 1. Hiroya Kawanabe with Eiji Harada (left) on the occasion of the International Symposium for Ecology in Shimane (September 1990).

Okada-san, vice-president of the International Union of Biological Sciences, to express my opinion on a prospective for such a study, and subsequently a well-planned project was proposed and worked out by young ecologists. I do not think I ever conceived beforehand that it was something that I ought to do. That came out in the course of freely considering the problem. You may say I am not discipline-guided.

EH: I know you have been observing ayu populations in the U-kawa in northern Kyoto Prefecture every summer since 1955, when you first engaged in research there.

HK: Yes. It is a small river, as you know, just about 10 m wide and about 10 km long, and it housed an estimated population of between a few tens of thousands and half a million ayu in the 1960s, with tenfold fluctuations in number. Then, the villages around it and the river itself changed; cattle disappeared, banks were altered into concrete dykes, and the number of dams increased from 4 to 10; the ayu population decreased and remained at around a few tens of thousands. The roads constructed in the weathered granite region upstream collapsed almost every summer and sand was washed out into the river to be deposited in its bed; some river pools that had been several metres deep previously became shallower, about 20 cm at most in some years, and the population of ayu reached a low of several thousand. Ayu fingerlings had been introduced annually from Biwa-ko and the fingerlings of some other fishes were inevitably among them. Benthic fishes, such as *Pseudogobio esocinus*, that prefer sandy bottoms and had been rare, showed a remarkable increase. More than 10 years passed and the sand was gradually swept away, and the ayu population recovered to the level of a few tens of thousands. Then once again, a land reformation scheme was started, that induced large-scale landscape alterations along the upper reaches of the river, and the ayu population decreased. Now, it is recovering.

My annual survey in the U-kawa is only for a week, but the component of the fluctuations in the population of ayu fish that is not due merely to random factors, but to environmental changes, can be sorted out, I think. More than one's lifetime might be necessary to come to a sound conclusion, and the continued observation and record-keeping should properly be executed by institutional bodies; 'long-term ecological study' is a phrase currently in fashion, although its virtue has long been realized. We know that it happens frequently that, when a problem arises concerning nature conservation, for instance, it turns out that little relevant information exists beforehand. People born earlier could no doubt have initiated long-term surveys earlier.

EH: I suppose the surveys in Naka-umi and adjacent regions during 1958–1962 and in Biwa-ko during 1962–1965 provided you with chances to think about and deal with communities.

HK: As you know, Miyadi-san was requested to organize and conduct those surveys, the former by the Shimane Prefectural Government and the latter by the Ministry of Construction. There were lots of arguments about whether the requests should be accepted. I understand you keep the records of early meetings of the latter survey. To be honest, it was really good to work together with scientists who differed in their way of study from mine. It was also fruitful for a student of pure science to engage in work that had planned objectives to be answered by the survey. However, we did not take the approach of pursuing immediate answers for the requests. For example, we did not address the direct effects on fisheries of the land reclamation or of the regulated fluctuation of water level for the benefit of the people. In contrast, we had some understanding of the importance of taking a broad view of these issues as biological and environmental problems, and argued with the officials of the Ministry of Construction about this. In the end, I think our understanding was correct, but not deep enough.

EH: The list of fishes of the Naka-umi region, compiled by Iwai & Asano (1962) and printed in the survey report edited by Professor Miyadi, has been frequently cited. In contrast, despite the advent of quantitative approaches, densities of animals, for example, are not so often referred to for comparison.

HK: That is true, but that is not, of course, because of the futility of quantitative data. I know you do not mean to belittle the importance of the quantitative aspect of ecological research. The long-term fluctuations of ayu in the U-kawa, as I have mentioned, could only be discerned because the surveys were conducted quantitatively. However, populations in nature usually vary greatly, more than ten-fold in time and space, and I admit it is a reasonable criticism that the quantitative results of a small-scale survey conducted only once or twice are misleading. There is also a point of view that a definite conclusion can only be drawn from qualities scored as present or absent, or in a modern way '1' or '0'. This may be a natural outcome of the human desire for a clear and distinct answer, or the natural human need for certainty, as Elton-san remarked early in 1940.

From the point of view of a man intrigued by the relationship, a citation is a relation between the matter to be cited and the person who cites it, no doubt an interactive issue. If nobody cites a paper, naturally no citation exists, no matter what value the report represents. Someone may argue: lack of citations may be taken to imply no value; but it is not necessarily so, and obviously, if no one is currently interested and working in the field concerned, a citation may not be made in the time span of the present. I do not intend to stress the absolute value of the work; it is as ambiguous as a relative evaluation, which I do not like. I mean that the relationship should be looked at from various sides.

EH: I remember about 30 years ago at a gathering held in a rural hot-spring resort, you expressed, or rather were asked by young scientists to explain, your concept of the ecological community by proposing a definition for it.

HK: You mean my advocacy of 'the community as the whole of the interrelationships'. I think I was interested in the relationship itself. For instance, if I were considering psychiatry, it would not be the disease itself but its background, the relationships causing it, that I would wish to see. I was not a boy who enthusiastically watched and collected insects or fishes. At present, children like me are said to be increasing, and it is a pity. I was an exception in my time. I tried to make myself love existing actual things like animals and plants and to find them attractive, but my concern remained on abstract forms; so I studied relationships, and not relationships that were rigid and persisting but those that were ever fluctuating. Recently, I mentioned that importance lies in equivocality or ambiguity, and was criticized for the equivocally or ambiguity of my words.

It is now becoming clear that a community cannot be understood by simply summing up the relationships analysed and elucidated between pairs of species/individuals, but by considering interrelationships as a whole. It is also gradually becoming clear that flexible, relative, conditional relationships are far more common and important in the organic world. The rigid, dichotomous way of categorizing relationships (negative or positive; black or white) has limitations. For instance, competition and cooperation should not always be regarded simply as opposites since a relationship that could be described as competitive cooperation is more common than previously thought. Furthermore, it appears to me that interspecific relationships (behaviour or state) contribute to create the characteristics of species (organisms or matter).

EH: Paine (1994) writes: 'The natural world is filled with uncertainty'. Well, as a scientist, you have involved yourself in several social problems, which are also fraught with uncertainty. Students of ecology in particular are supposed to be conscious of environmental problems.

HK: Maybe or maybe not. People somewhat apart from me regarded me as actively working, but among the people close to me, there must be some who do not think so. It is true that I testified in the courts trying the suits of Minamata disease in Niigata and damming of the Nagara-gawa. I also wrote articles critical of pollution problems around 1970. Much earlier than that, I was speaking against the way governmental agencies modified river channels, and surely they looked upon me as a nuisance. Attitudes have utterly changed now, and the urgency of solving global environmental problems is widely accepted. However, there are few really global affairs; they may be global in the sense that they affect everywhere in the world, but they must be solved locally, probably inevitably by changing our way of life in some way. To be honest, I rather doubt the possibility of 'sustainable development', particularly for the whole of the globe; a culture believing in 'no development' or prosperous reduction might need to be created, denoted by Wilson-san (1984) as the third ethical phase.

EH: You have published numerous scientific papers, books and essays on a wide range of subjects.

HK: A lot of people say many of them are essays. The first one I wrote of that kind was on the territory of ayu in a popular science journal 'Shizen (Nature)' when I was in the third year of the postgraduate course. On seeing this article, a professor in our department asked me, 'Did Professor Miyadi permit you to write this? You must know that writing an article of this kind at your age is the same as publishing a paper of a minus point', warning that it would have a negative influence on my reputation. Believe it or not, I myself made a distinction between these articles and my scientific papers. However, sometime later when I read them again, I realized that the distinction appeared to be obscured. Moreover, if any of my works displayed creative thinking it was these articles. I remember my family complaining that I was in a bad mood whenever I was writing an article of that kind, but not while preparing a scientific paper. Writing these articles is not an easy task, and I may not be suited for it. Imanashi-san repeatedly urged me to write a large article, scientific or popular, but I have not done it yet. Some day, I hope.

EH: I know you were born into the family of a Buddhist priest who inherited custodianship of a temple with a long history in Kyoto. Did you ever have contemplative ideas which go far beyond ecology, if I may repeat the same question put in the previous interviews?

HK: Do you mean something like the Taoism of Balon-san (e.g., 1988), the editor of this journal? I do not think I understand it well. My family belonged to a unique sect of Japanese Buddhism that officially allowed its priests to marry and admitted their children to be their successors since the middle ages. Thus, I was destined to become a priest. So, differing from most Japanese people, I was in a special situation that, if I wished to be secular, I needed to make greater efforts; and I did so, but I am afraid it was impossible to deny religious ties completely. I was often told that my way of thinking had some similarity to that of the founder priest of the sect, Shinran-san, about seven hundred years ago, although I was utterly unaware of him. My father died when I was a child and left a vast holding of books, including Japanese classics of all sorts. I love reading literature, and it has become my habit to carry a book in my hand to read in cars and on streets. I love listening to music, and in an extreme case I attended concerts 10 times a week. However, I must confess I have no hobbies actually.

EH: What value do you place on loving music, or literature?

HK: I do not think they have any direct bearing on my scientific activities; I mean reading novels and plays is not for the purpose of writing papers, and I do not see literature in that way. Studying languages is another thing and essential, and cavilling at the examination on Japanese language for the students of natural sciences is absurd.

EH: How about hot spring baths?

HK: Thas has, in contrast, a direct and practical effect – securing time for myself, telling nobody where I am staying. I may call it 'the importance of being 'solitary'', which has been your favourite.

EH: 'The truth is rarely pure and never simple'.

HK: 'Animal' life would be tedious if it were either, and modern 'ecology' a complete impossibility'.

EH: Now, would you like to add any comments?

HK: You asked me too much and I talked too much. It happens often when we are chatting. I wish I could hear more of your own opinions, but anyway I thank you for all the trouble you have taken. Next time, I would like to interview you.

References cited

Balon, E.K. 1988. Tao of life: universality of dichotomy in biology. 1. The mystic awareness. Rivista di Biologia/Biology Forum 81: 185–230 (in English and Italian).

Elton, C. 1940. Scholasticism in ecology. J. Anim. Ecol. 9: 151–152.

Huxley, T.H. 1863. On our knowledge of the causes of the phenomena of organic nature: being six lectures to working men, delivered at the Museum of Practical Geology. Robert Hardwicke, London. 156 pp.

Imanishi, K. 1949. The logic of organic society. Mainichi Newspaper Co., Tokyo. 256 pp. (in Japanese).

Ishikawa, C. 1934. Ishikawa's general zoology and botany with illustrations. Gakugei-shoin, Tokyo, 3 + 9 + 280 + 13 pp., 72 plates (in Japanese).

Iwai, T. & H. Asano. 1962. 3, List of organisms, 7, fishes. pp. 50–57. *In:* D. Miyadi (ed.) Ecological Report of Fishes and Other Organisms in Relation to Freshening and Land Construction in Naka-umi and Its Adjacent Area, Shimane Prefectural Government, Matsue (in Japanese).

Kitazawa, Y. 1952. Bio-economics. pp. 104–155. *In:* N. Yagi & K. Nomura (ed.) General Ecology, Yokendo, Tokyo (in Japanese).

Paine, R.T. 1994. Marine rocky shores and community ecology: an experimentalist's perspective. pp. 1–152. *In:* O. Kinne (ed.) Excellence in Ecology 4, Ecology Institute, Oldendorf.

Wilson, E.O. 1984. Biophilia. Harvard University Press, Cambridge. 157 pp.

Part 2: Invited reviews

Environmental Biology of Fishes **52**: 75–95, 1998.

Research on fish ecology in Japan: a brief history and selected review

Kurt D. Fausch[1] & Shigeru Nakano[2]
[1]*Department of Fishery and Wildlife Biology, Colorado State University, Fort Collins, CO 80523, U.S.A. (e-mail: kurtf@cnr.colostate.edu)*
[2]*Tomakomai Experimental Forest, Hokkaido University, Takaoka, Tomakomai, Hokkaido 053, Japan*

Received 7.7.1996 Accepted 10.3.1997

Key words: stream fish ecology, Lake Biwa, life history, behavior, Lake Tanganyika, marine reef fishes

Synopsis

Fish play an important role in the culture of Japan, supplying the primary source of food protein. Research on fish ecology in Japan began developing about 1955, and paralleled that in North America and Europe, but has a number of unique differences. Two main research foci, stream fish ecology and ecology of Lake Biwa, were founded by H. Kawanabe and his colleagues while graduate students in D. Miyadi's laboratory at Kyoto University in the 1950s. The focus on studies of individual differences in ecology, life history, and behavior by direct observation that were developed in this laboratory had a pervasive influence on research that has continued to the present. In the 1980s, Kawanabe, who had succeeded Miyadi as laboratory head, also co-organized a major research project on the complex communities of cichlid fishes in Lake Tanganyika, which has involved nearly every major Japanese investigator in fish ecology since. A review of research conducted by Japanese investigators in stream fish ecology, ecology of fishes in lakes Biwa and Tanganyika, and marine reef fish ecology revealed a wealth of literature of which most English-speaking fish ecologists are unaware. The most striking difference between Western and Japanese research is the focus on detailed observations and experiments in natural habitats of individual differences in behavior, life history, and ecology, and how these ultimately affect processes at the population and community level, as amply demonstrated by Kawanabe's early work on ayu, *Plecoglossus altivelis*. This approach predated by 20–30 years the current interest in individual differences and individual-based models in North America and Europe.

Introduction

Fish have long played a very important role in Japan, representing the predominant source of food protein. They also provide important symbols for art and culture, and recently, valued recreation. Fishing was an important source of food as early as 8000 B.C., some 7000 years before rice was introduced to the Japanese archipelago (Abe 1986). The rapid development of commercial fisheries during the last century has led to a concurrent increase in research on fisheries science and technology, and aquaculture.

In comparison to the history of research on ichthyology and fisheries science (Abe 1986), the study of fish ecology in Japan, that is, of the relationship between fish and their environment, is more recent. It paralleled the development in North America and Europe, but emerged somewhat later, increasing in depth and breadth after 1955. Moreover, there are a number of important features of the research that are unique to Japan, in methods, approaches, and problems considered important for study.

Our purpose here is to briefly review the history and scope of research on fish ecology in Japan, focusing on its development from 1955 to 1994. Okada (1959, 1960a, 1960b) provides an extensive review of fish biology from 1709 to 1957, and other papers in this volume (Yuma et al. 1998) discuss the most recent research. The goal is to provide researchers from other countries who read English an entree into this rich body of literature. We emphasize unique differences between Japanese and Western research, as well as discussing similarities, and take an historical approach to show linkages among investigators and ideas through time.

A brief history

The study of fish ecology in Japan largely grew out of research headed by two ecologists, K. Imanishi and D. Miyadi. During the 1950s, Imanishi worked at Kyoto University and studied social systems ('animal sociology') of primates (e.g., Imanishi 1951, 1960, Itani 1954, 1959, Kawamura 1959, Kawai 1965). Imanishi had a profound influence on the development of ecology in Japan (see Itô 1991 and Ohgushi 1992 for reviews), being well known for the habitat segregation theory he developed to explain distributions of lotic mayfly (Ephemeroptera) nymphs (Imanishi 1941), and a unique theory of evolution (Imanishi 1977, see Halstead 1985, 1987, 1988, Rossiter 1986, Balon 1989a, Asquith 1992).

During this same period, Miyadi was the professor of the Laboratory of Animal Ecology in the Department of Zoology at Kyoto University. Laboratories in Japanese academia are discrete subunits of departments, usually consisting of one full professor that heads the laboratory, one associate professor, one or two assistant professors, and their graduate students, all of whom conduct research toward a common goal in a general area. During the post-war period in Japan, Miyadi developed or administered several important research initiatives, including stream ecology, ecology of Lake Biwa, marine biology, and experimental evolution (e.g., Mori 1986, Mori &Suzuki 1986). For example, the 'Research Group on Ecology of Lotic Communities' was formed in the early 1950s by H. Kawanabe, N. Mizuno, and others, then graduate students, who actively pursued several research topics through about 1970 (e.g., Kawanabe 1957, 1969b, Mizuno 1960a, Nagoshi et al. 1962). Similarly, the 'Research Group for Biotic Resources of Lake Biwa', formed in the early 1960s, later became part of the International Biological Programme and focused on biological production of this large ancient lake, located near Kyoto in south-central Japan (e.g., Nagoshi 1966, 1969, Miura et al. 1976, see review by Kawanabe 1969a).

During the late 1970s, M. Kawabata and H. Kawanabe (the latter having been promoted to professor of the Laboratory of Animal Ecology at Kyoto University) jointly developed a research project on fish communities in Lake Tanganyika (e.g., Kawanabe 1981), one of the three large ancient African Rift lakes. This work stemmed from African studies begun by primatologists from Imanishi's group (e.g., Kawai & Mizuhara 1962, Itani & Suzuki 1967). The overall goal of the project, which is ongoing, has been to understand how species interactions combine and interact to shape complex communities of the highly diverse species flock of cichlid fishes. Since 1980, this project has, in some way, involved nearly every major Japanese investigator in fish ecology.

The general research initiatives described above form the historical basis for most investigators and research foci pursued at present. Below we describe the current research being conducted in each area.

The primacy of Kyoto University

Although the current group of fish ecologists emanates from many universities in Japan, the largest single lineage extends from Kyoto University. As described above, the laboratory of D. Miyadi, and H. Kawanabe who later succeeded him, produced many ecologists who study fish. These investigators and their students include a large number, but by no means all, of the fish ecologists now working in Japanese universities and government (federal and prefectural) laboratories.

Perhaps more important than the long lineage

from Kyoto has been the influence of ecologists Imanishi and Miyadi on the research ideas and approaches used in Japan (Ikeda & Sibatani 1995). For example, the individual identification and detailed behavioral observations of Japanese macaque, *Macaca fuscata*, in their social systems conducted by Imanishi and his colleagues influenced early work by Miyadi and Kawanabe and their contemporaries on how changes in social behavior of the salmon-like ayu, *Plecoglossus altivelis*, affected production (e.g., Kawanabe 1957, 1958, 1969b, 1970, 1976, Kawanabe et al. 1956, 1957a, 1957b, 1957c, 1959, Mizuno & Kawanabe 1957, Miyadi 1960). This focus on detailed study of individual fish behavior from the viewpoint of animal sociology was later modified and blended with ideas from the field of behavioral ecology, but nevertheless continues to be one of the primary approaches used by Japanese fish ecologists today. For example, Katano (1983, 1985, 1987, 1990, 1994) made an exhaustive study of the individual behavior of dark chub, *Zacco temmincki* (similar in morphology to North American creek chub *Semotilus atromaculatus*), and wrote an entire book in Japanese on the ecology of individuality (Katano 1991). Similarly, Japanese ecologists who study stream salmonids (Nakano et al. 1992, Nakano & Furukawa-Tanaka 1994, Nakano 1995a, b) and both marine and Lake Tanganyikan reef fishes (Kohda 1984, 1991, Kohda & Yanagisawa 1992, Kuwamura 1992, Ochi 1986a, 1993a, Yanagisawa 1993) have also conducted studies based on detailed observations of individual behavior.

Another important contribution made by Kawanabe was the view that communities can best be understood by detailed investigations of the interrelationships among individuals of the animal and plant species (Kawanabe 1960). This view apparently stemmed from his early work showing that production of ayu cannot be understood without elucidating how territoriality and modes of feeding change with density (Kawanabe 1969b, 1970). Similarly, his first research in Lake Tanganyika showed that feeding territories of one cichlid were defended against a different set of species, and had a very different function, than breeding territories. This led Kawanabe (1981) to suggest that more detailed study than had been conducted to date was needed on how these differences influence population regulation and community processes. This idea appears to have formed the basis for subsequent research conducted on reef fishes (e.g., Nakano & Nagoshi 1990, Kohda 1984, 1991, Nakai 1993).

Innovative methods are often required for research to advance. In 1955, Kawanabe and his colleagues began censusing ayu and observing their behavior by direct underwater observation in rivers via snorkeling (e.g., Kawanabe et al. 1956, Kawanabe 1970, Yuma & Harada 1998 this volume), a method rarely used in North America until later (e.g., Keenleyside 1962, Northcote & Wilkie 1963). Moreover, underwater observation has continued as the primary method for many Japanese fish ecologists who study fish ecology and behavior (e.g., Kohda & Hori 1993, Nakano 1995). Highly detailed underwater mapping of individual behavior is facilitated by the use of base maps and overlays drawn on plastic paper (e.g., Nakano & Furukawa-Tanaka 1994), a method borrowed from coral reef ecologists and applied during early research at Lake Tanganyika (Hori 1983, Hori et al. 1983, Yanagisawa 1986). Both methods continue to play an important role in research on fish ecology in Japan.

Areas of emphasis in Japanese fish ecology

Stream fish ecology

As described above, research on stream fish ecology was primarily founded by Kawanabe, Mizuno, and others while graduate students in Miyadi's laboratory at Kyoto University in the early 1950s. This group focused primarily on the study of ayu behavior, ecology, and production (Kawanabe 1957, 1958, 1966, 1969b, 1970, 1972, 1976, 1977, Kawanabe et al. 1956, 1957a, 1957b, 1957c, Mizuno & Kawanabe 1957, see also Azuma 1964, 1970, 1973a, 1973b, 1973c, Miura 1965, for ayu in Lake Biwa). Ayu are amphidromous fishes that live only one year. They spawn in the downstream reaches of rivers in fall, and the embryos hatch within a few weeks and drift to the ocean, ascending rivers as juveniles in the spring (Kawanabe 1969b). Schools of juveniles graze algae and diatoms from rock surfaces in riffles, where they may subsequently set up territories as adults. Research to 1970 showed that social behavior of adult ayu changed from territorial to schooling at high population density, and that this behavioral shift regulated their production more than algal food supply.

Ayu is an important food fish caught by both recreational anglers and commercial fishermen in Japan (Kawanabe 1969b). Recently, experiments on their upstream migratory behavior (Tsukamoto & Uchida 1990, 1992, Uchida et al. 1990a, 1990b) and aggressive behavior, mate choice, egg-size variation, and competition of larvae (Nishida 1986, Uchida et al. 1989, Iguchi 1991, 1993, Iguchi et al. 1991, Iguchi & Maekawa 1993, Iguchi & Yamaguchi 1994) have been conducted. This research is primarily to aid management of the remaining wild stocks in the face of many dams, and the outplanting of hatchery ayu. A unique study by Tsukamoto et al. (1987) revealed switching in life-history pattern among three types of ayu spawning in tributaries of Lake Biwa, which they proposed was primarily due to hatching time as determined from otolith daily growth increments.

Benthic stream fishes have been another topic of research since the 1950s. Studies by N. Mizuno and his students and colleagues on the taxonomy, distribution, and life history of amphidromous freshwater gobies (*Rhinogobius* spp.) in mountain streams of central Japan (e.g., Mizuno 1960a, 1960b, 1963a, 1963b, 1976) led naturally to research on their ecology (Mizuno et al. 1979, Mizuno 1980, Iguchi & Mizuno 1990, 1991, see also Sawara 1978, Sawara & Ogawa 1984) and zoogeography and evolution (Mizuno et al. 1982, Shimizu et al. 1993). Studies of interactive habitat segregation by species above and below waterfalls in the Ryukyu Islands (Kawanabe 1973, Kawanabe et al. 1974), research on ecology of four color morphs considered to be incipient species (Mizuno et al. 1979), and experiments identifying mechanisms of drift of larvae of an entirely fluvial form (Iguchi & Mizuno 1990, 1991) are particularly interesting.

Mizuno's work on sculpins and gobies (e.g., Mizuno & Niwa 1961, Mizuno 1963a) also influenced A. Goto, who began studying sculpins in streams of Hokkaido Island (the northern island in the Japan Archipelago) in the 1970s (e.g., Goto 1974, 1975a, 1975b, 1977). Goto has continued his detailed and painstaking studies of life history and reproductive behavior (e.g., Goto 1982, 1983, 1984a, 1987a, 1987b, 1988, 1989a, 1990, 1993a, 1993b, 1993c), as well as other work on ecology and distribution (Goto 1980, 1981, 1984b, 1986, 1989b, 1989c, 1994). He uses a unique method of carefully removing combinations of individual dorsal spines and rays (Goto 1985) to individually identify up to 500 sculpins within a watershed. Goto's work has revealed interesting reproductive life-history variation between closely related species that adapt them to living in upstream versus downstream reaches of Hokkaido mountain streams (e.g., Goto 1990). Isozyme analysis of these two forms also suggested that the more derived fluvial one inhabiting upstream reaches may have differentiated from the more ancestral amphidromous form in downstream reaches at several different times (i.e., polypatric speciation; Goto & Andoh 1990, see also Mizuno 1963a, Katoh & Nishida 1994).

Japanese fish biologists have long studied various aspects of salmonid biology (e.g., Onodera 1958, 1962, Onodera & Ueno 1961), but few studies of their ecology were conducted until relatively recently. K. Ishigaki's pioneering work on the distribution and ecology of two congeneric charrs (*Salvelinus*) throughout Hokkaido (Ishigaki 1969, 1987; see also Fausch et al. 1994), culminating in a widely-read book on the subject (Ishigaki 1984), served as a catalyst to influence several other investigators. For example, K. Maekawa and his colleagues have studied the systematics, reproductive life history, and evolution of a unique form of Dolly Varden charr, *Salvelinus malma miyabe*, landlocked in Lake Shikaribetsu, and of other Dolly Varden populations, since the 1970s (Maekawa 1977a, 1977b, 1978, 1983, 1984, 1985, 1989b, Maekawa & Hino 1986, 1987, 1990, Maekawa & Onozato 1986, Hino et al. 1990, Maekawa et al. 1993a, 1993b). These researchers are particularly interested in alternative reproductive life-history strategies of salmonids, and how they change via evolution under different environmental conditions (e.g., Maekawa & Goto 1982, Goto & Maekawa 1989, Gross & Maekawa 1989, Maekawa 1989a, Nakano et al. 1990, Yamamoto et al. 1992,

Maekawa et al. 1994, Kitano et al. 1994, Goto et al. 1994).

Other research on stream salmonid ecology includes investigations by M. Nagoshi and his students of many aspects of the population ecology of red-spotted masu salmon, *O. masou ishikawai* (formerly *O. m. rhodurus*), a landlocked form termed 'amago', in a small mountain stream in south-central Japan (Nagoshi & Sakai 1980, Nagoshi et al. 1982, 1988, Nakano & Nagoshi 1985, Nagoshi & Kurita 1986; see also Maekawa & Nakano 1994). T. Furukawa-Tanaka, who studied with Kawanabe and others at Kyoto, investigated interspecific interactions between masu salmon and white-spotted charr, *S. leucomaenis*, in central Japan (Furukawa-Tanaka 1985, 1988; see also Maruyama 1981), and of white-spotted charr and Dolly Varden in Hokkaido (Furukawa-Tanaka 1989). Tanaka's early work greatly influenced the junior author, who made detailed studies of individual behavior in the amago population studied by Nagoshi to elucidate the mechanisms structuring intraspecific dominance hierarchies (Nakano 1994, 1995b).

Furukawa-Tanaka's work in Hokkaido was also followed by a series of studies to understand mechanisms of foraging shifts by Dolly Varden and interspecific competition with its sympatric congener, the white-spotted charr (Nakano & Furukawa-Tanaka 1994, Fausch et al. 1994, 1997). The group of investigators that conducted this work recently collaborated on other studies of salmonids in Japan and Montana as well (Nakano et al. 1992, Kitano et al. 1993, 1994). Other research addressing fish ecology in Hokkaido is ongoing (see Goto & Nakano 1993 for a review), such as the response of fish communities to habitat changes at a variety of scales (e.g., Inoue & Nakano 1994; see also Nakamura et al. 1994 for work in central Japan). There is also renewed interest in studying the large Japanese huchen, *Hucho perryi*, in Hokkaido (Yamashiro 1965, Kimura 1966, Nakano 1992, Fukushima 1994), a species endangered by habitat loss and overfishing.

Despite the commercial importance of anadromous masu salmon, which has a juvenile life history similar to coho salmon, *O. kisutch* (i.e., usually rearing one or more years in fresh water) there has been relatively little research on its ecology in natural systems (but see Kato 1973, Kubo 1980). This is partly due to the lack of suitable study areas, because most masu salmon and chum salmon, *O. keta*, are captured in weirs as they ascend rivers so little natural spawning is allowed (Mayama 1990). However, some recent research on life history and growth has been conducted (e.g., Hirata et al. 1988, Kiso 1990, Kiso et al. 1992, Kiso & Matsumiya 1992, Mayama 1992, Nagata et al. 1994). Also of great importance are investigations showing the gradual decline in length, and increasing age at maturity, of returning adult chum salmon, concomitant with the rapid increase in the number of smolts released to the north Pacific Ocean after 1970 (Kaeriyama 1989, Kaeriyama & Urawa 1992; see Kaeriyama 1986 for freshwater early life history).

South-central Japan has a relatively rich cyprinid fauna, the basic ecology of which was first studied by Kawanabe and his colleagues (Mizuno et al. 1958, Kawanabe 1959, Nagoshi et al. 1962). More recent work by O. Katano on the behavioral ecology of dark chub (e.g., Katano 1987, 1990, 1994) has shown that the social behavior of this species is both complex and interesting, which led Katano (1993) to suggest that researchers in North America, Europe, and Japan may profit greatly from studying this familiar group of fishes. This example also illustrates that a new generation of Japanese fish ecologists has been reviving the study of stream fish ecology. After an energetic beginning in the 1950s, this field generally lay dormant during about 1970–1980 except for a few investigators (e.g., Maki 1972, 1973, 1976a, 1976b, Maki et al. 1970, Maki & Kakiuchi 1971, Hirai & Hikido 1976, Nagata & Nishiyama 1976, Takamura 1979; see also Kondo et al. 1984).

Sticklebacks are a group of fishes that occupy many different environments, ranging from lotic to lentic and freshwater to marine. Goto and his students have studied the biology and ecology of nine-spine stickleback, *Pungitius pungitius*, in Hokkaido (Goto et al. 1979, Takata et al. 1984, 1987; see also Tanaka & Hoshino 1979 for central Japan), and S. Mori (1984, 1985, 1987a, 1987b, 1987c, 1988, 1990, 1993, Mori & Nagoshi 1987, Nagata et al. 1988) studied reproductive behavior and ecology of threespine stickleback, *Gasterosteus aculeatus*, in rivers and ponds of central Japan.

Lakes Biwa and Tanganyika, and other lentic systems

Lake Biwa, a large, old lake with a diverse fauna of endemic cyprinids, silurids, ayu, and masu salmon (Kawanabe & Tokui 1984, Mori & Miura 1990) was the focus of one of the first coordinated studies of lentic fish ecology by Japanese ecologists (Yuma & Harada 1998 this volume). This research project, started in the early 1960s, later became part of the International Biological Programme (Kawanabe 1969a, 1978, Nakajima & Nakai 1994). M. Nagoshi and his colleagues made extensive studies of the population ecology, diet, and growth of the goby, *Chaenogobius isaza*, in the lake (Nagoshi 1966, 1969, 1982, Nagoshi & Kojima 1976, Takahashi 1981, Nakanishi & Nagoshi 1984, Hidaka & Takahashi 1987), and measured production of the endemic masu salmon (e.g., Miura et al. 1976). Others made detailed studies of the population ecology of various endemic cyprinids and other native fishes (e.g., Hirai 1964, 1969, 1970, 1971, 1972, Maki 1964, 1966a, 1966b, 1967, 1968a, 1968b, Nagoshi & Miura 1964, Sunaga 1964, 1970, Tanaka 1964, 1970a, 1970b, Nagoshi 1965, 1977, Miura 1966a, 1966b). Eutrophication and explosive increases in populations of introduced bluegill, *Lepomis macrochirus*, and largemouth bass, *Micropterus salmoides*, have caused dramatic changes in the ecosystem in the last 30 years (Terashima 1980, Nakajima & Nakai 1994). Much of the research on Lake Biwa was conducted under the direction of T. Miura, from the Otsu Hydrobiological Station at the south end near Kyoto. Investigators there also collaborated with Chinese fish biologists to study the effect of phytoplanktivorous fishes on the food web of an eutrophic lake in China (Miura 1989, 1990).

One lake in eastern Hokkaido originally had an endemic stock of kokanee, *O. nerka*, the resident form of sockeye salmon. In the 1890s this stock was transplanted to Lake Shikotsu, near Sapporo in western Hokkaido, and was subsequently extirpated from its original habitat (Kaeriyama personal communication). Kokanee are now stocked annually into Lake Shikotsu and several other Hokkaido lakes from hatchery sources (Kaeriyama et al. 1992). Kaeriyama (1991) made a detailed study of the factors that influenced population dynamics of kokanee in Lake Shikotsu based on data from 1898 to 1990, which probably represents one of the most extensive sets of data on this species anywhere in the world.

Research on fish communities in Lake Tanganyika grew out of interest by M. Kawabata of Shinshyu University in adaptive radiation of cichlids (e.g., Kawabata 1975). Kawabata interested Kawanabe in the fishes of the lake, and together they were successful in gaining funding to start work in 1979 (Kawanabe & Takamura 1979, Yuma & Harada 1998). The research has focused on community ecology of cichlids on rocky reef shores at various locations in Zaire, Zambia, Burundi, and Tanzania, and to date has included about 40 Japanese researchers from many fields of study (Y. Yanagisawa personal communication) as well as some international collaboration (Barel et al. 1985). The work also involved Zaire scientists (e.g., Mbomba 1985, Kawanabe 1986), three of whom earned doctoral degrees in Japan (e.g., Mbomba 1986, Gashagaza & Nagoshi 1988, Nshombo et al. 1985, Gashagaza 1991, Nshombo 1991, 1994a, 1994b).

The large body of research conducted in Lake Tanganyika has recently been aptly reviewed (Nakai 1993, Yuma 1993, Nakai et al. 1994) so only a brief synopsis of selected papers will be attempted here. Much of the research has focused on interspecific territoriality and mating systems of cichlids, following Kawanabe's ideas about the complex mechanisms of interspecific territoriality for feeding and mating (Kawanabe 1981, Hori et al. 1993). Investigators have discovered a great diversity of behavior, even among closely related congeneric species, in mouthbrooding (e.g., Yanagisawa 1986, Kuwamura 1988, Kuwamura et al. 1989, Yanagisawa & Sato 1990, Yanagisawa & Ochi 1991; see also Balon 1985), other kinds of parental care (Nagoshi 1985, 1987, Yanagisawa 1985, 1987, Nakano & Nagoshi 1990, Ochi et al. 1995), mating systems (Kuwamura 1986, 1987b, Yanagisawa & Nishida 1991, Ochi 1993a, 1993b, Yanagisawa 1993, Sato 1994), and interspecific territoriality (Takamura 1984, Kohda 1991, Kohda & Yanagisawa 1992). In general, similar species appear to segregate along depth strata along the shores of the lake (Kuwamura 1987a, Kohda & Yanagisawa 1992), and defend feeding territories most vigorously against conspecifics and heterospecifics that have similar diet (Kuwamura 1987b, Kohda 1991, Yanagisawa 1993).

The project in Lake Tanganyika has also resulted in exciting findings on complex interactions among fishes, especially the cichlids. For example, Hori (1993) and Takahashi & Hori (1994) recently showed that right- and left-handedness in a scale-eating cichlid is a polymorphism maintained by frequency-dependent selection, and can be explained based on simple genetics. Kohda & Hori (1993) and Kohda (1994) reported that two color morphs of another scale-eating cichlid had distinctly different predatory behavior, and proposed that the dichromatism served to optimize crypsis during foraging. Sato (1986) documented brood parasitism by a catfish on cichlids in the lake. Nakai et al. (1990) found that eight closely-related cichlids timed their spawning for moonlit nights to promote efficient egg guarding, and to allow their young to disperse two weeks later during moonless nights to enhance survival.

Many apparently commensal and mutualistic relationships in foraging and territoriality have emerged as well (Hori 1987, Nakai 1993, Yuma 1993). For example, five benthivorous cichlids that use different methods to prey on abundant atyid shrimp feed in groups, apparently because disturbance of shrimp by heterospecifics increases their foraging success in a mutualistic fashion (Yuma 1994). Similarly, Takamura (1983) reported that grazing activity by an algivorous cichlid improved the quality of filamentous algae for the dominant browsing cichlid, which was thought to account for the commensalistic overlap in their feeding territories. The prevalence of such indirect beneficial relationships suggests that, along with resource partitioning resulting from competition, mutualism and commensalism may be important mechanisms promoting the development and maintenance of high cichlid diversity in the African Rift lakes (Kawanabe et al. 1993, Matsuda et al. 1993, 1994, Nakai et al. 1994).

Ecology of marine reef fishes

The study of ecology of marine reef fishes has diverse origins in Japan (e.g., Mori 1956, Okuno 1971), and many investigators who worked on Lake Tanganyikan fishes also conducted research in marine systems. For example, the Seto Marine Biological Laboratory, founded in 1922, is a center for research on biology and taxonomy of crustaceans. However, E. Harada of Miyadi's laboratory, who became professor and head of the Seto Laboratory in 1977, also trained two ecologists now actively studying ecology of both Lake Tanganyikan and marine reef fishes, Y. Yanagisawa and T. Kuwamura. Yanagisawa's first work followed up on Harada's earlier research on shrimp-goby mutualistic interactions (e.g., Yanagisawa 1990). The shrimp digs a burrow that affords refuge for the goby, while the goby alerts the nearsighted shrimp of predators (Yanagisawa 1982, 1984). Kuwamura began his research by studying the aggressive mimicry and complex social structure of a cleaner wrasse (Kuwamura 1983, 1984).

Overall, the interests of Japanese reef fish ecologists have focused on behavioral ecology, especially mating systems and parental care. It is interesting that an American fish biologist, J. Moyer, who has studied reproductive behavior and social organization of coral reef fishes on Miyake-jima, an island south of Tokyo, for more than 20 years (e.g., Moyer 1975, 1976, 1977, 1979, 1987, 1989, 1990, Moyer & Sawyers 1973, Moyer & Bell 1976, Moyer & Nakazono 1978a, Moyer & Steene 1979, Moyer & Zaiser 1981), has strongly influenced research by Japanese fish ecologists. For example, Moyer and his collaborator, A. Nakazono, discovered sex change in the anemonefish, *Amphiprion clarkii* (Moyer & Nakazono 1978b), which was subsequently studied by Ochi, Yanagisawa, and Hattori (Ochi 1985, 1986b, 1989a, 1989b, Yanagisawa & Ochi 1986, Hattori & Yanagisawa 1991, Hattori 1994, 1995, Hattori & Yamamura 1994). All anemonefish begin life as males, but the pathways of sex change for individuals are apparently different for tropical versus more temperate locations due to ecological conditions. Other marine reef fishes change sex too (e.g., Kuwamura 1984), and Japanese coral reef fish ecologists recently contributed to a book on sex change in fishes

(Nakazono & Kuwamura 1987). A high point in the careers of many Japanese reef fish ecologists was the Second International Conference on Indo-Pacific Fishes (Uyeno et al. 1986), held in Tokyo in 1985, during which several of the younger scientists gave their first oral presentations in English, and Moyer introduced them to English-speaking coral reef fish ecologists from around the world (Yanagisawa personal communication).

Other current research topics of Japanese reef fish ecologists include population dynamics of a goby and its host coral in the Ryukyu Islands (Kuwamura et al. 1993, 1994), and intraspecific and interspecific territoriality in temperate damselfish (Kohda 1981, 1984, 1988). Sano has studied the effects of devastation of coral reefs by the crown-of-thorns starfish, *Acanthaster planci*, on populations and communities of coral polyp-feeding fishes, using both observational (Sano et al. 1987) and experimental approaches (Sano et al. 1984). He has also reported on diet and foraging behavior of marine fishes (e.g., Sano & Moyer 1985, Baba & Sano 1987, Moyer & Sano 1987, Sano 1989, 1993), and carried out an innovative field experiment testing habitat selection and interspecific competition between two mugiloidid reef fishes (Sano 1990). Noda et al. (1992) studied the importance of zooplankton patches as a food resource for a pomacentrid planktivore in southern Japan. Munehara (1988, 1991, 1992) investigated spawning behavior in marine cottids of coastal Hokkaido, which although not coral reef fishes, have internal fertilization and display complex reproductive behavior. Munehara et al. (1994) used DNA fingerprinting to show that males of one species guard eggs spawned by females that were previously inseminated by unrelated males, apparently to gain access to future reproduction with these females.

Other marine fishes

Japanese fish biologists and fishery scientists have gathered a wealth of biological data on a wide variety of marine fishes important to their commercial fisheries. This work has contributed much to understanding the basic ecology of species about which little is known, largely because they inhabit vast marine systems. A good example is the work by K. Tsukamoto and his many Japanese and international colleagues, who recently discovered the spawning grounds of the Japanese eel, *Anguilla japonica*, in the equatorial Pacific (Tsukamoto 1992), after years of careful research based on analysis of hatching dates using otolith microstructure (e.g., Tsukamoto 1990, Tsukamoto & Umezawa 1990, Umezawa & Tsukamoto 1991). Similarly, T. Taniuchi and M. Shimizu and their colleagues have reported on elasmobranch biology (e.g., Pratt et al. 1990, Taniuchi et al. 1993), including recent work on freshwater elasmobranchs of the Australian region (Shimizu & Taniuchi 1991), providing needed data for developing further hypotheses to test about the ecology of these species.

Unique differences and constraints

Differences in both language and culture between Japanese and Western fish ecologists have had important effects on the development of Japanese fish ecology and its current status. For example, writing papers in English is very difficult for any scientist for whom English is not a native language, but even more so for those like Japanese who must learn a completely different character set. As a result, few papers by Japanese fish ecologists were written in English before about 1960, and many lacked English abstracts. Some scientists began writing most of their work in English during the 1970s and 1980s (e.g., see papers by Goto, Maekawa, Nagoshi) and publishing them in international journals, a goal now pursued by most investigators.

Another unique difference is that, until recently, most Japanese investigations were based on observations of natural systems, rather than controlled laboratory or field experiments (but see Onodera 1958, 1962, Onodera & Ueno 1961, Yamagishi 1962, 1963, 1964, 1969, Sunaga 1971, 1976, Yamagishi et al. 1974, Nagoshi 1978, Nagoshi et al. 1978 for early experimental work). In many cases this stemmed from a lack of facilities for experiments, a persistent problem in the chronically under funded Japanese university system (e.g., Arima 1992, Hirano 1992).

However, the focus on observational natural history has also maintained a certain primacy in Japan, in part because many investigators express concern over artifacts that could be produced when large mobile organisms like fish are studied under controlled laboratory conditions (Fausch personal observation). This situation is changing, and many researchers have recently used experimental approaches, both in the laboratory (e.g., Uchida et al. 1990a, 1990b, 1991, Tsukamoto & Uchida 1990, Iguchi & Maekawa 1993, Iguchi & Yamaguchi 1994, Munehara et al. 1994, Nagata et al. 1994) and field (e.g., Yanagisawa & Ochi 1986, Tsukamoto et al. 1989, Sano 1990, Iguchi & Mizuno 1991, Kuwamura 1992, Yanagisawa 1993). Clearly, an ideal approach to guard against the artifacts described above would be to develop hypotheses based on careful observations of natural systems, test these via appropriate field or laboratory experiments, and then validate predictions from the experiments by further field observations.

Perhaps the most striking difference between Western and Japanese fish ecology is the focus on detailed observations and experiments of individual differences in behavior and life history, and how these ultimately affect processes at the population and community level (e.g., Yamagishi 1962, 1969, Maekawa & Onozato 1986, Yanagisawa & Ochi 1986, Tsukamoto et al. 1987, 1989, Katano 1990, Hattori & Yanagisawa 1991, Kuwamura 1992, Goto 1993c, Kohda & Hori 1993, Iguchi & Maekawa 1993, Mori 1993, Nakano 1994, 1995a, b). In the view of the senior author, Japanese scientists have studied important differences among individuals that many Western scientists tended to overlook in their attempts to fit organisms into categories (i.e., size or age classes, or groups based on behavior). This emphasis on proximate individuality and its effects on ultimate higher-order processes grew out of early work on primates by Miyadi and Imanishi, was applied to stream fishes by Kawanabe and his colleagues in the mid-1950s, and has been continued by many of the new generation of fish ecologists in Japan. It is interesting that this approach predated by 20–30 years the current interest in the ecology of individuals in Europe and North America (e.g., Łomnicki 1978, 1988, Ebenman & Persson 1988), including the burgeoning field of 'individual based modeling' (DeAngelis & Gross 1992).

The present and the future

In the late 1980s and early 1990s, Hiroya Kawanabe and his many colleagues organized and held several important international meetings in Japan that incorporated fish ecology, including the International Charr and Masu Salmon Symposium (when the senior author first traveled to Japan) in 1988 (Kawanabe et al. 1989), the Vth International Ecological Congress (INTECOL) in 1990 and its two satellite meetings titled Ecology for Tomorrow (Kawanabe et al. 1990) and Mutualism, Co-operation, and Organization in Natural Communities (Kawanabe et al. 1993), and a satellite meeting on Ecology and Ethology of Freshwater Fishes held in conjunction with the International Ethological Congress in 1991. Like the Second International Conference on Indo-Pacific Fishes in 1985 (Uyeno et al. 1986), these meetings succeeded in markedly increasing the international visibility of Japanese research on ecology and behavior of fishes and other organisms, and served to highlight the important contributions of many Japanese biologists. Moreover, the interactions and friendships forged fostered productive research collaboration between Japanese and Western scientists (e.g., Balon 1989b, Balon & Goto 1989, Gross & Maekawa 1989, Kawanabe et al. 1989, 1993, Hino et al. 1990, Nakano et al. 1992, Maekawa et al. 1993, Fausch et al. 1994, 1997, Kitano et al. 1994). Like Itô (1991), we hope that this trend continues, and leads to further international collaboration and global understanding.

Acknowledgements

This paper was written while the senior author was a Visiting Research Fellow of the Japan Society for the Promotion of Science (JSPS) during summer 1994. Much of the interaction with Japanese fish ecologists that led to the ideas presented here occurred during collaborative research funded jointly by the U.S. National Science Foundation (INT

9016512 to K. Fausch) and JSPS (to H. Kawanabe, M. Nagoshi, K. Maekawa, S. Nakano, K. Nakai, and T. Hino) through the U.S.-Japan Cooperative Science Program. This research was also partly supported by grants-in-aid from the Ministry of Education, Science, Sport, and Culture (Nos. 07308042, 07306022, 07740592, and 08454246). The senior author is grateful to his many hosts during his visit to Japan, including M. Nagoshi (Nara Women's University), Y. Yanagisawa (Ehime University), K. Maekawa, O. Katano, and K. Iguchi (National Institute of Fisheries Science, Ueda), A. Goto, F. Nakamura, and K. Ishigaki (Hokkaido University), H. Kawamura and M. Nagata (Hokkaido Fish Hatchery), M. Kaeriyama (National Hokkaido Salmon Hatchery), and M. Sano and K. Tsukamoto (Tokyo University). We thank Y. Yanagisawa and several anonymous reviewers for constructive comments that improved the manuscript.

References cited

Abe, T. 1986. A brief history of Japanese ichthyology. pp 1–6. *In:* T. Uyeno, R. Arai, T. Taniuchi & K. Matsuura (ed.) Indo-Pacific Fish Biology: Proceedings of the Second International Conference on Indo-Pacific Fishes, The Ichthyological Society of Japan, Tokyo.

Arima, A. 1992. Underfunding of basic science in Japan. Science 258: 590–591.

Asquith, P. J. 1992. Kinji Imanishi and ecology in Japan. Trends Ecol. Evol. 7: 285–286.

Azuma, M. 1964. The life history of the landlocked ayu-fish (*Plecoglossus altivelis*) in Lake Biwa, with special reference to their developmental stages. Physiol. Ecol. Japan 12: 55–71 (in Japanese with English summary).

Azuma, M. 1970. Studies on the variability of the landlocked ayu-fish *Plecoglossus altivelis* T. et S., in Lake Biwa. I. On the mode of distribution and some body form variation at early phases of development. Japan. J. Ecol. 20: 63–76.

Azuma, M. 1973a. Studies on the variability of the landlocked ayu-fish *Plecoglossus altivelis* T. et S., in Lake Biwa. II. On the segregation of population and the variations in each population. Japan. J. Ecol. 23: 126–139.

Azuma, M. 1973b. Studies on the variability of the landlocked ayu-fish *Plecoglossus altivelis* T. et S., in Lake Biwa. III. On the differences in the process of maturation, spawning habits, morphological features of each population. Japan. J. Ecol. 23: 147–159.

Azuma, M. 1973c. Studies on the variability of the landlocked ayu-fish *Plecoglossus altivelis* T. et S., in Lake Biwa. IV. Considerations on the grouping and features of variability. Japan. J. Ecol. 23: 255–265.

Baba, O. & M. Sano. 1987. Diel feeding patterns of the congiopodid fish *Hypodytes rubripinnis* in Aburatsubo Bay, Japan. Japan. J. Ichthyol. 34: 209–215.

Balon, E. K. 1985. The theory of saltatory ontogeny and life history models revisited. pp. 13–30. *In:* E.K. Balon (ed.) Early Life Histories of Fishes: New Developmental, Ecological, and Evolutionary Perspectives, Dev. in Env. Biol. Fish. 5, Dr W. Junk Publishers, Dordrecht.

Balon, E. K. 1989a. The Tao of life: from the dynamic unity of polar opposites to self-organization. pp. 7–40. *In:* M. N. Bruton (ed.) Alternative Life-History Styles of Animals, Perspectives in Vertebrate Science 6, Kluwer Academic Publishers, Dordrecht.

Balon, E.K. 1989b. Kariba-dam: kyodai dam ha onkei wo motarashi taka (Kariba: the dubious benefits of large dams). Tansui-gyo hogo (Freshwater Fish Protection) 1989: 69–79 (in Japanese).

Balon, E. K. & A. Goto. 1989. Styles in reproduction and early ontogeny. pp. 1–47. *In:* A. Goto & K. Maekawa (ed.) Reproductive Behavior in Fishes: Styles and Strategies, Tokai University Press, Tokyo (in Japanese).

Barel, C. D. N., R. Dorit, P. H. Greenwood, G. Fryer, N. Hughes, P. B. N. Jackson, H. Kawanabe, R. H. Lowe-McConnell, M. Nagoshi, A. J. Ribbink, E. Trewavas, F. Witte & K. Yamaoka. 1985. Destruction of fisheries in Africa's lakes. Nature 315: 19–20.

DeAngelis, D. L. & L. J. Gross. 1992. Individual-based models and approaches in ecology. Chapman & Hall, New York. 525 pp.

Ebenman, B. & L. Persson (ed.). 1988. Size-structured populations: ecology and evolution. Springer-Verlag, Berlin. 282 pp.

Fausch, K. D., S. Nakano & K. Ishigaki. 1994. Distribution of two congeneric charrs in streams of Hokkaido Island, Japan: considering multiple factors across scales. Oecologia 100: 1–12.

Fausch K. D., S. Nakano & S. Kitano. 1997. Experimentally induced foraging mode shift by sympatric charrs in a Japanese mountain stream. Beh. Ecol. 8: 414–420.

Fukushima, M. 1994. Spawning migration and redd construction of Sakhalin taimen, *Hucho perryi* (Salmonidae) on northern Hokkaido Island, Japan. J. Fish Biol. 44: 877–888.

Furukawa-Tanaka, T. 1985. The ecology of salmonid fishes in Japanese mountain streams I. Food condition and feeding habit of Japanese charr, *Salvelinus leucomaenis* (Pallas). Japan. J. Ecol. 35: 481–504.

Furukawa-Tanaka, T. 1988. The ecology of salmonid fishes in Japanese mountain streams. III. Interactive food segregation between red-spot masu salmon, *Salmo (Parasalmo) masou macrostomus*, and Japanese charr, *Salvelinus leucomaenis*, in relation to their social structure. Doctoral Dissertation, Kyoto University, Kyoto. 96 pp. (in Japanese and English).

Furukawa-Tanaka, T. 1989. Interactive focal point shift between Japanese and Dolly Varden charrs, *Salvelinus leucomaenis* and *S. malma*, in Hokkaido Island, Japan. Physiol. Ecol. Japan Spec. Vol. 1: 295–302.

Gashagaza, M. M. 1991. Diversity of breeding habits in lamprologine cichlids in Lake Tanganyika. Physiol. Ecol. Japan 28: 29–65.

Gashagaza, M. M. & M. Nagoshi. 1988. Growth of the larvae of a Tanganyikan cichlid, *Lamprologus attenuatus*, under parental care. Japan. J. Ichthyol. 35: 392–395.

Goto, A. 1974. Hypertrophy of the kidney of the male freshwater sculpin, *Cottus hangiongensis* Mori in the spawning period. Bull. Fac. Fish. Hokkaido Univ. 25: 211–214 (in Japanese with English summary).

Goto, A. 1975a. Ecological and morphological divergence of the freshwater sculpin, *Cottus nozawae* Snyder – I. Spawning behavior and process of development in the post-hatching stage. Bull. Fac. Fish. Hokkaido Univ. 26: 31–37 (in Japanese with English summary).

Goto, A. 1975b. Ecological and morphological divergence of the freshwater sculpin, *Cottus nozawae* Snyder – II. Morphological comparison of adult fishes of small-egg and large-egg types and their distribution. Bull. Fac. Fish. Hokkaido Univ. 26: 39–48 (in Japanese with English summary).

Goto, A. 1977. Some considerations on speciation and adaptation of the freshwater sculpins in Hokkaido, Japan. J. Michurin Biology 13: 39–47 (in Japanese with English summary).

Goto, A. 1980. Geographical distribution and variation of two types of *Cottus nozawae* in Hokkaido, and morphological characteristics of *C. amblystomopsis* from Sakhalin. Japan. J. Ichthyol. 27: 97–105.

Goto, A. 1981. Life history and distribution of a river sculpin, *Cottus hangiongensis*. Bull. Fac. Fish. Hokkaido Univ. 32: 10–21 (in Japanese with English summary).

Goto, A. 1982. Reproductive behavior of a river sculpin, *Cottus nozawae* Snyder. Japan. J. Ichthyol. 28: 453–457.

Goto, A. 1983. Spawning habits and reproductive isolating mechanism of two closely related river-sculpins, *Cottus amblystomopsis* and *C. nozawae*. Japan. J. Ichthyol. 30: 168–175.

Goto, A. 1984a. Comparative ecology of young-of-the-year between two amphidromous species of *Cottus* in Hokkaido – I. Upstream migration and growth. Bull. Fac. Fish. Hokkaido Univ. 35: 133–143.

Goto, A. 1984b. Sexual dimorphism in a river sculpin, *Cottus hangiongensis*. Japan. J. Ichthyol. 31: 161–166.

Goto, A. 1985. Individual identification by spine and ray clipping for freshwater sculpins. Japan. J. Ichthyol. 32: 359–362.

Goto, A. 1986. Movement and population size of the river sculpin *Cottus hangiongensis* in the Daitobetsu River of southern Hokkaido. Japan. J. Ichthyol. 32: 421–430.

Goto, A. 1987a. Life history variation in males of the river sculpin, *Cottus hangiongensis*, along the course of a river. Env. Biol. Fish. 19: 82–92.

Goto, A. 1987b. Polygyny in the river sculpin, *Cottus hangiongensis* (Pisces: Cottidae), with special reference to male mating success. Copeia 1987: 32–40.

Goto, A. 1988. Reproductive behavior and homing after downstream spawning migration in the river sculpin, *Cottus hangiongensis*. Japan. J. Ichthyol. 34: 488–496.

Goto, A. 1989a. Effects of parental care on egg survival estimates estimated from an experimental removal of nest-guarding males in the river-sculpin, *Cottus amblystomopsis*. Japan. J. Ichthyol. 36: 281–284.

Goto, A. 1989b. Growth differences in males of the river-sculpin *Cottus hangiongensis* along the river course, a correlate of the life-history variation. Env. Biol. Fish. 24: 241–249.

Goto, A. 1989c. Individual growth, age at maturity and longevity of females of the river-sculpins, *Cottus hangiongensis* in the Daitobetsu River, with special reference to their life-history variation. Japan. J. Ichthyol. 36: 90–99.

Goto, A. 1990. Alternative life-history styles of Japanese freshwater sculpins revisited. Env. Biol. Fish. 28: 101–112.

Goto, A. 1993a. Clinal life-history variation in the river-sculpin *Cottus hangiongensis*: an example of phenotypic plasticity. Japan. J. Ichthyol. 39: 363–370.

Goto, A. 1993b. Duration of male mating activity and male mate choice in the river sculpin, *Cottus hangiongensis*. Japan. J. Ichthyol. 40: 269–272.

Goto, A. 1993c. Male mating success and female mate choice in the river sculpin *Cottus nozawae* (Cottidae). Env. Biol. Fish. 37: 347–353.

Goto, A. 1994. Freshwater sculpins in the world and the Baikalian sculpins. pp. 165–195. *In:* H. Morino & N. Miyazaki (ed.) The Lake Baikal, Tokyo University Press, Tokyo (in Japanese).

Goto, A. & T. Andoh. 1990. Genetic divergence between the sibling species of the river-sculpins, *Cottus amblystomopsis* and *C. nozawae*, with special reference to their speciation. Env. Biol. Fish. 28: 257–266.

Goto, A. & K. Maekawa (ed.) 1989. Breeding behavior of fishes: their modes and strategies. Tokai University Press, Tokyo. 201 pp (in Japanese).

Goto, A. & S. Nakano. 1993. Distribution and ecology of freshwater fishes in Hokkaido. pp. 113–126. *In:* S. Higashi, A. Osawa & K. Kanagawa (ed.) Biodiversity and Ecology in the Northernmost Japan. Hokkaido Univ. Press, Sapporo.

Goto, A., K. Oishi & K. Takata. 1979. Breeding habits, growth and food of the ninespine stickleback, *Pungitius pungitius* (L.) in an artificial raceway of the Kunebetsu River system, Hokkaido. Bull. Fac. Fish. Hokkaido Univ. 30: 239–251 (in Japanese with English summary).

Goto, A., K. Tsukamoto & K. Maekawa (ed.) 1994. Diadromous fishes: life history and evolution. Tokai University Press, Tokyo. 279 pp. (in Japanese).

Gross, M. R. & K. Maekawa. 1989. Evolution of breeding strategies in fish. pp. 161–201. *In:* A. Goto & K. Maekawa (ed.) Breeding Behavior of Fishes: Their Modes and Strategies, Tokai University Press, Tokyo (in Japanese).

Halstead, B. 1985. Anti-Darwinism theory in Japan. Nature 317: 587–589.

Halstead, B. 1987. Imanishi's influence on evolution theory in Japan. Nature 326: 21.

Halstead, L. B. 1988. Kinji Imanishi – the view from the mountain top: a critique of Imanishi's theory of evolution. Tsukiji Shokan, Tokyo. 282 pp. (in Japanese).

Hattori, A. 1994. Inter-group movement and mate acquisition

tactics of the protandrous anemonefish, *Amphiprion clarkii*, on a coral reef, Okinawa. Japan. J. Ichthyol. 42: 159–165.

Hattori, A. 1995. Coexistence of two anemonefishes, *Amphiprion clarkii* and *A. perideraion*, which utilize the same host sea anemone. Env. Biol. Fish. 42: 345–353.

Hattori, A. & N. Yamamura. 1994. Coexistence of subadult males and females of the monogamous and protandrous anemonefish. Evol. Ecol. 8: 1–12.

Hattori, A. & Y. Yanagisawa. 1991. Life-history pathways in relation to gonadal sex differentiation in the anemonefish, *Amphiprion clarkii*, in temperate waters of Japan. Env. Biol. Fish. 31: 139–155.

Hidaka, T. & S. Takahashi. 1987. Reproductive strategy and interspecific competition in the lakeliving gobiid fish isaza, *Chaenogobius isaza*. J. Ethol. 5: 185–196.

Hino, T., K. Maekawa & J. B. Reynolds. 1990. Alternative male mating behaviors in landlocked Dolly Varden (*Salvelinus malma*) in south-central Alaska. J. Ethol. 8: 13–20.

Hirai, K. 1964. Comparative studies on ecology of four species of bitterlings in the Lake Biwa. Physiol. Ecol. Japan 12: 72–81 (in Japanese with English summary).

Hirai, K. 1969. Comparative study on food habits of kawachibuna *Carassius cuvieri*, gengoro-buna *C. cuvieri*, and nigorobuna *C. carassius grandoculis* larvae in the aquatic plant zone of Lake Biwa. Bull. Fac. Educ. Kanazawa Univ. 18: 123–131 (in Japanese with English summary).

Hirai, K. 1970. Ecological studies on fry and juvenile of fishes at aquatic plant area in a bay of Lake Biwa. I. On the distribution of fish larvae. Bull. Fac. Educ. Kanazawa Univ. 19: 93–105 (in Japanese with English summary).

Hirai, K. 1971. Ecological studies on fry and juvenile of fishes at aquatic plant areas in a bay of Lake Biwa. II. On the food habits of fish larvae. Bull. Fac. Educ. Kanazawa Univ. 20: 59–71 (in Japanese with English summary).

Hirai, K. 1972. Ecological studies on fry and juvenile of fishes at aquatic plant areas in a bay of Lake Biwa. III. Relationship of the food habits to the habitat of nigorobuna (*Carassius carassius glandoculis*) larvae. Japan. J. Ecol. 22: 69–92 (in Japanese with English summary).

Hirai, K. & T. Hikido. 1976. Morphological and ecological characters of the fishes of genus *Phoxinus* (Pisces: Cyprinidae) in Hokuriku district. Physiol. Ecol. Japan 17: 365–372. (in Japanese with English summary).

Hirano, Y. 1992. Public and private support of basic research in Japan. Science 258: 582–583.

Hirata, T., A. Goto & F. Yamazaki. 1988. Individual growth and smoltification of juvenile masu salmon, *Oncorhynchus masou* Brevoort, under rearing conditions. J. Fish Biol. 32: 77–84.

Hori, M. 1983. Feeding ecology of thirteen species of *Lamprologus (Teleostei; Cichlidae)* coexisting at a rocky shore of Lake Tanganyika. Physiol. Ecol. Japan 20: 129–149.

Hori, M. 1987. Mutualism and commensalism in a fish community in Lake Tanganyika. pp. 219–239. *In:* S. Kawanao, J.H. Connell, & T. Hidaka (ed.) Evolution and Coadaptation in Biotic Communities, University of Tokyo Press, Tokyo.

Hori, M. 1993. Frequency-dependent natural selection in the handedness of scale-eating cichlid fish. Science 260: 216–219.

Hori, M., M. M. Gashagaza, M. Nshombo & H. Kawanabe. 1993. Littoral fish communities in Lake Tanganyika: irreplaceable diversity supported by intricate interactions among species. Cons. Biol. 7: 657–666.

Hori, M., K. Yamaoka & K. Takamura. 1983. Abundance and micro-distribution of cichlid fishes on a rocky shore of Lake Tanganyika. African Study Monogr. 3: 25–38.

Iguchi, K. 1991. Visual stimuli causing the aggressive behavior of ayu *Plecoglossus altivelis*. Bull. National Res. Inst. Fish. Sci. 2: 15–23 (in Japanese with English summary).

Iguchi, K. 1993. Latitudinal variation in ayu egg size. Nipp. Suisan Gakk. (Bull. Japan. Soc. Sci. Fish.) 59: 2087.

Iguchi, K., F. Ito, K. Ikuta & M. Yamaguchi. 1991. Sexual dimorphism in the anal fin of ayu *Plecoglossus altivelis.* Nipp. Suisan Gakk. (Bull. Japan. Soc. Sci. Fish.) 57: 1501–1505.

Iguchi, K & K. Maekawa. 1993. Female mate preference and male mating success of ayu fish, *Plecoglossus altivelis* (Osmeridae) under a promiscuous mating system. Ethology 95: 193–201.

Iguchi, K. & N. Mizuno. 1990. Diel changes of larval drift among amphidromous gobies in Japan, especially *Rhinogobius brunneus*. J. Fish Biol. 37: 255–264.

Iguchi, K. & N. Mizuno. 1991. Mechanisms of embryonic drift in the amphidromous goby, *Rhinogobius brunneus*. Env. Biol. Fish. 31: 295–300.

Iguchi, K. & M. Yamaguchi. 1994. Adaptive significance of inter- and intrapopulational egg size variation in ayu *Plecoglossus altivelis* (Osmeridae). Copeia 1994: 184–190.

Ikeda, K. & A. Sibatani. 1995. Kinji Imanishi's biological thought. pp. 71–89. *In:* D.M. Lambert & H.G. Spencer (ed.) Speciation and the Recognition Concept, Theory and Application, The Johns Hopkins University Press, Baltimore.

Imanishi, K. 1941. Mayflies from Japanese torrents. IX. Life forms and life zones of mayfly nymphs. II. Ecological structure illustrated by life zone arrangement. Mem. Coll. Sci. Kyoto Imp. Univ. Ser. B 14: 1–35.

Imanishi, K. 1951. Natural society of Japanese macaques. Sizen (Nature) 6: 76–79 (in Japanese).

Imanishi, K. 1960. Social organization of subhuman primates in their natural habitat. Curr. Anthropol. 1: 393–407.

Imanishi, K. 1977. Theory of evolution from the subjectivism. Tyûôkôronsya, Tokyo. 189 pp. (in Japanese).

Inoue, M. & S. Nakano. 1994. Physical environment structure of a small stream with special reference to fish microhabitat. Japan. J. Ecol. 44: 151–160 (in Japanese with English summary).

Ishigaki, K. 1969. Ecological and morphological studies on chars in Hokkaido. Doctoral Dissertation, Hokkaido University, Sapporo. 249 pp. (in Japanese with English summary).

Ishigaki, K. 1984. Exploring the mystery of charrs. Iwanami-Shoten, Tokyo. 216 pp. (in Japanese).

Ishigaki, K. 1987. Studies on the biology in the early stages of two types of chars in Hokkaido. Res. Bull. Coll. Exp. Forests Hokkaido Univ. 44: 1121–1141.

Itani, J. 1954. Monkeys on Takasakiyama. Kobunsya, Tokyo. 284 pp. (in Japanese).

Itani, J. 1959. Parental care in the wild monkey, *Macaca fuscata*. Primates 2: 61–93.

Itani, J. & A. Suzuki. 1967. The social unit of chimpanzees. Primates 8: 355–381.

Itô, Y. 1991. Development of ecology in Japan, with special reference to the role of Kinji Imanishi. Ecol. Res. 6: 139–155.

Kaeriyama, M. 1986. Ecological study on early life of the chum salmon *Oncorhynchus keta* (Walbaum). Sci. Rep. Hokkaido Salmon Hatch. 40: 31–92.

Kaeriyama, M. 1989. Aspects of salmon ranching in Japan. Physiol. Ecol. Japan Spec. Vol. 1: 625–638.

Kaeriyama, M. 1991. Dynamics of the lacustrine sockeye salmon population in Lake Shikotsu, Hokkaido. Sci. Rept. Hokkaido Salmon Hatch. 45: 1–24 (in Japanese with English summary).

Kaeriyama, M. & S. Urawa. 1992. Future research by the Hokkaido Salmon Hatchery for the proper maintenance of Japanese salmonid stocks. pp. 57–62. *In:* Y. Ishida, K. Nagasawa, D. W. Welch, K. W. Myers & A. P. Shershnev (ed.) Proceedings of the International Workshop on Future Salmon Research in the North Pacific Ocean, National Res. Inst. Far Seas Fisheries, Shimizu.

Kaeriyama, M., S. Urawa & T. Suzuki. 1992. Anadromous sockeye salmon (*Oncorhynchus*) derived from nonanadromous kokanees: life history in Lake Toro. Sci. Rep. Hokkaido Salmon Hatch. 46: 157–174.

Katano, O. 1983. Spawning of the dark chub, *Zacco temmincki* (Temminck et Schlegel) (Pisces: Cyprinidae) with special reference to male social status. Physiol. Ecol. Japan 20: 33–52.

Katano, O. 1985. Aggressive behaviour and dominance relationships of the dark chub, *Zacco temmincki*, with special reference to their individual recognition. Japan. J. Ichthyol. 32: 225–238.

Katano, O. 1987. Social structure of the dark chub, *Zacco temmincki*, in a small pond in relation to individual differences. Physiol. Ecol. Japan 24: 1–132.

Katano, O. 1990. Dynamic relationships between the dominance of male dark chub, *Zacco temmincki*, and their acquisition of females. Anim. Beh. 40: 1018–1034.

Katano, O. 1991. The ecology of individuality. Kyoto Univ. Press, Kyoto. 286 pp.

Katano, O. 1993. Styles of general description. Env. Biol. Fish. 36: 325–327 (book critique).

Katano, O. 1994. Aggressive interactions between the dark chub, *Zacco temmincki*, and the pale chub, *Z. platypus*, in relation to their feeding behaviour. Japan. J. Ichthyol. 40: 441–449.

Kato, F. 1973. Ecological study on the sea-run form of *Oncorhynchus rhodurus*, found in Ise Bay, Japan. Japan. J. Ichthyol. 20: 225–234 (in Japanese with English summary).

Katoh, M. & M. Nishida. 1994. Biochemical and egg size evolution of freshwater fishes in the *Rhinogobius brunneus* complex (Pisces: Gobiidae) in Okinawa, Japan. Biol. J. Linnean Soc. 51: 325–335.

Kawabata, M. 1975. Fish community near Mkuyu, the east coast of Lake Tanganyika. Kyoto Univ. Afr. Studies 9: 21–31.

Kawai, M. 1965. Newly acquired pre-cultural behavior of the natural troop of Japanese monkeys on Koshima Islet. Primates 6: 1–30.

Kawai, M. & H. Mizuhara. 1962. An ecological study on the wild mountain gorilla (*Gorilla gorilla beringeri*). Primates 5: 1–42.

Kawamura, S. 1959. The process of subhuman culture propagation among Japanese macaques. Primates 2: 43–60.

Kawanabe, H. 1957. Social behaviour and production of a salmon-like fish, *Plecoglossus altivelis*, or ayu, with special reference to its population density. Japan. J. Ecol. 7: 131–137 (in Japanese with English summary).

Kawanabe. H. 1958. On the significance of the social structure for the mode of density effect in a salmon-like fish, 'ayu', *Plecoglossus altivelis*, Temminck et Schlegel. Mem. Coll. Sci. Univ. Kyoto, Ser. B 25: 171–180.

Kawanabe, H. 1959. Food competition among fishes in some rivers of Kyoto Prefecture, Japan. Mem. Coll. Sci. Univ. Kyoto, Ser. B 26: 253–268.

Kawanabe, H. 1960. An opinion on the study of lotic animal ecology. Physiol. Ecol. Japan 9: 1–10 (in Japanese with English summary).

Kawanabe, H. 1966. A short historical sketch of investigations on ayu (*Plecoglossus altivelis*). Japan. J. Michurin Biol. 2: 235–244 (in Japanese).

Kawanabe, H. 1969a. Fishes in rivers and lakes. Tyûôkôronsya, Tokyo. 196 pp. (in Japanese).

Kawanabe, H. 1969b. The significance of social structure in production of the 'ayu', *Plecoglossus altivelis*. pp. 243–251. *In:* T.G. Northcote (ed.) The Symposium on Salmon and Trout in Streams, H. R. MacMillan Lectures in Fisheries, University of British Columbia, Vancouver.

Kawanabe, H. 1970. Social behavior and production of ayu-fish, in the River Ukawa between 1955 and 1969, with reference to the stability of its territoriality. Japan. J. Ecol. 20: 144–151 (in Japanese with English summary).

Kawanabe, H. 1972. An evolutionary aspect of the territoriality of ayu-fish, *Plecoglossus altivelis*, with the social structure at the southern end of its distribution. Japan. J. Ecol. 22: 141–149 (in Japanese with English summary).

Kawanabe, H. 1973. Food and feeding habits of a freshwater goby, *Rhinogobius brunneus* in Okinawa Island, the Ryukyus, with some discussion on the interactive food segregation. Biol. Mag. Okinawa 10: 1–6 (in Japanese with English summary).

Kawanabe, H. 1976. A note on the territoriality of ayu, *Plecoglossus altivelis* Temminck et Schlegel (Pisces: Osmeridae) in the Lake Biwa stock, based on the 'relic social structure' hypothesis. Physiol. Ecol. Japan 17: 395–399 (in Japanese with English summary).

Kawanabe, H. 1977. Relic social structure hypothesis on the territoriality of ayu, *Plecoglossus altivelis* (Pisces: Osmeridae). Proc. Japan. Acad. Ser. B 53: 74–77.

Kawanabe, H. 1978. Some biological problems (in deep core of Lake Biwa). Verh. Intern. Verein. Limnol. 20: 2674–2677.

Kawanabe, H. 1981. Territorial behavior of *Tropheus moorii* (Osteichthyes: Cichlidae) with a preliminary consideration of the territorial form in animals. African Study Monogr. 1: 101–108.

Kawanabe, H. 1986. Cooperative study on the ecology of Lake Tanganyika between Japanese and African scientists, with special reference to mutual interactions among fishes. Physiol. Ecol. Japan 23: 119–128.

Kawanabe, H., J. E. Cohen & K. Iwasaki (ed.) 1993. Mutualism and community organization: behavioural, theoretical, and food-web approaches. Oxford University Press, Oxford. 426 pp.

Kawanabe, H., D. Miyadi, S. Mori, E. Harada, H. Mizuhara & R. Ohgushi. 1956. Ecology of natural stock of ayu, *Plecoglossus altivelis*. I. Contrib. Physiol. Ecol. Kyoto Univ. 79: 1–37 (in Japanese).

Kawanabe, H., N. Mizuno, D. Miyadi, S. Mori, R. Ohgushi & N. Nishimura. 1957. Ecology of natural stock of ayu, *Plecoglossus altivelis*. II. Physiol. Ecol. Japan 7: 145–167 (in Japanese with English summary).

Kawanabe, H., N. Mizuno & N. Nishimura. 1957. Population density of a salmon-like fish, *Plecoglossus altivelis*, or ayu, in different river-beds. Bull. Japan. Soc. Sci. Fisheries 23: 430–434 (in Japanese with English summary).

Kawanabe, H., S. Mori & N. Mizuno. 1957. Modes of utilizing the river-pools by a salmon-like fish *Plecoglossus altivelis*, or ayu, in relation to its population density. Japan. J. Ecol. 7: 22–26 (in Japanese with English summary).

Kawanabe, H., S. Mori & N. Mizuno. 1959. On the food economy of 'ayu' fish in relation to the production of algae. Physiol. Ecol. Japan 8: 117–123 (in Japanese with English summary).

Kawanabe, H., T. Ohgushi & M. Higashi (ed.) 1990. Ecology for tomorrow. Physiol. and Ecol. Japan, Spec. Vol. 27, 205 pp.

Kawanabe, H., T. Itai, S. Nishijima & S. Shokita. 1974. On the structure of the stream community in the Ryukyus with reference to food-chain relationships (preliminary report). Ecol. Stud. Natur. Cons. Ryukyu Isl. 1: 187–193 (in Japanese with English summary).

Kawanabe, H. & K. Takamura. 1979. Fish fauna near Katana, southwest coast of L. Kivu. pp. 7–8. *In:* M. Kawabata (ed.) Preliminary Report on the Aquatic Community with Special Reference to the Fish in Lake Tanganyika and Lake Kivu, Kyoto University, Kyoto.

Kawanabe, H. & T. Tokui. 1984. Fishes in Lake Biwa. pp. 339–360. *In:* S. Horie (ed.) Lake Biwa, Dr W. Junk Publishers, The Hague.

Kawanabe, H., F. Yamazaki & D. L. G. Noakes (ed.) 1989. Biology of charrs and masu salmon. Physiol. Ecol. Japan Spec. Vol. 1. 711 pp.

Keenleyside, M. H. A. 1962. Skin-diving observations of Atlantic salmon and brook trout in the Miramichi River, New Brunswick. J. Fish. Res. Board Can. 19: 625–634.

Kimura, S. 1966. On the life history of the salmonid fish, *Hucho perryi* (Brevoort), found in Nemuro, Hokkaido. Japan. J. Ichthyol. 14: 17–27 (in Japanese with English summary).

Kiso, K. 1990. Polymorphism of life form in masu salmon (*Oncorhynchus masou*) in the rivers of southern Sanriku district, Honshu, Japan. Bull. Inst. Zool. Acad. Sinica 29(3, Suppl.): 27–39.

Kiso, K., T. Akamine, S.Ohnishi & Y. Matsumiya. 1992. Mathematical examinations of the growth of sea-run and fluviatile forms of female masu salmon, *Oncorhynchus masou* in rivers of the southern Sanriku district, Honshu, Japan. Nipp. Suisan Gakk. (Bull. Japan. Soc. Sci. Fish.) 58: 1779–1784.

Kiso, K. & Y. Matsumiya. 1992. Growth of fluviatile form masu salmon *Oncorhynchus masou* in southern Sanriku district, Honshu, Japan. Nipp. Suisan Gakk. (Bull. Japan. Soc. Sci. Fish.) 58: 9–13.

Kitano, S., S. Nakano, M. Inoue, K. Shimoda & S. Yamamoto. 1993. Feeding and reproductive ecology of exotic rainbow trout *Oncorhynchus mykiss* in the Horonai Stream in Hokkaido, northern Japan. Nipp. Suisan Gakk. (Bull. Japan. Soc. Sci. Fish.) 59: 1837–1843 (in Japanese with English summary).

Kitano, S., K. Maekawa, S. Nakano & K. D. Fausch. 1994. Spawning behavior of bull trout in the upper Flathead drainage, Montana, with special reference to hybridization with brook trout. Trans. Amer. Fish. Soc. 123: 988–992.

Kohda, M. 1981. Interspecific territoriality and agonistic behavior of a temperate pomacentrid fish, *Eupomacentrus altus*. Z. Tierpsychol. 56: 205–216.

Kohda, M. 1984. Intra- and interspecific territoriality of a temperate damselfish *Eupomacentrus altus*, (Teleostei: Pomacentridae). Physiol. Ecol. Japan 21: 35–52.

Kohda, M. 1988. Diurnal periodicity of spawning activity of permanently territorial damselfishes (Teleostei: Pomacentridae). Env. Biol. Fish. 21: 91–100.

Kohda, M. 1991. Intra- and interspecific social organization among three herbivorous cichlid fishes in Lake Tanganyika. Japan. J. Ichthyol. 38: 147–163.

Kohda, M. 1994. Individual specialized foraging repertoires in the piscivorous cichlid fish, *Lepidiolamprologus profundicola*. Anim. Behav. 48: 1123–1131.

Kohda, M. & Y. Yanagisawa. 1992. Vertical distributions of two herbivorous cichlid fishes of the genus *Tropheus* in Lake Tanganyika, Africa. Ecol. Freshwater Fish 1: 99–103.

Kohda, M. & M. Hori. 1993. Dichromatism in relation to the trophic biology of predatory cichlid fishes in Lake Tanganyika, East Africa. J. Zool. Lond. 229: 447–455.

Kondo, T., J. Yamashita & M. Kano. 1984. Breeding ecology of five species of bitterling (Pisces: Cyprinidae) in a small creek. Physiol. Ecol. Japan 21: 53–62.

Kubo, T. 1980. Studies on the life history of the 'masu' salmon (*Oncorhynchus masou*) in Hokkaido. Sci. Rep. Hokkaido Salmon Hatch. 34: 1–95 (in Japanese with English summary).

Kuwamura, T. 1983. Reexamination on the aggressive mimicry of the cleaner wrasse *Labroides dimidiatus* by the blenny *Aspidontus taeniatus* (Pisces: Perciformes). J. Ethol. 1: 22–33.

Kuwamura, T. 1984. Social structure of the protogynous fish *Labroides dimidiatus*. Publ. Seto Mar. Biol. Lab. Kyoto Univ. 29: 117–177.

Kuwamura, T. 1986. Parental care and mating systems of cichlid fishes in Lake Tanganyika: a preliminary field survey. J. Ethol. 4: 129–146.

Kuwamura, T. 1987a. Distribution of fishes in relation to the depth and substrate at Myako, east-middle coast of Lake Tanganyika. Afr. Study Monogr. 7: 1–14.

Kuwamura, T. 1987b. Male mating territory and sneaking in a maternal mouthrooder, *Pseudosimochromis curvifrons* (Pisces: Cichlidae). J. Ethol. 5: 203–206.
Kuwamura, T. 1988. Biparental mouthbrooding and guarding in a Tanganyikan cichlid *Haplotaxodon microlepis*. Japan. J. Ichthyol. 35: 62–68.
Kuwamura, T. 1992. Overlapping territories of *Pseudosimochromis curvifrons* males and other herbivorous cichlid fishes in Lake Tanganyika. Ecol. Res. 7: 43–53.
Kuwamura, T., M. Nagoshi & T. Sato. 1989. Female-to-male shift of mouthbrooding in a cichlid fish, *Tanganicodus irsacae*, with notes on breeding habits of two related species in Lake Tanganyika. Env. Biol. Fish. 24: 187–198.
Kuwamura, T., Y. Yogo & Y. Nakashima. 1993. Size-assortative monogamy and parental egg care in a coral goby *Paragobiodon echinocephalus*. Ethology 95: 65–75.
Kuwamura, T., Y. Yogo & Y. Nakashima. 1994. Population dynamics of goby *Paragobiodon echinocephalus* and host coral *Stylophora pistillata*. Mar. Ecol. Prog. Ser. 103: 17–23.
Łomnicki, A. 1978. Individual differences among animals and natural regulation of their numbers. J. Anim. Ecol. 47: 461–475.
Łomnicki, A. 1988. Population ecology of individuals. Princeton University Press, Princeton. 223 pp.
Maekawa, K. 1977a. Studies on the variability of the land-locked miyabe charr, *Salvelinus malma miyabei* Oshima. I. Development and early life history. Japan. J. Ecol. 27: 91–102 (in Japanese with English summary).
Maekawa, K. 1977b. Studies on the variability of the land-locked miyabe charr, *Salvelinus malma miyabei*. III. Geographical variations of the Dolly Varden, *Salvelinus malma*, and morphological characters of the miyabe charr. Japan. J. Ichthyol. 24: 49–56 (in Japanese with English summary).
Maekawa, K. 1978. Growth and development of *Salvelinus malma miyabei* compared with other forms of *S. malma*. Japan. J. Ichthyol. 25: 9–18.
Maekawa, K. 1983. Streaking behaviour of mature male parr of the Miyabe charr, *Salvelinus malma miyabei*. Japan. J. Ichthyol. 30: 227–234.
Maekawa, K. 1984. Life history patterns of the Miyabe charr in Shikaribetsu Lake, Japan. pp. 233–250. *In:* L. Johnson & B.L. Burns (ed.) Biology of the Arctic Charr, University of Manitoba Press, Winnipeg.
Maekawa, K. 1985. Homing of lacustrine charr in a small lake with a few inlet creeks. Japan. J. Ichthyol. 32: 355–358.
Maekawa, K. 1989a. Breeding strategies of male salmon. pp. 50–60. *In:* A. Goto & K. Maekawa (ed.) Breeding Behavior of Fishes: Their Modes and Strategies, Tokai University Press, Tokyo (in Japanese).
Maekawa, K. 1989b. Occurrence of stream resident females in Miyabe charr. Japan. J. Ichthyol. 36: 141–143.
Maekawa, K. & A. Goto. 1982. Life history of freshwater fishes. Tyûkôronsya, Tokyo. 212 pp. (in Japanese).
Maekawa, K. & T. Hino. 1986. Spawning behaviour of Dolly Varden in southeastern Alaska, with special reference to the mature male parr. Japan. J. Ichthyol. 32: 454–458.
Maekawa, K. & T. Hino. 1987. Effect of cannibalism on alternative life histories in charr. Evolution 41: 1120–1123.
Maekawa, K. & T. Hino. 1990. Spawning tactics of female Miyabe charr (*Salvelinus malma miyabei*) against egg cannibalism. Can. J. Zool. 68: 889–894.
Maekawa, K., T. Hino, S. Nakano & W. W. Smoker. 1993. Mate preference of female anadromous and land-locked Dolly Varden in two Alaska streams. Can. J. Fish. Aquat. Sci. 50: 2375–2379.
Maekawa, K., S. Kitano & S. Takeyama. 1993. Type-assortative mating of Miyabe charr. Japan. J. Ichthyol. 39: 401–403.
Maekawa, K. & S. Nakano. 1994. Non-oviposition of mature eggs by female fluvial red-spotted masu salmon. Fish. Sci. (Japan) 60: 37–39.
Maekawa, K., S. Nakano & S. Yamamoto. 1994. Spawning behaviour and size-assortative mating of Japanese charr in an artificial lake-inlet stream system. Env. Biol. Fish. 39: 109–117.
Maekawa, K. & H. Onozato. 1986. Reproductive tactics and fertilization success of mature male Miyabe charr, *Salvelinus malma miyabei*. Env. Biol. Fish. 15: 119–129.
Maki, I. 1964. The relationship between fishes and their food in a bay of the Lake Biwa, Japan. Physiol. Ecol. Japan 12: 259–271 (in Japanese with English summary).
Maki, I. 1966a. Population studies of honmoroko, *Gnathopogon caerulescens* Sauvage, in Lake Biwa, Japan. I. On the critical life-cycle stages related to the annual fluctuation of the population. Japan. J. Ecol. 16: 183–190 (in Japanese with English summary).
Maki, I. 1966b. Studies on the population dynamics of *Gnathopogon caerulescens* Sauvage (Pisces), in Lake Biwa, Japan. II. An attempt to analyse the factores [sic.] regulating the yearly fluctuation of the population in winter. Japan. J. Ecol. 16: 254–263 (in Japanese with English summary).
Maki, I. 1967. Population studies of honmoroko, *Gnathopogon caerulescens*, Pisces in Lake Biwa, Japan. III. On the seasonal fluctuation of the fat content in the fish body of the under-yearling and the 'over-wintering' of them. Japan. J. Ecol. 17: 199–205 (in Japanese with English summary).
Maki, I. 1968a. Studies on the population dynamics of *Gnathopogon caerulescens* Sauvage (Pisces), in Lake Biwa, Japan. IV. Further analysis on the critical life cycle stages related to the annual fluctuation of the population. Japan. J. Ecol. 18: 112–119 (in Japanese with English summary).
Maki, I. 1968b. Studies on the population dynamics of *Gnathopogon caerulescens* Sauvage (Pisces), in Lake Biwa, Japan. V. Relation between the annual fluctuation of growth of the under-yearling fish and that of the population abundance in winter. Japan. J. Ecol. 18: 158–166 (in Japanese with English summary).
Maki, I., K. Naka & A. Uramoto. 1970. Seasonal and developmental changes of the food intake of oikawa (*Zacco platypus*, Cyprinidae) in the River Kishi-gawa, Wakayama, Japan. Bull. Fac. Educ. Wakayama Univ. 20: 43–54 (in Japanese with English summary).
Maki, I. & Y. Kakiuchi. 1971. An estimation of the net production rate of the oikawa (*Zacco platypus*, Cyprinidae) population in

the River Kishigawa, Wakayama, Japan. Bull. Fac. Educ. Wakayama Univ. 21: 33–43 (in Japanese with English summary).

Maki, I. 1972. On the production rate of the kawamutu (*Zacco temmincki*, Cyprinidae) population in the River Kishigawa, Wakayama, Japan. Bull. Fac. Educ. Wakayama Univ. 22: 25–38 (in Japanese with English summary).

Maki, I. 1973. Seasonal changes of the young and the adult fish production of oikawa (*Zacco platypus*, Cyprinidae) in the River Kishigawa, Wakayama, Japan. Bull. Fac. Educ. Wakayama Univ. 23: 37–50 (in Japanese with English summary).

Maki, I. 1976a. Population studies of takahaya, *Moroco steindachneri jouyi* (Pisces: Cyprinidae) at two small areas in the upper region of River Kinokawa. Physiol. Ecol. Japan 17: 247–260.

Maki, I. 1976b. Seasonal changes of population numbers, growth, and productivity of kawamutu (*Zacco temmincki*, Pisces, Cyprinidae) in a small tributary of the River Kishigawa, Wakayama. Bull. Fac. Educ. Wakayama Univ. 25: 5–14 (in Japanese with English summary).

Maruyama, T. 1981. Comparative ecology on the fluvial forms of *Salmo (Oncorhynchus) masou masou* (Brevoort) and *Salvelinus leucomaenis* (Pallas) (Pisces, Salmonidae) I. Structure of spawning redds and spawning sites in Kamidani, River Yura. Japan. J. Ecol. 31: 269–284 (in Japanese with English summary).

Matsuda, H., P. A. Abrams & M. Hori. 1993. The effect of adaptive anti-predator behavior on exploitative competition and mutualism between predators. Oikos 68: 549–559.

Matsuda, H., M. Hori & P. A. Abrams. 1994. Effect of predator-specific defence on community complexity. Evol. Ecol. 8: 628–638.

Mayama, H. 1990. Masu salmon propagation in Hokkaido, Japan. Bull. Inst. Zool. Acad. Sinica 29 (3: Suppl.): 95–104.

Mayama, H. 1992. Studies on the freshwater life and propagation technology of masu salmon, *Oncorhynchus masou* (Brevoort). Sci. Rep. Hokkaido Salmon Hatch. 46: 1–156 (in Japanese with English summary).

Mbomba, N. B. 1985. Comparative feeding ecology of 11 sympatric species of cichlid algae feeders in Lake Tanganyika. Verh. Internat. Verein. Limnol. 22: 2662.

Mbomba, N. B. 1986. Comparative feeding ecology of aufwuchs eating cichlid fishes in Lake Tanganyika with reference to their developmental changes. Physiol. Ecol. Japan 23: 79–108.

Miura, T. 1965. Population studies based on relative abundance of five different life history stages of ayu, *Plecoglossus altivelis* (Pisces: Plecoglossidae), in Lake Biwa. Res. Pop. Ecol. 7: 87–98.

Miura, T. 1966a. Competitive influence of isaza, *Chaenogobius isaza*, on ayu, *Plecoglossus altivelis*, in Lake Biwa. Res. Pop. Ecol. 8: 37–50.

Miura, T. 1966b. Ecological notes on the fishes and the interspecific relations among them in Lake Biwa. Japan. J. Limnol. 27: 1–24.

Miura, T. 1989. East Lake: a phytoplanktivorous fishes dominated lake ecosystem. Ohtsu Hydrobiological Station, Kyoto University, Kyoto, Foreign Language Ser. 334. 142 pp.

Miura, T. 1990. The effects of planktivorous fishes on the plankton community in a eutrophic lake. Hydrobiol. 200/201: 567–579.

Miura, T., N. Suzuki, M. Nagoshi & K. Yamamura. 1976. The rate of production and food consumption of the biwamasu, *Oncorhynchus rhodurus*, population in Lake Biwa. Res. Popul. Ecol. 17: 135–154.

Miyadi, D. 1960. The narrative of ayu. Iwanami, Tokyo. 226 pp (in Japanese).

Mizuno, N. 1960a. Description of a new freshwater goby from Japan. Mem. Coll. Sci. Univ. Kyoto Ser. B 27: 117–119.

Mizuno, N. 1960b. Study on a freshwater goby, *Rhinogobius similis* Gill, with a proposition on the relationships between landlocking and speciation of some freshwater gobies in Japan. Mem. Coll. Sci. Univ. Kyoto, Ser. B 27: 97–115.

Mizuno, N. 1963a. Distributions of *Cottus japonicus* Okada (Cottidae) and *Tukugobius flumineus* Mizuno (Gobiidae), with special reference to their peculiarities in both the land-locking and the speciation from amphidromous ancestors. Mem. Osaka Gakugei Univ. 11: 129–161 (in Japanese with English summary).

Mizuno, N. 1963b. Distribution of two gobiid fishes, *Rhinogobius brunneus* and *Tukugobius flumineus*. I. Distributions of them in and near stagnant waters. Japan. J. Ecol. 13: 242–247.

Mizuno, N. 1976. Studies on a freshwater fish, *Rhinogobius brunneus*. III. Distribution of four colour types in Shikoku and Kyushu islands. Physiol. Ecol. Japan 17: 373–381.

Mizuno, N. & H. Kawanabe. 1957. Behaviour of salmon-like fish, 'ayu', in an area with closely established territories. Japan. J. Ecol. 7: 26–30 (in Japanese with English summary).

Mizuno, N., H. Kawanabe, D. Miyadi, S. Mori, H. N. Kodoma, R. Ohgushi, A. Kusakabe & Y. Y. Furuya. 1958. Life history of some stream fishes with special reference to four cyprinid species. Contr. Physiol. Ecol. Kyoto Univ. 81: 1–48 (in Japanese).

Mizuno, N., M. Mukai, A. Goto & K. Hamada. 1982. Studies of the freshwater fish in Hokkaido, Japan – II. Distribution of the two types of a goby, *Rhinogobius brunneus*. Bull. Fac. Fish. Hokkaido Univ. 33: 115–125 (in Japanese with English summary).

Mizuno, N. & H. Niwa. 1961. Two ecological types of the freshwater scorpion, *Cottus pollux* Gunther. Zool. Mag. (Japan) 70: 267–275 (in Japanese with English summary).

Mizuno, H., S. Uehara & M. Maki. 1979. Studies on a freshwater fish, *Rhinogobius brunneus* (Gobiidae) IV. Habitat segregation among sympatric populations of 4 colour types. Japan. J. Ecol. 29: 137–147.

Mori, S. 1956. Social organization of a group of young girelloid fish, *Girella punctata* Gray, confined in a tide-pool, with special reference to the relation between social hierarchy and territorial system. Japan. J. Ecol. 5: 145–150 (in Japanese with English summary).

Mori, S. 1984. Sexual dimorphism of the landlocked three-spined stickleback *Gasterosteus aculeatus microcephalus* from Japan. Japan. J. Ichthyol. 30: 419–425 (in Japanese with English summary).

Mori, S. 1985. Reproductive behaviour of the landlocked three-spined stickleback, *Gasterosteus aculeatus microcephalus*, in Japan. Behaviour 93: 21–35.

Mori, S. 1986. Change of characters of *Drosophila melanogaster* brought about during the life in constant darkness and consideration on the process through which these changes are induced. Zool. Sci. 3: 945–957.

Mori, S. 1987a. Divergence in reproductive ecology of the three-spined stickleback, *Gasterosteus aculeatus*. Japan. J. Ichthyol. 34: 165–175.

Mori, S. 1987b. Geographical variation in freshwater populations of the three-spined stickleback, *Gasterosteus aculeatus*, in Japan. Japan. J. Ichthyol. 34: 33–46.

Mori, S. 1987c. Multinesting behaviour by the freshwater three-spined stickleback *Gasterosteus aculeatus*. J. Ethol. 5: 199–202.

Mori, S. 1988. The upright nesting behaviours on a vertical shore-wall in the three-spined stickleback, *Gasterosteus aculeatus* (*leiurus* form). J. Ethol. 6: 59–62.

Mori, S. 1990. Two morphological types in the reproductive stock of three-spined stickleback, *Gasterosteus aculeatus*, in Lake Harutori, Hokkaido Island. Env. Biol. Fish. 27: 21–31.

Mori, S. 1993. The breeding system of the three-spined stickleback, *Gasterosteus aculeatus* (forma *leiura*) with reference to spatial and temporal patterns of nesting activity. Behaviour 126: 97–124.

Mori, S. & T. Miura. 1990. List of plant and animal species living in Lake Biwa. Mem. Fac. Sci. Kyoto Univ., Ser. Biol. 14: 13–32.

Mori, S. & M. Nagoshi. 1987. Growth and maturity size of the three-spined stickleback *Gasterosteus aculeatus* in rearing pool. Bull. Fac. Fish. Mie Univ. 14: 1–10.

Mori, S. & N. Suzuki. 1986. Change of *Drosophila's* hereditary nature of behavior, reacting to illumination, caused by selection and changes in other characters seen in these selected flies. Zool. Sci. 3: 151–165.

Moyer, J. T. 1975. Reproductive behavior of the damselfish *Pomacentrus nagasakiensis* at Miyake-jima, Japan. Japan. J. Ichthyol. 22: 151–163.

Moyer, J. T. 1976. Geographical variation and social dominance in Japanese populations of the anemonefish *Amphiprion clarkii*. Japan. J. Ichthyol. 23: 12–22.

Moyer, J. T. 1977. Aggressive mimicry between juveniles of the snapper *Lutjanus bohar* and species of the damselfish genus *Chromis* from Japan. Japan. J. Ichthyol. 24: 218–222.

Moyer, J. T. 1979. Mating strategies and reproductive behavior of ostraciid fishes at Miyake-jima, Japan. Japan. J. Ichthyol. 26: 148–160.

Moyer, J. T. 1987. Quantitative observations of predation during spawning rushes of the labrid fish *Thalassoma cupido* at Miyake-jima, Japan. J. Ichthyol. 34: 76–81.

Moyer, J. T. 1989. Reef channels as spawning sites for fishes on the Shiraho coral reef, Ishigaki Island, Japan. Japan. J. Ichthyol. 36: 371–375.

Moyer, J. T. 1990. Social and reproductive behavior of *Chaetodontoplus mesoleucus* (Pomacanthidae) at Bantayan Island, Philippines, with notes on pomacanthid relationships. Japan. J. Ichthyol. 36: 459–467.

Moyer, J. T. & L. J. Bell. 1976. Reproductive behavior of the anemonefish *Amphiprion clarkii* at Miyake-jima, Japan. Japan. J. Ichthyol. 23: 23–32.

Moyer, J. T., & A. Nakazono. 1978a. Population structure, reproductive behavior and protogynous hermaphroditism in the angelfish *Centropyge interruptus* at Miyake-jima, Japan. Japan. J. Ichthyol. 25: 25–39.

Moyer, J. T., & A. Nakazono. 1978b. Protandrous hermaphroditism in six species of the anemonefish genus *Amphiprion* in Japan. Japan. J. Ichthyol. 25: 101–106.

Moyer, J. T. & M. Sano. 1987. Feeding habits of two sympatric ostraciid fishes at Miyake-Jima, Japan. Japan. J. Ichthyol. 34: 108–112.

Moyer, J. T. & R. C. Steene. 1979. Nesting behavior of the anemonefish *Amphiprion polymnus*. Japan. J. Ichthyol. 26: 209–214.

Moyer, J. T. & C. E. Sawyers. 1973. Territorial behavior of the anemonefish *Amphiprion xanthurus* with notes on the life history. Japan. J. Ichthyol. 20: 85–93.

Moyer, J. T. & M. J. Zaiser. 1981. Social organization and spawning behavior of the pteroine fish *Dendrochirus zebra* at Miyake-jima, Japan. Japan. J. Ichthyol. 28: 52–69.

Munehara, H. 1988. Spawning and subsequent copulating behavior of the elkhorn sculpin *Alcicthys alcicornis* in an aquarium. Japan. J. Ichthyol. 35: 358–364.

Munehara, H. 1991. Utilization and ecological benefits of a sponge as a spawning bed by the little dragon sculpin *Blepsias cirrhosus*. Japan. J. Ichthyol. 38: 179–184.

Munehara, H. 1992. Utilization of polychaete tubes as spawning substrate by the sea raven *Hemitripterus villosus* (Scorpaeniformes). Env. Biol. Fish. 33: 395–398.

Munehara, H., A. Takenaka & O. Takenaka. 1994. Alloparental care in the marine sculpin *Alcicthys alcicornis* (Pisces: Cottidae): copulating in conjunction with parental care. J. Ethol. 12: 115–120.

Nagata, M., M. Nakajima & M. Fujiwara. 1994. Dispersal of wild and domestic masu salmon fry (*Oncorhynchus masou*) in an artificial channel. J. Fish Biol. 45: 99–109.

Nagata, Y. & K. Nishiyama. 1976. Reproductive behaviour of a bitterling, *Rhodeus ocellatus* (Kner.). Physiol. Ecol. Japan 17: 85–90 (in Japanese with English summary).

Nagata, Y., T. Yoshimura & S. Mori. 1988. Courtship behavior of the freshwater three-spined stickleback, *Gasterosteus aculeatus* (*leiurus* form) in Japan. Mem. Osaka Kyoiku Univ. Ser. III 37: 29–36.

Nagoshi, M. 1965. Ecological studies on the population of gengorobuna, *Carassius cuvieri*, in Lake Biwa. – II. Mortality, size of spawning population and number of emigrants estimated with tagging method. Rep. Fac. Fish. Mie Pref. Univ. 5: 307–317.

Nagoshi, M. 1966. Ecological studies on the population of isaza, *Chaenogobius isaza* Tanaka, in Lake Biwa, with special reference to the effects of population density upon its growth. Res. Popul. Ecol. 8: 20–36.

Nagoshi, M. 1969. Ecological studies on the population of isaza, *Chaenogobius isaza*, in Lake Biwa, with consideration on the factors of annual fluctuation of the population density. Rep. Fac. Fish. Mie Univ. 6: 87–96.

Nagoshi, M. 1977. Ecological studies on the population of gengorobuna, *Carassius cuvieri*, in Lake Biwa – III. Yearly fluctuation of age composition and growth. Bull. Fac. Fish. Mie Univ. 4: 31–44 (in Japanese with English summary).

Nagoshi, M. 1978. The effects of food and size hierarchy on the growth in the fry of crucian carp, *Carassius cuvieri*. Bull. Fac. Fish. Mie Univ. 5: 1–11.

Nagoshi, M. 1982. Diel vertical migration of zooplankters and fish larvae in Lake Biwa. Bull. Fac. Fish. Mie Univ. 9: 1–10.

Nagoshi, M. 1985. Growth and survival in larval stage of the genus *Lamprologus* (Cichlidae) in Lake Tanganyika. Verh. Internat. Verein. Limnol. 22: 2663–2670.

Nagoshi, M. 1987. Survival of broods under parental care and parental roles of the cichlid fish, *Lamprologus toae*, in Lake Tanganyika. Japan. J. Ichthyol. 34: 71–75.

Nagoshi, M., T. Kanda & K. Horiya. 1978. Relation between exploitation and production in experimental populations of *Poecilia reticulata*. Res. Pop. Ecol. 20: 33–49.

Nagoshi, M., H. Kawanabe, N. Mizuno, D. Miyadi, S. Mori, Y. Sugiyama & I. Maki. 1962. Studies on the life of river fish. III. The life history of pale chub, 'oikawa', *Zacco platypus*. Contrib. Physiol. Ecol. Kyoto Univ. 82: 1–19 (in Japanese with English summary).

Nagoshi, M. & T. Kojima. 1976. The annual changes of growth and the critical size for maturation of isaza, *Chaenogobius isaza*, in Lake Biwa. Physiol. Ecol. Japan 17: 297–301 (in Japanese with English summary).

Nagoshi, M. & H. Kurita. 1986. Relationship between population density and production of the redspot masu-trout *Oncorhynchus rhodurus* in Japanese mountain stream. Nipp. Suisan Gakk. (Bull. Japan. Soc. Sci. Fish.) 52: 1875–1879 (in Japanese with English summary).

Nagoshi, M. & T. Miura. 1964. Population ecological studies of gengorobuna, *Carassius cuvieri* Temminck et Schlegel, in Lake Biwa. I. Daily estimation of the spawning population, immigrants and emigrants at Hayasaki Lagoon by tagging method. Physiol. Ecol. Japan 12: 93–98 (in Japanese with English summary).

Nagoshi, M., S. Nakano & Y. Tokuda. 1988. Changes in the utilization of microhabitat and food with growth of amago *Oncorhynchus rhodurus* in a Japanese mountain stream. Nipp. Suisan Gakk. (Bull. Japan. Soc. Sci. Fish.) 54: 33–38 (in Japanese with English summary).

Nagoshi, M. & T. Sakai. 1980. Relation between body size and food of the amago, *Oncorhynchus rhodurus*, from Hirakura Stream, Mie Prefecture, Japan. Japan. J. Ichthyol. 26: 342–350 (in Japanese with English summary).

Nagoshi, M, S. Suzuki, H. Semi, K. Sakaguchi, Y. Iwamoto & Y. Okamoto. 1982. Production of the amago, *Oncorhynchus rhodurus*, from mountain stream. Bull. Fac. Fish. Mie Univ. 9: 69–88 (in Japanese with English summary).

Nakai, K. 1993. Foraging of brood predators restricted by territoriality of substrate-brooders in a cichlid fish assemblage. pp. 84–108. *In:* H. Kawanabe, J. E. Cohen & K. Iwasaki (ed.) Mutualism and Community Organization: Behavioural, Theoretical, and Food-web Approaches, Oxford Univ. Press, Oxford.

Nakai, K., H. Kawanabe & M. M. Gashagaza. 1994. Ecological studies on the littoral cichlid communities of Lake Tanganyika: the coexistence of many endemic species. Arch. Hydrobiol. Beih. Ergebn. Limnol. 44: 375–391.

Nakai, K., Y. Yanagisawa, T. Sato, Y. Niimura & M. M. Gashagaza. 1990. Lunar synchronization of spawning in cichlid fishes of the tribe Lamprologini in Lake Tanganyika. J. Fish Biol. 37: 589–598.

Nakajima, T. & K. Nakai. 1994. Lake Biwa. Arch. Hydrobiol. Beih. Ergebn. Limnol. 44: 43–54.

Nakamura, T., T. Maruyama & E. Nozaki. 1994. Seasonal abundance and the re-establishment of iwana charr *Salvelinus leucomaenis* f. *pluvius* after excessive sediment loading by road construction in the Hakusan National Park, central Japan. Env. Biol. Fish. 39: 97–107.

Nakanishi, N. & M. Nagoshi. 1984. Yearly fluctuation of food habits of the isaza, *Chaenogobius isaza*, in Lake Biwa. Japan. J. Limnol. 45: 279–288 (in Japanese with English summary).

Nakano, S. 1992. A preliminary report on food habit of Japanese huchen *Hucho perryi* in northern Hokkaido. Biol. Inland Waters 7: 20–23 (in Japanese with English summary).

Nakano, S. 1994. Variation in agonistic encounters in a dominance hierarchy of freely interacting red-spotted masu salmon (*Oncorhynchus masou ishikawai*). Ecol. Freshwater Fish. 3: 153–158.

Nakano, S. 1995a. Competitive interactions for foraging microhabitats in a size-structured interspecific dominance hierarchy of two sympatric stream salmonids in a natural habitat. Can J. Zool. 73: 1845–1854.

Nakano, S. 1995b. Individual differences in resource use, growth and emigration under the influence of a dominance hierarchy in fluvial red-spotted masu salmon in a natural habitat. J. Anim. Ecol. 64: 75–84.

Nakano, S., K. D. Fausch, T. Furukawa-Tanaka, K. Maekawa & H. Kawanabe. 1992. Resource utilization by bull char and cutthroat trout in a mountain stream in Montana, U.S.A. Japan. J. Ichthyol. 39: 211–217.

Nakano, S. & T. Furukawa-Tanaka. 1994. Intra- and interspecific dominance hierarchies and variation in foraging tactics of two species of stream-dwelling chars. Ecol. Res. 9: 9–20.

Nakano, S., K. Maekawa & S. Yamamoto. 1990. Change of life cycle of *Salvelinus leucomaenis* following artificial lake by damming. Nipp. Suisan Gakk. (Bull. Japan. Soc. Sci. Fish.) 56: 1901–1905.

Nakano, S. & M. Nagoshi. 1985. Density regulation and growth of a redspot masu-trout, *Oncorhynchus rhodurus*, in a mountain stream. Physiol. Ecol. Japan 22: 1–16.

Nakano, S. & M. Nagoshi. 1990. Brood defence and parental roles in a biparental cichlid fish *Lamprologus toae* in Lake Tanganyika. Japan. J. Ichthyol. 36: 468–476.

Nakazono, A. & T. Kuwamura (ed.) 1987. Sex change in fishes. Tokai Univ. Press, Tokyo (in Japanese).

Nishida, M. 1986. Geographic variation in the molecular, morphological and reproductive characters of the ayu *Plecoglossus altivelis* (Plecoglossidae) in the Japan-Ryukyu Archipelago. Japan. J. Ichthyol. 33: 232–248.

Noda, M., K. Kawabata, K. Gushima & S. Kakuda. 1992. Importance of zooplankton patches in foraging ecology of the planktivorous reef fish *Chromis chrysurus* (Pomacentridae) at Kuchinoerabu Island, Japan. Mar. Ecol. Prog. Ser. 87: 251–263.

Northcote, T. G. & D. W. Wilkie. 1963. Underwater census of stream fish populations. Trans. Amer. Fish. Soc. 92: 146–151.

Nshombo, M. 1991. Occasional egg-eating by the scale-eater *Plecodus straeleni* (Cichlidae) of Lake Tanganyika. Env. Biol. Fish. 31: 207–212.

Nshombo, M. 1994a. Foraging behavior of the scale-eater *Plecodus straeleni* (Cichlidae, Teleostei) in Lake Tanganyika. Env. Biol. Fish. 39: 59–72.

Nshombo, M. 1994b. Polychromatism of the scale-eater *Perissodus microlepis* (Cichlidae, Teleostei) in relation to foraging behavior. J. Ethol. 12: 141–161.

Nshombo, M., Y. Yanagisawa & M. Nagoshi. 1985. Scale-eating in *Perissodus microlepis* (Cichlidae) and change of its food habits with growth. Japan. J. Ichthyol. 32: 66–73.

Ochi, H. 1985. Temporal patterns of breeding and larval settlement in a temperate population of the tropical anemonefish, *Amphiprion clarkii*. Japan. J. Ichthyol. 32: 248–257.

Ochi, H. 1986a. Breeding synchrony and spawning intervals in the temperate damselfish, *Chromis notata*. Env. Biol. Fish. 17: 117–123.

Ochi, H. 1986b. Growth of anemonefish *Amphiprion clarkii* in temperate waters, with special reference to the influence of settling time on the growth of 0-year olds. Marine Biol. 92: 223–229.

Ochi, H. 1989a. Acquisition of breeding space by nonbreeders in the anemonefish *Amphiprion clarkii* in temperate waters of southern Japan. Ethology 83: 279–294.

Ochi, H. 1989b. Mating behavior and sex change of the anemonefish, *Amphiprion clarkii*, in the temperate waters of southern Japan. Env. Biol. Fish. 26: 257–275.

Ochi, H. 1993a. Maintenance of separate territories for mating and feeding by males of a maternal mouthbrooding cichlid, *Gnathochromis pfefferi*, in Lake Tanganyika. Japan. J. Ichthyol. 40: 173–182.

Ochi, H. 1993b. Mate monopolization by a dominant male in a multi-male social group of a mouthbrooding cichlid, *Ctenochromis horei*. Japan. J. Ichthyol. 40: 209–218.

Ochi, H., Y. Yanagisawa & K. Omori. 1995. Intraspecific brood-mixing of the cichlid fish *Perissodus microlepis* in Lake Tanganyika. Env. Biol. Fish. 43: 201–206.

Ohgushi, R. 1992. Japanese ecology – K. Imanishi and his surroundings. Tokai Univ. Press, Tokyo. 212 pp. (in Japanese).

Okada, Y. 1959. Studies on the freshwater fishes of Japan. I. General part. Jour. Faculty of Fisheries, Prefectural University of Mie 4: 1–265 (plates I–XII).

Okada, Y. 1960a. Studies on the freshwater fishes of Japan. II. Special part. Jour. Faculty of Fisheries, Prefectural University of Mie 4: 266–588 (plates XIII–XLIX).

Okada, Y. 1960b. Studies on the freshwater fishes of Japan. II. Special part. Jour. Faculty of Fisheries, Prefectural University of Mie 4: 589–860 (plates L–LKI).

Okuno, R. 1971. Ecology of reef fishes. Sougen-sya, Osaka. 204 pp. (in Japanese).

Onodera, K. 1958. On the survival of trout fingerlings stocked in a mountain brook. I. The estimation of survival rate by observation. Bull. Japan. Soc. Sci. Fish. 24: 428–434.

Onodera, K. 1962. Carrying capacity in a trout stream. Bull. Freshwat. Fish. Res. Lab. 12: 1–42.

Onodera, K. & T. Ueno. 1961. On the survival of trout fingerlings stocked in a mountain brook – II. Survival rate measured and scouring effect of flood as a cause of mortality. Bull. Japan. Soc. Sci. Fish. 27: 530–557.

Pratt, H. L. Jr., S. H. Gruber & T. Taniuchi (ed.) 1990. Elasmobranchs as living resources: advances in the biology, ecology, systematics, and the status of the fisheries. U.S. Dept. of Commerce, NOAA, Tech. Rep. NMFS 90. 520 pp.

Rossiter, A. 1986. Evolutionary 'classics' may self-destruct. Nature 322: 315–316.

Sano, M. 1989. Feeding habits of Japanese butterflyfishes (Chaetodontidae). Env. Biol. Fish. 25: 195–203.

Sano, M. 1990. Patterns of habitat and food utilization in two coral-reef sandperches (Mugiloididae): competitive or noncompetitive coexistence? J. Exp. Mar. Biol. Ecol. 140: 209–223.

Sano, M. 1993. Foraging activities and diets of males and females in a haremic sandperch (Pisces: Pinguipedidae). Mar. Ecol. Prog. Ser. 98: 55–59.

Sano, M. & J. T. Moyer. 1985. Bathymetric distribution and feeding habits of two sympatric cheilodactylid fishes at Miyake-Jima, Japan. Japan. J. Ichthyol. 32: 239–247.

Sano, M., M. Shimizu & Y. Nose. 1984. Changes in structure of coral reef fish communities by destruction of hermatypic corals: observational and experimental views. Pacific Sci. 38: 51–79.

Sano, M., M. Shimizu & Y. Nose. 1987. Long-term effects of destruction of hermatypic corals by *Acanthaster planci* infestation on reef fish communities at Iriomote Island, Japan. Marine Ecol. Prog. Ser. 37: 191–199.

Sato, T. 1986. A brood parasitic catfish of mouthbrooding cichlid fishes in Lake Tanganyika. Nature 323: 58–59.

Sato, T. 1994. Active accumulation of spawning substrate: a determinant of extreme polygyny in a shell-brooding cichlid fish. Anim. Beh. 48: 669–678.

Sawara, Y. 1978. Ecological studies on the common freshwater goby, *Rhinogobius brunneus*, especially on the growth, food habits and feeding activity. J. Fac. Sci. Univ. Tokyo Sec. IV 14: 201–236.

Sawara, Y. & T. Ogawa. 1984. Presence of food as an entraining factor of rhythymic activity in *Chaenogobius laevis* in captivity. Japan. J. Ichthyol. 31: 93–96.

Shimizu, T., N. Taniguchi & N. Mizuno. 1993. An electrophoretic study of genetic differentiation of a Japanese freshwater goby, *Rhinogobius flumineus*. Japan. J. Ichthyol. 39: 329–343.

Shimizu, M & T. Taniuchi. 1991. Studies on elasmobranchs collected from seven river systems in northern Australia and Papau New Guinea. Univ. of Tokyo Museum, Nature and Culture 3. 109 pp.

Sunaga, T. 1964. On the seasonal change of food habits of several fishes in the Lake Biwa. Physiol. Ecol. Japan 12: 252–258 (in Japanese with English summary).

Sunaga, T. 1970. Relation between food consumption and food selectivity in fishes I. Food selectivity of juvenile hasu (*Opsariichthys uncirostris* Teminck et Schlegel), in Lake Biwa. Japan. J. Ecol. 20: 129–137 (in Japanese with English summary).

Sunaga, T. 1971. Relation between food consumption and food selectivity in fishes II. Experimental study on the food selection by juvenile guppy (*Poecilia reticulata*). Japan. J. Ecol. 21: 67–70 (in Japanese with English summary).

Sunaga, T. 1976. Relation between food consumption and food selectivity in fishes IV. Experimental study on the food selection by the juvenile guppy (*Poecilia reticulata*) 2. Physiol. Ecol. Japan 17: 217–220 (in Japanese with English summary).

Takahashi, S. 1981. Vertical distribution and diel migration of isaza, *Chaenogobius isaza*, Pisces, in Lake Biwa. Zool. Mag. 90: 145–151.

Takahashi, S. & M. Hori. 1994. Unstable evolutionarily stable strategy and oscillation: a model on lateral assymetry in scale-eating cichlids. Amer. Nat. 144: 1001–1020.

Takamura, K. 1979. Utilization of habitat by young pale chub: *Zacco platypus* (Temminck et Schlegel) (Pisces: Cyprinidae). Japan. J. Ecol. 29: 295–306 (in Japanese with English summary).

Takamura, K. 1983. Interspecific relationship between two aufwuchs eaters *Petrochromis polyodon* and *Tropheus moorei* (Pisces: Cichlidae) of Lake Tanganyika, with a discussion on the evolution and functions of a symbiotic relationship. Physiol. Ecol. Japan 20: 59–69.

Takamura, K. 1984. Interspecific relationship of aufwuchs-eating fishes in Lake Tanganyika. Env. Biol. Fish. 10: 225–241.

Takata, K., A. Goto & K. Hamada. 1984. Geographic distribution and variation of three species of ninespine sticklebacks (*Pungitius tymensis, P. pungitius* and *P. sinensis*) in Hokkaido. Japan. J. Ichthyol. 31: 312–326.

Takata, K., A. Goto & F. Yamazaki. 1987. Biochemical identification of a brackish water type of *Pungitius pungitius*, and its morphological and ecological features in Hokkaido, Japan. Japan. J. Ichthyol. 34: 176–183.

Tanaka, S. 1964. The food and growth of 'hasu', *Opsariichthys uncirostris* (T. & S.) in the Lake Biwa. Physiol. Ecol. Japan 12: 106–114 (in Japanese with English summary).

Tanaka, S. 1970a. Studies on the growth of 'hasu', *Opsariichthys uncirostris*, in Lake Biwa. I. On the body length at each age and the growth curve estimated from the spawning populations. Japan. J. Ecol. 19: 13–25 (in Japanese with English summary).

Tanaka, S. 1970b. Studies on the growth of 'hasu', *Opsariichthys uncirostris*, in Lake Biwa. II. On the growth variation of individuals and the growth compensation observed in the process of growth. Japan. J. Ecol. 20: 80–90 (in Japanese with English summary).

Tanaka, S. & M. Hoshino. 1979. Growth and maturity of ninespine sticleback, *Pungitius sinensis* Guichenot, in the Kamokawa, a streamlet in Toyama Prefecture. Bull. Toyama Sci. Mus. 1: 19–29.

Taniuchi, T., H. Tachikawa, M. Shimizu & Y. Nose. 1993. Geographical variation in reproductive parameters of shortspine spurdog in the north Pacific. Nipp. Suisan Gakk. (Bull. Japan. Soc. Sci. Fish.) 59: 45–51.

Terashima, A. 1980. Bluegill – vacant niche remained also in Lake Biwa. pp. 63–70. *In:* T. Kawai, H. Kawanabe & N. Mizuno (ed.) Japanese Freshwater Organisms – Ecology of Invasion and Disturbance, Tokai University Press, Tokyo (in Japanese).

Tsukamoto, K. 1990. Recruitment mechanism of the eel, *Anguilla japonica*, to the Japanese coast. J. Fish Biol. 36: 659–671.

Tsukamoto, K. 1992. Discovery of the spawning area for Japanese eel. Nature 356: 789–791.

Tsukamoto, K., R. Ishida, K. Naka & T. Kajihara. 1987. Switching of size and migratory pattern in successive generations of landlocked ayu. Amer. Fish. Soc. Symp. 1: 492–506.

Tsukamoto, K., H. Kuwada, J. Hirokawa, M. Oya, S. Sekiya, H. Fujimoto & K. Imaizumi. 1989. Size-dependent mortality of red sea bream, *Pagrus major*, juveniles released with fluorescent otolith-tags in News Bay, Japan. J. Fish Biol. 35(Suppl A): 59–69.

Tsukamoto, K. & K. Uchida. 1990. Spacing and jumping behaviour of the ayu *Plecoglossus altivelis*. Nipp. Suisan Gakk. (Bull. Japan. Soc. Sci. Fish.) 56: 1383–1392.

Tsukamoto, K. & K. Uchida. 1992. Migration mechanism of the ayu. pp. 145–172. *In:* V. I. Ilyichev & V. V. Anikiev (ed.) Oceanic and Anthropogenic Controls of Life in the Pacific Ocean. Kluwer Academic Publishers, Dordrecht.

Tsukamoto, K & A. Umezawa. 1990. Early life history and oceanic migration of the eel, *Anguilla japonica*. La mer (Soc. Franco-Japonaise d'Oceanographie) 28: 188–198.

Uchida, K., K. Tsukamoto, R. Ishida, S. Ishii & T. Kajihara. 1990a. Effect of parental origin on jumping and spacing behaviours of juvenile ayu *Plecoglossus altivelis* with special reference to their upstream migration. Nipp. Suisan Gakk. (Bull. Japan. Soc. Sci. Fish.) 56: 1401–1405.

Uchida, K., K. Tsukamoto, S. Ishii, R. Ishida & T. Kajihara. 1989. Larval competition for food between wild and hatchery-reared ayu, *Plecoglossus altivelis* Temminck et Schlegel, in culture ponds. J. Fish Biol. 34: 399–407.

Uchida, K., K. Tsukamoto & T. Kajihara. 1990b. Effects of environmental factors on jumping behavior of the juvenile ayu *Plecoglossus altivelis* with special reference to their upstream migration. Nipp. Suisan Gakk. (Bull. Japan. Soc. Sci. Fish.) 56: 1393–1399.

Umezawa, A. & K. Tsukamoto. 1991. Factors influencing otolith increment formation in Japanese eel, *Anguilla japonica* T. & S., elvers. J. Fish Biol. 39: 211–223.

Uyeno, T., R. Arai, T. Taniuchi & K. Matsuura (ed.) 1986. Indo-Pacific fish biology: proceedings of the second international conference on Indo-Pacific fishes. The Ichthyological Soc. of Japan, Tokyo. 990 pp.

Yamagishi, H. 1962. Growth relation in some small experimental populations of rainbow trout fry, *Salmo gairdneri* Richardson with special reference to social relations among individuals. Japan. J. Ecol. 12: 43–53.

Yamagishi, H. 1963. Some observations on growth variation and feeding behaviour in the fry of two races of Japanese crucian carp, *Carassius carassius* Linne. Japan. J. Ecol. 13: 156–161.

Yamagishi, H. 1964. An experimental study on the effect of aggressiveness to the variability of growth in the juvenile rainbow trout, *Salmo gairdneri* Richardson. Japan. J. Ecol. 14: 228–232.

Yamagishi, H. 1969. Postembryonal growth and its variability of the three marine fishes with special reference to the mechanism of growth variation in fishes. Res. Pop. Ecol. 11: 14–33.

Yamagishi, H., T. Maruyama & K. Mashiko. 1974. Social relation in a small experimental population of *Odontobutis obscurus* (Temminck et Schlegel) as related to individual growth and food intake. Oecologia 17: 187–202.

Yamamoto, S., S. Nakano & Y. Tokuda. 1992. Variation and divergence of the life-history of Japanese charr *Salvelinus leucomaenis* in an artificial lake-inlet stream system. Japan. J. Ecol. 42: 149–157 (in Japanese with English summary).

Yamashiro, S. 1965. Age and growth of the ito (*Hucho perryi*) in northeastern Hokkaido. Japan. Soc. Sci. Fish. 31: 1–7 (in Japanese with English summary).

Yanagisawa, Y. 1982. Social behavior and mating system of a gobiid fish *Amblyeleotris japonica*. Japan. J. Ichthyol. 28: 401–422.

Yanagisawa, Y. 1984. Studies on the interspecific relationship between gobiid fish and snapping shrimp II. Life history and pair formation of snapping shrimp *Alpheus bellulus*. Publ. Seto Mar. Biol. Lab. Kyoto Univ. 29: 93–116.

Yanagisawa, Y. 1985. Parental strategy of the cichlid fish *Perissodus microlepis*, with particular reference to intraspecific brood 'farming out'. Env. Biol. Fish. 12: 241–249.

Yanagisawa, Y. 1986. Parental care in a monogamous mouthbrooding cichlid *Xenotilapia flavipinnis* in Lake Tanganyika. Japan. J. Ichthyol. 33: 249–261.

Yanagisawa, Y. 1987. Social organization of a polygynous cichlid *Lamprologus furcifer* in Lake Tanganyika. Japan. J. Ichthyol. 34: 82–90.

Yanagisawa, Y. 1990. Strange seabed fellows. Natural History 8: 46–50.

Yanagisawa, Y. 1993. Long-term territory maintenance by female *Tropheus duboisi* (Cichlidae) involving foraging during the mouth-brooding period. Ecol. Freshwater Fish 2: 1–7.

Yanagisawa, Y. & M. Nishida. 1991. The social and mating system of the maternal mouthbrooder *Tropheus moorii* (Cichlidae) in Lake Tanganyika. Japan. J. Ichthyol. 38: 271–282.

Yanagisawa, Y. & H. Ochi. 1986. Step-fathering in anemonefish *Omphiprion clarkii*: a removal study. Anim. Beh. 34: 1769–1780.

Yanagisawa, Y. & H. Ochi. 1991. Food intake by mouthbrooding females of *Cyphotilapia frontosa* (Cichlidae) to feed both themselves and their young. Env. Biol. Fish. 30: 353–358.

Yanagisawa, Y. & T. Sato. 1990. Active browsing of mouthbrooding females of *Tropheus duboisi* and *Tropheus moorii* (Cichlidae) to feed the young and/or themselves. Env. Biol. Fish. 27: 43–50.

Yuma, M. 1993. Competitive and co-operative interactions in Tanganyikan fish communities. pp. 213–227. *In:* H. Kawanabe, J. E. Cohen & K. Iwasaki (ed.) Mutualism and Community Organization: Behavioural, Theoretical, and Food-web Approaches, Oxford University Press, Oxford.

Yuma, M. 1994. Food habits and foraging behaviour of benthivorous cichlid fishes in Lake Tanganyika. Env. Biol. Fish. 39: 173–182.

Yuma, M. & E. Harada. 1998. The life and work of Hiroya Kawanabe: the priest ecologist. Env. Biol. Fish. 52: 11–35 (this volume).

Yuma, M., I. Nakamura & K.D. Fausch (ed.) 1998. Fish biology in Japan: an anthology in honour of Hiroya Kawanabe. Dev. Env. Biol. Fish. 18, Kluwer Academic Publishers, Dordrecht. 405 pp.

Environmental Biology of Fishes **52**: 97–124, 1998.

Distribution of the freshwater fishes of Japan: an historical overview

Masahide Yuma[1], Kazumi Hosoya[2] & Yoshikazu Nagata[3]
[1] *Center for Ecological Research, Kyoto University, Otsu 520-0105, Japan*
(e-mail: yuhma@ecology.kyoto-u.ac.jp)
[2] *National Research Institute of Fisheries Science, Ueda 386-0031, Japan*
[3] *Department of Biology, Osaka Kyoiku University, Kashiwara 582-8582, Japan*

Received 25.2.1997 Accepted 28.11.1997

Key words: indigenous fishes, exotic fishes, historical processes, extinct fishes, endangered fishes, current distribution patterns, Lake Biwa, artificial water system

Synopsis

Japanese freshwater fishes, including lampreys, comprise 15 orders, 35 families, and 96 genera, with 211 species and subspecies. Most belong to the families Cyprinidae (29% of species and subspecies), Gobiidae (21%), Salmonidae (10%), and Cobitidae (8%). Cyprinids and cobitids presumably originated from east Asia, gobiids from southeast Asia, and cottids and salmonids from the north Pacific. Japanese freshwater fishes include 88 endemic species and subspecies, of which three have been extirpated. Fishes introduced into natural rivers and lakes for inland commercial fisheries and sport fishing, and by accident, include many exotic species, of which 23 now inhabit natural freshwaters. These often have destroyed the local fish fauna by predation, and caused genetic pollution by hybridization with local strains. Destruction of freshwater environments by land development also poses a threat to Japanese freshwater fish communities. In addition Japanese freshwater systems have been markedly altered by development of rice paddy fields which have caused some species to decline but others to flourish, and changed the distribution patterns of fishes between upstream and downstream areas. To conserve endangered species and declining communities of Japanese freshwater fishes, we need to clarify the characteristics of their original habitats and the effects of developing paddy fields, from both the ecological and historical points of view.

Introduction

The Japanese Archipelago extends from northeast to southwest and consists of four major islands (Hokkaido, Honshu, Shikoku, and Kyushu) and other smaller island chains to the south, such as Ryukyu Islands, Ogasawara Islands and numerous other smaller islands (Figure 1). The archipelago is located close to the eastern margin of the Eurasian continent and spans from the subarctic in the north to the subtropical in the south, with a wide range of seasonal temperature changes and a large amount of rainfall. There are numerous small mountain rivers; the longest, the Shinano, is 367 km, and the second longest, the Tone, is 322 km. There are also many small lakes and a few large ones; the largest is Lake Biwa (670 km^2 in surface area), and the second largest is lake Kasumigaura (168 km^2).

The geologic history and geography of Japan has had profound effects on the freshwater fish fauna. In this paper, the composition and distribution of Japanese freshwater fish are reviewed with respect

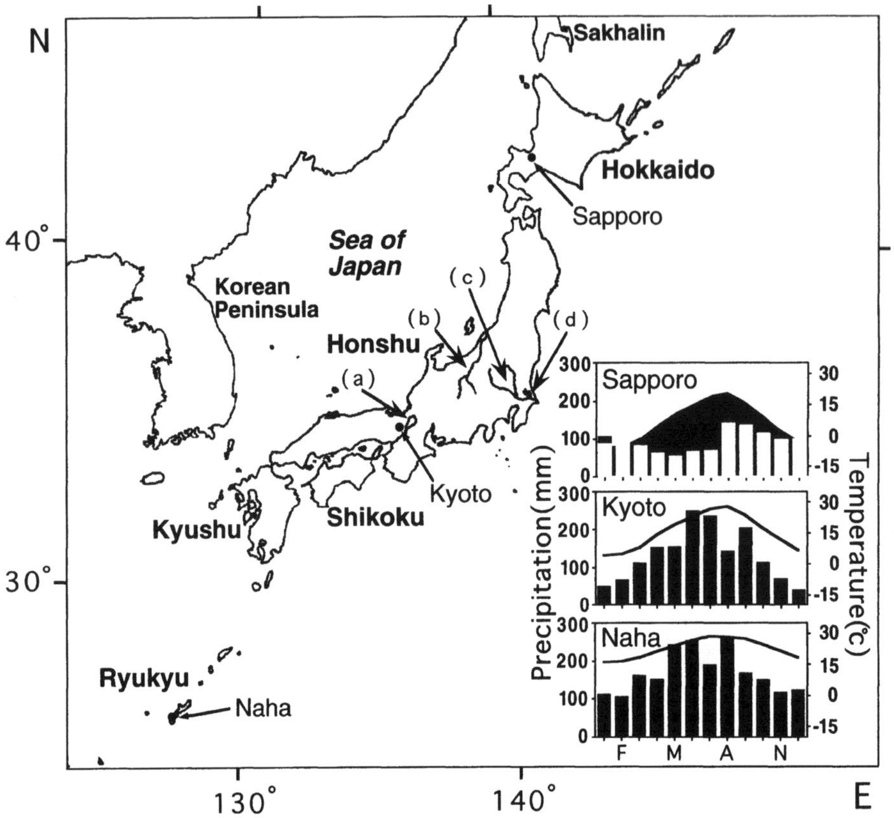

Figure 1. Map of Japan. Average temperature (lines) and precipitation (bars) between 1961 and 1990 for three cities are shown: a – Lake Biwa, b – River Shinano, c – River Tone, d – Lake Kasumigaura.

to their centers of origin, plate tectonics of the Japanese archipelago, and the introduction of exotic fishes. Their distribution is also discussed in relation to expansion of artificial freshwater systems due to development of rice paddy cultivation in Japan.

Freshwater fish fauna

Okada (1959–1960) reviewed 1949 articles on Japanese freshwater fishes published from 1709 to 1959, and synthesized their information. These papers refer to 147 species and subspecies belonging to 10 orders, 32 families, and 82 genera. Later Kawanabe & Mizuno (1989) and Nakabo (1993) listed 211 species and subspecies of Japanese freshwater fishes belonging to 15 orders, 35 families, and 96 genera (Table 1, Appendix). These fishes comprise 6% of about 3800 fish species that have been recorded from marine and freshwater habitats of Japan, and include some taxonomically undetermined forms. The freshwater fish fauna of Japan consists mostly of the families Cyprinidae (29% of species and subspecies), Gobiidae (21%), Salmonidae (10%), and Cobitidae (8%). In general, the cyprinids are widely distributed, while the salmonids and cottids are confined to the northern area and the gobiids to the south.

These freshwater fishes include 134 fluvial and lacustrine forms (63% of species and subspecies) and 77 diadromous ones (37%). These include 88 endemic species and subspecies (41%; 32 cyprinids, 17 gobiids, 14 cobitids, 7 salmonids, 4 bagrids, 4 gasterosteids, 3 cottids, 2 silurids, 1 osmerid, 1 plecoglossid, 1 engraulid, 1 amblycipid and 1 salangid, and 23 exotic species, mostly cyprinids and salmonids (11%) (Table 1).

The freshwater fish fauna of Japan is composed of freshwater fishes like cyprinids and cobitids that

Table 1. Number of freshwater fish species and subspecies in Japan after Kawanabe & Mizuno (1989) and Nakabo (1993). Fishes living mainly in brackish and marine waters were excluded. Scientific names follow Nakabo (1993).

Taxon	Number of genera (%)	Number of species & subspecies (%)	Primary freshwater fishes	Diadromous species	Fishes endemic to Japan	Fishes only in Hokkaido & north	Fishes only in Ryukyu Is. & south	Exotic fishes
Petromyzontiformes								
Petromyzontidae	2 (2.1)	4 (1.9)	2	2	–	–	–	–
Acipenseriformes								
Acipenseridae	1 (1.0)	1 (0.5)	0	1	–	–	–	–
Anguilliformes								
Anguillidae	1 (1.0)	2 (0.9)	0	2	–	–	–	–
Clupeiformes								
Engraulididae	1 (1.0)	1 (0.5)	1	0	1	–	–	–
Cypriniformes								
Cyprinidae	24 (25.0)	61 (28.9)	60	1	32	1	–	5
Cobitidae	6 (6.3)	16 (7.6)	16	0	14	2	–	–
Siluriformes								
Bagridae	3 (3.1)	4 (1.9)	4	0	4	–	–	–
Siluridae	1 (1.0)	3 (1.4)	3	0	2	–	–	–
Amblycipitidae	1 (1.0)	1 (0.5)	1	0	1	–	–	–
Clariidae	1 (1.0)	1 (0.5)	1	0	–	–	1	1
Salmoniformes								
Osmeridae	3 (3.1)	4 (1.9)	1	3	1	2	–	–
Plecoglossidae	1 (1.0)	2 (0.9)	0	2	1	–	1	–
Salangidae	2 (2.1)	1 (0.5)	1	0	1	–	–	
Salmonidae	5 (5.2)	21 (10.0)	10	11	7	3	–	5
Atheriniformes								
Atherinidae	1 (1.0)	1 (0.5)	1	0	–	–	–	1
Cyprinodontiformes								
Poeciliidae	2 (2.1)	2 (0.9)	2	0	–	–	–	2
Beloniformes								
Adrianichthyidae	1 (1.0)	1 (0.5)	1	0	–	–	–	–
Hemiramphidae	1 (1.0)	1 (0.5)	1	0	–	–	–	–
Gasterosteiformes								
Gasterosteidae	2 (2.1)	8 (3.8)	7	1	4	1	–	–
Syngnathiformes								
Syngnathidae	1 (1.0)	2 (0.9)	0	2	–	–	2	–
Synbranchiformes								
Synbranchidae	1 (1.0)	1 (0.5)	1	0	–	–	–	–
Scorpaeniformes								
Scorpaenidae	1 (1.0)	1 (0.5)	0	1	–	–	1	–
Cottidae	2 (2.1)	7 (3.3)	3	4	3	1	–	–
Perciformes								
Ambassidae	1 (1.0)	2 (0.9)	0	2	–	–	2	–
Percichthyidae	1 (1.0)	1 (0.5)	1	0	–	–	–	–
Terapontidae	1 (1.0)	3 (1.4)	0	3	–	–	3	–
Kuhliidae	1 (1.0)	2 (0.9)	0	2	–	–	–	–
Centrarchidae	2 (2.1)	2 (0.9)	2	0	–	–	–	2
Scatophagidae	1 (1.0)	1 (0.5)	0	1	–	–	–	–
Cichlidae	2 (2.1)	3 (1.4)	3	0	–	–	–	3
Mugilidae	1 (1.0)	1 (0.5)	0	1	–	–	1	–
Rhyancichthyidae	1 (1.0)	1 (0.5)	0	1	–	–	1	–
Gobiidae	19 (19.8)	44 (20.9)	7	37	17	–	18	–
Belontiidae	1 (1.0)	2 (0.9)	2	0	–	–	1	1
Channidae	1 (1.0)	3 (1.4)	3	0	–	–	1	3
Total	96 (100.0)	211 (100.0)	134	77	88	10	32	23

originated from east Asia, and diadromous ones such as gobiids from southeast Asia and cottids and salmonids from the north Pacific (Aoyagi 1957, Okada 1959–1960, Mizuno 1987). According to Nakabo (1993), 21% of the 188 indigenous freshwater fishes in Japan are also found on the Korean Peninsula, 28% of them are widely distributed in southeast Asia (southern China, Taiwan and the western Pacific), and 15% of them are also found in eastern Russia (farther north than Sakhalin).

The distribution of Japanese cyprinids indicates a strong affinity with those of Korea, inasmuch as 23% of the 56 indigenous Japanese cyprinids are also found on the Korean Peninsula, as compared with 13% in the Far East (Russia) and 4% in southeast Asia. As far as the freshwater gobiids are concerned, 52% of the 56 indigenous Japanese species and subspecies are widely distributed from Japan to southeast Asia, while 23% are found on the Korean Peninsula and only 9% in the Far East of Russia. Among salmonids and cottids, 48% of the 23 species and subspecies also occur in the Far East of Russia, 13% on the Korean Peninsula, and only 8% in southeast Asia.

The distribution of the freshwater fishes also depends on the geological history of the Japanese Archipelago and climatic conditions, primarily temperature (Goto 1987). Until the early Miocene (15–20 million years ago) the Japanese Archipelago was a part of the Eurasian Mainland, at which time it became separated from the mainland by tectonic plate movements. Until the mid-Pliocene (ca. 2.8–3 million years aso), a huge freshwater lake (the Paleo-Japan Sea) is thought to have existed at the location of the present Sea of Japan (Nishimura 1980). The primary freshwater fishes are considered to have been widely distributed around the Paleo-Japan Sea. The Paleo-Japan Sea then became a bay with a narrow mouth at the location of the present Ishikari Plain in western Hokkaido near Sapporo. During the Quaternary, there was a series of at least four marine transgressions and regressions. Freshwater fishes might have migrated between the Eurasian Mainland and the Japanese Archipelago by means of a land bridge during the times of marine regression, and speciated there during the marine transgressions (Aoyagi 1957, Lindberg 1972, Nish-

Table 2. Freshwater fishes in 'The list of endangered wild animals in Japan (Red Data Book in Japan): vertebrates' published by the Environment Agency of Japan (1991). The list covers the fishes living in fresh and brackish waters, and includes undescribed species, subspecies, and forms. Scientific names follow Nakabo (1993). See Appendix for Japanese and English names.

Extinct species
- *Oncorhynchus nerka kawamurae*, Salmonidae
- *Pungitius kaibarae*, Gasterosteidae

Endangered species
- A form of *Salvelinus leucomanis japonicus*, Salmonidae
- *Oncorhynchus masou ishikawae*, Salmonidae
- A color variant of *Oncorhynchus masou masou*, Salmonidae
- *Plecoglossus altivelis ryukyuensis*, Plecoglossidae
- *Salanx ariakensis*, Salangidae
- *Neosalanx reganius*, Salangidae
- *Aphyocypris chinensis*, Cyprinidae
- *Pseudorasbora pumila* subsp., Cyprinidae
- *Acheilognathus longipinnis*, Cyprinidae
- *Rhodeus ocellatus kurumeus*, Cyprinidae
- *Rhodeus atremius suigensis*, Cyprinidae
- *Tanakia tanago*, Cyprinidae
- *Leptobotia curta*, Cobitidae
- *Coreobagrus ichikawai*, Bagridae
- *Pseudobagrus tokiensis*, Bagridae
- *Pungitius* sp., Gasterosteidae

Vulnerable species
- *Hucho perryi*, Salmonidae
- *Salvelinus leucomaenis imbrius*, Salmonidae
- *Tribolodon* sp., Cyprinidae
- *Gasterosteus microcephalus*, Gasterosteidae
- *Boleophthalmus pectinirostris*, Gobiidae
- *Trachidermus fasciatus*, Cottidae

Rare species
- *Entosphenus tridentata*, Petromyzontidae
- *Lethenteron kessleri*, Petromyzontidae
- *Coilia nasus*, Engraulididae
- *Salvelinus malma miyabei*, Salmonidae
- *Salvelinus malma malma*, Salmonidae
- *Oncorhynchus masou* subsp., Salmonidae
- *Pseudorasbora pumila pumila*, Cyprinidae
- *Acheilognathus typus*, Cyprinidae
- *Pungitius tymensis*, Gasterosteidae
- *Macropodus opercularis*, Belontiidae
- *Coreoperca kawamebari*, Percichthyidae
- *Lates japonicus*, Centropomidae
- *Hypseleotris cyprinoides*, Gobiidae
- *Chaenogobius* sp. 3, Gobiidae
- *Luciogobius pallidus*, Gobiidae
- *Luciogobius albus*, Gobiidae
- *Rhyacichthys aspro*, Rhyncichthyidae

Local populations
- *Coilia nasus* Engraulididae, in the River Rokkaku, Saga Pref.
- *Hemigrammocypris rasborella*, Cyprinidae, in Shizuoka Pref.
- *Liobagrus reini*, Amblycipitidae, in the Kyushu District
- *Oryzias latipes*, Adrianichthyidae, in Okinawa Pref.
- *Monopterus albus*, Synbranchidae, in Okinawa Pref.
- *Gasterosteus* sp., Gasterosteidae, in the Aizu District, Fukushima Pref.
- *Gasterosteus* sp., Gasterosteidae, in the Ohno Basin, Fukui Pref.

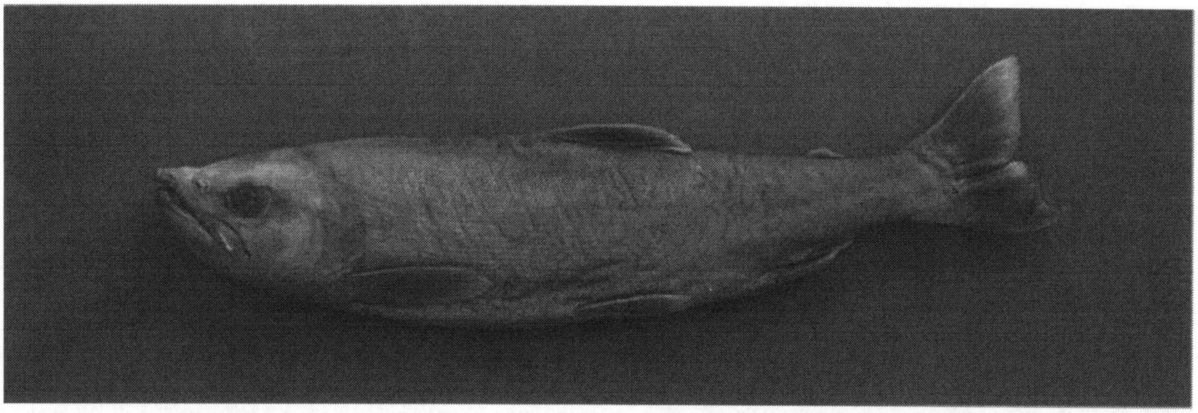

Figure 2. Kuni-masu, black kokanee, *Oncorhynchus nerka kawamurai,* specimen from the Center for Ecological Research, Kyoto University. Photograph by H. Akiyama.

imura 1980, Mizuno 1987). Thus, the species and genera endemic to Japan are considered to have immigrated from the mainland in earlier times of marine regression, while the species and genera held in common with those of the mainland are considered to have immigrated during subsequent marine regressions (Lindberg 1972, Nishimura 1980).

There may have been two different routes of migration over the land bridge during the times of marine regression, one through the Korean Peninsula and the other through Sakhalin. The latter route is indicated by the distribution patterns of the Sakhalin lake minnow, *Phoxinus percnurus sachalienesis*, the Siberian stone loach, *Noemacheilus barbatulus toni* and the Hokkaido eight-barbel loach, *Lefua nikkonis,* for each of which the southern limit is located in the Ishikari Plain (Goto et al. 1978). The former route is indicated by many fishes commonly distributed in southwestern Japan and the Korea-China Region (Aoyagi 1957, Mizuno 1987).

Some fishes might have spread to northern Japan with the propagation of paddy-field cultivation during historical times, such as the topmouth gudgeon, *Pseudorasbora parva,* the field gudgeon, *Gnathopogon elongatus elongatus,* the pike gudgeon, *Pseudogobio esocinus esocinus,* the steed barbel, *Hemibarbus barbus,* the common carp, *Cyprinus carpio,* the silver crucian carp, *Carassius auratus langsdorfi,* the slender bitterling, *Tanakia lanceolata,* the Asian pond loach, *Misgurnus anguillicaudatus,* the sand loach, *Cobitis biwae,* and the Far Eastern catfish, *Silurus asotus* (Nishimura 1980, Balon 1995). The original distributions of Japanese freshwater fishes are, therefore, becoming obscured due to human activities including the transplantation of fishes and the alteration of the natural environment.

Endangered species

In 1991 the Environment Agency of Japan published 'The list of endangered wild animals in Japan (Red Data Book of Japan): vertebrates'. The list included 2 extinct, 16 endangered, 6 vulnerable, and 17 rare species of freshwater fishes, and 7 declining local populations of widely distributed species (Table 2).

One of the extinct species is the black kokanee, *Oncorhynchus nerka kawamurai* (Salmonidae) (Figure 2). The fish, called kuni-masu or kinosiri-masu in Japanese, once occurred in Lake Tazawa, Akita Prefecture. Kawanabe & Watanabe (1993a) cited a work entilted 'History of Akita Region', or 'Akita fudoki' in Japanese, written during the mid-Edo Period (ca. 1700 AD), which notes that 'there are fish called kuni-masu. They provide the usual food resource for the people around the lake'. The kuni-masu seems to have exhibited a black nuptial coloration throughout the year, which deepened particularly in January-March and July-October (Oshima 1941). Among the salmonid fishes, such reproductive changes are unique.

Lake Tazawa is the deepest lake in Japan with a surface area of 25.8 km^2, a maximum depth of

Figure 3. Minami-tomiyo, Kyoto ninespine stickleback, *Pungitius kaibarae*; holotype ZUMT 8197 (Zoology, University Museum, the University of Tokyo). Photograph by K. Hosoya.

Figure 4. The paddy field of Japanese parsley where *Pungitius kaibarae* survived until the 1960s. Photograph by K. Hosoya.

423.4 m, and once had transparency to 30 m (recorded in 1926; it is the second clearest record for lakes in Japan, but dropped to 4 m in 1993 by recent eutrophication). However, in 1940, strong acid water was diverted into Lake Tazawa from the adjacent River Tama to work a hydro-power station, at which time the pH of the lake abruptly dropped to 4.6 and almost all of organisms in the lake including the kuni-masu were extirpated, thus preventing any study of their biological and ecological features. Apart from the type specimens in the Field Museum of Natural History, Chicago, only a few specimens remain, eight at the Center for Ecological Research (formerly the Otsu Hydrobiological Station) of Kyoto University, three at the Fisheries Research Station, Faculty of Agriculture, Kyoto University, two at the Lake Tazawa Regional Museum, and several at Akita University and the Akita Prefectural Museum. The Japanese dace, *Tribolodon hakonensis,* is the only species of fish now living in the lake.

Another extinct species is the Kyoto ninespine stickleback (minami-tomiyo), *Pungitius kaibarae* (Gasterosteidae) (Figure 3). This species inhabited springs and small streams around Kaibara Town, Hyogo Prefecture, and the southwestern part of Kyoto City. Tan (1928) and Kobayashi (1933) reported that the fish was found in paddies of Japanese parsley with cool spring water (Figure 4) and in shallow spring ponds. However, these populations declined after reclamation of paddies and ponds for land development. An attempt to transplant a population into the Kyoto City Zoo in 1959 was unsuccessful, and the fish disappeared after the early 1960s (Kawanabe & Watanabe 1993b). Type specimens have been preserved at the University Museum, University of Tokyo, and there are more than 20 specimens at the Center for Ecological Research, Kyoto University. In addition to these two fishes, the Suwa gudgeon, *Gnathopogon elongatus suwae* (Cyprinidae), which was endemic to Lake Suwa in Nagano Prefecture, must have become extinct following its replacement by *G. elongatus elongatus* (Miyadi et al. 1976; Hosoya in Nakabo 1993).

The endangered, vulnerable, and rare fishes listed in the Red Data Book include (1) those confined to restricted areas which are referred to as local species and subspecies such as the swampy bullhead, *Coreobagrus ichikawai,* the cuttailed bullhead, *Pseudobagrus tokiensis,* the Ryukyu-ayu, *Plecoglossus atlivelis ryukyuensis,* the Miyabe charr, *Salvelinus malma miyabei,* and the Biwa trout, *Oncorhynchus masou* subsp.; (2) those living in small streams and shallow ponds such as the Venus fish, *Aphyocypris chinensis,* the dwarf topmouth gudgeon, *Pseudorasboro pumila pumila*, an undescribed gudeon, *P. pumila* subsp., the Japanese rose bitterling, *Rhodeus ocellatus kurumeus,* the Suigen rose bitterling, *R. atremius suigensis,* the netted bitterling, *Acheilognathus typus*, and the kissing loach, *Leptobotia curta;* as well as (3) diadromous fishes in large rivers such as the red-spotted masu trout, *Oncorhynchus masou ishikawae.*

Following the publication of many full-colored books and introductory essays on endangered species, the appearance of the Red Data Book further publicized the threatened status of organisms and their natural environments. It also may have contributed to enactment of the Basic Environmental Law of 1993 and to the increased use of improved techniques to conserve natural organisms and their environments associated with public works such as channelization of rivers and revetment of their banks.

Table 3. Fishes registered and protected as National Natural Monuments of Japan. Scientific names follow Nakabo (1993).

Species registered and protected in all of Japan
Acheilognathus longipinnis, Cyprinidae
Designated in 1974
Tanakia tanago, Cyprinidae
Designated in 1974
Leptobotia curta, Cobitidae
Designated in 1977
Coreobagrus ichikawai, Bagridae
Designated in 1977
Local populations
Anguilla japonica, Anguillidae
In Kayugawa, Minami-mura, Gujo-gun, Gifu Pref.
Designated in 1924
Anguilla japonica, Anguillidae
In Kashiko-numa, Hiranuma-no-uchi, Iwaki City, Fukushima Pref.
Designated in 1939
Anguilla marmorata, Anguillidae
In Shirahama-cho, Nishimuro-gun, Wakayama Pref.
Designated in 1923
Anguilla marmorata, Anguillidae
In Hahagawa, Kaifu-cho, Kaifu-gun, Tokushima Pref
Designated in 1923
Anguilla marmorata, Anguillidae
In Nomozaki-cho, Nishisonogi-gun, Nagasaki Pref.
Designated in 1923
Tribolodon hakonensis, Cyprinidae
In Yokoyama, Tsuyama-cho, Motoyoshi-gun, Miyagi Pref.
Designated in 1935
Tribolodon hakonensis, Cyprinidae
In Yanaizu, Yanaidu-cho, Kawanuma-gun, Fukushima Pref.
Designated in 1940
Carassius auratus langsdorfi, scarlet morph, Cyprinidae
In Lake Harutori, Harutori, Kushiro City, Hokkaido
Designated in 1937
Carassius auratus subsp. var., Cyprinidae
In Yutori-numa, Miyazaki-cho & Onoda-cho, Kami-gun, Miyagi Pref.
Designated in 1933
Gasterosteus aculeatgus, Gasterosteidae
In Hongan-shouzu, Itoyo-cho, Ohno City, Fukui Pref.
Designated in 1934
Cottus kazika, Cottidae
In River Kuzuryu, Ohono City and Fukui City, Fukui Pref.
Designated in 1935

Figure 5. Ita-sen-para, deepbodied bitterling, *Acheilognathus longipinnis*. Photograph by Y. Nagata.

Before the Red Data Book was published, four species of freshwater fishes confined to restricted areas had been designated as National Natural Monuments (tennen-kinenbutsu) and were protected by the national government under the Cultural Properties Protection Law (Table 3, Kato et al. 1995). One is the deepbodied bitterling, *Acheilognathus longipinnis* (Cyprinidae, Figure 5), which was designated a National Natural Monument in 1974. This bitterling is confined to the River Yodo system and its tributaries (Osaka and Kyoto prefectures), rivers in the Noubi Plains (Aichi and Gifu prefectures), and the Toyama Plains (Toyama Prefecture). In the Yodo, it is found in small inlets between the spur dikes, which were constructed 110 years ago to facilitate boat passage. The metropolitan bitterling, *Tanakia tanago* (Cyprinidae), also was designated a National Natural Monument in 1974. It is now confined to small streams near springs and small ponds in the Tochigi, Saitama, and Chiba prefectures, although it seems to have become extinct in the Tokyo and Kanagawa prefectures. The third is the kissing loach, *Leptobotia curta* (Cobitidae, Figure 6), which was designated a National Natural Monument in 1977. It is distributed only in the River Yodo system through the Shiga, Kyoto, and Osaka prefectures and the Yoshii, Asa-

Figure 6. Ayu-modoki, kissing loach, *Leptobotia curta*. Photograph by H. Akiyama.

hi, and Takahashi rivers in Okayama Prefecture. The last species is the stumpy bullhead, *Coreobagrus ichikawai* (Bagridae), designated a National Natural Monument in 1977. Its distribution is confined to the mid-reaches of rivers around the Ise Bay and Mikawa Bay in Mie and Aichi prefectures.

In addition, 11 local populations of 7 species of freshwater fishes have been registered and protected as National Natural Monuments, although these seem to have been nominated more in response to regional or social demands than for biological or ecological reasons (Table 3). Many other populations of freshwater fishes have also been registered and protected as natural monuments at the prefectural, city, town, or village level. Thus, the endangered and protected fishes described above are those confined to restricted areas either because they are local species or subspecies in a biological sense, or because their distribution was reduced by human activities.

Human activities and freshwater fishes

Inland fisheries, fish stocking, and culture of indigenous species

Japanese freshwater fishes, as well as marine ones, have served as an important source of food. For example, the abundant populations of salmon in rivers supported many people during the Jomon Period (ca. 10 000–300 BC) (Ichikawa 1977), and bones of freshwater fishes have been found in ruins of their shell mounds. The oldest Japanese literature, 'Record of ancient matters' (Ko-ji-ki) (712 AD) and 'Chronicles of Japan' (Nihon-shoki) (720 AD), refer to ayu, common carp, and crucian carp together with marine fishes such as sea bream, tuna, perch, and shark.

Among the protocol manuals for the Shinto religion, 'Engi-shiki' (905–927 AD), the eighth volume describes dishes of salmon, trout, ayu, common carp, and crucian carp. The traditional cooking style since the early Heian Period (ca. 800–1200 AD), Shijo-ryu, transmits a typical traditional preparation of common carp called 'Carp at the gate of the dragon' (Ryumon no koi) derived from Chinese culture, even though the Japanese, during the Nara and early Heian periods, generally had inhibitions about eating fishes and mammals because of Buddhist precepts (Yano 1981).

Thus, Japanese freshwater fishes were highly exploited as a source of food. Since 676 AD, on the other hand, they were also stocked into rivers and ponds in order to compensate for their consumption as food, according to the protocol of shrines and temples called 'Houjou-e' (Masutani & Kanaoka 1994). 'Chronicles of Japan' (Nihon-shoki) (720 AD) mentions how the Emperors Tenmu (reign 673–686 AD) and Jitou (690–697 AD) ordered the local people to release living organisms. This protocol has been continued up to the present with people releasing common carp, crucian carp, eel, and soft-shelled turtles. In addition, the aquaculture of common carp in paddy fields has been carried on from at least the early Edo Period (16th–17th centuries) (Matsui 1948).

At present, the harvest of inland water fishes is much less than that of marine fishes. In 1994, the annual harvest of inland water fishes was 138 506 metric tons, but this was only 2% of the total fish harvest in Japan including marine fishes and aquaculture. Among freshwater harvest in 1994, 36% was from rivers (61 388 mt), 18% from lakes (31 122 mt) and 45% from aquaculture facilities (76 579 mt).

In 1984 and 1994 respectively, the major fish species taken from inland waters were ayu (22% and 23% of the annual fish harvest), anadromous salmon and trout (13% and 28%), crucian carp (12% and

7%), and common carp (11% and 8%) (Table 4). The major species produced by aquaculture were eel (40% and 39% of freshwater fish culture yield in 1984 and 1994, respectively), common carp (22% and 18%), rainbow trout (18% and 18%), and ayu (12% and 15%).

To support these inland water fisheries, many eggs and juvenile fishes have been transplanted into natural rivers and lakes (Table 5). From 200 to 250 million juvenile ayu per year were released into rivers between 1973 and 1988. The ayu, *Plecoglossus altivelis altivelis,* is an amphidromous fish, young of which swim upstream to grow in the middle reaches of rivers. The adults spawn downstream in the lower reaches, and the early juveniles move downstream to grow in a littoral marine environment. A population of ayu in Lake Biwa and its tributaries is, however, a fluvial-lacustrine one. In the spring, some fish move from the lake into its tributaries where they graze on the algae attached to stones (amphidromous life history), while the others remain in the lake, feed on zooplankton, and mature

Table 4. Fish landings (in metric tons, mt) from inland water fisheries and fish culture. Data are from the 'Annual report of fisheries and culture (1996)' by the Statistics and Information Department, Japan Ministry of Agriculture, Forestry and Fisheries.

		1984		1994	
		mt	%	mt	%
Inland water fisheries					
Anadromous salmon and trout	Salmon	8639	(12.9)	17460	(27.8)
	Pink salmon	317	(0.5)	3598	(5.7)
	Masu salmon	45	(0.1)	91	(0.1)
	Subtotal	9001	(13.5)	21149	(33.6)
Landlocked salmon and trout	Kokanee	114	(0.2)	83	(0.1)
	Rainbow trout	530	(0.8)	567	(0.9)
	Masu salmon	645	(1.0)	761	(1.2)
	Japanese charr	256	(0.4)	403	(0.6)
	Others	554	(0.8)	703	(1.1)
	Subtotal	2099	(3.1)	2517	(4.0)
Japanese smelt		3872	(5.8)	2082	(3.3)
Ayu		14919	(22.3)	14272	(22.7)
Japanese icefish		1196	(1.8)	1119	(1.8)
Carp		7594	(11.4)	4968	(7.9)
Crucian carp		8034	(12.0)	4402	(7.0)
Japanese dace		5216	(7.8)	3216	(5.1)
Dark chub		3248	(4.9)	1672	(2.7)
Japanese eel		1573	(2.4)	949	(1.5)
Asian pond loach		647	(1.0)	223	(0.4)
Mugilid species		2011	(3.0)	973	(1.5)
Gobiid species		1919	(2.9)	1338	(2.1)
Others		5556	(8.3)	3983	(6.3)
	Total	66885	(100.0)	62863	(100.0)
Fish culture					
Trout	Rainbow trout	16773	(17.5)	13280	(17.6)
	Others	3047	(3.2)	4628	(6.1)
Ayu		11705	(12.2)	11554	(15.3)
Carp		21071	(22.0)	13284	(17.6)
Crucian carp		1492	(1.6)	1014	(1.3)
Eel		38030	(39.6)	29431	(38.9)
Tilapia		3544	(3.7)	2125	(2.8)
Others		280	(0.3)	327	(0.4)
	Total	95942	(100.0)	75643	(100.0)

at a smaller size (lacustrine life history) (Miyadi 1960).

In 1913, C. Ishikawa discovered that when juveniles of the lacustrine ayu population in Lake Biwa were transplanted to other rivers they grew as large as amphidromous ayu (Ishikawa 1932, Miyadi 1960). Inasmuch as juvenile ayu in the lake are easy to catch in large quantities, the transplantation of ayu from Lake Biwa to other Japanese freshwaters has become well established. Moreover, because the ayu in Lake Biwa are more aggressive in guarding their territories than those from any other area (Kawanabe 1976, 1977), they are favored as a game fish and recently have been transplanted more for sport fishing than for the commercial fishery.

Ayu in Lake Biwa are genetically different from other populations in Japan (Nishida 1985, Tsujimura & Taniguchi 1995). However, because Lake Biwa ayu reproduce earlier than sea-run populations, transplanting ayu from Lake Biwa does not seem to cause any genetic contamination of the sea-run populations (Nishida 1985). Furthermore, because Lake Biwa ayu show only a weak tolerance to marine salinity, they are unlikely to spread to other freshwaters accessible only via the ocean (Azuma 1980). On the other hand, the transplantation of ayu from Lake Biwa has resulted in the chance introduction of other fishes such as *Opsariichthys uncrostris uncrostris* and *Zacco platypus* which are captured with the juvenile ayu (Sunaga 1980, Mori & Nagoshi in Kawanabe & Mizuno 1989).

Table 5. Fish stocking for Japanese inland fisheries. Data are from the 'Fisheries Census in Japan' by the Statistics and Survey Division, Ministry of Agriculture and Forestry, Government of Japan. (+ = fewer than 1 million eggs were transplanted).

	1973	1978	1983	1988
Young of fishes (millions of indiv.)				
Anadromous salmon and trout	47	113	489	766
Landlocked salmon and trout	20	130	31	59
Rainbow trout	–	–	–	20
Red-spotted masu salmon	–	–	–	13
Masu salmon	–	–	–	13
Japanese charr	–	–	–	6
Others	–	–	–	8
Ayu	230	204	219	251
Common carp	43	64	26	29
Crucian carp	23	74	16	24
Eel	10	16	10	7
Other fishes	469	960	113	498
Total	841	1561	905	1633
Eggs of fishes (millions of eggs)				
Anadromous salmon and trout	17	–	–	–
Landlocked salmon and trout	+	–	–	–
Ayu	341	–	–	–
Common carp	–	–	–	–
Crucian carp	13	–	–	–
Japanese smelt	–	–	8679	17886
Other fishes	16021	–	–	1009
Total	16392	86436	8679	18894

Another fish that is becoming popular for sport fishing is the deepbodied crucian carp which is a variety developed from the *Carassius cuvieri* endemic to Lake Biwa. This species lives in the offshore water of the lake and feeds on phytoplankton (Tokui & Kawanabe 1984), so it is transplanted into artificial ponds and reservoirs with open waters. The fish is both easy for beginners to catch and of angling interest to veterans, so it has been introduced all over Japan and has often hybridized with the local crucian carps.

The number of stocked juveniles of anadromous salmon and trout, mainly the chum salmon, *Oncorhynchus keta*, has greatly increased, from 47 million fish in 1973 to 766 million in 1988. Since 1877, transplanting salmon and trout has been recommended as public projects for promotion of inland fisheries by national and local governments. In addition to indigenous salmonids, exotic ones such as the rainbow trout, *Oncorhynchus mykiss,* have been stocked widely in natural rivers and lakes (see below).

The number of transplanted eggs of the Japanese smelt, *Hypomesus transpacificus niponensis,* has increased markedly, to nearly 18 billion in 1988. The original purpose of this stocking was to improve the inland fisheries but it now has changed to serve for sport fishing.

Japanese freshwater fishes have also played a vital role in the hobby aquarium trade. The first record of this pasttime can be found in 'Nihon-shoki' (Chronicles of Japan), when people in the 7th century enjoyed keeping common carp in garden ponds. Although the colored carp 'nishikigoi' are now popular throughout the world, these color variants of the common carp appeared only in later years (Balon 1995). The primitive forms of colored carp with orange, white, and dark-blue colors are mutant forms of the common carp that first appeared in Niigata Prefecture during the Bunka and Bunsei eras (1804–1830) (Kuroki 1986, Kataoka 1989), Kataoka (1989) suggested that these forms were produced from wild common carp, because Suzuki (1836–1842) reported that common carp, ayu, trout, and salmon were common in the River Uono, a tributary of the Shinano. The white and red form, 'kouhaku', was claimed to have been crossbred during the Tempou Era (1830–1843) and treated as one of the important local products from the domain of Nagaoka (now in Niigata Prefecture) (Kuroki 1986). The beautiful colored carp, nishikigoi, of today were produced by also crossbreeding with the German common carp by the people of Ojiya City, Nagaoka City, and Yamakoshi Village in Niigata Prefecture during the late-Meiji and Taisho periods (ca. 1910–1925) (Kuroki 1986). More beautiful colored strains were developed after World War II from selected strains which were saved from consumption as food during the acute food shortages of the war (personal communication with residents of Yamakoshi Village in 1994).

Fancy goldfish are another renowned home aquarium fish that was developed in Japan. The ancestor of the goldfish is considered to have been a red form of *Carassius auratus auratus* (Balon 1995) in China during the Chinese Jin Dynasty (265–419 AD), which was introduced into Japan in 1502 (Ojima 1988, Ojima & Takai 1995). Since then, many different forms of goldfish, like the 'wakin' and 'ranchuu', have been produced in Japan, as described by Matsui (1935, 1971).

Introduction of exotic species

Many species of foreign freshwater fishes have been introduced into Japan. The records of introduction of freshwater fishes after 1877 have been reviewed in detail by Maruyama et al. (1987). The introduction of 103 species and subspecies of freshwater fishes into Japan has been recorded, excluding those introduced only for hobby aquaria, but including those accidentally introduced into the natural environment as a result. These belong mainly to the Cyprinidae (33 species and subspecies), Salmonidae (18), Cichlidae (11), Acipenseridae (9), Anguillidae (5), Centrarchidae (5), and some others. They include fishes important for inland fisheries, and some have been intensively used in attempts to stock ponds, lakes, and rivers. As a result, at least 23 species and subspecies are now established in natural freshwaters in Japan. These belong to the Salmonidae (5 species and subspecies), Cyprinidae (5), Channidae (3), Cichlidae (3), Poeciliidae (2), Centrarchidae (2), Atherinidae (1), Clariidae (1), and Belontiidae (1).

Suzuki (1992) has categorized the various reasons for fish introduction: (1) to promote commercial fishing from the 1880s to the 1940s, (2) to increase food/protein resources during the 1940s, and (3) for culture or multiplication of fish stocks after the 1950s. The purpose of introduction of mosquitofish, *Gambusia affinis*, was unique, that is, to control mosquito populations. In recent years, the importation of aquarium fishes has increased sharply, which probably increases the chance of escape of exotic aquarium fishes. In addition, stocking fishes for sport angling also has increased, as sport fishing has become a booming leisure activity. Stocked fishes include indigenous species such as ayu and trout but also exotic species such as the rainbow trout, *Oncorhynchus mykiss,* and largemouth bass, *Micropterus salmoides.* The largemouth bass has become a predatory piscine pest that has likely destroyed native fish communities in may Japanese freshwaters (Ohtani & Kikukawa 1980).

The introduction of subspecies or other genetically different strains often pollutes the genome of the indigenous form of the same species by hybridization. A typical case is the interaction between the Japanese rose bitterling, *Rhodeus ocellatus kurumeus,* and an exotic rose bitterling, *R. ocellatus ocellatus* (Nagata et al. 1996). The latter subspecies was introduced accidentally with the 1942 introduction of the grass carp, *Ctenopharyngodon idellus,* from China. In later years, it was introduced probably for aquarium purposes, and became widespread in natural habitats of the indigenous Japanese rose bitterling. As a result of hybridization between the two subspecies or replacement of the native subspecies by exotic one, the indigenous subspecies now faces extinction (Nagata 1980, Nagata et al. 1969).

Another case of widespread introduction concerns the common carp. At least nine subspecies or forms of this species have been introduced into Japan from Indonesia, Taiwan, Israel, and Germany (Maruyama et al. 1987). Hitherto, the common carp in Japan was thought to be the form brought with the technique of carp-culture in paddy fields as one of the tools of rice farming, and Jordan & Fowler (1903) claimed that the common carp in Japan was an introduced species (see also Balon 1995). However, a native species of common carp might have existed in Japan because many fossil species of *Cyprinus* have been found from Tertiary formations in the Fukui, Gifu, Shiga, and Nagasaki prefectures (Hosoya, in Kawanabe & Mizuno 1989). Nakajima (1994) demonstrated the process of evolution in species of *Cyprinus* from ancient Lake Biwa (ca. 3 million years ago) to recent Lake Biwa, based on the fossil record. Pharyngeal teeth of the common carp were found in ruins of shell mounds from the early Jomon Period (ca. 10 000–6000 BC) in Ibaragi Prefecture (Kafuku 1980) which was long before rice culture appeared in Japan. Based on this evidence, some form of common carp probably existed in Japan, but a detailed comparative analysis with the common carp in recent Japan has yet to be carried out.

Lake fish fauna

There are numerous lakes, created by volcanic and tectonic activity, especially in the middle and eastern mountainous areas of Japan. Originally, few fishes inhabited such lakes because they were isolated from rivers by falls, but various species, especially salmonids and cyprinids, have been introduced to provide inland fisheries.

A typical example is Lake Chuzenji in Tochigi Prefecture. The lake is situated 1271 m above sea level and has a surface area of 11.6 km^2 and a maximum depth of 161.5 m. The Kegon Fall (97 m) inhibited upstream migration of fishes from the River Daiya. Tanaka (1936) mentioned that there had been no fish in the lake. However, Tochigi Prefecture (1980, 1993) recently reported more than 17 species and subspecies of fishes from the lake. Of course, these fishes are considered all to be introduced species, such as the Japanese eel, *Anguilla japonica,* Japanese charr, *Salvelinus leucomaenis,* brook charr, *S. fontinalis*, lake charr, *S. namaycush,* brown trout, *Salmo trutta,* rainbow trout and steelhead, *Oncorhynchus mykiss,* Biwa trout, *O. masou subsp.,* kokanee, *O. nerka nerka*, Japanese smelt, *Hypomesus transpacificus nipponensis*, Japanese dace, *Tribolodon hakonensis,* common carp, *Cyprinus carpio*, silver crucian carp, *Carassius auratus langsdorfi*, a species of cobited loach, *Cobitis* sp., a cottid, *Cottus* sp., and the common freshwater goby, *Rhinogobius* sp.

Another example is Lake Ashinoko in Kanagawa Prefecture. It is 724.5 m above sea level with a surface area of 7.0 km^2 and a maximum depth of 40.6 m. Ishihara et al. (1986) presented a list of fishes in the lake (Table 6), including eight species thought to be indigenous, five introduced Japanese ones, three introduced exotic ones, and nine which were probably accidentally introduced without any commercial purpose. The year of introduction was included when known and records of unsuccessful introductions were also listed.

Fishes in Lake Biwa

Lake Biwa is the largest lake in Japan with a maximum depth of 104 m. The catchment area is 3848 km^2 with about 120 rivers flowing into the lake, one natural outlet, the River Seta, flowing into Osaka Bay and one artificial outlet, the Kyoto Canal, constructed in 1890 to provide water for neighboring Kyoto City.

There are two outstanding characteristics of the fish fauna in the lake; first, the large number of species, more than in any other lake in Japan, and second the high number of endemic species and subspecies (Ueno 1984, Kawanabe 1996). At present, there are 71 species and subspecies of freshwater fishes found in Lake Biwa and its tributaries (Table 7), which represents more than half of the primary freshwater fish fauna of Japan. More than half of the fishes of Lake Biwa belong to the family Cyprinidae (37 species and subspecies), and the rest to the Cobitidae (6), Gobiidae (6), Salmonidae (5), Siluridae (3), and others. This accounting includes nine exotic species and subspecies and four introduced species from other Japanese waters, but excludes three extinct species (the deepbodied bitterling, the Japanese rose bitterling, and the kissing loach). Also excluded from the list are species that seem to have appeared accidentally, e.g., having escaped from private aquaria.

The greater part of the Lake Biwa fauna is of boreal origin, believed to be derived from the Eurasian freshwater fauna and some brackish water and marine stocks (Ueno 1984). Fishes like the piscivorous chub, *Opsariichthys uncirostris uncirostris,* and the lakeweed chub, *Ischikauia steenackeri,* are considered to be relics of an old freshwater fauna that was once distributed widely in Eurasia before the most recent ice age.

There are 13 species and subspecies of fishes endemic to lake Biwa (Table 7). These represent 22% of the 58 species and subspecies indigenous to the lake and 15% of the fishes endemic to Japan. The ratio of endemism is the highest of any lake in Japan. In Lake Biwa there is also an endemic genus or subgenus of snail, *Biwamelania,* which consists of at

Table 6. Fish fauna in Lake Ashinoko (after Ishihara et al. 1986). Scientific names follow Nakabo (1993). ◎= successful introduction, × = unsuccessful introduction, △ = uncertain record.

Taxon		Japanese name	Native indigenous fish	Introduced Japanese fish	Introduced exotic fish	Accidentally introduced fish	Year of introduction
1 Anguillidae	*Anguilla japonica*	Unagi	△	×		△	1919
2 Cyprinidae	*Cyprinus carpio*	Koi	△	◎			1938
3 Cyprinidae	*Tribolodon hakonensis*	Ugui	△	◎			1981
4 Salmonidae	*Oncorhynchus masou masou*	Yamame	△	◎			1975
5 Salmonidae	*Oncorhynchus masou ishikawae*	Amago	△	◎			1985
6 Siluridae	*Silurus asotus*	Namazu	△			△	
7 Gobiidae	*Rhinogobius* sp.	Yoshinobori sp.	△				
8 Gobiidae	*Chaenogobius urotaenia*	Uki-gori	△				
9 Cyprinidae	*Carassius auratus langsdorfi*	Gin-buna		◎			1940
10 Cyprinidae	*Carrasius cuvieri*	Gengorou-buna		◎			1967
11 Osmeridae	*Hypomesus transpacificus nipponensis*	Wakasagi		◎			1918
12 Salmonidae	*Oncorhynchus masou* subsp.	Hon-masu		◎			1880
13 Salmonidae	*Oncorhynchus nerka nerka*	Hime-masu		◎			1909
14 Osmeridae	*Plecoglossus altivelis altivelis*	Ayu		×			1940
15 Salmonidae	*Salvelinus leucomaenis* subsp.	Iwana		×		◎	1889
16 Salmonidae	*Oncorhynchus masou* subsp.	Biwa-masu		×			1907
17 Salmonidae	*Oncorhynchus nerka nerka*	Beni-masu		×			1928
18 Salmonidae	*Oncorhynchus keta*	Sake		×			1938
19 Salmonidae	*Salmo trutta*	Brown trout			◎		1972
20 Salmonidae	*Oncorhynchus mykiss*	Niji-masu			◎		1910
21 Centrarchidae	*Micropterus salmoides*	Ookuchi-bass			◎		1925
22 Salmonidae	*Salvelinus fontinalis*	Kawa-masu			×		1928
23 Atherinidae	*Odonthestes bonariensis*	Peherei			×		1974
24 Cyprinidae	*Carassius auratus* subsp. 2	Kin-buna				◎	
25 Cyprinidae	*Zacco platypus*	Oikawa				◎	1977
26 Centrarchidae	*Lepomis macrochirus*	Bluegill				◎	1984
27 Cyprinidae	*Sarcocheilichthys variegatus* subsp.	Higai sp.				△	
28 Cyprinidae	*Pseudorasbora parva*	Motsugo				△	
29 Cyprinidae	*Phoxinus lagowski steindachneri*	Abura-haya				△	
30 Cobitididae	*Misgurnus anguillicaudatus*	Dojou				△	
31 Gobiidae	*Tridentiger obscurus*?	Chichibu?				△	
32 Cyprinidae	*Rhodeus ocellatus ocellatus*	Tairiku-bara-tanago				×	1977

Table 7. Fishes in Lake Biwa and its tributaries. Scientific names follow Nakabo (1993). Naka (1991) shows the year of introduction of Japanese and alien fishes. Symbols referring to origin of taxa are defined in the footnote. Our assessment of the origin of each species is shown by symbols to the left of scientific names.

Taxon			Japanese name	Mori (1970)	Mori & Miura (1980)	Tokui & Kawanabe (1984)	Kawanabe & Mizuno (1989)	Biwako bunkakan (1991)	Naka (1991)	Nakabo (1993)
Petromyzontidae										
○	1	*Lethenteron reissneri*	Suna-yatsume	○	○	○		○	○	
Anguillidae										
○	2	*Anguilla japonica*	Unagi	○	○	○		○	○	
Cyprinidae										
○	3	*Zacco temminckii*	Kawamutsu B							
○	4	*Zacco* sp.	Kawamutsu A	○	○	○		○	○	
○	5	*Zacco platypus*	Oikawa	○	○	○	○	○	○	
○	6	*Opsariichthys uncirostris uncirostris*	Hasu	★	★	○	○	○	○	○
○	7	*Hemigrammocypris rasborella*	Kawabata-moroko	○	○	○	○	○	○	
○	8	*Tribolodon hakonensis*	Ugui	○	○	○		○	○	
○	9	*Phoxinus lagowski steindachneri*	Abura-haya	○	○	○	○	○		
○	10	*Phoxinus oxycephalus jouvi*	Taka-haya	○	○	○	○	○		
+	11	*Ctenopharyngodon idellus*	Sougyo	○	○	+	○	○	+ 1916	
+	12	*Mylopharyngodon piceus*	Ao-uo					○	+ 1956?	
★	13	*Ischikauia steenackeri*	Wataka	★	★	○	★	○	★	★
+	14	*Hypophthalmichthys molitrix*	Hakuren			○		○	+ 1957?	
○	15	*Gnathopogon elongatus elongatus*	Ta-moroko	○	○	○	○	○	○	
★	16	*Gnathopogon caerulescens*	Hon-moroko	★	★	★	★	○	★	★
○	17	*Pungtungia herzi*	Mugitsuku	○	○	○	○	○		
○	18	*Pseudorasbora parva*	Motsugo	○	○	○	○	○	○	
★	19	*Sarcocheilichthys variegatus microoculus*	Biwa-higai	○	○	○	★	○	★	★
★	20	*Sarcocheilichthys biwaensis*	Abura-higai				★	○	★	★
○	21	*Pseudogobio esocinus esocinus*	Kamatsuka	○	○	○		○	○	
△	22	*Abbottina rivularis*	Tsuchifuki	○	○	◎		○	△1937?	
○	23	*Biwia zezera*	Zezera	○	○	○	○	○	○	○
★	24	*Squalidus chankaensis biwae*	Sugo-moroko	○	○	○	★	○	○	★
○	25	*Squalidus japonicus japonicus*	Deme-moroko	○	○	○	○	○	○	○
○	26	*Squalidus gracilis gracilis*	Ito-moroko			○	○	○		
○	27	*Hemibarbus barbus*	Nigoi	○	○	○	○	○	○	
○	28	*Hemibarbus labeo*	Kourai-nigoi							
○	29	*Hemibarbus longirostris*	Zunaga-nigoi			○	○	○		
○	30	*Cyprinus carpio*	Koi	○	○	○	○	○	○	○
★	31	*Carassius auratus grandoculis*	Nigoro-buna	○	○	★	★	○	★	★
○	32	*Carassius auratus langsdorfi*	Gin-buna	○	○	○	○	○	○	
★	33	*Carrasius cuvieri*	Gengorou-buna	★	★	★	★	○	★	★
○	34	*Tanakia lanceolata*	Yari-tanago	○	○	○	○	○	○	
○	35	*Tanakia limbata*	Abura-bote	○	○	○	○	○	○	
+	36	*Rhodeus ocellatus ocellatus*	Tairiku-bara-tanago		○	+	○	○	+1960s	
●?	37	*Rhodeus ocellatus kurumeus*	Nippon-bara-tanago	○	○		○	○	○	○
○	38	*Acheilognathus cyanostigma*	Ichimonji-tanago	○	○	○	○	○	○	
○	39	*Acheilognathus tabira tabira*	Shiro-hire-tabira	○	○	○	○	○	○	○
○	40	*Acheilognathus rhombeus*	Kanehira	○	○	○	○	○	○	
●	41	*Acheilognathus longipinnis*	Ita-sen-para							
Cobitidae										
●	42	*Leptobotia curta*	Ayu-modoki	○	○	○	○	○	○	○
○	43	*Misgurnus anguillicaudatus*	Dojou	○	○	○		○	○	
★	44	*Cobitis* sp. 1	Suji-sima-dojou (large)	○	○	○	★	○	?	★
★	45	*Cobitis* sp.2 subsp. 4	Suji-sima-dojou (small-Biwako)				★		?	★
○	46	*Cobitis biwae*	Shima-dojou	○	○	○	○	○		
○	47	*Niwaella delicata*	Ajime-dojou			○	○	○		○
○	48	*Lefua echigonia*	Hotoke-dojou			○		○		

Table 7. Continued.

Taxon	Japanese name	Mori (1970)	Mori & Miura (1980)	Tokui & Kawanabe (1984)	Kawanabe & Mizuno (1989)	Biwako bunkakan (1991)	Naka (1991)	Nakabo (1993)
Bagridae								
○ 49 *Pelteobagrus nudiceps*	Gigi	○	○	○	○	○	○	
Siluridae								
○ 50 *Silurus asotus*	Namazu	○	○	○	○	○	○	
★ 51 *Silurus biwaensis*	Biwako-oo-namazu	★	★	★	★	○	★	★
★ 52 *Silurus lithophilus*	Iwatoko-namazu	★	★	★	★	○	★	★
Amblycipitidae								
○ 53 *Liobagrus reini*	Akaza	○	○	○		○		
Osmeridae								
△ 54 *Hypomesus transpacificus nipponensis*	Wakasagi	○	○	○		○	△1910	
Plecoglossidae								
○ 55 *Plecoglossus altivelis altivelis*	Ayu	○	○	○	○	○	○	
Salmonidae								
○ 56 *Salvelinus leucomaenis japonicus*	Yamato-iwana			○	○	○		○
+ 57 *Oncorhynchus mykiss*	Niji-masu	○	○	+		○	+1911	
△ 58 *Oncorhynchus masou masou*	Yamame					○		
○ 59 *Oncorhynchus masou ishikawae*	Amago	○	○		○	○	○	
★ 60 *Oncorhynchus masou* subsp.	Biwa-masu	★	★	○	○	★	★	★
Adrianichthyidae								
○ 61 *Oryzias latipes*	Medaka	○	○	○		○	○	
Gasterosteidae								
○ 62 *Gasterosteus microcephalus*	Hariyo	○	○	○	○	○		○
Cottidae								
○ 63 *Cottus pollux*	Kajika	○	○	○		○		
○ 64 *Cottus reinii*	Utsusemi-kajika			○	★	○	★	
Centrarchidae								
+ 65 *Micropterus salmoides*	Ookuchi-bass		○	+	○	○	+1974	
+ 66 *Lepomis macrochirus*	Bluegill		○	+	○	○	+1965	
Cichlidae								
+ 67 *Tilapia zillii*	Zilli tilapia							
Gobiidae								
○ 68 *Odontobutis obscura obscura*	Donko	○	○	○	○	○		
○ 69 *Rhinogobius* sp. OR	Tou-yoshinobori	○	○	○	○	○	○	
○ 70 *Rhinogobius flumineus*	Kawa-yoshinobori			○		○		
△ 71 *Tridentiger brevispinis*	Numa-chichibu						△1989	
★ 72 *Chaenogobius isaza*	Isaza	★	★	★	★	○	★	★
○ 73 *Chaenogobius urotaenia*	Uki-gori	○	○	○	○	○	○	
Channidae								
+ 74 *Channa argus*	Kamuruchii	○	○	+	○	○	+1933	

★ = species or subspecies endemic to Lake Biwa.
○ = indigenous species or subspecies of Lake Biwa.
● = extinct species or subspecies.
△ = introduced Japanese species or subspecies.
+ = exotic species or subspecies.

least 14 nominal endemic species (Watanabe & Nishino 1995). Such high endemism in the animals of Lake Biwa is evidence that the current lake is ancient, perhaps as old as 300 000 to 500 000 years (Horie 1984, Yokoyama 1995, Kawanabe 1996). However, also there was an old lake Biwa in the Ueno Basin in Mie Prefecture 4 to 6 million years ago.

The endemic species and subspecies of fishes in Lake Biwa contain those living in offshore water of the lake, such as the piscivorous chub, the Biwa gudgeon, *Gnathopogon caerulescens,* the deepbodied crucian carp, *Carassius cuvieri,* the Biwa catfish, *Silurus biwaensis,* the Biwa salmon, and the Biwa goby, *Chaenogobius isaza*. No other fish is found in

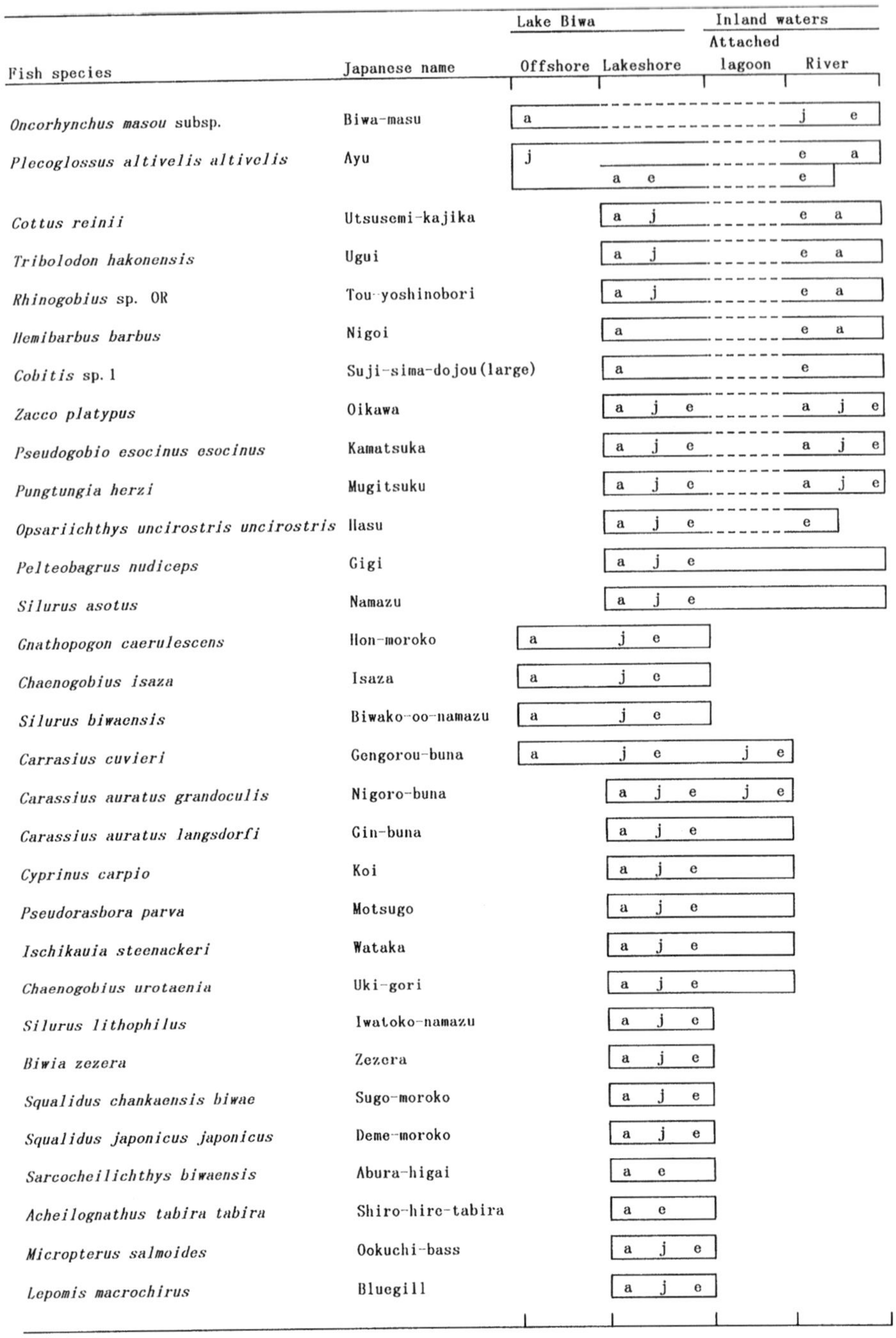

Figure 7. Habitat utilization by fishes in Lake Biwa: a – adults, j – juveniles, e – embryos. After information in Kawanabe & Mizuno (1989), Biwako Bunkakan (1990) and Naka (1991).

offshore water of the lake except the ayu, which has a unique subpopulation adapted to offshore waters of the lake (see above). This suggests that these fishes speciated in response to the unexploited habitat offshore (Tokui & Kawanabe 1984, Kawanabe 1996).

However, all fishes in Lake Biwa, including those living offshore, need the shallow littoral zone to reproduce (Figure 7). For example, numerous individuals of the deepbodied crucian carp, the round crucian carp, *C. auratus grandoculis* and the common carp, are known to migrate to the lakeshore and associated lagoons from spring to early summer for reproduction (e.g. Kawanabe 1996). The spawning behavior of Lake Biwa fishes can be categorized into four types: those attaching eggs to weeds near the surface as do common carp, crucian carps, and gudgeons; those attaching eggs to the undersides of stones as do gobiids and cottids; those scattering eggs on the bottom or covering them over as do silurids, chubs, ayu, and salmonids; and those inserting the eggs inside mussels as do bitterlings. No Lake Biwa fish spawns in open water or on the deep bottom, and all spawning takes place in shallow waters. Many fish species also migrate into the tributaries to reproduce and their eggs and juveniles develop there (Figure 7).

People, however, have disturbed the lake's environment to satisfy the demands of an increasing population with its need for more cultivated fields. For example, they have reconstructed the lakeshore and reclaimed the estuaries and lagoons connected with Lake Biwa, reducing the area of the lagoons from about 3440 ha before 1944 to about 570 ha in 1991,[1,2] and the area of reeds beds along the lakeshore from 260 ha in 1953 to 130 ha in 1991 (Itoh 1992). These areas served as spawning and nursery sites for Lake Biwa fishes (Kawanabe 1996). The Amagase Dam was constructed for hydropower on the lower reach of the River Seta (Uji City, Kyoto Prefecture) in 1964, and this completely blocked upstream migration of catadromous fishes like the Japanese eel and the ayu. In 1905, the Seta Weir was constructed 5 km downstream from the outlet of Lake Biwa to control the water level of the lake for flood prevention and water supply for downstream cities. It was rebuilt in 1961. However, water level control is disadvantageous for fishes because the lake is kept at low level during the rainy season when the fishes migrate into the shallow shores and adjoining waters to spawn.

The environmental alteration of the lake and surrounding areas has increased since 1972 due to a national comprehensive development project for Lake Biwa. A significant change in the aquatic environment resulted from installing a drinking-water system for more than 90% of the houses around the lake in 1975. Previously, in 1955 only 5% had been so provided (Kada 1995). Dams were constructed in the lake tributaries to provide drinking-water, and the number and area of artificial ponds decreased. The development of this water supply system also reduced the need for water in small canals and ditches that had played an important role in the daily lives of the people, so these were covered over or even diverted. Water systems around paddies were also made more effective by using concrete bottoms and water gates. Many steps were also constructed in the river beds to moderate their gradient.

These changes in the lake and its tributaries threaten the freshwater organisms, especially fishes that migrate between the lake and inland waters. For example, the Biwa salmon has been forced to spawn in rivermouths because its upstream migration was interrupted by weirs and steps (Y. Fujioka personal communication), and Shiga Prefecture was forced to construct artificial streams to provide spawning sites for the ayu.

Some changes in the fish community of Lake Biwa can be surmised from a study of the fish harvested from the lake. Although fishery activities have gradually decreased, the catch of ayu has increased, and the catches of oikawa, hon-moroko, and hasu have remained unchanged (Table 8). Since 1985, catch of the exotic largemouth bass has increased. The species was discovered in the lake in 1964 and eight years later it had spread over the entire lake (Biwako Bunkakan 1991). The population of the Bi-

1 [1] Lake Institute, Shiga University (ed.) 1974. Lake Biwa II. The future development. Sankyo Shuppan, Tokyo. 190 pp. (in Japanese).

2 [2] Shiga Prefecture. 1993. Report on lake environment at the 4th basic survey for conservation of natural environment. Environment Agency of Japan 25: 1–124 (in Japanese).

wa goby (isaza) has been known to fluctuate with the population of ayu, which competes with it for food, both being zooplankton-feeders in the offshore water (Miura 1996). Meanwhile, the catch of crucian carps (funa) decreased abruptly after 1985, especially of the nigoro-buna which is an important rural food fish. In the inland waters around Lake Biwa, fishes such as eel, chub, bitterling, loach, and catfish were reported to have decreased during the last 20–30 years (Yuma et al. 1995).

Another problem faced by the lake fish fauna is the introduction of exotic species; as of now, there are 14 exotic species and subspecies in Lake Biwa. Fishermen have paid much attention to the predatory largemouth bass, because of its effect on the populations of commercially important fishes like the ayu. As mentioned earlier, many attempts have been made to introduce species of fishes into Lake Biwa. According to Furukawa & Awano (1969), since 1883, 24 potentially important fishes have been introduced into the lake, but only the rainbow trout and the grass carp have become established.

In addition to these two deliberately introduced species, some fishes have been accidentally but successfully introduced into the lake. One is the bluegill, *Lepomis macrochirus*. This species was cultured in ponds to serve as a host fish for the glochidia of the Biwa pearly mussel, *Hyriopsis schlegeli*. It appeared in the lake unnoticed in mid 1960s and spread throughout the lake in the early 1970s (Furukawa & Awano 1969, Terashima 1980). Terashima (1980) suggested that the bluegill occupied a vacant niche in the lake fish community due to its unique breeding habit and wide range of prey. Another is the short-spined Japanese trident goby. *Tridentiger brevispinis,* which became common in Lake Biwa after 1989 (Takahashi 1990, Ohno 1994). Ohno (1994) suggested that the algal feeding habit of this

Table 8. Fish harvest (annual mean, metric tons) in Lake Biwa. Data are from the Shiga Statistics Information Office, Bureau of Agriculture and Fisheries of Kinki District. Japanese names for fishes are employed (see Table 7 and Appendix) because the classification in fisheries literature is often not consistent with the accepted scientific one.

	1954–1955	1959–1963	1969–1974	1975–1979	1980–1984	1985–1989	1990–1994
Masu-salmon	53.4	36.0	32.9	16.4	25.6	25.0	32.4
Ayu	377.7	327.7	842.2	1050.0	1285.0	1614.6	1580.0
juveniles	–	(327.7)	(527.3)	(614.2)	(699.1)	(920.8)	(921.8)
subadults	–	–	(314.9)	(435.8)	(585.9)	(693.8)	(658.1)
Koi	80.5	92.2	166.3	162.7	132.7	86.2	62.4
Funa-group	653.9	810.3	616.2	628.8	702.0	476.5	167.6
Nigoro-buna	–	–	–	–	–	177.6	82.0
Ugui	33.0	27.3	53.6	60.6	56.7	27.6	17.9
Oikawa	10.9	15.0	30.8	39.3	48.7	31.8	31.8
Unagi	27.0	23.6	29.3	10.5	7.9	8.4	9.0
Isaza	24.7	113.7	420.1	397.4	422.4	293.4	21.1
Moroko-group	–	297.2	404.3	227.2	271.5	291.7	365.9
Hon-moroko	–	(178.6)	(330.0)	(200.8)	(222.8)	(242.0)	(223.1)
Hasu	–	75.8	139.4	106.4	110.1	102.7	100.8
Other fishes	732.0	375.2	512.5	397.2	256.2	361.8	330.4
Wataka	–	(47.8)	–	–	–	–	–
Gori (gobiids)	–	(99.7)	–	–	–	–	–
Gigi	–	(19.2)	–	–	–	–	–
Higai-group	–	(3.6)	–	–	–	–	–
Largemouth bass	–	–	–	–	–	(76.5)	(88.7)
Total	1993	2194	3248	3097	3319	3320	2719
No. of fisherman	–	1704	1536	1597	1377	1242	1113
No. of fishing boats	–	1961	2058	2272	2167	2006	1750
Cumulative days of fishing	–	211939	209967	227975	185619	180013	156134

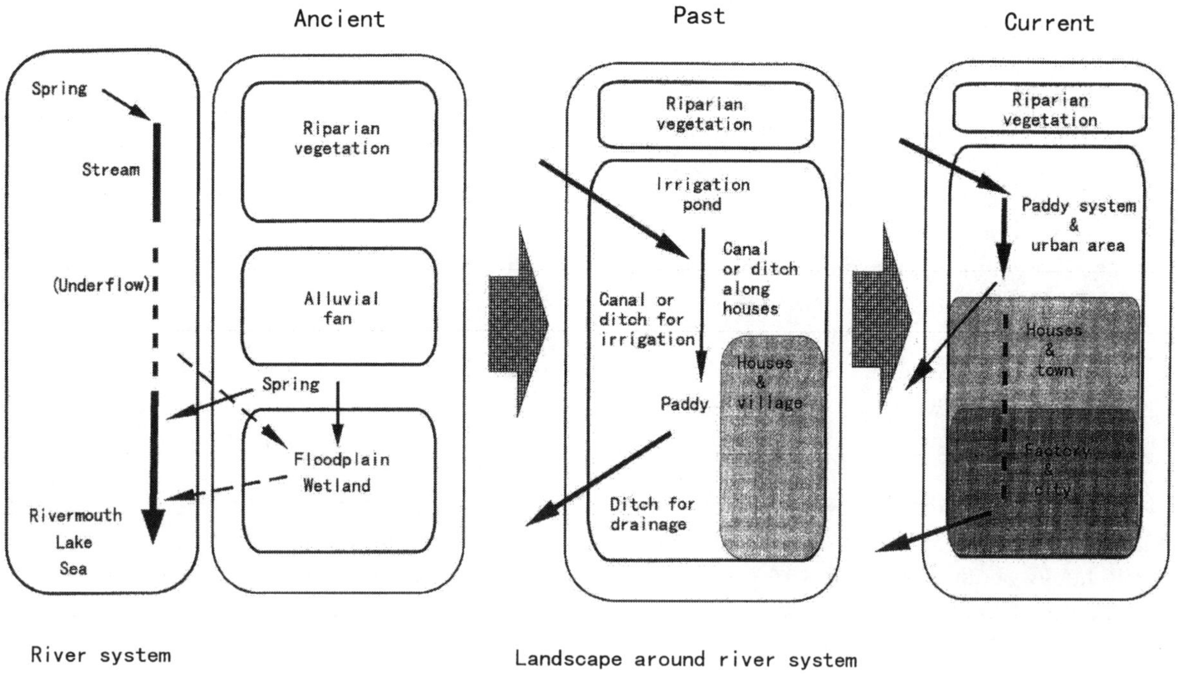

Figure 8. A schematic of changes in landscapes and water systems in Japan from ancient times to the present.

species might have helped it become established in the fish community along the lakeshore where no fish used the attached algae. It is noteworthy that a cichlid, *Tilapia zillii*, is confined to a restricted area of a lake tributary polluted with warm water from a factory. More recently, the Japanese smelt was found alive in the lake. This species had been introduced repeatedly from 1910 to 1919 and from 1939 to 1953 (Furukawa & Awano 1969), but only a few fish were ever found. However, its population increased abruptly from 1994 (The Kyoto Newspaper 17.3.1996, an interview in Kohoku Town), despite the fact that no more introductions of this species had been carried out since 1954. This may suggest some unknown changes are taking place in the lake ecosystem.

Development of paddy-fields and the distribution of fishes

To understand the distributions of Japanese freshwater fishes, we need information on the geological influences on migration and isolation of organisms as well as the patterns of ecological and physiological adaptation. We also have to understand changes in the natural environment during historical times because humans have greatly modified the landscape, including water systems. Especially in Japan, the role of rice paddies is very important for understanding the current distributions of freshwater organisms because the Japanese have operated artificial water systems for irrigation on a huge scale. Such artificial water systems included canals, ditches, and ponds near the paddies, and similar systems developed around each family's house for daily use.

In Japan, paddy fields occupied a large part of the land, with a maximum area of 3.4 million ha in 1969 and 2.8 million ha in 1990. This is more than twice the surface area of natural freshwater lakes and rivers, which accounted for 1.3 million ha in 1989.[3] Paddy field cultivation started as long ago as the Jomon Period (ca. 1000 BC) (Kubodera 1993). The area of paddy fields was estimated to reach 1 million ha during the Nara Period (610–784 AD), and was en-

3 [3] Yano Kataro Memorial Foundation (ed.) 1991. A hundred years of Japan, showing by numerals, 3rd ed. Kokuseisha, Tokyo. 542 pp. (in Japanese).

larged during the Kamakura and early Edo periods and the early Meiji Period (Furushima 1967).

Paddy fields were developed on valley slopes at the foot of the mountains, and on the plains surrounding the estuary near the sea. They replaced natural environments like riparian forests in the valleys, the dry land on alluvial fans near the foothills, and the plains and marshes around estuaries (Figure 8). Paddy fields were constructed on alluvial fans by diverting water from mountain streams and irrigation ponds, changing the dry land into artificial wetlands. Surface waters had originally been scarce on alluvial fans because they were well drained. Deep marshes around the lower reaches of rivers were changed into shallow wetlands by eliminating water from the marshes. As a result, the relatively homogenous environment of paddy fields with their artificial water systems became continuous from the estuaries all the way to the foothills, including the area that was once dry land on alluvial fans and plains.

Following these environmental changes, freshwater organisms inhabiting shallow waters expanded their distribution through the paddy fields and artificial water systems, while freshwater organisms in deeper waters likely decreased or become extinct. To analyze and understand the current distribution of freshwater organisms in Japan, it is essential to remember that some species living in the lower reaches or rivers have expanded their distributions through the alluvial fans up to the foothills (Yuma 1995).

In mountains, small, rapid streams often have been changed into significantly slower ones by constructing small weirs for irrigation and flood control. This may cause organisms adapted to rapid streams to move into more mountainous areas, and result in their replacement by organisms adapted to slower streams.

Recent land development creating towns, cities, and factories has decreased the paddy fields and artificial water systems, especially in the lower and middle reaches of rivers, and altered the land use from wet to dry. Such human activity forced freshwater organisms out of artificial wetlands around the lower and middle reaches of rivers, and may cause the loss of freshwater organisms, especially those living in stagnant waters on the plains. It also might bring about the decline of aquatic biota that inhabit stagnant or shallow water during certain parts of their life cycles, such as during reproduction. The latter organisms include fishes like cyprinids and cobitids which constitute the major part of the freshwater fish fauna of Japan. The same might occur for the lacustrine-riverine fishes around Lake Biwa, as mentioned above.

With regard to the present distribution of freshwater organisms, it can be said that some species that originated in wetlands around the lower and middle reaches of rivers can maintain their populations in small artificial wetlands such as paddy fields, canals, ditches, and irrigation ponds in the upper zone of the alluvial fans and the lower part of the valley (Yuma 1995). Examples of fish with such distribution patterns are the Japanese rose bitterling, the topmouth gudgeon *Pseudorasbora parva,* and the golden Venus fish, *Hemigrammocypris rasborella,* which are found in irrigation ponds, and the kissing loach, and the metropolitan bitterling, which occur in irrigation canals (Kawanabe & Mizuno 1989). These patterns suggest that original habitats of such species have already disappeared and that they inhabit artificially managed freshwaters which may resemble their original habitats.

The distribution pattern of Japanese freshwater fishes was originally established by migration and speciation of fishes throughout the geological history of the Japanese Archipelago. Recent human activities such as destruction of freshwater environments by land development, and fish introductions in natural habitats, apparently caused a decline in Japanese freshwater fish communities. In addition, the historical changes of Japanese freshwater systems by development of paddy fields has caused some species to decline but others to flourish, and changed the distribution patterns of fishes between upstream and downstream areas. To conserve endangered species and declining communities of freshwater fishes, we need to clarify characteristics of their original habitats and also the ecological advantages and disadvantages of modification and maintenance of freshwater systems by humans.

Acknowledgements

We thank James Atz and Kurt Fausch for a thorough revision of language and style, as well as Eugene Balon, Ed Crossman and David Noakes for detailed review of earlier drafts of this manuscript. We also thank Nagatoshi Hayase, Ibaragi Prefectural Natural Museum, Hiroshi Ikeda, Agency for Cultural Affairs, Hiroya Kawanabe, Yasuhiro Fujioka, and Masakazu Maehata, Lake Biwa Museum, Toshio Yokoyama, Institute of Humanities, Kyoto University, and Seiro Kimura, Fukuoka City, for their useful information. The discussions in the Kyoto International Seminar at the Kyoto Seminar House, Keihoku-cho, in 1994, including those with Eugene K. Balon who was invited by the Japan Society for the Promotion of Science in 1994 were most helpful for us. This study was supported in part by grants-in-aid for the International Scientific Research Program (No. 04044088, 07044194), for Cooperative Scientific Research (No. 05304004), for Scientific Research (No. 05804004, 06640814, 07458120) from the Japanese Ministry of Education, Science, Sports, and Culture, and by the Nippon Life Insurance Foundation (1995–1996).

References cited

Aoyagi, H. 1957. General notes on the freshwater fishes of the Japanese Archipelago. Daishuukan Shoten, Tokyo. 272 pp. (in Japanese).

Azuma, M. 1980. Small ayu of Lake Biwa: only the invader of one generation? pp. 154–161. *In:* T. Kawai, H. Kawanabe & N. Mizuno (ed.) Freshwater Organisms in Japan: Their Ecology of Invasion and Disturbance, Tokai University Press, Tokyo (in Japanese).

Balon, E.K. 1995. The common carp, *Cyprinus carpio:* its wild origin, domestication in aquaculture, and selection as colored nishikigoi. Guelph Ichthyol. Rev. 3: 1–55.

Biwako Bunkakan (ed.) 1991. Fishes in Lake Biwa, 3rd ed. Daiichi-Houki Publ., Tokyo. 190 pp (in Japanese).

Environment Agency of Japan. 1991. A list of endangered wild animals in Japan (Red Data Book in Japan): vertebrates. Research Center for Wildlife of Japan, Tokyo. 331 pp. (in Japanese).

Furukawa, Y. & K. Awano. 1969. Records of introduction of freshwater animals and plants into Lake Biwa. Sci. Rep. Shiga Pref. Fish. Exper. St. 22: 245–350.

Furushima, T. 1967. History found in landscape. Iwanami Shoten, Tokyo. 222 pp. (in Japanese).

Goto, A. 1987. Freshwater fishes: their grouping by life history strategy and their distribution pattern. pp. 231–244. *In:* N. Mizuno & A. Goto (ed.) Freshwater Fishes in Japan: Their Distribution, Variation and Speciation, Tokai University Press, Tokyo (in Japanese).

Goto, A., T. Nakanishi, H. Utoh & K. Hamada. 1978. A preliminary study of the freshwater fish fauna of rivers in southern Hokkaido. Bull. Fac. Fish. Hokkaido Univ. 29: 118–130 (in Japanese).

Horie, S. (ed.) 1984. Lake Biwa. Dr. W. Junk Publishers, Dordrecht. 654 pp.

Ichikawa, T. 1977. Salmon of Japan: its cultural aspect and fishery. Nihon Housou Publ., Tokyo. 242 pp. (in Japanese).

Ishihara, T., M. Kikkawa, K. Kurimoto & N. Uetsuma. 1986. Fishes in Hakone. Kanagawa Shinbunsha, Yokohama. 259 pp. (in Japanese).

Ishikawa, C. 1932. 'Koayu' and marine young 'ayu'. Kagaku Chishiki 25:8 (in Japanese).

Itoh, M. 1992. The reed beds conservation ordinance in Lake Biwa. pp. 145–149. *In:* H. Isozaki, M. Ando & Y. Natori (ed.) Towards Wise Use of Asian Wetlands, International Lake Environmental Committee Foundation, Kusatsu.

Jordan, D.S. & H.W. Fowler. 1903. A review of the cyprinoid fishes of Japan. Proc. U.S. National Mus. 26(1334): 811–862.

Kada, Y. 1995. Environmental science in human life: messages from Lake Biwa. Nousan Gyoson Bunka Kyoukai, Tokyo. 320 pp. (in Japanese).

Kafuku, T. 1980. Sociology of fishes. Kyoritsu Shuppan, Tokyo. 214 pp. (in Japanese).

Kataoka, M. 1989. Talking on nishikigoi. Musashi-token, Kumagaya City. 76 pp. (in Japanese).

Kato, M., M. Numata, K. Watanabe & M. Hata. 1995. Natural monuments of Japan. Kodansha, Tokyo. 1101 pp. (in Japanese).

Kawanabe, H. 1976. A note on the territoriality of ayu, *Plecoglossus altivelis* Temminck et Schlegel (Pisces: Osmeridae) in the Lake Biwa stock, based on the 'relic social structure'. Physiol. Ecol. Japan 17: 395–399 (in Japanese).

Kawanabe, H. 1977. Relic social structure hypothesis on the territoriality of ayu, *Plecoglossus altivelis* (Pisces: Osmeridae). Proc. Japan Acad. 53 (Ser. B): 74–77.

Kawanabe, H. 1996. Asian great lakes, especially Lake Biwa, Env. Biol. Fish. 47: 219–234.

Kawanabe, H. & N. Mizuno (ed.) 1989. Freshwater fishes of Japan. Yama-to-Keikoku-sha, Tokyo. 720 pp. (in Japanese).

Kawanabe, H. & K. Watanabe. 1993a. Black kokanee, *Oncorhynchus nerka kawamurai*. pp. 130–132. *In:* S. Ueno (ed.) Japanese Animals Going Extinct, Tsukiji Shokan, Tokyo (in Japanese).

Kawanabe, H. & K. Watanabe. 1993b. Kyoto ninespine stickleback, *Pungitius kaibarae.* pp. 133–135. *In:* S. Ueno (ed.) Japanese Animals Going Extinct, Tsukiji Shokan, Tokyo (in Japanese).

Kobayashi, J. 1933. Ecology of a stickleback, *Pungitius sinensis* var. *kaibarae* (Tanaka). J. Sci. Hiroshima Univ., ser. B. Div. 1: 2(5): 1–19.

Kubodera, K. 1993. Folklore and culture of rice. Sekai Seiten Kankou Kyoukai, Tokyo. 252 pp. (in Japanese).

Kuroki, T. 1986. Introduction to nishikigoi. Shin Nihon Kyouiku Tosho Co., Shimonoseki. 304 pp. (in Japanese).

Lindberg, G.U. 1972. Large-scale fluctuation of sea level in the Quaternary Period. Izd. Nauka, Moscow (in Russian); Japanese translation by T. Shinbori & F. Kanemitsu, 1981, Tokai University Press, Tokyo. 548 pp.

Maruyama, T., K. Fujii, T. Kijima & H. Maeda. 1987. Process of introduction of exotic fishes. Section of Resources at Japan Fisheries Department and National Research Institute of Aquaculture. 157 pp. (in Japanese).

Masutani, F. & H. Kanaoka. 1994. Causal dictionary of Buddhism. Taiyo Shuppan, Tokyo. 596 pp. (in Japanese).

Matsui, Y. 1935. Study of goldfish from the viewpoints of science and hobby. Koudoukaku, Tokyo. 421 pp. (in Japanese).

Matsui, Y. 1948. Paddy field fish culture. Tomi Shoten, Kyoto. 166 pp. (in Japanese).

Matsui, Y. 1971. Goldfish. Hoikusha, Osaka. 123 pp.

Miura, T. 1966. Competitive influence of isaza, *Chaenogobius isaza*, on ayu, *Plecoglossus altivelis*, in Lake Biwa. Res. Popul. Ecol. 8: 37–50.

Miyadi, D. 1960. Story of ayu-fish. Iwanami Shoten, Tokyo. 226 pp. (in Japanese).

Miyadi, D., H. Kawanabe & N. Mizuno. 1976. Colored illustrations of the freshwater fishes of Japan (revised edition). Hoikusha, Osaka. 462 pp. (in Japanese).

Mizuno, N. 1987. Establishment of freshwater fish fauna in Japan. pp. 231–244. *In:* N. Mizuno & A. Goto (ed.) Freshwater Fishes in Japan: Their Distribution, Variation and Speciation, Tokai University Press, Tokyo (in Japanese).

Mori, S. 1970. List of animal and plant species living in or on Lake Biwa, Mem. Fac. Sci., Kyoto Univ., Ser. Biol. 3: 22–46.

Mori, S. & T. Miura. 1980. List of plant and animal species living in Lake Biwa (revised). Mem. Fac. Sci., Kyoto Univ., Ser. Biol. 8: 1–33.

Nagata, Y. 1980. Rose bitterling, *Rhodeus ocellatus ocellatus:* a crisis of indigenous strain. pp. 147–153. *In:* T. Kawai, H. Kawanabe & N. Mizuno (ed.) Freshwater Organisms in Japan: Their Ecology of Invasion and Distribution, Tokai University Press, Tokyo (in Japanese).

Nagata, Y., T. Tetsukawa, T. Kobayashi & K. Numachi. 1996. Genetic markers distinguishing between two subspecies of the rosy bitterling, *Rhodeus ocellatus* (Cyprinidae). Ichthyol. Res. 43: 117–124.

Naka, K. 1991. Changes of fish fauna in Lake Biwa. pp. 55–64. *In:* Lake Biwa Institute (ed.) Report of the 9th Symposium at Lake Biwa Institute, Otsu (in Japanese).

Nakabo, T. (ed.) 1993. Fishes of Japan with pictorial keys to the species. Tokai University Press, Tokyo. 1474 pp. (in Japanese).

Nakajima, T. 1994. Cyprinid fishes and their pharyngeal teeth. pp. 168–203. Cyprinid fishes. pp. 235–275. *In:* Research Group for Natural History of Lake Biwa (ed.) The Natural History of Lake Biwa, Yasaka Shobo, Tokyo (in Japanese).

Nishida, M. 1985. Substantial genetic differentiation in ayu *Plecoglossus altivelis* of the Japan and Ryukyu Islands. Bull. Japan. Soc. Sci. Fish. 51: 1269–1274.

Nishimura, S. 1980. Establishment of the Japan Sea, 2nd ed. Tsukiji Shokan, Tokyo. 228 pp. (in Japanese).

Ohno, Y. 1994. Resource utilization by an invader to Lake Biwa, *Tridentiger brevispinis.* Masters Thesis, Kyoto University, Kyoto. 43 pp.

Ohtani, K. & Y. Kikukawa. 1980. Largemouth bass: a typical harmful fish? pp. 20–29. *In:* T. Kawai, H. Kawanabe & N. Mizuno (ed.) Freshwater Organisms in Japan: Their Ecology of Invasion and Disturbance, Tokai University Press, Tokyo (in Japanese).

Ojima, Y. 1988. Stories of fishes. Gihoudo Shuppan, Tokyo. 191 pp. (in Japanese).

Ojima, Y. & A. Takai. 1995. World of fishes: from microbiology to macrobiology. Shoukabou Co., Tokyo. 160 pp. (in Japanese).

Okada, Y. 1959–1960. Studies on the freshwater fishes of Japan. J. Fac. Fish. Mie Prefect Univ. 4: 1–860.

Oshima, M. 1941. Studies on a rare salmonid fish 'Kuni-masu' in Lake Tazawa, Akita Prefecture. Nihon Gakujutsu Kyoukai Koukoku (Report of Japanese Scientific Society) 16: 254–259 (in Japanese).

Sunaga, T. 1980. *Opsariichthys uncirostris uncirostris:* ecological conditions for piscivorous fish. pp. 30–36. *In:* T. Kawai, H. Kawanabe & N. Mizuno (ed.) Freshwater Organisms in Japan: Their Ecology of Invasion and Disturbance, Tokai University Press, Tokyo (in Japanese).

Suzuki, B. 1836–1842. Natural features in Hokuriku region of Japan. Iwanami Shoten, Tokyo. 348 pp. (in Japanese, corrected and revised by T. Okada in 1936).

Suzuki, S. 1992. Some problems of introduced exotic fishes. pp. 5–9. *In:* Union of Inland Fisheries Cooperation (ed.) All of the Largemouth Bass and Bluegill. Report of Discussion for Introduced Fishes, Tokyo (in Japanese).

Takahashi, S. 1990. *Tridentiger kuroiwae brevispinis* caught in Lake Biwa. Ann. Rep. Biwako Bunkakan 11: 7 (in Japanese).

Tan, N. 1928. Ecology of *Pungitius kaibarae* Tanaka. Published by himself, Kyoto. 41 pp. (in Japanese).

Tanaka, S. 1936. Fishes in Nikko. pp. 346–354. *In:* Nikko Toshogu Shrine (ed.) Plants and Animals in Nikko, Yokendo, Tokyo (in Japanese).

Terashima, A. 1980. Bluegill: a vacant ecological niche in Lake Biwa. pp. 63–70. *In:* T. Kawai, H. Kawanabe & N. Mizuno (ed.) Freshwater Organisms in Japan: Their Ecology of Invasion and Disturbance, Tokai University Press, Tokyo (in Japanese).

Tochigi Prefecture. 1980. Report on the lake environments in Tochigi Prefecture (The second survey for conservation of natural environment). Environment Agency of Japan, Tokyo. 64 pp. (in Japanese).

Tochigi Prefecture. 1993. Report on the lake environments in Tochigi Prefecture (The fourth survey for conservation of natural environment). Environment Agency of Japan, Tokyo. 62 pp. (in Japanese).

Tokui, T. & H. Kawanabe. 1984. Fishes. pp. 339–360. *In:* S. Horie (ed.) Lake Biwa, Dr W. Junk Publishers, Dordrecht.

Tsujimura, A. & N. Taniguchi. 1995. Genetic differences in some reproductive characters found between landlocked and amphidromous forms of ayu, *Plecoglossus altivelis altivelis*. Nippon Suisan Gakkaishi 61: 165–172.

Ueno, M. 1984. Biogeography of Lake Biwa. pp. 625–633. *In:* S. Horie (ed.) Lake Biwa, Dr W. Junk Publishers, Dordrecht.

Watanabe, N. & M. Nishino. 1995. A study on taxonomy and distribution of the freshwater snails, genus *Semisulcospira* in Lake Biwa, with descriptions of eight new species. Lake Biwa Study Monographs 6: 1–36.

Yano, K. 1981. Folklore of fishes. Yuzankaku Shuppan, Tokyo. 190 pp. (in Japanese).

Yokoyama, T. 1969. Paleography of ancient Lake Biwa in late Pliocene deduced from paleocurrent directions of Kobiwako group at Sanagu area in Mie Prefecture, Japan. Sci. Eng. Rev. Doshisha University 15: 37–44.

Yokoyama, T. 1995. Lake Biwa and its geohistorical migration. Housei Shuppan Yawata City. 312 pp. (in Japanese).

Yuma, M. 1995. Familiar animal communities in artificial water systems, in relation to paddy-field cultivation. Environmental Technology 24: 695–700 (in Japanese).

Yuma, M., Y. Kada & Y. Fujioka. 1995. Historical changes of aquatic fauna and childhood-play in familiar freshwaters – through a three-generational interviewing survey. Res. Environ. System 23: 20–31 (in Japanese).

Appendix. Freshwater fishes shown in Kawanabe & Mizuno (1989) and Nakabo (1993). Fishes living mainly in brackish and marine waters are excluded. Scientific names follow Nakabo (1993). English names follow Miyadi et al. (1976). Habitat: F = freshwater; B = brackish water; M = marine; ◎= main habitat; ○ = common habitat. Distribution: H, only in Hokkaido and north; R, only in Ryukyu Islands and south.

Taxon	English name	Japanese name	Habitat			Distri-bution		Endemic to Japan	Alien fish (origin, year of intro-duction)
			F	B	M	H	R		
Petromyzontiformes									
Petromyzontidae									
1 *Entosphenus tridentata*	Pacific lamprey	Mitsuba-yatsume	○	○	○				
2 *Lenthenteron japonicum*	Arctic lamprey	Kawa-yatsume	○	○	○				
3 *Lenthenteron kessleri*	Landlocked Siberian lamprey	Siberia-yatsume	○			▲			
4 *Lenthenteron reissneri*	Sand lamprey	Suna-yatsume	○						
Acipenseriformes									
Acipenseridae									
5 *Acipenser medirostris*	Sakhalin green sturgeon	Chouzame	○	○	○				
Anguilliformes									
Anguillidae									
6 *Anguilla japonica*	Japanese eel	Unagi	○	○	○				
7 *Anguilla marmorata*	Marbled eel	Oo-unagi	○	○	○				
Clupeiformes									
Engraulididae									
8 *Coilia nasus*		Etsu	○	○				★	
Cypriniformes									
Cyprinidae									
9 *Zacco temminckii*	Dark chub type B	Kawamutsu B	◎						
10 *Zacco* sp.	Dark chub type A	Kawamutsu A	◎						
11 *Zacco platypus*	Pale chub	Oikawa	◎						
12 *Opsariichthys uncirostris uncirostris*	Piscivorous chub	Hasu	◎						
13 *Aphyocypris chinensis*	Venus fish	Hina-moroko	◎						
14 *Hemigrammocypris rasborella*	Golden venus fish	Kawabata-moroko	◎					★	
15 *Tribolodon hakonensis*	Japanese dace	Ugui	◎	○	○				
16 *Tribolodon* sp.		Ukekuchi-ugui	◎	○				★	
17 *Tribolodon brandti*	Far Eastern dace	Maruta-ugui	◎						
18 *Tribolodon ezoe*	Rosyface dace	Ezo-ugui	◎						
19 *Phoxinus percnurus sachalinensis*	Sakhalin lake minnow	Yachi-ugui	◎			▲			
20 *Phoxinus lagowski steindachneri*	Japanese fat minnow	Abura-haya	◎						
21 *Phoxinus lagowski yamanotis*	Yamanaka minnow	Yamanaka-haya	◎					★ (subsp)	
22 *Phoxinus oxycephalus jouvi*	Chinese minnow	Taka-haya	◎						
23 *Ctenopharyngodon idellus*	Grass carp	Sougyo	◎						● (China)
24 *Mylopharyngodon piceus*	Black Chinese roach	Ao-uo	◎						● (China, 1943)
25 *Ishikauia steenackeri*	Lakeweed chub	Wataka	◎					★	
26 *Hypophthalmichthys nobilis*	Striped bighead	Kokuren	◎						● (China)
27 *Hypophthalmichthys molitrix*	Silver bighead	Hakuren	◎						● (China, 1943)
28 *Gnathopogon elongatus elongatus*	Field gudgeon	Ta-moroko	◎					★ (subsp)	
29 *Gnathopogon elongatus suwae*	Suwa gudgeon	Suwa-moroko	◎					★ (subsp)	
30 *Gnathopogon caerulescens*	Biwa gudgeon	Hon-moroko	◎					★	
31 *Pungtungia herzi*	Black stripe gudgeon	Mugitsuku	◎						
32 *Pseudorasbora parva*	Topmouth gudgeon	Motsugo	◎						
33 *Pseudorasbora pumila pumila*	Dwarf topmouth gudgeon	Shinai-motsugo	◎					★ (subsp)	
34 *Pseudorasbora pumila* subsp.		Ushi-motsugo	◎					★ (subsp)	
35 *Sarcocheilichthys variegatus variegatus*	River gudgeon	Kawa-higai	◎					★ (subsp)	
36 *Sarcocheilichthys variegatus microoculus*	Biwa gudgeon	Biwa-higai	◎					★ (subsp)	
37 *Sarcocheilichthys biwaensis*	Oily gudgeon	Abura-higai	◎					★	
38 *Pseudogobio esocinus esocinus*	Pike gudgeon	Kamatsuka	◎						
39 *Abbottina rivularis*	Chinese false gudgeon	Tsuchifuki	◎						
40 *Biwia zezera*	Biwa gudgeon	Zezera	◎					★	
41 *Squalidus chankaensis biwae*		Sugo-moroko	◎					★ (subsp)	
42 *Squalidus chankaensis* subsp.		Kourai-moroko	◎						

Appendix. Continued.

Taxon	English name	Japanese name	Habitat F	Habitat B	Habitat M	Distribution H	Distribution R	Endemic to Japan	Alien fish (origin, year of introduction)
43 *Squalidus japonicus japonicus*		Deme-moroko	◎					★ (subsp)	
44 *Squalidus gracilis gracilis*	Slender gudgeon	Ito-moroko	◎					★ (subsp)	
45 *Hemibarbus barbus*	Steed barbel	Nigoi	◎	◎				★	
46 *Hemibarbus labeo*		Kourai-nigoi	◎	◎					
47 *Hemibarbus longirostris*	Longnose barbel	Zunaga-nigoi	◎					★	
48 *Cyprinus carpio*	Common carp	Koi	◎	◎					
49 *Carassius auratus* subsp. 1		Naga-buna	◎					★ (subsp)?	
50 *Carassius auratus* subsp. 2	Golden crucian carp	Kin-buna	◎					★ (subsp)	
51 *Carassius auratus buergeri*		Oo-kin-buna	◎						
52 *Carassius auratus grandoculis*	Round crucian carp	Nigoro-buna	◎					★ (subsp)	
53 *Carassius auratus langsdorfi*	Silver crucian carp	Gin-buna	◎						
54 *Carassius cuvieri*	Deepbodied crucian carp	Gengorou-buna	◎					★	
55 *Tanakia lanceolata*	Slender bitterling	Yari-tanago	◎						
56 *Tanakia limbata*	Oily bitterling	Abura-bote	◎						
57 *Tanakia tanago*	Metropolitan bitterling	Miyako-tanago	◎					★	
58 *Rhodeus ocellatus ocellatus*	Rose bitterling	Tairiku-bara-tanago	◎						◎(China, 1940s)
59 *Rhodeus ocellatus kurumeus*	Japanese rose bitterling	Nippon-bara-tanago	◎					★ (subsp)	
60 *Rhodeus atremius atremius*	Kyushu rose bitterling	Kazetoge-tanago	◎					★ (subsp)	
61 *Rhodeus atremius suigensis*	Suigen rose bitterling	Suigen-zeni-tanago	◎						
62 *Acheilognathus melanogaster*	Broadstriped bitterling	Tanago	◎					★	
63 *Acheilognathus cyanostigma*	Striped bitterling	Ichimonji-tanago	◎					★	
64 *Acheilognathus tabira tabira*	White rock bitterling	Shiro-hire-tabira	◎					★ (subsp)	
65 *Acheilognathus tabira* subsp. 1	Red rock bitterling	Aka-hire-tabira	◎					★ (subsp)	
66 *Acheilognathus tabira* subsp. 2	Spotted rock bitterling	Seboshi-tabira	◎					★ (subsp)	
67 *Acheilognathus rhombeus*	Flat bitterling	Kanehira	◎						
68 *Acheilognathus longipinnis*	Deepbodied bitterling	Ita-sen-para	◎					★	
69 *Acheilognathus typus*	Netted bitterling	Zeni-tanago	◎					★	
Cobitidae									
70 *Leptobotia curta*	Kissing loach	Ayu-modoki	◎					★	
71 *Misgurnus anguillicaudatus*	Asian pond loach	Dojou	◎						
72 *Cobitis matsubarai*	Spined loach	Yamato-sima-dojou	◎					★	
73 *Cobitis* sp. 1	Striped spined loach (large)	Suji-sima-dojou (large)	◎					★?	
74 *Cobitis* sp. 2 subsp. 1	Striped spined loach (small-San'you)	Suji-sima-dojou (small-San'you)	◎					★ (subsp)?	
75 *Cobitis* sp. 2 subsp. 2	Striped spined loach (small-Toukai)	Suji-sima-dojou (small-Toukai)	◎					★ (subsp)?	
76 *Cobitis* sp. 2 subsp. 3	Striped spined loach (small-spotted)	Suji-sima-dojou (small-tenkogata)	◎					★ (subsp)?	
77 *Cobitis* sp. 2 subsp. 4	Striped spined loach (small-Biwako)	Suji-sima-dojou (small-Biwako)	◎					★ (subsp)?	
78 *Cobitis* sp. 3	Striped spined loach (middle)	Suji-sima-dojou (middle)	◎					★?	
79 *Cobitis biwae*	Sand loach	Shima-dojou	◎					★	
80 *Cobitis takatsuensis*	Secret loach	Ishi-dojou	◎					★	
81 *Niwaella delicata*	Delicate loach	Ajime-dojou	◎					★	
82 *Noemacheilus barbatulus toni*	Siberian stone loach	Fuku-dojou	◎			▲			
83 *Lefua echigonia*	Japanese eight barbel loach	Hotoke-dojou	◎					★	
84 *Lefua nikkonis*	Hokkaido eight barbel loach	Ezo-hotoke-dojou	◎			▲		★	
85 *Lefua* sp.		Nagare-hotoke-doyou	◎					★	
Siluriformes									
Bagridae									
86 *Pelteobagrus nudiceps*	Folktail bullhead	Gigi	◎					★	
87 *Pseudobagrus tokiensis*	Cuttailed bullhead	Gibachi	◎					★	
88 *Pseudobagrus aurantiacus*	Kyushu bullhead	Ariake-gibachi	◎					★	
89 *Coreobagrus ichikawai*	Swampy bullhead	Neko-gigi	◎					★	
Siluridae									
90 *Silurus asotus*	Far Eastern catfish	Namazu	◎						
91 *Silurus biwaensis*	Biwa catfish	Biwako-oo-namazu	◎					★	
92 *Silurus lithophilus*	Rock catfish	Iwatoko-namazu	◎					★	

Appendix. Continued.

Taxon	English name	Japanese name	Habitat			Distribution		Endemic to Japan	Alien fish (origin, year of introduction)
			F	B	M	H	R		
Amblycipitidae									
93 *Liobagrus reini*	Reddish bullhead	Akaza	◎					★	
Clariidae									
94 *Clarias fuscus*		Hire-namazu	◎						● (Taiwan)
Salmoniformes									
Osmeridae									
95 *Spirinchus lanceolatus*	Shishamo smelt	Shishamo	○	○	○			★	
96 *Osmerus eperlanus mordax*	Olive rainbow smelt	Kyuuri-uo	○	○	○	▲			
97 *Hypomesus transpacificus nipponensis*	Japanese smelt	Wakasagi	◎	○	○				
98 *Hypomesus olidus*	Pond smelt	Ishikari-wakasagi	◎			▲			
Plecoglossidae									
99 *Plecoglossus altivelis altivelis*	Ayu	Ayu	◎	○	○				
100 *Plecoglossus altivelis ryukyuensis*	Ryukyu ayu	Ryuukyuu-ayu	○	○	○		▼	★ (subsp)	
Salangidae									
101 *Neosalanx reganius*	Regan icefish	Ariake-hime-sirauo	◎	○				★	
Salmonidae									
102 *Hucho perryi*	Sakhalin huchen	Itou	◎	○	○	(▲)			
103 *Salvelinus malma malma*	Dolly Yarden	Oshorokoma	◎	○	○	▲			
104 *Salvelinus malma miyabei*	Miyabe charr	Miyabe-iwana	◎			▲		★ (subsp)	
105 *Salvelinus leucomaenis leucomaenis*	Whitespotted charr	Ame-masu, Ezo-iwana	◎	○	○				
106 *Salvelinus leucomaenis pluvius*	Japanese charr	Nikkou-iwana	◎	○	○			★ (subsp)	
107 *Salvelinus leucomaenis japonicus*	Japanese charr	Yamato-iwana	◎					★ (subsp)	
108 *Salvelinus leucomaenis imbrius*	Head spotted charr	Gogi	◎					★ (subsp)	
109 *Salvelinus fontinalis*	Brook trout	Kawa-masu	◎						● (USA, 1902)
110 *Salvelinus namaycush*	Lake trout	Lake trout	◎						● (Canada, 1966)
111 *Salmo trutta*	Brown trout	Brown trout	◎						● (USA?)
112 *Oncorhynchus mykiss*	Rainbow trout	Niji-masu	◎						● (USA, after 1877)
113 *Oncorhynchus masou masou*	Masu salmon, Masu trout	Sakura-masu, Yamame	◎	○	○				
114 *Oncorhynchus masou ishikawae*	Red-spotted masu trout	Satsuki-masu, Amago	◎	○	○			★ (subsp)	
115 *Oncorhynchus masou* subsp.	Biwa trout	Biwa-masu	◎					★ (subsp)	
116 *Oncorhynchus kisutch*	Coho salmon	Gin-zake	○	○	○				
117 *Oncorhynchus tshawytscha*	Chinook salmon	Masu-no-suke	○	○	○				
118 *Oncorhynchus keta*	Chum salmon	Sake	○	○	○				
119 *Oncorhynchus gorbuscha*	Pink salmon	Karafuto-masu	○	○	○				
120 *Oncorhynchus nerka nerka*	Sockeye salmon, Kokanee	Beni-zake, Hime-masu	◎	○	○	▲			
121 *Oncorhynchus nerka kawamurai*	Black kokanee	Kuni-masu	◎					★ (subsp)	
122 *Coregonus lavaretus maraena*	Sig	Shinano-yuki-masu	◎						● (Europe?)
Atheriniformes									
Atherinidae									
123 *Odonthestes bonariensis*	Peherei	Peherei	◎						● (Argentina, 1966)
Cyprinodontiformes									
Poeciliidae									
124 *Gambusia affinis*	Common gambusia	Ka-dayashi	◎	◎					● (through Taiwan, 1916)
125 *Poecilia reticulata*	Guppy	Guppy	◎	◎					● (Aquarium trade)
Beloniformes									
Adrianichthyidae									
126 *Oryzias latipes*	Medaka	Medaka	◎	◎	◎				
Hemiramphidae									
127 *Hyporhamphus intermedius*	Japanese brackish halfbeak	Kurume-sayori	◎	◎					

Appendix. Continued.

Taxon	English name	Japanese name	Habitat F	Habitat B	Habitat M	Distribution H	Distribution R	Endemic to Japan	Alien fish (origin, year of introduction)
Gasterosteiformes									
Gasterosteidae									
128 *Gasterosteus aculeatus*	Threespine stickleback (anadromous)	Itoyo; anadromous form	◎	○	○				
129 *Gasterosteus* sp.	Threespine stickleback (landlocked)	Itoyo; landlocked form	◎					★	
130 *Gasterosteus microcephalus*	Naked threespine stickleback	Hariyo	◎					★	
131 *Pungitius tymensis*	Short-spined ninespine stickleback	Ezo-tomiyo	◎			▲			
132 *Pungitius pungitius*	Ninespine stickleback	Ibara-tomiyo, Kita-no-tomiyo	◎	◎					
133 *Pungitius* sp.	Musashi ninespine stickleback	Musashi-tomiyo	◎					★	
134 *Pungitius sinensis*	Chinese ninespine stickleback	Tomiyo	◎						
135 *Pungitius kaibarae*	Kyoto ninespine stickleback	Minami-tomiyo	◎					★	
Syngnathiformes									
Syngnathidae									
136 *Microphis (Oostethus) brachyurus brachyurus*		Tengu-youji	○	○	○		▼		
137 *Microphis (Coelonotus) leiaspis*	Coelonotous pipefish	Issen-youji	○	○	○		▼		
Synbranchiformes									
Synbranchidae									
138 *Monopterus albus*	Swamp eel	Ta-unagi	◎						
Scorpaeniformes									
Scorpaenidae									
139 *Tetraroge niger*		Higesori-okoze	○	○	○		▼		
Cottidae									
140 *Trachidermus fasciatus*	Roughskin sculpin	Yama-no-kami	○	○					
141 *Cottus kazika*	Fourspine sculpin	Ayu-kake, Kamakiri	○	○	○			★	
142 *Cottus hangiongensis*	Tyuman-river sculpin	Kankyou-kajika	○	○	○				
143 *Cottus amblyostomopsis*		Ezo-hana-kajika	○	○	○	▲			
144 *Cottus nozawae*	Wrinklehead sculpin	Hana-kajika	◎	○	○			★	
145 *Cottus pollux*	Japanese sculpin	Kajika	◎						
146 *Cottus reinii*		Utsusemi-kajika	◎					★	
Perciformes									
Ambassidae									
147 *Ambassis miops*		Sesuji-takasago-ishimochi	○	○	○		▼		
148 *Ambassis urotaenia*		Takasago-ishimochi	○	○	○		▼		
Percichthyidae									
149 *Coreoperca kawamebari*	Japanese aucha perch	Oya-nirami	◎						
Terapontidae									
150 *Mesopristes* sp.		Misuji-sima-isaki	○	○	○		▼		
151 *Mesopristes argenteus*		Nise-sima-isaki	○	○	○		▼		
152 *Mesopristes cancellatus*		Yoko-sima-isaki	○	○	○		▼		
Kuhliidae									
153 *Kuhlia marginata*	Spotted flagtail	Yugoi	○	○	○		(▼)		
154 *Kuhlia rupestris*	Rock flagtail	Ookuchi-yugoi	○	○	○		(▼)		
Centrarchidae									
155 *Micropterus salmoides*	Largemouth bass	Ookuchi-bass	◎						● (USA, 1925)
156 *Lepomis macrochirus*	Bluegill	Bluegill	◎						● (USA, 1960)
Scatophagidae									
157 *Scatophagus argus*		Kurohoshi-manjuudai	○	○	○				
Cichlidae									
158 *Oreochromis mossambicus*	Mozambique tilapia	Mozambique tilapia	◎	◎	◎				● (Thailand, 1954)
159 *Oreochromis niloticus*	Nile tilapia	Nile tilapia	◎	◎					● (Egypt, 1962)
160 *Tilapia zillii*	Zilli tilapia	Zilli tilapia	◎	◎	◎				● (Egypt, 1962)
Mugilidae									
161 *Crenimugil heterocheilos*		Nagare-fuurai-bora	○	○	○		▼		

Appendix. Continued.

Taxon	English name	Japanese name	Habitat			Distribution		Endemic to Japan	Alien fish (origin, year of introduction)
			F	B	M	H	R		
Rhyncichthyidae									
162 *Rhyacichthys aspro*	Rhyacichthyid fish	Tsubasa-haze	○	○	○		▼		
Gobiidae									
163 *Eleotris oxycephala*	Spined sleeper	Kawa-anago	○	○	○				
164 *Eleotris fusca*	Indian spined sleeper	Tenjiku-kawa-anago	○	○	○		▼		
165 *Eleotris acanthopoma*		Chichibu-modoki	○	○	○				
166 *Ophieleotris* sp.		Tametomo-haze	○	○	○		▼		
167 *Odontobutis obscura obscura*	Dark sleeper	Donko	◎					★ (subsp)	
168 *Hypseleotris cyprinoides*	Carp-like sleeper	Tanago-modoki	○	○	○		▼		
169 *Schismatogobius roxasi*		Eso-haze	○	○	○		▼		
170 *Schismatogobius* sp.		Shima-eso-haze	○	○	○		▼		
171 *Awaous melanocephalus*		Kuro-minami-haze	○	○	○		▼		
172 *Awaous ocellaris*	Ocellated river goby	Minami-haze	○	○	○		▼		
173 *Redigobius bikolanus*	Blackspotted goby	Hina-haze	○	○	○				
174 *Rhinogobius giurinus*	Paradise goby	Gokuraku-haze	◎	○	○				
175 *Rhinogobius* sp. cross-band CB	Common freshwater goby (CB)	Shima-yoshinobori	◎	○	○				
176 *Rhinogobius* sp. dark DA	Common freshwater goby (DA)	Kuro-yoshinobori	◎	○	○			★?	
177 *Rhinogobius* sp. depressed large dark DL	Common freshwater goby (DL)	Hira-yoshinobori	○	○	○		▼	★?	
178 *Rhinogobius* sp. large dark LD	Common freshwater goby (LD)	Oo-yoshinobori	◎	○	○			★?	
179 *Rhinogobius* sp. cobalt CO	Common freshwater goby (CO)	Ruri-yoshinobori	◎	○	○			★?	
180 *Rhinogobius* sp. orange OR	Common freshwater goby (OR)	Tou-yoshinobori	◎	○	○			★?	
181 *Rhinogobius* sp. mosaic MO	Common freshwater goby (MO)	Aya-yoshinobori	◎	○	○		▼	★?	
182 *Rhinogobius* sp. yellow belly YB	Common freshwater goby (YB)	Kibara-yoshinobori	◎				▼	★?	
183 *Rhinogobius* sp. blue belly BB	Common freshwater goby (BB)	Aobara-yoshinobori	◎				▼	★?	
184 *Rhinogobius flumineus*	Lizard goby	Kawa-yoshinobori	◎					★	
185 *Stenogobius* sp.	Filletcheek goby	Tanekawa-haze	○	○	○				
186 *Tridentiger obscurus*	Japanese trident goby	Chichibu	◎	◎	○				
187 *Tridentiger brevispinis*	Short-spined Japanese trident goby	Numa-chichibu	◎	◎	○				
188 *Tridentiger kuroieae*		Nagano-gori	◎	○	○			★	
189 *Glossogobius celebius*		Iwa-haze	○	○	○		▼		
190 *Glossogobius aureus*		Konjiki-haze	○	○	○				
191 *Pandaka trimaculata*		Mitsuboshi-goma-haze	○	○	○		▼		
192 *Chaenogobius laevis*		Juzukake-haze	◎					★?	
193 *Chaenogobius isaza*	Biwa goby	Isaza	◎					★	
194 *Chaenogobius urotaenia*	Floating goby	Uki-gori	◎	○	○				
195 *Chaenogobius* sp. 1		Sumi-uki-gori	○	○	○				
196 *Chaenogobius* sp. 2		Shima-uki-gori	○	○	○				
196 *Luciogobius guttatus*	Flathead goby	Mimizu-haze	○	○	○				
197 *Luciogobius pallidus*	Well goby	Ido-mimizu-haze	○	○	○			★	
198 *Luciogobius albus*	Cave goby	Doukutsu-mimizu-haze	○	○				★	
199 *Leucopsarion petersii*	Ice goby	Shiro-uo	○	○	○				
200 *Sicyopus leprurus*		Kaeru-haze	○	○	○		▼	★?	
201 *Sicyopus zosterophorum*		Fude-haze, Aka-bouzu-haze	○	○	○		▼		
202 *Lentipes armatus*		Yoroi-bouzu-haze	○	○	○		▼	★	
203 *Stiphodon elegans*		Nan'you-bouzu-haze	○	○	○				
204 *Stiphodon stevensoni*		Hayase-bouzu-haze	○	○	○		▼		
205 *Sicyopterus macrostetholepis*		Ruri-bouzu-haze	○	○	○		▼		
206 *Sicyopterus japonicus*	Parrot goby	Bouzu-haze	○	○	○				
Belontiidae									
207 *Macropodus chinenis*	Roundtailed paradise fish	Chousen-buna	◎						● (Korea, ca. 1914)
208 *Macropodus opercularis*	Paradise fish	Taiwan-kingyo	◎				▼		
Channidae									
209 *Channa argus*	Northern snakehead	Kamuruchii	◎						● (Korea?, 1923–24)
210 *Channa maculata*	Formosan snakehead	Taiwan-dojou	◎						● (Taiwan, 1906)
211 *Channa asiatica*		Koutai	◎				▼		● (Taiwan)

Environmental Biology of Fishes **52**: 125–135, 1998.

New aspects of sex change among reef fishes: recent studies in Japan

Tetsuo Kuwamura[1] & Yasuhiro Nakashima[2]
[1] *Biological Laboratory, Faculty of Liberal Arts, Chukyo University, Yagoto, Nagoya 466, Japan*
[2] *School of Nursing, Miyagi University, Taiwa-cho, Miyagi 981-36, Japan (e-mail: PXB11261@niftyserve.or.jp)*

Received 16.10.1996 Accepted 22.1.1997

Key words: hermaphroditism, protandry, protogyny, 2-way sex change, size-advantage model, social status

Synopsis

New aspects of sex change in reef fishes are reviewed with special emphasis on recent studies in Japan. For protogyny, studies on both monandric and diandric species have been conducted, but the distinction of primary males from prematurational secondary males seems to need further examination. For protandry, detailed field studies on anemonefishes have revealed alternative life-history styles associated with movements between hosts before or after maturation. The most interesting new aspect has been the discovery of 2-way sex change within a species. Conditions for evolution of 2-way sex change are examined in relation to the size-advantage model and social control mechanisms. A fish may change sex when it becomes dominant in a mating group, but a dominant fish may also change sex in the reverse direction when its social status changes to subordinate through inter-group movement. Two-way sex change has hitherto been reported only from basically protogynous fishes (e.g., Gobiidae, Pomacanthidae, Cirrhitidae, Epinephelinae). Possibilities of the reverse sex change in the protandrous anemonefishes are discussed with data from some unpublished studies.

Introduction

Sex change from female to male or vice versa is well known among teleost fishes. The evolution of sex change, as a life-history style alternative to gonochorism, has been explained by the size-advantage model (Gishelin 1969, Warner 1975, 1988a, Charnov 1982). If the product of the probability of survival to a particular age and expected fecundity at that age increases with age or size faster for one sex than the other, then an individual that changes sex will have a higher lifetime reproductive success than one that does not. This model predicts that the direction of sex change, protogyny or protandry, will be determined by the mating system of each species, as has been confirmed in many reef fishes (Warner 1984, 1988b).

Sex change has also been one of the main interests of Japanese ichthyologists. Theoretical and empirical studies on sex change among fishes were reviewed by Nakazono & Kuwamura (1987). The book was one of the products of the research project 'Optimal strategy and social structure of vertebrates' (1983–1986, supported by the Japan Ministry of Education, Science and Culture), which was headed by Hiroya Kawanabe. An English translation of the book was prepared by K. Asoh, T. Yoshikawa and D. Shapiro, but unfortunately it has not yet been published.

The purpose of this essay is to review recent studies on sex change of fishes conducted in Japan and to evaluate new aspects in this field. First, we give a brief summary of each chapter of the above-mentioned book (Nakazono & Kuwamura op. cit.). Then we review the recent studies in Japan, focusing on ecological and evolutionary aspects of sex change. The most interesting new aspect in this field seems to be the finding of sex change in either direction within a species, which has been confirmed in several species of fishes by Japanese researchers (e.g., Kuwamura et al. 1994). Therefore, we put special emphasis on this aspect and discuss requisites and possibilities of the reverse-directional sex change among other hermaphroditic fishes including protandrous anemonefishes.

Recent studies on sex change of fishes in Japan

Until 1987

First, we briefly introduce each chapter of Nakazono & Kuwamura (1987). Chapter 1 'Hermaphroditism and the evolutionary aspects of its occurrences in fishes' (by Y. Yogo) is a phylogenetic review of hermaphroditic fishes with lists of species. By that time, functional hermaphroditism had been reported from at least 350 species of 34 families in eight orders of teleost fishes. Chapter 2 'Histology and physiology of sex change in fishes' (by M. Nakamura) reviews histological and hormonal changes in sex-changing fishes, e.g., a protogynous wrasse *Thalassoma duperrey*. Chapter 3 'Theories for sex change' (by Y. Yanagisawa) reviews the evolutionary models of Ghiselin (1969), Warner (1975), Charnov (1982), etc.

Chapter 4 'Social control of sex change in harems of *Labroides dimidiatus*' (by T. Kuwamura) reviews the well-known example of the protogynous cleaner wrasse, mainly referring to Robertson (1974) and Kuwamura (1984). Chapter 5 'Social organization and protogynous hermaphroditism in marine angelfishes (Pomacanthidae)' (by J.T. Moyer) is a review with special emphasis on the occurrence of early sex change (Moyer & Zaiser 1984, also see Moyer 1990). Chapter 6 'Group structure and mechanism of sex change in *Anthias squamipinnis*' (by Y. Yogo) reviews its mating system (Yogo 1985) and the sex-ratio threshold model (e.g., Shapiro & Lubbock 1980). Chapter 7 'Diandry in Japanese labrid fishes and alternative strategies of primary males' (by A. Nakazono) compares three Japanese wrasses (Nakazono 1979) and the bluehead wrasse *Thalassoma bifasciatum* (e.g., Warner & Hoffman 1980).

Chapter 8 'Sexual patterns in damselfishes (Pomacentridae) with special reference to protandry in anemonefishes' (by H. Ochi) compares social systems of protogynous *Dascyllus* (e.g., Coates 1982) and protandrous *Amphiprion* (e.g., Fricke & Fricke 1977, also see Ochi 1989a, b). Chapter 9 'Partial protandry in the alpheid shrimp *Athanas kominatoensis*: a case of sex change in Crustacea' (by Y. Nakashima; also see Nakashima 1987) was included, though it is not a fish, because examples of protandrous fishes whose social systems were well known were very limited. Thus, the book is composed of these nine chapters with more than 400 references cited.

Since 1987

Since Nakazono & Kuwamura (1987), the study of sex change has been attracting many Japanese ichthyologists. Hermaphroditism in fishes was reviewed by Nakazono (1991) in various aspects, and by Suzuki (1989) from the viewpoint of comparative gonad histology. Physiological aspects of sex change, such as on function of sex hormones, have also been studied (e.g., Nakamura et al. 1994). For theoretical aspects Iwasa (1991) developed a model of sex-change evolution by dynamic programming, and proposed that the evolutionary stable sexual style may include an extended nonreproductive period that intervenes between male and female phases, if costs of reproduction are concerned. This may correspond to early sex change. Iwasa (op. cit.) also showed that the difference between sexes either in mortality (mortality advantage) or in growth rate (growth-rate advantage) favors the evolution of sex change even if the size advantage (size-fecundity advantage) is the same between the sexes. For ecological and histological aspects of hermaphroditic fishes living in Japanese waters, more than 30 papers have been published since 1987, summarized as follows.

Protogynous fishes

Among protogynous fishes, labrids have been most

intensively studied. Three sexual pathways in individual life histories are known (Figure 1). Males derived without sex change are called primary males and those through sex change secondary males; species containing both primary and secondary males are called diandric and those containing only secondary males monandric (e.g., Warner & Robertson 1978). Dichromatism (i.e. two color phases) related to body size is often referred to as initial phase (IP) and terminal phase (TP).

Jack Moyer, who conducted much of the pioneer work on reproductive behavior of hermaphroditic fishes in Japan, reported spawning behavior and mating systems of nine protogynous labrids from Miyake-jima Island, Izu Islands (Moyer 1991). Fukui et al. (1991) studied a temperate wrasse *Halichoeres poecilopterus* in Seto Inland Sea, western Japan. The population showed higher density, higher percentage of primary males (about 40% in both IP and TP fish) and smaller size at sex change (or change from IP to TP) than the conspecific population of northern Kyushu studied by Nakazono (1979). Fukui et al. (1991) observed only group spawning by a female and IP and TP males, suggesting that TP males could not monopolize females in such a high-density population. They also suggested that the fishing pressure on large individuals may have reduced the size at sex change to a smaller size in fishes in their study area.

Halichoeres poecilopterus was also studied by Kobayashi & Suzuki (1994) in Suruga Bay, Central Japan. The percentage of primary males and the size at sex change were similar to those in the Seto Inland Sea. Immature IP fish had either ovaries, intersexual gonads with many oocytes and seminal lobules, or primary testes with markedly developed seminal lobules and a few generating oocytes. This suggests that primary testes may have also experienced ovarian features in their early stages, although primary males have been distinguished from 'prematurational secondary males' whose gonads are derived from immature ovaries through prematurational sex change (Robertson & Warner 1978). Further examination seems to be needed for the distinction of primary males and prematurational secondary males to confirm whether they differ genetically or represent a conditional style.

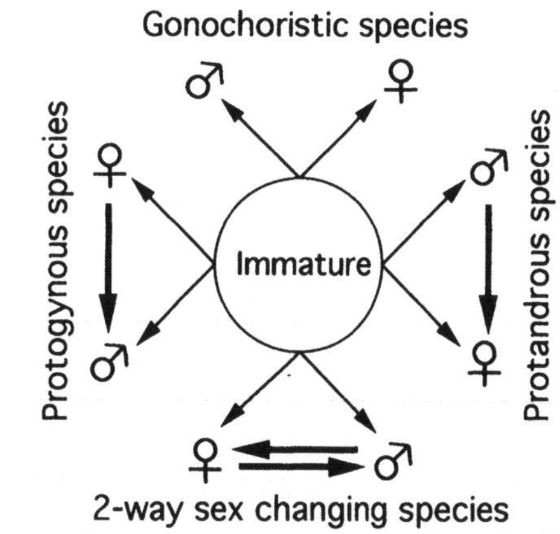

Figure 1. Possible sexual pathways of individual life histories within a species of each category. Arrows indicate sexual maturation (thin line) or functional sex change (thick line).

A congener *H. marginatus* was studied at Kuchierabu-jima Island, south of Kyushu, by Shibuno et al. (1993a). This species is also diandric, but primary males made up only 5% of the population. TP males establish territories in the late afternoon, to which females migrate to spawn either by pair-mating with the TP males (sometimes accompanied by sneaking or streaking IP/TP males) or by group mating with IP and TP males. The spawning sites that a female had used were subsequently used for its mating after it had changed sex (Shibuno et al. 1993b). We (K. Karino, T. Kuwamura, Y. Nakashima, and A. Nema) have also been studying sex change and sexual selection of another diandric labrid *H. melanurus* at Sesoko Island, Okinawa (unpublished).

In *Cirrhilabrus temmincki* from Suruga Bay, no primary (and IP) males were found, all TP fish being secondary males (Kobayashi & Suzuki 1990). The same was found in *Choerodon schoenleinii* from Okinawa Island (Ebisawa et al. 1995). Its congener *C. azurio* in northern Kyushu was also suggested to be monandric (Nakazono & Kusen 1991), as was a temperate parrotfish *Calotomus japonicus* from Makurazaki, southern Kyushu (Kusen & Nakazono 1991). In the latter two species, however, the number of specimens used for the gonad examination was so small (15–17) that it seems difficult to consider them monandric, because the proportion of primary males may sometimes be very small (< 5%; e.g., Shibuno et al. 1993a).

Prematurational sex change may occur even in monandric species (Robertson & Warner 1978, Kuwamura 1981, Tanaka et al. 1990). Alternative mating and life-history style of the monandric wrasse *Labroides dimidiatus* have been studied in Uwa Sea, western Shikoku, by Y. Sakai (1995, p. 77, in 24th International Ethological Conference Abstracts), who has also studied conditions for early sex change in the angelfish *Centropyge ferrugatus* at Sesoko Island, Okinawa (Sakai 1992, 1995).

Studies have also been done for protogynous fishes of other families besides Labridae. The sand-diver *Trichonotus filamentosus* in Hakata Bay, northern Kyushu, was found to be protogynous with all males having secondary testes (Kusen et al. 1991). This is the first evidence of protogyny from the family Trichonotidae.

The sandperch *Parapercis snyderi* (Parapercidae) has been intensively studied in recent years. In Suruga Bay it is monandric and, different from the labrid secondary males, the ovarian cavity clearly remains only in the early stage and later disappears in the secondary testis (Kobayashi et al. 1993a). The secondary males maintained territories including a harem of 3–10 females, and not only pair-spawning but also streaking by other males were observed. When females were kept together in a tank, not only the largest fish changed sex to male but also the remaining came to have intersexual gonads. However, when females were kept singly no gonadal change occurred (Kobayashi et al. 1993b). Mating system and sex change of *P. snyderi* have been studied also in other localities: Uwa Sea, western Shikoku [N. Onishi & Y. Yanagisawa 1992, p. 47 in Advance Abstracts for 25th Annual Meeting of the Ichthyological Society of Japan (in Japanese; hereafter abbreviated to AMIJ); N. Onishi 1995, p. 53 in 28th AMIJ], and two localities in Kagoshima, southern Kyushu (M. Kobayakawa, K. Baba and A. Shinomiya 1992, p. 23 in 25th AMIJ; K. Baba, M. Kobayakawa and A. Shinomiya 1995, p. 52 in 28th AMIJ). Among these localities, differences have been found in longevity and size at sex change.

Protandrous fishes

Detailed studies on protandry are still limited to those on the anemonefishes. Females are larger than males in monogamous pairs, and it was believed that a male would always change sex after death of his mate (e.g., Fricke & Fricke 1977). However, alternative life-history styles have been revealed by intensive field work both in temperate waters (Uwa Sea, western Shikoku) and on the subtropical coral reefs of southern Japan (Sesoko Island, Okinawa).

In *Amphiprion clarkii* living in the temperate water where host sea anemones occur in high density, Ochi (1989a) found that after mate loss, males often moved between hosts to acquire new mates, usually without sex change. He suggested that sex change is a style rather like the BBS (best of a bad situation) in this habitat, because it may take a longer time to change sex than to acquire a new heterosexual mate by movement. This was confirmed by the experimental removal of females from breeding pairs (Hattori & Yanagisawa 1991a).

In this high host-density habitat (Uwa Sea), breeding pairs establish almost contiguous territories, and nonbreeders shelter in relatively small hosts in the fringes of the pairs' territories. The nonbreeders may mature into either sex according to their social status, or body size relative to their new mates (Ochi 1989b, Hattori & Yanagisawa 1991b). The examination of gonad histology in relation to body-color change revealed three sexual pathways in life histories of the anemonefish: (1) from immature male to functional male, (2) and then to functional female by sex change, or (3) from immature male to immature female and then to functional female (prematurational sex change or femininity differentiation in nonbreeder state; Hattori & Yanagisawa 1991b; see Figure 1).

Even in coral reefs where the host sea anemones occur in much lower densities (Sesoko Island), *A. clarkii*, especially nonbreeders, often move between hosts to acquire new mates and breeding sites, and the same three life-history pathways were detected (Hattori 1994). Hattori & Yamamura (1995) developed an ESS model for the co-existence of immature males and females under frequency-dependent selection, suggesting that the alternative life-history pathways in *A. clarkii* have evolved as style of breeding site acquisition.

Amphiprion frenatus living in the same coral

reefs with lower host-densities sometimes moves between hosts (Hattori 1991, Hirose 1995). By contrast, *A. perideraion*, a smaller species inhabiting the same host species as *A. clarkii*, seldom moves (Hattori 1995, Hirose 1995). Since dominant females strongly suppress growth of their mates in *A. frenatus*, the size difference between the sexes in pairs of *A. frenatus* is much larger than in *A. clarkii*, although females sizes are similar between the two species (Hattori 1991, 1994). It is suggested that the difference in mobility among the three species of anemonefishes may affect the patterns of mate acquisition and the size composition of breeding groups (Hirose 1995).

These studies on monogamous anemonefishes indicate that if movement between hosts is strictly limited, protandrous life-history may be optimal as predicted by the size-advantage model, but the ability of movement may determine the frequencies of sex change and other life-history styles. In addition, social systems and gonad histology of *A. ocellaris*, which is believed to be most strictly host-attached, have been studied at Iriomote Island, Okinawa (Y. Inagaki 1992, p. 3 in 25th AMIJ; A. Yamashita, K. Koike and A. Hattori 1993, p. 35 in 26th AMIJ).

Two-way sex change

The occurrence of reverse-directional sex change was first, but only briefly, reported from a protogynous grouper *Epinephelus akaara*. Males larger than 30 cm TL were observed to change sex in captivity (Tanaka et al. 1990). After about 30 males were kept together in a tank for nine months, a few of them, larger than the average size, were found to have mature oocytes in their gonads (K. Nogami & H. Tanaka personal communication).

Similarly, although protogyny is common in polygynous angelfishes (Pomacanthidae; Moyer 1990), the reverse sex change was suggested in *Apolemichthys trimaculatus*. When two functional males were kept in an aquarium, the smaller came to have an intermediate gonad at the end of the experiment (Hioki & Suzuki 1995). Later, spawning after male-to-female sex change was observed in 2-male experiments of *Centropyge flavissimus, C. fisheri,* and *C. acanthops* (Hioki & Suzuki 1996).

In the hawkfish *Cirrhitichthys aureus* (Cirrhitidae), spawning was observed after female-to-male sex change in a 2-female experiment, and also after male-to-female sex change in a 2-male experiment in a tank (Kobayashi & Suzuki 1992). Specimens collected from Suruga Bay had either immature ovaries, secondary testes, or in most cases ambisexual gonads. The smaller ambisexual fish showed active oogenetic processes in the major ovarian zones, while the larger ambisexual ones had slender androgenic gonads with active spermatogenesis in various testicular zones. Similar ambisexual gonads were also found in other hawkfishes, *Cyprinocirrhites polyactis* from Suruga Bay and *Cirrhitichthys aprinus, C. falco* and *Cirrhitopus hubbardi* from Hachijo-jima Island, Izu Islands. Kobayashi & Suzuki (1992) regarded these bisexual gonadal aspects as an intermediate between sequential (protogynous) and simultaneous hermaphroditism. However, simultaneously functional hermaphrodites have not been reported from the hawkfishes, and it is uncertain whether the gonad structure represents an evolutionarily intermediate interval. The ambisexual gonads may be maintained because of the advantage in changing sex in either direction.

Two-way sex change has been also confirmed by aquarium experiments in a small goby *Trimma okinawae*, whose males have ovarian tissue within a functional testis (Sunobe & Nakazono 1993). In Kagoshima, southern Kyushu, its mating system was resource defense polygyny, with paternal egg care. The largest female of a social unit changed sex following the removal of the dominant male, indicating socially controlled protogyny (Sunobe & Nakazono 1990). In aquarium experiments, however, when males were kept together, the smaller changed back to female and spawned within 4–12 days (Sunobe & Nakazono 1993). Also, in a congener *T. grammistes*, the larger changed sex in a 2-female experiment and the smaller in a 2-male experiment (Y. Shiobara, G. Nishi, N. Okutomi, M. Takahashi and H. Takeoka 1992, p. 24 in 25th AMIJ).

Evidence for 2-way sex change in natural populations was obtained for the first time in the coral goby *Paragobiodon echinocephalus* at Sesoko Island, Okinawa (Kuwamura et al. 1994). In this monogamous fish inhabiting *Stylophora* corals, sex change occurs according to the relative size of new partners; usually the larger becomes male and the smaller female. This was confirmed by monitoring marked individuals in the field (Kuwamura et al. 1994), and by a mate removal experiment in the field as well as an aquarium experiment with 2 consexuals (Nakashima et al. 1995). Both sexes of gobies showed equal ability to change sex both ways in the experiments (within 3–4 weeks at the shortest). In the natural population, however, maturation as female and subsequent female-to-male sex change occurred much more frequently than the other way. In any event, the sexual life-history pathways of the gobies were more flexible than protandrous or protogynous species (Figure 1).

Two-way sex change has been confirmed also in other monogamous gobies (*Gobiodon micropus, G. oculolineatus, G. quinquestrigatus* and *G. rivulatus rivulatus*), inhabiting *Acropora* corals, by the field and aquarium experiments at Sesoko Island (Nakashima et al. 1996).

Thus during the last few years, sex change in either direction, or in the reverse direction to the former sex, has been revealed by Japanese researchers among several groups of reef fishes. However, in most cases at present it has been confirmed only in captive conditions.

Social control of 2-way sex change

Social control of sex change is well known among reef fishes. The dominant fish becomes male in polygynous mating groups, while the dominant becomes female in monogamous anemonefishes (Robertson 1972, Fricke & Fricke 1977, Warner 1984, Nakazono & Kuwamura 1987, Ross 1990). The size-advantage model predicts that the dominant should do so to increase its fitness. It is also evident that the 2-way sex change is socially controlled at least in the gobies *Trimma, Paragobiodon,* and *Gobiodon* (Sunobe & Nakazono 1993, Kuwamura et al. 1994, Nakashima et al. 1995, 1996). In these species, the dominant fish becomes male, although the evolutionary factors differ between *Trimma* and the others.

Trimma has polygynous mating systems, and so the dominant fish takes advantages of becoming male as in many other protogynous fishes. By contrast, mating occurs monogamously in *Paragobiodon* and *Gobiodon*, so the dominant should be a female, if the size-fecundity advantage exists as in the anemonefishes. However, no size-fecundity advantage was detected in *Paragobiodon* (Kuwamura et al. 1993, 1994). Breeding pairs are formed by a male and a female of similar size, and reproductive success is positively correlated with body size of the male and female almost equally. In other words, females' fecundity and males' ability of egg care may change with body size in similar fashion. In such size-assortative mating, which was also the case in *Gobiodon* (Nakashima et al. 1996), gonochorism should be favored if other factors than the size-fecundity advantage are not acting (Warner 1984). The major factor was, however, the growth-rate advantage in *Paragobiodon* and at least two species of *Gobiodon* (Kuwamura et al. 1994, Nakashima et al. 1996). That is, because females grow faster than their mates after new pair formation until the start of breeding, and because the smaller sex limits the pair's reproductive success, the smaller individual should become female. Thus, the larger, dominant fish in new pairs should be male in these monogamous gobies.

If the dominant fish becomes a male, protogyny will be favored, as observed in the natural population of *Trimma* (Sunobe & Nakazono 1990) and *Paragobiodon* (Kuwamura et al. 1994). This is because when small, a fish may usually reproduce as female with a larger mate, and when it becomes large and dominant, it had better change sex to male to mate with smaller fish. Sex change in the reverse direction, i.e., male to female, will occur only when the social status of the dominant fish (male) changes by inter-group movement of itself or its new mate (Figure 2). This was confirmed by field observations and experiments in *Paragobiodon* (Kuwamura et al. 1994, Nakashima et al. 1995) and *Gobiodon* (Nakashima et al. 1996). Inter-coral movement often occurred to acquire new mates in these gobies, and unmated fish preferred pairing with nearby consexuals over searching for heterosexuals from farther distances.

Change of social status of the dominant fish has not been observed in the field in *Trimma*. Sunobe & Nakazono (1993) suggested that if the largest female, whose size was sometimes similar to the dominant male, grows faster than the male, change of their status within a social group would occur. In *Paragobiodon*, however, even if social status changed within a pair due to the faster growth in females, simultaneous sex change in both mates was never observed, probably because the cost of sex change may exceed the growth-rate advantage (Kuwamura et al. 1994).

It is noteworthy here that the subordinate fish changes sex in the 'returning' sex change: in 2-male pairs the smaller changes sex in the gobies (Figure 2). The dominant fish will increase its reproductive success by forcing the subordinate to change sex. This is quite different from the suppression theory of social control (i.e., the dominant fish changes sex and socially suppresses sex change of the subordinate) in the former one-way sex change theory (e.g., Ross 1990). Sex change of subordinate males as BBS was reported in a shrimp *Anthanas kominatoensis*, in which secondary sex change (female-to-male) when becoming dominant was also suggested but not confirmed (Nakashima 1987). In case of the polygynous *Trimma*, the subordinate male may change sex as BBS when it has lost mates or has been taken over by the larger male; the same may occur in the angelfishes. By contrast, in the monogamous *Paragobiodon* and *Gobiodon*, the subordinate male would increase its reproductive success by pairing with a larger male and changing sex to grow faster as a female.

The case of the hawkfish *Cirrhitichthys aureus* appears to be contradictory. The larger fish (90.0 mm vs. 74.5 mm SL) became male in a 2-female experiment, while the larger (93.6 mm vs. 89.2 mm) became female in a 2-male experiment (Kobayashi & Suzuki 1992). However, it seems to be the rule that the larger becomes male, because the larger ambisexual specimens had male-biased functions (Kobayashi & Suzuki 1992) and also because males were larger in monogamous pairs in the field (Yogo 1987). Further field observations are needed to confirm whether this species is strictly monogamous or polygynous with facultative monogamy. It is also unknown at present why the relatively large males of the protogynous grouper *Epinephelus akaara* developed oocytes in multi-male

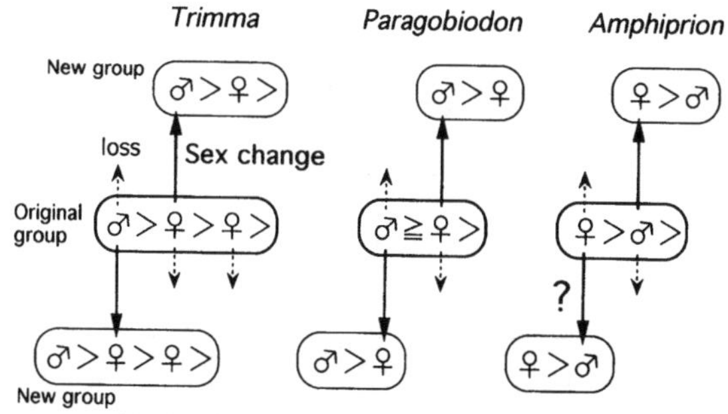

Figure 2. Social dominance and control of 2-way sex change in polygynous (*Trimma*) and monogamous (*Paragobiodon* and *Amphiprion*) groups. Sex change in different directions may occur after loss (shown by dashed arrows) of the dominant mate (top) or after loss of the subordinate mate(s) (bottom). In newly formed pairs of *Paragobiodon* males are initially larger than their mates, but later become the same size due to the growth-rate advantage of females.

experiments (Tanaka et al. 1990, K. Nogami & H. Tanaka personal communication). It would be the smaller subordinate males that should change sex, if there is a social control of functional sex.

In addition to the fishes of the Japanese waters, another two gobies, *Lythrypnus dalli* and *L. zebra*, inhabiting the rocky shore of southern California and Mexico, have been reported to exhibit 2-way sex change in laboratory experiments (St. Mary 1993, 1994, 1996). These gobies have only vitellogenic eggs, only tailed sperm, or both active gametes of various proportions in the gonad. The behavioral (functional) sex can be altered within a few weeks by reallocation of the gonad. Although St. Mary called these species as simultaneous hermaphrodites based on gonadal sex (also see Fishelson 1989), it sounds confusing to call such sex-changing (i.e., sequential) hermaphrodites as 'simultaneous' even if they often have bisexual gonads. The gonad structure of *Lythrypnus*, separated into ovarian and testicular parts, is similar to *Trimma*, but different from *Paragobiodon* and *Gobiodon* (Fishelson 1989, Cole 1990, St. Mary 1993, Sonobe & Nakazono 1993). As in *Trimma*, the mating systems of *Lythrypnus* are polygynous with paternal egg care (St. Mary 1994, 1996). Egg-guarding males tend to be larger than females, but with a large overlap within a stock. When consexuals were kept together, larger fish tended to become males, although no such tendency was found in female-group experiment of *L. zebra* (St. Mary 1994, 1996). Because they occur in high densities, social units of *Lythrypnus* may not be so fixed as they are in *Trimma*. Further investigations are needed in the field to know whether, or how, social conditions such as change of social status induce sex change in individuals of *Lythrypnus*.

Nakashima et al. (1995) suggested that change of social status by inter-group movement may also occur, even if only infrequently, in other hermaphroditic fishes living in social groups, and therefore sex change in the reverse direction, or 2-way sex change, will be detected in many of them. In the case of the protogynous sandperch *Parapercis snyderi*, however, no sex change was observed within 37 days when five males were kept in an aquarium (Kobayashi et al. 1993b). The authors suggested that the gonad structure might constrain the reverse sex change; in the secondary testis the ovarian cavity gradually degenerates and disappears in the sandperch. However, it is also likely that the conditions in the aquarium were not suitable for inducing sex change. Four of five males died after 33–37 days, although no aggressive interactions were observed, and all fish were reduced in size (Kobayashi et al. 1993b).

Reverse sex change has been reported only from basically protogynous fishes (e.g., Gobiidae, Pomacanthidae, Cirrhitidae, Epinephelinae). Is this simply because protogynous species are predominant over protandrous ones among reef fishes (Yogo 1987), or do protandrous species never exhibit the reverse sex change due to some unknown constraints? In the anemonefish *Amphiprion bicinctus*, Fricke & Fricke (1977) reported that forcible pairing of functional females resulted in death or severe injury to the subdominants and that the reversal of sex change from female to male could not be induced.

During our preliminary experiments on *A. clarkii* at Sesoko Island (Kuwamura, Nakashima & Karino unpublished), the smaller of two females kept in a tank changed the shape of its urogenital papilla into one that was intermediate between the sexes within six months. In the field experiment during more than 18 months, however, two unmated females in 15 m distance never moved to form a 2-female pair; they accepted juveniles but never resumed breeding.

In the female-dominance system of the monogamous anemonefishes, the reproductive success of a pair depends on the female body size, which affects fecundity, but not on male size (e.g., Fricke & Fricke 1977, Warner 1984). This is because males of the anemonefishes can take care of a clutch deposited by much larger mates, which is different from *Paragobiodon* males. Therefore, if a female anemone fish could acquire a new mate larger than herself, sex change to male would be more advantageous than remaining a female (Figure 2). For the larger female, however, the size of her mate would not affect her own reproductive success.

In such situations as the anemonefish, a female should accept the smaller female and force it to change sex, but only when the duration needed for the sex change is expected to be shorter than that for maturation of recruited juveniles or their immigration. Although it may take a few years for juveniles to mature, late juveniles can be expected to immigrate and mature within a few months in *A. clarkii* at Sesoko Island (Hattori 1994, Hirose 1995). If sex change from female to male takes more than a few months (cf., about two months in male-to-female change; Hattori 1994), the larger female should wait for immigration of late juveniles. It seems that the two unmated females of *A. clarkii* in our field experiment adopted this style, but unfortunately no immigration of late juveniles occurred for more than a year. Further experiments are needed to study the possibility of reverse sex change in the anemonefishes, although it is suggested that mobility of adults and late juveniles may affect its evolution.

Closing comments

Recent studies on sex change among fishes have revealed various life-history pathways in hermaphroditic species (Figure 1). The discovery of 2-way sex change suggests that the difference between protandrous and protogynous species may be much smaller than has been believed from histological studies. If there is any social condition that favors evolution of sex change in one direction, the reverse sex change will also be likely to evolve, so far as the social status of the dominant fish can change (Figure 2). This can be explained by the size-advantage model (sensu Warner 1988a, Kuwamura et al. 1994, Nakashima et al. 1995) incorporated with the social-control system. Further studies should focus on finding evolutionary constraints preventing hermaphroditic species from conducting the reverse sex change. The gonad structure, however, should not be regarded as a constraint, but should be treated as an expression of reproductive styles.

Acknowledgements

I (T.K.) would like to express my sincere thanks to H. Kawanabe, who stimulated my interest in behavioral ecology of fishes, including sex change, after his critical reading of my doctoral thesis, although he himself has never been a behavioral ecologist. We thank H. Tanaka and K. Nogami for providing unpublished information on the grouper experiments, and A. Hattori and an anonymous reviewer for comments on the manuscript. This work was partly supported by funding for T.K. to be a visiting professor of Sesoko Station of Tropical Biosphere Research Center, University of the Ryukyus (1995) and by a Grant-in-Aid (# 05640718) from the Japan Ministry of Education, Science, Sports and Culture. This is a contribution from the Sesoko Station.

References cited

Charnov, E.L. 1982. The theory of sex allocation. Princeton University Press, Princeton. 355 pp.

Coates, D. 1982. Some observations on the sexuality of humbug damselfish, *Dascyllus aruanus* (Pisces, Pomacentridae) in the field. Z. Tierpsychol. 59: 7–18.

Cole, K.S. 1990. Patterns of gonad structure in hermaphroditic gobies (Teleostei : Gobiidae). Env. Biol. Fish. 28: 125–142.

Ebisawa, A., K. Kanashiro, T. Kyan & F. Motonaga. 1995. Aspects of reproduction and sexuality in the back-spot tuskfish, *Choerodon schoenleinii*. Japan. J. Ichthyol. 42: 121–130.

Fishelson, L. 1989. Bisexuality and pedogenesis in gobies (Gobiidae : Teleostei) and other fish, or, why so many little fish in tropical seas? Seckenbergiana Marit. 20: 147–169.

Fricke, H. & S. Fricke. 1977. Monogamy and sex change by aggressive dominance in coral reef fish. Nature 266: 830–832.

Fukui, Y., S. Gushima, S. Kakuda & H. Hashimoto. 1991. Growth-related changes in color and sex in *Halichoeres poecilopterus*. Japan. J. Ichthyol. 37: 395–401. (in Japanese).

Ghiselin, M.T. 1969. The evolution of hermaphroditism among animals. Quart. Rev. Biol. 44: 189–208.

Hattori, A. 1991. Socially controlled growth and size-dependent sex change in the anemonefish *Amphiprion frenatus* in Okinawa, Japan. Japan. J. Ichthyol. 38: 165–177.

Hattori, A. 1994. Inter-group movement and mate acquisition tactics of the protandrous anemonefish, *Amphiprion clarkii*, on a coral reef, Okinawa. Japan. J. Ichthyol. 41: 159–165.

Hattori, A. 1995. Coexistence of two anemonefishes, *Amphiprion clarkii* and *A. perideraion*, which utilize the same host sea anemone. Env. Biol. Fish. 42: 345–353.

Hattori, A. & N. Yamamura. 1995. Co-existence of subadult males and females as alternative tactics of breeding post acquisition in a monogamous and protandrous anemonefish. Evol. Ecol. 9: 292–303.

Hattori, A. & Y. Yanagisawa. 1991a. Sex change of the anemonefish *Amphiprion clarkii* in a habitat of high host density: a removal study. Japan. J. Ecol. 41: 1–8. (in Japanese).

Hattori, A. & Y. Yanagisawa. 1991b. Life-history pathways in relation to gonadal sex differentiation in the anemonefish, *Amphiprion clarkii*, in temperate waters of Japan. Env. Biol. Fish. 31: 139–155.

Hioki, S. & K. Suzuki. 1995. Spawning behavior, eggs, larvae, and hermaphroditism of the angelfish, *Apolemichthys trimaculatus*, in captivity. Bull. Inst. Oceanic Res. & Develop., Tokai Univ. 16: 13–22. (in Japanese).

Hioki, S. & K. Suzuki. 1996. Sex changing from male to female on the way of protogynous process in three *Centropyge* angelfishes (Pomacanthidae: Teleostei). Bull. Inst. Oceanic Res. & Develop., Tokai Univ. 17: 27–34 (in Japanese).

Hirose, Y. 1995. Patterns of pair formation in protandrous anemonefishes, *Amphiprion clarkii*, *A. frenatus* and *A. perideraion*, on coral reefs of Okinawa, Japan. Env. Biol. Fish. 43: 153–161.

Iwasa, Y. 1991. Sex change evolution and cost of reproduction. Behav. Ecol. 2: 56–68.

Kobayashi, K. & K. Suzuki. 1990. Gonadogenesis and sex succession in the protogynous wrasse, *Cirrhilabrus temmincki*, in Suruga Bay, Central Japan. Japan. J. Ichthyol. 37: 256–264.

Kobayashi, K. & K. Suzuki. 1992. Hermaphroditism and sexual function in *Cirrhitichthys aureus* and the other Japanese hawkfishes (Cirrhitidae : Teleostei). Japan. J. Ichthyol. 38: 397–410. (in Japanese).

Kobayashi, K. & K. Suzuki. 1994. Intersexuality and reproduction in *Halichoeres poecilopterus*, in Suruga Bay, Central Japan. J. Fac. Mar. Sci. Technol., Tokai Univ. 38: 233–256. (in Japanese).

Kobayashi, K., K. Suzuki & Y. Shiobara. 1993a. Reproduction and hermaphroditism in *Parapercis snyderi* (Teleostei, Parapercidae) in Suruga Bay, Central Japan. J. Fac. Mar. Sci. Technol., Tokai Univ. 35: 149–168. (in Japanese).

Kobayashi, K., K. Suzuki & Y. Shiobara. 1993b. Studies on the gonadal sex succession in *Parapercis snyderi* cross examined in captivity. Bull. Inst. Oceanic Res. & Develop., Tokai Univ. 14: 83–91. (in Japanese).

Kusen, J.D., K. Nakagawa, Y. Yogo & A. Nakazono. 1991. Protogynous hermaphroditism in the sand diver *Trichonotus filamentosus*. Nippon Suisan Gakkaishi 57: 35–40.

Kusen, J.D. & A. Nakazono. 1991. Protogynous hermaphroditism in the parrotfish, *Calotomus japonicus*. Japan. J. Ichthyol. 38: 41–45.

Kuwamura, T. 1981. Life history and population fluctuation in the labrid fish, *Labroides dimidiatus*, near the northern limit of its range. Publ. Seto Mar. Biol. Lab. 26: 95–117.

Kuwamura, T. 1984. Social structure of the protogynous fish *Labroides dimidiatus*. Publ. Seto Mar. Biol. Lab. 29: 117–177.

Kuwamura, T., Y. Nakashima & Y. Yogo. 1994. Sex change in either direction by growth-rate advantage in the monogamous coral goby, *Paragobiodon echinocephalus*. Behav. Ecol. 5: 434–438.

Kuwamura, T., Y. Yogo & Y. Nakashima. 1993. Size-assortative monogamy and paternal egg care in a coral goby *Paragobiodon echinocephalus*. Ethology 95: 65–75.

Moyer, J.T. 1990. Social and reproductive behavior of *Chaetodontoplus mesoleucus* (Pomacanthidae) at Bantayan Island, Philippines, with notes on pomacanthid relationships. Japan. J. Ichthyol. 36: 459–467.

Moyer, J.T. 1991. Comparative mating strategies of labrid fishes. Monogr. No. 1, Watanabe Ichthyological Institute, Tokyo. 90 pp.

Moyer, J.T. & M.J. Zaiser. 1984. Early sex change: a possible mating strategy of *Centropyge* angelfishes (Pisces : Pomacanthidae). J. Ethol. 2: 63–67.

Nakamura, M., T. Mariko & Y. Nagahama. 1994. Ultrastructure and *in vitro* steroidogenesis of the gonads in the protandrous anemonefish *Amphiprion frenatus*. Japan. J. Ichthyol. 41: 47–56.

Nakashima, Y. 1987. Reproductive strategies in partially protandrous shrimp, *Athanas kominatoensis* (Decapoda, Alpheidae): sex change as the best of a bad situation for subordinates. J. Ethol. 5: 145–159.

Nakashima, Y., T. Kuwamura & Y. Yogo. 1995. Why be a both-ways sex changer? Ethology 101: 301–307.

Nakashima, Y., T. Kuwamura & Y. Yogo. 1996. Both-ways sex change in monogamous coral gobies, *Gobiodon* spp. Env. Biol. Fish. 46: 281–288.

Nakazono, A. 1979. Studies on the sex reversal and spawning behavior of five species of Japanese labrid fishes. Rep. Fish. Res. Lab. Kyushu Univ. 4: 1–64. (in Japanese).

Nakazono, A. 1991. Functional hermaphroditism. pp. 327–361. *In:* Y. Itazawa & I. Hanyu (ed.) Fish Physiology, Kouseisha-Kouseikaku, Tokyo. (in Japanese).

Nakazono, A. & J.D. Kusen. 1991. Protogynous hermaphroditism in the wrasse *Chaerodon azurio*. Nippon Suisan Gakkaishi 57: 417–420.

Nakazono, A. & T. Kuwamura (ed.). 1987. Sex change in fishes. Tokai University Press, Tokyo. 284 pp. (in Japanese).

Ochi, H. 1989a. Mating behavior and sex change of the anemonefish, *Amphiprion clarkii*, in temperate waters of southern Japan. Env. Biol. Fish. 26: 257–275.

Ochi, H. 1989b. Acquisition of breeding space by nonbreeders in the anemonefish *Amphiprion clarkii* in temperate waters of southern Japan. Ethology 83: 279–294.

Robertson, D.R. 1972. Social control of sex reversal in a coral-reef fish. Science 177: 1007–1009.

Robertson, D.R. 1974. A study of the ethology and reproductive biology of the labrid fish, *Labroides dimidiatus*, at Heron Island, Great Barrier Reef. Ph.D. Thesis, University of Queensland, St. Lucia. 295 pp.

Robertson, D.R. & R.R. Warner. 1978. Sexual patterns in the labroid fishes of the western Caribbean, II: the parrotfishes (Scaridae). Smithsonian Contrib. Zool. 255: 1–26.

Ross, R.M. 1990. The evolution of sex-change mechanisms in fishes. Env. Biol. Fish. 29: 81–93.

Sakai, Y. 1992. Early sex change of an angelfish *Centropyge ferrugatus* (Pomacanthidae). Galaxea 11: 69.

Sakai, Y. 1995. Studies on the ecology of the polygynous angelfish *Centropyge ferrugatus*. Ph.D. Thesis, Osaka City University, Osaka. 116 pp.

Shapiro, D.Y. & R. Lubbock. 1980. Group sex ratio and sex reversal. J. Theor. Biol. 82: 411–426.

Shibuno, T., I. Chiba, K. Gushima, S. Kakuda & H. Hashimoto. 1993a. Reproductive behavior of the wrasse, *Halichoeres marginatus*, at Kuchierabu-jima. Japan. J. Ichthyol. 40: 351–359.

Shibuno, T., K. Gushima & S. Kakuda. 1993b. Female spawning migrations of the protogynous wrasse, *Halichoeres marginatus*. Japan. J. Ichthyol. 39: 357–362.

St. Mary, C.M. 1993. Novel sexual patterns in two simultaneously hermaphroditic gobies, *Lythrypnus dalli* and *Lythrypnus zebra*. Copeia 1993: 1062–1072.

St. Mary, C.M. 1994. Sex allocation in a simultaneous hermaphrodite, the blue-banded goby (*Lythrypnus dalli*): the effects of body size and behavioral gender and the consequences for reproduction. Behav. Ecol. 5: 304–313.

St. Mary, C.M. 1996. Sex allocation in a simultaneous hermaphrodite, the zebra goby *Lythrypnus zebra*: insights gained through a comparison with its sympatric congener, *Lythrypnus dalli*. Env. Biol. Fish. 45: 177–190.

Sunobe, T. & A. Nakazono. 1990. Polygynous mating system of *Trimma okinawae* (Pisces : Gobiidae) at Kagoshima, Japan with a note on sex change. Ethology 84: 133–143.

Sunobe, T. & A. Nakazono. 1993. Sex change in both directions by alteration of social dominance in *Trimma okinawae* (Pisces : Gobiidae). Ethology 94: 339–345.

Suzuki, K. 1989. Hermaphroditism in teleost fishes. Saibai-gyogyo-gijutsu-kaihatsu-kenkyu (Rep. Technol. Develop. Fish Culture) 18: 45–55 (in Japanese).

Tanaka, H., K. Hirose, K. Nogami, K. Hattori & N. Ishibashi. 1990. Sexual maturation and sex reversal in red spotted grouper, *Epinephelus akaara*. Bull. Nat. Res. Inst. Aquaculture 17: 1–15. (in Japanese).

Warner, R.R. 1975. The adaptive significance of sequential hermaphroditism in animals. Amer. Nat. 109: 61–82.

Warner, R.R. 1984. Mating behavior and hermaphroditism in coral reef fishes. Amer. Scientist 72: 128–136.

Warner, R.R. 1988a. Sex change and the size-advantage model. Trends Ecol. Evol. 3: 133–136.

Warner, R.R. 1988b. Sex change in fishes: hypotheses, evidence, and objections. Env. Biol. Fish. 22: 81–90.

Warner, R.R. & S.G. Hoffman. 1980. Local population size as a determinant of a mating system and sexual composition in two tropical reef fishes (*Thalassoma* spp.). Evolution 34: 508–518.

Warner, R.R. & D.R. Robertson. 1978. Sexual patterns in the labroid fishes of the western Caribbean, I: the wrasses (Labridae). Smithsonian Contrib. Zool. 254: 1–27.

Yogo, Y. 1985. Studies on the sexual maturation and reproductive ecology in three protogynous fishes. Rep. Fish. Res. Lab. Kyushu Univ. 7: 37–83 (in Japanese).

Yogo, Y. 1987. Hermaphroditism and the evolutionary aspects of it occurrences in fishes. pp. 1–47. *In:* A. Nakazono & T. Kuwamura (ed.) Sex Change in Fishes, Tokai University Press, Tokyo (in Japanese).

Part 3: Full papers and brief communications

Environmental Biology of Fishes **52**: 139–148, 1998.

Evolution of freshwater eels of the genus *Anguilla*: a probable scenario

Katsumi Tsukamoto & Jun Aoyama
Ocean Research Institute, University of Tokyo, 1-15-1, Minamidai, Nakano, Tokyo, 164, Japan
(e-mail: ktpc@ori.u-tokyo.ac.jp)

Received 10.7.1996 Accepted 26.5.1997

Key words: migration loop, Tethys Sea, speciation, catadromy

Synopsis

We present a molecular phylogeny of freshwater eels from three oceans and give hypotheses to address major questions about the evolution and geographic distribution of this group. A phylogenetic tree obtained from mitochondrial cytochrome b sequences of eight species of *Anguilla* suggests that the African species *A. mossambica* and Australian species *A. australis* form a clade together with the two Atlantic species, the European eel, *A. anguilla*, and American eel, *A. rostrata* , whereas *A. marmorata* in the Indo-Pacific Ocean, *A. reinhardti* in northeastern Australia and the Japanese eel, *A. japonica*, in the northwestern Pacific are placed in another. Most speciation among the lineages is proposed to have occurred during the Eocene to Oligocene (45–30 million years ago, Ma). However, the two Atlantic species are estimated to have separated much later, approximately 10 Ma. The following evolutionary scenario for the dispersal and speciation of these species of anguillid eels is proposed based on general global paleogeography and paleo-circulation. Ancestral eels evolved during the Eocene or earlier, in the western Pacific Ocean near present-day Indonesia. A group derived from this ancestor dispersed westward, by transport of larvae in the global circum-equatorial current through the northern edge of the Tethys Sea. This group split into the ancestor of the European and American eels, which entered the Atlantic Ocean, and a second group, which dispersed southward and split into the east African species and Australian species.

Introduction

Some information has emerged in the last 80 years about the unusual life history of the freshwater anguillid eels, but very little has been learned about most of the 15 species. Questions about the phylogenetic relationships of this group and implications of their long spawning migrations and larva period on their global distributions have not been addressed. Freshwater eels of the genus *Anguilla* (family Anguillidae) are catadromous and have a relatively long larva period (the larva called leptocephalus) in the open ocean. All other Anguilliform fishes (congers, morays, snake eels) have the same marine larva period, but have entirely marine life histories (Nelson 1984). For this reason eels are considered to be derived from a marine species. Freshwater eels might have originated in the tropics, because two-thirds of the recognized 15 *Anguilla* species inhabit the tropics while only five species are from temperate regions. Seven of the 15 species occur in the western Pacific around Indonesia, and *A. celebesensis*, which mainly inhabits Sulawesi Island (Celebes) has been supposed to be one of the possible ancestral species of the Anguillidae (Ege 1939, Catsle & Williamson 1974). All these considerations strongly suggest that the western Pacific in and around Indonesia could be the place of origin for freshwater eels.

The world wide geographic distribution of freshwater eels appears to be related to the subtropical circulation of the oceans, with most species being located on the west side of the Atlantic, Pacific and Indian oceans (Ege 1939, Tesch 1977). Warm westward flowing equatorial currents reach and flow along the east coast of each continent, whereas the west coasts are swept by cold currents which originate in high latitudes. The species for which there is some information about spawning areas and distributions of larvae, are clearly associated with warm water or the flow of the subtropical gyres in the north-Atlantic and Pacific (Bertin 1956). This association with specific hydrographic conditions may explain why eels inhabit east coasts of Eurasia, Australia, Africa and North America, but no eels inhabit the west coasts of North and South America, Australia and Africa. Although the European eel, *A. anguilla*, is able to inhabit Europe and the Mediterranean because of the relatively warm eastward flow of the Gulf Stream and North Atlantic drift, it is not known why there are no freshwater eels along the east coast of South America despite the existence of the warm Brazil Current. The geographic distribution and unique life history traits of anguillid eels give rise to many questions about when these eels evolved to form the 15 present-day species and how they achieved their world-wide distribution. In particular, how did the two Atlantic species become completely isolated far from the majority of other eel species in the Indo-Pacific, and why are there no eels in the South Atlantic (e.g. Tucker 1959). To answer these questions, the phylogenetic relationships of these eels need to be clarified and the evolutionary history of *Anguilla* should be considered in light of global paleogeographical events.

In this paper, we used mitochondrial DNA sequence analysis to determine a phylogenetic tree among eight species of *Anguilla* distributed in six different regions of the world. We used these phylogenetic relationships and information about global paleo-geography and paleo-circulation to develop a hypothesis for an evolutionary scenario of speciation and dispersal to account for the unusual geographic distributions and great migrations of anguillid eels.

Materials and methods

Eight species of freshwater eel, *A. celebesensis, A. marmorata, A. mossambica, A. japonica, A. anguilla, A. reinhardti, A. australis* (*A. australis australis*), and *A. rostrata*, were collected from various localities during March 1979 and September 1994 (Figure 1).

Direct extraction of DNA was performed in the field using a specially designed buffer for tropical regions: 8M Urea, 10mM Tris-HCl pH8.5, 125mM NaCl, 50mM EDTA, 1%v/w SDS. Liver tissues were taken from live eels, and minced in the buffer. These samples were transported to the laboratory at ambient temperatures. This buffer was also used for DNA extraction from ethanol preserved tissues. The DNA was purified by extracting twice with phenol/chloroform/isoamyl alcohol (25 : 24 : 1vol/vol), twice with diethyl ether, and concentrated by ethanol precipitation.

The mitochondrial cytochrome b gene was amplified with PCR in order to sequence, using two oligonucleotide primers, GLU-L (Palumbi et al.1991) and H15149 (Kocher 1989). Sequencing of the amplified cytochrome b gene was performed using a Uniplex Cycle DNA Sequence kit (Millipore MBBLUNP02).

In the phylogenetic analysis, we adopted two different methods: neighbor-joining (Saitou & Nei 1987) and maximum parsimony by means of the software package PHYLIP Ver.3.5 (Felsenstein[1]). We selected *A. celebesensis* as an out-group based on the morphological study by Ege (1939), which found this species to have ancestral characteristics among the genus.

Phylogenetic relationships

The sequence divergence between the two Atlantic species, *A. anguilla* and *A. rostrata*, in the 410 base pairs of the cytochrome b gene determined, was lower (0.024) than any other combination among

1 Felsenstein, J. 1993. Phylogeny inference package, PHYLIP Version 3.5. University of Washington, Seattle (also on the web site: http://evolution.genetics.washington.edu/phylip.html).

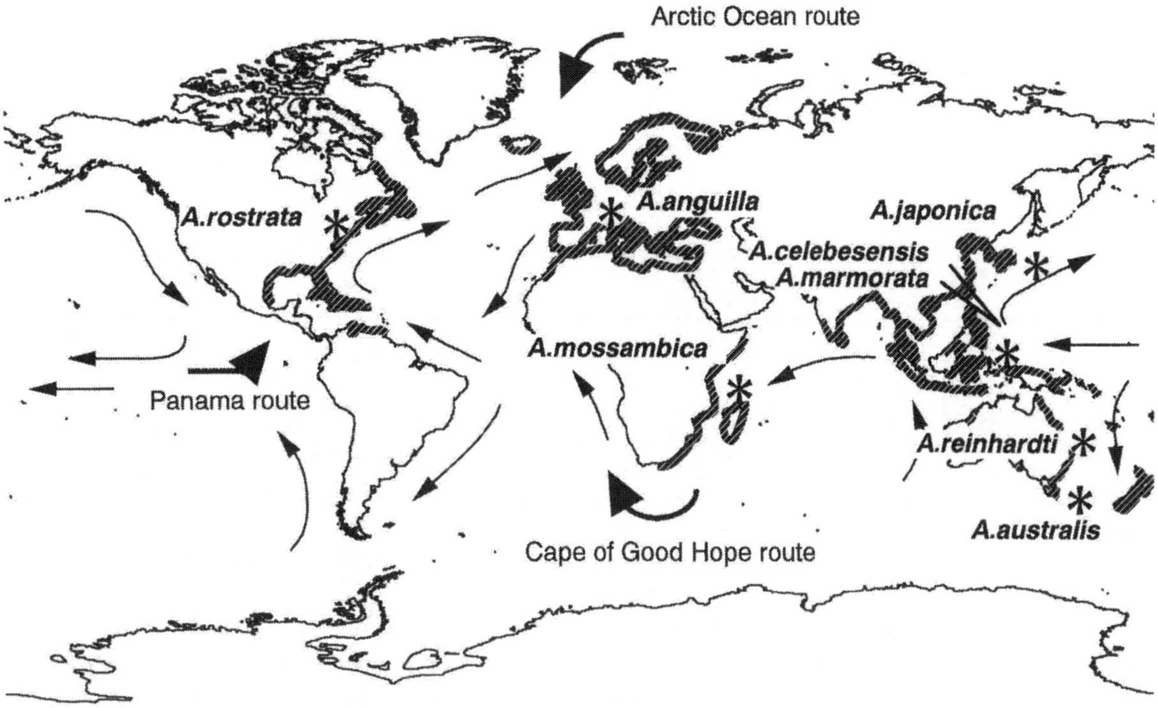

Figure 1. World distribution of all freshwater eels and the sampling locations of eight species studied (asterisks). The geographic ranges of the eight species are as follows: *A. marmorata* in tropical regions of the Indian and Pacific oceans, *A. mossambica* on the eastern coast of Africa, *A. japonica* on the eastern coast of the Eurasian continent, *A. anguilla* on the western coast of the Eurasian continent, *A. rostrata* on the eastern coast of North America, *A. reinhardti* on the eastern coast of Australia, *A. australis australis* in south-eastern Australia and *A. celebesensis* in the Sulawesi Island region. We identified these specimens according to the morphological key presented by Ege(1939).

the species examined (range 0.051-0.085, Aoyama et al. 1996). Phylogenetic analysis with the neighbor-joining method suggested that *A. mossambica, A. australis, A. rostrata* and *A. anguilla* are members of the same clade (Figure 2). Interestingly, *A. reinhardti* and *A. australis* were indicated to be members of different clades in spite of their sympatry in eastern Australia. Since tree lengths in neighbor-joining analysis are a representation of relative evolutionary time, it appears that most of the speciation of *Anguilla* occurred relatively early in comparison with that of the two Atlantic species.

The parsimony method presented three equally parsimonious trees in 75 steps (Aoyama et al. 1996). Each of these trees showed similarity with the neighbor-joining tree in Figure 2. Furthermore, the bootstrap consensus tree based on the results from both methods (Figure 3) was in agreement with one of the parsimonious trees, and with the neighbor-joining tree (except for the position of *A. japonica*). Thus the consensus tree was accepted as a phylogeny of eight *Anguilla* species because it represented both the parsimonious and neighbor-joining trees. An alternate analysis excluding the third codons, which might have more frequent multiple-substitution than the other two codons, showed about the same result as the consensus tree. The only difference was that the positions of *A. australis* and *A. mossambica* were switched. This suggested the possibility that *A. australis* may have branched earlier than *A. mossambica*.

The one previous study of the phylogeny of freshwater eels recognized 18 species including three subspecies, and divided them into four groups. If the scheme of species relationships proposed by Ege(1939) is to be interpreted as a phylogeny and the node species are considered to be the most likely ancestral species, then the scheme can be explained as follows: the group including the ancestral species *A. celebesensis* originated first, next the sec-

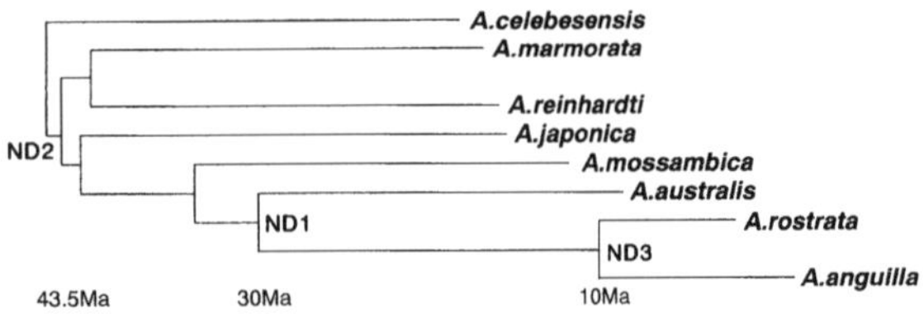

Figure 2. Phylogenetic tree for eight *Anguilla* species examined in the present study, constructed by the neighbor-joining method (Saitou & Nei 1987). The pairwise distance was calculated according to Kimura's two parameter method (Kimura 1980) with weighted modification of 5:1 for unequal rates of transitions and transversion by the DNAdist program in the software package PHYLIP Ver.3.5 (Felsenstein[1]). The node (ND1) indicating a split between ancestors of south Australian species and Atlantic species must be dated before closure of the Tethys Sea, 25-30 Ma, assuming that the Tethys Corridor Hypothesis applies to the expansion of *Anguilla* distributions.

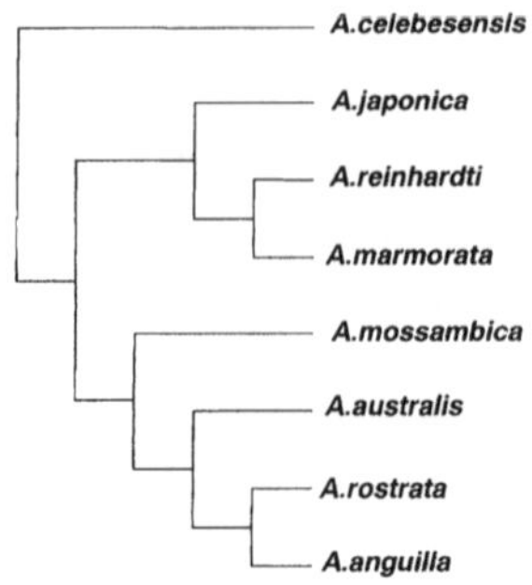

Figure 3. Bootstrap consensus tree for eight *Anguilla* species in the present study. We accepted the consensus tree of the corresponding maximum parsimony and neighbor-joining analysis by means of 1000 replications of bootstrap (Felsenstein 1985), as a phylogeny for the eight *Anguilla* species.

ond group containing *A. marmorata* and *A. reinhardti* appeared, then the third group consisting of *A. japonica, A. anguilla* and *A. rostrata*, and finally the fourth group including *A. australis* appeared. Except for the position of *A. japonica*, the phylogenetic tree obtained from genetic analysis in this study agreed with that resulting from the morphological analysis of Ege (1939).

Ege (1939) concluded that the two Atlantic species, *A. anguilla* and *A. rostrata*, were closely related to *A. japonica* of the North Pacific. However, despite these morphological similarities, their present geographical distributions (separated in different oceans) and the apparent lack of close genetic similarity suggest that these eel species do not have a particularly close phylogenetic relationship. The morphological similarity between Atlantic species and *A. japonica* may be parallel or convergent developments in response to their temperate habitats and long oceanic spawning migrations, which may require common morphological characteristics and functions.

Speciation and dispersal

In principle, each diadromous fish has a 'migration loop' peculiar to each species (Figure 4a). The migration loop consists of different habitats for spawning and growth, and two routes connecting these habitats (e.g. McDowall 1987). Once either of these two habitats of the migration loop is shifted, speciation begins to occur. Although both of the two Atlantic eels, *A. anguilla* and *A. rostrata* breed in the Sargasso Sea, their growth habitats are segregated in Europe and North America, respectively (Figure 4b). Different growth habitats would cause different routes for spawning migration and thus different timing of arrival to the breeding place. The two Atlantic species, despite their roughly common spawning habitat, are reproductively isolated in time in the Sargasso Sea. The peak season is in February for *A. rostrata* and April for *A. anguilla* (McCleave et al. 1987). In contrast, New Zealand long fin eel *A. dieffenbachi* and short fin eel *A. australis* are sympatric in their growth habitat (Figure 4c). However, it has generally been considered that they have different spawning habitats since the maturation interval of silver eels migrating to the sea are quite different and thus, distance to each spawning habitat seems to be different (Jellyman 1987). Therefore, it seems quite possible that a shift of migration loop is a key mechanism for speciation of freshwater eels.

When the ancestral eels dispersed worldwide, their peculiar larvae, the leptocephali (transparent, willow leaf-like), probably played a more important role than the adult/juvenile form because the latter probably do not move much between river systems once they recruit to freshwater. Leptocephali have long duration and appear to be highly adapted to a planktonic life in the sea. Premetamorphic leptoce-

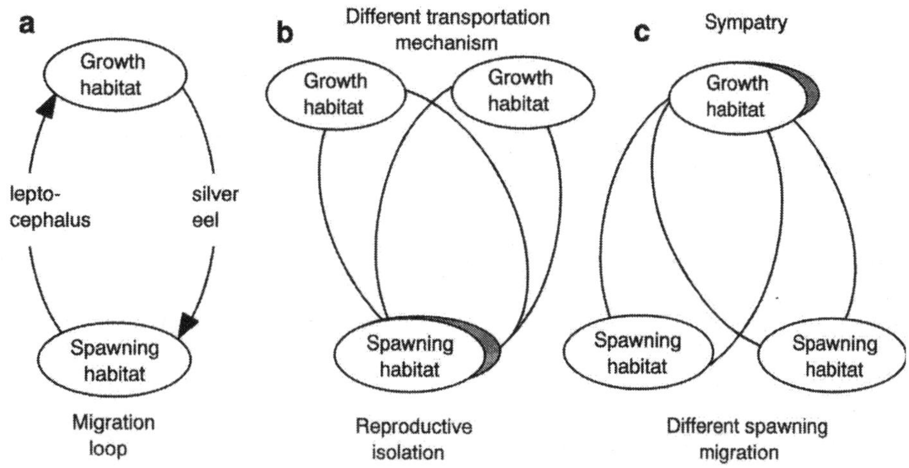

Figure 4. Migration loop of freshwater eels and the process of speciation: a - migration loop, b - two species with common spawning habitats but inhabiting different growth habitats; e.g. *A. anguilla* and *A. rostrata* in the Atlantic Ocean, c - two species sympatric but with different spawning habitats; e.g. *A. dieffenbachi* and *A. australis schmidti* in New Zealand.

phali have a 3.5 times larger body surface area (Tsukamoto & Umezawa 1994) and 1.2 times higher water content (Callamand 1943) than glass eels. This should facilitate depth regulation by leptocephali and enable them to passively drift for long distances in current systems.

The leptocephalus interval is unique in elopomorph fishes such as Elopiformes, Anguilliformes, Saccopharyngiformes, and Notacanthiformes. It seems likely that the ancestral eels already possessed this interval when they appeared and could consequently disperse widely on ocean currents during their leptocephalus phase.

Invasion of the Atlantic ocean

Assuming a western Pacific origin for the ancestor of the Atlantic species, what might have been its invasion route? Although it is impossible to rule out the possibility of changes in the world wide distribution of eels due to extinctions and global climate change in the past, we will discuss four possible routes of entry into the Atlantic Ocean based on present knowledge of zoogeography, paleogeography and paleo-circulation : (1) Panama route, (2) Cape of Good Hope route, (3) Arctic Ocean route and (4) Tethys Sea route (Figure 1).

For a Panama route, the ancestral eels would have had to move across the much wider paleo-Pacific Ocean from the western end to the eastern end. However, the eastern limit of *Anguilla* distribution in the Pacific is presently the Tuamotu Islands (Ege 1939). Freshwater eels do not occur on Easter Island and Galapagos Islands which are far to the east in the Pacific, nor on the west coasts of the North and South American continents. These facts suggest that the Panama route is unlikely. The second route via the Cape of Good Hope (southern tip of Africa) is also unlikely because of the absence of recent freshwater eels along both the east coast of South America and the west coast of Africa (except in the far north). In the third route via the Arctic Ocean, the low temperature and a great distance would act as a barrier to prevent dispersal of ancestral eels from tropical regions. As far as is known, all eels have their breeding places in small areas of tropical or subtropical waters. Even the European eel, which is most adapted to low temperatures, continues to breed at low latitude, in the Sargasso Sea. This suggests that these eels have retained their breeding place at low latitudes throughout their evolution. Even if the Arctic was warm enough in the past, this route is also unlikely because the migratory distance for one life cycle would be too long for these eels which are very likely to have dispersed as larvae from their breeding place in low latitudes.

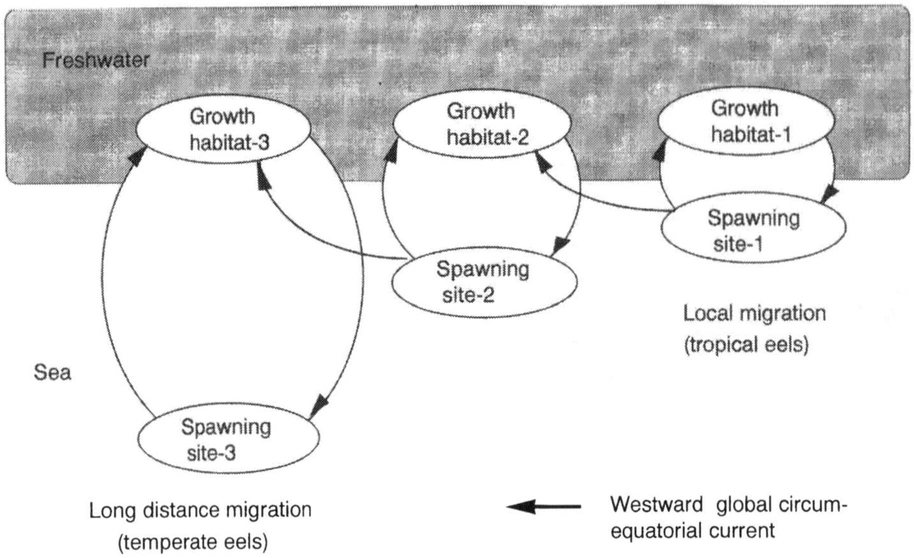

Figure 5. Evolution of migratory behaviour in freshwater eels.

The most likely route would seem to have been the Tethys Sea that separated Laurasia (North American and Eurasian continents) from Gondwana (Africa, South America, and India) after the break up of Pangea in the Triassic (200 Ma). The Tethys Sea existed during the Mesozoic to the beginning of the Tertiary period (approximately 200 to 30 Ma). There was a global circum-equatorial current which flowed from east to west encircling the earth along the equator (Haq 1984, Obata 1993). Therefore it is likely that the leptocephali of ancestral eels dispersed westward in this current through the Tethys Sea to enter the paleo-Atlantic Ocean from the western Pacific, via the present-day Indian Ocean.

Age estimation

Based on such a Tethys Corridor Hypothesis for the dispersal of ancestral eels, they must have entered the Atlantic before the Tethys Sea closed near the middle to late Oligocene (25-30 Ma: Haq 1984, Andel 1985). This suggests that the node between the ancestor of the two Atlantic species and *A. australis* in the neighbor-joining tree (ND1 in Figure 2), should occur more than 25-30 Ma, namely before the closure of the Tethys Sea. Thus, we conventionally dated this node (ND1) as 30 Ma. On the basis of this dating, it is suggested that the split of *A. celebesensis* and the ancestor of the other seven species (ND2 in Figure 2) was approximately 43.5 Ma, i.e. in the Eocene. Similarly, the split of *A. anguilla* and *A. rostrata* was estimated at approximately 10.2 Ma (ND3 in Figure 2). This aging may be underestimated because the speciation between the Atlantic species and *A. australis* may have already proceeded by reproductive isolation of adults in different oceans even though the Tethys Sea was opened and there were some supply of leptocephali from the Indian Ocean. This means that the timing of the closure of the Tethys Sea (30 Ma) would be possibly located much later in the phylogenetic tree, suggesting much earlier speciation of anguillid eels, for instance, in the Cretaceous.

Fossils placed in the order Anguilliformes have been described from the Upper Cretaceous (Robins 1989). Moreover, Regan (1912) concluded that the Anguillidae is primitive among the order Anguilliformes. This suggests that the Anguilliformes appeared at the latest in the Cretaceous, approximately 100 Ma, and that the Anguillidae evolved at an early time in the history of the order.

The sequence divergence between *A. anguilla* and *A. rostrata* was 0.024 (Aoyama et al. 1996), which is approximately congruent with the 0.03 val-

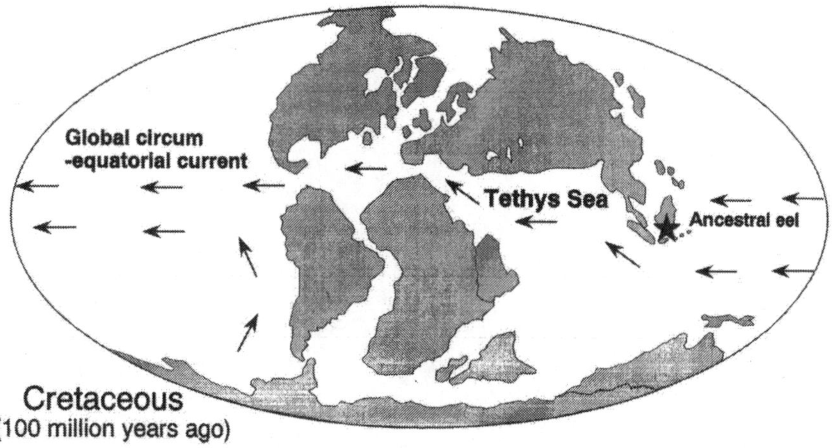

Figure 6. Diagrammatic representation of the proposed scenario for the dispersal and speciation of anguillid eels. The map shows a Cretaceous global paleogeographic reconstruction, approximately 100 Ma (Haq 1984). The Tethys Sea was present between the Eurasian and African continents. The ancestral *Anguilla* species originated in the western Pacific, in present-day Indonesia.

ue calculated by mitochondrial DNA restriction-site polymorphism (Avise et al. 1986, 1990). These authors estimated the separation time between *A. anguilla* and *A. rostrata* as 1.5 million years, using a conventional mitochondrial DNA clock calibration of 2% sequence divergence per million years. Applying this value of separation time to the phylogenetic tree obtained in this study, the oldest branch (ND2 in Figure 2) was calculated to be only 6.4 million years ago. This would indicate that the speciation events of the *Anguilla* occurred mainly during late Miocene to early Pliocene. However, this period was characterized by the accentuation of a cooling climate associated with the development of the Antarctic ice-sheet (Haq 1984). If this time estimation is adopted, the westward invasion of eels into the Atlantic would be impossible since the Tethys Corridor would have already closed much earlier and as an alternative the arctic region would have been too cold. Although the Isthmus of Panama was still opened in that period, the current was in the wrong direction to transport larvae into the Atlantic, even if some larvae drifted across the Pacific Ocean. Therefore, it seems reasonable to conclude that molecular evolution for eels may be much slower than for other animals, since it is now generally known that different animals have different molecular clocks (Meyer 1992).

Evolution of migratory behaviour

Tropical eels such as those inhabiting Sumatra Island are considered to spawn in waters comparatively close to the coast; i.e. leptocephali of four species were collected at the Mentawai Trench only 300 km east of Sumatra Island (Jespersen 1942), suggesting their local migration with a small migration loop (Figure 5). Similar to these present tropical species it is likely that the ancestral eels also showed a local migration. Accidental drift of larvae by a global circum-equatorial current might have caused a westward expansion of their growth habitat. After growing in the new habitat, if some of the adults found a new spawning site, a new migration loop would have been established and new species would appear (Figure 5). As the result of such sequential shifts of the migration loop, it is likely that the speciation and global dispersal of freshwater eels might have occurred simultaneously. For eels such as the two Atlantic species and *A. japonica* which inhabit temperate regions, retaining their spawning habitat in the tropics would mean migrating thousands of kilometers for spawning; they developed long distance migrations according to the expansion of the Atlantic plate and the Philippine plate, respectively.

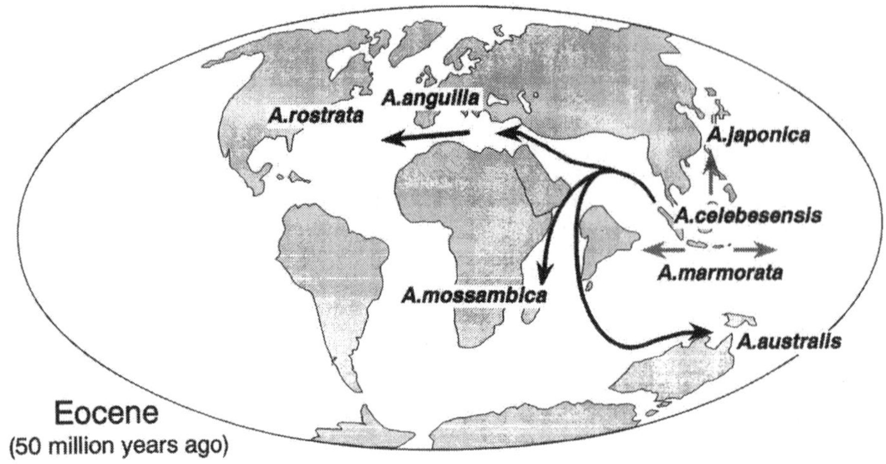

Figure 7. Diagrammatic representation of the proposed scenario for the dispersal and speciation of anguillid eels. The map shows a Eocene global paleogeographic reconstruction, approximately 50 Ma (Haq 1984). The range of the ancestral eels expanded westward by transport of their larvae in the Tethys Sea, via the paleo-global circum-equatorial current, eventually reaching the north Atlantic. At that time, the south Atlantic probably had not opened sufficiently to have a subtropical gyre suitable for the development of freshwater eel migration.

Evolutionary scenario

Based on our estimate of the timing of divergence for the *Anguilla* species, we compared the molecular phylogenetic trees (Figures 2, 3) and propose the following evolutionary scenario for the speciation and dispersal of freshwater eels of the genus *Anguilla*. In the Eocene or earlier, the ancestral eel originated in the western Pacific, around present-day Indonesia (Figure 6) and then split into two groups. One group dispersed westward by the global circum-equatorial current along the northern margin of the Tethys Sea, while the other group did not disperse so widely. In the Eocene, according to this scenario, the former group which may have spread throughout the Tethys Sea split into two lin-

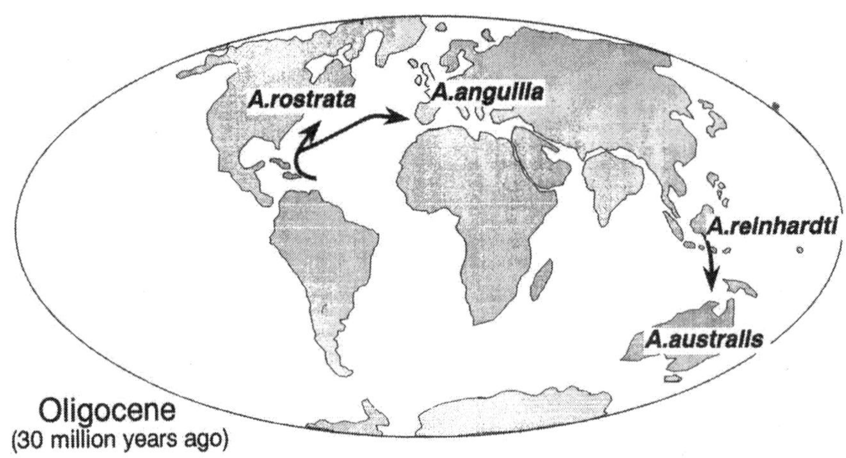

Figure 8. Diagrammatic representation of the proposed scenario for the dispersal and speciation of anguillid eels. The map shows a Oligocene global paleogeographic reconstruction, approximately 30 Ma (Haq 1984). Two Atlantic species began to differentiate from each other. The Australian Continent had moved north after its separation from Antarctica, and was populated by the lineage of *A. reinhardti* from nearby Indonesia.

eages: one being the ancestor of the two Atlantic species, which was carried farther westward by the current into the Atlantic Ocean, and the other being the ancestor of *A. mossambica* and *A. australis* which dispersed southward in the current that flows along the east coast of the African Continent (Figure 7). The ancestor of *A. anguilla* and *A. rostrata* that entered the Atlantic Ocean may have populated most of the coastal regions of the North Atlantic, as a result of dispersal of larvae in the paleo-North Atlantic gyre, since the exit of the Tethys Sea opened directly to the North Atlantic. However, this species may not have been able to spread into the South Atlantic because the paleo-South Atlantic had not opened sufficiently to have a warm subtropical gyre suitable for development of larvae. Even after a gyre was formed, the circulation of the northern and southern gyres probably did not overlap. That may be why no eels now occur in the South Atlantic.

In the western Pacific, according to this scenario, the other group may have remained at its place of origin for a while before part of the group dispersed northward by the North Pacific gyre resulting in the lineage of *A. japonica* (Figure 7). The others that remained near Indonesia evolved into the lineages of *A. reinhardti* and *A. marmorata*. The former dispersed south and the latter extended east and west.

In the latest part of the Eocene and early Oligocene, the Australian continent had moved north after its separation from Antarctica, and was populated by the lineage of *A. reinhardti* from nearby Indonesia (Figure 8). Eels that inhabit the Australian continent thus now consist of two lineages that arrived there by different routes at different times. This evolutionary scenario is consistent with the two Australian species, *A. reinhardti* and *A. australis*, being resolved as members of two separate clades in the phylogenetic tree (Figure 3).

The novel findings included in the above scenario indicate quite different phylogenetic relationships than might be expected based on available morphological data. The Japanese eel, *A. japonica*, is apparently not closely related to the Atlantic eels, as has previously been supposed, even though recent findings have shown them to have very similar migratory and recruitment styles (Tsukamoto 1992). Instead, it is the southern species *A. mossambica* and *A. australis* that are phylogenetically closer to the North Atlantic eels, all of which appear to have diverged from a common Indonesian ancestor. Keys to this divergence were the availability of the Tethys corridor and dispersal of larvae westwards and southwards on the paleo-circumglobal current. If we accept the Tethys Corridor hypothesis presented here, the rate of sequence divergence for *Anguilla* is slower than the previously accepted standard rate. These results provide a new basis for further evaluation of the evolutionary implications of the remarkable life history style of freshwater eels, *Anguilla*.

Acknowledgements

We thank T. Kobayashi for his cooperation in molecular analysis, and M. Miller and P.H.J. Castle for constructive comments and for improving the manuscript at an early stage. We also thank M. Gross and J.C. Avise for critical reading of the manuscript and comments.

References cited

Andel, T.H.V. 1985. New views on an old planet – continental drift and the history of earth. Tsukiji shokan, Tokyo. 352 pp. (in Japanese)

Aoyama, J., T. Kobayashi & K. Tsukamoto. 1996. Phylogeny of eels suggested by mitochondrial DNA sequences. Nippon Suisan Gakkaishi 63: 370–375 (in Japanese).

Avise, J.C., G.S. Helfman, N.C. Saunders & L.S. Hales. 1986. Mitochondrial DNA differentiation in North Atlantic eels, population genetic consequences of an unusual life history pattern. Proc. Natl. Acad. Sci. USA 83: 4350–4354.

Avise, J.C., W.S. Nelson, J. Arnold, R.K. Koehn, G.C. Williams & V. Thorsteinsson. 1990. The evolutionary genetic status of Icelandic eels. Evolution 44: 1254–1262.

Bertin, L. 1956. Eels. A biological study. Cleaver Hume Press, London. 192 pp.

Castle, P.H.J. & G.R. Williamson. 1974. On the validity of the fresh water eel species *Anguilla ancestralis* Ege, from Celebes. Copeia 1974: 569–570.

Callamand, O. 1943. L'Anguille Europeenne. Les bases physiologique de sa migration. Annll. Inst. Oceanogr. 21: 361–440.

Ege, V. 1939. A revision of the genus *Anguilla* Shaw, a systematic, phylogenetic and geographical study. Dana Rep. 16: 1–256.

Felsenstein, J.1985. Confidence limits on phylogenies: an approach using the bootstrap. Evolution 39: 783–791.

Haq, B.U. 1984. Paleoceanography: a synoptic overview of 200 million years of ocean history. pp. 201–234. *In*: B.U. Haq & J.D. Milliman (ed.) Marine Geography and Oceanography of Arabian Sea and Coastal Pakistan, Van Nostrand Reinhold, New York.

Jellyman, D.J. 1987. Review of the marine life history of Australian temperate species of *Anguilla*. pp. 276–285. *In*: M.J. Dadswell, R.J. Klauda, C.M. Moffitt, R.L. Saunders, R.A. Rulifson & J.E. Cooper (ed.) Common Strategies of Anadromous and Catadromous Fishes, Proceedings of an International Symposium, American Fisheries Society, Bethesda.

Jespersen, P. 1942. Indo-Pacific leptocephalids of the genus *Anguilla*. Dana Rep. 22: 1–128.

Kimura, M. 1980. A simple method for estimating evolutionary rate of base substitution through comparative studies of nucleotide sequences. J. Mol. Evol. 16: 111–120.

Kocher, T.D., W.K. Thomas, A. Meyer, S.V. Edwards, V. Paabo, F.X. Villablanca & A.C. Wilson. 1989. Dynamics of mitochondrial DNA evolution in animals: amplification and sequencing with conserved primers. Proc. Natl. Acad. Sci. USA 86: 6196–6200.

McCleave, J.D., R.C. Kleckner & M. Castonguay. 1987. Reproductive sympatry of American and European eels and implications for migration and taxonomy. pp. 286–297. *In*: M.J. Dadswell, R.J. Klauda, C.M. Moffitt, R.L. Saunders, R.A. Rulifson & J.E. Cooper (ed.) Common Strategies of Anadromous and Catadromous Fishes, Proceedings of an International Symposium, American Fisheries Society, Bethesda.

McDowall, R.M. 1987. The occurrence and distribution of diadromy among fishes. pp. 1–13. *In*: M.J. Dadswell, R.J. Klauda, C.M. Moffitt, R.L. Saunders, R.A. Rulifson & J.E. Cooper (ed.) Common Strategies of Anadromous and Catadromous Fishes, Proceeding of an International Symposium, American Fisheries Society, Bethesda.

Meyer, A. 1992. Evolution of mitochondial DNA in fishes. pp. 1–38. *In*: P.W. Hochachka & T.P. Mommsen (ed.) Biochemistry and Molecular Biology of Fishes, vol. 2, Elsevier Science Publishers, New York.

Nelson, J.S. 1984. Fishes of the world, 2nd ed. Wiley-Interscience, New York. 523 pp.

Obata, I. 1993. Natural history of the Cretaceous. University of Tokyo Press, Tokyo. 200 pp. (in Japanese).

Palumbi, S., A. Martin, S. Romano, W.O. McMillan, L. Stice & G. Grabowski. 1991. The simple fool's guide to PCR Ver.2.0. University of Hawaii, Honolulu. 40 pp.

Regan, C.T. 1912. The osteology and classification of the teleostean fishes of the order Apodes. Ann.Mag.Nat.Hist. 8 (10): 377–387.

Robins, C.R. 1989. The phylogenetic relationships of the Anguilliform fishes. pp. 9–23. *In*: E.B. Böhlke (ed.) Fishes of the Western North Atlantic, Sears Foundation for Marine Research, Yale University, New Haven.

Saitou, N. & M. Nei. 1987. The neighbor-joining method: a new method for reconstructing phylogenetic trees. Mol. Biol. Evol. 4: 406–425.

Tesch, F.W. 1977. The eel. Chapman and Hall, London. 434 pp.

Tsukamoto, K. 1992. Discovery of the spawning area for Japanese eel. Nature 356: 789–791.

Tsukamoto, K. & A. Umezawa. 1994. Metamorphosis: a key factor of larval migration determining geographic distribution and speciation of eels. pp. 231–248. *In*: Faculty of Fishes, Kasetart University (ed.) Proceedings Fourth Indo-Pacific Fish Conference, Kasetart University, Bangkok.

Tucker, D.W. 1959. A new solution to the Atlantic eel problem. Nature 183: 495–501.

Environmental Biology of Fishes **52**: 149–161, 1998.

Genetic structure and differentiation of four *Lethenteron* taxa from the Far East, deduced from allozyme analysis

Yuji Yamazaki & Akira Goto
Laboratory of Aquatic Breeding Science, Faculty of Fisheries, Hokkaido University, 3-1-1 Minatocho, Hakodate 041, Japan (e-mail: akir@pop.fish.hokudai.ac.jp)

Received 20.1.1997 Accepted 27.5.1997

Key words: lamprey, Petromyzontidae, electrophoresis, genetic variability, genetic relationship, satellite species

Synopsis

Genetic structure and differentiation among four *Lethenteron* taxa, *L. japonicum, L. kessleri,* and two groups of *L. reissneri*, collected from Japan and the Far Eastern region of Russia, were investigated by electrophoretic analysis. Several complete-allele substitutions were found between all possible pairs of taxa in regions of sympatry, strongly suggesting the existence of reproductive isolation between them. Therefore, four *Lethenteron* taxa should be regarded as discrete species, respectively. In each taxon, the genetic variability within each population (H=0.062 – 0.127 for *L. japonicum*, H = 0.062– 0.128 for *L. kessleri*, H = 0.026 – 0.148 for the northern group of *L. reissneri,* and H = 0.015 – 0.102 for the southern group of *L. reissneri*) was considerable, suggesting large effective sizes for most populations. The sample of *L. japonicum* collected from Kyiya, the basin of Amur, was somewhat divergent from the other intraspecific samples. This may have resulted mainly because it does not migrate to the sea. In contrast *L. kessleri* has a fluvial life style, but the genetic differentiation between populations (G_{ST} = 0.117, D = 0.000 – 0.061) was less than that for each group of *L. reissneri* (G_{ST} = 0.493, D = 0.000 – 0.226 for the northern group and G_{ST} = 0.660, D = 0.008 – 0.422 for the southern group), probably meaning more recent dispersal of *L. kessleri* than the two groups of *L. reissneri*. The parasitic and anadromous *L. japonicum* appeared to be closely related to *L. kessleri* (D = 0.042 – 0.090) and the northern group of *L. reissneri* (D = 0.163 – 0.355), which have nonparasitic and fluvial life styles, whereas the southern group of *L. reissneri* was greatly divergent from the other three *Lethenteron* species (D = 0.559 – 0.926), suggesting that the former three species might be monophyletic.

Introduction

Many lamprey genera are composed of species which have different life styles such as parasitic or nonparasitic. Such nonparasitic species which morphologically resemble the parasitic species are generally called paired species (Zanandrea 1959, Hardisty & Potter 1971). Vladykov & Kott (1979), however, have appropriately called these satellite species because some genera are composed of one parasitic species and two or more nonparasitic species. The nonparasitic (satellite) species were probably derived from the parasitic species belonging to the same genus (Zanandrea 1959, Hardisty & Potter 1971, Hubbs & Potter 1971, Vladykov & Kott 1979, Potter 1980). To test this hypothesis based on only the morphological and biogeographical features (Hardisty & Potter 1971, Potter 1980), will require

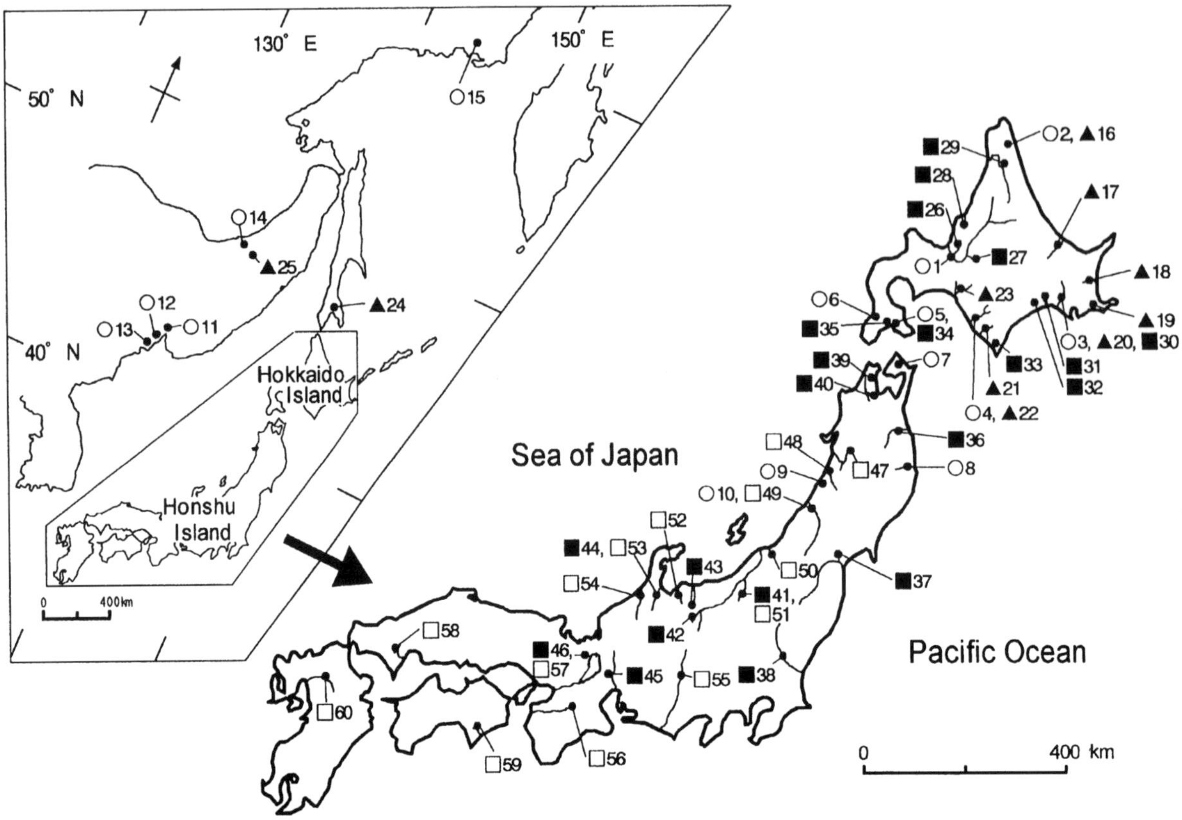

Figure 1. Collection sites of four *Lethenteron* taxa around Japan. Open circle = *L. japonicum*; solid triangle = *L. kessleri*; solid square = northern group of *L. reissneri*; open square = southern group of *L. reissneri* (rivers sampled were: 1 Ishikari, 2 Sarufutsu, 3 Hororo, 4 Mena, 5 Ohno, 6 Uzura, 7 Oh-hata, 8 Omoto, 9 Nishime, 10 Mogami, 11 Gryaznaya, 12 Barbashovka, 13 Narba, 14 Kyiya, 15 Ola, 16 Sarufutsu, 17 Tokoro, 18 Chashikotsu, 19 Bekanbeushi, 20 Hororo, 21 Horobetsu, 22 Mena, 23 Abira, 24 Byeraya, 25 Arychy, 26 Tobetsu, 27 Uenbetsu, 28 Nobusha, 29 Otoineppu, 30 Hororo, 31 Charo, 32 Okoppe, 33 Utabetsu, 34 Ohno, 35 Hekiriji, 36 Kumahara, 37 Abukuma, 38 Naka-gawa, 39 Kanita, 40 Iwaki, 41 Uono, 42 Sai-kawa, 43 Hime-kawa, 44 Shou-gawa, 45 Tsuya, 46 Shoushu, 47 Hinokinai, 48 Koyoshi, 49 Mogami, 50 Kaji-gawa, 51 Uono, 52 Kado-kawa, 53 Shou-gawa, 54 Tedori, 55 Tenryu, 56 Kino-kawa, 57 Shoushu, 58 Nishiki, 59 Ogawa-dani, 60 Tade).

clarifying phylogenetic relationships. Phylogenetic relationships both between and within genera have been based on dentition (Hubbs & Potter 1971, Potter 1980) and the number of trunk myomeres (Vladykov & Kott 1979), but genetic relationships have never been examined.

In the Far East, three *Lethenteron* species have been reported: the Arctic lamprey, *L. japonicum*, characterized by a parasitic, anadromous life style, and the Siberian lamprey, *L. kessleri* and the Far Eastern brook lamprey, *L. reissneri*, both of which are nonparasitic and fluvial (Berg 1931, 1948, Sato 1951, Hubbs & Potter 1971, Choi et al. 1984, Iwata et al. 1985). Recently, two groups, northern and southern, were reported in *L. reissneri* collected from nearly all of Japan, following allozyme analyses (Yamazaki & Goto 1996). Yamazaki & Goto (1996) concluded that they should be classified as different species, owing to their large genetic divergence and genetical evidence showing the lack of their hybridization in regions of sympatry.

According to the biological species concept (Mayr 1963), the existence of reproductive isolation between sympatric populations, at least in animals, suggests that they should be regarded as different species. While *L. japonicum* and *L. kessleri* have different life-history styles, they have similar morphological characteristics (Iwata et al. 1985, Iwata & Hamada 1986), suggesting that they are not distinct species, but are ecotypes within the same species, as is observed in some salmonid fishes (Svärdson 1971, Johnson 1980, Kawanabe 1989, Yamamoto et al. 1992). It is important, therefore, to clarify the genetic structure and differentiation of these taxa.

Vladykov & Kott (1979) have regarded two nonparasitic *Lethenteron* species from Japan, *L. matsubarai* and *L. reissneri*, as satellite species of parasitic *L. japonicum*. Subsequently, Iwata et al. (1985) reported that *L. matsubarai* should be synonymized with *L. kessleri*. However, the specimens of *L. reissneri*, examined by Vladykov & Kott (1979) might have included examples of both groups of *L. reissneri*. Therefore, the phylogenetic relationships among four *Lethenteron* taxa remain to be solved. In the present study, allozyme analyses were performed to determine the genetic characteristics of each *Lethenteron* taxon and to estimate the degree of genetic differentiation among four taxa occurring in Japan and the Far Eastern region of Russia.

Table 1. Enzymes and proteins examined, loci detected, and the buffers and tissues used. Enzyme commission numbers are given in parentheses. ([1] A = citrate-N-(3-aminopropyl)-diethanolamine pH 6.9, B = Ridgway buffer, [2] M = muscle, L = liver).

Enzyme and protein	Abbreviation	Locus	Buffer[1]	Tissue[2]
Aspartate aminotransferase	AAT	*AAT-1**	A	M
(EC 2.6.1.1)		*AAT-2**	A	M
Acid phosphatase	ACP	*ACP**	A	L
(EC 3.1.3.2)				
Aconitate hydrogenase	AH	*AH-1**	A	M, L
(EC 4.2.1.3)		*AH-2**	A	M, L
Esterase	EST	*EST**	B	L
(EC 3.1.1.-)				
Fumarate hydrogenase	FH	*FH**	B	M
(EC 4.2.1.2)				
Glycerol-3-phosphate dehydrogenase	G3PDH	*G3PDH**	A	M
(EC 1.1.1.8)				
Glycerol-6-phoshate dehydrogenase	G6PDH	*G6PDH**	B	M
(EC 1.1.1.49)				
Glucose-6-phosphate isomerase	GPI	*GPI-1**	B	M, L
(EC 5.3.1.9)		*GPI-2**	B	M
$_L$-Iditol dehydrogenase	IDDH	*IDDH**	B	L
(EC 1.1.1.14)				
Isocitrate dehydrogenase (NADP+)	IDHP	*IDHP-1**	A	M, L
(EC 1.1.1.42)		*IDHP-2**	A	M, L
		*IDHP-3**	A	M, L
		*IDHP-4**	A	M, L
$_L$-Lactate dehydrogenase	LDH	*LDH**	A	M, L
(EC 1.1.1.27)				
Malate dehydrogenase	MDH	*MDH-1**	A	M, L
(EC 1.1.1.37)		*MDH-2**	A	M, L
		*MDH-3**	A	M, L
		*MDH-4**	A	M, L
Malic enzyme (NADP+)	MEP	*MEP-1**	A	M
(EC 1.1.1.40)		*MEP-2**	A	M
Phosphogluconate dehydrogenase	PGDH	*PGDH**	A	M
(EC 1.1.1.44)				
Phosphoglucomutase	PGM	*PGM**	A	M, L
(EC 5.4.2.2)				
General protein	PROT	*PROT-1**	B	M
		*PROT-2**	B	M

Materials and methods

Samples of the four *Lethenteron* taxa were collected from Hokkaido and Honshu Islands, Japan, and the Far Eastern region of Russia from 1993 to 1996 (Figure 1). All the specimens were classified into *L. japonicum, L. kessleri,* and *L. reissneri* based on morphological features reported by Iwata et al. (1985). Because it is difficult to distinguish between the northern and southern groups of *L. reissneri* based on morphological characters (Yamazaki & Goto 1997), all the specimens of the two groups were identified by diagnostic allozymic alleles on 11 loci, the *AAT-1*, G6PDH*, GPI-2*, IDHP-1*, 2*, 3*, 4*, MDH-3*, 4*, MEP-1** and *PGM**, according to Yamazaki & Goto (1996). After morphological assessment, all samples were immediately frozen on dry ice or liquid nitrogen and stored in the laboratory at −30°C until electrophoretic analysis was done.

Horizontal starch gel electrophoresis in Citrate-N-(3-aminopropyl)-diethanolamine (C-APE) at pH 6.9 (Clayton & Tretiak 1972) and Ridgway buffers (Ridgway et al. 1970) of extracts of isolated aqueous homogenates of skeletal muscle and liver samples was used to resolve fifteen enzymes and one general protein encoded by 27 loci (Table 1). Staining procedures followed Shaw & Prasad (1970), with slight modifications. Identification of loci and alleles, and genetic nomenclature, followed Yamazaki & Goto (1996) and Shaklee et al. (1990), respectively.

Allele frequencies were calculated directly from observed genotypes. The distribution of observed genotypes was compared with that expected from Hardy-Weinberg equilibrium using chi-square contingency tests. The number of alleles per locus, the percentage of polymorphic loci (P), and the observed and expected mean heterozygosities (H) were calculated for each population. The coefficient of gene differentiation (G_{ST}; Nei 1975) was calculated to estimate interpopulational diversity. Nei's (1978) unbiased genetic distance (D) was calculated between all possible pairs of all samples from allele frequencies. A dendrogram based on the D values was generated using the UPGMA method (Sneath & Sokal 1973) to estimate the genetic relationships among populations. Allozyme data were analyzed using BIOSYS-1 vers. 1.7 (Swofford & Selander 1981).

Results

Allele and genotypic frequencies

Twenty-seven gene loci encoding 15 enzymes and one general protein were assumed in the present study. Four of the 27 loci were monomorphic ($p > 0.95$) or fixed for the same allele in all samples: *IDDH*, LDH*, PROT-1** and *2**. The other 23 loci were polymorphic ($p < 0.95$): *AAT-1*, 2*, ACP*, AH-1*, 2*, EST*, FH*, G3PDH*, G6PDH*, GPI-1*, 2*, IDHP-1*, 2*, 3*, 4*, MDH-1*, 2*, 3*, 4*, MEP-1*, 2*, PGDH** and *PGM**.

Comparisons of observed genotypic frequencies with Hardy-Weinberg expectations showed signif-

Table 2. Individual frequency distribution of the four *Lethenteron* taxa from eight sites in which they were sympatrically distributed (n = total sample size).

River	n	*L. japonicum*	*L. kessleri*	the northern group of *L. reissneri*	the southern group of *L. reissneri*
Sarufutsu	115	68	47	–	–
Hororo	103	26	56	21	–
Mena	92	18	74	–	–
Ohno	55	14	–	41	–
Mogami	61	30	–	–	31
Uono	37	–	–	26	11
Shou	86	–	–	52	34
Shoushu	33	–	–	24	9

Table 3. Estimates of genetic variability in *Lethenteron* taxa on 27 loci: * a locus is considered polymorphic if the frequency of the most common allele did not exceed 0.95, ** based on direct count of heterozygotes, *** a biased estimate based on Hardy-Weinberg expectations (Nei 1978).

Taxon Population	Mean sample size per locus (mean ± SE)	Mean number of alleles per locus (mean ± SE)	Percentage of loci polymorphic*	Proportion of heterozygotes	
				Observed** (mean ± SE)	Expected*** (mean ± SE)
L. japonicum					
1. Ishikari	50.8 ± 0.1	1.6 ± 0.2	29.6	0.090 ± 0.029	0.099 ± 0.031
2. Sarufutsu	66.6 ± 0.6	1.7 ± 0.2	29.6	0.080 ± 0.026	0.096 ± 0.030
3. Hororo	25.6 ± 0.3	1.7 ± 0.2	37.0	0.105 ± 0.032	0.107 ± 0.029
4. Mena	18.0 ± 0.0	1.4 ± 0.1	29.6	0.099 ± 0.030	0.114 ± 0.037
5. Ohno	13.8 ± 0.2	1.4 ± 0.1	29.6	0.070 ± 0.022	0.075 ± 0.022
6. Uzura	40.0 ± 0.0	1.6 ± 0.2	33.3	0.096 ± 0.029	0.114 ± 0.035
7. Oh-hata	41.8 ± 0.2	1.7 ± 0.2	33.3	0.082 ± 0.023	0.100 ± 0.028
8. Omoto	36.0 ± 0.0	1.6 ± 0.2	29.6	0.090 ± 0.026	0.103 ± 0.030
9. Nishime	44.4 ± 0.4	1.6 ± 0.2	25.9	0.062 ± 0.021	0.075 ± 0.025
10. Mogami	29.5 ± 0.3	1.5 ± 0.1	25.9	0.079 ± 0.029	0.081 ± 0.029
11. Gryaznaya	23.5 ± 0.3	1.6 ± 0.2	25.9	0.098 ± 0.032	0.107 ± 0.035
12. Barbashovka	25.5 ± 0.4	1.5 ± 0.2	29.6	0.102 ± 0.033	0.117 ± 0.037
13. Narba	37.3 ± 0.5	1.6 ± 0.2	37.0	0.114 ± 0.032	0.110 ± 0.031
14. Kyiya	31.4 ± 0.5	1.4 ± 0.1	25.9	0.099 ± 0.032	0.107 ± 0.035
15. Ola	34.6 ± 0.3	1.6 ± 0.2	37.0	0.127 ± 0.037	0.116 ± 0.032
L. kessleri					
16. Sarufutsu	46.6 ± 0.2	1.9 ± 0.2	29.6	0.097 ± 0.028	0.114 ± 0.033
17. Tokoro	41.5 ± 0.3	1.4 ± 0.1	22.2	0.062 ± 0.020	0.064 ± 0.021
18. Chashikotsu	51.6 ± 0.3	1.6 ± 0.2	33.3	0.092 ± 0.029	0.090 ± 0.028
19. Bekanbeushi	29.6 ± 0.3	1.4 ± 0.1	29.6	0.117 ± 0.039	0.112 ± 0.035
20. Hororo	55.0 ± 0.7	1.8 ± 0.2	33.3	0.102 ± 0.031	0.108 ± 0.033
21. Horobetsu	26.4 ± 0.3	1.4 ± 0.1	37.0	0.105 ± 0.032	0.106 ± 0.031
22. Mena	73.1 ± 0.5	1.6 ± 0.2	29.6	0.108 ± 0.034	0.117 ± 0.036
23. Abira	35.7 ± 0.2	1.4 ± 0.1	25.9	0.100 ± 0.037	0.094 ± 0.033
24 Byeraya	28.0 ± 0.6	1.6 ± 0.2	33.3	0.128 ± 0.039	0.133 ± 0.041
25. Arychy	39.0 ± 0.0	1.4 ± 0.1	29.6	0.123 ± 0.039	0.122 ± 0.037
northern group of *L. reissneri*					
26. Tobetsu	40.2 ± 0.4	1.4 ± 0.1	33.3	0.093 ± 0.027	0.104 ± 0.030
27. Uenbetsu	40.0 ± 0.0	1.6 ± 0.2	37.0	0.123 ± 0.035	0.129 ± 0.035
28. Nobusha	40.7 ± 0.9	1.5 ± 0.1	29.6	0.073 ± 0.023	0.087 ± 0.027
29. Otoineppu	41.0 ± 0.0	1.6 ± 0.1	40.7	0.128 ± 0.032	0.134 ± 0.033
30. Hororo	21.0 ± 0.0	1.6 ± 0.2	25.9	0.086 ± 0.033	0.108 ± 0.038
31. Charo	40.8 ± 0.2	1.7 ± 0.2	29.6	0.098 ± 0.029	0.114 ± 0.035
32. Okoppe	41.1 ± 0.4	1.7 ± 0.2	29.6	0.114 ± 0.037	0.128 ± 0.040
33. Utabetsu	17.0 ± 0.0	1.3 ± 0.1	18.5	0.083 ± 0.035	0.082 ± 0.034
34. Ohno	40.0 ± 0.6	1.7 ± 0.2	40.7	0.148 ± 0.040	0.151 ± 0.040
35. Hekiriji	23.8 ± 0.2	1.6 ± 0.1	40.7	0.134 ± 0.037	0.140 ± 0.038
36. Kumahara	34.7 ± 0.2	1.3 ± 0.1	22.2	0.064 ± 0.030	0.065 ± 0.028
37. Abukuma	14.0 ± 0.0	1.3 ± 0.1	22.2	0.066 ± 0.025	0.087 ± 0.033
38. Naka-gawa	21.5 ± 0.5	1.4 ± 0.1	25.9	0.109 ± 0.037	0.110 ± 0.037
39. Kanita	44.5 ± 0.3	1.6 ± 0.2	29.6	0.105 ± 0.035	0.118 ± 0.040
40. Iwaki	37.7 ± 0.2	1.5 ± 0.1	25.9	0.090 ± 0.031	0.094 ± 0.031
41. Uono	26.0 ± 0.0	1.2 ± 0.1	14.8	0.057 ± 0.030	0.057 ± 0.029
42. Sai-kawa	49.7 ± 0.3	1.3 ± 0.1	14.8	0.055 ± 0.025	0.058 ± 0.026
43. Hime-kawa	49.1 ± 0.5	1.3 ± 0.1	11.1	0.026 ± 0.017	0.034 ± 0.020
44. Shou-gawa	51.6 ± 0.3	1.6 ± 0.1	22.2	0.090 ± 0.032	0.101 ± 0.035
45. Tsuya	50.0 ± 0.0	1.6 ± 0.1	18.5	0.061 ± 0.026	0.067 ± 0.026
46. Shoushu	24.0 ± 0.0	1.4 ± 0.1	25.9	0.079 ± 0.031	0.085 ± 0.030
southern group of *L. reissneri*					
47. Hinokinai	55.6 ± 1.3	1.3 ± 0.1	18.5	0.052 ± 0.026	0.058 ± 0.028
48. Koyoshi	39.6 ± 0.3	1.3 ± 0.1	18.5	0.073 ± 0.031	0.076 ± 0.032
49. Mogami	31.0 ± 0.0	1.4 ± 0.1	18.5	0.060 ± 0.025	0.060 ± 0.025
50. Kaji-gawa	25.4 ± 0.3	1.5 ± 0.2	22.2	0.102 ± 0.040	0.105 ± 0.041
51. Uono	11.0 ± 0.0	1.3 ± 0.1	14.8	0.071 ± 0.035	0.069 ± 0.030
52. Kado-kawa	38.6 ± 0.3	1.4 ± 0.1	29.6	0.096 ± 0.032	0.103 ± 0.034
53. Shou-gawa	33.8 ± 0.1	1.4 ± 0.1	14.8	0.059 ± 0.027	0.067 ± 0.029
54. Tedori	50.4 ± 0.4	1.4 ± 0.2	11.1	0.049 ± 0.025	0.051 ± 0.027
55. Tenryu	49.0 ± 0.0	1.2 ± 0.1	7.4	0.019 ± 0.012	0.024 ± 0.016
56. Kino-kawa	49.6 ± 0.2	1.4 ± 0.1	29.6	0.093 ± 0.032	0.099 ± 0.033
57. Shoushu	9.0 ± 0.0	1.3 ± 0.1	29.6	0.066 ± 0.028	0.073 ± 0.028
58. Nishiki	15.9 ± 0.1	1.2 ± 0.1	7.4	0.025 ± 0.023	0.031 ± 0.024
59. Ogawa-dani	15.0 ± 0.0	1.1 ± 0.1	7.4	0.015 ± 0.011	0.013 ± 0.010
60. Tade	8.0 ± 0.0	1.4 ± 0.1	25.9	0.088 ± 0.033	0.085 ± 0.033

icant deviations at the *PGM** (p = 0.009) and *AH-1** (p = 0.031) in the Bekanbeushi and Mena samples of *L. kessleri*, respectively. In the Tsuya samples of the northern group of *L. reissneri* a single significantly divergent locus, *AH-2** (p = 0.032) was detected. The southern group of *L. reissneri* showed significant deviations from genotype frequencies expected for Hardy-Weinberg equilibrium in two populations, the Kino-kawa (*ACP**, p = 0.019) and Nishiki (*EST**, p = 0.043) samples. Since the remaining 22 polymorphic loci examined were not significantly divergent, the deviations mentioned above were considered unavoidable, and samples were treated as random mating populations. In the other samples, the observed genotypic frequencies of all polymorphic loci conformed with the Hardy-Weinberg expectations.

Diagnostic alleles among taxa

Four *Lethenteron* taxa could be distinguished from each other by having diagnostic alleles at 1 to 11 loci (see Appendix). The samples of *L. japonicum* were distinguished from *L. kessleri* samples by the *MDH-3*-122* allele, from the northern group of *L. reissneri* by the *IDHP-1*97* and *2*92* alleles, and from the southern group of *L. reissneri* by the *AAT-1*100, G6PDH*100, GPI-2*100* and **48, IDHP-1*97, 2*92, 3*-100, 4*-100, MDH-3*-122, 4*-100, MEP-1*53,* and *PGM*-43* and **-100* alleles. The samples of *L. kessleri* were distinguished from the northern group of *L. reissneri* by the *IDHP-1*97, 2*92*, and *MDH-3*-93* alleles, and from the southern group of *L. reissneri* by the *AAT-1*100, G6PDH*100, GPI-2*100* and **48,*

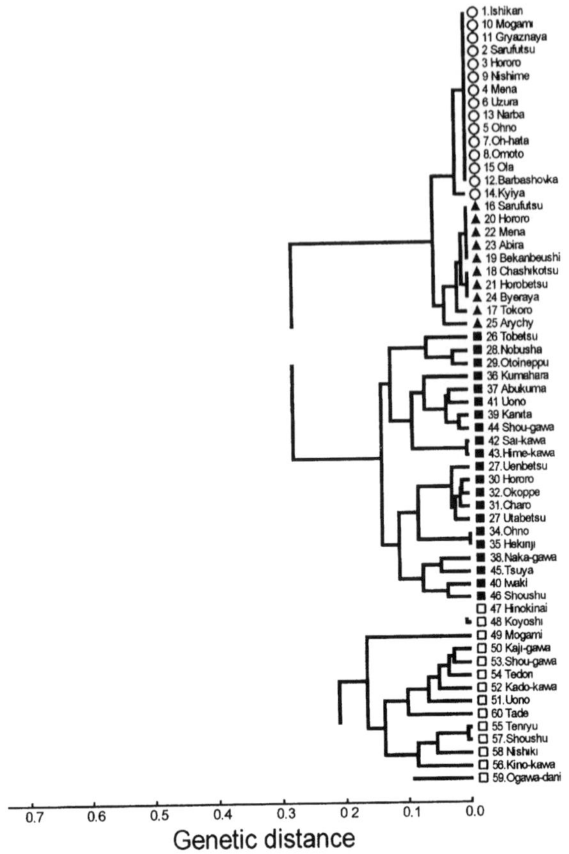

Figure 2. An UPGMA dendrogram using Nei's (1978) unbiased genetic distance for 60 samples of *Lethenteron*, based on 27 loci of biochemical genetic markers. Symbols and figures refer to taxa and sample sites listed in Figure 1.

Table 4. Average values of Nei's (1978) unbiased genetic distance between pairs of the four *Lethenteron* taxa. Ranges are shown in parentheses.

	L. japonicum	*L. kessleri*	*L. reissneri*	
			northern group	southern group
L. japonicum	0.006 (0.000–0.025)			
L. kessleri	0.055 (0.042–0.090)	0.016 (0.000–0.061)		
northern group of *L. reissneri*	0.273 (0.163–0.355)	0.294 (0.211–0.394)	0.116 (0.000–0.226)	
southern group of *L. reissneri*	0.684 (0.603–0.830)	0.691 (0.591–0.846)	0.707 (0.559–0.926)	0.158 (0.008–0.422)

*IDHP-1*97, 2*92, 3*-100, 4*-100, MDH-3*-93, 4*-100, MEP-1*53,* and *PGM*-43* and **-100* alleles. The northern-group samples of *L. reissneri* were distinguished from the southern-group samples of *L. reissneri* by the *AAT-1*100, G6PDH*100, GPI-2*100* and **48, IDHP-1*100, 2*100, 3*-100, 4*-100, MDH-3*-100* and **-122, 4*-100* and **-129, MEP-1*100,* and *PGM*-43* and **100* alleles.

All possible pairs of taxa, except *L. kessleri* and the southern group of *L. reissneri*, were found in at least one river. Three taxa, *L. japonicum, L. kessleri,* and the northern group of *L. reissneri,* were collected from the Hororo River in Hokkaido Island, Japan (Table 2). No hybrids between the taxa were found in any sympatric samples.

Genetic variability and differentiation

Intrapopulational variability is given in Table 3. The G_{ST} values were 0.064, 0.117, 0.493 and 0.660 for *L. japonicum, L. kessleri*, and the northern and southern groups of *L. reissneri* overall, respectively. To estimate the genetic relationships, Nei's (1978) unbiased genetic distances (D) were calculated between all possible pairs of samples, and are shown in Table 4.

The four taxa each made up a separate cluster on a dendrogram using the UPGMA clustering algorithm (Figure 2). Each group of *L. reissneri* was divided into some subgroups, as already shown in Yamazaki & Goto (1996). In *L. japonicum* and *L. kessleri*, however, samples were almost homogenous, with the exception of samples collected from the Amur basin, the Kyiya in *L. japonicum* and the Arychy in *L. kessleri.* On the UPGMA dendrogram, the cluster consisting of *L. japonicum* samples was closely connected with the cluster of *L. kessleri* samples, and the cluster of these two taxa was connected somewhat distantly with the cluster of the northern-group samples of *L. reissneri.* The cluster of these three taxa was most distantly connected with that of the southern-group samples of *L. reissneri.*

Discussion

The low genetic distances (D = 0.042 – 0.090, average: 0.055) detected between *L. japonicum* and *L. kessleri*, are similar to those found among intraspecific populations in fish (D = 0.002 – 0.065; Shaklee et al. 1982). Although *L. japonicum* can be distinguished from *L. kessleri* by the existence of a dark dot on the second dorsal fin, there is considerable overlap in morphometric and meristic characters (Iwata et al. 1985). Their life-history styles are also different; *L. japonicum* is parasitic and anadromous, while *L. kessleri* is nonparasitic and fluvial (Iwata et al. 1985). Iwata & Hamada (1986), however, found a population of *L. japonicum* from southern Hokkaido in which one nonparasitic fluvial specimen was included. Similarly, Beamish (1987) reported parasitic and nonparasitic populations of *Lampetra richardsoni.* These findings suggest that *L. japonicum* and *L. kessleri* are not distinct species but variations within a single species. However, a complete-allele substitution for the *MDH-3** locus was found between populations of *L. japonicum* and *L. kessleri* in the Sarufutsu, Hororo and Mena where the two taxa were sympatrically distributed: the former possesses the *MDH-3*-122* allele, whereas the latter have the *MDH-3*-93* allele. In addition, hybrids between the taxa were not detected in any of the three sites, suggesting that the sympatric taxa in each site do not exchange genes, and are thus reproductively isolated (Dowling & Moore 1984).

Although sympatric populations were not recognized between *L. kessleri* and the southern group of *L. reissneri*, a large genetic divergence (D = 0.591 – 0.846, average: 0.691) was found between them. Yamazaki & Iwata (1997) reported that there is also a considerable divergence in morphological characteristics between the two taxa. Consequently, the present data strongly suggest that each of the four taxa should be recognized as fully discrete species.

The genetic variability of diadromous species, at least in fishes, is generally larger than that of related fluvial species (Gyllensten 1985, Taniguchi et al. 1986, Goto & Andoh 1990). However, mean heterozygosity of *L. japonicum* from the Hororo was almost equal to or slightly larger than that of *L. kess-*

leri and the northern group of *L. reissneri* from the same site, respectively. Kijima & Fujio (1984) showed a significant positive correlation between mean heterozygosity and the effective size of fluvial populations in the chum salmon, *Oncorhynchus keta*. The results mentioned above suggest that the effective population sizes of the three fluvial *Lethenteron* taxa are as large as that of anadromous *L. japonicum*. Similar phenomena were found in *Lampetra* species (Beamish & Withler 1986) and *Tribolodon* species (Sakai 1995).

Although a high gene flow among intraspecific populations in *L. japonicum* probably occurs in the period of sea life, the Kyiya population was somewhat genetically divergent from the others (see Figure 2), meaning low gene flow between this population and the other populations. Iwata & Hamada (1986) reported a fluvial and dwarf male of *L. japonicum* from southern Hokkaido, Japan. Some lamprey species such as *Petromyzon marinus* show such variation of life-history styles between populations, in that one is the sea-run form and the other is the lake-run form (Hubbs & Potter 1971, Smith 1971, Potter 1980). Probably the population from Kyiya, located over 800 km upstream from the sea, has never migrated to the sea. This finding might support the hypothesis proposed by Beamish (1985, 1987) that some nonparasitic species evolved through an intermediate fluvial and parasitic interval, not directly from anadromous species.

Since it is extremely difficult for fluvial fishes to migrate from river to river, probably little or no gene flow occurs between the river populations or local populations (Mizuno 1963, Grudzien & Turner 1984, Goto & Andoh 1990, Shimizu et al. 1993, Okumura et al. 1996). However, the genetic differentiation between the populations of *L. kessleri* was less than that of each group of *L. reissneri* (see Table 4 and Figure 2) and other fluvial species of fishes (Gyllensten 1985, Mitsuboshi et al. 1992, Shimizu et al. 1993), but similar to that of *Tribolodon ezoe* (Sakai 1995).

It is well known that some glacial epochs occurred intermittently even in the Pleistocene. During these glacial epochs, several solely freshwater fishes, such as the Siberian stone loach, *Noemacheilus barbatulus toni,* and the Sakhalin lake minnow, *Phoxinus percunurus sachalinensis*, originated in the Siberian region and subsequently dispersed into the northern part of Japan (the present Hokkaido Island), via a landbridge connection with the Eurasian Continent (Aoyagi 1957, Goto et al. 1978, Maekawa & Goto 1982). Both *L. kessleri* and the northern group of *L. reissneri* were supposed to have the same historical experience as mentioned above (Yamazaki & Goto 1996, Yamazaki & Iwata 1997). The dispersal of *L. kessleri* from the Siberian region to the northern part of Japan might have occurred more recently than that of the northern group of *L. reissneri*, judging from Nei's (1975) divergence time of animals.

Based on the number of trunk myomeres (TM), Vladykov & Kott (1979) proposed that *L. japonicum* (TM = 65–72) was more closely related to *L. kessleri* (TM = 66–70) than to *L. reissneri* (TM = 57–63). Recently, Yamazaki & Goto (1997) reported that the two groups of *L. reissneri* have a similar number of trunk myomeres (northern group 51–66; southern group 50–62). This suggests that, based on morphology, both groups of *L. reissneri* are somewhat distantly related to *L. japonicum* at about the same distance, compared with the closer phylogenetic relationship between *L. japonicum* and *L. kessleri*.

On the other hand, our data show that the values of Nei's unbiased genetic distance (D = 0.042 – 0.394) among *L. japonicum, L. kessleri,* and the northern group of *L. reissneri* fell into the range from congeneric fish species (D = 0.025 – 0.609; Shaklee et al. 1982). The values (D = 0.559 – 0.926) between the three *Lethenteron* taxa and the southern group of *L. reissneri* were almost equivalent to those found among intergeneric fish species (D = 0.580 – 1.21; Shaklee et al. 1982, D = 0.495 – 1.42; Taniguchi et al. 1986). Judging from the somewhat low genetic divergence, the following three *Lethenteron* species, *L. japonicum, L. kessleri,* and the northern group of *L. reissneri*, may be monophyletic, suggesting the latter two species are satellite species of the former species. Because of the large genetic divergence between the southern group of *L. reissneri* and the other three *Lethenteron* species, however, it is difficult to determine whether the former has evolved monophyletically from the latter species. Further comparative studies of the molecular genetic and ecological features among the four *Lethenteron* species, using other genera as outgroups, are necessary to resolved the polarity of phylogeny and the process of speciation in lampreys.

Acknowledgements

We thank F. Yamazaki for his invaluable suggestions in the course of this study. We are also grateful to H. Sakai, K. Takata, and A. Iwata for providing valuable suggestions and helping with the field collections. Thanks are also due to S. Mori, S. Nakano, Y. Suzawa, I. Hosoya, S. Konno, and S. Takeyama for field collections.

References cited

Aoyagi, H. 1957. General notes on the freshwater fishes of the Japanese Archipelago. Daisyukan, Tokyo. 272 pp. (in Japanese).

Beamish, R.J. 1985. Freshwater parasitic lamprey on Vancouver Island and a theory of the evolution of the freshwater parasitic and nonparasitic life history types. pp. 123–140. *In*: R.E. Foreman, A. Gorbman, J.M. Dodd & R. Olsson (ed.) Evolutionary Biology of Primitive Fishes, Plenum Publishing Corporation, New York.

Beamish, R.J. 1987. Evidence that parasitic and nonparasitic life history types are produced by one population of lamprey. Can. J. Fish. Aquat. Sci. 44: 1779–1782.

Beamish, R.J. & R.E. Withler. 1986. A polymorphic population of lampreys that may produce parasitic and nonparasitic varieties. pp. 31–49. *In*: T. Uyeno, R. Arai, T. Taniguchi & K. Matsuura (ed.) Indo-Pacific Fish Biology, Proceedings of the Second International Conference on Indo-Pacific Fishes, Ichthyological Society of Japan, Tokyo.

Berg, L.S. 1931. A review of the lampreys of the northern hemisphere. Ann. Mus. Zool. Acad. Sci. USSR. 31: 87–116.

Berg, L.S. 1948. The freshwater fishes of the USSR and adjacent countries, 4th edition, Part 1. Akademia Nauk USSR, Moskow/Leningrad. 466 pp. (in Russian, English translation published by Israel Program for Scientific Translations, Jerusalem, 1965. 505 pp.)

Choi, K.C., S.R. Jeon & I.S. Kim. 1984. The atlas of Korean freshwater fishes. Korean Inst. Fresh. Biol. 1–95.

Clayton, J.W. & D.N. Tretiak. 1972. Amine-citrate buffers for pH control in starch gel electrophoresis. J. Fish. Res. Board Can. 29: 1169–1172.

Dowling, T.E. & W.S. Moore. 1984. Level of reproductive isolation between two cyprinid fishes, *Notropis cornutus* and *N. chrysocephalus*. Copeia 1984: 617–628.

Goto, A. & T. Andoh. 1990. Genetic divergence between the sibling species of river-sculpins, *Cottus amblystomopsis* and *C. nozawae*, with special reference to speciation. Env. Biol. Fish. 28: 257–266.

Goto, A., T. Nakanishi, H. Utoh & K. Hamada. 1978. A preliminary study of the freshwater fish fauna of rivers in southern Hokkaido. Bull. Fac. Fish. Hokkaido Univ. 29: 118–130 (in Japanese).

Grudzien, T.A. & B.J. Turner. 1984. Genetic identity and geographic differentiation of trophically dichotomous *Ilyodon* (Teleostei: Goodeidae). Copeia 1984: 102–107.

Gyllensten, U. 1985. The genetic structure of fish: differences in the intraspecific distribution of biochemical genetic variation between marine, anadromous, and freshwater species. J. Fish Biol. 26: 691–699.

Hardisty, M.W. & I.C. Potter. 1971. Paired species. pp. 249–277. *In*: M.W. Hardisty & I.C. Potter (ed.) The Biology of Lampreys, Volume 1, Academic Press, London.

Hubbs, C.L. & I.C. Potter. 1971. Distribution, phylogeny and taxonomy. pp. 1–65. *In*: M.W. Hardisty & I.C. Potter (ed.) The Biology of Lampreys, Volume 1, Academic Press, London.

Iwata, A. & K. Hamada. 1986. A dwarf male of the arctic lamprey, *Lethenteron japonicum* from the Assabu River, Hokkaido, Japan. Bull. Fac. Fish. Hokkaido Univ. 37: 17–22.

Iwata, A., A. Goto & K. Hamada. 1985. A review of the Siberian lamprey, *Lethenteron kessleri*, in Hokkaido, Japan. Bull. Fac. Fish. Hokkaido Univ. 36: 182–190.

Johnson, L. 1980. The arctic charr, *Salvelinus alpinus*. pp. 15–98. *In*: E.K. Balon (ed.) Charrs, Salmonid Fishes of the Genus *Salvelinus*, Dr W. Junk Publishers, Dordrecht.

Kawanabe, H. 1989. Japanese char(r(r))s and masu-salmon problems: a review. Physiol. Ecol. Japan, Spec. 1: 13–23.

Kijima, A. & Y. Fujio. 1984. Relationship between average heterozygosity and river population size in chum salmon. Bull. Japan. Soc. Sci. Fish. 50: 603–608.

Mayr, E. 1963. Animal species and evolution. Harvard University Press, Cambridge. 797 pp.

Maekawa, K. & A. Goto. 1982. Natural history of freshwater fishes in Japan. Chuo-koronsha, Tokyo. 212 pp. (in Japanese).

Mitsuboshi, T., A. Goto & F. Yamazaki. 1992. Genetic differentiation of the dolly varden *Salvelinus malma* in Hokkaido, Japan. Bull. Fac. Fish. Hokkaido Univ. 43: 153–161.

Mizuno, N. 1963. Distribution of *Cottus japonicus* Okada (Cottidae) and *Tukogobius flumineus* Mizuno (Gobiidae), with special references to their peculiarities in both the land-locking and the speciation from amphidromous ancestors. Bull. Osaka Gakugei Univ. 11: 129–161 (in Japanese).

Nei, M. 1975. Molecular population genetics and evolution. North-Holland Pub. Co., Amsterdam. 238 pp.

Nei, M. 1978. Estimation of average heterozygosity and genetic distance from a small number of individuals. Genetics 89: 583–590.

Okumura, N., Y. Hasebe & A. Goto. 1996. Genetic variations of *Cottus nozawae* populations from five tributaries of the Shiribetsu River of southern Hokkaido, Japan. Ichthyol. Res. 43: 65–71.

Potter, I.C. 1980. The Petromyzoniformes with particular reference to paired species. Can. J. Fish. Aquat. Sci. 37: 1595–1615.

Ridgway, G.J., S.W. Sherburne & R.D. Lewis. 1970. Polymorphism in the esterases of Atlantic herring. Trans. Amer. Fish. Soc. 99: 147–151.

Sakai, H. 1995. Life-histories and genetic divergence in three species of *Tribolodon* (Cyprinidae). Mem. Fac. Fish. Hokkaido Univ. 42: 1–98.

Sato, S. 1951. Discussion of lampreys on Hokkaido, Japan. Bull. Fac. Fish. Hokkaido Univ. 1: 54–62 (in Japanese).

Shaklee, J.B., C.S. Tamaru & R.S. Waples. 1982. Speciation and evolution of marine fishes studied by the electrophoretic analysis of proteins. Pacific Science 36: 141–157.

Shaklee, J.B., F.W. Allendorf, D.C. Morizot & G.S. Whitt. 1990. Gene nomenclature for protein-coding loci in fish. Trans. Amer. Fish. Soc. 119: 2–15.

Shaw, C.R. & R. Prasad. 1970. Starch gel electrophoresis of enzymes, a compilation of recipes. Biochem. Genet. 4: 297–320.

Shimizu, T., N. Taniguchi & N. Mizuno. 1993. An electrophoretic study of genetic differentiation of a Japanese freshwater goby, *Rhinogobius flumineus*. Japan. J. Ichthyol. 39: 329–343.

Smith, B.R. 1971. Sea lamprey in the Great Lakes of North America. pp. 207–248. *In*: M.W. Hardisty & I.C. Potter (ed.) The Biology of Lampreys, Volume 1, Academic Press, London.

Sneath, P.H.A. & R.R. Sokal. 1973. Numerical taxonomy. W.H. Freeman and Co., San Francisco. 573 pp.

Svärdson, G. 1979. Speciation of Scandinavian *Coregonus*. Rep. Inst. Freshw. Res. Drottningholm 23: 1–95.

Swofford, D.L. & R.B. Selander. 1981. BIOSYS-1: A FORTRAN program for the comprehensive analysis of electrophoretic data in population genetics and systematics. J. Hered. 72: 281–283.

Taniguchi, N., M. Fujita & M. Akazaki. 1986. Genetic divergence and systematics in sparid fish from Japan. pp. 849–858. *In*: T. Uyeno, R. Arai, T. Taniguchi & K. Matsuura (ed.) Indo-Pacific Fish Biology, Proceedings of the Second International Conference on Indo-Pacific Fishes, Ichthyological Society of Japan, Tokyo.

Vladykov, V.D. & E. Kott. 1979. Satellite species among the holarctic lampreys (Petromyzonidae). Can. J. Zool. 57: 860–867.

Yamamoto, S., S. Nakano & Y. Tokuda. 1992. Variation and divergence of the life- history of Japanese charr *Salvelinus leucomaenis* in an artificial lake-inlet stream system. Japan. J. Ecol. 42: 149–157 (in Japanese).

Yamazaki, Y. & A. Goto. 1996. Genetic differentiation of *Lethenteron reissneri* populations, with reference to the existence of discrete taxonomic entities. Ichthyol. Res. 43: 283–299.

Yamazaki, Y. & A. Goto. 1997. Morphometric and meristic characteristics of two groups of *Lethenteron reissneri*. Ichthyol. Res. 44: 15–25.

Yamazaki, Y. & A. Iwata. 1997. First record of the Siberian lamprey, *Lethenteron kessleri*, from Honshu Island, Japan. Japan. J. Ichthyol. 44: 51–55 (in Japanese).

Zanandrea, G. 1959. Speciation among lampreys. Nature 184: 380.

Appendix. Genotypic distributions at the 27 loci for 60 samples of *Lethenteron* taxa. Numbers refer to sample sites listed in Figure 1.

Locus	Genotype	1	2	3	4	5	6	7	8	9	10	11	12	13	14	15	16	17	18	19	20	21	22	23	24	25	26	27	28	29	30	31	32	33	34	35	36	37	38	39	40	41	42	43	44	45	46	47	48	49	50	51	52	53	54	55	56	57	58	59	60
AAT-1*	*173/173	-	1	1	-	-	-	-	-	-	-	-	-	-	11	1	1	-	2	-	-	-	-	-	1	13	-	-	-	-	-	-	-	-	-	-	-	-	-	-	-	-	-	-	-	-	-	-	-	-	-	-	-	-	-	-	-	-	-	-	-
	*173/100	4	3	5	-	-	2	1	2	-	-	-	-	4	12	12	-	2	16	4	7	3	2	-	9	20	-	-	-	-	1	-	2	-	5	1	-	-	-	1	-	-	-	-	-	3	-	-	-	-	-	-	-	-	-	-	-	-	-	-	-
	*135/135	-	-	-	-	-	-	-	-	-	-	-	-	-	-	-	-	-	-	-	-	-	-	-	-	-	-	-	-	-	-	-	-	-	-	-	-	-	-	-	-	-	-	-	-	-	-	-	-	-	-	-	-	-	-	-	-	-	-	15	-
	*123/123	-	-	-	-	-	-	-	-	-	-	-	-	-	-	-	-	-	-	-	-	-	-	-	-	-	-	-	-	-	-	-	-	-	-	-	-	-	-	-	-	-	-	-	-	-	-	57	40	31	26	11	18	34	51	49	50	9	16	-	8
	*123/39	-	-	-	-	-	-	-	-	-	-	-	-	-	-	-	-	-	-	-	-	-	-	-	-	-	-	-	-	-	-	-	-	-	-	-	-	-	-	-	-	-	-	-	-	-	-	-	-	-	-	-	15	-	-	-	-	-	-	-	-
	*100/100	47	64	20	18	14	38	41	34	45	30	24	26	34	9	22	46	40	34	26	49	24	72	36	19	6	41	40	42	41	20	41	40	17	36	23	35	14	22	44	28	26	50	50	52	47	24	-	-	-	-	-	-	-	-	-	-	-	-	-	-
	*100/42	-	-	-	-	-	-	-	-	-	-	-	-	-	-	-	-	-	-	-	-	-	-	-	-	-	-	-	-	-	-	-	-	-	-	-	-	-	-	-	10	-	-	-	-	-	-	-	-	-	-	-	-	-	-	-	-	-	-	-	-
	*39/39	-	-	-	-	-	-	-	-	-	-	-	-	-	-	-	-	-	-	-	-	-	-	-	-	-	-	-	-	-	-	-	-	-	-	-	-	-	-	-	-	-	-	-	-	-	-	-	-	-	-	-	4	-	-	-	-	-	-	-	-
AAT-2*	*-74/-74	-	-	-	-	-	-	-	-	-	-	-	-	-	-	-	-	-	-	-	-	-	-	-	-	-	1	-	-	-	-	-	-	-	-	-	-	-	-	-	-	-	-	-	-	-	-	-	-	-	-	-	-	-	-	-	-	-	-	-	-
	*-74/-100	-	-	-	-	-	-	-	-	-	-	-	-	-	-	-	-	-	-	-	-	-	-	-	-	-	10	-	-	-	-	-	-	-	-	-	-	-	-	-	1	-	-	-	-	-	-	-	-	-	-	-	-	-	-	-	-	-	-	-	-
	*-100/-100	51	68	26	18	14	40	42	36	45	30	24	26	38	32	35	47	42	52	30	56	27	74	36	29	39	30	40	42	41	21	41	42	17	41	24	35	14	22	45	37	26	-	-	37	50	24	57	40	31	26	11	39	33	51	49	50	9	16	15	8
	*-100/-113	-	-	-	-	-	-	-	-	-	-	-	-	-	-	-	-	-	-	-	-	-	-	-	-	-	-	-	-	-	-	-	-	-	-	-	-	-	-	-	-	-	-	-	14	-	-	-	-	-	-	-	-	1	-	-	-	-	-	-	-
	*-111/-111	-	-	-	-	-	-	-	-	-	-	-	-	-	-	-	-	-	-	-	-	-	-	-	-	-	-	-	-	-	-	-	-	-	-	-	-	-	-	-	-	-	50	48	-	-	-	-	-	-	-	-	-	-	-	-	-	-	-	-	-
	*-113/-113	-	-	-	-	-	-	-	-	-	-	-	-	-	-	-	-	-	-	-	-	-	-	-	-	-	-	-	-	-	-	-	-	-	-	-	-	-	-	-	-	-	-	-	1	-	-	-	-	-	-	-	-	-	-	-	-	-	-	-	-
	*-121/-121	-	-	-	-	-	-	-	-	-	-	-	-	-	-	-	-	-	-	-	-	-	-	-	-	-	-	-	-	-	-	-	-	-	-	-	-	-	-	-	-	-	-	2	-	-	-	-	-	-	-	-	-	-	-	-	-	-	-	-	-
ACP*	*209/209	-	-	-	-	-	-	-	-	-	-	1	-	-	-	-	-	-	-	-	-	-	-	-	-	-	-	-	-	-	-	-	-	-	-	-	-	-	2	-	-	-	-	-	-	-	-	-	-	-	-	-	-	-	2	-	-	-	-	-	-
	*209/71	-	1	-	-	-	-	1	1	-	-	1	1	-	-	-	-	-	-	-	-	-	-	-	-	-	-	-	-	-	-	-	-	-	-	-	-	-	-	-	-	-	-	-	-	-	-	1	-	-	-	-	-	-	3	-	-	-	-	-	-
	*209/-100	-	1	-	-	-	-	-	-	1	-	-	-	-	-	-	1	-	-	-	-	-	1	-	-	-	-	-	-	-	-	-	-	-	-	1	-	-	2	-	-	-	-	-	-	-	-	-	-	-	-	-	-	-	-	-	-	-	-	-	-
	*71/71	6	5	4	3	-	6	3	4	2	3	2	1	2	1	1	4	-	1	-	3	3	11	3	1	2	-	-	-	-	2	-	-	2	1	-	-	-	-	-	-	1	4	-	-	-	-	49	39	-	-	-	-	-	3	-	-	-	-	-	4
	*71/-100	14	17	6	8	4	10	8	5	7	10	4	3	12	10	6	13	3	4	1	12	12	23	4	4	8	-	1	-	1	1	2	1	2	1	-	-	-	-	4	1	2	10	-	-	-	-	5	1	-	1	1	9	-	15	-	-	-	-	2	3
	*71/-286	-	-	-	-	-	1	-	-	-	-	-	-	-	1	-	-	-	-	-	-	-	-	-	-	-	-	1	-	-	-	7	2	-	1	-	-	-	-	-	-	-	-	-	-	-	-	-	-	-	1	-	-	-	2	-	-	-	-	-	-
	*71/-386	-	-	-	-	-	-	-	1	-	-	-	-	-	-	-	-	-	-	-	-	-	-	-	1	-	-	-	-	-	-	2	-	-	-	-	-	-	-	-	-	-	-	-	-	-	-	-	-	-	-	-	-	-	1	-	-	-	-	-	-
	*-100/-100	31	43	15	7	9	23	30	25	34	16	15	19	24	20	28	28	39	47	29	41	12	39	27	19	29	15	3	-	1	-	2	4	9	5	5	14	2	5	38	37	23	36	49	51	12	22	2	-	31	7	-	26	14	17	-	33	1	-	13	-
	*-100/-286	-	1	-	-	1	-	-	-	1	1	1	2	-	-	-	1	-	-	-	-	-	-	2	1	-	17	12	-	11	1	4	14	4	20	9	15	4	8	3	-	-	-	1	-	22	2	-	-	-	6	7	4	13	5	-	11	4	-	-	-
	*-100/-386	-	-	1	-	-	-	-	-	-	-	-	-	-	-	-	-	-	-	-	-	-	-	-	3	-	-	-	-	1	1	-	6	-	-	-	-	-	-	-	-	-	-	-	-	-	-	-	-	-	3	-	-	-	-	-	-	-	-	-	-
	*-286/-286	-	-	-	-	-	-	-	-	-	-	-	-	-	-	-	-	-	-	-	-	-	-	-	-	-	9	23	32	23	2	13	10	-	13	9	5	8	5	-	-	-	-	-	1	16	-	-	-	-	4	3	-	7	3	49	6	4	16	-	1
	*-286/-386	-	-	-	-	-	-	-	-	-	-	-	-	-	-	-	-	-	-	-	-	-	-	-	-	-	-	-	9	3	10	5	5	-	-	-	-	-	-	-	-	-	-	-	-	-	-	-	-	-	1	-	-	-	-	-	-	-	-	-	-
	*-386/-386	-	-	-	-	-	-	-	-	-	-	-	-	-	-	-	-	-	-	-	-	-	-	-	-	-	-	-	1	1	4	6	-	-	-	-	-	-	-	-	-	-	-	-	-	-	-	-	-	-	1	-	-	-	-	-	-	-	-	-	-
AH-1*	*132/132	1	2	-	2	-	3	2	-	-	1	-	-	-	-	-	3	-	-	2	3	-	4	2	2	1	-	-	-	-	-	-	-	-	-	-	-	-	-	3	-	-	-	-	1	-	-	-	-	-	1	1	-	-	-	2	-	-	-	-	-
	*132/100	8	5	1	2	-	4	3	2	1	3	-	-	1	1	2	5	1	2	6	6	-	9	4	2	1	-	-	-	-	-	-	-	-	2	-	-	-	-	7	1	-	-	-	1	-	-	7	5	5	8	-	-	2	-	9	-	4	-	-	-
	*132/72	-	2	-	-	-	1	-	-	1	-	3	-	-	-	-	1	-	-	-	3	-	-	2	-	1	-	1	1	-	1	-	-	-	-	-	-	-	-	5	-	-	-	-	-	-	-	-	-	-	-	-	-	-	-	-	-	-	-	-	-
	*100/100	19	33	14	7	6	12	26	25	22	12	9	4	6	8	6	22	29	45	17	25	14	34	16	5	8	39	4	2	5	2	2	-	-	23	21	-	8	1	10	27	26	-	-	8	1	12	50	11	26	6	9	-	22	51	29	35	4	-	-	-
	*100/72	13	14	7	4	2	12	4	6	6	5	3	9	15	3	19	10	4	-	-	7	4	16	6	3	15	1	14	5	13	5	8	-	-	5	3	1	5	4	10	6	-	1	-	15	6	8	-	8	-	5	1	4	5	-	6	8	1	-	-	-
	*72/72	9	4	1	3	-	8	1	3	4	3	2	9	9	6	6	4	1	-	-	5	3	4	-	1	13	-	21	16	23	13	28	36	17	3	-	34	1	4	8	4	-	41	38	21	43	4	-	8	-	5	-	32	3	-	3	2	-	16	15	8
AH-2*	*244/244	-	1	-	-	-	2	-	1	1	-	-	2	-	-	-	1	2	2	2	2	-	2	1	1	1	2	7	10	23	1	1	7	1	4	6	1	2	2	4	-	-	-	-	-	3	-	-	-	-	-	-	-	-	-	-	-	-	-	-	-
	*244/100	-	12	9	5	2	15	9	14	10	1	4	5	7	-	7	15	11	16	7	14	8	15	15	6	6	10	23	7	14	2	11	12	3	16	9	2	3	4	6	-	-	1	3	-	7	-	-	-	-	-	-	-	-	-	-	-	-	-	-	-
	*244/-34	-	-	-	-	-	-	-	-	-	-	-	-	-	-	-	-	-	-	-	1	-	1	-	2	-	-	-	-	-	-	-	-	-	-	-	-	-	1	4	-	-	-	-	-	-	-	-	-	-	-	-	-	-	-	-	-	-	-	-	-
	*100/100	49	46	17	12	12	23	33	21	27	23	13	8	18	32	18	29	29	26	16	36	19	48	20	13	32	27	10	7	4	18	29	17	13	12	9	26	9	12	13	10	25	20	18	19	40	-	-	-	-	-	-	-	-	-	-	-	-	-	-	-
	*100/-34	2	2	-	1	-	-	-	-	3	-	2	1	1	-	1	1	-	1	-	3	-	1	-	-	-	-	-	-	-	-	-	-	-	-	-	-	-	3	11	14	-	21	19	24	-	1	-	-	-	-	-	-	-	-	-	-	-	-	-	-
	*100/-122	-	-	-	-	-	-	-	-	-	-	-	-	-	-	-	-	-	-	-	-	-	-	-	-	-	-	-	-	-	-	-	-	-	-	-	-	-	-	-	-	1	-	-	-	-	-	-	-	-	-	-	-	-	-	-	-	-	-	-	-
	*-34/-34	-	-	-	-	-	-	-	-	1	-	1	1	-	-	-	-	-	-	-	-	-	-	-	-	-	-	-	-	-	-	-	-	-	1	-	-	-	-	5	9	-	7	10	13	-	17	9	16	-	-	-	-	-	-	47	44	4	16	15	-
	*-34/-122	-	-	-	-	-	-	-	-	-	-	-	-	-	-	-	-	-	-	-	-	-	-	-	-	-	-	-	-	-	-	-	-	-	-	-	-	-	-	-	-	-	-	-	-	-	5	8	14	-	-	-	-	1	-	2	5	3	-	-	-
	*-122/-122	-	-	-	-	-	-	-	-	-	-	-	-	-	-	-	-	-	-	-	-	-	-	-	-	-	-	-	-	-	-	-	-	-	-	-	-	-	-	-	-	-	-	-	-	-	1	6	7	22	26	11	39	33	51	-	1	2	-	-	8
	*-122/-315	-	-	-	-	-	-	-	-	-	-	-	-	-	-	-	-	-	-	-	-	-	-	-	-	-	-	-	-	-	-	-	-	-	-	-	-	-	-	-	-	-	-	-	-	-	-	-	-	9	-	-	-	-	-	-	-	-	-	-	-
EST*	*105/105	-	-	-	-	1	3	4	5	-	-	-	-	-	-	1	1	2	1	1	-	1	-	-	-	-	3	4	1	5	-	-	1	-	-	-	-	-	-	-	-	-	-	-	-	-	2	-	-	-	-	-	-	-	-	-	-	-	1	-	-
	*105/100	-	-	3	-	3	8	10	10	-	-	-	4	12	-	4	-	13	6	10	4	4	2	-	-	3	9	13	5	10	1	2	7	-	1	-	1	-	-	-	2	-	-	-	3	1	-	1	-	-	-	-	-	-	-	-	-	-	-	-	-
	*100/100	51	68	23	18	10	29	28	20	45	30	24	22	26	32	30	46	27	45	19	52	22	72	36	29	36	28	22	36	26	20	39	34	17	40	24	34	14	22	45	36	26	50	50	49	49	22	56	40	31	26	11	39	34	51	49	50	9	15	15	8
FH*	*100/100	51	67	25	18	14	40	41	36	44	30	24	26	38	32	35	47	41	52	30	51	27	74	36	29	39	41	-	42	33	-	3	1	-	2	1	35	14	22	42	34	26	50	50	52	47	24	-	-	31	26	11	39	34	51	49	50	9	16	15	8
	*100/86	-	1	-	-	-	-	-	-	-	-	-	-	-	-	-	-	1	-	-	-	-	-	-	-	-	-	-	-	6	-	9	9	1	5	2	-	-	-	2	-	-	-	-	-	-	-	1	-	-	-	-	-	-	-	-	-	-	-	-	-
	*100/78	-	-	-	-	-	-	-	-	-	-	-	-	-	-	-	-	-	-	-	-	-	-	-	-	-	-	-	-	-	-	-	-	-	-	-	-	-	-	-	-	-	-	-	-	3	-	-	-	-	-	-	-	-	-	-	-	-	-	-	-
	*86/86	-	-	1	-	-	-	1	-	1	-	-	-	-	-	-	-	-	-	-	1	-	-	-	-	-	-	39	-	2	21	29	32	16	34	21	-	-	-	1	-	-	-	-	-	-	-	52	40	-	-	-	-	-	-	-	-	-	-	-	-
	*86/78	-	-	-	-	-	-	-	-	-	-	-	-	-	-	-	-	-	-	-	-	-	-	-	-	-	-	1	-	-	-	-	-	-	-	-	-	-	-	-	-	-	-	-	-	-	-	-	-	-	-	-	-	-	-	-	-	-	-	-	-
G3PDH*	*54/-100	-	-	-	-	-	-	-	-	-	-	-	-	-	-	-	-	-	-	-	-	-	-	-	-	-	-	-	-	-	-	-	-	-	-	-	-	-	-	-	-	-	-	-	-	-	-	-	-	-	-	-	-	-	-	-	-	-	-	-	3
	*-100/-100	51	68	26	18	14	40	42	36	45	30	24	26	38	32	35	46	42	52	30	56	27	74	36	29	39	41	40	42	30	21	41	42	17	2	1	4	14	21	45	38	26	50	50	51	50	11	57	40	28	26	11	39	34	51	49	50	8	15	15	5
	*-100/-526	-	-	-	-	-	-	-	-	-	-	-	-	-	-	-	-	-	-	-	-	-	-	-	-	-	-	-	-	11	-	-	-	-	23	11	13	-	-	-	-	-	-	-	1	-	11	-	-	-	-	-	-	-	-	-	-	1	1	-	-
	*-100/-942	-	-	-	-	-	-	-	-	-	-	-	-	-	-	-	1	-	-	-	-	-	-	-	-	-	-	-	-	-	-	-	-	-	-	-	-	-	1	-	-	-	-	-	-	-	-	-	-	3	-	-	-	-	-	-	-	-	-	-	-
	*-526/-526	-	-	-	-	-	-	-	-	-	-	-	-	-	-	-	-	-	-	-	-	-	-	-	-	-	-	-	-	-	-	-	-	-	16	12	18	-	-	-	-	-	-	-	-	-	2	-	-	-	-	-	-	-	-	-	-	-	-	-	-
G6PDH*	*148/148	-	-	-	-	-	-	-	-	-	-	-	-	-	-	-	-	-	-	-	-	-	-	-	-	-	-	-	-	-	-	-	-	-	-	-	-	-	-	-	-	-	-	-	-	-	-	57	40	31	26	11	39	34	51	49	50	9	16	15	8
	*100/100	51	68	26	18	14	40	42	36	45	30	24	26	38	32	35	47	42	52	30	56	27	74	36	29	39	41	40	42	41	21	41	42	17	41	24	35	14	22	45	38	26	50	50	52	50	24	-	-	-	-	-	-	-	-	-	-	-	-	-	-

Appendix. Continued.

Locus	Genotype	1	2	3	4	5	6	7	8	9	10	11	12	13	14	15	16	17	18	19	20	21	22	23	24	25	26	27	28	29	30	31	32	33	34	35	36	37	38	39	40	41	42	43	44	45	46	47	48	49	50	51	52	53	54	55	56	57	58	59	60
GPI-1*	*196/196	-	1	-	-	-	-	1	-	-	-	-	-	-	2	1	-	-	-	-	-	-	-	-	-	6	-	-	-	-	-	-	-	-	-	-	-	-	-	-	-	-	-	-	-	-	-	-	-	-	-	-	-	-	-	-	-	-	-	-	-
	*196/186	-	-	-	-	-	-	-	-	-	-	-	-	-	-	-	-	-	-	-	-	-	-	-	-	-	-	-	-	-	1	-	-	-	-	-	-	-	-	-	-	-	-	-	-	-	-	-	-	-	-	-	-	-	-	-	-	-	-	-	-
	*196/100	1	1	1	-	-	-	2	-	-	1	1	-	2	17	10	4	-	15	-	3	-	-	-	9	22	-	-	-	-	-	-	1	-	-	-	-	-	-	-	-	-	-	-	-	-	-	-	-	-	-	-	-	-	-	-	-	-	-	-	-
	*186/124	-	-	-	-	-	-	-	-	-	-	-	-	-	-	-	-	-	-	-	-	-	-	-	-	-	-	-	-	-	1	-	-	-	-	-	-	-	-	-	-	-	-	-	-	-	-	-	-	-	-	-	-	-	-	-	-	-	-	-	-
	*134/134	1	-	-	-	-	-	1	-	-	1	-	-	-	-	-	-	-	-	-	-	-	-	-	-	-	-	-	-	-	3	1	-	-	5	2	-	-	-	-	1	2	1	-	-	-	1	57	22	-	5	4	7	33	1	1	-	1	-	-	-
	*134/124	-	-	-	-	-	-	-	-	-	-	-	-	-	-	-	-	-	-	-	-	-	-	-	-	-	-	-	-	-	2	-	1	-	-	-	-	1	-	-	-	4	-	-	-	-	-	-	-	-	-	-	-	-	-	-	-	-	-	-	-
	*134/100	2	3	1	-	1	1	2	2	1	-	-	-	3	-	-	4	-	1	-	4	-	1	2	1	-	1	-	-	6	6	4	-	-	12	10	-	1	-	-	2	2	15	-	4	-	4	-	17	2	11	-	5	-	9	6	1	-	-	-	-
	*134/48	-	-	-	-	-	-	-	-	-	-	-	-	-	-	-	-	-	-	-	-	-	-	-	-	-	-	-	-	-	-	-	1	-	6	-	-	-	-	-	-	-	-	-	-	-	-	-	-	-	3	6	15	1	-	-	-	-	-	-	1
	*124/124	-	-	-	-	-	-	-	-	-	-	-	-	-	-	-	-	-	-	-	-	-	-	-	-	-	-	4	-	-	-	17	8	5	-	-	10	-	-	2	-	2	-	-	-	1	-	-	-	-	-	-	-	-	-	-	-	-	-	-	3
	*124/100	-	-	-	-	-	-	-	-	-	-	-	-	-	-	-	-	-	-	-	-	-	-	-	-	-	-	12	-	10	2	10	13	9	2	-	22	-	-	8	-	9	-	-	-	-	-	-	-	-	-	-	-	-	-	-	-	-	-	-	2
	*124/90	-	-	-	-	-	-	-	-	-	-	-	-	-	-	-	-	-	-	-	-	-	-	-	-	-	-	-	-	-	-	-	-	-	-	-	-	-	-	-	-	-	-	-	-	6	-	-	-	-	-	-	-	-	-	-	-	-	-	-	-
	*124/48	-	-	-	-	-	-	-	-	-	-	-	-	-	-	-	-	-	-	-	-	-	-	-	-	-	-	-	-	-	-	-	-	-	1	-	-	-	-	-	-	-	-	-	-	-	-	-	-	-	-	-	-	-	-	-	-	-	-	-	1
	*100/100	43	55	23	14	13	36	33	33	43	28	23	26	33	13	24	37	42	36	30	48	21	51	34	19	11	26	24	37	25	5	6	17	3	11	6	3	11	7	30	35	7	34	43	44	-	18	-	1	29	1	-	-	-	33	42	29	8	16	15	-
	*100/48	-	-	-	4	-	2	2	1	1	-	-	-	-	-	-	-	-	-	-	1	3	11	-	-	-	8	-	4	-	-	-	1	-	4	2	-	1	14	-	-	-	-	1	-	-	-	-	-	-	-	-	3	-	1	-	16	-	-	-	-
	*100/-12	2	-	-	-	-	1	1	-	-	-	-	-	-	-	-	-	-	-	-	-	2	-	-	-	-	-	-	-	-	-	-	-	-	-	3	-	-	-	-	-	-	-	-	-	-	1	-	-	-	-	-	-	-	-	-	-	-	-	-	-
	*90/90	-	-	-	-	-	-	-	-	-	-	-	-	-	-	-	-	-	-	-	-	-	-	-	-	-	-	-	-	-	-	-	-	-	-	-	-	-	-	-	-	-	-	-	-	42	-	-	-	-	-	-	-	-	-	-	-	-	-	-	-
	*90/38	-	-	-	-	-	-	-	-	-	-	-	-	-	-	-	-	-	-	-	-	-	-	-	-	-	-	-	-	-	-	-	-	-	-	-	-	-	-	-	-	-	-	-	-	1	-	-	-	-	-	-	-	-	-	-	-	-	-	-	-
	*48/48	-	-	1	-	-	-	-	-	-	-	-	-	-	-	-	-	-	-	-	-	-	1	-	-	-	-	-	1	-	-	-	-	-	-	-	-	-	1	-	-	-	-	-	-	-	-	-	-	-	-	1	9	-	-	-	2	-	-	-	1
	*48/-12	-	-	-	-	-	-	-	-	-	-	-	-	-	-	-	-	-	-	-	-	1	-	-	-	-	-	-	-	-	-	-	-	-	-	-	-	-	-	-	-	-	-	-	-	-	-	-	-	-	-	-	-	-	-	-	-	-	-	-	-
	*-12/-12	-	-	-	-	-	-	-	-	-	-	-	-	-	-	-	-	-	-	-	-	-	-	-	-	-	-	-	-	-	-	-	-	-	-	1	-	-	-	-	-	-	-	-	-	-	-	-	-	-	-	-	-	-	-	-	-	-	-	-	-
GPI-2*	*196/196	-	-	-	-	-	-	-	-	-	-	-	-	-	-	-	-	-	-	-	-	-	-	-	-	-	-	-	-	-	-	-	-	-	-	-	-	-	-	-	-	-	-	-	-	-	-	-	-	-	-	-	-	4	-	-	-	-	-	-	-
	*196/168	-	-	-	-	-	-	-	-	-	-	-	-	-	-	-	-	-	-	-	-	-	-	-	-	-	-	-	-	-	-	-	-	-	-	-	-	-	-	-	-	-	-	-	-	-	-	-	-	-	-	-	4	16	13	-	-	-	1	-	-
	*196/124	-	-	-	-	-	-	-	-	-	-	-	-	-	-	-	-	-	-	-	-	-	-	-	-	-	-	-	-	-	-	-	-	-	-	-	-	-	-	-	-	-	-	-	-	-	-	-	-	-	-	-	-	1	-	-	-	-	-	-	-
	*168/168	-	-	-	-	-	-	-	-	-	-	-	-	-	-	-	-	-	-	-	-	-	-	-	-	-	-	-	-	-	-	-	-	-	-	-	-	-	-	-	-	-	-	-	-	-	-	2	1	-	11	-	24	10	24	-	-	-	3	-	-
	*168/124	-	-	-	-	-	-	-	-	-	-	-	-	-	-	-	-	-	-	-	-	-	-	-	-	-	-	-	-	-	-	-	-	-	-	-	-	-	-	-	-	-	-	-	-	-	-	11	7	4	1	-	10	1	6	-	-	-	4	-	-
	*168/72	-	-	-	-	-	-	-	-	-	-	-	-	-	-	-	-	-	-	-	-	-	-	-	-	-	-	-	-	-	-	-	-	-	-	-	-	-	-	-	-	-	-	-	-	-	-	-	-	4	6	-	-	-	-	-	-	-	3	-	-
	*134/134	-	-	-	-	-	-	-	-	-	-	-	-	-	-	-	-	-	-	-	-	-	2	-	-	-	-	-	-	-	-	-	1	-	-	-	-	-	-	-	-	-	-	-	-	-	-	-	-	-	-	-	-	-	-	-	-	-	-	-	-
	*134/100	-	-	-	1	2	1	-	1	-	-	-	-	-	-	-	1	-	-	-	-	-	4	3	-	-	-	-	1	-	-	-	-	-	-	-	-	-	-	-	-	-	-	-	1	-	-	-	-	-	-	-	-	-	-	-	-	-	-	-	-
	*134/48	-	-	-	-	-	-	-	-	-	-	-	-	-	-	-	-	-	-	-	-	-	1	-	-	-	-	-	-	-	-	-	1	-	-	-	-	-	-	-	-	-	-	-	3	-	-	-	-	-	-	-	-	-	-	-	-	-	-	-	-
	*124/124	-	-	-	-	-	-	-	-	-	-	-	-	-	-	-	-	-	-	-	-	-	-	-	-	-	-	-	-	-	-	-	-	-	-	-	-	-	-	-	-	-	-	-	-	-	-	44	32	-	-	10	1	-	-	-	-	-	-	-	7
	*124/72	-	-	-	-	-	-	-	-	-	-	-	-	-	-	-	-	-	-	-	-	-	-	-	-	-	-	-	-	-	-	-	-	-	-	-	-	-	-	-	-	-	-	-	-	-	-	-	-	9	-	1	-	-	-	-	-	-	2	-	1
	*100/100	35	45	21	15	9	33	31	24	37	24	16	16	26	30	16	28	31	19	12	46	10	55	26	6	38	-	14	2	3	2	-	2	5	34	19	1	5	-	8	-	11	50	44	8	-	6	-	-	-	-	-	-	-	-	-	-	-	-	-	-
	*100/48	11	7	3	2	3	5	10	11	7	6	6	8	10	2	14	7	-	11	1	8	-	11	6	6	1	11	20	17	16	3	4	15	10	7	5	2	5	8	20	2	13	-	-	19	-	15	-	-	-	-	-	-	-	-	-	-	-	-	-	-
	*100/-12	-	-	2	-	-	-	1	-	-	-	2	-	-	-	-	3	-	3	-	1	-	-	-	2	-	-	-	-	-	-	2	-	-	-	-	-	-	-	2	-	-	-	-	-	2	-	-	-	-	-	-	-	-	-	-	-	-	-	-	-
	*100/-69	-	1	-	-	-	-	-	-	-	-	-	1	2		2	2	11	13	14	1	15	-	-	9	-	-	-	-	-	-	-	-	-	-	-	-	-	-	-	-	-	-	-	-	-	-	-	-	-	-	-	-	-	-	-	-	-	-	-	-
	*72/72	-	-	-	-	-	-	-	-	-	-	-	-	-	-	-	-	-	-	-	-	-	-	-	-	-	-	-	-	-	-	-	-	-	-	-	-	-	-	-	-	-	-	-	-	-	-	-	-	12	2	-	-	-	-	-	-	-	3	-	-
	*72/0	-	-	-	-	-	-	-	-	-	-	-	-	-	-	-	-	-	-	-	-	-	-	-	-	-	-	-	-	-	-	-	-	-	-	-	-	-	-	-	-	-	-	-	-	-	-	-	-	1	-	-	-	-	-	-	-	-	-	-	-
	*48/48	3	3	-	-	-	1	-	-	1	-	-	1	-	-	-	2	-	2	-	-	-	1	-	3	-	24	5	22	22	16	28	23	2	-	-	32	4	14	8	17	2	-	-	17	4	3	-	-	-	-	-	-	-	-	-	-	-	-	-	-
	*48/-12	-	-	-	-	-	-	-	-	-	-	-	-	-	-	-	-	-	-	-	-	-	-	-	-	-	-	1	-	-	-	7	-	-	-	-	-	-	-	2	19	-	-	-	-	25	-	-	-	-	-	-	-	-	-	-	-	-	-	-	-
	*48/-69	-	-	-	-	-	-	-	-	-	-	-	-	-	-	2	1	-	2	1	-	-	-	-	-	-	-	-	-	-	-	-	-	-	-	-	-	-	-	-	-	-	-	-	-	1	-	-	-	-	-	-	-	-	-	-	-	-	-	-	-
	*0/0	-	-	-	-	-	-	-	-	-	-	-	-	-	-	-	-	-	-	-	-	-	-	-	-	-	-	-	-	-	-	-	-	-	-	-	-	-	-	-	-	-	-	-	-	-	-	-	-	1	-	-	-	-	-	-	-	-	-	-	-
	*0/-54	-	-	-	-	-	-	-	-	-	-	-	-	-	-	-	-	-	-	-	-	-	-	-	-	-	-	-	-	-	-	-	-	-	-	-	-	-	-	-	-	-	-	-	-	-	-	-	-	-	-	-	-	-	-	-	1	1	-	-	-
	*-12/-12	-	-	-	-	-	-	-	-	-	-	-	-	-	-	-	-	-	-	-	-	-	-	-	-	-	-	-	-	-	-	-	-	-	-	-	-	-	-	-	-	-	-	-	-	17	-	-	-	-	-	-	-	-	-	-	-	-	-	-	-
	*-54/-54	-	-	-	-	-	-	-	-	-	-	-	-	-	-	-	-	-	-	-	-	-	-	-	-	-	-	-	-	-	-	-	-	-	-	-	-	-	-	-	-	-	-	-	-	-	-	-	-	-	-	-	-	-	-	49	10	8	-	15	-
	*-54/-100	-	-	-	-	-	-	-	-	-	-	-	-	-	-	-	-	-	-	-	-	-	-	-	-	-	-	-	-	-	-	-	-	-	-	-	-	-	-	-	-	-	-	-	-	-	-	-	-	-	-	-	-	-	-	-	21	-	-	-	-
	*-69/-69	-	-	-	-	-	-	-	-	-	-	-	-	-	-	1	1	-	2	2	-	2	-	-	3	-	-	-	-	-	-	-	-	-	-	-	-	-	-	-	-	-	-	-	-	-	-	-	-	-	-	-	-	-	-	-	-	-	-	-	-
	*-100/-100	-	-	-	-	-	-	-	-	-	-	-	-	-	-	-	-	-	-	-	-	-	-	-	-	-	-	-	-	-	-	-	-	-	-	-	-	-	-	-	-	-	-	-	-	-	-	-	-	-	-	-	-	-	-	-	16	-	-	-	-
IDDH*	*100/100	51	68	26	18	14	40	42	36	45	30	24	26	38	32	35	47	42	52	30	56	27	74	36	29	39	41	40	42	41	21	41	42	17	41	24	35	14	22	45	38	26	50	50	52	50	24	57	40	31	26	11	39	34	51	49	50	9	16	15	8
IDHP-1*	*129/129	-	-	-	-	-	-	-	-	-	-	-	-	-	-	-	-	-	-	-	-	-	-	-	-	-	-	-	-	-	-	-	-	-	-	-	-	-	-	-	-	-	-	-	-	-	-	57	40	31	26	11	39	34	51	49	50	9	16	15	8
	*100/100	-	-	-	-	-	-	-	-	-	-	-	-	-	-	-	-	-	-	-	-	-	-	-	-	-	41	40	42	41	21	41	42	17	41	24	35	14	22	45	38	26	50	50	52	50	24	-	-	-	-	-	-	-	-	-	-	-	-	-	-
	*97/97	51	68	26	18	14	40	42	36	45	30	24	26	38	32	35	47	42	52	30	56	27	74	36	29	39	-	-	-	-	-	-	-	-	-	-	-	-	-	-	-	-	-	-	-	-	-	-	-	-	-	-	-	-	-	-	-	-	-	-	-
IDHP-2*	*155/155	-	-	-	-	-	-	-	-	-	-	-	-	-	-	-	-	-	-	-	-	-	-	-	-	-	-	-	-	-	-	-	-	-	-	-	-	-	-	-	-	-	-	-	-	-	-	57	40	31	25	11	39	34	47	49	50	9	16	15	8
	*155/127	-	-	-	-	-	-	-	-	-	-	-	-	-	-	-	-	-	-	-	-	-	-	-	-	-	-	-	-	-	-	-	-	-	-	-	-	-	-	-	-	-	-	-	-	-	-	-	-	-	1	-	-	-	4	-	-	-	-	-	-
	*100/100	-	-	-	-	-	-	-	-	-	-	-	-	-	-	-	-	-	-	-	-	-	-	-	-	-	41	40	42	41	21	40	42	17	41	24	35	14	18	41	38	26	50	50	51	49	24	-	-	-	-	-	-	-	-	-	-	-	-	-	-
	*100/72	-	-	-	-	-	-	-	-	-	-	-	-	-	-	-	-	-	-	-	-	-	-	-	-	-	-	-	-	-	-	1	-	-	-	-	-	-	4	3	-	-	-	-	1	1	-	-	-	-	-	-	-	-	-	-	-	-	-	-	-
	*92/92	51	66	25	17	14	40	42	34	45	29	23	25	36	31	34	46	42	51	30	55	27	72	36	29	39	-	-	-	-	-	-	-	-	-	-	-	-	-	-	-	-	-	-	-	-	-	-	-	-	-	-	-	-	-	-	-	-	-	-	-
	*92/63	-	2	1	1	-	-	-	2	-	1	1	1	2	1	1	1	-	1	-	1	-	2	-	-	-	-	-	-	-	-	-	-	-	-	-	-	-	-	-	-	-	-	-	-	-	-	-	-	-	-	-	-	-	-	-	-	-	-	-	-
	*72/72	-	-	-	-	-	-	-	-	-	-	-	-	-	-	-	-	-	-	-	-	-	-	-	-	-	-	-	-	-	-	-	-	-	-	-	-	-	-	1	-	-	-	-	-	-	-	-	-	-	-	-	-	-	-	-	-	-	-	-	-
IDHP-3*	*99/31	-	-	-	-	-	-	-	-	-	-	-	-	-	-	-	-	-	-	-	-	-	-	-	-	-	-	-	-	-	-	-	-	-	-	-	-	-	-	-	-	-	-	-	-	-	-	-	-	-	-	-	1	-	4	-	-	-	-	-	-
	*31/31	-	-	-	-	-	-	-	-	-	-	-	-	-	-	-	-	-	-	-	-	-	-	-	-	-	-	-	-	-	-	-	-	-	-	-	-	-	-	-	-	-	-	-	-	-	-	57	40	31	26	11	38	34	47	49	50	9	16	15	8
	*-100/-100	51	68	26	18	14	40	42	36	45	30	24	26	38	32	35	47	42	52	30	56	27	74	36	29	39	41	40	42	41	21	41	42	17	41	24	35	14	22	45	38	26	50	50	52	50	24	-	-	-	-	-	-	-	-	-	-	-	-	-	-
IDHP-4*	*-51/51	-	-	-	-	-	-	-	-	-	-	-	-	-	-	-	-	-	-	-	-	-	-	-	-	-	-	-	-	-	-	-	-	-	-	-	-	-	-	-	-	-	-	-	-	-	-	57	40	31	26	11	39	34	51	49	50	9	16	15	8
	*-100/-100	51	68	26	18	14	40	42	36	45	30	24	26	38	32	35	47	42	52	30	56	27	74	36	29	39	41	40	42	41	21	41	42	17	41	24	35	14	22	45	38	26	50	50	52	50	24	-	-	-	-	-	-	-	-	-	-	-	-	-	-

Appendix. Continued.

Locus	Genotype	1	2	3	4	5	6	7	8	9	10	11	12	13	14	15	16	17	18	19	20	21	22	23	24	25	26	27	28	29	30	31	32	33	34	35	36	37	38	39	40	41	42	43	44	45	46	47	48	49	50	51	52	53	54	55	56	57	58	59	60
LDH*	*-55/-100	-	-	-	-	-	-	-	-	-	-	-	-	-	-	-	-	-	-	-	-	-	-	-	-	-	-	-	-	-	-	-	1	-	-	-	-	-	-	-	-	-	-	-	-	-	-	-	-	-	-	-	-	-	-	-	-	-	-	-	-
	*-100/-100	51	68	26	18	14	40	42	36	45	30	24	26	38	32	35	47	42	52	30	56	27	74	36	29	39	41	40	40	41	21	41	41	17	41	24	35	14	22	45	38	26	50	50	52	50	24	57	40	31	26	11	39	34	51	49	50	9	16	15	8
	*-100/-140	-	-	-	-	-	-	-	-	-	-	-	-	-	-	-	-	-	-	-	-	-	-	-	-	-	-	-	2	-	-	-	-	-	-	-	-	-	-	-	-	-	-	-	-	-	-	-	-	-	-	-	-	-	-	-	-	-	-	-	-
MDH-1*	*152/100	-	-	-	-	-	-	-	-	-	-	-	-	-	-	-	-	-	-	-	-	-	-	-	-	-	-	-	-	-	-	-	-	-	-	-	-	-	-	-	-	-	-	-	1	-	-	-	-	-	1	-	-	-	-	-	-	-	-	-	1
	*100/100	51	68	26	18	14	40	42	36	45	30	24	26	38	32	35	47	42	52	30	56	27	74	36	29	39	41	40	42	41	21	41	42	17	41	24	35	14	22	45	38	26	50	50	51	50	24	57	40	31	25	11	39	34	51	49	50	9	16	15	7
MDH-2*	*174/174	-	-	-	-	-	-	-	-	-	-	-	-	-	-	-	-	-	-	-	-	-	-	-	-	-	-	-	-	-	-	-	-	-	-	-	-	-	-	-	-	-	-	-	-	-	-	-	-	-	-	-	-	-	-	-	10	-	-	15	-
	*174/100	-	-	-	-	-	-	-	-	-	-	-	-	-	-	-	-	-	-	-	-	-	-	-	-	-	-	-	-	-	-	-	-	-	-	-	-	-	-	-	-	-	-	-	-	-	-	-	-	-	-	-	-	-	-	-	23	-	-	-	-
	*100/100	50	68	26	18	14	40	42	36	45	30	24	26	38	32	35	46	42	52	30	56	27	74	36	29	39	41	40	42	41	21	41	42	17	41	24	35	14	22	45	7	26	50	50	52	50	24	57	40	31	26	11	39	34	51	49	17	9	16	-	8
	*100/42	1	-	-	-	-	-	-	-	-	-	-	-	-	-	-	1	-	-	-	-	-	-	-	-	-	-	-	-	-	-	-	-	-	-	-	-	-	-	-	18	-	-	-	-	-	-	-	-	-	-	-	-	-	-	-	-	-	-	-	-
	*42/42	-	-	-	-	-	-	-	-	-	-	-	-	-	-	-	-	-	-	-	-	-	-	-	-	-	-	-	-	-	-	-	-	-	-	-	-	-	-	-	13	-	-	-	-	-	-	-	-	-	-	-	-	-	-	-	-	-	-	-	-
MDH-3*	*-56/-56	-	-	-	-	-	-	-	-	-	-	-	-	-	-	-	-	-	-	-	-	-	-	-	-	-	-	-	-	-	-	-	-	-	-	1	-	-	-	-	-	-	-	-	-	-	-	-	-	-	-	-	-	-	-	-	-	-	-	-	-
	*-56/-100	-	-	-	-	-	-	-	-	-	-	-	-	-	-	-	-	-	-	-	-	-	-	-	-	-	-	-	-	-	-	-	-	-	3	1	-	-	-	-	-	-	-	-	-	-	-	-	-	-	-	-	-	-	-	-	-	-	-	-	-
	*-84/-84	-	-	-	-	-	-	-	-	-	-	-	-	-	-	-	-	-	-	-	-	-	-	-	-	-	-	-	-	-	-	-	-	-	-	-	-	-	-	-	-	-	-	-	-	-	-	-	-	-	-	-	-	-	-	-	-	-	-	-	-
	*-93/-93	-	-	-	-	-	-	-	-	-	-	-	-	-	-	-	47	42	52	30	56	27	74	36	29	39	-	-	-	-	-	-	-	-	-	-	-	-	-	-	-	-	-	-	-	-	-	-	-	-	26	11	39	34	51	49	50	9	16	15	8
	*-100/-100	-	-	-	-	-	-	-	-	-	-	-	-	-	-	-	-	-	-	-	-	-	-	-	-	-	-	40	-	-	21	41	42	17	38	22	35	14	22	45	38	26	50	50	52	50	24	-	-	-	-	-	-	-	-	-	-	-	-	-	-
	*-113/-113	-	-	-	-	-	-	-	-	-	-	-	-	-	-	-	-	-	-	-	-	-	-	-	-	-	-	-	-	-	-	-	-	-	-	-	-	-	-	-	-	-	-	-	-	-	-	57	40	31	-	-	-	-	-	-	-	-	-	-	-
	*-122/-122	51	68	26	18	14	40	42	36	45	30	24	26	38	32	35	-	-	-	-	-	-	-	-	-	-	41	-	42	41	-	-	-	-	-	-	-	-	-	-	-	-	-	-	-	-	-	-	-	-	-	-	-	-	-	-	-	-	-	-	-
MDH-4*	*-64/-64	-	-	-	-	-	-	-	-	-	-	-	-	-	-	-	-	-	-	-	-	-	-	-	-	-	-	-	-	-	-	-	-	-	-	-	-	-	-	-	-	-	-	-	-	-	-	-	-	-	-	-	-	-	-	-	-	-	-	-	-
	*-64/-93	-	-	-	-	-	-	-	-	-	-	-	-	-	-	-	-	-	-	-	-	-	-	-	-	-	-	-	-	-	-	-	-	-	-	-	-	-	-	-	-	-	-	-	-	-	-	-	-	-	-	-	7	-	-	-	-	-	-	-	-
	*-71/-100	-	-	-	-	-	-	-	-	-	-	-	-	-	-	-	1	-	-	-	-	-	-	-	-	-	-	-	-	-	-	-	-	-	-	-	-	-	-	-	-	-	-	-	-	-	-	-	-	-	-	-	11	-	-	-	-	-	-	-	-
	*-93/-93	-	-	-	-	-	-	-	-	-	-	-	-	-	-	-	-	-	-	-	-	-	-	-	-	-	-	-	-	-	-	-	-	-	-	-	-	-	-	-	-	-	-	-	-	-	-	-	-	-	-	-	-	-	-	-	-	-	-	-	-
	*-93/-122	-	-	-	-	-	-	-	-	-	-	-	-	-	-	-	-	-	-	-	-	-	-	-	-	-	-	-	-	-	-	-	-	-	-	-	-	-	-	-	-	-	-	-	-	-	-	57	40	31	3	10	21	33	51	-	3	-	-	-	-
	*-100/-100	51	68	26	18	14	39	40	36	45	29	23	26	37	32	35	44	42	52	29	56	27	74	36	29	39	5	32	1	-	20	39	28	17	37	22	-	-	21	8	25	-	4	-	-	49	24	-	-	-	13	1	-	1	-	1	25	-	-	4	3
	*-100/-114	-	-	-	-	-	-	-	-	-	-	-	-	-	-	-	-	-	-	-	-	-	-	-	-	-	-	1	-	-	-	-	-	-	1	-	-	-	-	-	-	-	-	-	-	-	-	-	-	-	-	-	-	-	-	-	-	-	-	-	-
	*-100/-129	-	-	-	-	-	1	2	-	-	1	1	-	1	-	-	2	-	-	1	-	-	-	-	-	-	10	7	10	7	1	2	10	-	3	2	-	-	1	19	11	-	22	-	-	1	-	-	-	-	-	-	-	-	-	-	-	-	-	-	-
	*-122/-122	-	-	-	-	-	-	-	-	-	-	-	-	-	-	-	-	-	-	-	-	-	-	-	-	-	-	-	-	-	-	-	-	-	-	-	-	-	-	-	-	-	-	-	-	-	-	-	-	-	10	-	-	-	-	48	22	9	16	11	5
	*-129/-129	-	-	-	-	-	-	-	-	-	-	-	-	-	-	-	-	-	-	-	-	-	-	-	-	-	26	-	31	34	-	-	4	-	-	-	31	14	-	18	2	26	24	50	52	-	-	-	-	-	-	-	-	-	-	-	-	-	-	-	-
	*129/-147	-	-	-	-	-	-	-	-	-	-	-	-	-	-	-	-	-	-	-	-	-	-	-	-	-	-	-	-	-	-	-	-	-	-	-	4	-	-	-	-	-	-	-	-	-	-	-	-	-	-	-	-	-	-	-	-	-	-	-	-
MEP-1*	*100/100	-	-	-	-	-	-	-	-	-	-	-	-	-	-	-	-	-	-	-	-	-	-	-	-	-	21	35	40	40	21	41	42	17	36	15	35	14	22	45	37	26	50	50	52	50	24	-	-	-	-	-	-	-	-	-	-	-	-	-	-
	*100/53	7	3	-	-	-	1	-	-	1	-	-	-	-	-	-	2	-	2	-	1	-	1	-	-	-	15	4	2	1	-	-	-	-	5	9	-	-	-	-	1	-	-	-	-	-	-	-	-	-	-	-	-	-	-	-	-	-	-	-	-
	*78/78	-	-	-	-	-	-	-	-	-	-	-	-	-	-	-	-	-	-	-	-	-	-	-	-	-	-	-	-	-	-	-	-	-	-	-	-	-	-	-	-	-	-	-	-	-	-	-	-	-	-	-	-	-	-	-	-	-	-	-	-
	*53/53	44	65	26	18	14	39	42	36	44	30	24	26	38	32	35	45	42	50	30	55	27	73	36	29	39	5	1	-	-	-	-	-	-	-	-	-	-	-	-	-	-	-	-	-	-	-	57	40	31	26	11	39	34	51	49	50	9	16	15	8
MEP-2*	*-49/-49	1	3	-	2	-	1	-	-	-	-	-	-	3	-	-	1	-	1	3	1	-	9	-	-	4	-	-	-	-	-	-	-	-	-	-	-	2	-	-	-	-	-	2	-	-	-	-	-	-	-	-	-	-	-	-	-	-	-	-	-
	*-49/-100	13	10	7	6	3	6	7	7	6	4	7	12	10	12	8	11	3	9	12	17	4	34	6	3	14	-	2	-	9	1	2	-	-	-	-	-	5	-	-	-	-	2	6	-	1	-	-	-	-	-	-	-	-	-	-	-	-	-	-	4
	*-100/-100	37	55	19	10	11	33	35	29	39	26	17	11	28	20	27	35	39	42	15	38	23	31	30	26	21	41	38	42	32	20	39	42	17	41	24	35	7	22	45	38	26	48	42	52	49	24	57	40	31	26	11	39	34	51	49	50	8	16	15	4
	*-100/-140	-	-	-	-	-	-	-	-	-	-	-	-	-	-	-	-	-	-	-	-	-	-	-	-	-	-	-	-	-	-	-	-	-	-	-	-	-	-	-	-	-	-	-	-	-	-	-	-	-	-	-	-	-	-	-	-	1	-	-	-
PGDH*	*184/100	-	-	-	-	-	-	-	-	-	-	-	-	-	-	1	-	-	-	-	1	-	-	-	-	-	-	-	-	-	-	-	-	-	-	-	-	-	-	-	-	-	-	-	-	-	-	-	-	-	-	-	-	-	-	-	-	-	-	-	-
	*137/137	2	7	-	4	-	4	6	5	1	-	1	4	2	-	1	3	28	-	6	3	3	10	9	2	1	-	1	1	-	1	-	-	-	4	3	-	-	-	-	-	-	-	-	1	-	-	-	-	-	-	-	-	-	-	-	-	-	-	-	-
	*137/100	21	24	11	7	1	17	15	14	14	13	11	12	17	12	14	17	13	16	13	17	11	35	23	13	20	-	7	9	2	1	4	-	-	14	8	-	-	-	3	-	-	-	-	5	-	-	-	-	-	2	-	-	-	-	-	-	-	-	-	-
	*137/41	-	1	-	-	-	-	-	-	-	-	-	1	1	-	-	1	-	-	-	-	-	-	-	-	-	-	-	-	-	-	1	-	-	-	-	-	-	-	-	-	-	-	-	-	-	-	-	-	-	-	-	-	-	-	-	-	-	-	-	-
	*100/100	28	34	7	7	11	19	21	17	29	15	9	8	17	19	18	26	1	36	11	14	8	29	4	9	17	41	32	32	39	19	29	35	17	12	7	35	14	22	41	38	26	50	50	41	49	24	57	40	3	23	11	20	21	51	49	50	9	16	15	8
	*100/82	-	-	-	-	-	-	-	-	-	-	-	-	-	-	-	-	-	-	-	-	-	-	-	-	-	-	-	-	-	-	-	-	-	-	-	-	-	-	-	-	-	-	-	-	-	-	-	-	7	-	-	-	-	-	-	-	-	-	-	-
	*100/41	-	2	1	-	1	-	-	-	-	-	-	-	-	1	1	-	-	-	-	4	-	-	-	1	1	-	-	-	-	-	7	1	-	-	-	-	-	-	1	-	-	-	-	5	1	-	-	-	-	1	-	10	9	-	-	-	-	-	-	-
	*82/82	-	-	-	-	-	-	-	-	-	-	-	-	-	-	-	-	-	-	-	-	-	-	-	-	-	-	-	-	-	-	-	-	-	-	-	-	-	-	-	-	-	-	-	-	-	-	-	-	21	-	-	-	-	-	-	-	-	-	-	-
	*41/41	-	-	-	-	1	-	-	-	-	-	-	-	1	-	-	-	-	-	-	-	-	-	-	-	-	-	-	-	-	-	-	-	-	-	-	-	-	-	-	-	-	-	-	-	-	-	-	-	-	-	-	3	3	-	-	-	-	-	-	-
PGM*	*0/-43	-	-	-	-	-	-	-	-	-	-	-	-	-	-	-	-	-	-	-	-	-	-	-	-	-	-	-	-	-	-	-	-	-	-	-	-	-	-	-	-	-	-	-	1	-	2	-	-	-	-	-	-	-	-	-	-	-	-	-	-
	*0/-100	-	-	-	-	-	-	-	-	-	-	-	-	-	-	-	-	-	-	-	-	-	-	-	-	-	-	2	-	-	-	-	-	-	-	-	-	-	-	-	-	-	-	-	-	-	-	-	-	-	-	-	-	-	-	-	-	-	-	-	-
	*-22/-39	-	-	-	-	-	-	-	-	-	-	-	-	-	-	-	-	-	-	-	-	-	-	-	-	-	-	-	-	-	-	-	-	-	-	-	-	-	-	-	-	-	-	-	-	-	-	-	-	1	-	-	-	-	-	1	-	-	-	-	-
	*-39/-39	-	-	-	-	-	-	-	-	-	-	-	-	-	-	-	-	-	-	-	-	-	-	-	-	-	-	-	-	-	-	-	-	-	-	-	-	-	-	-	-	-	-	-	-	-	-	2	-	25	26	2	31	33	51	48	-	8	12	15	8
	*-39/-95	-	-	-	-	-	-	-	-	-	-	-	-	-	-	-	-	-	-	-	-	-	-	-	-	-	-	-	-	-	-	-	-	-	-	-	-	-	-	-	-	-	-	-	-	-	-	18	17	5	-	4	7	1	-	-	11	-	-	-	-
	*-39/-144	-	-	-	-	-	-	-	-	-	-	-	-	-	-	-	-	-	-	-	-	-	-	-	-	-	-	-	-	-	-	-	-	-	-	-	-	-	-	-	-	-	-	-	-	-	-	3	1	-	-	-	-	-	-	-	-	1	-	-	-
	*-43/-43	4	6	2	2	1	4	3	1	4	1	6	4	3	2	-	7	1	-	1	6	-	12	8	2	19	-	1	41	2	5	27	4	2	4	1	-	-	10	-	36	-	-	-	11	48	20	-	-	-	-	-	-	-	-	-	-	-	-	-	-
	*-43/-100	23	26	9	7	2	16	14	7	12	15	11	6	10	10	12	14	6	8	22	27	5	39	22	14	17	4	11	1	21	7	12	17	9	16	8	-	-	9	11	2	9	2	-	26	1	2	-	-	-	-	-	-	-	-	-	-	-	-	-	-
	*-43/-158	-	-	-	-	-	-	-	-	-	-	-	-	-	-	-	-	-	-	-	-	-	-	-	-	-	-	-	-	-	-	1	2	-	-	-	-	-	-	-	-	-	-	-	-	-	-	-	-	-	-	-	-	-	-	-	-	-	-	-	-
	*-95/-95	-	-	-	-	-	-	-	-	-	-	-	-	-	-	-	-	-	-	-	-	-	-	-	-	-	-	-	-	-	-	-	-	-	-	-	-	-	-	-	-	-	-	-	-	-	-	20	18	-	-	5	1	-	-	-	37	-	-	-	-
	*-95/-144	-	-	-	-	-	-	-	-	-	-	-	-	-	-	-	-	-	-	-	-	-	-	-	-	-	-	-	-	-	-	-	-	-	-	-	-	-	-	-	-	-	-	-	-	-	-	13	4	-	-	-	-	-	-	-	-	-	-	-	-
	*-100/-100	23	33	15	9	11	20	25	27	29	14	7	16	25	19	23	21	29	44	7	23	17	23	6	13	3	30	26	-	18	9	1	9	6	21	15	35	14	3	34	-	17	48	45	14	1	-	-	-	-	-	-	-	-	-	-	-	-	-	-	-
	*-100/-121	-	-	-	-	-	-	-	-	-	-	-	-	-	-	-	-	-	-	-	-	-	-	-	-	-	-	-	-	-	-	-	-	-	-	-	-	-	-	-	-	-	-	5	-	-	-	-	-	-	-	-	-	-	-	-	-	-	-	-	-
	*-100/-158	1	1	-	-	-	-	-	1	-	-	-	-	-	-	-	-	-	-	-	-	-	-	-	-	-	2	-	-	-	-	-	2	-	-	-	-	-	-	-	-	-	-	-	-	-	-	-	-	-	-	-	-	-	-	-	-	-	-	-	-
	*-144/-144	-	-	-	-	-	-	-	-	-	-	-	-	-	-	-	-	-	-	-	-	-	-	-	-	-	-	-	-	-	-	-	-	-	-	-	-	-	-	-	-	-	-	-	-	-	-	1	-	-	-	-	-	-	-	-	-	-	-	-	-
	*-158/-158	-	-	-	-	-	-	-	-	-	-	-	-	-	-	-	-	-	-	-	-	-	-	-	-	-	-	-	-	-	-	-	2	-	-	-	-	-	-	-	-	-	-	-	-	-	-	-	-	-	-	-	-	-	-	-	-	-	-	-	-
PROT-1*	*100/100	51	68	26	18	14	40	42	36	45	30	24	26	38	32	35	47	42	52	30	56	27	74	36	29	39	41	40	42	41	21	41	42	17	41	24	35	14	22	45	38	26	50	50	52	50	24	57	40	31	26	11	39	34	51	49	50	9	16	15	8
PROT-2*	*100/100	51	68	26	18	14	40	42	36	45	30	24	26	38	32	35	47	42	52	30	56	27	74	36	29	39	41	40	42	41	21	41	42	17	41	24	35	14	22	45	38	26	50	50	52	50	24	57	40	31	26	11	39	34	51	49	50	9	16	15	8

Environmental Biology of Fishes **52**: 163–171, 1998.

Osteology of *Zacco pachycephalus*, sensu Jordan & Evermann (1903), with special reference to its systematic position

Hiroshi Ashiwa[1] & Kazumi Hosoya[2]
[1] *Department of Zoology, Faculty of Science, Kyoto University, Kitashirakawa, Oiwake-cho, Sakyo-ku, Kyoto 606, Japan (e-mail: ashiwa@ecol.zool.kyoto-u.ac.jp)*
[2] *National Research Institute of Fisheries Science, 1088 Komaki, Ueda, Nagano 386, Japan*

Received 15.5.1997 Accepted 5.12.1997

Key words: Formosan cyprinid, skeletal system, *Opsariichthys*, fish

Synopsis

The skeletal system of *Zacco pachycephalus*, sensu Jordan & Evermann (1903), is described and compared with that of other related barilines to clarify its systematic position. Osteological characters fall generally into two categories in terms of apomorphic features. The first category is exemplified by autapomorphies such as lower jaw bone shape and the number of branchiostegal rays and hypurals. The latter characters are highly unusual among Cyprinidae. The second category consists of synapomorphies linking *Z. pachycephalus* and *Opsariichthys uncirostris*. Specialization of jaw bones for a piscivorous habit is also notable. The quadrate-pterygoid fenestra of *Z. pachycephalus* and *O. uncirostris* is considered derived. We conclude that *Zacco pachycephalus* should be transferred from *Zacco* to *Opsariichthys* based on the present osteological analysis.

Introduction

The cyprinid genus *Zacco* comprises eight species (Bănărescu 1968): *Z. platypus* (Temminck & Schlegel), *Z. temminckii* (Temminck & Schlegel), *Z. pachycephalus* (Günther), *Z. spilurus* (Günther), *Z. barbatus* (Regan), *Z. chengtui* Kimura, *Z. macrophthalmus* Kimura and *Z. macrolepis* Wu. *Z. temminckii* includes two genetically distinctive species, so-called A and B types (Okazaki et al. 1991).

Zacco occurs in freshwater throughout the region of southeast China, the Korean Peninsula, Japan, Taiwan and Hainan Islands (Oshima 1919, Uchida 1939, Harada 1943, Wu 1964, Nakamura 1969, Bănărescu 1991). Among *Zacco, Z. pachycephalus* is unique biogeographically because it is confined to Taiwan Island. Although this species has general *Zacco* features of gross morphology, it presents some evolutionary trends towards *Opsariichthys,* such as a mouth with a large gape and slightly undulated jaws; *Z. pachycephalus* seems to link *Zacco* with *Opsariichthys*. Its true systematic position is uncertain, however, because there has been no detailed osteological description of it (see Howes 1980, 1983, for partial descriptions). And, it has been treated erratically with regard to generic taxonomy, as *Zacco* or *Opsariichthys.* The purpose of this study is to describe the osteological features of *Z. pachycephalus,* focusing on its relationship with ad hoc taxa.

Materials and methods

The skeletal system was examined in cleared and counter-stained specimens prepared by an improved method following Kawamura & Hosoya

Figure 1. Zacco pachycephalus, sensu Jordan & Evermann (1903), FRLM 11127 (74.9 mm SL) from Taipei, Taiwan.

(1991). Osteological terminology basically follows Harrington (1955). Nelson (1969), Kusaka (1974), Howes (1980) and Fujita (1989) also referred to some anatomical units (i.e. infraorbital, urohyal, cephalic portion and caudal complex). Character polarity for phylogenetic analysis was decided by comparison with the osteological information of generalized cyprinids (i.e. Ramaswami 1955, Howes 1980, 1983, Fink & Fink 1981, Hosoya 1986, 1989).

Specimens examined in this study are deposited in the following institutions: Department of Fisheries, Faculty of Agriculture, Kyoto University, Kyoto (FAKU); Fisheries Research Laboratory, Mie University, Mie (FRLM).

Material examined

Z. pachycephalus, sensu Jordan & Evermann (1903); FAKU 115723, 1 specimen, 80.7 mm standard length (SL), Taipei, Taiwan; FRLM 11127–11128, 2 specimens, 74.6–74.9 mm SL, Taipei, Taiwan (Figure 1).

Comparative materials:

O. uncirostris uncirostris: FAKU 115901, 1 specimen, 81.6 mm SL, Lake Biwa, Shiga Pref., Japan; FAKU 115902, 1 specimen, 162.2 mm SL, Hasu River, Fukui Pref., Japan.

Z. platypus: FAKU 115897, 1 specimen, 85.2 mm SL, Lake Biwa, Shiga Pref., Japan; FAKU 115757, 1 specimen, 79.3 mm SL, Heisou-gun, Kougen-do, Korea.

Z. temminckii, A type, sensu Okazaki et al. (1991): FAKU 115715–115716, 2 specimens, 83.6–107.9 mm SL, Maruyama River, Okayama Pref., Japan; FAKU 56842, 1 specimen, 91.8 mm SL, Tokita River, Watarai-gun, Mie Pref., Japan.

Z. temminckii, B type, sensu Okazaki et al. (1991): FAKU 56840, 115899, 115900, 3 specimens, 79.1–103.2 mm SL, Tokita River, Watarai-gun Mie Pref., Japan; FAKU 115898, 1 specimen, 62.1 mm SL, Iroha River, Oita Pref., Japan.

Barilius baruna: FAKU 115707, 115712, 2 specimens, 54.3–96.5 mm SL, Nepal.

Osteological description

We provide a description of the osteology of *Z. pachycephalus*. Derived features, judged by comparison with comparative materials in this study and other general cyprinids, are labeled with numbers in parentheses.

Neurocranium

Neurocranium (Figure 2) subtriangular in dorsal view, rather oblong with spinous pterotic projections. Roof of cranium without fontanels. Cranium narrow. Supraethmoid (SE) broad. Ethmoid cartilage (ec) slightly attached to anterior part of vomer (VO). Mesethmoid (ME) in contact with lateral ethmoid (LE) posteriorly, with deeply notched olfactory foramen. Lateral ethmoid relatively thin. Parasphenoid (PS) large and connected to vomer at notched anterior border. In lateral view, parasphenoid curved upward from midpoint, and sutured with orbitosphenoid septum (oss). Posterior part of orbitosphenoid septum deeply indented. Carotid foramen (cf) very small. Prootic (PRO) contacts pterosphenoid (PTS) anteriorly and medially via lateral commissure (lc, see Howes 1980), and contacts epiotic (EPO), exoccipital (EXO) and basioccipital (BO) posteriorly. Lateral jugular bridge (ljb, see Howes 1980) thin. Pterosphenoid contacts orbitosphenoid (OS) anteriorly. Subtemporal fossa (stf) connected with posttemporal fossa (ptf). Nasal (N) tubular in shape. Frontal (F) elongate, narrow at midpoint along lateral edge. Frontals in contact

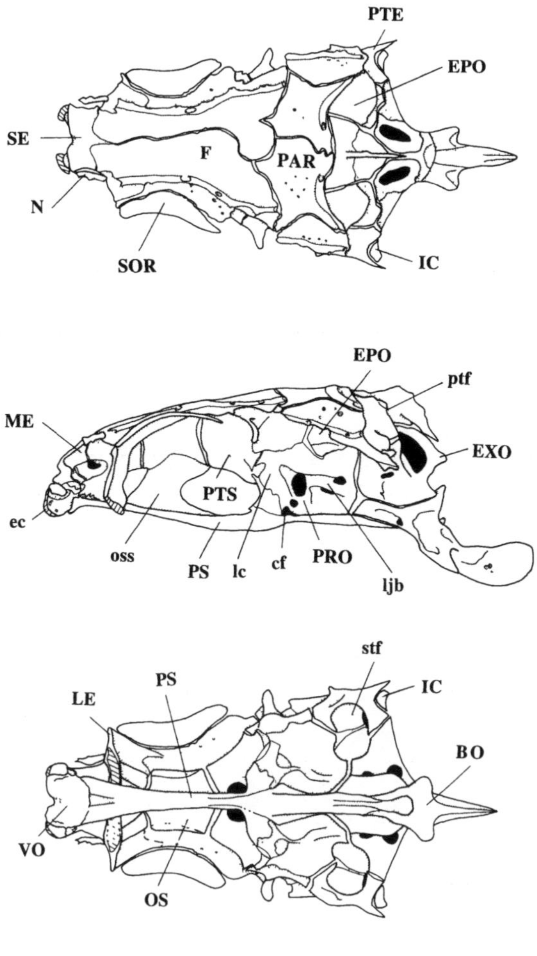

Figure 2. Neurocranium of *Zacco pachycephalus*, sensu Jordan & Evermann (1903). Dorsal, lateral and ventral views, from top to bottom. Cartilage is cross-hatched. Scale indicates 2 mm (BO = basioccipital; cf = carotid foramen; ec = ethmoid cartilage; EPO = epiotic; EXO = exoccipital; F = frontal; IC = intercalar; lc = lateral commissure; LE = lateral ethmoid; ljb = lateral jugular bridge; ME = mesethmoid; N = nasal; OS = orbitosphenoid; oss = orbitosphenoid septum; PAR = parietal; PRO = prootic; PS = parasphenoid; PTE = pterotic; ptf = posttemporal fossa; PTS = pterosphenoid; SE = supraethmoid; SOR = supraorbital; stf = subtemporal fossa; VO = vomer).

with each other in midline. Parietals (PAR) broad and short, in contact in midline. Supraorbital (SOR) long and narrow. Intercalar (IC) relatively large, not pointed posteriorly, free and articulated with posteromarginal concavity of pterotic (PTE) by chondral suture. Supratemporal canal partly membranous, running along posterior margin of parietal.

Jaws

(1) Jaws along oral surface slightly undulated. Maxillary (MAX) having both anterior and medial cavities on its surface (Figure 3a). Maxillary medial cavity well developed, supporting outer division of adductor mandibular muscle. Premaxillary (PMX) thin, flat, with vertical anterior edge. (2) Dentary (D) slightly undulate along dorsal surface, making a deep uncinate gape with the premaxillary (Figure 3b). Mentomeckelian (MM) capping tip of meckelian cartilage (mc) and attached onto dentary. An arrowhead shaped coronomeckelian (CM) suspended above cartilage. Retroarticular (RA) positioned on lower margin of angular (A). Angular housing meckelian cartilage on its oral surface. (3) Anterior part of angular fused with dentary on its posterior margin.

Circumorbital and opercular series

Infraorbital series (Figure 3c) composed of lachrymal (IO 1 (L)) and four infraorbital bones (IO 2–5). Lachrymal the largest in series, pointed anterodorsad. Second and third infraorbital bones large and deep, while fifth, small and shallow.

Opercular series (Figure 3d) composed of opercle, preopercle, interopercle and subopercle. Opercle (OP) the largest, triangular shaped and extending caudad. Preopercle (PO) crescentic, interopercle (IOP) bullet shaped, subopercle (SOP) horn shaped.

Suspensorium

Suspensorium amphystilically articulating with neurocranium through palatine (PAL) and hyomandibular (HM, Figure 3e). Posterior edge of palatine cartilaginous, articulating with entopterygoid (ENT). Ectopterygoid (ECT) long and slender, irregularly curved at anterior edge. (4) Metapterygoid (MET) with dorsal and ventral borders deeply concave, 'H' shaped, and articulating posteriorly with both hyomandibular and symplectic (SY). Anterior margin of metapterygoid extending dorsad to

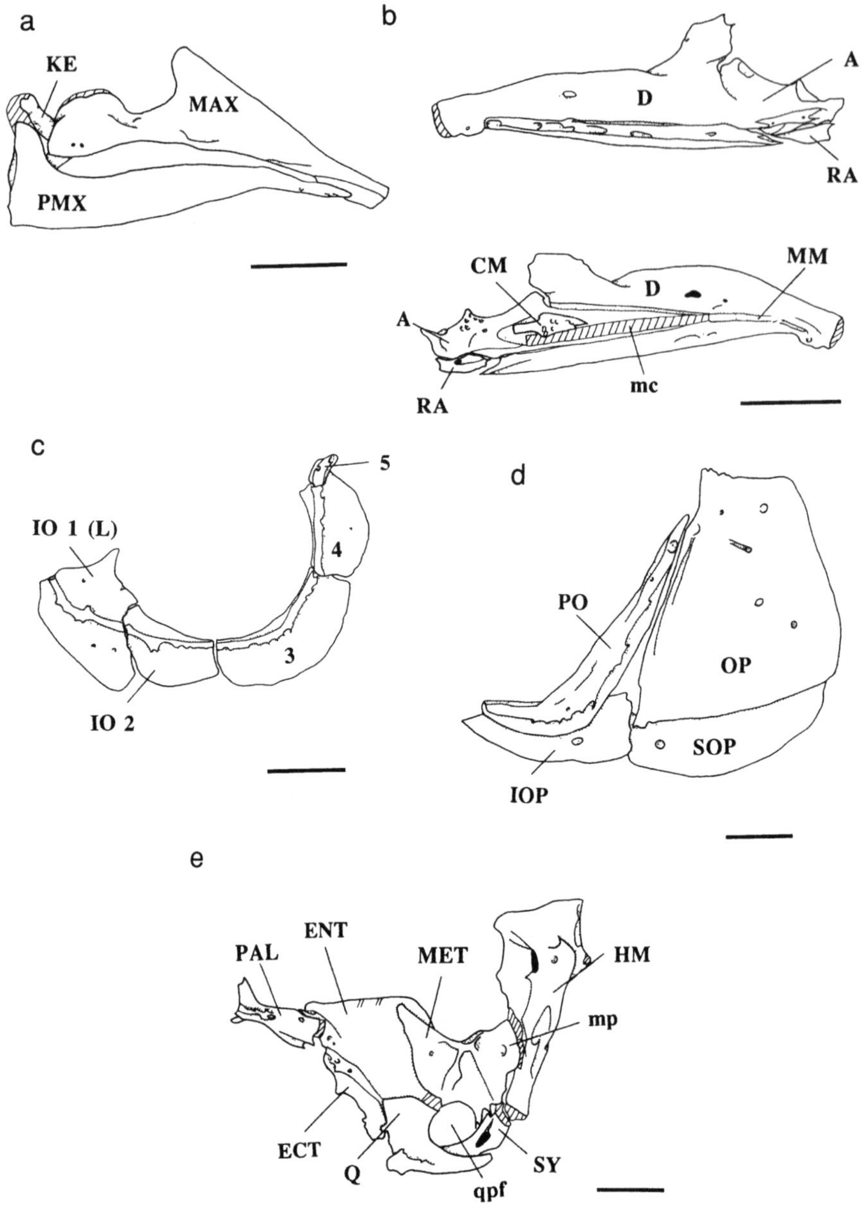

Figure 3. Skeletal systems of the head region in *Zacco pachycephalus,* sensu Jordan & Evermann (1903): a – upper jaw; b – lower jaw; c – infraorbital series; d – opercles; e – suspensorium; all in lateral view. Cartilage is cross-hatched. Scale indicates 2 mm (A = angular; cf = carotid foramen; CM = coronomeckelian; D = dentary; ECT = ectopterygoid; ENT = entopterygoid; HM = hyomandibular; IO = infraorbital; IOP = interopercle; KE = kinethmoid; L = lachrymal; MAX = maxillary; MET = metapterygoid; MM = mentomeckelian, mc = meckelian cartilage; mp = metapterygoid process; OP = opercle; PAL = palatine; PMX = premaxillary; PO = preopercle; Q = quadrate, qpf = quadrate-pterygoid fenestra; RA = retroarticular; SOP = subopercle; SY = symplectic).

cover whole posterior margin of entopterygoid. (5) Metapterygoid process (mp, see Howes 1980) dorso-posteriorly developed. (6) Quadrate (Q) dorso-posteriorly deeply concave, 'L' shaped. (7) Quadrate-pterygoid fenestra (qpf) well developed, surrounded by metapterygoid, quadrate and symplectic.

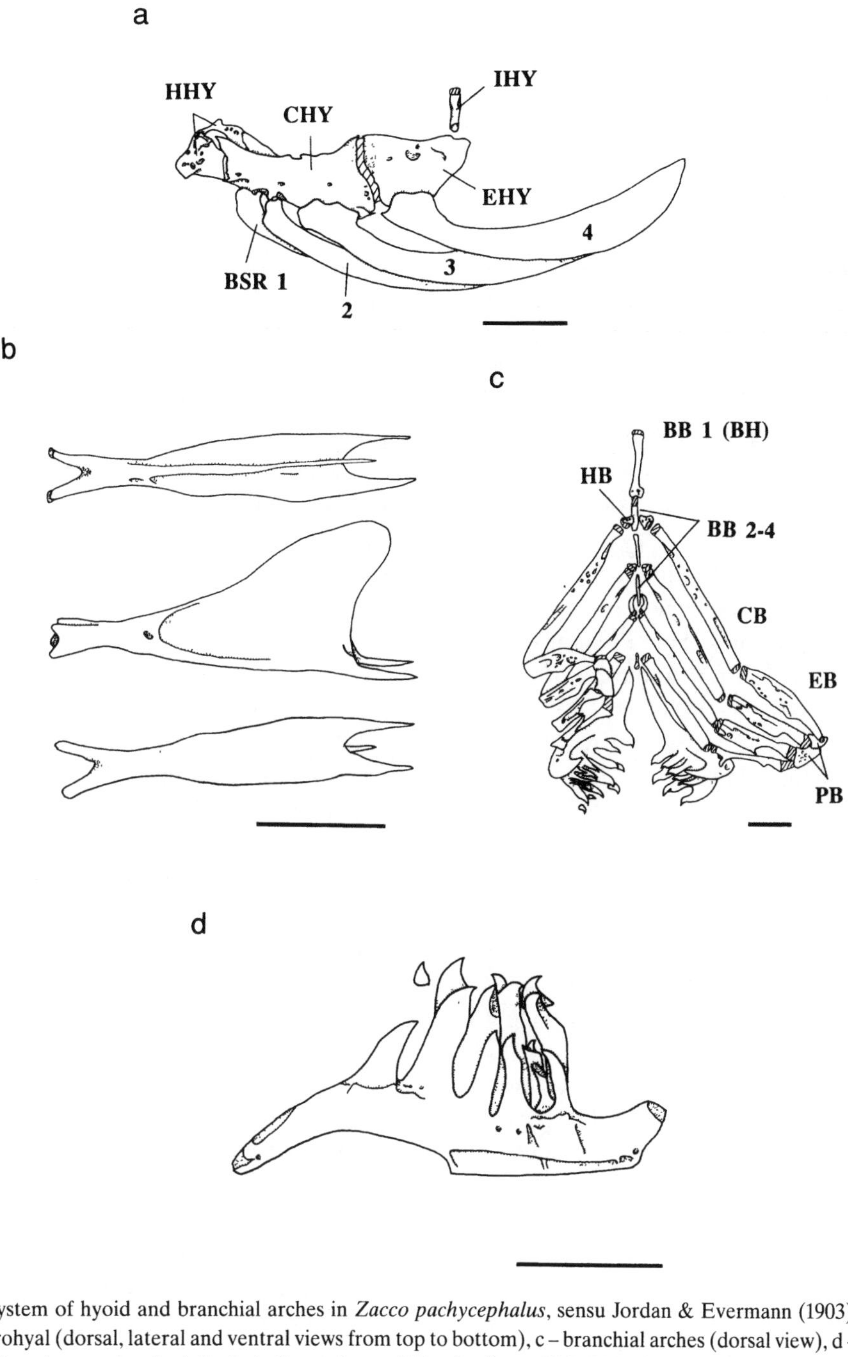

Figure 4. Skeletal system of hyoid and branchial arches in *Zacco pachycephalus*, sensu Jordan & Evermann (1903): a – hyoid arches (lateral view), b – urohyal (dorsal, lateral and ventral views from top to bottom), c – branchial arches (dorsal view), d – pharyngeal teeth (lateral view). Cartilage is cross-hatched. Scale indicates 2 mm (BB = basibranchial; BH = basihyal; BSR = branchiostegal ray; CB = ceratobranchial; CHY = ceratohyal; EB = epibranchial, EHY = epihyal; HB = hypobranchial; HHY = hypohyal; IHY = interhyal; PB = pharyngobranchial).

Hyoid and branchial arches

Ceratohyal (CHY) articulating anteriorly with upper and lower hypohyals (HHY), and posteriorly with triangular epihyal (EHY, Figure 4a). Posterodorsal margin of epihyal concave for interhyal (IHY) articulation. (8) Branchiostegal rays (BSR 1–4) three or, notably, four pairs.

Urohyal (Figure 4b) arrow-tail shape (see Kusaka 1974 on topography). Vertical plate triangular in lateral view. Horizontal plate rather slender, making a wing-like shape in ventral view, anterior end pointed and posterior end fork sharped.

Four rod-like basibranchials (BB 1(BH) and BB 2–4, Figure 4c) and a small cartilaginous copula posteriorly followed. Articulation between each basibranchial held by a pair of hypobranchials (HB). Four elongate ceratobranchials. Four epibranchials; fourth epibranchial thin. First pharyngobranchial (PB) absent, second and third present.

Pharyngeal teeth (Figure 4d) in three rows with a dental formula 2.3.5–5.3.2. Each tooth with serrated cutting edge or decurved tip.

Anterior vertebral region

Weberian apparatus (Figure 5a) composed of claustrum (CLA), scaphium (SCA), intercalarium (INC), supraneurals (SN), and tripus (TR). Neural arch (NA) and synchondral joint (scj) without any modification. Second centrum not fused to third. Os suspensorium (os) long and ventrally tapered, to meet its fellow curved inward at attachment onto anterior portion of swimbladder.

Fin supports

Pectoral girdle, composed of so-called primary shoulder girdle (coracoid (CR) and scapula), secondary shoulder girdle (posttemporal, supracleithrum (SC), cleithrum (CL) and elongate postcleithrum (PC) and radials (R, Figure 5b, also see Starks 1930). Extrascapular (ES) attached to parietal without fusion. In pelvic girdle, pelvic bone deeply forked anteriorly; ischiac process roundish caudad (Figure 5c). Dorsal fin support composed of eight pterygiophores (PP) and stay (STY). Middle radials absent from three anterior pterygiophores (DP, Figure 5d). Anal fin support composed of 10 pterygiophores and a stay; only five or six posterio-most pterygiophore with middle radials. Anal fin elongate in mature male. In caudal complex, posterior margin of elements fringed with cartilage (Figure 5e). Epural (EPU) long. Pleurostyle (PUS, see Dunn 1983) elongate, laterally with a pair of free uroneurals (URN) on its distal side; the bone posteriorly displaced by opisthural cartilage (cop, see Fujita 1989). (9) Hypurals (HP) five. First hypural fused to parhypural (PHP). Sixth hypural reduced in size to a small ossicle, often disappearing or remaining as a small ossicle.

Discussion

Osteological comparison

Many researchers have discussed the systematic position of *Z. pachycephalus* (i.e. Bănărescu 1968, Howes 1980), but its true systematic position is still uncertain. This is because of a shortage of anatomical information, not only in *Z. pachycephalus* but also in other related barilines. In this study, skeletal systems of *Zacco pachycephalus* were described in detail from head to tail. First, we compared it with *O. uncirostris* and other related cyprinid fishes (general Teleostei. Fujita 1989, Kusaka 1974, Otophysi. Fink & Fink 1981, Cyprinidae. Ramaswami 1955, Barilinae. Howes 1980, 1983, Gobioninae. Hosoya 1989). Based on those comparisons, we can divide the derived features of *Z. pachycephalus* into the following two categories:

1. Characters unique to *Zacco pachycephalus*: Specializations in jaws, branchial arch and caudal complex, (3), (8) and (9) were confirmed to be unique to *Z. pachycephalus*. Also, these three characters seem to be unique ones even among cyprinid fishes; the angular of cyprinids generally does not fuse with the basal dentary as has been reported in Barilinae (Howes 1980) and Gobioninae (Hosoya 1986). Cyprinids normally have only three branchiostegal rays (McAllister 1968) and six hypurals.

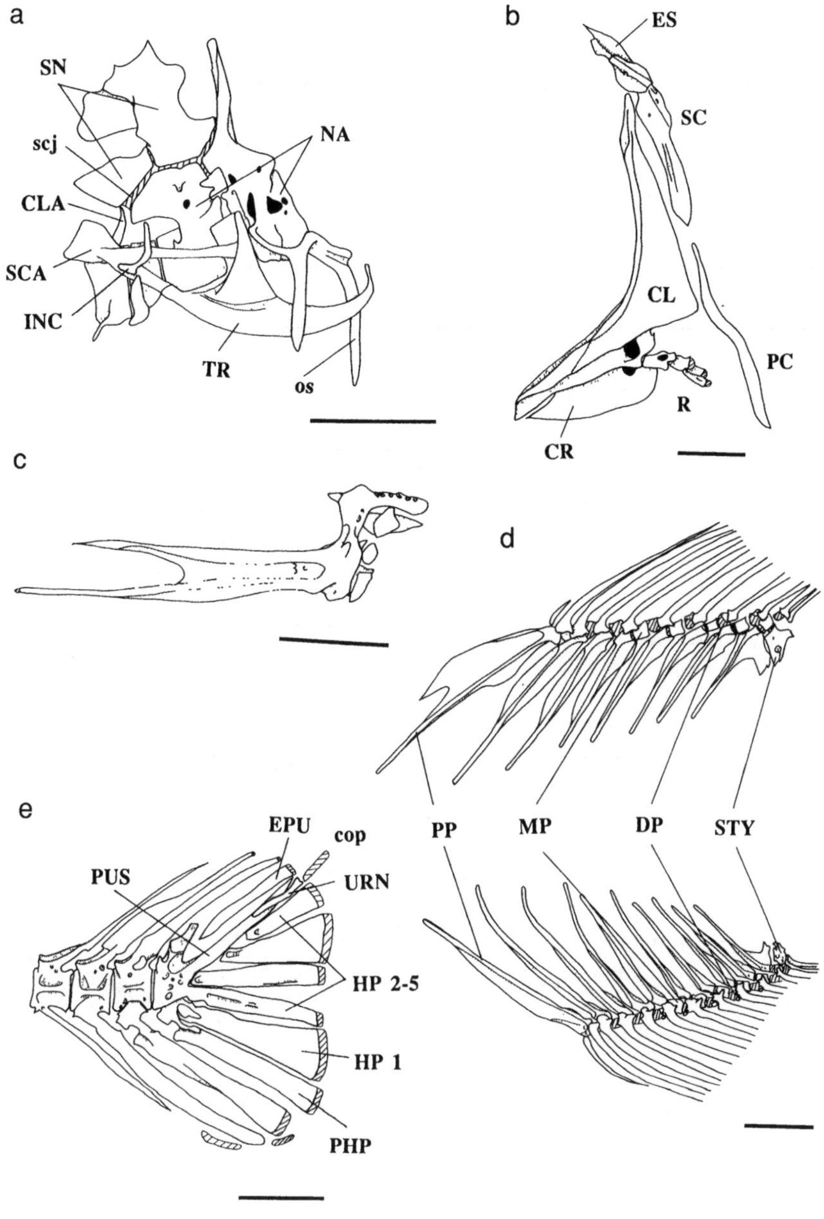

Figure 5. Vertebral and appendicular skeletons of *Zacco pachycephalus*, sensu Jordan & Evermann (1903): a – anterior vertebrae; b – pectoral girdle; c – pelvic girdle; d – fin supports; e – caudal complex (a, c, d and e = lateral view; b = dorsal view). Cartilage is cross-hatched. Scale indicates 2 mm (CL = cleithrum; CLA = claustrum; cop = opisthural cartilage; CR = coracoid; DP = distal pterygiophore; EPU = epural; ES = extrascapular; HP = hypural; INC = intercalarium; MP = median pterygiophore; NA = neural arch; os = os suspensorium; PC = postcleithrum; PHP = parhypural; PP = proximal pterygiophore; PUS = pleurostyle; R = radial; SCA = scaphium; scj = synchrondral joint; SN = supraneural; STY = stay; TR = tripus; URN = uroneural).

Zacco pachycephalus, with four branchiostegals and five hypurals is thus highly unusual as a cyprinid.

2. Synapomorphies of *Zacco pachycephalus* with *Opsariichthys uncirostris: Z. pachycephalus* and *O. uncirostris* share (1), (2), (4), (5), (6) and (7). All the characters are related to specializations of the jaw bones, pharyngeal teeth, infraorbital bones, and suspensorium attending with the development of the adductor mandibulae muscle A_1 for a piscivo-

rous habit (Gosline 1973, Howes 1980). We take those characters as evolutionary trends to *O. uncirostris*. Though quadrate-pterygoid fenestra are present in other *Zacco* congeners too, they are not so well developed as in *Z. pachycephalus* and *Opsariichthys*. This character state, the presence of fenestra, has traditionally been regarded by many researchers as a primitive character that implies the relationship between Cyprinoidea and Carasinoidea (i.e. Regan 1911, Ramaswami 1955). We agree with them only in that the presence of fenestra is primitive. However, we also stress that enlargement of fenestra is rather derivative. From the above observations, we reconfirmed that *Z. pachycephalus* and *O. uncirostris* are strongly linked by a sister species relationship to each other.

Systematic position of Zacco pachycephalus *sensu Jordan & Evermann (1903)*

The generic placement of *Zacco pachycephalus* has been erratic. Günther (1868) originally described the new species *Opsariichthys pachycephalus*, but it was transferred by Jordan & Evermann (1903) from the genus *Opsariichthys* to *Zacco*. Since then this taxonomic treatment has been accepted by many investigators (i.e. Oshima 1919, Wu 1964, Bănărescu 1968, Tzeng 1986, Min & Yang 1991), but Howes (1980) concluded that the species belongs to *Opsariichthys*. When Jordan & Evermann (1903) erected the genus *Zacco* (type species *Z. platypus*), the discriminating criterion for both genera was only that '*Opsariichthys* is present of peculiar notched jaws, but *Zacco* is absent'. However, we found that the jaws of *Z. pachycephalus* are similar to *O. uncirostris*. In fact, all the characters, including jaw morphology, that *Z. pachycephalus* and *O. uncirostris* share are specializations for piscivory. These facts lead us to support Howes' (1980) proposal that '*Zacco pachycephalus*' should be changed to *Opsariichthys pachycephalus*.

Acknowledgements

We cordially thank to K.H. Chang and C.S. Tzeng for providing every possible facility during the junior author's (K.H.) stay in Taiwan. The senior author (H.A.) greatly appreciates the generous guidance of Y. Nagata and S.S. Kimura. We are also grateful to K. Kawamura for technical advice and to K. Hoshino for sampling the fish for comparative study.

References cited

Bănărescu, P. 1968. Revision of the genera *Zacco* and *Opsariichthys* (Pisces, Cyprinidae). Věst. Čs. zool. spol. 32: 305–311.

Bănărescu, P. 1991. Zoogeography of fresh waters, vol. 2. Aula-Verlag, Wiesbaden. 571 pp.

Dunn, J.R. 1983. The utility of developmental osteology in taxonomic and systematic studies of teleost larvae: a review. NOAA Tech. Rep., NMFS Circ. 450, Department of Commerce, Washington, D.C. 19 pp.

Fink, S. & W.L. Fink. 1981. Interrelationships of the ostaryophysan fishes (Teleostei). Zool. J. Lin. Soc. London 72: 297–353.

Fujita, K. 1989. Nomenclature of cartilaginous elements in the caudal skeleton of teleostean fishes. Japan. J. Ichthyol. 36: 22–29.

Gosline, W.A. 1973. Considerations regarding the phylogeny of cypriniform fishes, with special reference to structures associated with feeding. Copeia 1973: 761–776.

Günther, A. 1868. Report on a collection of fishes made by Dr. Kirk in the Zambezi and Nyasse regions. Proc. Zool. Soc. London: 303–314.

Harada, I. 1943. Freshwater fishes of Hainan Island. Political Affairs Bureau Special Service Agency, The Imperial Navy of Japan, Tokyo. 114 pp. (in Japanese).

Harrington, R.W. 1955. The osteocranium of American cyprinid fish *Notropis bifrenatus*, with an annotated synonymy of teleost skull bones. Copeia 1955: 267–290.

Hosoya, K. 1986. Interrelationships of the Gobioninae (Cyprinidae). pp. 484–501. *In:* T. Ueno, R. Arai, T. Taniuchi & K. Matsuura (ed.) Indo-Pacific Fish Biology, Proc. Second Int. Conf. Indo-Pac. Fishes, Japan. J. Ichthyol, Tokyo.

Hosoya, K. 1989. Osteology of *Squalidus multimaculatus* (Teleostei, Cyprinidae), with comments on Korean *Squalidus* zoogeography. Bull. Mus. Nat. Hist. Nat., Paris 11: 234–245.

Howes, G.J. 1980. The anatomy, phylogeny and classification of the bariliine cyprinid fishes. Bull. Br. Mus. Nat. Hist. (Zool.) 37: 129–198.

Howes, G.J. 1983. Additional notes on bariliine cyprinid fishes. Bull. Br. Mus. Nat. Hist. (Zool.) 45: 95–101.

Jordan, D.S. & B.W. Evermann. 1903. Note on a collection of fishes from the island of Formosa. Proc. U.S. Nat. Mus. 25: 315–368.

Kawamura, K. & K. Hosoya. 1991. A modified double staining technique for making a transparent fish-skeletal specimen. Bull. Nat. Inst. Aquaculture 20: 11–18 (in Japanese).

Kusaka, T. 1974. The urohyal of fishes. University of Tokyo Press, Tokyo. 320 pp. (in Japanese).

McAllister, D.E. 1968. Evolution of branchiostegals and classification of teleostome fishes. Bull. Nat. Mus. Can. (Biol.) 77: 1–239.

Min, M.S. & S.Y. Yang. 1991. Systematic study on the genus *Zacco* (Pisces, Cyprinidae). II. Phylogenetic relationships of the genera *Zacco* and *Candidia.* Korean J. Zool. 34: 571–584 (in Korean).

Nakamura, M. 1969. Cyprinid fishes of Japan. Spec. Publ. Res. Inst. Nat. Resources 4: viii + iv + 455 pp. (in Japanese).

Nelson, G.J. 1969. Gill arches and the phylogeny of fishes, with notes on the classification of vertebrates. Bull. Amer. Mus. Nat. Hist. 141: 479–552.

Okazaki, T., M. Watanabe, K. Mizuguchi & K. Hosoya. 1991. Genetic differentiation between two types of dark chub, *Zacco temminckii*, in Japan. Japan. J. Ichthyol. 38: 133–140.

Oshima, M. 1919. Contribution of study of the fresh water fishes of the Island of Formosa. Ann. Carnegie. Mus. 196–328.

Ramaswami, L.S. 1955. Skeleton of cyprinid fishes in relation to phylogenetic studies. vii, Skull and weberian apparatus of Cyprininae (Cyprinidae). Acta. Zool. 36: 199–242.

Regan, C.T. 1911. The classification of the teleostean fishes of order Ostariophysi: I. Cyprinoidea. Ann. Mag. Nat. Hist. 8: 12–577.

Starks, E.C. 1930. The primary shoulder girdle of bony fishes. Stanford Univ. Publ., Univ. Ser. Biol. Sci. 6: 149–239.

Tzeng, C.S. 1986. The freshwater fishes of Taiwan. Taiwan Prov. Educ. Dept. Pres, Taipei. 66 pp. (in Chinese).

Uchida, K. 1939. Freshwater fishes of Chosen (Korea). Nematognathii and Eventognathii. Chosen Sotokufu Suisan Sikenjyo Houkoku, Pusan 6: 8 + 458, 47 pl. (in Japanese).

Wu, H.W. 1964. The cyprinid fishes of China. Volume 1. Sci. Tech. Press, Shanghai. 288 pp. (in Chinese).

Environmental Biology of Fishes **52**: 173–179, 1998.

A study of the systematics of cyprinid fishes by two-dimensional gel electrophoresis

Jun-Ichi Miyazaki[1], Tamio Hirabayashi[1], Kazumi Hosoya[2] & Tetsuo Iwami[3]
[1] *Institute of Biological Sciences, University of Tsukuba, Tsukuba, Ibaraki 305, Japan (e-mail: junichi@sakura.cc.tsukuba.ac.jp)*
[2] *National Research Institute of Aquaculture, Nansei, Mie 516-01, Japan*
[3] *Laboratory of Biology, Tokyo Kasei Gakuin University, 2600 Aihara, Machida, Tokyo 194-02, Japan*

Received 13.11.1995 Accepted 20.1.1997

Key words: genetic distance, Cyprininae, Acheilognathinae, Leuciscinae, Gobioninae, Cyprinidae, phenogram

Synopsis

This study was carried out to shed light on confused subfamilial groupings in the Cyprinidae from the biochemical viewpoint at the molecular level, specifically by using two-dimensional gel electrophoresis of liver proteins. Six pairs of cypriniform fishes, which are different from one another at familial, subfamilial, generic, specific, subspecific, and individual levels, were compared. The genetic distances between pairs of fishes increased as taxonomic ranks of the pairs became higher, confirming the reliable usefulness of this technique. Four species representing the different subfamilies, Cyprininae, Gobioninae, Acheilognathinae, and Leuciscinae, were compared to give new insight into relationships at the subfamilial level. *Cyprinus carpio* (Cyprininae) and *Pseudogobio esocinus esocinus* (Gobioninae) gave the smallest genetic distance and the largest values were obtained between either one of the above species and *Acheilognathus melanogaster* (Acheilognathinae), suggesting that the former two subfamilies compose the most closely related group that is in turn distantly related to Acheilognathinae. *Tribolodon hakonensis* (Leuciscinae) had almost equal genetic distances to the three other species.

Introduction

The family Cyprinidae is one of the largest groups of freshwater fishes, and there has been confusion on taxonomic groupings within the family. For example, according to Hensel (1970) who reviewed the history of cyprinid classification, Kryzhanovsky (1947) grouped cyprinids into four subfamilies, mainly based on embryonic and zoogeographical data. Nikolsky (1954) classified them into nine subfamilies mainly by morphological traits, such as the number of rows of pharyngeal teeth. Berg (1955) grouped the Cyprinidae into four subfamilies and Wu (1964) classified Chinese cyprinids into 10 subfamilies. Nelson (1994), after reviewing works of several authors, grouped the Cyprinidae into eight subfamilies.

We find inconsistency in the classification of Japanese cyprinids as well. Kobayasi (1952) classified Japanese cyprinids from the lepidological viewpoint into five subfamilies. Matsubara (1955) and Okada (1959) did not accept the distinction at the familial level between the Cyprinidae and the Cobitidae because of the presence of their intermediate form, i.e. Chinese *Gobiobotia*, and thus they concluded that Cyprinidae consists of two subfamilies,

Cyprininae and Cobitinae. Miyadi et al. (1976) grouped Japanese cyprinids into seven subfamilies.

Phylogenetic studies at the subfamilial level other than those mentioned above have been carried out by inter-subfamilial hybridization (Suzuki 1968) and karyological comparison (Arai 1982). However, no biochemical study has been carried out at the subfamilial level of cyprinids.

We have two electrophoretic techniques for systematic study; one is a conventional technique to mainly analyze enzymes, and the other is a two-dimensional technique to analyze protein constituents of tissues. We employed the latter in this study for investigating phylogenetic relationships of cyprinids at the subfamilial level. A previous study (Miyazaki et al. 1987) suggested that the latter technique is more suitable than the former for evaluating genetic differences between differentiated animals, i.e. those different at higher taxonomic levels.

First, to confirm usefulness of this method, protein components of livers from cypriniform fishes were analyzed and compared pairwise. Six pairs of fishes were used that are different from each other at the familial, subfamilial, generic, specific, subspecific, or population level. Secondly, four additional species of cyprinids were used for investigating phylogenetic relationships at the subfamilial level. Based on the results of this study, applicability of two-dimensional gel electrophoresis for systematic study is demonstrated, and phylogenetic relationships of cyprinids are discussed.

Materials and methods

Materials

Eight species including four subspecies of the Cyprinidae and one species of the Cobitidae were used in this study. *Acheilognathus tabira* subsp. (so-called 'Akahire-tabira' sensu Nakamura 1963), *A. rhombeus, A. melanogaster, Rhodeus ocellatus ocellatus, Pseudorasbora parva, Cyprinus carpio, Tribolodon hakonensis* and *Misgurnus anguillicaudatus* were

Table 1. Correlation of genetic distances[1] with taxonomic ranks.

Taxonomic rank	Freshwater fish[2]	Mouse[3]
	Genetic distance to *A. tabira* subsp.	Genetic distance to *P. maniculatus* GA
Individual	0.027 *A. tabira* subsp.	0.049 *P. maniculatus* VT 0.084 *P. maniculatus* OR
Subspecies	0.088 *A. tabira tabira*	
Species	0.237 *A. rhombeus*	0.147 *P. leucopus* 0.188 *P. boylii*
Genus	0.284 *R. ocellatus ocellatus*	0.401 *O. nuttalli* 0.453 *S. hispidus*
Subfamily	0.422 *Ps. parva*	
Family	0.517 *Mi. anguillicaudatus*	0.496 *M. musculus*

[1] Genetic distances were calculated according to Aquadro & Avise (1981).

[2] Generic names of fishes are abbreviated as follows; *A.* = *Acheilognathus*; *R.* = *Rhodeus*; *Ps.* = *Pseudorasbora*; *Mi.* = *Misgurnus*.

[3] Data were cited from Aquadro & Avise (1981) for comparison. Generic names of mice are abbreviated as follows; *P.* = *Peromyscus*; *O.* = *Ochrotomys*; *S.* = *Sigmodon*; *M.* = *Mus*.

collected in Lake Kasumigaura and in rivers emptying into the lake at Ibaraki. Two specimens of *A. tabira* subsp. for intraspecific comparison were collected at Omuro, the west side of the lake, and at Okijuku, the east side of the lake. *Pseudogobio esocinus esocinus* was collected in the Kinu River in Ibaraki. *Acheilognathus tabira tabira* was collected in Lake Biwa at Shiga.

According to Nikolsky (1954), the cyprinids used are included in the following subfamilies: Acheilognathinae – *A. tabira* subsp., *A. tabira tabira, A. rhombeus, A. melanogaster* and *R. ocellatus ocellatus*; Gobioninae – *P. esocinus esocinus*; Cyprininae – *C. carpio*; Leuciscinae – *T. hakonensis*. These four subfamilies have been recognized consistently by many taxonomists. The position of *P. parva* is controversial and tentatively included in the Barbinae (Hosoya 1988).

Electrophoresis

Livers from the above mentioned species were used for electrophoretic analysis. Two-dimensional gel electrophoresis was carried out as described previously (Hirabayashi 1981, Hirabayashi et al. 1983, Oh-ishi & Hirabayashi 1988). In brief, extraction of proteins was done by homogenizing the liver in 20 volumes of a medium containing 8 M guanidine-HCl, which inhibits protease activities thoroughly.

Table 2. Comparison among cyprinids at the subfamilial level[1].

Combination	Genetic distance[2]
Cyprinus carpio	
vs. *Pseudogobio esocinus esocinus*	0.445
vs. *Tribolodon hakonensis*	0.468
vs. *Acheilognathus melanogaster*	0.525
Pseudogobio esocinus esocinus	
vs. *Tribolodon hakonensis*	0.467
vs. *Acheilognathus melanogaster*	0.539
Tribolodon hakonensis	
vs. *Acheilognathus melanogaster*	0.473

[1] The subfamilies that the examined species represent are as follows; *C. carpio*, Cyprininae; *P. esocinus esocinus*, Gobioninae; *T. hakonensis*, Leuciscinae; *A. melanogaster*, Acheilognathinae.
[2] Genetic distances were calculated according to Aquadro & Avise (1981).

After centrifugation at 60000 × g, the supernatant was subjected to isoelectric focusing of the first dimension for 12500–13500 V-h. SDS-polyacrylamide gel electrophoresis of the second dimension was performed by using a running gel of a concentration gradient of 12–20% acrylamide and a stacking gel of 3% acrylamide. Proteins were stained with Coomassie brilliant blue in picrate as described by Stephano et al. (1986). In earlier experiments, a focusing gel with slightly different composition was used and proteins were stained with Coomassie brilliant blue in 50% trichloroacetic acid. These changes did not lead to significant effects when comparing electrophoretic patterns.

Analysis

Electrophoretic patterns were compared visually by the triplet method in which two different samples (60 μl) and their mixture (40 + 40 μl) were focused at the same time (Miyazaki et al. 1987). For confirmation of usefulness of this electrophoresis technique, pairwise comparison was made on each of six triplet patterns. The combination of six pairs of fishes (Table 1) was designed to show genetic distances at familial, subfamilial, generic, specific, subspecific, and individual levels. *Acheilognathus tabira* subsp., one of the most common cyprinids in Lake Kasumigaura, was used as a standard counterpart for comparison. Six pairs of cyprinids (Table 2) were compared in the next experiments for investigation of phylogenetic relationships at the subfamilial levels.

Genetic distances were calculated according to the formula of Aquadro & Avise (1981); $D = 1 - 2N_{xy}/(N_x + N_y)$, where D is the genetic distance between specimens x and y, N_{xy} is the number of protein spots shared by x and y, and N_x and N_y are the numbers of protein spots scored for x and y, respectively. The phenogram was depicted by the unweighted pair-group clustering method (UPGM) according to Sneath & Sokal (1973).

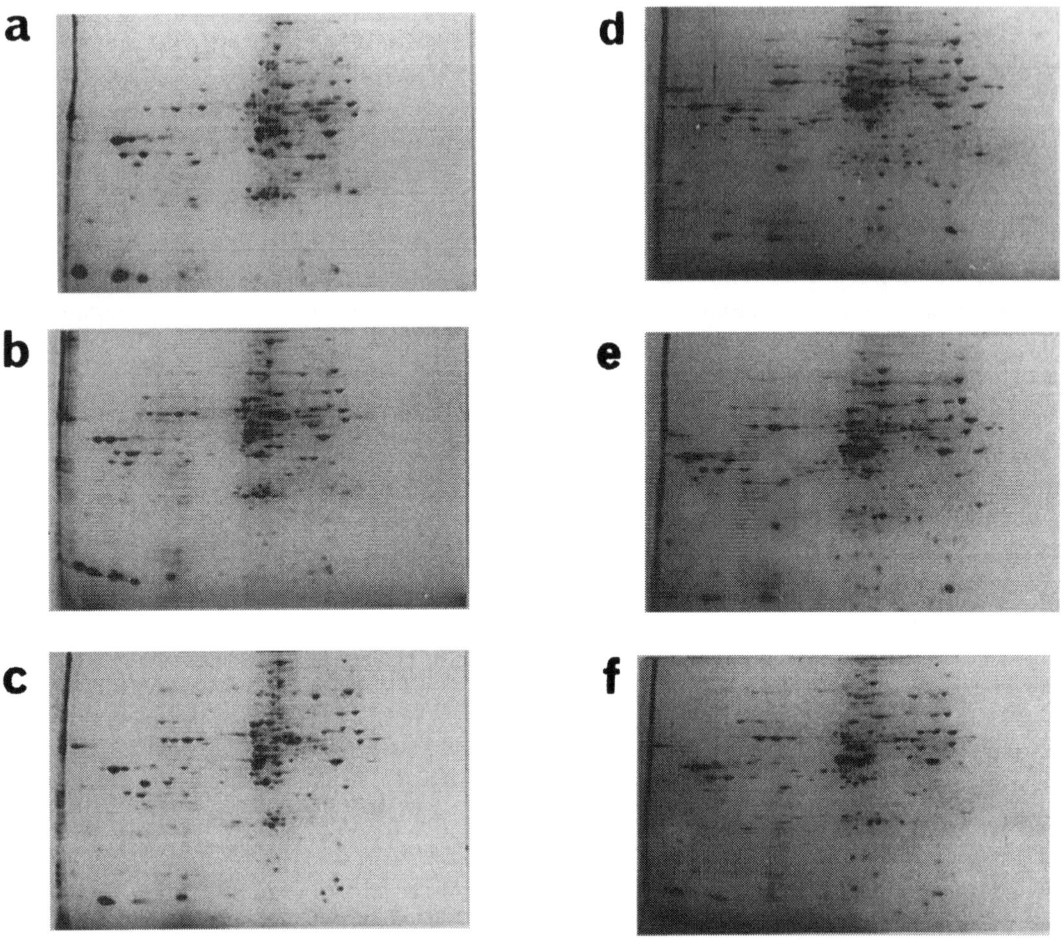

Figure 1. Two-dimensional gel electrophoresis patterns of fish livers for intersubfamilial comparison of the Cyprinidae. Triplet patterns on the left are for comparison between the Cyprininae and the Leuciscinae: a – *Cyprinus carpio* (Cyprininae); c – *Tribolodon hakonensis* (Leuciscinae); b – their mixture. Triplet patterns on the right are for comparison between the Gobioninae and the Acheilognathinae: d – *Pseudogobio esocinus esocinus*; f – *Acheilognathus melanogaster*; e – their mixture.

Results

Two-dimensional gel electrophoresis pattern

Representative two-dimensional gel electrophoresis patterns of fish livers are shown in Figure 1. Triplet patterns on the left are for comparison between *Cyprinus carpio* (a) and *Tribolodon hakonensis* (c). *Cyprinus carpio* represents the Cyprininae, while *T. hakonensis* is a member of the Leuciscinae. A mixture of samples of both species was also analyzed for precise comparison (b). Triplet patterns on the right are for comparison between *Pseudogobio esocinus esocinus* (d) and *Acheilognathus melanogas-*

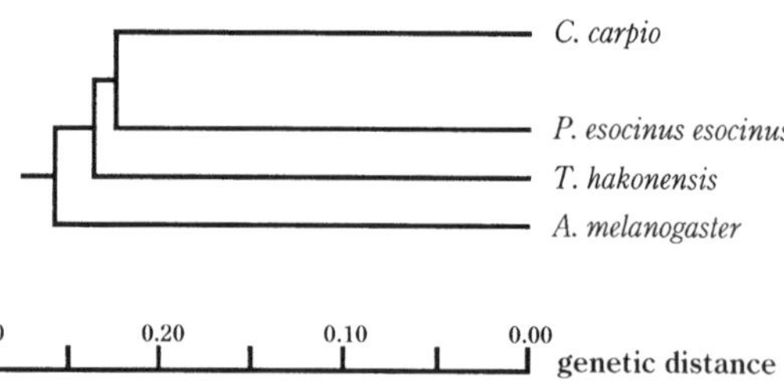

Figure 2. Relationships among four cyprinid species. Phenogram of four species representing different subfamilies was depicted by using UPGM (Sneath & Sokal 1973) based on the genetic distances shown in Table 2.

ter (f). The former represents the Gobioninae, and the latter the Acheilognathinae. The middle (e) is the pattern of their mixed sample. Twelve sets of triplet patterns were prepared and an average of 322 (255 to 476) protein spots on those patterns was compared.

Usefulness of two-dimensional gel electrophoresis

Six pairs of fishes were compared pairwise to confirm usefulness of two-dimensional gel electrophoresis in systematic study of freshwater fishes (the left column in Table 1). The smallest value (0.027) was obtained between two different specimens of *A. tabira* subsp., and the largest one (0.517) between *A. tabira* subsp. and *Misgurnus anguillicaudatus*, the former of which belongs to the family Cyprinidae and the latter to the family Cobitidae. The results also showed that pairs of fishes give larger genetic distances when they are different at higher taxonomic levels, clearly presenting a perfect correlation between taxonomic rank and extent of genetic differentiation evaluated by the two-dimensional gel electrophoresis method. Therefore, the results confirmed that this method could be useful for systematic study of freshwater fishes.

Phyletic analysis of cyprinids at the subfamilial level

Four species, which belong to different subfamilies, were compared pairwise to give new insight into phylogenetic relationships at the subfamilial level (Table 2). *Cyprinus carpio* and *P. esocinus esocinus* gave the smallest genetic distance (0.445) in mutual comparison, and the largest values (0.525 of the former and 0.539 of the latter) in comparison with *A. melanogaster*. Interestingly enough, *T. hakonensis* gave almost equal values to the other three species (0.467–0.473). All values from comparisons of six pairs of fishes (0.445–0.539) were considerably larger than that of the pair of consubfamilial genera (0.284 of *A. tabira* subsp. vs. *Rhodeus ocellatus ocellatus* in Table 1). Those values are, however, very close to each other and also close to that of the pair of confamilial subfamilies (0.422 of *A. tabira* subsp. vs. *Pseudorasbora parva* in Table 1), suggesting close relationships. Such relationships are shown well in the phenogram (Figure 2) depicted by the unweighted pair-group clustering method (Sneath & Sokal 1973).

Discussion

Applicability of two-dimensional gel electrophoresis

Comparison was made between fishes at six different taxonomic levels. The taxonomic status of those fishes is accepted by many taxonomists. Results showed a correlation between taxonomic rank and genetic distance (Table 1), confirming the reliable usefulness of two-dimensional gel electrophoresis to the systematic study of freshwater fishes as was expected from previous studies (Miyazaki et al. 1987, Miyazaki et al. 1988). Usefulness of this technique was also demonstrated by a study of rodents (Aquadro & Avise 1981), which showed a good correlation between taxonomic rank and genetic distance as presented in the right column of Table 1. It is interesting to note that the genetic distances given in each taxonomic level are similar between the two different taxa, fishes and rodents, with the exception of the generic values (considerably larger in rodents), suggesting the possibility that taxonomic ranks in these groups can be roughly defined by genetic distances.

Phylogenetic relationships of cyprinids at the subfamilial level

Comparison between four species representing different subfamilies gave the smallest genetic distance (0.445) between *Cyprinus carpio* and *Pseudogobio esocinus esocinus*, and the largest values (0.525 and 0.539) between either one of the above species and *Acheilognathus melanogaster* (Table 2). Therefore, these results indicate that the Cyprininae and the Gobioninae subfamilies are the most closely related to each other and distantly related to the Acheilognathinae.

This is supported by both intersubfamilial hybridization (Suzuki 1968) and karyological and morphological studies (reviewed by Arai 1982). Results of hybridization showed that hybrids between cyprinine and gobionine fishes could hatch and in part develop until feeding. On the other hand, attempts to hybridize either cyprinine or gobionine with acheilognathine fishes did not result in successful fertilization when eggs of acheilognathine fishes were inseminated with sperm from cyprinine or gobionine fishes. When acheilognathine sperm were used for insemination of cyprinine or gobionine eggs, fertilization was successful, but hybrids could not hatch nor develop until feeding. Based on karyological and morphological studies, Arai (1982) showed a diagram of provisional relationships of Eurasian cyprinid subfamilies. In his diagram, the Cyprininae and the Gobioninae compose the most closely related group, which is distantly related to the Acheilognathinae. Furthermore, Kafuku (1958) showed that an acheilognathine pattern of intestinal coiling was largely different from those of other subfamilies. Although Chen et al. (1984) suggested, based on cladistic morphological analysis, that the Gobioninae are more closely related to the Acheilognathinae than to the Cyprininae, Hosoya (1988) showed the Cyprininae and the Gobioninae compose a sister group by sharing some derived characters. Based on the above data, it can be stated with confidence that the Cyprininae and the Gobioninae are more closely related to each other than either one is to the Acheilognathinae.

Arai (1982) showed in his diagram that the Leuciscinae were positioned distantly from both the Acheilognathinae and the Cyprininae-Gobioninae. The present study showed that *Tribolodon hakonensis* had nearly equal genetic distances to three other species (Table 2). Therefore, it is not obvious at present whether the Leuciscinae are more closely related to the Cyprininae-Gobioninae than to the Acheilognathinae or vice versa. In any case, closeness of branching points among the four subfamilies (see Figure 2) and their great diversification lead to lack of reliable criteria for grouping cyprinids into subfamilies, as pointed out by Hensel (1970), consequently resulting in taxonomic confusion at the subfamilial level.

Remark on the systematic position of the Cobitidae

The genetic distances between *A. melanogaster* and either one of *C. carpio* and *P. esocinus esocinus* (0.525 and 0.539 in Table 2) are somewhat larger than that between *A. tabira* subsp. and *Misgurnus anguillicaudatus* (0.517 in Table 1). Many taxonomists consider that cobitids descended from a cyprinid ancestor (reviewed by Hensel 1970), suggesting that cyprinids might have already diversified before the first cobitid arose. Therefore, it is conceivable that some cyprinids may have larger genetic distances to other cyprinids than to cobitids. The problems of the systematic position of the Cobitidae and the paraphyletic status of the Cyprinidae remain to be studied.

Prospect of the phylogenetic study of cyprinids

We investigated four species representing different subfamilies that are accepted by many taxonomists, and suggested possible phylogenetic relationships of the subfamilies. The relationships, however, are not conclusive, because only one species in each subfamily was used in this study. The phylogeny of cyprinids at the subfamilial level should be deduced by examining more species in each subfamily and including species from other, less well-established subfamilies. Nevertheless, this is the first report which suggests intersubfamilial phylogenetic relationships of cyprinids from a biochemical viewpoint, prompting further investigation on the phylogeny of cyprinids.

References cited

Aquadro, C.F. & J.C. Avise. 1981. Genetic divergence between rodent species assessed by using two-dimensional electrophoresis. Proc. Nat. Acad. Sci. USA 78: 3784–3788.

Arai, R. 1982. A chromosome study on two cyprinid fishes, *Acrossocheilus labiatus* and *Pseudorasbora pumila pumila*, with notes on Eurasian cyprinids and their karyotypes. Bull. Nat. Sci. Mus. Tokyo Ser. A 8: 131–152.

Berg, L.S. 1955. Classification of fishes, both recent and fossil. Trud. Zool. Inst. Akad. Nauk S.S.S.R. 20: 1–286. (in Russian).

Chen, X.-L., P.-Q. Yue & R.-D. Lin. 1984. Major groups within the family Cyprinidae and their phylogenetic relationships. Acta Zootaxonomica Sinica 9: 424–440.

Hensel, K. 1970. Review of the classification and of the opinions on the evolution of Cyprinoidei (Eventognathi) with an annotated list of genera and subgenera described since 1921. Annot. Zool. Bot. 57: 1–45.

Hirabayashi, T. 1981. Two-dimensional gel electrophoresis of chicken skeletal muscle proteins with agarose gels in the first dimension. Anal. Biochem. 117: 443–451.

Hirabayashi, T., R. Tamura, I. Mitsui & Y. Watanabe. 1983. Investigation of actin in *Tetrahymena* cells. A comparison with skeletal muscle actin by a devised two-dimensional gel electrophoresis method. J. Biochem. 93: 461–468.

Hosoya, K. 1988. Interrelationships of the Gobioninae (Cyprinidae). Ph.D. Dissertation, University of Kyoto, Kyoto. 103 pp. (in Japanese).

Kafuku, T. 1958. Speciation in cyprinid fishes on the basis of intestinal differentiation, with some references to that among catostomids. Bull. Freshwater Fish. Res. Lab. (Tokyo) 8: 45–78.

Kobayasi, H. 1952. Phylogeny and evolution of Japanese freshwater fishes based on the comparative morphology of scale characters (preliminary report). Mem. Aiti Gakugei Univ. 1: 80–83. (in Japanese).

Kryzhanovsky, S.G. 1947. Classification of the Cyprinidae. Zool. Zhurn. 26: 53–64. (in Russian).

Matsubara, K. 1955. Fish morphology and hierarchy. Part I. Ishizaki-shoten, Tokyo. 789 pp. (in Japanese).

Miyadi, D., H. Kawanabe & N. Mizuno. 1976. Colored illustrations of the freshwater fishes of Japan. Hoikusha Publishing, Osaka. 462 pp. (in Japanese).

Miyazaki, J.-I., K. Sekiguchi & T. Hirabayashi. 1987. Application of an improved method of two-dimensional electrophoresis to the systematic study of horseshoe crabs. Biol. Bull. 172: 212–224.

Miyazaki, J.-I., R. Ueshima & T. Hirabayashi. 1988. Application of a two-dimensional electrophoresis method to the systematic study of land snails of subgenus *Luchuphaedusa* from southwestern Japan islands. Biol. Bull. 175: 372–377.

Nakamura, M. 1963. Keys to the freshwater fishes of Japan fully illustrated in colors. Hokuryukan, Tokyo. 262 pp.

Nelson, J.S. 1994. Fishes of the world. John Wiley & Sons, New York. 600 pp.

Nikolsky, G.V. 1954. Special ichthyology. Sovetskaya Nauka, Moscow. 458 pp. (in Russian).

Oh-ishi, M. & T. Hirabayashi. 1988. Micro-two-dimensional gel electrophoresis with agarose gel in the first dimension. Physiochem. Biol. 32: 113–120.

Okada, Y. 1959. Studies on the freshwater fishes of Japan. Pref. Univ. Mie, Tsu. 860 pp. (in Japanese).

Sneath, P.H.A. & R.R. Sokal. 1973. Numerical taxonomy. W.H. Freeman, San Francisco. 573 pp.

Stephano, J.L., M. Gould & L. Rojas-Galicia. 1986. Advantages of picrate fixation for staining polypeptides in polyacrylamide gels. Anal. Biochem. 152: 308–313.

Suzuki, R. 1968. Hybridization experiments in cyprinid fishes, XI. Survival rate of F1 hybrids with special reference to the closeness of taxonomical position of combined fishes. Bull. Freshwater Fish. Res. Lab. 18: 113–155.

Wu, H.-W. 1964. Chinese cyprinid fishes. Vol. 1. Sci. Techn. Print House, Shanghai. 228 pp. (in Chinese).

Environmental Biology of Fishes **52**: 181–186, 1998.

Fertilization environment of the non-copulating marine sculpin, *Hemilepidotus gilberti*

Youichi Hayakawa[1] & Hiroyuki Munehara[2]
[1] *Research Institute of North Pacific Fisheries, Faculty of Fisheries, Hokkaido University, 3-1-1 Minato-Cho, Hakodate 041-082, Japan (e-mail: youichi.hayakawa@f2.hines.hokudai.ac.jp)*
[2] *Usujiri Fisheries Laboratory, Hokkaido University, Minami-Kayabe, Hokkaido 041-1613, Japan*

Received 23.4.1997 Accepted 1.12.1997

Key words: sperm motility, oviparous, ovarian fluid, external insemination

Synopsis

To clarify the extracellular environment for external fertilization in the non-copulating marine sculpin *Hemilepidotus gilberti*, sperm motility was measured in NaCl, KCl, mannitol solutions, seawater, and ovarian fluid. Spermatozoa of *H. gilberti* actively moved in seminal plasma the moment they were removed from the genital papilla. Spermatozoa showed higher motility in NaCl solution at osmolalities between 300–400 mOsmol kg^{-1}. In KCl and in mannitol solutions, spermatozoa actively moved at osmolalities between 500 and 800 mOsmol kg^{-1}, and at osmolality 300 mOsmol kg^{-1}, respectively. The ovarian fluid was a transparent and viscous gelatinous material, rich in sodium with an osmolality of 340 mOsmol kg^{-1}. Sperm motility in the ovarian fluid lasted more than 90 min, which was six times longer than in seawater. This sperm motility under conditions isotonic to body fluid is similar to that of copulating marine sculpins rather than to other non-copulating marine fishes. In addition, eggs of *H. gilberti* could be fertilized in the ovarian fluid. This suggests that external fertilization takes place under physiological conditions similar to the internal conditions of the ovary provided by the ovarian fluid, which isolates the eggs from sea water for several hours after spawning. This manner of fertilization is thought to be one of the evolutionary pre-adaptations allowing copulation among marine sculpins.

Introduction

Most teleost fishes do not copulate, unlike other vertebrates. The typical reproductive behavior of non-copulating fishes consists of egg release and simultaneous sperm emission. Eggs released in the extracellular aquatic environment must become associated with spermatozoa within a short time, or else they will lose their capacity for fertilization due to auto-activation induced by being exposed to conditions anisotonic to body fluid (Yamamoto 1961). The situation seems to be true of spermatozoa as well. In addition, because the sperm of teleost fishes are not replenished with motive energy after emission, they must reach the eggs using only the energy source supplied during spermatogenesis. Sperm are immotile in the male reproductive organ, preserving the energy source for motility, but become mobile in the aquatic environment so that they can penetrate into eggs. The aquatic environment has several properties that affect sperm motility, including osmolality, ion composition, concentration of principal ions, and pH. It is known that the sperm of freshwater and marine fishes initiate motility when exposed to hypotonic freshwater and hypertonic seawater in which they spawn their eggs, respectively. Sperm of salmonid fishes initiate motility in media with lower potassium concentration than that in the seminal plasma (Morisawa 1985, Billard & Cosson 1992).

Hemilepidotus gilberti is a non-copulating sculpin (Hayakawa & Munehara 1996), although many marine sculpins are known to copulate. Female *H. gilberti* very slowly deposit eggs enclosed in viscous ovarian fluid onto a substrate. Spawning continues without interruption for approximately 20–50 min. Males emit semen against the deposited eggs several minutes after the beginning of spawning. Ovarian fluid covers the whole egg mass during spawning, although it dissolves in seawater within a few hours. This observation suggests that gamete association of *H. gilberti* might occur not in seawater but in the ovarian fluid. Females of the copulating sculpins release eggs that have been associated internally with sperm (Munehara et al. 1989, 1991). Sperm of such copulating sculpins show higher motility when exposed to certain properties that the ovarian fluid provides (Koya et al. 1993). In fact, sperm retain their movement far longer in ovarian fluid than seawater (Koya et al. 1993). This study was undertaken to clarify whether gamete association and subsequent fertilization of *H. gilberti* occur in seawater or ovarian fluid, in comparison with copulating species.

Materials and methods

Hemilepidotus gilberti were collected by angling and gill nets in the coastal waters (50–90 m depth) off Usujiri (41°57′ N, 140°, 58′ E), southern Hokkaido, Japan, from June to September 1992. Males and females were kept separately in 1000 l tanks until maturity.

Semen was collected by sucking it into capillary tubes from the tip of the genital papilla while guarding against contamination with urine. The seminal plasma was obtained by centrifuging the semen in microtubes at 390 g for 15 min. Ovarian fluid was collected by squeezing the egg masses from ripe females and centrifuging the masses at 270 g for 15 min. Blood was collected from the caudal vessel with a syringe and centrifuged at 270 g for 15 min to obtain the serum. All samples were frozen and stored at – 25 °C until analysis.

Sodium, potassium, calcium, and magnesium were analyzed in all samples using a flame spectrophotometer (Hitachi 170–30). The osmolality of each sample was measured by the freezing point method using a Knauer semimicro-osmometer.

The motility of spermatozoa in different test solutions was determined by the duration of their forward movement. Observations were carried out under a light microscope with 100 μl of each test solution placed on a glass slide at room temperature. The measurement started when the semen was quickly added to the test solution and mixed using a glass rod. Test solutions applied were: NaCl, KCl, and mannitol solutions ranging from 100 to 1000 mOsmol kg^{-1} and buffered to pH = 7.4 using HEPES-NaOH in distilled water. In addition, ovarian fluid, seawater and various mixtures of both (ovarian fluid: sea water = 1:3, 1:1 and 3:1) were applied. All chemicals were of reagent grade.

Artificial inseminations were doen *in vitro* and *in vivo* to confirm whether fertilization occurred in ovarian fluid. For the *in vitro* experiment, a small egg mass was placed into a beaker (300 ml) and inseminated with 300 μl semen without being immersed in seawater. For the *in vivo* experiment, the same volume of semen was injected into the ovary and gently mixed with eggs using a capillary tube. For the control, eggs inseminated *in vitro* were immersed in seawater. Control and *in vitro* eggs were incubated at 5 °C and the fish inseminated artificially were kept in a tank at 5 to 8 °C. The number of developing eggs in both experiments were counted 7 and 30 h after insemination.

Results

Chemical characteristics of seminal plasma, ovarian fluid and serum

Inorganic electrolyte concentrations and osmolalities of seminal plasma, ovarian fluid, and serum are shown in Table 1. In both seminal plasma and male serum, sodium concentrations were the highest among the four ions measured. Potassium and mag-

Table 1. Osmolalities and inorganic electrolyte concentrations of male serum, seminal plasma, female serum, and ovarian fluid. Values are mean ± SE. Means that differ from serum of fish of the same sex are marked with an asterisk (t-test, $p < 0.01$). Numbers in parentheses shows sample size.

	Na^+ (mM kg^{-1})	K^+ (mM kg^{-1})	Ca^{2+} (mM kg^{-1})	Mg^{2+} (mM kg^{-1})	Osmolality (mOsmol kg^{-1})
male serum	204 ± 9.5 (8)	4.5 ± 0.68 (8)	2.10 ± 0.16 (8)	2.06 ± 0.13 (8)	364.7 ± 22.4 (7)
seminal plasma	162 ± 12.1 (5)	20.4 ± 2.60* (5)	0.97 ± 0.16* (4)	2.52 ± 0.21 (4)	328.0 ± 13.2 (3)
female serum	176 ± 2.7 (9)	7.5 ± 1.47 (9)	2.32 ± 0.22 (9)	1.55 ± 0.04 (9)	358.2 ± 7.9 (9)
ovarian fluid	129 ± 5.7* (5)	6.68 ± 0.76 (5)	1.42 ± 0.21 (5)	1.56 ± 0.11 (5)	340.0 ± 3.9 (5)

nesium concentrations were higher in the seminal plasma than in the serum, especially for potassium, which was about 4–5 times higher ($p < 0.01$ by t-test). Calcium concentrations in the seminal plasma were significantly lower than in the serum ($p < 0.01$). The osmolality of the serum was higher than that of the seminal plasma.

The osmolality of the ovarian fluid was similar to that of the serum. Sodium concentration was the highest among the four ions measured in both ovarian fluid and female serum, and was significantly higher in the serum than in the ovarian fluid ($p < 0.01$). Serum potassium concentrations was higher than in the ovarian fluid. The calcium concentration of the ovarian fluid was 1.42 mM kg^{-1}. Concentration of magnesium in both ovarian fluid and serum was almost identical.

Sperm motilities

Spermatozoa of *H. gilberti* actively moved in seminal plasma the moment they were removed from genital papilla, and their movement continued for at least one week in a microtube.

Spermatozoa of *H. gilberti* could swim in all solutions with wide ranges of osmolalities ranging from hypotonic solutions (200–300 mOsmol kg^{-1}) through seminal plasma (328.0 mOsmol kg^{-1}) to solutions isotonic with seawater (1000 mOsmol kg^{-1}) (Figure 1). In NaCl solution, longer motilities occurred at osmolalities between 300 and 700 mOsmol kg^{-1}, with a peak of about 50 min activity occur-

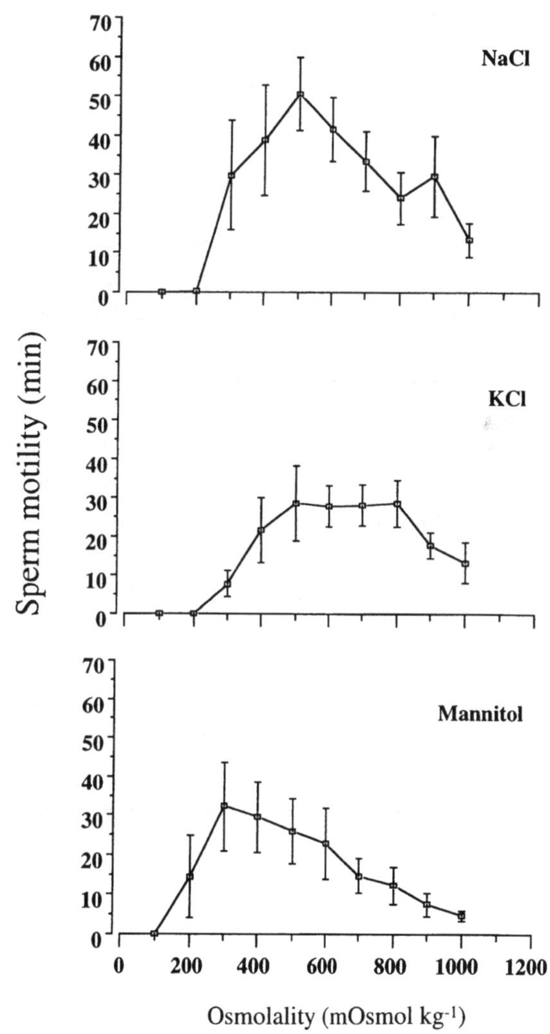

Figure 1. Sperm motility of *H. gilberti* in solutions of various osmolalities. Values are means ± SE (N = 5).

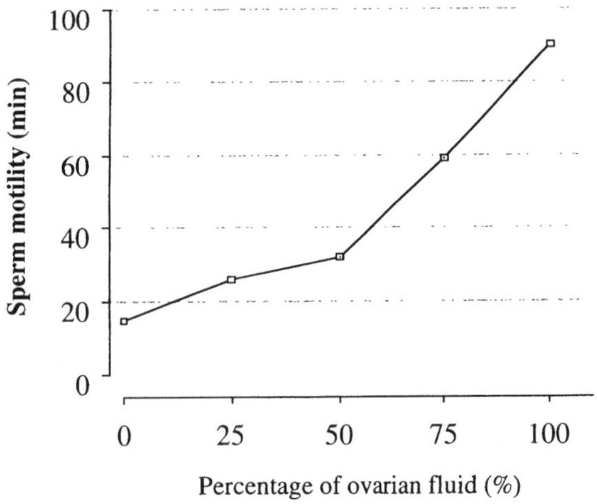

Figure 2. Sperm motility of *H. gilberti* in solutions of various volume ratios of ovarian fluid to sea water (N = 1).

ring at 500 mOsmol kg^{-1}. While sperm motility was 13.4 min at an osmolality of 1000 mOsmol kg^{-1} isotonic with seawater, spermatozoa could move for 30 to 40 min at osmolalities between 300 and 400 mOsmol kg^{-1}, which is isotonic to ovarian fluid. In KCl solution, although spermatozoa moved for about 30 min at osmolalities between 500 and 800 mOsmol kg^{-1}, the duration of motility was shorter than that in NaCl. In non-electrolyte mannitol solution, the longest motility occurred at an osmolality of 300 mOsmol kg^{-1}. In mixed solutions of ovarian fluid and seawater, the higher the proportion of ovarian fluid, the longer the spermatozoa were motile (Figure 2). In ovarian fluid, sperm motility lasted more than 90 min, which was six times longer than the motility in seawater.

Artificial insemination

Fertilization in ovarian fluid occurred both *in vitro* and *in vivo* (Table 2). Eggs placed in seawater (control) and *in vitro* developed to the 2–4 cell stages 7 h after insemination, and reached the morula stages 30 h after insemination. The *in vitro* fertilization rate (94.1%) was higher than the *in vivo* fertilization rate (65.9%). The development was so delayed *in vivo* that no eggs were observed to have been fertilized 7 h after insemination, and few eggs (34.1%) had developed to the morula stages 30 h after insemination.

Discussion

The inorganic composition of seminal plasma has been reported for some teleost species (Morisawa 1985, Koya et al. 1993). The seminal plasma of salmonid fishes contains as much sodium as the blood plasma, whereas potassium levels are markedly higher in the seminal plasma than in the blood plasma (Morisawa 1985). In cyprinid fishes, sodium concentration in the seminal plasma is lower than that in the blood plasma, while potassium concentration is markedly higher than in blood plasma (Morisawa 1985). In marine species, such as the puffer *Fugu niphobles* and the black seabream *Acanthopagrus schlegele,* sodium concentrations in the seminal plasma and blood plasma are similar, but potassium concentration in the seminal plasma is lower than that in the blood plasma (Morisawa 1985). In the

Table 2. Occurrence of fertilized eggs and developmental stages of artificial insemination in the ovarian fluid both externally and internally (* are observations 7 h after insemination, and † 30 h after insemination). The percentage of eggs fertilized in vivo that had reached each stage 30 h after fertilization is shown below.

	n	Percentage of fertilized eggs						Percentage of unfertilized eggs
control*	100	99 (99)						1 (1)
*in vitro**	102	94.1 (96)						5.9 (6)
in vivo†	167	65.9 (110)						34.1 (57)
		stages	2 cell	4~8 cell	16~32 cell	morula	unidentified #	
			0.6	3.0	13.2	34.1	15.0	

# indicates that development showed deformity such as eggs with loose cell aggregations.

seminal plasma of copulating marine sculpins, such as *Alcichthys alcicornis* and *Blepsias cirrhosus*, the potassium concentration is 4–10 times higher than that in serum, and sodium concentration is lower than in the serum (Koya et al. 1993). In the present study, the seminal plasma of *H. gilberti* contained lower sodium levels than the serum did, but potassium levels were approximately five times higher than in the serum. Thus, the seminal ion composition of *H. gilberti* is similar to that of copulating marine sculpins.

In general, the spermatozoa of teleosts are immotile in the male reproductive organ, and motility is initiated when the spermatozoa are ejaculated into the spawning medium (Morisawa 1985). In salmonid fishes, sperm motility is inhibited by the high concentration of potassium in seminal plasma, and contact with a medium containing lower potassium levels at spawning induces sperm motility (Morisawa et al. 1983, Stoss 1983). The sperm motility of freshwater species is suppressed in seminal plasma and is initiated by a decrease in osmolality at spawning (Billard 1978, Morisawa et al. 1983, Lahnsteiner et al. 1995). On the other hand, the sperm motility of marine species is initiated by an increase in osmolality (Billard 1978, Morisawa & Suzuki 1980). In contrast, the spermatozoa of copulating marine sculpins belonging to Scorpaeniformes are motile when they are removed from the sperm duct, and neither a high potassium concentration nor an isotonic osmolality in the seminal plasma inhibits sperm motility in the sperm duct (Koya et al. 1993). *H. gilberti* is a non-copulating species, but their sperm showed movement similar to that of sperm of the copulating sculpins. Recently, it has been reported that spermatozoa of turbot, *Scophthalmus maximus*, also show a similar phenomenon, and it was suggested that sperm motility is controlled by the oxygen concentration (Chauvaud et al. 1995). Although the mechanisms that inhibit sperm motility in the sperm duct of Scorpaeniformes are unknown, sperm of sculpins may also be of this type.

Spermatozoa of *H. gilberti* showed longer motility in NaCl solution. The longest motility was observed at an osmolality of 500 mOsmol kg^{-1}, but spermatozoa were motile for about 30 to 40 min at osmolalities of 300 and 400 mOsmol kg^{-1}, which is isotonic to body fluid. In contrast, sperm longevity at an osmolality of 1000 mOsmol kg^{-1} was less than half of that between 300 and 400 mOsmol kg^{-1}. These results suggest that solutions which are isotonic to body fluid and which contain high sodium levels are suitable for maintaining sperm movement. Furthermore, the spermatozoa of *H. gilberti* exhibited far higher motility in the ovarian fluid than in seawater. The ovarian fluid of *H. gilberti* has an osmolality of about 340 mOsmol kg^{-1}, and is rich in sodium. Therefore we conclude that the most suitable extracellular environment for movement of spermatozoa of this species is ovarian fluid, not seawater. High sperm motility under conditions isotonic to the body fluid is similar to that for copulating marine sculpins, such as *A. alcicornis* and *B. cirrhornis* (Koya et al. 1993). Thus, spermatozoa of *H. gilberti* have characteristics similar to copulating marine sculpins, rather than to other non-copulating marine species.

The eggs of *H. gilberti* were fertilized in ovarian fluid under both *in vitro* and *in vivo* conditions, although embryonic development *in vivo* was markedly delayed compared to *in vitro* development. In copulating marine sculpins, fertilization occurs not internally but externally after spawning (Munehara et al. 1989, Munehara et al. 1991, Koya et al. 1993). The inhibition of internal fertilization of those fishes is thought to be due to a deficiency of calcium ions in the ovarian fluid (= 0.41 mM kg^{-1} in *A. alcicornis*: 0.31 mM kg^{-1} in *B. cirrhosus*) which is slightly lower than the threshold concentration required for fertilization. Fertilization is triggered by the rise in calcium concentration caused by the surrounding seawater after egg deposition (Munehara et al. 1994). The calcium concentration in the ovarian fluid of *H. gilberti* is 1.42 mM kg^{-1}, which is approximately three to four times greater than that in the copulating sculpins. Thus, the high calcium concentrations may allow fertilization to occur in the ovarian fluid of *H. gilberti*.

It has been proposed that during the long period of egg release in *H. gilberti* the deposited eggs are unlikely to be directly exposed to seawater, because the egg mass is enclosed in gelatinous ovarian fluid for a few hours (Hayakawa & Munehara 1996). Under these conditions, eggs may avoid auto-activation due to exposure to seawater (unpublished data). The present study demonstrated that spermatozoa of *H. gilberti* are well adapted for movement in the ovarian fluid, suggesting that gametic association of *H. gilberti* takes place in the ovarian fluid. In addition, the results of artificial insemination experiments indicated that fertilization can also occur in the ovarian fluid. Since the ovarian fluid provides a chemical environment similar to internal physiological conditions, external fertilization of *H. gilberti* likely corresponds physiologically to internal fertilization. It is possible that this manner of fertilization may be an evolutionary pre-adaptation allowing copulation in sculpins.

Acknowledgements

We thank K. Shimazaki, H. Ogi,and Y. Sakurai, Faculty of Fisheries, Hokkaido University, for their suggestions and Shirley E. Hughes, Zoology Department, University of Aberdeen and John Richard Bower, Faculty of Fisheries, Hokkaido University, for critically reading the manuscript. We are also indebted to the staff of Usujiri Fisheries Laboratory, Hokkaido University, and the Usujiri Fisheries Cooperative Society, for the collection of specimens. This study was supported in part by Research Fellowships of the Japan Society for the Promotion of Science for Young Scientists. This paper is contribution number 135 from the Usujiri Fisheries Laboratory, Faculty of Fisheries, Hokkaido University.

References cited

Billard, R. 1978. Changes in structure and fertilization ability of marine and freshwater fish spermatozoa diluted in media of various salinities. Aquaculture 14: 187–198.

Billard, R. & M.P. Cosson. 1992. Some problems related to the assessment of sperm motility in freshwater fish. J. Exp. Zool. 261: 122–131.

Chauvaud, L., J. Cosson, M. Suquet & R. Billard. 1995. Sperm motility in turbot, *Scophthalmus maximus*: initiation of movement and change with time of swimming characteristics. Env. Biol. Fish. 43: 341–349.

Hayakawa, Y. & H. Munehara. 1996. Non-copulatory spawning and female participation during early egg care in a marine sculpin *Hemilepidotus gilberti.* Ichthyol. Res. 43: 73–78.

Koya, Y., H. Munehara, K. Takano & H. Takahashi. 1993. Effects of extracellular environments on the motility of spermatozoa in several marine sculpins with internal gametic association. Comp. Biochem. Physiol. 106: 25–29.

Lahnsteiner, F., B. Berger, T. Weismann & R. Patzner. 1995. Fine structure and motility of spermatozoa and composition of the seminal plasma in the perch. J. Fish Biol. 47: 492–508.

Morisawa, M. 1985. Initiation mechanism of sperm motility at spawning in teleosts. Zool. Sci. 2: 605–615.

Morisawa, M. & K. Suzuki. 1980. Osmolality and potassium ions: their roles in initiation of sperm motility in teleosts. Science 210: 1145–1146.

Morisawa, M., K. Suzuki & S. Morisawa. 1983. Effects of potassium and osmolality on spermatozoan motility of salmonid fishes. J Exp. Biol. 107: 105–113.

Munehara, H., K. Takano & Y. Koya. 1989. Internal gametic association and external fertilization in the elkhorn sculpin, *Alcichthys alcicornis.* Copeia 1989: 673–678.

Munehara, H., K. Takano & Y. Koya. 1991. The little dragon sculpin *Blepsias cirrhosus,* another case of internal gametic association and external fertilization. Japan. J. Ichthyol. 37: 391–394.

Munehara, H., Y. Koya & K. Takano. 1994. Conditions for initiation of fertilization of eggs in the copulating elkhorn sculpin. J. Fish Biol. 45: 1105–1111.

Stoss, J. 1983. Fish gamete preservation and spermatozoan physiology. pp. 305–350. *In:* W.S. Hoar, J.D. Randall & E.M. Donaldson (ed.) Fish Physiology, Vol. 9, Academic Press, New York.

Yamamoto, T. 1961. The physiology of fertilization in fish eggs. Int. Rev. Cytol. 12: 361–405.

Environmental Biology of Fishes **52**: 187–201, 1998.

Comparison of laboratory-reared eggs, embryos and larvae of five labrid fishes

Seishi Kimura, Yoshinori Nakayama[1] & Takanari Kiriyama[2]
Fisheries Research Laboratory, Mie University, P.O. Box 11, Wagu, Shima, Mie, 517–07, Japan
(e-mail: kimura-s@bio.mie-u.ac.jp)
[1] *Present address: 6–39 Kamoike-cho, Tokoname, Aichi, 479, Japan*
[2] *Present address: Aquaculture Research Laboratory, Nagasaki Prefectural Institute of Fisheries, Nomozaki, Nagasaki, 851–05, Japan*

Received 11.9.1995 Accepted 3.3.1997

Key words: embryonic development, morphology of larvae, identification of larvae

Synopsis

Eggs, embryos and larvae of five labrid fishes, *Thalassoma cupido, Pteragogus flagellifer, Pseudolabrus japonicus, Halichoeres tenuispinnis,* and *H. poecilopterus*, reared in the laboratory are described and compared. The eggs were buoyant and spherical, with a single, spherical oil globule. *P. japonicus* eggs were unique in lacking melanophores on the oil globule. Eggs of the remaining species closely resembled each other, except in diameter. Incubation periods were short, ranging from ca. 19 h in *H. poecilopterus* to ca. 31 h in *P. japonicus*. The newly-hatched embryos also resembled each other, having a short tail and large oval or pear-shaped yolksac, the anterior tip of which extended beyond the snout. The single oil globule was located at the anterior tip of the yolk. As the yolksac diminished with growth, its anterior tip moved posteriorly. The yolk and oil globule were completely absorbed 3 or 4 days after hatching. In all free embryos and larvae except for *Pteragogus flagellifer*, needle-like projections appeared on both the dorsal and anal finfold margins 12 h to 1 day after hatching. Although morphology of free embryos and larvae of all five species was very similar, differences in pigmentation, location of the anus, and the needle-like projections were apparent. Artificial keys to the newly-hatched embryos and larvae are given.

Introduction

Labrid fishes are commonly distributed in southern Japan, being caught by both commercial and recreational fishing. More than 120 species belonging to the family Labridae have been recorded from Japanese waters (Shimada 1993), 13 of these from Ago Bay and adjacent waters, Mie Prefecture, central Japan (Kimura & Suzuki 1980, 1982), where the present study was carried out. Of the latter, *Thalassoma cupido, Pteragogus flagellifer, Pseudolabrus japonicus, Halichoeres tenuispinnis,* and *H. poecilopterus* are common. Although early developmental stages of Japanese labrids, including the above five species, have already been reported by Kamiya (1925), Kubo (1939), Mito (1962), Masuda & Tanaka (1962), Kojima (1988), Mito & Ikeda (1988) and Kimura & Kiriyama (1993), some accounts were brief or gave only fragmented descriptions of eggs, embryos, and larvae. Because spawning periods of the above five labrids overlap (Kamiya 1925, Nakazono & Tsukahara 1974, Meyer 1977, Nakazono

Table 1. Collection intervals and rearing tanks for the parental fishes.

Species	Collection intervals	No. of individuals	Rearing tanks
Thalassoma cupido	26 May–19 Aug 1988	51	1 m^3 black polyethylene tank
Pteragogus flagellifer	5 May–20 Jul 1988	33	1.5 m^3 concrete tank
Pseudolabrus japonicus	9 Aug–19 Oct 1987	242	10 m^3 concrete tank
	23 Oct–11 Nov 1987*	96	10 m^3 concrete tank
Halichoeres tenuispinnis	5 May–19 Aug 1988	107	10 m^3 concrete tank
H. poecilopterus	30 Apr–31 May 1987	35	10 m^3 concrete tank

* 9 IU of gonadotropin (HCG) per gram of body weight was injected in the epaxial musculature.

1979, Kimura & Kiriyama unpublished data), their eggs, embryos and larvae appear sympatrically and synchronous, and specific identification of their early life stages has been difficult.

The aims of this paper are to describe the morphology of laboratory-reared eggs, embryos, and larvae of the above labrids and to clarify their diagnostic characters.

Material and methods

All parental fishes were collected by angling in Ago Bay (34°16–18′N, 136°45–49′E) and reared at the Fisheries Research Laboratory, Mie University (FRLM), situated on Zaga Island in the bay. Collection and rearing data are shown in Table 1. Each rearing tank was equipped with a net to collect spawned eggs from the overflow. The net was examined for eggs every 30 minutes from 6:00 to 20:00 h and emptied.

After spawning, the eggs collected were transferred to a gauze net cage (45 × 45 × 45 cm) suspended in a rearing tank (0.5 m^3 capacity black polyethylene tank) containing weakly aerated sea water. Just after hatching, the embryos (sensu Balon 1980, 1990) were released from the net cage and reared in still water for 2 or 3 days. Subsequently, sea water was supplied at rates of 30–100 ml min^{-1}. Cultured marine chlorella, *Nannochloropsis oculata* (ca. 2×10^5 cells ml^{-1}), was added to the rearing tank just after hatching. Two or three days after hatching, rotifers, *Brachionus plicatilis* (ca. 5–30 individuals ml^{-1}) , and trochophore larvae of the oyster, *Crassostrea gigas* (ca. 60 individuals ml^{-1}), were introduced in tanks containing embryos as potential food items. The incubating and rearing temperatures were 21.9–25.0 °C for *Thalassoma cupido*, 20.9–23.0 °C for *Pteragogus flagellifer*, 14.1–24.7 °C for *Pseudolabrus japonicus*, 21.7–23.0 °C for *Halichoeres tenuispinnis,* and 19.8–26.8 °C for *H. poecilopterus*.

Free embryos and larvae were sampled periodically from the rearing tank, being anesthetized with 10–100 ppm ethylene glycol monophenyl ether prior to morphological observations and measurements. Thereafter, they were preserved in 5 % buffered formalin (diluted with sea water) and deposited in FRLM.

Table 2. Spawning intervals and times of the five labrids.

Species	Spawning intervals	Spawning time (h)
Thalassoma cupido	17 Jun–29 Aug 1988	8:00–12:00
Pteragogus flagellifer	9 Jun–24 Sep 1988	15:00–18:00
Pseudolabrus japonicus	8 Oct 1987–13 Jan 1988	8:00–12:00
Halichoeres tenuispinnis	1 Jun–23 Aug 1988	7:00–11.00 & 14:00–17:00
H. poecilopterus	2 Jun–15 Sep 1987	7:00–11:00

Results and discussion

Spawning

Spawning periods and times for the five labrids are shown in Table 2. *Thalassoma cupido, Pteragogus flagellifer, Halichoeres tenuispinnis,* and *H. poecilopterus* are type 1 (warm season) spawners and *Pseudolabrus japonicus*, a type 2 (cold season) spawner. Other type 1 Japanese labrids include *Stethojulis interrupta terina*, which is found in Ago Bay, and *Cirrhilabrus temminckii* (Nakazono 1979, Suzuki et al. 1981, Bell 1983). *Labroides dimidiatus* is a type 2 spawner (Suzuki et al. 1981).

Spawning times could be divided into three categories, category 1 spawning in the morning (*T. cupido, P. japonicus* and *H. poecilopterus*), category 2 afternoon (*Pteragogus flagellifer*) and category 3 twice a day (*H. tenuispinnis*). *Cirrhilabrus temminckii* belongs to category 2 (Bell 1983).

Eggs

Morphology and incubation data of the eggs are shown in Figure 1 and Tables 3 and 4. Eggs of the five species were buoyant and spherical, with a smooth, colorless, transparent envelope, narrow perivitelline space, colorless yolk, and a single oil globule. Eggs of *Thalassoma cupido* were smallest, with those of *Halichoeres tenuispinnis* and *H. poecilopterus* being medium-sized, and those of *Pteragogus flagellifer* and *Pseudolabrus japonicus* large. The oil globule of *H. tenuispinnis* was colorless, whereas that of the remaining species was pale yellow. Eggs of *P. japonicus* were unique in lacking melanophores on the oil globule. Eggs of the other species closely resembled each other, except in diameter. Among other labrid fishes, eggs of *Labroides dimidiatus* are characteristic in having 50–70 small oil globules before closure of the germ ring (Suzuki et al. 1981). The oil globule of embryos of *Cirrhilabrus temminckii* changes from spherical to oval just before hatching (Bell 1983).

Incubation periods of the species studied were all short, those of *H. poecilopterus* (ca. 19 h) and *P. ja-*

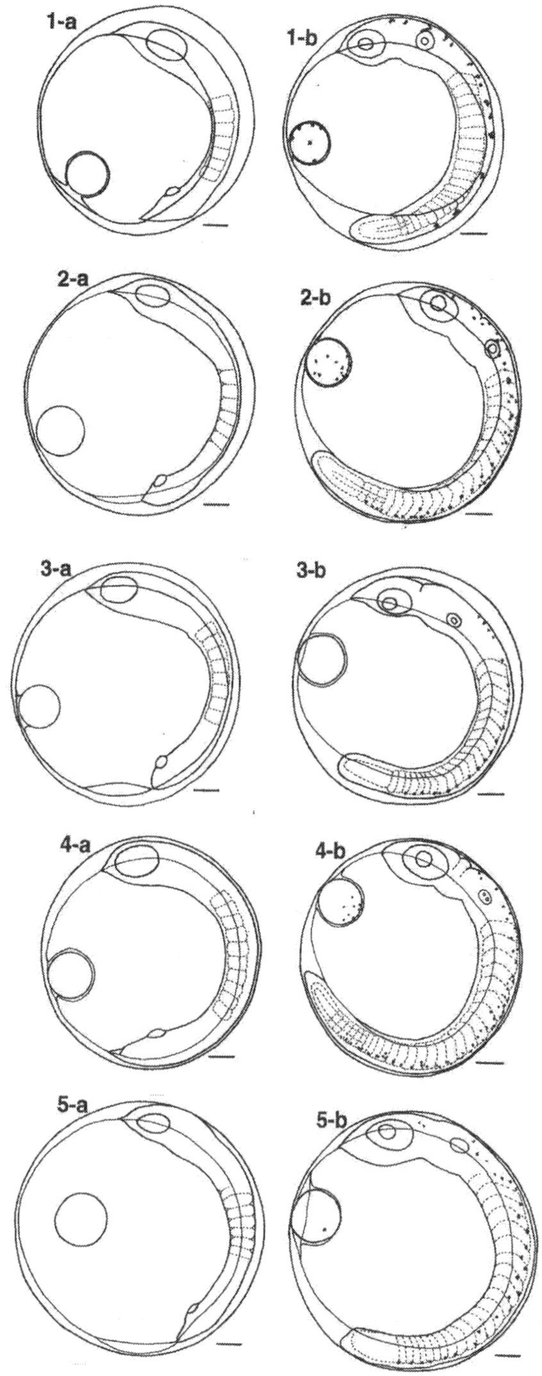

Figure 1. Embryos of five labrid fishes: 1 – *Thalassoma cupido*, 17 h 16 min (a) and 23 h 48 min (b) after fertilization; 2 – *Pteragogus flagellifer*, 18 h 43 min (a) and 26 h 20 min (b); 3 – *Pseudolabrus japonicus*, 15 h 19 min (a) and 26 h 2 min (b); 4 – *Halichoeres tenuispinnis*, 14 h 20 min (a) and 24 h 30 min (b); 5 – *H. poecilopterus*, 10 h 36 min (a) and 16 h 14 min (b) (after Kimura & Kiriyama 1993). Scales indicate 0.1 mm.

Table 3. Egg measurements for the five labrids. Figures in parentheses are mean values.

Species	Egg diameter (mm)	Diameter of oil globule (mm)
Thalassoma cupido	0.54–0.65 (0.58)	0.11–0.12 (0.11)
Pteragogus flagellifer	0.75–0.78 (0.76)	0.12–0.15 (0.14)
Pseudolabrus japonicus	0.76–0.83 (0.79)	0.12–0.16 (0.15)
Halichoeres tenuispinnis	0.64–0.67 (0.66)	0.13–0.15 (0.14)
H. poecilopterus	0.67–0.72 (0.68)	0.13–0.16 (0.14)

ponicus (ca. 31 h in almost the same water temperature) being the shortest and longest, respectively.

Description of Thalassoma cupido *embryos and larvae*

Morphology. — External morphology and measurements are shown in Figure 2 and Table 5, respectively. The newly-hatched embryos had a short tail, a large oval yolksac 60% of the total length (TL) including a spherical oil globule at its tip, and 25 to 29 myomeres. The anterior tip of the yolksac extended beyond the snout, its posterior end being well forward of the anus. With growth, the body became slender, the tail elongated, and the yolksac diminished, its anterior tip moving posteriorly. The yolk and oil globule were completely absorbed 4 days after hatching. Needle-like projections appeared on both the dorsal and anal finfold margins 12 h after hatching. Subsequently, the projections increased in both number and size up to 4 days after hatching, thereafter decreasing gradually. The anlage of the pectoral fin appeared 2 days after hatching.

Pigmentation. — Newly-hatched embryos had 7–12 branched melanophores on the dorsal contour. The melanophores subsequently moved towards the ventral surface, decreasing in number. Whereas 1 day-old embryos had 1–4 melanophores on the head, 4–7 on the trunk and 1–4 on the dorsal aspect of the tail, only 1–3 and 1–2 melanophores remained

Table 4. Incubation times (h:min) and developmental stage after insemination for the five labrids. Figures in parentheses indicate ranges of water temperature during embryonic development.

Embryonic stages	Species				
	Thalassoma cupido (23.8–24.9 °C)	*Pteragogus flagellifer* (20.9–22.1 °C)	*Pseudolabrus japonicus* (22.3–24.7 °C)	*Halichoeres tenuispinnis* (21.7–22.2 °C)	*Halichoeres poecilopterus* (21.9–24.7 °C)
2-cell	0:40	0:57	0:51	0:40	0:50
4-cell	1:07	1:11	1:18	1:05	1:03
8-cell	1:52	1:31	1:44	1:31	1:16
16-cell	2:12	1:48	2:09	1:36	1:36
Blastula	4:41	3:36	4:57	3:40	3:58
Gastrula	8:20	6:38	7:55	6:35	7:49
Formation of embryo	11:12	9:46	9:58	9:44	8:48
Formation of myomeres	12:11	13:15	14:06	12:20	9:29
Formation of optic vesicles	15:03	16:35	14:39	12:32	10:36
Closure of blastopore	15:52	18:16	17:52	13:13	11:29
Appearance of Kupffer's vesicles	17:03	18:43	15:19	14:20	10:36
Formation of otic capsules	20:05	22:00	21:11	17:05	16:14
Appearance of melanophores on the embryo	17:16	22:48	22:50	19:30	13:39
Formation of optic lens	22:42	24:50	21:52	20:45	15:29
Disappearance of Kupffer's vesicle	21:35	26:11	22:27	22:10	16:09
Beginning of heart pulse	23:48	26:20	26:02	23:13	16:39
Beginning of hatching	26:12	30:18	31:01	26:05	18:48
End of hatching	27:23	32:00	33:27	27:35	20:25

on the head and dorsal aspect of the tail, respectively, in 2 day-old embryos. The dorsal melanophores on the tail decreased to 1, moving to the dorsal finfold in 3 day-old embryos. Melanophores on the head completely disappeared 3 days after hatching. One to three branched melanophores, which appeared on the ventral contour of the tail 12 h after hatching, decreased to 1 in 1 day-old embryos, and subsequently disappeared in larvae 4 days after hatching. Five to ten branched melanophores scattered over the entire surface of the oil globule just after hatching, subsequently aggregated on the anterior hemisphere in 1 day-old embryos, and extended onto the yolk in 2 day-old embryos. These melanophores disappeared with yolk and oil globule consumption. One or two melanophores were present on the central gut in 1 day-old embryos, increasing in number to 2–3 in 2 day-old embryos, before further developing and becoming interconnected on the dorsal surface of the gut in 4 day-old larvae. Melanophores were deposited on the eyes 2

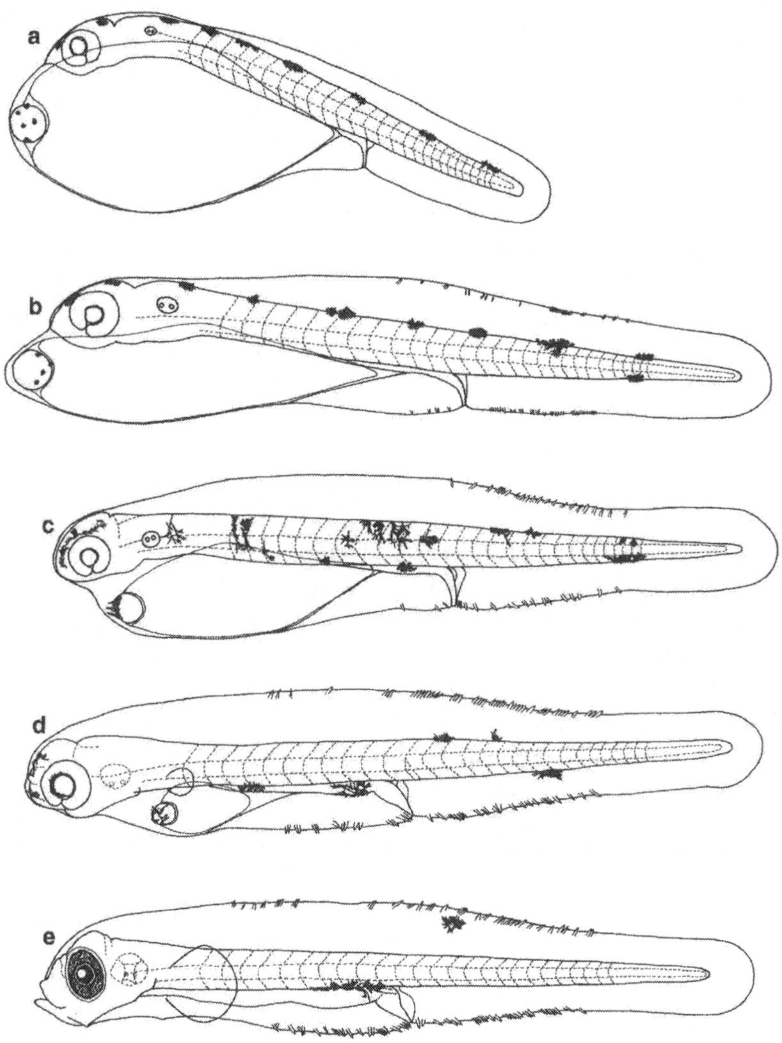

Figure 2. Free embryos and larvae of *Thalassoma cupido:* a – just after hatching at 1.28 mm notochord length (NL); b – 12 h, 1.69 mm; c – 1 day, 1.89 mm; d – 2 days, 2.10 mm; and e – 5 days, 2.15 mm.

Table 5. Measurements of *Thalassoma cupido* embryos and larvae. Figures in parentheses indicate mean values and sample sizes.

Time elapsed after hatching	Developmental interval	Total length (mm)	Notochord length (mm)	Yolksac length (mm)
Just after hatching	Free embryo	1.45–1.52 (1.49, 6)	1.27–1.32 (1.30, 6)	0.86–0.91 (0.89, 6)
1 day	Free embryo	2.01–2.10 (2.04, 6)	1.89–1.98 (1.93, 6)	0.58–0.67 (0.76, 6)
2 days	Free embryo	2.17–2.23 (2.20, 6)	2.04–2.12 (2.08, 6)	0.22–0.41 (0.30, 6)
3 days	Free embryo	2.05–2.30 (2.19, 5)	1.98–2.15 (2.06, 5)	0.09–0.19 (0.13, 5)
4 days	Larva	2.20–2.27 (2.23, 5)	2.07–2.15 (2.09, 5)	
6 days	Larva	2.28–2.32 (2.30, 5)	2.12–2.19 (2.15, 5)	

days after hatching, with iridophores appearing on the eyes the following day.

Remarks. — Kubo (1939) described the artificially-inseminated eggs and embryos of this species, but his illustration of the embryos 16 h after hatching lacked needle-like projections on the finfold margins (possibly disregarded by Kubo, since they were present in all specimens examined here). The embryos described by Mito (1962) as Labrina No. 7 were determined to be those of *T. cupido*, owing to their close resemblance to the latter in initial egg diameter, diameter and color of the oil globule, seasonal occurrence, embryo pigmentation, and myomere count.

Description of Pteragogus flagellifer *embryos and larvae*

Morphology. — External morphology and measurements are shown in Figure 3 and Table 6, respectively. The general appearance of embryos was similar to that of *Thalassoma cupido*. The newly-hatched embryos had a large pear-shaped yolksac, 64% of TL, and 24 to 27 myomeres. The anterior tip of the yolksac extended well forward, its posterior end being situated just before the anus. The yolk and oil globules were completely absorbed 3 days after hatching. No needle-like projections were present on the finfold. The anlage of the pectoral fin appeared 1 day after hatching.

Pigmentation. — The newly-hatched embryos had two rows of branched melanophores on the dorsal contour, each row including 45–55 pigment cells, which subsequently moved towards the ventral surface, and decreased in number. Whereas a few melanophores were distributed on the head, margin of the eye, trunk, tail, and dorsal surface of the gut in 2 day-old embryos, only those on the gut remained in 4 day-old larvae. One to three branched melanophores appeared on the ventral contour of the tail 12 h after hatching, subsequently developing and becoming interconnected in 4 day-old larvae. An obvious melanophore appeared on the snout tip of embryos 2 days after hatching, but completely disappeared 4 days after hatching. Seven to ten branched melanophores scattered over the entire surface of the oil globule just after hatching subsequently aggregated on the anterior hemisphere 1 day after hatching and extended onto the yolk 2 days after hatching. These melanophores disappeared with yolk and oil globule consumption. One or two melanophores were present on the central gut in 1 day-old embryos, increasing to 2–3 in 2 day-old embryos, before further developing and becoming interconnected on the dorsal surface of the gut in 4 day-old larvae. Some newly-hatched embryos had a branched melanophore on the posterior end of the yolksac, but it completely disappeared several hours after hatching. A melanophore appearing on the corner of the gut 12 h after hatching subsequently increased in size, and became connected with melanophores on the gut dorsal surface. Melanophores were deposited on the eyes 2 days after hatching, with iridophores appearing on the eyes the following day.

Remarks. — The embryo described by Kamiya (1925) as *Labrus* sp. No. 4 was similar to the above in egg and oil globule diameter, seasonal occurrence, and embryo pigmentation. It should be considered as *P. flagellifer*, as suggested by Mito (1962).

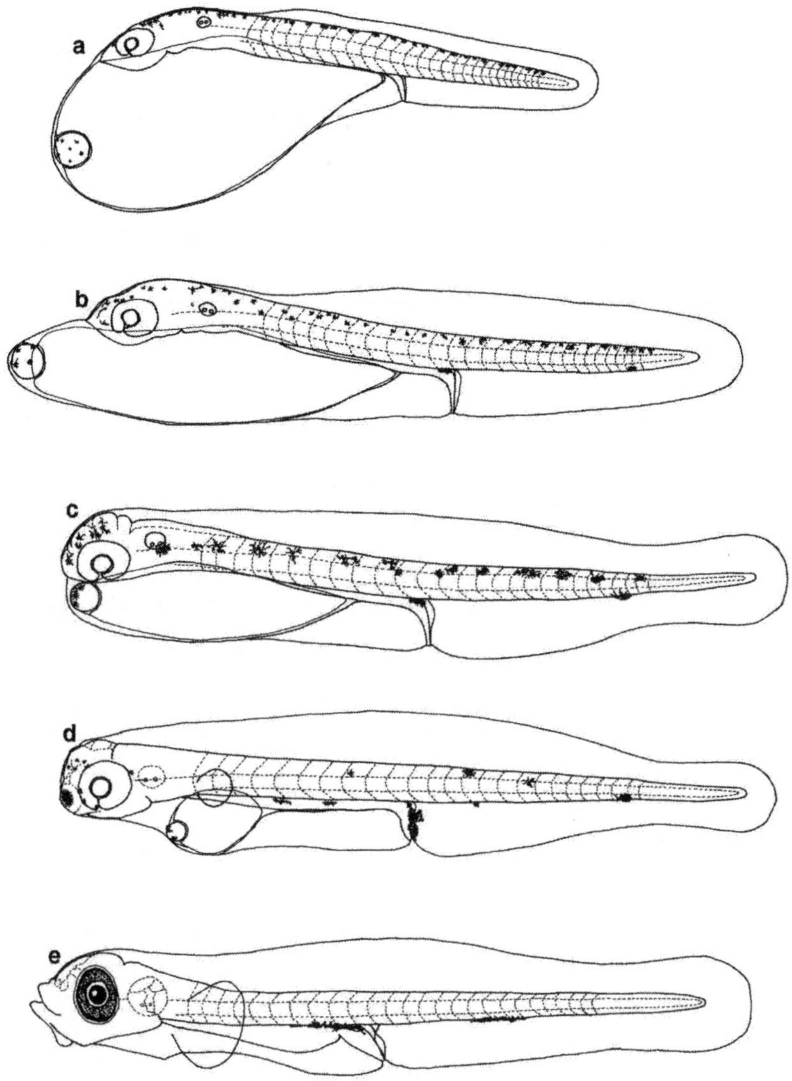

Figure 3. Free embryos and larvae of *Pteragogus flagellifer:* a – just after hatching at 1.98 mm NL; b – 12 h, 2.12 mm; c – 1 day, 2.48 mm; d – 2 days, 2.61 mm; and e – 4 days, 2.65 mm.

Table 6. Measurements of *Pteragogus flagellifer* embryos and larvae. Figures in parentheses indicate mean values and sample sizes.

Time elapsed after hatching	Developmental interval	Total length (mm)	Notochord length (mm)	Yolksac length (mm)
Just after hatching	Free embryo	1.66–1.93 (1.74, 6)	1.62–1.87 (1.67, 6)	1.02–1.26 (1.11, 6)
1 day	Free embryo	2.60–2.71 (2.66, 6)	2.48–2.63 (2.54, 6)	0.72–0.87 (0.84, 6)
2 days	Free embryo	2.54–2.82 (2.69, 6)	2.41–2.66 (2.55, 6)	0.16–0.21 (0.18, 6)
3 days	Larva	2.77–2.83 (2.80, 6)	2.60–2.68 (2.65, 6)	
4 days	Larva	2.77–2.83 (2.80, 4)	2.62–2.65 (2.64, 4)	

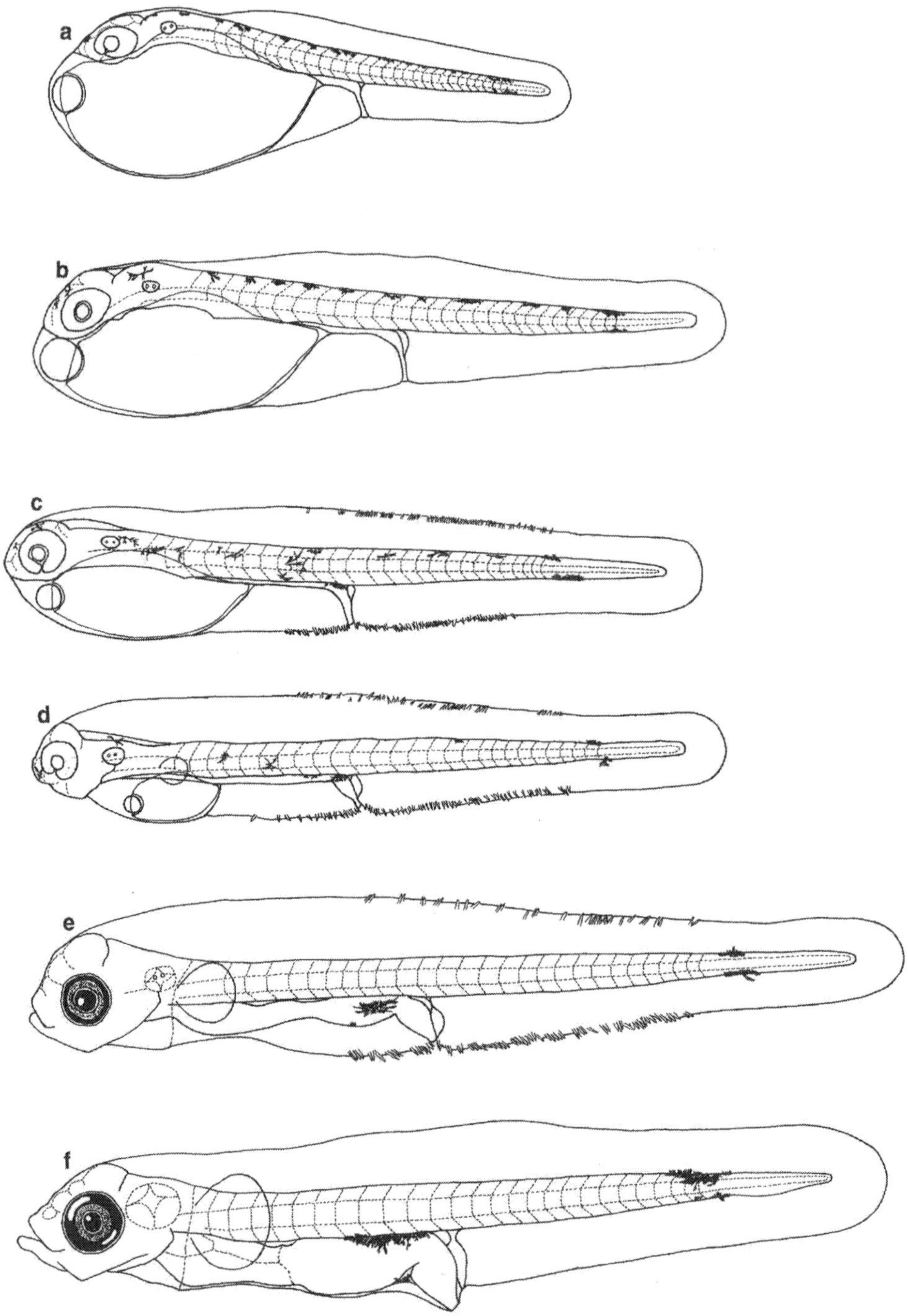

Figure 4. Free embryos and larvae of *Pseudolabrus japonicus:* a – just after hatching at 1.70 mm NL; b – 8 h, 1.91 mm; c – 1 day, 2.48 mm; d 2 days, 2.61 mm; e – 4 days, 2.80 mm; and f – 22 days, 4.02 mm.

Table 7. Measurements of *Pseudolabrus japonicus* embryos and larvae. Figures in parentheses indicate mean values and sample sizes.

Time elapsed after hatching	Developmental interval	Total length (mm)	Notochord length (mm)	Yolksac length (mm)
Just after hatching	Free embryos	1.76–1.90 (1.82, 6)	1.70–1.76 (1.74, 6)	0.96–1.10 (1.05, 6)
1 day	Free embryos	2.42–2.66 (2.53, 6)	2.34–2.48 (2.42, 6)	0.88–1.10 (0.95, 6)
2 days	Free embryos	2.52–2.76 (2.66, 6)	2.36–2.64 (2.51, 6)	0.44–0.74 (0.62, 6)
3 days	Free embryos	2.80–3.00 (2.86, 6)	2.58–2.82 (2.68, 6)	0.22–0.40 (0.34, 6)
4 days	Larva	2.80–2.98 (2.91, 6)	2.64–2.80 (2.74, 6)	
7 days	Larva	2.80–3.22 (3.04, 5)	2.58–3.02 (2.83, 5)	
10 days	Larva	3.00–3.12 (3.07, 5)	2.80–2.92 (2.87, 5)	
15 days	Larva	3.12–3.54 (3.33, 4)	2.96–3.28 (3.10, 4)	
20 days	Larva	3.92–4.32 (4.18, 3)	3.74–4.14 (3.97, 3)	
22 days	Larva	4.22–4.60 (4.39, 3)	4.02–4.32 (4.16, 3)	

Description of Pseudolabrus japonicus *embryos and larvae*

Morphology. — External morphology and measurements are shown in Figure 4 and Table 7, respectively. The general appearance of the embryos and larvae was similar to that of *Thalassoma cupido*. The newly-hatched embryos had a large oval yolksac, 57% of TL, and 25 to 27 myomeres. The anterior tip of the yolksac extended beyond the snout, its posterior end being well forward of the anus. The yolk and oil globule were completely absorbed 4 days after hatching. Needle-like projections appeared on both the dorsal and anal finfold margins 12 h after hatching. Subsequently, these projections increased in both number and size up to 18 days after hatching, thereafter decreasing gradually and disappearing completely in 22 day-old larvae. The anlage of the pectoral fin appeared 2 days after hatching.

Pigmentation. — The newly-hatched embryos had 2 rows of branched melanophores on the dorsal contour, each row including 10–16 pigment cells, which subsequently moved towards the ventral surface and decreased in number. One day-old embryos had 1–3 melanophores on the head, 4–6 on the trunk and 3–5 on the dorsal aspect of the tail. Of these, only a single melanophore remained dorsally on the tail in 3 day-old embryos. One to three branched melanophores on the ventral surface of the tail in newly-hatched embryos subsequently decreased in number to a single melanophore in 2 day-old embryos. A single, large, branched melanophore and 1–2 small ones appeared on the gut 1 day after hatching, with the latter disappearing in 2-day old embryos. A branched melanophore appeared on the corner of the gut in 4 day-old larvae. Melanophores were deposited on the eyes 3 days after hatching, with iridophores appearing on the eyes the following day.

Description of Halichoeres tenuispinnis *embryos and larvae*

Morphology. — External morphology and measurements are shown in Figure 5 and Table 8, respectively. The general appearance of the embryos and larvae was similar to that of *Pseudolabrus japonicus*. The newly-hatched embryos had a large, oval yolksac, 56% of TL, and 25 to 27 myomeres. The anterior tip of the yolksac extended beyond the snout, its posterior end being well forward of the anus. The yolk and oil globule were completely absorbed 3 days after hatching. Needle-like projections appeared on both the dorsal and anal finfold margins 12 h after hatching, subsequently increasing in both number and size up to 1 day after hatching, thereafter gradually disappearing. The anlage of the pectoral fin appeared 1 day after hatching.

Pigmentation. — The newly-hatched embryos had 2 rows of branched melanophores on the dorsal contour, each row including 19–25 pigment cells, which subsequently moved towards the ventral surface, and decreased in number. In contrast, 1 day-old embryos had 5–7 melanophores on the sides of the

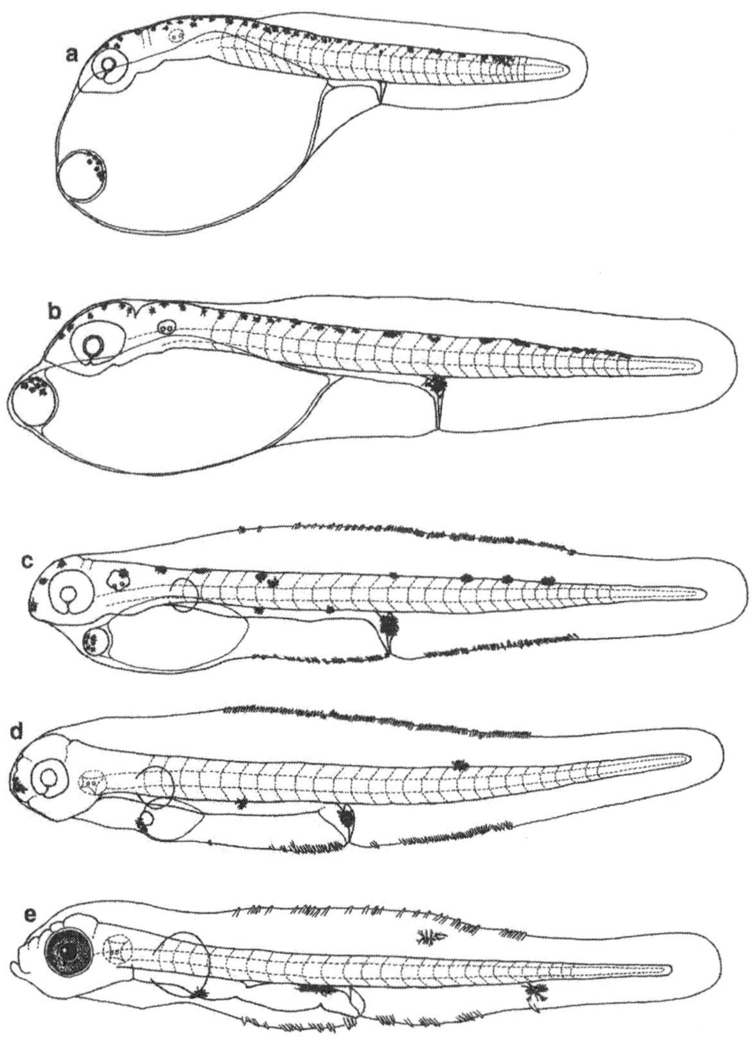

Figure 5. Free embryos and larvae of *Halichoeres tenuispinnis:* a – just after hatching at 1.54 mm NL; b – 8 h, 1.90 mm; c – 1 day, 2.29 mm; d – 2 days, 2.43 mm; and e – 4 days, 2.49 mm.

Table 8. Measurements of *Halichoeres tenuispinnis* embryos and larvae. Figures in parentheses indicate mean values and sample sizes.

Time elapsed after hatching	Developmental interval	Total length (mm)	Notochord length (mm)	Yolksac length (mm)
Just after hatching	Free embryo	1.60–1.67 (1.63, 6)	1.51–1.57 (1.54, 6)	0.89–0.94 (0.92, 6)
1 day	Free embryo	2.38–2.54 (2.46, 6)	2.28–2.40 (2.35, 6)	0.62–0.68 (0.64, 6)
2 days	Free embryo	2.53–2.60 (2.57, 6)	2.35–2.44 (2.40, 6)	0.26–0.28 (0.27, 6)
3 days	Larva	2.59–2.69 (2.63, 5)	2.44–2.55 (2.48, 5)	
4 days	Larva	2.50–2.70 (2.61, 5)	2.38–2.54 (2.47, 5)	
5 days	Larva	2.65–2.72 (2.68, 3)	2.52–2.60 (2.55, 3)	

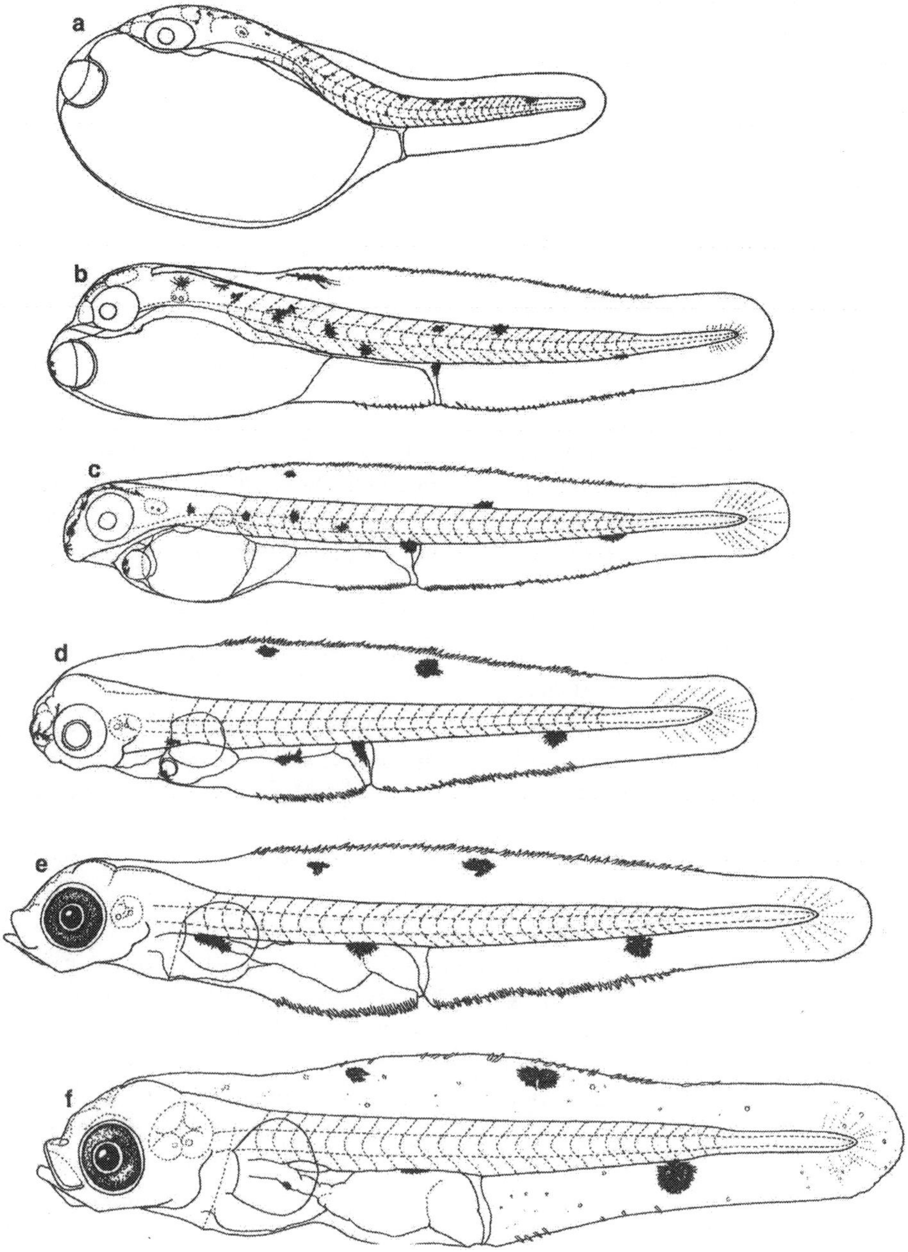

Figure 6. Free embryos and larvae of *Halichoeres poecilopterus:* a – just after hatching at 1.43 mm NL; b – 8 h, 1.80 mm; c – 1 day, 2.30 mm; d – 2 days, 2.43 mm; e – 4 days, 2.46 mm; and f – 14 days, 3.61 mm (a, c, d and f, after Kimura & Kiriyama 1993).

head and trunk, and 2–3 on the dorsal aspect of the tail. Some trunk melanophores subsequently moved to the gut and the overall total number decreased until only a single melanophore remained dorsally on the tail in 2 day-old embryos. This melanophore moved to the dorsal finfold the following day. Branched melanophores appeared on the snout tip 1 day after hatching, developing further in 2 day-old embryos, but disappeared in 4 day-old larvae. Five to seven branched melanophores were scattered on the posterior surface of the oil globule just after hatching, and subsequently moved to the

Table 9. Measurements of *Halichoeres poeclilopterus* embryos and larvae. Figures in parentheses indicate mean values and sample sizes.

Time elapsed after hatching	Developmental interval	Total length (mm)	Notochord length (mm)
Just after hatching	Free embryo	1.46–1.70 (1.61, 6)	1.30–1.50 (1.45, 6)
1 day	Free embryo	2.34–2.71 (2.48, 6)	2.22–2.54 (2.32, 6)
2 days	Free embryo	2.46–2.74 (2.58, 6)	2.34–2.58 (2.45, 6)
3 days	Larva	2.52–2.72 (2.61, 6)	2.38–2.58 (2.45, 6)
4 days	Larva	2.50–2.68 (2.57, 6)	2.34–2.48 (2.40, 6)
8 days	Larva	2.92–3.00 (2.95, 4)	2.64–2.82 (2.76, 4)
12 days	Larva	2.68–3.58 (3.12, 4)	2.50–3.38 (2.94, 4)
14 days	Larva	3.40–3.82 (3.56, 4)	3.22–3.68 (3.45, 4)
21 days	Larva	5.18 (1)	4.48 (1)

anterior hemisphere before extending onto the yolk 2 days after hatching. These melanophores disappeared with yolk and oil globule consumption. A single branched melanophore appeared on the corner of the gut 6 h after hatching. Two or three melanophores were present on the central gut 1 day after hatching, and decreased to a single cell in 3 day-old larvae. The latter had a single well-developed melanophore on both the dorsal and anal finfold. Melanophores were deposited on the eyes 2 days after hatching, with iridophores appearing on the eyes the following day.

Description of Halichoeres poecilopterus *embryos and larvae*

Because development of this species has already been described in detail by Kimura & Kiriyama (1993), it is described only briefly here. External morphology and measurements are shown in Figure 6 and Table 9, respectively. In general appearance, the embryos and larvae closely resembled those of *H. tenuispinnis*. Myomeres numbered 25 to 27. The yolk and oil globule were completely absorbed 3 or 4 days after hatching. Needle-like projections on the finfold margins were present in 2 day-old embryos.

Pigmentation. — The newly-hatched embryos had 20 or more punctate melanophores on the dorsal contour. A branched melanophore appeared on the anterior part of the dorsal finfold 8 h after hatching, with a similar pigment cell appearing on the posterior part of the finfold 2 days after hatching. A branched melanophore present on the ventral contour of the tail in 1 day-old embryos subsequently moved onto the anal finfold.

Comparison of embryo and larva morphology

Although external morphologies of the embryos and larvae of the five labrid species were very similar, a few differences were obvious, as shown in Tables 10–12. In newly-hatched embryos, differences in pigmentation on the dorsum and location of the anus were significant. In free embryos also a few differences in pigmentation were recognized. However, the pattern of melanophore distribution varied and overlapped among species making identification of free embryos very difficult. In larvae, some differences were apparent in pigmentation and the needle-like projections on the finfold margins.

The following artificial keys to newly-hatched embryos and preflexion larvae include some related species (viz. *Labroides dimidiatus*, see Suzuki et al. 1981, *Stethojulis interrupta terina*, see Mito 1962, and *Cirrhilabrus temminckii*, see Suzuki et al. 1981)

Key to newly-hatched embryos

1a. Melanophores present on yolksac 2

2a. Melanophores present on anterior tips of yolksac and oil globule; oil globule ellipsoid . . . . *Cirrhilabrus temminckii*

2b. Melanophores present on posterior tip of

Table 10. Comparison of newly-hatched embryos of Japanese labrids.

Species	Melanophores on dorsum	Melanophores on oil globule	Other melanophores	Position of anus	Source
Thalassoma cupido	Single row, 7–12 cells	Whole surface	Absent	Well posterior to yolksac	Present study
Pteragogus flagellifer	2 rows, 45–55 cells each	Whole surface	Absent or posterior end of yolksac	Just behind posterior end of yolksac	Present study
Pseudolabrus japonicus	2 rows, 10–16 cells each	Absent	Ventral aspect of tail (1–3)	Well posterior to yolksac	Present study
Halichoeres tenuispinnis	2 rows, 19–25 cells each	Posterior hemisphere	Absent	Well posterior to yolksac	Present study
H. poecilopterus	20 or more scattered cells	Whole surface	Absent	Well posterior to yolksac	Present study
Stethojulis interrupta terina	Single row, 8–9 cells	Present*	Absent	Just behind posterior end of yolksac	Mito (1962)
Cirrhilabrus temminckii	Present*	Anterior hemisphere	Anterior tip of yolksac	Well posterior to yolksac	Suzuki et al. (1981)
Labroides dimidiatus	14 cells*	Absent	Absent	Well posterior to yolksac	Suzuki et al. (1981)

* Distribution pattern unknown.

Table 12. Melanophore distribution patterns in larvae of Japanese labrids.

Species	Finfolds	Gut	Body	Source
Thalassoma cupido	1 cell posteriorly on dorsal finfold	1 elongated cell along dorsal surface	Absent	Present study
Pteragogus flagellifer	Absent	1 elongated cell along dorsal surface	1 cell on ventral contour of tail	Present study
Pseudolabrus japonicus	Absent	Single cells on dorsal and ventral surfaces of corner	Single cells on dorsal and ventral contours of tail	Present study
Halichoeres tenuispinnis	Single cells posteriorly on dorsal and anal finfolds	Single cells on anterior part and corner	Absent	Present study
H. poecilopterus	Single cells anteriorly and posteriorly on dorsal finfold, and 1 cell on anal finfold	Single cells on anterior part and corner	Absent	Present study
Stethojulis interrupta terina	1–4 cells on ventral finfold margin just before anus, and 0–1 cell on central dorsal finfold	Single cells on dorsal and ventral surfaces of corner	Absent	Mito (1962)
Cirrhilabrus temminckii	Absent	Absent	Single cells on dorsal surface of trunk and tail	Suzuki et al. (1981)

Table 11. Melanophore distribution patterns in free embryos of Japanese labrids.

Species	No. of melanophores on gut	Ventral contour of tail	Tip of snout	Oil globule	Yolksac	Source
Thalassoma cupido	1–2	Present	Absent	Present	Absent or present	Present study
Pteragogus flagellifer	1–2	Present	Present	Present	Absent or present	Present study
Pseudolabrus japonicus	1–3	Present	Absent	Absent	Absent	Present study
Halichoeres tenuispinnis	1–4	Absent	Present	Present	Absent or present	Present study
H. poecilopterus	1–3	Present	Absent	Present	Absent or present	Present study
Tethojulis interrupta terina	3–6	Absent	Present	Present	Present	Mito (1962)
Cirrhilabrus temminckii	1	Absent	Present	Present	Present	Suzuki et al. (1981)
Labroides dimidiatus	0	Present	Absent	Absent	Absent	Suzuki et al. (1981)

yolksac and over entire surface of oil globule; oil globule spherical . *Pteragogus flagellifer*

1b. Melanophores absent on yolksac 3

3a. Melanophores present on oil globule 4

4a. Melanophores arranged dorsally in a single or 2 rows 5

5a. Melanophores arranged dorsally in 2 rows 6

6a. Each dorsal pigment row including 45–55 melanophores; melanophores distributed over entire surface of oil globule ...... *Pteragogus flagellifer*

6b. Each dorsal pigment row including 19–25 melanophores; melanophores on oil globule restricted to posterior hemisphere *Halichoeres tenuispinnis*

5b. Melanophores arranged dorsally in a single row 7

7a. Anus situated well posterior to yolksac; melanophores present over entire surface of the oil globule .... *Thalassoma cupido*

7b. Anus situated just behind posterior end of the yolksac; melanophores on oil globule restricted to anterior hemisphere .... *Stethojulis interrupta terina*

4b. Melanophores not arranged dorsally in row; 20 or more melanophores scattered dorsally on body .... *Halichoeres poecilopterus*

3b. Melanophores absent on oil globule 8

8a. One to three melanophores on ventral contour of tail .. *Pseudolabrus japonicus*

8b. No melanophores on ventral contour of tail .... *Labroides dimidiatus*

Key to the preflexion larvae

1a. Needle-like projections present on finfold margins 2

2a. Melanophore(s) present on finfold(s) ... 3

3a. Melanophore(s) present on anal finfold 4

4a. Single melanophore posteriorly on dorsal finfold ... *Halichoeres tenuispinnis*

4b. Two melanophores on dorsal finfold .. .. *Halichoeres poecilopterus*

3b. Melanophores absent on anal finfold .. 5

5a. No melanophores on ventral finfold; single melanophore present posteriorly on dorsal finfold .. *Thalassoma cupido*

5b. One to four melanophores on ventral finfold margin just before anus; single melanophore present or absent centrally on dorsal finfold .... *Stethojulis interrupta terina*

2b. Melanophores absent on finfolds .... *Pseudolabrus japonicus*

1b. Needle-like projections absent 6

6a. Gut melanophores extending anterio-posteriorly .... *Pteragogus flagellifer*

6b. Gut melanophores absent .... *Cirrhilabrus temminckii*

Acknowledgement

We are sincerely grateful to K. Fukusho and M. Okauchi, National Research Institute of Aquaculture, for advice regarding the culture of food organisms. Thanks are also due to Y. Tsukamoto, National Research Institute of Fisheries Science, for his help during the study and G. S. Hardy, Thames, New Zealand, for valuable suggestions and a critical review of the manuscript. This work was partially supported by a Grant-in-Aid for Scientific Research from the Ministry of Education, Science, and Culture.

References cited

Balon, E.K. 1980. Early ontogeny of the lake charr, *Salvelinus (Cristivomer) namaycush*. pp. 485–562. *In*: E.K. Balon (ed.) Charrs: Salmonid Fishes of the Genus Salvelinus, Perspectives in Vertebrate Science 1, Dr W. Junk Publishers, The Hague.

Balon, E.K. 1990. Epigenesis of an epigeneticist: the development of some alternative concepts on the early ontogeny and evolution of fishes. Guelph Ichthyol. Reviews 1: 1–42.

Bell, L.J. 1983. Aspects of the reproductive biology of the wrasse, *Cirrhilabrus temminckii*, at Miyake-jima, Japan. Japan. J. Ichthyol. 30: 158–167.

Kamiya, T. 1925. The pelagic eggs and larvae of fishes in Tateyama Bay (Pref. Chiba), Report No. 3. J. Imp. Fish. Inst. 21: 27–29, 71–85 (in Japanese).

Kimura, S. & T. Kiriyama. 1993. Development of eggs, larvae and juveniles of the labrid fish, *Halichoeres poecilopterus*, reared in the laboratory. Japan. J. Ichthyol. 39: 371–377.

Kimura, S. & K. Suzuki. 1980. Fish fauna of Ago Bay and its adjacent waters, Mie Prefecture, Japan. Rep. Fish. Res. Lab. Mie Univ. 2: 1–58.

Kimura, S & K. Suzuki. 1982. Fish fauna of Ago Bay and its adjacent waters, Mie prefecture, Japan, Supplement I. Rep. Fish. Res. Lab. Mie Univ. 3: 1–20.

Kojima, J. 1988. Labridae. pp. 575–591. *In*: M. Okiyama (ed.) An Atlas of the Early Stage Fishes in Japan, Tokai University Press, Tokyo (in Japanese).

Kubo, I. 1939. Notes on the development of a teleost, *Thalassoma cupido* (Temminck & Schlegel). Bull. Soc. Sci. Fish. Japan 8: 165–167 (in Japanese).

Masuda, T. & K. Tanaka. 1962. Young of labroid and scaroid fishes from the central Pacific coast of Japan. J. Tokyo Univ. Fish. 48: 1–98.

Meyer, K.A. 1977. Reproductive behavior and patterns of sexuality in the Japanese labrid fish *Thalassoma cupido*. Japan. J. Ichthyol. 24: 101–112.

Mito, S. 1962. Pelagic fish eggs from Japanese waters – IV Labrina. Sci. Bull. Fac. Agr. Kyushu Univ. 19: 493–502 (in Japanese).

Mito, S. & T. Ikeda. 1988. Key to the eggs and hatched larvae. pp. 999–1083. *In*: M. Okiyama (ed.) An Atlas of the Early Stage Fishes in Japan, Tokai University Press, Tokyo (in Japanese).

Nakazono, A. 1979. Studies on the sex reversal and spawning behavior of five species of Japanese labrid fishes. Rep. Fish. Res. Lab. Kyushu Univ. 4: 1–64 (in Japanese).

Nakazono, A. & H. Tsukahara. 1974. Underwater observation of the spawning behavior of the wrasse *Duymaeria flagellifera* (Cuvier et Valenciennes). Rep. Fish. Res. Lab. Kyushu Univ. 2: 1–11 (in Japanese).

Shimada, K. 1993. Labridae. pp. 848–885. *In*: T. Nakabo (ed.) Fishes of Japan with Pictorial Keys to the Species, Tokai University Press, Tokyo (in Japanese).

Suzuki, K., K. Hioki, K. Kobayashi & Y. Tanaka. 1981. Developing eggs and early larvae of the wrasses, *Cirrhilabrus temminckii* and *Labroides dimidiatus*, with a note on their spawning behaviors. J. Fac. Mar. Sci. Technol. Tokai Univ. 14: 369–377 (in Japanese).

Environmental Biology of Fishes **52**: 203–212, 1998.

Life-history variations in the fluvial sculpin, *Cottus nozawae* (Cottidae), along the course of a small mountain stream

Akira Goto
Department of Biology, Faculty of Fisheries, Hokkaido University, 3-1-1 Minatocho, Hakodate 041, Japan (e-mail: akira@pop.fish.hokudai.ac.jp)

Received 27.11.1995 Accepted 7.2.1997

Key words: mark-and-recapture, size at maturity, age at maturity, restricted movement, individual growth, sex ratio, population density, phenotypic plasticity, fish

Synopsis

Life-history variations in male and female fluvial sculpins, *Cottus nozawae,* were studied in a small mountain stream in Hokkaido, Japan, primarily by using capture-mark-recapture methods. At three study areas established along the stream course, the majority of marked sculpins were recaptured in their original location over one or more years, indicating their long-term occupation of each restricted habitat area. Sculpin densities increased toward the upstream habitats, whereas individual growth rates were more rapid downstream. In both sexes, sculpins distributed downstream matured at a larger body size and later in life than upstream sculpins, clearly demonstrating a clinal variation in these respects. A comparison of life-history variations in *C. nozawae* with those in amphidromous *C. hangiongensis* suggests that intrapopulational life-history variations in the former might be environmentally induced, and that one of the most important determinants for the variations in *Cottus* species might be population density.

Introduction

A species inhabiting a variety of habitats should exhibit different life-history patterns as a result of different selection pressures. Numerous examples exist of geographical variation in life-history styles of freshwater fish species (Schaffer & Elson 1975, Fox 1978, Leggett & Carscadden 1978, Reznick & Endler 1982, Mann et al. 1984, Jonsson et al. 1991). In the American shad, *Alosa sapidissima,* for example, the age at sexual maturity and the frequency of repeat spawners increases with latitude along the east coast of North America (Leggett & Carscadden 1978). This was thought to be an adaptation to differences in water temperature (Glebe & Leggett 1981).

Although such intraspecific life-history variations in fishes may be evident on a broad geographic scale, there are few studies that have compared the life-history patterns among several stocks of a single population inhabiting different habitats. Recently, Goto (1987, 1989a, b, 1993a) reported the occurrence of clinal life-history variations within a population of the amphidromous sculpin, *Cottus hangiongensis,* and suggested that these were environmentally induced.

The fluvial sculpin, *Cottus nozawae* Snyder, is a small, bottom-dwelling fish abundantly distributed in the rivers of Hokkaido Island and the most northeastern part of Honshu Island, Japan (Goto 1975b, 1980). On Hokkaido Island *C. nozawae* generally inhabits the middle and/or upper courses of rivers,

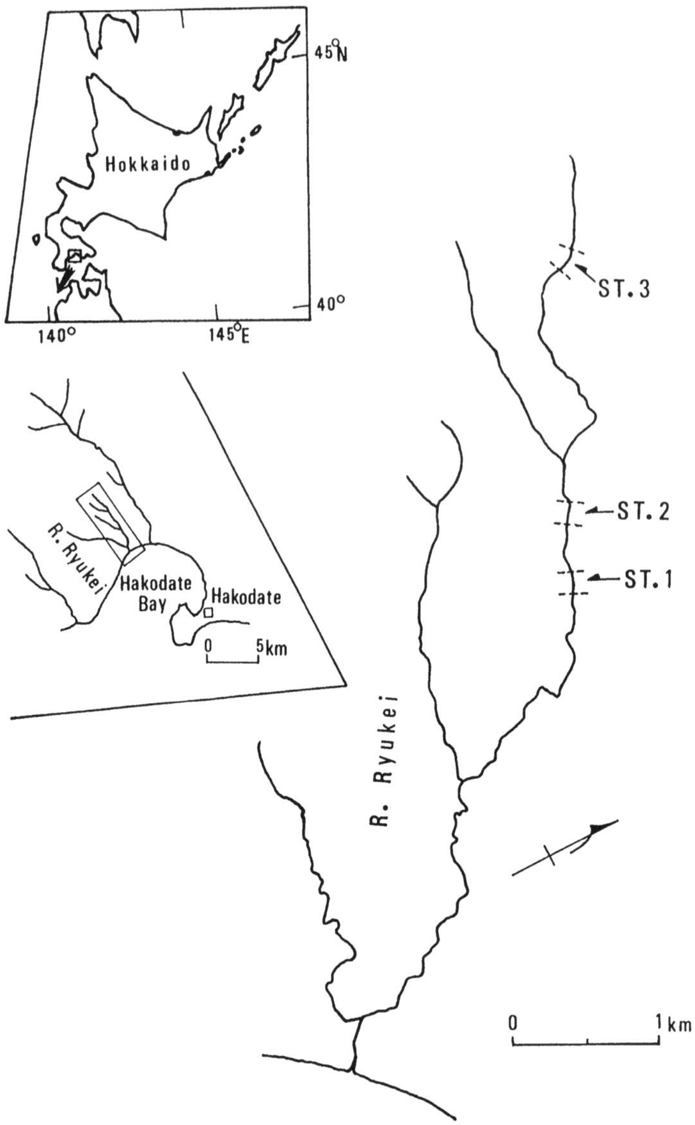

Figure 1. Location and map of the Ryukei River showing the study sites.

and breeds in early spring from late April to mid-May (Goto 1975a, 1982). The reproductive style is that of a hole nester, with paternal care as in congeneric species (Morris 1954, Savage 1963, Goto 1982, 1990, Balon & Goto 1989). Females of *C. nozawae* lay a small number of large eggs from which well-developed, benthic young hatch (Goto 1975a, 1977, 1983, 1990). The mating system is polygynous, females prefering to mate with males whose nests already contain eggs (Goto 1993b).

In the present study, the structure of adult *C. nozawae* stocks inhabiting three sections, situated along the course of a short mountain stream, was characterized by examining the population density, sex ratio, and age composition. The long-term spatial stability, growth patterns of individual fishes, and size at sexual maturity were examined primarily by using capture-mark-recapture methods, in order to clarify variations in male and female life histories.

Figure 2. A marked female *Cottus nozawae*, which was designated as IV, VII: 3, 6, 14, and recaptured 16 months after marking. The body length was 96.0 mm SL at the time of recapture, 5 December 1992. Arrows indicate positions of spines or rays that were removed.

Study area and methods

Fieldwork was conducted from June 1989 to December 1991 in the Ryukei River of southern Hokkaido, Japan (41 50′ N, 140 40′ E: Figure 1). The river originates on Oshima Mountain and drains into Hakodate Bay, being characterized by a relatively steep gradient and short length (9.6 km). In reaches about 4 km upstream from the river mouth, sites from 0.5 to 3 km apart from each other were chosen for the overall study area, owing to the relative abundance of the fluvial sculpins compared with that in other sites, and the low density of the only other resident fish species, white-spotted charr, *Salvelinus leucomaenis.*

Three study stations (St. 1–3) were established along the stream course (Figure 1). The most downstream station, St. 1, was 140 m long and 3.9 m wide on average. Stations 2 and 3 were 110 m long and 4.3 m wide, and 70 m long and 3.2 m wide, respectively. Water temperatures did not differ by more than 3 °C between stations in all seasons, being generally higher downstream in summer and upstream in winter. The greatest difference in water temperature was recorded between St. 1 and St. 3 in August of each year.

Throughout the study period, sculpins were captured monthly at each station, with 3 mm stretched-mesh dip nets, except from January to March when the stream was ice-covered. The sculpins captured were anesthetized with 0.0001% ethyl p-aminobenzoate solution and measured for standard length (SL) to the nearest 0.1 mm using a caliper. Sculpins larger than 50 mm SL were individually marked by removing dorsal-fin spines and rays (Goto 1985). Capture-mark-recapture methods enabled assessment of the population density, as well as individual spatial stability and sculpin growth at each station. Population size was estimated for each station using the Jolly-Seber method (Seber 1973). Sexual maturity of fish was determined by examination of the gonad condition, i.e. whether or not the fish oozed milt or had ripe eggs during the spawning period (April to May). Age at maturity was estimated by examination of otoliths or the growth pattern indicated by successive measurements of each fish, although the latter was only applicable for individuals smaller than 60 mm (age 1 +) at the time of marking. For consistency in aging, April was considered as the standard spawning time in each year.

Results

Spatial stability

During the entire study period, a total of 699 individuals were marked: 60 males and 158 females at St. 1, 88 males and 186 females at St. 2, and 75 males

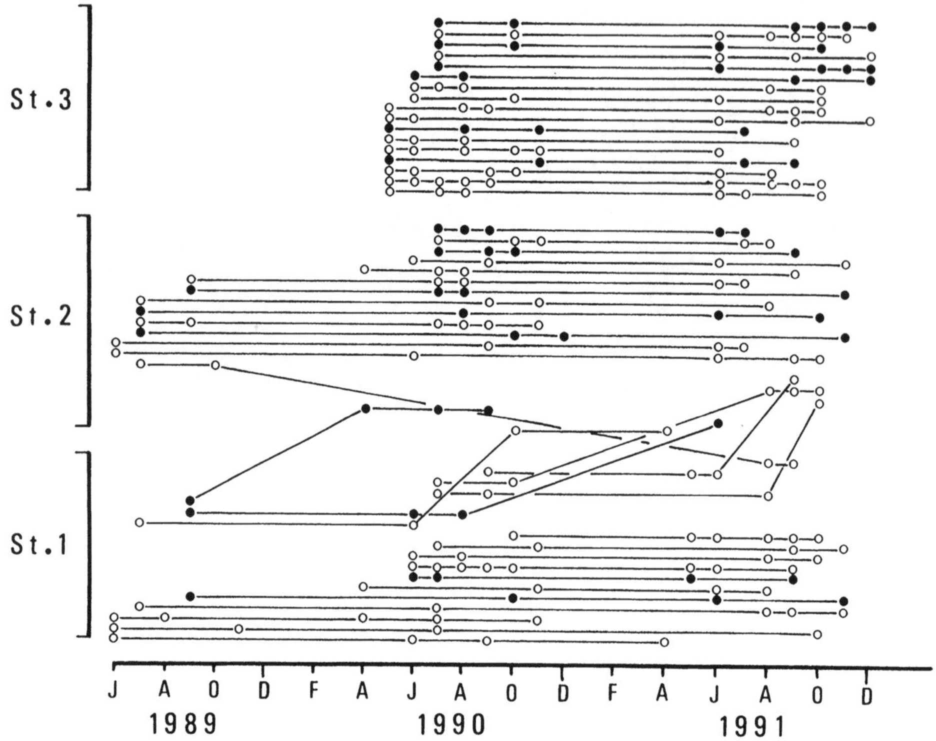

Figure 3. Long-term spatial stability, and movement between stations, of selected adult individuals of *Cottus nozawae* in the Ryukei River, from June 1989 to December 1991 (● = male; ○ = female).

and 132 females at St. 3. Of these, 638 fish were recaptured at least once.

Figure 2 shows a female which was recaptured 16 months after marking. The combination of spines and rays removed from its dorsal fins was so distinct that the specimen could be distinguished from all other marked individuals, demonstrating the suitability of such a marking technique for long-term capture-mark-recapture studies of this species.

Nearly all the marked sculpins that were recaptured successively for one or two years were at their original stations, i.e. the station in which each fish was marked (Figure 3). No individuals marked at St. 3 were found to have moved to the downstream stations. A single female, marked at St. 2, was found to have moved downstream to St. 1, and six fish out of those marked at St. 1 (2 males and 4 females) moved upstream to St. 2. Such results demonstrated that sculpins could be active but have limited home ranges, and suggested that fish inhabiting each station could be referred to as a particular habitat

Table 1. Characteristics of population structure in each habitat group of *Cottus nozawae* in the Ryukei River.

Station	Population size (fish)	Population density (fish m^{-2})	Sex ratio		Age composition	
			female	male	female	male
St. 1	110	0.17	1	0.37	0–VI	0–VII
St. 2	240	0.43	1	0.60	0–VI	0–VII
St. 3	180	0.69	1	0.34	0–V	0–VI

group, because of their complete or almost complete isolation from each other.

Population size and structure

The population sizes estimated from monthly capture-mark-recapture data from June to August 1990, were 110 individuals at St. 1, 240 at St. 2 and 180 at St. 3, giving densities of 0.17, 0.43 and 0.69 fish m^{-2}, respectively (Table 1).

The size composition of each habitat group captured in the prespawning season (early April) is shown in Figure 4. The standard lengths of mature fish at St. 1, 2 and 3 ranged from 83.6 to 102.3 mm, 81.0 to 108.3 mm, and 70.8 to 95.2 mm for males, and from 76.9 to 102.2 mm, 70.2 to 98.3 mm, and 65.5 to 83.4 mm for females, respectively. In all three habitats, sexual dimorphism was evident, with males being larger than females.

The age composition of males and females was almost the same in the two downstream habitat groups (St. 1 and 2), but differed between those and the upstream habitat group (St. 3, Table 1). The latter lacked the oldest age classes (7 years old, hereinafter abbreviated as age VII for males, and age VI for females). For all three habitat groups, the oldest males lived one year longer (age VI or VII) than females.

The sex ratio of adults collected from June to August 1990, was biased toward females at all three stations. The female bias was greater at St. 1 (0.37 males per females) and St. 3 (0.34) than St. 2 (0.60) (Table 1).

Individual growth

The individual growth of marked sculpins from each station was estimated from the standard length increases at successive recaptures (Figure 5). Although considerable variation in growth was found among individuals of the same habitat group, and smaller fish appeared to grow more rapidly at all stations, less growth occurred in fish from St. 3 compared with those from St. 1 and 2. For example, mean annual growth increments (± SD) for the

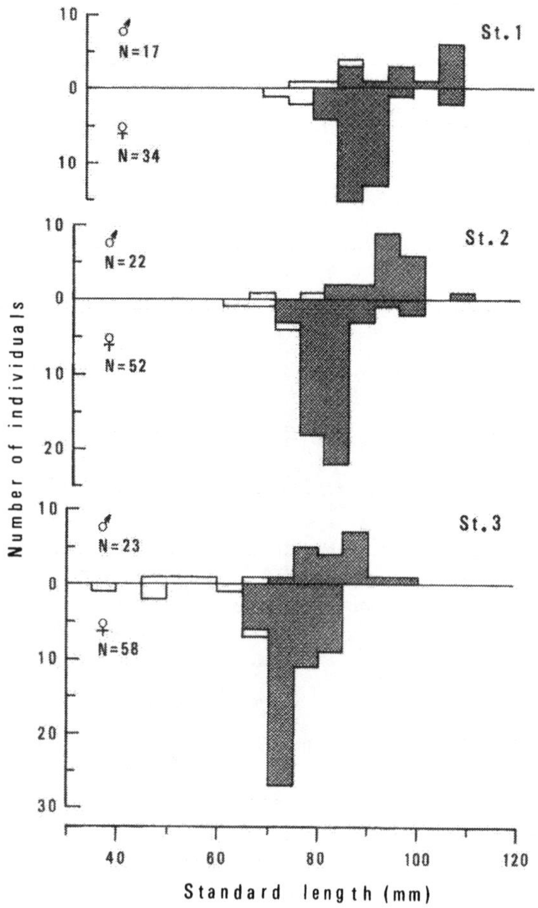

Figure 4. Length-frequency distribution of male and female *Cottus nozawae* captured at three stations in the Ryukei River during the prespawning season (early April) of 1990 (□ = immature fish; □ = mature fish).

sculpins of the 70–79 mm size class were 14.5 ± 4.5 mm at St. 1 (n = 21), 9.7 ± 3.5 mm at St. 2 (n = 20) and 3.1 ± 2.4 mm at St. 3 (n = 24) for males, and 6.7 ± 1.7 mm (n = 25), 5.9 ± 2.2 mm (n = 26) and 1.5 ± 0.9 mm (n = 32) for females, respectively. Growth increments were significantly smaller in fish from upstream habitats than downstream habitats (Mann-Whitney U-test, $p < 0.05$), except for females between St. 1 and St. 2. At St. 3, for example, female # 3040 (70 mm SL at marking in May 1990) had reached only 71.9 mm SL at the fifth recapture in December 1991, demonstrating growth of 1.9 mm over about 1.5 years. At St. 1, on the other hand, female # 1077 (82.4 mm SL at marking in April 1990) grew more rapidly, having reached 89.7 mm

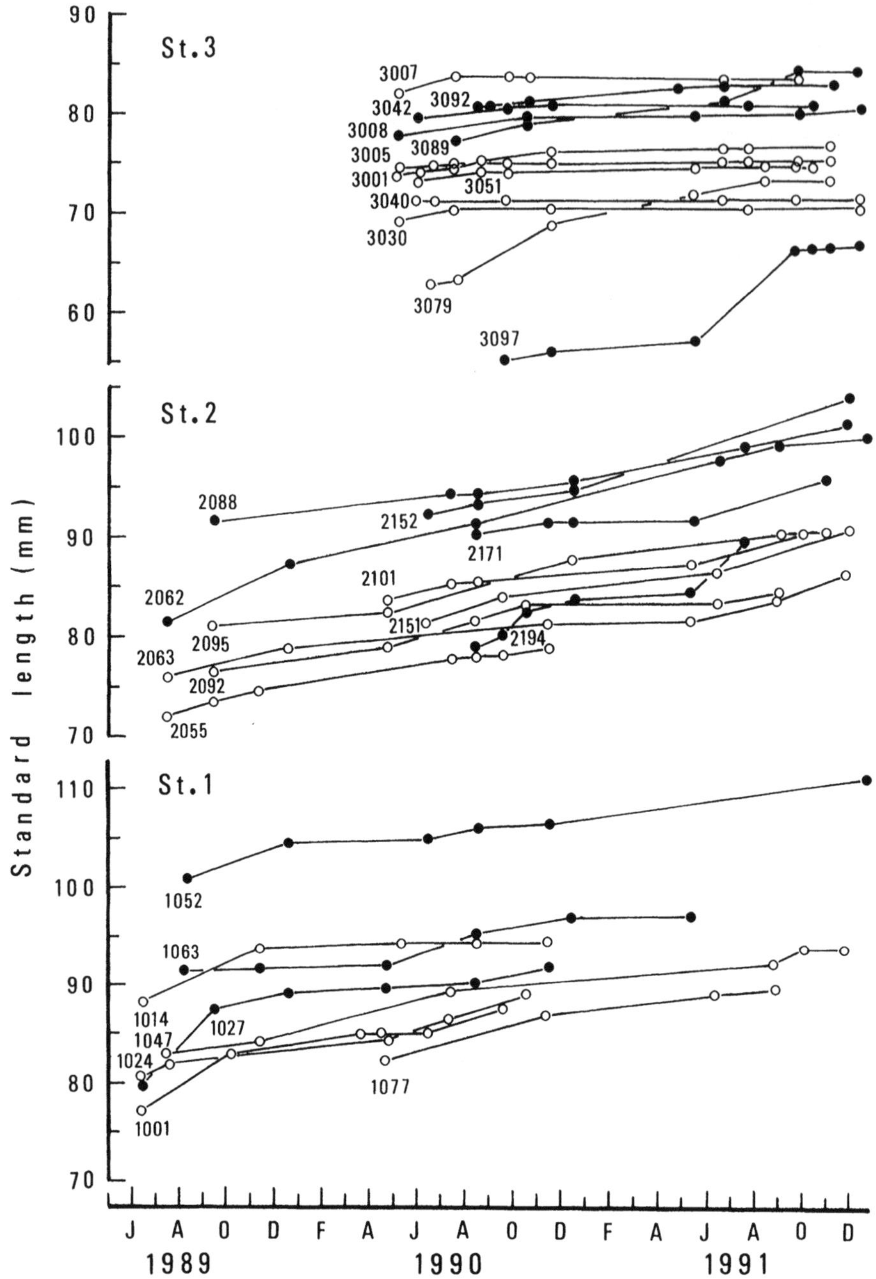

Figure 5. Individual growth patterns of adult *Cottus nozawae* at three stations in the Ryukei River from June 1989 to December 1991 (○ = male; ● = female).

SL in September 1991. Female # 2151 (81.2 mm SL in June 1990) from St. 2 also grew faster reaching 91.1 mm by December 1991. Similarly, slow growth was found in males from St. 3, as compared with those from St. 1 and 2 (Figure 5).

Size and age at maturity

In both sexes, the standard lengths of mature fish decreased toward upstream habitats: the mean length (± SD) at maturity was 94.3 ± 7.5 mm SL at

St. 1, 93.3 ± 6.1 mm SL at St. 2 and 83.6 ± 5.5 mm SL at St. 3 for males, and 85.7 ± 5.8 mm SL, 81.2 ± 5.0 mm SL and 74.5 ± 4.2 mm SL for females, respectively (Figure 4). Statistical analyses showed significant differences for both males and females between the habitat groups (Mann-Whitney U-test, $p < 0.01$), except for males between St. 1 and 2. The minimum size of mature fish, recorded as the size at first evidence of sexual maturity, was also smaller for both sexes in fishes from the upstream habitat.

Age at first maturity was slightly lower in the St. 3 habitat group (age 2 or 3 years for males and 2 years for females) than in the two downstream habitat groups (age 3 years for males, and 2 or 3 years for females).

Discussion

Previous studies have demonstrated that the fluvial, land-locked sculpin, *C. nozawae,* has greater morphological and genetic variation between populations from different river systems than the amphidromous, sibling species, *C. amblystomopsis* (Goto 1975b, 1977, 1980, Goto & Andoh 1990). The present study showed that within a single population, *C. nozawae* had considerable life-history variation between several habitat groups distributed separately along a river course.

The long-term limited home ranges of individual sculpins observed in this study may facilitate such life-history differences among habitat groups, because of the resultant reproductive isolation between the groups. Each habitat group of *C. nozawae*, which is separately distributed throughout several successive riffles, seems to be reproductively isolated from the others owing to the existence of large pools between the riffles, so that recruitment of a given habitat group may be dependent almost solely upon the offspring. Adult *C. nozawae* spawn a small number of large eggs in the middle or upper course of rivers, from which well-developed, benthic young hatch. These exhibit negative phototaxis and are able to settle on the gravel bottom in the spawning grounds (Goto 1975a, 1977, 1990). At two weeks of age they have completely absorbed their yolk, and are fully formed definitive phenotypes (sensu Balon 1975, 1985). Furthermore, our recent genetic study showed that significant heterogeneities of allele and mtDNA haplotype frequencies were detected among some tributary populations of *C. nozawae* from a single river system, suggesting genetic differentiation between them (Okumura et al. 1996).

Similar, strongly restricted movements have been observed in congeneric sculpins (McCleave 1964, Goto 1986, 1988, Morgan & Ringler 1992) and some stream-resident salmonids (Miller 1957, Maruyama 1981, Bachman 1984, Nakano et al. 1990). In an amphidromous sculpin, *C. hangiongensis*, adult

Table 2. Comparison of intrapopulational life-history variations in *Cottus nozawae* and *C. hangiongensis**.

	C. nozawae	*C. hangiongensis*
Life-history variations in upstream vs. downstream populations	Smaller body size and younger age at sexual maturity	Larger body size and greater age at sexual maturity
Cause of the life-history variations	Phenotypic plasticity?	Phenotypic plasticity
Gradient of biotic/abiotic environments	Similar to each other in water temperature, abundance of aquatic insects, etc.	
Life-cycle:	Fluvial	Amphidromous
Habitats:	Middle to upper reaches	Lower to middle reaches
Spawning grounds:	Throughout the distribution range	Restricted to lower reaches
Mating system:	Polygynous	Polygynous
Female mate preference:	Egg-guarding males	Larger males
Upstream population densities:	Higher	Lower
Upstream growth rates:	Lower	Higher

* Results presented in this table were from Goto (1975a, 1982, 1983, 1986, 1987, 1988, 1989a, b, 1993a) or are from the present study.

individuals remained within a narrow limited stream area, all movements being shorter than 100 m during the non-breeding season, although they migrated downstream to the spawning grounds in the breeding period (Goto 1988). For stream-resident cutthroat trout, *Oncorhynchus clarki*, Miller (1957) reported that subsequent to the early juveniles, individuals showed very restricted movements, probably remaining within one pool for long periods. In fluvial fish, restricted movement appears to be a behavioral characteristic, though the spatial separation between rich feeding habitats and suitable spawning sites might bring about extensive migrations (Gerking 1959, Northcote 1978). In the case of fluvial *C. nozawae,* there was no spatial separation between feeding and spawning sites, thereby enabling the fish to remain in a limited home range within short reaches of the stream.

Conspicuous differences in aspects of their life-history between the habitat groups of *C. nozawae* along the stream course can be expressed briefly as follows: in both sexes, sculpins distributed downstream mature later in life and at a larger body size, whereas maturity occurs at a younger age and smaller body size in the upstream area.

Although growth rates of individual sculpins from the same habitat varied considerably, they were generally slower in fishes from the upstream habitat. Such growth variations in populations along the stream course may result from differences in biotic and/or abiotic conditions in their respective habitats. In spite of their slow growth, the upstream sculpins exhibited earlier (precocious) maturity, which may counteract the relatively low clutch size resulting from their small body size at maturity. The opposite situation may occur in downstream individuals.

Regarding geographic variations in life-history aspects of freshwater sculpins, it is known that in the fluvial *Cottus gobio,* fast growth, early maturation, short life-span and multiple annual egg-batches occur in unproductive, northern streams of England, whereas the opposite traits occur in fishes from more productive, southern streams (Fox 1978, Mann et al. 1984). By reciprocal transfer experiments, Mann et al. (1984) demonstrated that the variation in the number of egg-batches and differences in growth rate at least, were not genotypic, but environmentally-induced (phenotypic) responses. Such findings suggest that most fluvial sculpins, including *C. nozawae*, have much phenotypic plasticity in their life-history characteristics, possibly resulting from their continual exposure to fluctuating environmental conditions, as demonstrated in several poeciliids (Trexler 1989, Meffe 1992) and freshwater snails (Russell-Hunter 1978, Brown 1985, Brown et al. 1985). As suggested by Brown (1985) for the pond snail, *Lymnaea elodes*, phenotypic plasticity in life-history may itself be adaptive for the inhabitants of heterogenous or unpredictable environments, such as vernal ponds and rivers. In the case of *C. nozawae*, if the population were characterized by a single, specific life-history pattern, the species might not exploit as broad a range of stream habitats.

Intrapopulational life-history variations similar to those observed in *C. nozawae* have been found also in amphidromous congeneric sculpin, *C. hangiongensis* (Goto 1987, 1989a, b, 1993a). In the latter, however, the life-history traits varied in the reverse direction to those in *C. nozawae,* sexual maturity occurring earlier in life and at a smaller size in the downstream habitats, but later in life and at a larger size in upstream habitats (Table 2).

How can such differences in responses to environmental gradients between the two *Cottus* species be explained? Based on the field survey and rearing experiments, the clinal life-history variations in *C. hangiongensis* are considered to be environmentally-induced responses to different habitat conditions, primarily differences in population density: rapid growth and early maturity occur in the downstream habitat where the density is high, whereas slow growth and delayed maturity take place in the low density, upstream habitat (Goto 1989a, b, 1993a). In the case of *C. nozawae*, the life-history variation along the stream course also appeared to be a response to different population densities, mainly because the population density was greater upstream. The slight differences in water temperature found between the stations probably was not responsible for the life-history variation, because the remarkable differences occurred only in summer and winter seasons when sculpins grew very little (Goto 1977). Thus, one of the most important determinants for life-history variation within a single population of species of *Cottus* might be the population density. Further field and experimental studies on intrapopulational life-history variations of other congeneric species are needed to test this hypothesis.

Acknowledgements

I wish to express my thanks to F. Yamazaki for his invaluable advice in the course of this study. Thanks are also due to N. Okumura, S. Takeyama and M. Higuchi for their field assistance. This work was partly supported by Grants-in-Aid (No. 02304002 and 05660193) from the Japan Ministry of Education, Science, Sports, and Culture.

References cited

Bachman, R.A. 1984. Foraging behavior of free-ranging wild and hatchery brown trout in a stream. Trans. Amer. Fish. Soc. 113: 1–32.

Balon, E.K. 1975. Terminology of intervals in fish development. J. Fish. Res. Board Can. 32: 1663–1670.

Balon, E.K. (ed.) 1985. Early life histories of fishes: new developmental, ecological and evolutionary perspectives. Developments in Env. Biol. Fish. 5, Dr W. Junk Publishers, Dordrecht. 280 pp.

Balon, E.K. & A. Goto. 1989. Styles in reproduction and early ontogeny. pp. 1–47. *In:* A. Goto & K. Maekawa (ed.) Reproductive Behaviour in Fish: Styles and Strategies, Tokai University Press, Tokyo (in Japanese).

Brown, K.M. 1985. Intraspecific life history variation in a pond snail: the roles of population divergence and phenotypic plasticity. Evolution 39: 387–395.

Brown, K.M., D.R. DeVries & K.B. Leathers. 1985. Causes of life history variation in the freshwater snail *Lymnaea elodes*. Malacologia 26: 191–200.

Fox, P.J. 1978. Preliminary observations on different reproductive strategies in the bullhead (*Cottus gobio* L.) in northern and southern England. J. Fish Biol. 12: 5–11.

Gerking, D.L. 1959. The restricted movement of fish populations. Biol. Rev. 34: 221–242.

Glebe, B.D. & W.C. Leggett. 1981. Latitudinal differences in energy allocation and use during the freshwater migration of American shad (*Alosa sapidissima*) and their life history consequences. Can. J. Fish. Aquat. Sci. 38: 806–820.

Goto, A. 1975a. Ecological and morphological divergence of the freshwater sculpin, *Cottus nozawae* Snyder-I. Spawning behavior and process of the development in the post-hatching stage. Bull. Fac. Fish. Hokkaido Univ. 26: 31–37 (in Japanese).

Goto, A. 1975b. Ecological and morphological divergence of the freshwater sculpin, *Cottus nozawae* Snyder-II. Morphological comparison of adult fishes of small-egg and large-egg types and their distribution. Bull. Fac. Fish. Hokkaido Univ. 26: 39–48 (in Japanese).

Goto, A. 1977. Ecological and morphological studies on the freshwater sculpins in Hokkaido, Japan. Doctoral Dissertation, Hokkaido University, Sapporo. 207 pp. (in Japanese).

Goto, A. 1980. Geographic distribution and variations of two types of *Cottus nozawae* in Hokkaido, and morphological characteristics of *C. amblystomopsis* from Sakhalin. Japan. J. Ichthyol. 27: 97–105.

Goto, A. 1982. Reproductive behavior of a river sculpin, *Cottus nozawae* Snyder. Japan. J. Ichthyol. 28: 453–457.

Goto, A. 1983. Spawning habits and reproductive isolating mechanisms of two closely related river-sculpins, *Cottus amblystomopsis* and *C. nozawae*. Japan J. Ichthyol. 30: 168–175.

Goto, A. 1985. Individual identification by spine and ray clipping for freshwater sculpins. Japan. J. Ichthyol. 32: 359–362.

Goto, A. 1986. Movement and population size of the river sculpin *Cottus hangiongensis* in the Daitobetsu River of southern Hokkaido. Japan J. Ichthyol. 32: 421–430.

Goto, A. 1987. Life-history variation in males of the river sculpin, *Cottus hangiongensis*, along the course of a river. Env. Biol. Fish. 19: 82–92.

Goto, A. 1988. Reproductive behavior and homing after downstream spawning migration in the river sculpin, *Cottus hangiongensis*. Japan. J. Ichthyol. 34: 488–496.

Goto, A. 1989a. Growth differences in males of the river-sculpin *Cottus hangiongensis* along a river course, a correlate of the life-history variation. Env. Biol. Fish. 24: 241–249.

Goto, A. 1989b. Growth patterns, age at maturity and longevity of females of the river-sculpin, *Cottus hangiongensis* in the Daitobetsu River, with special reference to their life-history variation. Japan. J. Ichthyol. 36: 90–99.

Goto, A. 1990. Alternative life-history styles of Japanese freshwater sculpins revisited. Env. Biol. Fish. 28: 101–112.

Goto, A. 1993a. Clinal life-history variation in the river-sculpin, *Cottus hangiongensis*; an example of phenotypic plasticity. Japan. J. Ichthyol. 39: 363–370.

Goto, A. 1993b. Male mating success and female mate choice in the river sculpin, *Cottus nozawae* (Cottidae). Env. Biol. Fish. 37: 347–353.

Goto, A. & T. Andoh. 1990. Genetic divergence between the sibling species of river-sculpin, *Cottus amblystomopsis* and *C. nozawae*, with special reference to speciation. Env. Biol. Fish. 28: 257–266.

Jonsson, N., L.P. Hansen & B. Jonsson. 1991. Variation in age, size and repeat spawning of adult Atlantic salmon in relation to river discharge. J. Anim. Ecol. 60: 937–947.

Leggett, W.C. & J.E. Carscadden. 1978. Latitudinal variation in reproductive characters of American shad (*Alosa sapidissi-*

ma): evidence for population specific life history strategies in fish. J. Fish. Res. Board Can. 35: 1469–1478.

Mann, R.H.K., C.A. Mills & D.T. Crisp. 1984. Geographical variation in the life-history tactics of some species of freshwater fish. pp. 171–186. *In:* G.W. Potts & R.J. Wootton (ed.) Fish Reproduction: Strategies and Tactics, Academic Press, London.

Maruyama, T. 1981. Comparative ecology on the fluvial forms of *Salmo (Oncorhynchus) masou masou* (Brevoort) and *Salvelinus leucomaenis* (Pallas) (Pisces, Salmonidae). I. Structure of spawning redds and spawning sites in Kamidani, River Yura. Japan. J. Ecol. 31: 269–284 (in Japanese).

McCleave, J.D. 1964. Movement and population of the mottled sculpin (*Cottus bairdi* Girard) in a small Montana stream. Copeia 1964: 506–513.

Meffe, G.K. 1992. Plasticity of life-history characters in eastern mosquitofish (*Gambusia holbrook*: Poeciliidae) in response to thermal stress. Copeia 1992: 94–102.

Miller, R.B. 1957. Permanence and size of home territory in stream-dwelling cutthroat trout. J. Fish. Res. Board Can. 14: 687–691.

Morgan, C.R. & N.H. Ringler. 1992. Experimental manipulation of sculpin (*Cottus cognatus*) populations in a small stream. J. Freshw. Ecol. 7: 227–232.

Morris, D. 1954. The reproductive behavior of the river bullhead (*Cottus gobio*) with special reference to the fanning activity. Behaviour 7: 1–32.

Nakano, S., T. Kachi & M. Nagoshi. 1990. Restricted movement of the fluvial form of red-spotted masu salmon, *Oncorhynchus masou rhodurus*, in a mountain stream, central Japan. Japan. J. Ichthyol. 37: 158–163.

Northcote, T.G. 1978. Migratory strategies and reproduction in freshwater fishes. pp. 326–359. *In:* S.G. Gerking (ed.) Ecology of Freshwater Fish Production, Halsted Press, New York.

Okumura, N., Y. Hasebe & A. Goto. 1996. Genetic variations of *Cottus nozawae* populations from five tributaries of the Siribetsu River of southern Hokkaido, Japan. Ichthyol. Res. 43: 65–71.

Reznick, D.N. & J.A. Endler. 1982. The impact of predation on life history evolution in Trinidadian guppies (*Poecilia reticulata*). Evolution 36: 160–177.

Russell-Hunter, W.D. 1978. Ecology of fresh water pulmonates. pp. 335–384. *In:* V. Fretter & J. Peake (ed.) Pulmonates, Vol. 2A, Systematics, Evolution, and Ecology, Academic Press, New York.

Savage, T. 1963. Reproductive behavior of the mottled sculpin, *Cottus bairdi* Girard. Copeia 1963: 317–325.

Schaffer, W.M. & P.F. Elson. 1975. The adaptive significance of variations in life history among local populations of Atlantic salmon in North America. Ecology 56: 577–590.

Seber, G.A.F. 1973. The estimation of animal abundance and related parameters. Griffin, London. 506 pp.

Trexler, J.C. 1989. Phenotypic plasticity in poeciliid life histories. pp. 201–214. *In:* G.K. Meffe & F.F. Snelson (ed.) Ecology and Evolution of Livebearing Fishes (Poeciliidae), Prentice Hall, Englewood Cliffs.

Environmental Biology of Fishes **52**: 213–223, 1998.

Spawning ecology of Japanese surf smelt, *Hypomesus pretiosus japonicus* (Osmeridae), in Otsuchi Bay, northeastern Japan

Taro Hirose[1] & Kouichi Kawaguchi[2]
[1] *Japan Sea National Fisheries Research Institute, 1-5939-22 Suido-cho, Niigata 951-8121, Japan (e-mail: taro@ss.jsnf.affrc.go.jp)*
[2] *Ocean Research Institute, University of Tokyo, 1-15-1 Minamidai, Nakano-ku, Tokyo 164-0014, Japan*

Received 13.11.1996 Accepted 15.12.1997

Key words: chika, breeding behavior, reproductive style, egg distribution, sex ratio of spawners, age composition of spawners

Synopsis

The spawning habits of Japanese surf smelt, *Hypomesus pretiosus japonicus* (Japanese name 'chika'), were studied from 1992 to 1995 at Akaiso Beach in Otsuchi Bay off the Pacific coast of northern Honshu. Investigations were conducted in relation to time, tidal rhythm, sand grain size on the spawning ground, and sex and age of spawners. Spawning occurred from late March to early May with its peak at the spring tide period during full moon in April. Before dark, fish gathered to school from 1–3 m depth, and 10–20 m off the shoreline of the spawning beach. Just before dark, they repeatedly approached the shoreline and stayed near the shoreline to spawn just after dark. In the spawning process, three intervals, i.e. aggregation, approach, and spawning, were recognized and these intervals are thought to be functionally connected. Mating occurred in a unit or group formed by one female and several males. During 1992–1995, except for 1994, spawning took place in a restricted region of the beach where backwash was stronger than other areas, and most sand grains were more than 1 mm in diameter and coarser than those of other areas. Although spawning was observed near the shoreline (34 to 120 cm tide level), most of the eggs were dispersed by backwash and each became attached to a few sand grains close to a step formed at 1–2 m depth off the spawning site. Spawners were composed mainly of age 0+ fish and some 1+ fish. Males usually outnumbered females on the spawning ground. Their spawning styles are discussed in relation to their ecology and habitat.

Introduction

Smelts (family Osmeridae) are distributed from temperate to subarctic areas of the Northern hemisphere (Nelson 1994). They show various life history patterns and spawning habits. Some of them, like longfin smelt, *Spirinchus thaleichthys*, and delta smelt, *Hypomesus transpacificus*, are anadromous or landlocked and spawn in fresh water, while others like capelin, *Mallotus villosus*, and surf smelt, *H. pretiosus pretiosus*, live in the sea throughout life (Hart 1973, Scott & Crossman 1973, Jangaard 1974, Ueno 1984). Japanese surf smelt, *H. pretiosus japonicus* (Japanese name 'chika'), occur in the coastal water off northern Japan and also on the Korean Peninsula and Russian coast of the Japan Sea up to the Kuril Islands (Ueno 1984). This species is a popular target species of coastal fisheries and recreational fishing off northern Japan.

North American surf smelt, *H. pretiosus pretiosus*, a subspecies of chika, was reported to spawn near the shoreline of sand beaches on the west coast

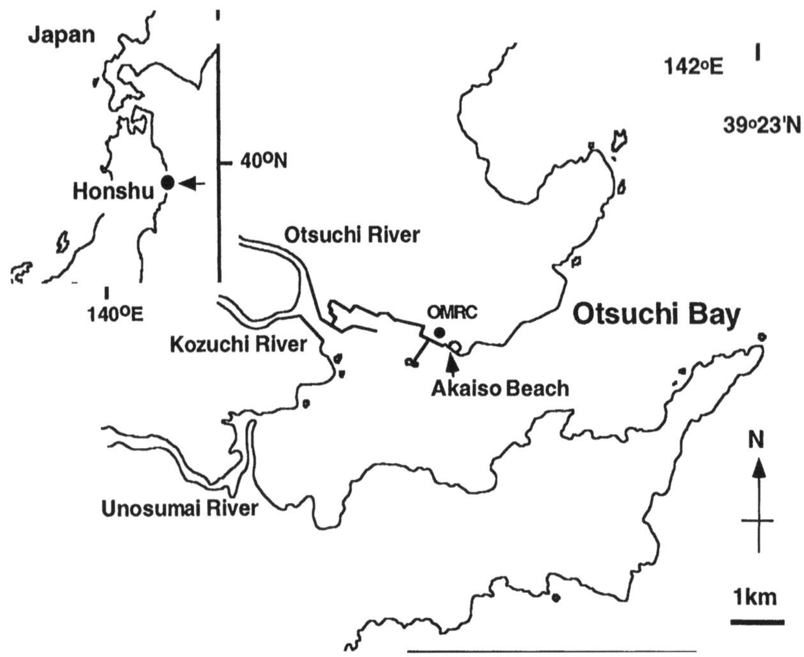

Figure 1. Location of study site, Akaiso Beach, Otsuchi Bay, off northeastern Honshu (OMRC = Otsuchi Marine Research Center, Ocean Research Institute, University of Tokyo). Hokkaido Island is shown, in part, north of Honshu.

of North America (Thompson et al. 1936, Loosanoff 1937, Penttila 1973, 1978). Yanagawa (1981) reported demersal adhesive eggs of chika sampled off a sand beach in Hokkaido. These facts suggest that chika also spawn on sand beaches like North American surf smelt, but detailed knowledge of its spawning habits is lacking. In addition, it is known that the spawning of some species like California grunion, *Lauresthes tenuis*, Atlantic silverside, *Menidia menidia*, and grass puffer, *Takifugu niphobles*, which are intertidal zone spawners, closely connect their spawning with the lunar cycle and tidal rhythms (Walker 1952, Middaugh 1981, Shiraish 1986). The purpose of our study was to describe the spawning behavior of chika based on observation of spawning and collection of spawners in the field in relation to time, tidal rhythm, and substrate, and also to discuss their spawning style.

Materials and methods

The study was done at Akaiso Beach in Otsuchi Bay on the Pacific coast of northeastern Honshu, the main island of Japan (Figure 1). Akaiso Beach is a middle sized sand beach (ca. 80 m long) in this bay and is surrounded by cliffs (Figure 2). Routine observations were made in the spawning season of four consecutive years, i.e., from 17 April to 8 May 1992, from 30 March to 8 May 1993, from 27 March to 7 May 1994, and from 25 March to 3 May 1995. Observation was started one hour before sunset and continued until the end of spawning, or for three hours after sunset if spawning did not occur. Twenty-four-hour observations were carried out on 5 April (spring tide) and 16 April (normal tide) 1993 to determine the relationship among tide, light level, and spawning. Observations right after flood tide at midnight were done on 9 and 10 April 1993 to determine the relationship between flood tide and the timing of spawning. We also collected information in the more offshore area (1–2 m deep) from a boat on 11 and 25 April 1994 to examine the possibility of offshore spawning. An electric light was used for the observation after dark. Spawning behavior was recorded by photograph and video tape recorder. The beginning and ending time of spawning was recorded when spawning was observed. Surface wa-

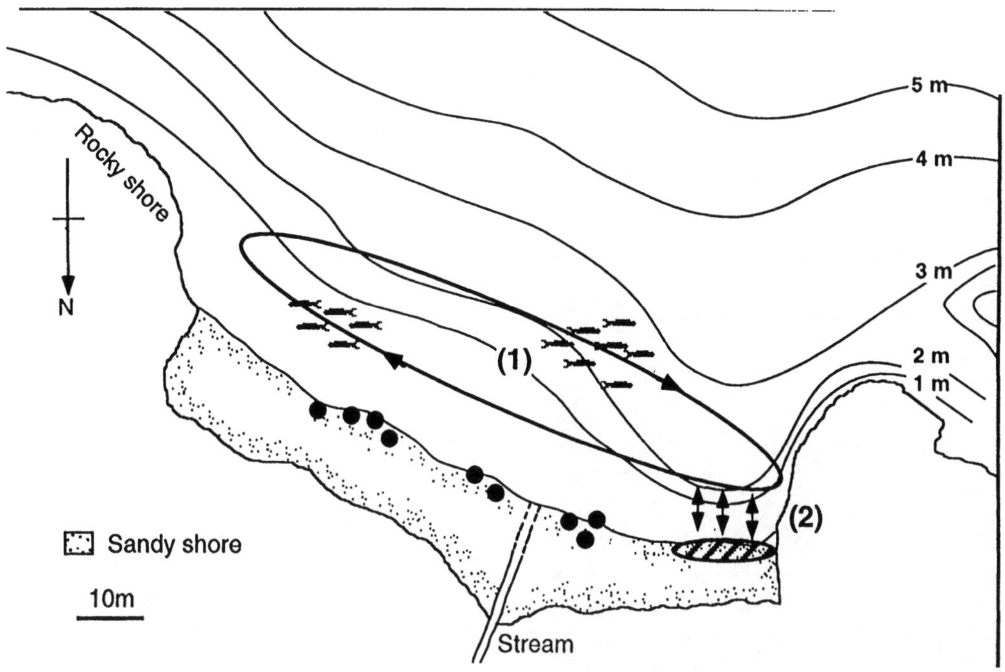

Figure 2. Diagrammatic representation of spawning habit of Japanese surf smelt at Akaiso Beach. Japanese surf smelt swam around in the area 1–2 m depth at dusk (1) and repeatedly approached and left the shoreline at the west side of the beach just before dark (2). Just after dark, they began to spawn near the shoreline. Striped area shows spawning ground in 1992, 1993 and 1995. Black dots indicate unusual spawning sites observed in 1994 when the usual spawning area was disturbed by a large fallen pine tree. Shoreline and isobathometric lines were measured at the time of mean sea level in Otsuchi Bay.

ter temperature was measured near shoreline during the spawning season of 1994 and 1995. Spawners (total 1075 individuals) were sampled with a dip net or a cast net during spawning to determine their age and sex ratio. Live samples were taken to our laboratory, and some were used for artificial insemination. All samples were then preserved in 10% buffered formalin. The fish were aged by scale reading following Hamada (1956), and sexed by examining gonads. Age of the fish was expressed as if 1 May was their birthday in Otsuchi Bay. Stomach contents of 853 individual spawners representative of those captured during all sampling days in 1994 and 1995 were also investigated to assess egg cannibalism.

We defined spawning magnitude as the number of fish in a spawning aggregation. Where possible this estimate was obtained from the photographs or videos. However, when environmental conditions precluded using photographic techniques, we estimated spawning magnitue visually. In this case, we counted 10–20 individuals in the spawning school and then estimated the area they occupied. We estimated total area covered by all spawners to calculate total number. The spawning group size was described as: XL (extremely large = more than 500 individuals gathered), L (large = 200–500 fish), M (medium = 50–200 fish) and S (small = less than 50 fish).

The distribution of eggs deposited was studied at 14 sampling stations by sampling the bottom sand with a Smith-McIntyre grab sampler at different depths along four parallel transect lines off Akaiso Beach on 27 April 1992. From each sample, four subsamples were collected from the surface to 10 cm depth with a cylindrical core-sampler of 7.2 cm in diameter. The sand collected was preserved in 10% formalin and eggs adhering to sand grains were counted to measure the egg density at each sampling station. Sand size composition was also analyzed as an index of habitat in the spawning area. Bottom sand was sampled with a cylindrical

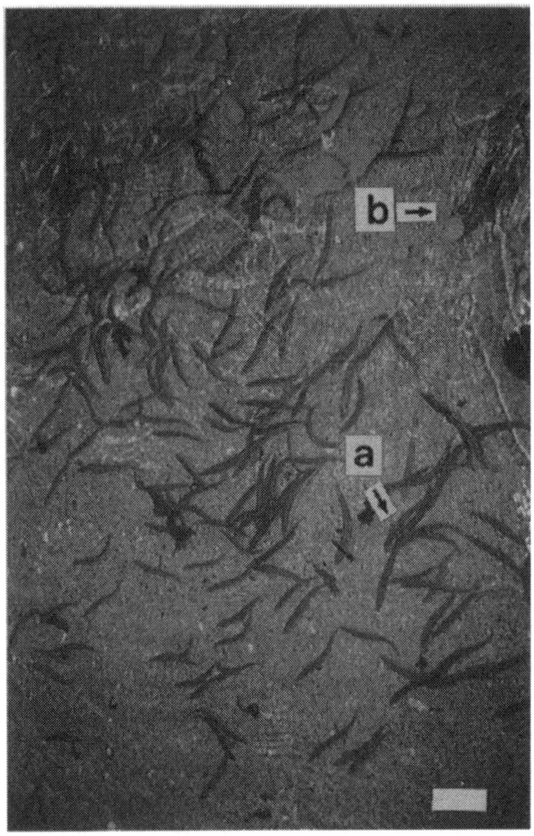

Figure 3. Spawning school of chika on the spawning ground in Akaiso Beach on 14 April 1995: A – entwining individuals, B – dashing individuals. White bar indicates 10 cm.

Table 1. Relation between spawning group size and the duration that spawners remained at the spawning site in 1992–1994, based on 37 observations.

Duration (h)	Spawning group size*				
	S	M	L	XL	Total
< 0.5	8	6			14
0.5–1	8	3	2	1	14
1–2	1	2	3	1	7
2–3			1	1	2
Total	17	11	6	3	37

* XL = more than 500 spawners.
L = 200–500 spawners; M = 50–200 spawners.
S = less than 50 spawners.

core-sampler 5 cm in diameter and 13 cm tall by diving on 28 April 1994. Sampling stations were set every 3 m from the shoreline at the highest tide to a point 18 m offshore, on four transect lines set at regular intervals. Sand samples were washed in fresh water, dried at 60° C for 24 hours, and the grain size measured by geological sieves of various mesh sizes.

Results

Spawning behavior

The spawning process of chika could be divided into three intervals (Figure 2). Soon after sunset, spawners aggregated 20–30 m off the beach and swam in an area 1–3 m in depth all along the beach (aggregation interval). As darkness increased, the school gradually moved toward the shoreline and began to repeatedly approach and leave the shoreline just before complete darkness (approach interval). This behavior was observed every year, but only at the west side of the beach where the spawning took place in 1992, 1993 and 1995. Just after dark, the fish stayed near the shoreline and began to spawn (spawning interval) (Figure 3). During spawning no individuals were seen in the deeper area which they had occupied during the aggregation interval at dusk. Spawning always occurred just after dark when the illumination was less than 0.1 lux. The act was not related to daily tidal rhythm. Spawning duration was different from day to day, ranging from 30 minutes to 2–3.5 hours. Spawning duration was longer in large spawning schools (Table 1).

The spawning school swam back and forth repeatedly near the shoreline according to the wave action, and some fish were washed ashore. On such occasions, 2 to more than 10 fish were often seen entwined with each other in the spawning school during the spawning interval (Figures 3, 4a). Afterwards, they aligned in parallel and made a quick dash for about 1 m (or 1 second), scattering sand (Figures 3, 4b, c). Sometimes other individuals which were not entwined also participated in the dash. The dashing individuals dip-netted were usually composed of one female and several males (Ta-

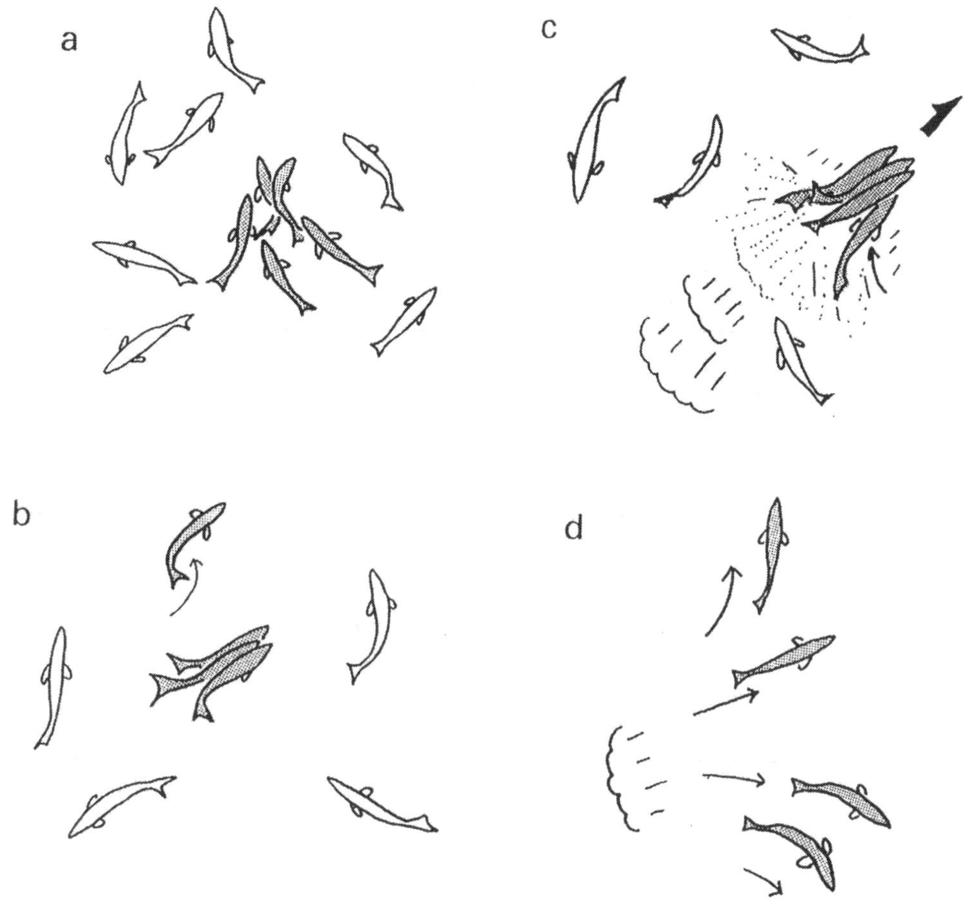

Figure 4. Spawning behaviour of chika: A – some individuals entwined with each other in a spawning school, B – they suddenly lined up in parallel, C – they made a very fast dash for 1 meter within 1 second, scattering sand. It was sometimes observed that some entwining individuals did not participate in the dash and some others not entwining did (fishes with narrow arrow in B, C). Wide arrow shows direction of the dash. D – After the dash they parted company and returned to the school separately. Shaded and unshaded fish show individuals participating in the dash and not participating, respectively.

Table 2. Sex composition of individuals which showed dashing behavior as a unit group, captured during the 1995 spawning season.

Date	Sample no.	Female	Male
26 Apr	1	1	4
	2	1	3
	3	1	3
	4	1	4
27 Apr	5	1	3
	6	1	1
	7	1	3
29 Apr	8	1	4
1 May	9	1	1
	10	1	3
	11	1	2

ble 2). After the dash, they scattered and joined the school again (Figure 4d). The dashes did not have a predictable direction. Dash behaviors were always observed in the backwash.

Fish were very wary during the aggregation and approach intervals, but once spawning started they ignored the presence of observers, flash photography, or lights for video cameras, and were easily dip-netted. In the rare case that dip-netting did disturb the school, they gathered again after a few minutes. Although they formed a dense school, no predators were observed during spawning, except humans. Stomachs of the spawners were almost empty (ratio of empty stomach was 93%), and no newly spawned eggs were found in their stomachs.

Table 3. Sex and age composition of spawners sampled randomly at Akaiso Beach during the spawning season from 1993 to 1995.

Date	Total no.	Female (%)	Age 1+ (%)	Spawning group size
1993				
2 Apr	24	7.7	7.7	M
3	13	0	0	M
5	7	0	0	XL
6	21	14.3	14.3	XL
7	12	0	0	L
8	25	0	0	L
13	8	2.5	12.5	M
15	41	0	9.8	M
24	4	0	0	M
25	3	63.7	0	S
26	6	0	16.7	M
2 May	1	0	0	S
Total	165	5.1	6.5	
1994				
31 Mar	5	40.0	40.0	S
1 Apr	18	5.6	0	M
4	12	0	0	M
5	21	14.3	23.8	M
6	15	0	13.3	L
19	21	0	4.8	S
23	45	2.5	0	XL
24	58	0	1.7	L
Total	195	6.2	3.1	
1995				
28 Mar	41	2.4	0	L
29	19	1.6	0	M
30	44	9.1	0	L
4 Apr	9	0	11.1	L
6	11	9.1	9.1	L
7	22	9.1	4.5	L
10	35	11.4	5.7	XL
11	76	0	0	XL
14	122	2.5	0.1	L
15	36	0	0	XL
16	20	25	0	L
17	45	0	0	XL
18	26	11.5	0	L
19	28	3.6	0	L
20	47	4.3	2.1	L
21	23	4.3	0	L
22	33	3.0	6.0	L
25	24	41.7	4.2	L
26	11	18.2	0	L
28	16	43.8	0	S
1 May	27	0	0	L
Total	715	5.7	1.4	

Age and sex composition of spawners

In all years, more than 90% of fish in the spawning schools were age 0+ (93.5% in 1993, 96.9% in 1994, 98.6% in 1995), the remainder being age 1+ (Table 3). Fish of both sexes captured on the spawning ground were mature enough for artificial insemination. In all but one case males outnumbered females, and in many samples no females were captured. Overall, less than 10% of the catch was females. The ratio of female and age 1+ fish in a spawning school did not change with date and spawning magnitude in any year (Kolmogorov-Smirnov test for ratios vs. date, and Kruskal-Wallis test for ratios vs. spawning magnitude, $p > 0.10$).

Spawning season and site

Spawning at Akaiso Beach was observed from 17 to 19 April 1992, from 2 April to 2 May 1993, from 31 March to 24 April 1994 and from 28 March to 1 May 1995 (Figure 5). The peak spawning was observed during the spring tide period at the full moon every year, although it appeared that more spawners were present in 1995 than other years. In 1995, the spawning did not relate closely to the phases of the moon, in contrast to other years. Surface water temperature during spawning season ranged from 5.3 to 9.8° C in 1994 and from 6.3 to 10.7° C in 1995. No significant relationship was observed between peak spawning and water temperature. Water temperatures ranged from 6.1 to 10.0° C when the largest groups (XL) spawned.

Spawning took place only at the west side of the beach without exception in 1992 and 1993 (Figure 2). The mean slope of this shore was 1 : 12 and the tide level during each spawning changed from 34 to 120 cm (mean sea level in Otsuchi Bay is 86 cm). In 1994 a large pine tree had fallen down from the cliff and covered the usual spawning area. In this year behavior during the two pre-spawning intervals, i.e., the aggregation and approach intervals, was the same as in the other years, but spawning was sometimes observed in unusual areas of the center or eastern part of the beach (Figure 2). When the tide level was low enough for the water to retreat more

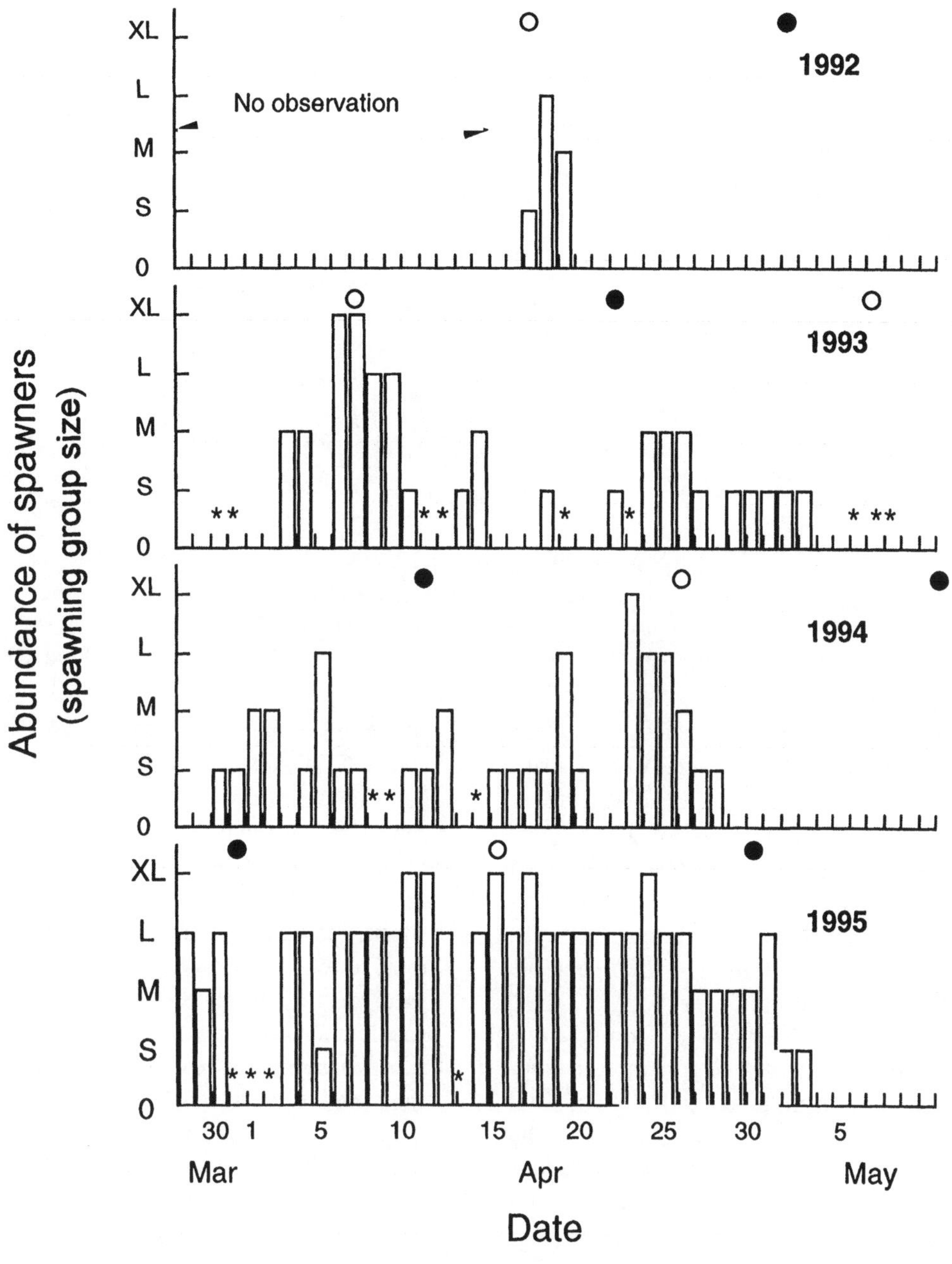

Figure 5. Spawning density of chika on Akaiso Beach in 1992–1995. The categories S, M, L, XL and 0 indicate that the spawning group was small (less than 50 individuals of spawners), medium (50–200 fish), large (200–500 fish), extremely large (more than 500 fish), or that no fish were present. Open and filled circles show full and new moon, respectively. Asterisks show dates when no observations were made due to rough sea conditions.

seaward than the fallen tree, fish spawned at the same site as in other years. After the tree was removed from the beach by a typhoon in autumn 1994, the spawning in 1995 occurred in the same place as in 1992 and 1993.

Distribution of eggs and sand grain size

Each deposited egg usually adhered to a few sand grains 1–2 mm in diameter. Eggs were scattered separately and did not stick together in clumps. Fish

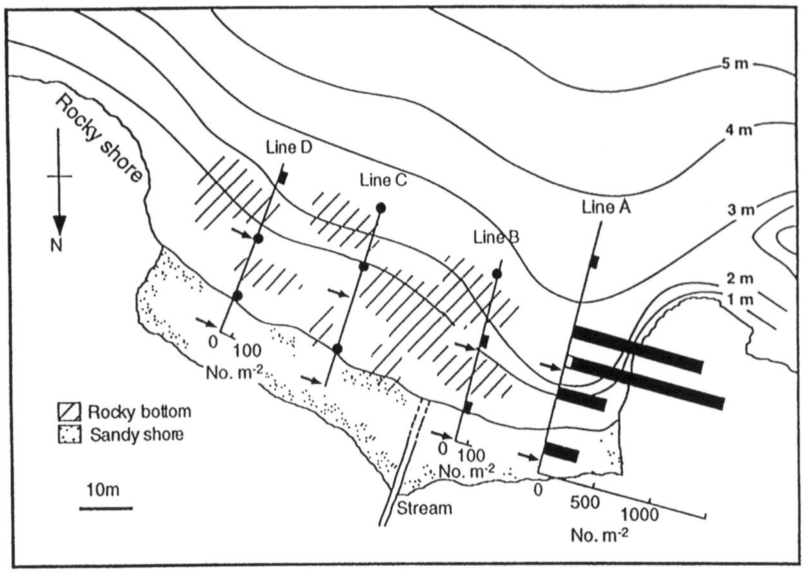

Figure 6. Egg density on the four transects at Akaiso Beach, 27 April 1992. Solid circles on transects are the sampling stations without eggs. In the bar charts, black and white parts indicate live and dead egg density, respectively. Shore line and isobathometric lines are at the time of mean sea level in Otsuchi Bay. Sediment size composition (Figure 7) was analyzed at seven points between the arrows on each transect.

spawned near the shoreline at the west side of the beach in 1992, but almost all the eggs were distributed around the 'step' at 1–2 m depth along transect A just off the spawning site (Figure 6). The highest density, about 1500 eggs m^{-2}, was observed at 2 m depth just below the step along transect A. No eggs were taken along transect C and numbers of eggs collected along transects B and D were negligible compared with those along transect A.

Sand in Akaiso Beach originates from granite. Bottom sand along transect A was composed mainly of grains < 1 mm in diameter (60–90% of total sand grains in weight), which were coarser than the sand sampled along the other transects. The center of egg distribution were composed of coarser sand than those of other areas in Akaiso Beach (Figure 7).

Discussion

Function of the three spawning intervals

Chika spawned only at a restricted site on the western side of Akaiso Beach in 1992, 1993 and 1995, but environmental change due to a fallen tree caused a change in the spawning sites in 1994. However, the 'aggregation' and 'approach' intervals were observed at the western site as in other years. These facts suggest that the aggregation interval, together with the early approach interval, have evolved to ensure high reproductive success through searching for a suitable spawning site, characterized by coarse sand grains for egg deposition and a strong backwash for egg dispersal. The three intervals, aggregation approach, and spawning, recognized in the spawning process are considered closely integrated with each other and ensure successful mating. Similar searching behavior for a proper spawning site just before initiation of spawning were reported in Atlantic silverside and California grunion (Middaugh et al. 1981, Walker 1952). It was suggested that they also await a suitable time and tide level. Chika spawned just after dark, but they showed complex prespawning behavior before dark, suggesting that they were assessing spawning habitat.

Sex and age composition of spawners

Males nearly always outnumbered females at the spawning ground on Akaiso Beach. Yanagawa

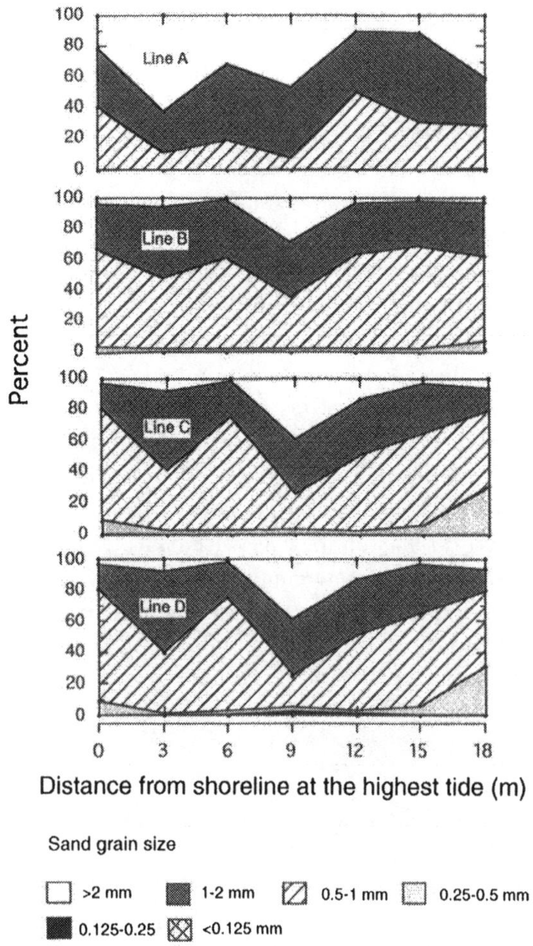

Figure 7. Size composition of the sand grains on the four transects sampled at regular intervals at Akaiso Beach. Seven sampling stations were set every 3 m from the uppermost shoreline on each transect (see Figure 6). The area from 0 to 13 m is the intertidal zone on each transect. Graphs exhibits change in the percentage by weight of each grain size on transects A–D.

(1981) reported that the sex ratio of chika schools in Hokkaido was almost equal in the nonbreeding season, but shifted to male groups or female groups at the beginning of spawning season. Based on histological examination of the gonads (Hirose 1996), mature females spawn all of their eggs during one spawning run, but males mature earlier in the spawning season and participated in spawning more than once. This sexual difference in the spawning mode would explain the dominance of males on the spawning ground. The spawning of chika is performed by one female and a number of males (Table 2). This type of spawning may cause sperm competition and uncertainty of fertilization for males, but males could compensate for these unstable factors by repeated participation in spawning.

Spawning groups of chika were composed mainly of age 0+ fish and some age 1+ fish. In Hokkaido, spawning schools of chika are also mainly composed of age 0+ fish, but the number of age 1+ fish is higher than in Otsuchi Bay (Yanagawa 1981). For example, the proportion of age 1+ and older females in daily samples during 1974 at Hakodate in southern Hokkaido ranged from 15.4% to 61.4%. Yamagawa (1981) also reported a change in age composition of spawning groups as the spawning season progressed. On the northwest coast of North America, age 1+ surf smelt *H. pretiosus pretiosus*, are more numerous than age 0+ fish in spawning schools (Loosanoff 1937, Penttila 1978). Otsuchi Bay is near the southern limit of distribution of chika (Ueno 1984), but the growth in this bay may usually be better than that in more northern distribution areas (Hirose unpublished data). So, fish may mature at age 0+ because of fast growth. On the other hand, the very low proportion of age 1+ fish on the spawning ground suggests that the life span of chika may be only one year in Otsuchi Bay.

Distribution of eggs and selection of spawning site

The center of egg distribution in Akaiso Beach was located off the spawning site where the sand grain size was the coarsest, most particles being larger than 0.5 mm (Figure 7). This indicates that water velocities were strongest there because sand grains larger than 0.1 mm require higher water velocities to be transported (Sverdrup et al. 1961). So, the average current velocity at this spawning site is thought to be the strongest on the beach. These facts also suggest that chika select spawning sites that are more suitable for offshore egg dispersal, because they spawn only in back wash. When artificial insemination was done in calm water, the eggs easily attached to each other and formed a clump (Hirose unpublished data), but eggs in the field are distributed separately. Separate egg distribution is also thought to be more efficient to avoid predation (see Bailey & House 1989). Coarser sand grains along transect A suggest that the current is stronger and interstices of sand grains are larger than in other areas. Eggs are apparently strong enough to withstand mechanical shock by sand, since they were not broken in hand-treated sand samples. Strong current and the larger interstices in larger sand grains would ensure a good oxygen supply to the eggs. Sand size composition and/or water current are known to be the key factors for American surf smelt, capelin, ayu, *Plecoglossus altivelis*, and many salmonids in selection of their spawning sites (Thompson et al. 1936, Shiraishi & Suzuki 1962, Nishida 1978, Frank & Leggett 1981, Kondolf et al. 1991, Tachihara & Kimura 1991). These fishes are considered to have evolved the ability to select suitable spawning sites in order to increase survival of eggs and embryos. Chika may also choose their spawning ground for the same reason.

Spawning time

The spawning peak for chika occurs during the spring tide at full moon. During the spawning season at Otsuchi Bay, flood tide occurs just before dark at the spring tide period. That is, chika spawn at the beginning of the ebb tide, which would also help their sinking eggs to disperse seaward efficiently. American surf smelt, California grunion, Atlantic silverside and grass puffer spawn near shore at spring tide at full or new moon (Loosanoff 1937, Baxter 1960, Nozaki et al. 1976, Johannes 1978, Middaugh 1981, Shiraishi 1986). This specific tide level at spring tide was selected according to their spawning habits and mode of egg distribution to increase the probability of their reproductive success and survival of their developing embryos (Baxter 1960, Middaugh 1981, Middaugh et al. 1981, Shiraishi 1986).

In the period without spring tide, high tide is not synchronous with sunset, but the spawning of chika was observed only just after dark regardless of tidal level. This suggests that sunset, namely change in light condition, is a cue that starts spawning. Many fishes which release pelagic eggs spawn at dusk when the numbers of planktivores are reduced (Johannes 1978), since both diurnal and nocturnal fishes are inactive at that time (Helfman 1993). During spawning, there were no predators of adult chika and their eggs except for fishermen. Spawning just after dark would be an effective method to avoid predation for both newly spawned eggs and spawning adults.

The sandy beach is easily affected by many environmental factors such as wave strength that changes daily, which may cause eggs and embryos of chika to encounter various environmental conditions. It would be advantageous for chika to have a long spawning period to adapt to this environmental variability. Although spring tide is the best condition for spawning, chika may adopt a 'bet hedging' style (Lambert 1984, Lambert & Ware 1984). This may be the reason that spawning of chika shows weaker synchroneity with tidal cycles than grass puffer and California grunion.

Acknowledgements

We thank John Paxton, of the Australian Museum, and Izumi Akagawa, University of Miyazaki for critically reading our manuscript. We are indebted to K. Morita, T. Kawamura and K. Hirano, and other staff of Otsuchi Marine Research Center, Ocean Research Institute, University of Tokyo for the field sampling. Special thanks are due to K. Takahashi and other staff of the Plankton Laboratory for help in the field work and fruitful discussion during this study.

References cited

Bailey, K.M. & E.D. Houde. 1989. Predation on eggs and larvae of marine fishes and the recruitment problem. Adv. Mar. Biol. 25: 1–83.

Baxter, J.L. 1960. Inshore fishes of California. California Department of Fish and Game, Sacramento. 80 pp.

Frank, K. & W.C. Leggett. 1981. Subtidal occurrence of larval capelin (*Mallotus villosus*) as related to variation in the intertidal distribution of their eggs. Rapp. P.-V Réun. Cons. Int. Explor. Mer. 178: 179–180.

Hart, J.L. 1973. Pacific fishes of Canada. Fish. Res. Board Can. Bull. 180. 740 pp.

Hamada, K. 1956. Ecological study of *Hypomesus japonicus*, I.

Conclusion about age. Bull. Fac. Fish. Hokkaido Univ. 7: 219–224 (in Japanese).

Hirose, T. 1996. Fisheries ecology of chika, *Hypomesus pretiosus japonicus*, in Otsuchi Bay, Northeastern Japan. Ph.D. Thesis, University of Tokyo, Tokyo. 157 pp. (in Japanese).

Helfman, G.S. 1993. Fish behavior by day, night and twilight. pp. 479–512. *In:* T.J. Pitcher (ed.) Behaviour of Teleost Fishes, 2nd edition, Chapman & Hall, London.

Jangaard, P.M. 1974. The capelin (*Mallotus villosus*). Fish. Res. Board Can. Bull. 186. 70 pp.

Johannes, R.E. 1978. Reproductive strategies of coastal marine fishes in the tropics. Env. Biol. Fish. 3: 65–84.

Kondolf, G.M., G.F. Cada, M.J. Sale & T. Felando. 1991. Distribution and stability of potential salmonid spawning gravels in steep boulder bed streams of eastern Sierra Nevada. Trans. Amer. Fish. Soc. 120: 177–186.

Lambert, T.C. 1984. Larval cohort succession in herring (*Clupea harengus*) and capelin (*Mallotus villosus*). Can. J. Fish. Aquat. Sci. 41: 1552–1564.

Lambert, T.C. & D.M. Ware. 1984. Reproductive strategies of demersal and pelagic spawning fish. Can. J. Fish. Aquat. Sci. 41: 1565–1569.

Loosanoff, V.L. 1937. The spawning run of the Pacific surf smelt, *Hypomesus pretiosus* (Girard). Int. Rev. Hydrobiol. Hydrogra. 36: 170–183.

Middaugh, D.P. 1981. Reproductive ecology and spawning periodicity of the Atlantic silverside, *Menidia menidia* (Pisces: Atherinidae). Copeia 1981: 766–776.

Middaugh, D.P., G.I. Scott & J.M. Dean. 1981. Reproductive behavior of the Atlantic silverside, *Menidia menidia* (Pisces: Atherinidae). Env. Biol. Fish. 6: 269–276.

Nelson, J.S. 1994. Fishes of the world, 3rd edition. John Wiley & Sons, New York. 600 pp.

Nishida, M. 1978. Spawning habit of the dwarf ayu-fish in Lake Biwa. Bull. Japan. Soc. Sci. Fish. 44: 577–585 (in Japanese).

Nozaki, M., T. Tsutsumi, H. Kobayashi, Y. Takei, T, Ichikawa, K. Tsuneki, K. Miyagawa, H. Uemura & Y. Tsutsumi. 1976. Spawning habit of the puffer, *Fugu niphobles* (Jordan & Snyder) I. Zool. Mag. 85: 156–168 (in Japanese).

Penttila, D. 1973. Observations on some Puget Sound spawning beaches of the surf smelt (*Hypomesus pretiosus*, Girard). Suppl. Prog. Rep. Wash. Dept. Fish., Seattle. 20 pp.

Penttila, D. 1978. Studies of the surf smelt (*Hypomesus pretiosus*) in Puget Sound. Wash. Dept. Fish. Tech. Rep. 42: 1–47.

Scott, W.B. & E.J. Crossman. 1973. Freshwater fishes of Canada. Fish. Res. Board Can. Bull. 184. 966 pp.

Shiraishi, M. 1986. Study on the spawning and early ecology of panther puffer *Takifugu pardalis* and grass puffer *Takifugu niphobles*. Ph.D. Thesis, University of Tokyo, Tokyo. 217 pp. (in Japanese).

Shiraishi, Y. & N. Suzuki. 1962. The spawning activity of ayu-fish, *Plecoglossus altivelis*. Bull. Freshwat. Fish. Res. Lab. 12: 83–107 (in Japanese).

Sverdrup, H.U., M.W. Johnson & R.H. Fleming. 1961. The oceans, their physics, chemistry, and general biology. Chales E. Tuttle Company, Tokyo. 1059 pp.

Tachihara, K. & S. Kimura. 1991. Maturity and spawning of land locked ayu *Plecoglossus altivelis* in Lake Ikeda, Kagoshima Prefecture. Nippon Suisan Gakkaishi 57: 2187–2192 (in Japanese).

Thompson, W.F., F.H. Bell, H.A. Dunlop, P. Schultz & R.V. Cleve. 1936. The spawning of the silver smelt, *Hypomesus pretiosus*. Ecology 17: 158–168.

Ueno, T. 1984. *Hypomesus pretiosus japonicus*. pp. 33. *In:* H. Masuda, K. Amaoka, C. Araga, T. Ueno & T. Yoshino (ed.) The Fishes of the Japanese Archipelago, 2nd edition, Tokai University Press, Tokyo.

Walker, B.W. 1952. A guide to the grunion. Calif. Fish. Game 38: 409–420.

Yanagawa, H. 1981. Studies on the local form and dispersal of the chika, *Hypomesus pretiosus japonicus* (Brevoort) in Japan. Mem. Fac. Fish. Hokkaido Univ. 27: 1–77.

Environmental Biology of Fishes **52**: 225–229, 1998.

Timing and sites of parturition of the viviparous scorpionfish, *Sebastiscus marmoratus*

Hiroshi Fujita & Masanori Kohda
Laboratory of Animal Sociology, Department of Biology, Faculty of Science, Osaka City University, Sumiyoshi-ku, Osaka 558, Japan (e-mail: fujita@sci.osaka-cu.ac.jp)

Received 6.6.1997 Accepted 4.12.1997

Key words: night spawning, larva predation, internal fertilization, Scorpaenidae

Synopsis

Timing and sites of parturition of the viviparous scorpionfish *Sebastiscus marmoratus* were investigated both in aquaria and natural habitats of a rocky coast in southern Japan. Females released 4 mm TL larvae during the early night in winter. Night spawning may be advantageous in avoiding predators of both larvae and spawners, because most such predators are inactive during the night. Females inhabiting shallow areas (< 6 m depth) moved to deeper sites for parturition, but ones inhabiting deeper areas (6–8 m depth) gave birth within their usual habitats. This migration may be advantageous in effective larval dispersal.

Introduction

The ecological significance of the timing and sites of spawning has been studied in many pelagically spawning, oviparous reef fishes (Thresher 1984). In many cases, spawning timing and sites seem to be related to avoidance of egg predation. For example, many fishes spawn at dusk when egg predators are inactive or at specific tidal cycles in the daytime when eggs can be effectively dispersed offshore where the predators are in low density (Johannes 1978, Taylor 1984). Some fishes spawn at certain sites with strong currents, usually removed from their foraging areas, where eggs are quickly dispersed offshore (Robertson & Hoffman 1977, Johannes 1978).

Many fishes with internal fertilization release pelagic larvae. However, in contrast to pelagic egg spawners, the timing and sites of parturition of pelagic larvae have not been studied.

The Scorpaenidae is a large family that includes many viviparous species (Wourms 1991). In many species of this family, general reproductive ecology (e.g. season, fecundity) has been described based on laboratory studies of field-sampled specimens (e.g. Moser 1967, Fujita 1958, Eldridge et al. 1991, Kusakari 1995). *Sebastiscus marmoratus* is a member of this family, being common on rocky coasts in southern Japan. This fish, which has strong site fidelity, usually appears within narrow home ranges in the daytime and hides in holes or crevices during the night (Fujita unpublished data). Copulation occurs in early winter (Fujita & Kohda 1996). Females release tens of thousands of pelagic larvae (4 mm TL) several times from mid-winter to early spring (Mizue 1959, Shiokawa 1962). We found that females of this fish released larvae during the early night, and females in shallow areas migrated to deeper sites for parturition. Here we describe the parturition of this fish, and discuss its adaptive significance.

Materials and methods

We observed parturitions of *S. marmoratus* at Morode Beach (33° 00′ N, 132° 30′ E), on the west coast of Shikoku Island, Japan, from mid-January to mid-March in 1993, 1995 and 1996, with the aid of SCUBA. We set up an 18 m × 40 m quadrat on a rocky bottom at a depth of 3–8 m. Before the parturition season, we captured most females of this species in the quadrat, marked them by removing one to three dorsal fin spines and measured standard length (SL) underwater. The fish were then released at the capture sites.

Direct observations of parturition in the study area were quite difficult. To estimate parturition time, we visually classified the abdomen size of 27 marked females into three categories (flat, distended, largely distended) during 16 observation periods in early morning, evening and at midnight during 9–15 February 1993.

When the largely distended abdomen of a female changed to flat during the next observation period, she was regarded as having given birth between them. We made focal observations on one female with a largely distended abdomen during 17:00–21:00 h on 1–3 March 1993.

To describe spawning behaviour and estimate more exact spawning time, we observed females just prior to parturition in aquaria in 1993, 1995, and 1996. Such females could be easily distinguished from others by their largely distended abdomen and darker body colour (see Results). We captured a total of 20 such females in the quadrat, and each was put in a lighted 63 l aquaria (30 × 60 × 35 cm) within 1 h of capture. Around sunset, we turned off the light and covered the aquaria with a black nylon sheet. To confirm when they gave birth, we checked the tank at 15–25 min intervals in 12 females and at 1 h or more in the other eight. To examine the ovary conditions after spawning, 10 of them were fixed in 10% formalin.

Results

Parturition behaviours were observed in four females in aquaria. From about one hour before par-

a

b

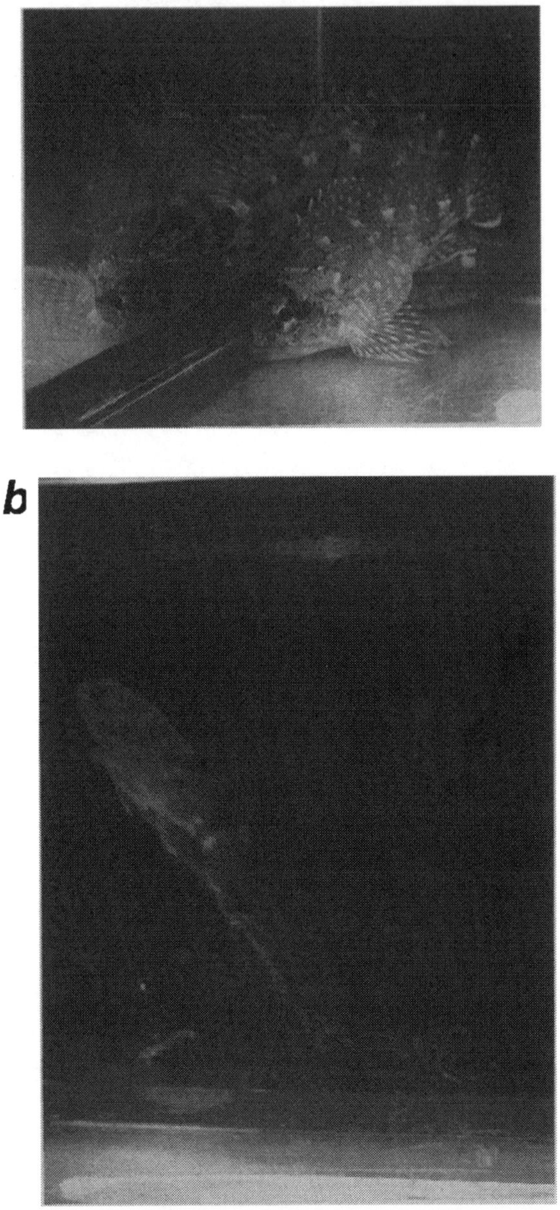

Figure 1. a– Pre-parturition female, and b– female at parturition of *Sebastiscus marmoratus*. Photographs by N. Ohnishi.

turition, they were usually stationary on the bottom in a tail-up position (Figure 1a), and sometimes turned and swam near the bottom. Thereafter, they suddenly swam upward and released larvae along with mucous material for several seconds (Figure 1b). After parturition, they quickly sawm down and were stationary on the bottom again. At that time, their abdomens were flat. Within 30 min after parturition, the mucus disappeared and almost all larvae swam and floated near the water surface. Gonadal examination of the females revealed that only unhatched embryos and a few larvae remained in their ovaries. This indicates that almost all larvae of one clutch are released during one parturition ascent.

In the study quadrat, 13 of the 27 females marked and observed appeared to have given birth between observation periods, based on shrinkage of their abdomens. Nine of the 13 females gave birth between successive periods separated by < 17 h intervals (Table 1), whereas the other four gave birth between periods > 22 h apart. Seven of the nine gave birth between evening and early the next morning, and the other two (F7, F8) did by early evening (Table 1). These females were found in their usual home ranges during these periods. The female on which we made focal observations gave birth within her home range around 20:30 h on 3 March 1993, although we missed her parturition. Before parturition, she was stationary on the bottom in the tail-up position as was observed in aquaria. After parturition, she was stationary on the bottom in her home range with a flat abdomen.

In aquaria, all 20 females released larvae before 24:00 h on the capture day. Of 12 females checked at intervals of 15–25 min, 10 females gave birth between 18:00–22:00 h, and the other two (F17, F18) near sunset before lights were turned off; F18 gave birth before entering an aquarium (Table 2).

In the study area, the resident marked females with largely distended abdomens usually remained stationary on the bottom within their home ranges at a depth of 6–8 m. A total of 20 non-resident (i.e. non-marked) females which had darker body color were observed in the study area in late afternoon. These darker non-residents always had largely distended abdomens. After they had been observed in the field for a while, they were captured for observation of parturition behaviors in aquaria. All gave birth within the day (see above). In contrast to resident females, all of the non-residents gradually shifted their position to deeper (i.e. more offshore) areas in the quadrat. They usually shifted about 10 m, and up to 20 m in some cases, before capture. The areas deeper than the quadrat had a sandy bottom, where this fish was rare, in contrast to the shallower areas with rocky bottom. This suggests that the non-resident females may inhabit shallower areas (i.e. < 6 m depth, more onshore) and move to areas deeper than their usual habitats for parturition. There was no difference in body size between the resident (mean SL ± SD = 104.3 mm ± 15.1, n = 13) and the non-resident females (107.8 mm ± 24.9, n = 20; Mann-Whitney U-test, U = 124.5, p > 0.1).

Table 1. Observation time before and after female *Sebastiscus marmoratus* had apparently given birth (based on reduced abdomen size) in the study quadrat in February 1993.

Female code	SL (mm)	Time, date	
		Before	After
F1	115	16:58, 9	08:01, 10
F2	95	16:58, 9	08:01, 10
F3	95	16:58, 9	08:01, 10
F4	97	17:03, 10	08:00, 11
F5	98	17:07, 11	08:00, 12
F6	98	17:07, 11	08:00, 12
F7	103	01:23, 14	17:46, 14
F8	110	07:34, 14	17:46, 14
F9	150	16:59, 14	05:47, 15

Table 2. Parturition time of female *Sebastiscus marmoratus* in aquaria, estimated as the median between observation times (at 15–25 min intervals) before and after parturition (* based on direct observation). Sunset was at about 17:00 to 18:00 h during the direct observations.

Female code	SL (mm)	Parturition date	Time placed in aquarium	Parturition time
F11	95	27 Feb 1993	17:20	20:35*
F12	100	9 Mar 1993	18:00	18:34*
F13	95	13 Mar 1993	17:35	19:35*
F14	105	12 Jan 1995	11:45	18:51
F15	133	25 Jan 1995	16:56	20:52
F16	158	26 Jan 1995	17:10	21:25
F17	100	6 Feb 1995	16:45	16:57
F18	84	10 Feb 1995	–	16:52
F19	103	14 Feb 1995	17:20	18:00*
F20	92	24 Feb 1995	17:25	21:22
F21	120	9 Mar 1995	17:41	21:17
F22	84	20 Jan 1996	16:50	18:19

Discussion

Some viviparous scorpionfishes have been suggested to give birth during the night (Eldridge et al. 1991, Kusakari 1995). Our observations indicate that female *S. marmoratus* gives birth between dusk and midnight, mainly during the early night. Because most predators of larvae (zooplanktivores) are inactive at night in the study areas (Fujita personal observation), night parturition would be advantageous in avoiding such predators. Moreover, early night parturition would be advantageous for effective dispersal of larvae, because it allows larvae the longest period to disperse under the cover of darkness.

In paternal brooding fishes, such as cardinalfishes or pipefishes, males release larvae during the night (N. Okuda personal communication, S. Watanabe personal communication). Embryos of many reef fishes with demersal eggs, such as damselfishes, also hatch during the early night (Thresher 1984, Kohda 1988). The adaptive significance of the timing of these reproductive events is likely to be the same as for scorpionfish.

Many coral reef fishes that release pelagic eggs also spawn at dusk (Helfman 1993). This may be advantageous in avoiding egg predators, because zooplanktivores are less active at dusk than during the day, although predators on spawners are active at this time (Helfman 1993). Although night spawning is also advantageous against egg predation and even spawner predation (Johannes 1978), many reef fishes that spawn pelagic eggs do not spawn at night. Pelagic egg spawners need mates for egg release, while pelagic livebearers can release larvae without mates. For pelagic egg spawners, synchronous release of gametes in close proximity of mates is essential for egg fertilization. Such synchronization may require visual contract, and therefore be difficult during the night. In contrast, pelagic livebearers are free from this restriction, such as female *S. marmoratus* which gave birth during the night. Similarly, cardinalfishes and pipefishes need no mates to release larvae, which occurs at night.

Many pelagic egg spawners in reef fishes conduct spawning ascent, which may be adaptive for effective egg dispersal (Robertson & Hoffman 1977, Johannes 1978). Spawning ascent of *S. marmoratus* may also have the same significance for larvae.

Some fishes release eggs offshore where predation pressure is low and/or water currents are strong, both of which are advantageous for effective egg dispersal (Johannes 1978). Females of *S. marmoratus* inhabiting shallow areas appeared to move to deeper sites (offshore) for parturition, while females inhabiting deeper areas gave birth in their usual habitats. Although the exact purpose of the parturition movements is not certain, one possibility is that the offshore location results in more effective dispersal of larvae.

Acknowledgements

We are grateful to members of the Laboratory of Animal Sociology of Osaka City University, Y. Yanagisawa and members of the UWA Marine Science Center of Ehime University for their considerable support, and N. Ohnishi for photography. This study was funded by a Research Fellowship for Young Scientists from the Japan Society for the Promotion of Science, and by a Grand-in-Aid for Scientific Research (no. 08640811) from the Japan Ministry of Education, Science, Sports, and Culture.

References cited

Eldridge, M.B., J.A. Whipple, M.J. Bowers, B.M. Jarvis & J. Gold. 1991. Reproductive performance of yellowtail rockfish, *Sebastes flavidus*. Env. Biol. Fish. 30: 91–102.

Fujita, H. & M. Kohda. 1996. Male mating effort in the viviparous scorpionfish, *Sebastiscus marmoratus*. Ichthyol. Res. 43: 247–255.

Fujita, S. 1958. On the egg development and larval stages of a viviparous scorpaenid fish, *Sebastes oblongus* Günther. Bull. Jap. Soc. Scient. Fish. 24: 475–479 (in Japanese).

Helfman, G.S. 1993. Fish behaviour by day, night and twilight. pp. 479–512. *In:* T.J. Pitcher (ed.) Behaviour of Teleost Fishes, 2nd ed., Chapman & Hall, London.

Johannes, R.E. 1978. Reproductive strategies of coastal marine fishes in the tropics. Env. Biol. Fish. 3: 65–84.

Kohda, M. 1988. Diurnal periodicity of spawning activity of permanently territorial damselfishes (Teleostei: Pomacentridae). Env. Biol. Fish. 21: 91–100.

Kusakari, M. 1995. Studies on the reproductive biology and artificial juvenile production of kurosoi *Sebastes schlegeli* (Hilgendrof). Scient. Rep. Hokkaido Fish. Exp. St. 47: 41–124 (in Japanese).

Mizue, K. 1959. Studies on a scorpaenous fish *Sebastiscus marmoratus* Cuvier et Valenciennes – V: on the maturation and the seasonal cycle of the ovaries of the marine ovoviviparous teleost. Bull. Fac. Fish. Nagasaki Univ. 8: 84–110 (in Japanese).

Moser, H.G. 1967. Reproduction and development of *Sebastodes paucispinis* and comparison with other rockfishes off southern California. Copeia 1967: 773–797.

Robertson, D.R. & S.G. Hoffman. 1977. The roles of female mate choice and predation in the mating systems of some tropical labroid fishes. Z. Tierpsychol. 45: 298–320.

Shiokawa, T. 1962. Growth and maturity of the common rockfish, *Sebastiscus marmoratus* Cuvier et Valenciennes. Rec. Oceanogr. W. Japan. 6: 91–102.

Taylor, M.H. 1984. Lunar synchronization of fish reproduction. Trans. Amer. Fish. Soc. 113: 484–493.

Thresher, R.E. 1984. Reproduction in reef fishes. T.F.H. Publications, Neptune City. 399 pp.

Wourms, J.P. 1991. Reproduction and development of *Sebastes* in the context of the evolution of piscine viviparity. Env. Biol. Fish. 30: 111–126.

Environmental Biology of Fishes **52**: 231–242, 1998.

Coexistence of permanently territorial cichlids of the genus *Petrochromis* through male-mating attack

Masanori Kohda
Laboratory of Animal Sociology, Faculty of Science, Osaka City University, Sumiyoshi-ku, Osaka 558, Japan (e-mail: maskohda@sci.osaka-cu.ac.jp)

Received 31.8.1995 Accepted 20.1.1997

Key words: coexistence mechanism, feeding territory, herbivorous fishes, mating, removal experiments, Lake Tanganyika

Synopsis

Coexistence mechanisms of permanently territorial fishes proposed hitherto have been mainly based on interspecific competition. To test the hypothesis of the coexistence through male-mating aggression (Kohda, 1995a), spacing patterns of feeding territories of three coexisting herbivorous cichlids, *Petrochromis polyodon, P. trewavasae* and *P. famula*, were investigated in a homogeneous habitat at a shallow rocky shore in Lake Tanganyika. All three species maintained individual feeding territories that were defended against both conspecifics and congenerics and rarely overlapped either intra- or interspecifically. Territories of three species were scattered and covered almost all rocky bottoms. Territorial attacks usually occurred near the border of feeding territories. But male *P. polyodon*, the largest and socially most dominant fish, attacked conspecific males far beyond their feeding territories, which were widely separated. Removal of some territory owners suggested that such attacks result in the separate distribution of male *P. polyodon* territories. The interstitial space between *P. polyodon* feeding territories resulting from their mating aggression can be occupied by subordinate congeneric fishes. Male-mating attack of dominant males of *Petrochromis* will facilitate the coexistence of other congeners.

Introduction

Many tropical herbivorous fishes have feeding territories. They usually coexist competitively and their territories are arranged in a mosaic with each being occupied by an individual fish (Sale 1974 1991, Kohda 1991, Kohda & Yanagisawa 1992). Much effort has been made to elucidate the mechanism allowing such fishes, usually related species, to coexist (damselfishes: Sale 1974, 1978, 1991, Robertson 1984, Ebersole 1985; surgeonfishes: Robertson et al. 1979, Robertson & Gaines 1986; cichlid fishes: Takamura 1984, Kohda 1991, Kohda & Yanagisawa 1992, Kuwamura 1992). Two main mechanisms for the coexistence have been proposed: lottery competition for space (Sale 1978, 1982, 1991, Chesson & Warner 1981) and micro-habitat segregation (Robertson & Lassig 1980, Takamura 1984, Ebersole 1985, Kohda & Yanagisawa 1992). All are based on interspecific competition.

It is widely accepted that related species can coexist where intraspecific competition is stronger than interspecific competition (Pianka 1978, Tilman 1982). Permanently territorial damselfishes attack conspecifics more strongly than heterospecifics (Myrberg & Thresher 1974, Thresher 1976). Some

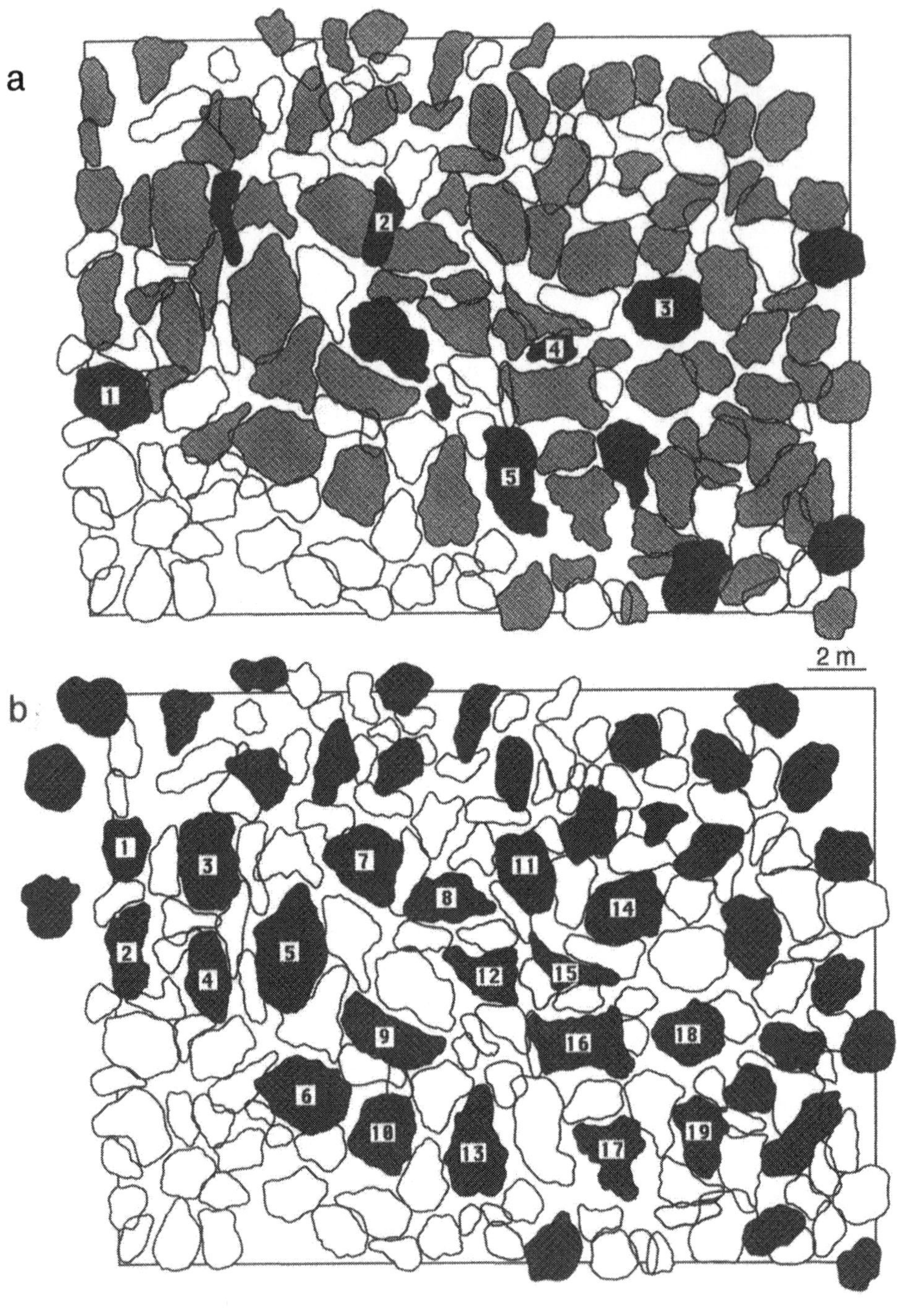

Figure 1. Arrangement of feeding territories of three species of *Petrochromis* in the study area. Solid lines are borders of feeding territories. A number in each territory is a code of its owner: a – shaded areas are territories of *P. polyodon*, solid areas territories of male *P. famula*, and open areas *P. trewavasae*. b – solid areas are territories of male *P. polyodon*, and the others open. Some territories of *P. polyodon* near the study area are added. Data were obtained 22 November to 7 December 1994.

damselfish males and cichlid males attack conspecific males far outside their feeding territories as compared with attacks against heterospecifics (Itzkowitz 1978, Kohda 1984, 1991, 1995b). Curiously, such stronger intraspecific competition for space has been ignored in explanation of the coexistence (Sale 1991). Recently, however, Kohda (1995a) proposed the hypothesis that such male-mating aggression may result in wider separation of the distribution of feeding territories of conspecific males, and provide space which congeneric fishes can inhabit. The coexistence mechanism through male-mating attack and mechanisms through interspecific competition are not mutually exclusive. The aim of this paper is to test the hypothesis of a coexistence mechanism through male-mating attack.

In Lake Tanganyika, six species of the genus *Petrochromis* that graze on unicellular and micro-filamentous algae usually cohabit in shallow rocky areas (Yamaoka 1982, 1983, 1991, Takamura 1984, Kohda 1991, 1995a). Both sexes of *Petrochromis* fishes defend a permanent feeding territory against conspecifics and congenerics, with territories arranged in a mosaic pattern (Takamura 1984, Kohda 1991, 1995a). Larger individuals are intra- and interspecifically dominant over neighbour territory owners (Kohda 1991, 1995a). Males are usually larger than conspecific females. Males of *Petrochromis* fishes usually attack conspecific males far beyound their feeding territories (Kuwamura 1987, Kohda 1991, 1995a, b). Species of *Petrochromis* are maternal mouthbrooders. Females deposit eggs in nests in male territories and incubate them for a few months (Kuwamura 1986, Kohda personal observation). Reproductive activities continue throughout the year. In this study, I will describe the spacing pattern and territorial behaviour of three cohabiting species of *Petrochromis*, and conduct removal experiments to examine the effect of male-mating attacks on the spacing of territories.

Materials and methods

Study area

Lake Tanganyika harbours more than 170 endemic cichlid species (Poll 1986, Brichard 1989, Coulter 1991). This study was conducted at a rocky shore of Kasenga Point (8° 43′ S, 31° 08′ E) at the southern end of the lake, 7 km north-east of Mpulungu, Zambia. At this shore, *Petrochromis* species maintained feeding territories. Removal of territorial fishes result in recolonization of their territories within a few days, suggesting a limitation of territory space in this area (Hori 1991, Kohda 1995a).

Petrochromis species exhibit micro-habitat segregation, according to rock types and water depth (Takamura 1984). To minimize these effects on species coexistence, a study site was selected in a homogeneous rocky substrate within a narrow depth zone. A 26 m (parallel to the shore line) × 18 m quadrat divided into 2 m × 2 m grids with string was set at a depth of 2–5 m, which included a study area of Kohda (1995a). A variety of rock sizes, 10 cm to 1 m in diameter, composed the rocky substrate that provided an abundance of foraging sites and shelter holes for fishes throughout the study area. The configuration of rocks more than 0.5 m in diameter in the quadrat was mapped on plastic paper (scale = 1/40), which was used to record fish behaviour.

Observations of territorial behaviour

Underwater observations of *Petrochromis* were made from October to December 1994. Eighty-two *P. polyodon*, 73 *P. trewavasae* and 12 *P. famula* were observed. Most fishes were sexed on the basis of the sexual dichromatism (Kuwamura 1986) and reproductive behaviour. Most males of *P. polyodon* and *P. famula* were identified by the number of 'egg spots' on anal or dorsal fins, specific body coloration and injury and/or body size. Total length was estimated for each fish by eye and measured to the nearest 5 mm using a scale placed near the fish. Juveniles were omitted from observations.

Swimming tracks made by each fish were observed during a 15 min period and were recorded twice on the plastic paper with an interval more than 1 month between observations. The outermost area of tracking during foraging was defined as the border of a feeding territory, and attack points usually coincided with the border. During the 15 min obser-

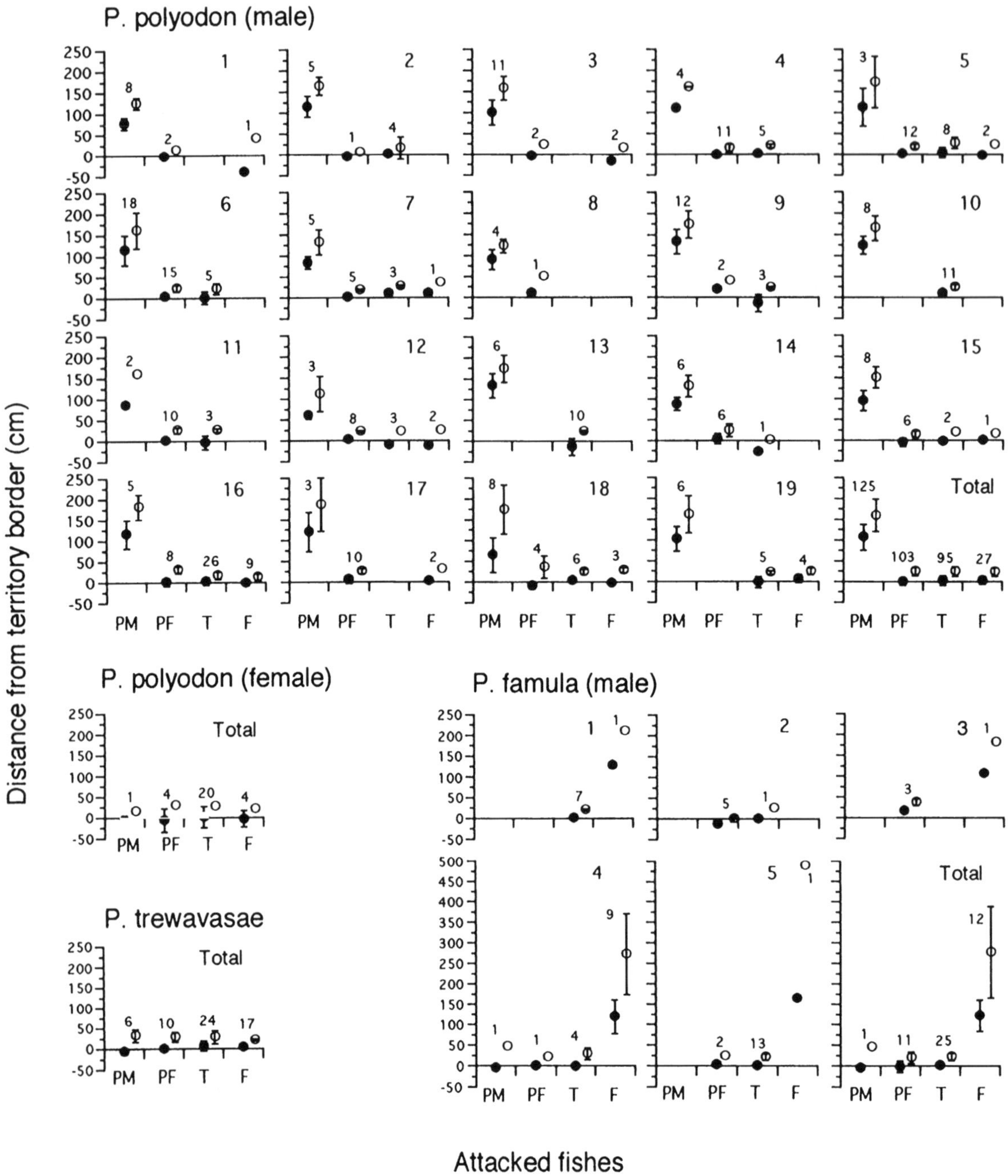

Figure 2. Distance between territory borders and (1) attack points (solid circle) and (2) end points of chasing (open circle) by *Petrochromis polyodon* male (PM) and female (PF), *P. trewavasae* (T) and *P. famula* (F). Numbers in the upper right of graph panels refer to territories in Figure 1. Vertical lines with sample sizes are SD.

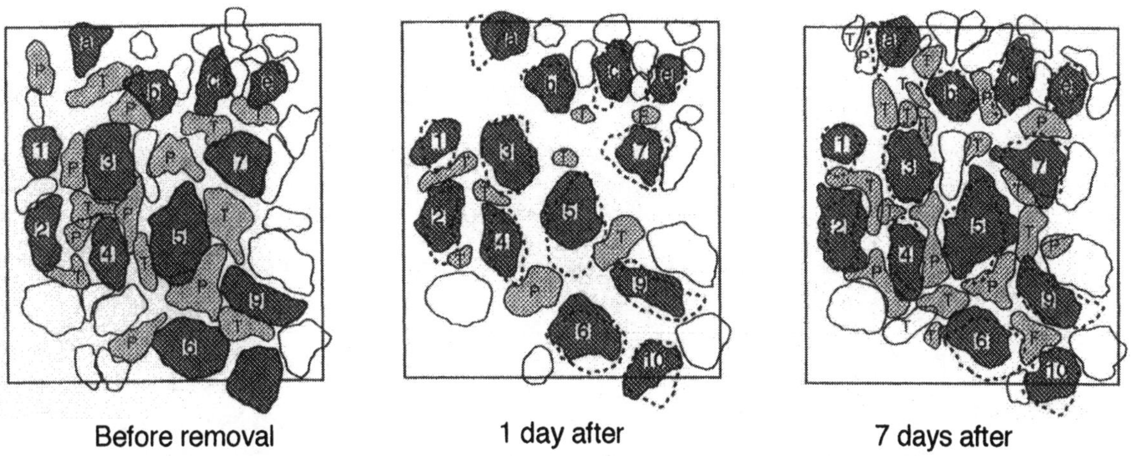

Figure 3. Territory arrangement of *Petrochromis polyodon* (P), *P. trewavasae* (T) and *P. famula* (F) before and after removal of residents other than *P. polyodon* males. Darker areas are territories of male *P. polyodon* of which codes are shown by numbers or small letters. Areas enclosed with dashed lines are territory sites of male *P. polyodon* before the experiment. Territories of fish removed and those immigrating are shown by light shading.

vation period, the locations of intra- and interspecific aggression, courtship behaviour and foraging sites were recorded on the map.

Removal experiments

According to the lottery hypothesis, a vacant area after removal of a territory owner of a species will be occupied by chance by a fish of the same species or another territorial species (Sale 1978, 1991). In contrast, 'male-mating attack' hypothesis predicts that a vacant area adjacent to a feeding territory of a reproductively active male will be occupied by fishes other than conspecific males (Kohda 1995a). To examine the pattern of occupation of vacant territories, a total of 17 territory owners, 7 *P. polyodon* females and 10 *P. trewavasae*, all adjacent to feeding territories of male *P. polyodon*, were removed and fish immigrating into the vacant areas were observed 1 and 7 days after the removal.

'Male-mating attack' hypothesis predicts that removal of a territorial male will affect the arrangement of the territories of conspecific males located nearby. To test this, five males of *P. polyodon* were removed, and territories of neighbours and conspecific males around them were observed after the removal.

Results

Territories and territory arrangement of the three cichlids

Individuals of *Petrochromis polyodon, P. trewavasae* and *P. famula* had a territory from which conspecifics and congeners were repelled. Each fish foraged exclusively within its territory. Almost all available substrate was covered with their feeding territories (Figure 1a). Each territory of male *P. polyodon* and *P. famula* was maintained at the same size, shape, and location over one month. Most other territories were similarly maintained by the same species of fish of the same size, and maybe the same individuals. Most territories abutted territories held by conspecifics and congenerics. However, territories of male *P. polyodon* were usually separated 1 m or more from each other (Figure 1b). The space between territories of male *P. polyodon* was usually occupied by conspecific females or congeneric fishes.

Territories of male *P. famula* were also separated. The lower left part of the study area was deeper than 5 m, which was occupied by *P. trewavasae* that were predominant on rocky bottoms deeper than the study area.

Mean total length ($\pm$ SD) of male *P. polyodon* (20.8 cm $\pm$ 0.8, n = 43) was larger than conspecific

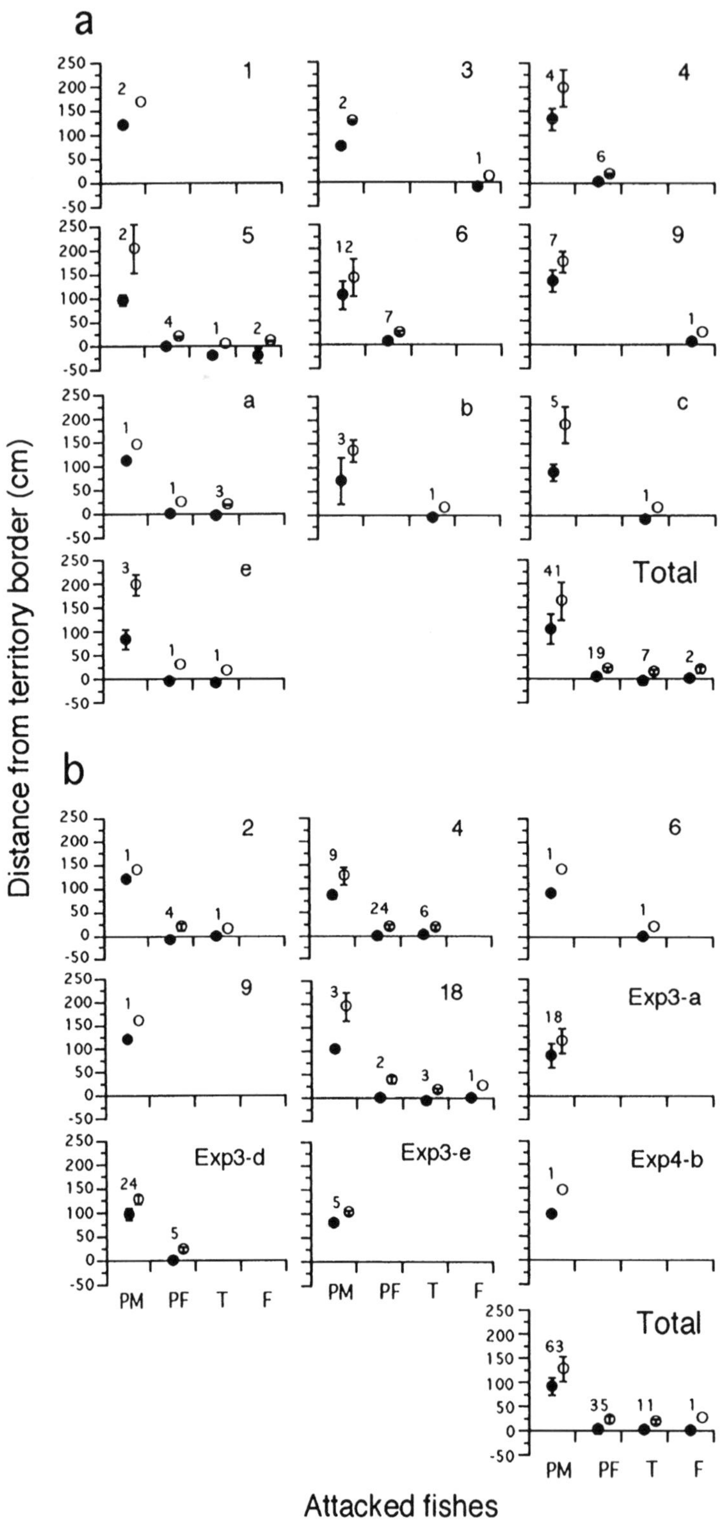
a
1
3
4
5
6
9
a
b
c
e
Total
b
2
4
6
9
18
Exp3-a
Exp3-d
Exp3-e
Exp4-b
Total
Distance from territory border (cm)
PM PF T F
Attacked fishes

←

Figure 4. Distance between territory borders and (1) attack points (solid circle) and (2) end points of chasing (open circle) by *Petrochromis polyodon*: a – after removal of fishes other than male *P. polyodon* and b – after removal of male *P. polyodon*. Numbers and small letters are male codes appearing in Figures 1 and 3. See Figure 2 for abbreviations.

females (17.6 cm ± 0.8, n = 39, Mann Whitney U-test, z = 7.75, $p < 0.01$), *P. trewavasae* (16.5 cm ± 3.1, n = 73, z = 7.93, $p < 0.01$) and *P. famula* (17.4 cm ± 1.4, n = 11, z = 5.23, $p < 0.01$). Each territorial male *P. polyodon* was larger than its immediate neighbours (by 4.9 cm ± 3.8, n = 158), and was dominant over them (Kohda 1995a).

Attack between territorial neighbours and between conspecific males

Petrochromis polyodon attacked territorial neighbours that were conspecific females and heterospecifics near their borders and chased them for several decimeters (Figure 2). In contrast, these males attacked conspecific males far beyond the borders and chased them over longer distances. When each male was inside his territory, aggressive interactions did not occur, even if they were in close proximity. But if a male went out his territory, the territorial 'neighbour' male attacked him at points usually over neighbouring conspecific females or heterospecifics. The attacked male usually fled back to his own territory.

As with *P. polyodon* males, *P. famula* males attacked conspecific males far outside their territories, but attacked congeneric fishes only near the border of their territories. Attacks between male *P. trewavasae* were not observed.

Female *P. polyodon* and *P. trewavasae* of both sexes attacked neighbours near the border of their territories.

Male mating activities

Within its territory, each male *P. polyodon* had one or two spawning sites. Females mostly from outside the study area visited males for mating, usually swimming more than 1 m above the substrate. Such swimming did not elicit aggressive responses from territory owners. When a male found a visiting female, he quickly approached and courted her, usually in spaces around his territory (89% of 83 courtship events of 17 males). Courted females usually followed him and arrived at the spawning site, though most left without mating. Male-male aggression occasionally occurred during the sexual affair. Males rarely courted neighbouring conspecific females, many of which seemed to incubate eggs.

Male *P. famula* also courted conspecific females usually outside their territories (86%, n = 28) and led them to the spawning site within their territories.

Removal experiments

After removal of territorial fish other than male *P. polyodon*, the resident male *P. polyodon* rarely enlarged or shifted their feeding territories to the vacant areas (Figure 3). Even when vacant areas were not fully occupied, males attacked conspecific males far beyond their feeding territories as they did before the removal experiment began (Figure 4a).Consequently, a total of 18 female *P. polyodon* or congenerics immigrated or shifted into the areas. New male *P. polyodon* never immigrated. This result favours the male-mating attack hypothesis. The immigrants ($\bar{x} \pm SD$ = 14.5 cm ± 2.7, n = 18) were smaller than removed fish (17.2 cm ± 1.1, n = 17, Mann Whitney U-test, U = 69.5, $p < 0.01$), and much smaller than resident male *P. polyodon* (20.9 cm ± 0.6, U = 0, $p < 0.001$). Shapes of newly established territories seemed to be formed on a pattern of the interstice, suggesting that territory formation of newcomers does not largely affect territory arrangement of resident male *P. polyodon*.

In contrast, removal of male *P. polyodon* resulted in the shift of territories of 'neighbouring' conspecific males (Figure 5). The rearrangement had already started by the next day's observation. Males that enlarged and/or shifted their territories to va-

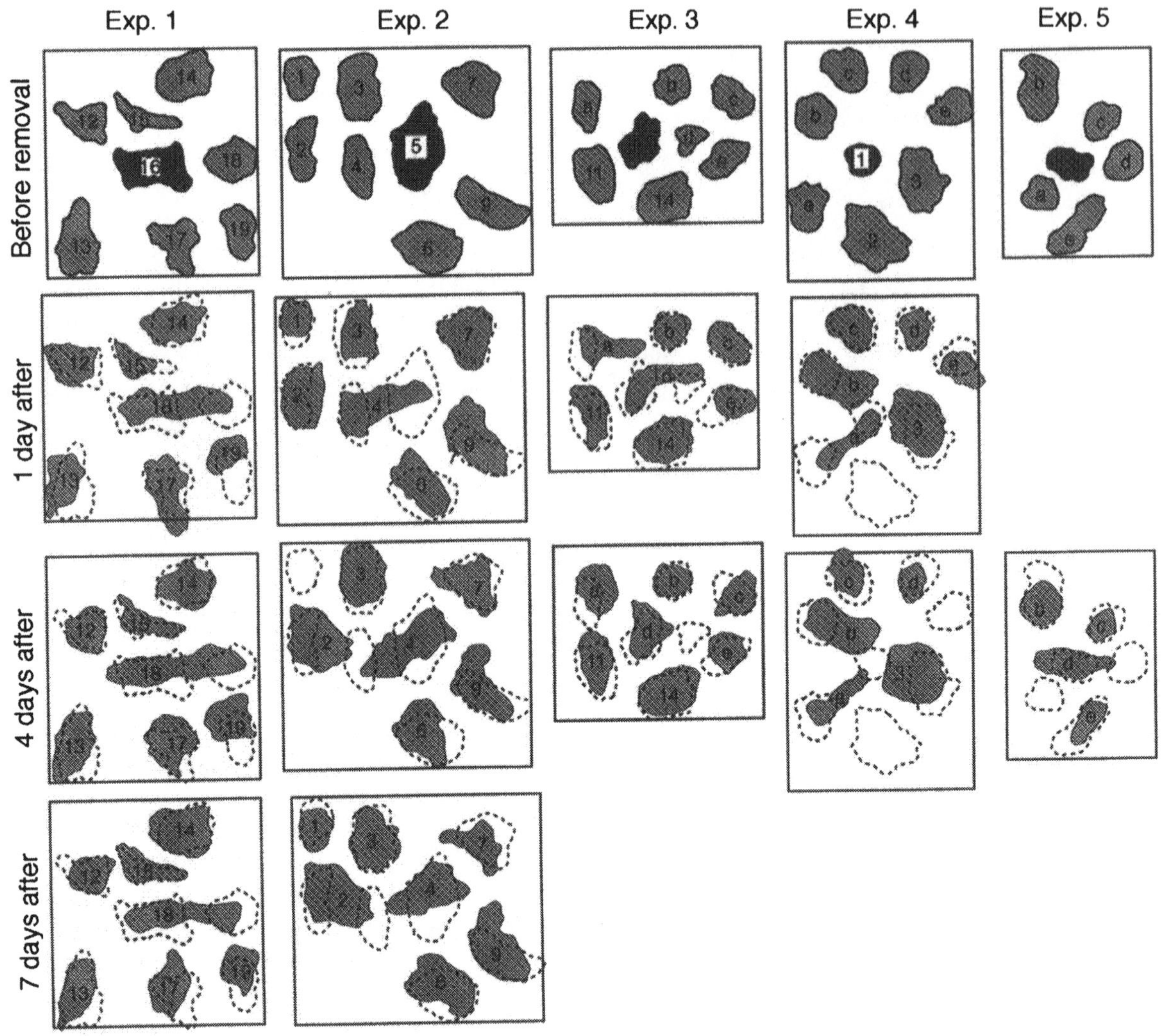

Figure 5. Territory arrangement of *P. polyodon* after removal of 5 male *P. polyodon*. Owners of black territories were removed. Dashed lines show territory sites before removal. Numbers in territories are codes of owners appearing in Figure 1. Small letters are also male codes. Territories other than male *P. polyodon* are omitted.

cant areas usually invaded one or two territories of immediate neighbours. These smaller neighbours, irrespective of species, could not stop the male invasion. After the removal, remaining males attacked conspecific males far outside their territories, but attacked the other fish only near the border of territories as occurred before the removal experiment (Figure 4b). Four and seven days after removal, rearranged territories of male *P. polyodon* were separated from each other as before (Figure 5).

Comparison in arrangement of territories of male P. polyodon *between 1992 and 1994*

In 1992, male *P. polyodon* territories were separated from each other (Figure 6). It is noteworthy that, out of 16 territories observed in 1992, six males could be re-identified in this study and they remained at sites where they were previously observed. The territory arrangement of male *P. polyodon* is likely to be continously stable for a long time.

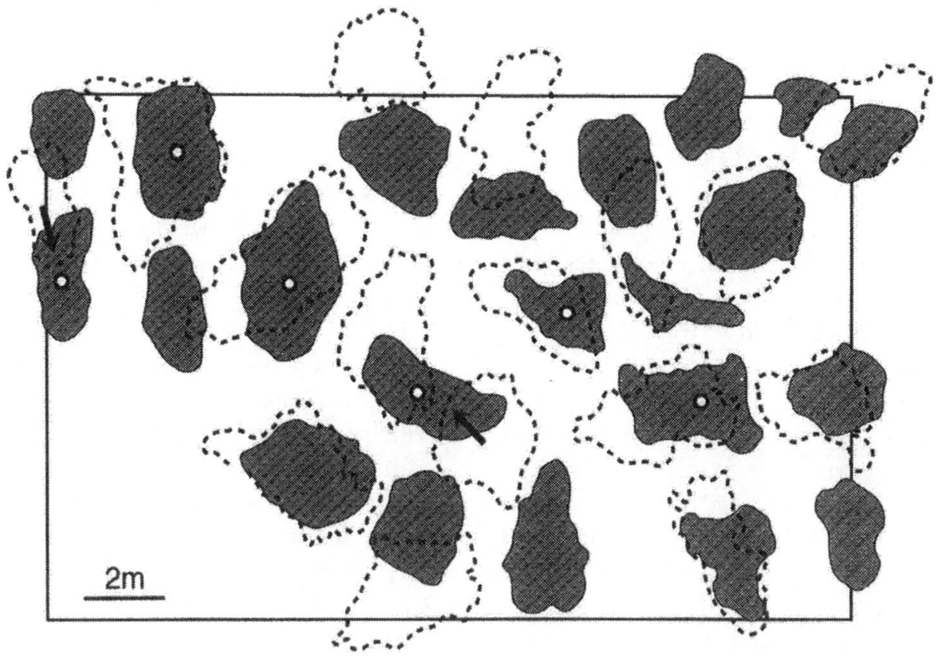

Figure 6. Territory arrangement of male *P. polyodon* in 1992 and 1994. Areas enclosed by broken lines are territories in 1992. Territories in this study are shaded. Owners of territories with dots are the same individuals that were identified at the same site in 1992. Arrows indicate shift of territories between the two years. The 1992 data are from Kohda (1995a).

Discussion

Herbivorous territorial cichlids are members of species-rich fish communities of rocky habitats in Lake Tanganyika (Kawanabe 1981, Takamura 1984, Hori 1987, 1991, Yanagisawa & Nishida 1991). Several herbivorous fishes of the same genus usually cohabit. The cohabiting mechanism proposed hitherto was differential competitive advantage in heterogeneous rocky substrate (Takamura 1984, Kohda & Yanagisawa 1992). The results of this study show that three territorial cichlids of the genus *Petrochromis* cohabit within a shallow homogeneous habitat where *P. polyodon* can be large, socially dominant and abundant. However, the coexistence could not be easily explained by the heterogeneity of habitat.

In the study area, territories of male *P. polyodon* were separated from them. Vacated territories adjacent to those of male *P. polyodon* were occupied by fishes other than male *P. polyodon*. Male *P. polyodon* did not expand territories into nearby vacated areas, probably because male-male interactions restrained them from intruding into these areas. If a new male *P. polyodon* came to the vacant area, he would be repelled. Thus, conspecific males could not occupy vacant areas between male *P. polyodon* territories. In contrast, after removal of male *P. polyodon*, neighbouring male(s) expanded and shifted territories and usually occupied the vacant areas. Rearranged territories of male *P. polyodon*, however, were also separated from each other, mainly because males attacked each other as they did before experimental removal of other members. The substrate of the study area is homogeneous, and the three species take similar foods in a similar manner (Kohda 1995a). Territories of male *P. polyodon* can shift after removal experiments. These suggest that micro-habitat differences could not explain the separate distribution of their territories. The attacks between male *P. polyodon* must be the mechanism primarily responsible for maintaining the separation of their territories.

Mating territories defended against conspecific males, where the owner males exclusively court visiting females, are common in maternal mouthbrooding cichlids, and increase male mating opportunity (McKaye 1984, Kuwamura 1987, 1992, Ochi 1993, Rossiter 1994, Hert 1995). Areas around feeding territories of male *P. polyodon* should be the mating territory from which conspecific males are repelled, and in which they court visiting females. Male *P. polyodon* that keep out rival males from this space and court females may increase their mating success. Thus, separated distribution of male feeding territories may inevitably result from their aggressive mating styles.

Available space on shallow rocky areas of this lake is usually occupied by territories of *Petrochromis* species (Takamura 1984, Hori 1991, Kohda 1995a). Although male *P. polyodon* are large and dominant over other congeners in shallow areas (Takamura 1984, Kohda 1995a), they do not monopolize the area, because of their mating competition. Between their feeding territories, male *P. polyodon* inevitably make space that other fishes can inhabit. Thus, mating attacks between dominant male *P. polyodon* regulate their density in the area, and facilitate the micro-habitat coexistence of congeneric fishes. Male *P. famula* attack conspecific males far beyond their feeding territories, and their feeding territories are also separated. Similar arguments will be applicable to these males.

In a deep rocky bottom in the nothern end of this lake (500 km apart), *Petrochromis orthognathus* males are large and socially dominant (Kohda 1991). They also maintain larger mating territories against conspecific males far beyond their feeding territories, which are also separated from each other. The interstices between the feeding territories are occupied by subordinate congeners including smaller *P. polyodon*. Males of *Petrochromis* species have mating territories far beyond their feeding territories (Kuwamura 1987, Kohda 1991, 1995b). Therefore, separated distribution of dominant male territories due to mating attacks between conspecific males should be common in *Petrochromis* fishes.

Such distribution patterns of territorial males are likely to be common in maternal mouthbrooding cichlid fishes. For example, in Malawian cichlids that have a lek-like mating system, territorial males are separated from conspecific males, but neighbour with heterospecific males (McKaye 1984, Hert 1995). These males do not have feeding territories but have mating territories against conspecific males. So, heterospecific neighbouring males are less likely to interfere in courting and mating, and therefore a multi-specific distribution pattern reflects intraspecific male-male competition avoidance (McKaye 1984).

On coral reefs, coexistence mechanisms of permanently territorial fishes are well studied in herbivorous damselfishes (Sale 1974, 1978, 1991, Itzkowitz 1977, Robertson 1984, Ebersole 1985). Female damselfishes visit and mate at nest sites in male territories. As in male *Petrochromis*, male damselfishes maintain a mating territory against conspecific rival males far beyond their feeding territories (Itzkowitz 1978, Kohda 1984, 1995b). Males court visiting females in these mating territories. Such territoriality of male damselfishes (Myrberg & Thresher 1974, Thresher 1976, Kohda 1984) suggests that male-mating attack is common and may result in separated distribution of their feeding territories in many species. The male-mating attack hypothesis should receive attention for appreciation of the effects on spatial organization and coexistence that occur through this behaviour.

The prevailing mechanisms to explain competitive coexistence of related fishes with permanent territories are based on interspecific competition for space (Sale 1977, 1978, 1991, Robertson & Lassig 1980, Takamura 1984, Ebersole 1985). Coexistence through male-male attack for mating space is not incompatible with these mechanisms. Both intra- and interspecific competition would play a role in coexistence of permanently territorial fishes in which males have mating territories larger than feeding ones.

Acknowledgements

The Lake Tanganyika Research Unit (L.T.R.U.) of the Department of Fisheries, Zambia provided facilities for the field research. This work was partly supported by a Grant-in-Aid for Overseas Scientific Research (No. 0404178 and 04044088) and for

Scientific Research on Priority Areas (#319), from the Japan Ministry of Education, Science, Sport and Culture.

References cited

Brichard, P. 1989. Cichlids and all the other fishes of Lake Tanganyika. T.F.H. Publications, Neptune City. 544 pp.

Chesson, P.L. & R.P. Warner. 1981. Environmental variability promotes coexistence in lottery competitive systems. Amer. Nat. 117: 923–943.

Coulter, G.W. 1991. Lake Tanganyika and its life. Oxford University Press, London. 354 pp.

Ebersole, J.P. 1985. Niche separation of two damselfish species by aggression and differential microhabitat utilization. Ecology 66: 14–20.

Hert, E. 1995. The impact of intralacustrine introductions with regard to space utilization and competition for territories to a cichlid fish community in Lake Malawa, Africa. Ecol. Res. 10: 117–124.

Hori, M. 1987. Mutualism and commensalism in a fish community in Lake Tanganyika. pp. 219–239. *In:* S. Kawano, J.H. Connell & T. Hidaka (ed.) Evolution and Coadaptation in Biotic Communities, University of Tokyo Press, Tokyo.

Hori, M. 1991. Feeding relationships among cichlid fishes in Lake Tanganyika: effects of intra- and interspecific variations of feeding behavior on their coexistence. INTECOL Bull. 19: 89–101.

Itzkowitz, M. 1977. Spatial organization of the Jamaican damselfish community. J. Exp. Mar. Biol. Ecol. 28: 217–242.

Itzkowitz, M. 1978. Group organization of a territorial damselfish, *Eupomacentrus planifrons*. Behaviour 65: 125–137.

Kawanabe, H. 1981. Territorial behaviour of *Tropheus moorei* (Osteichthyes : Cichlidae) with a preliminary consideration on the territorial forms in animals. Afr. Stud. Monogr. 1: 101–108.

Kohda, M. 1984. Intra- and interspecific territoriality of a temperate damselfish, *Eupomacentrus altus* (Teleostei : Pomacentridae). Physiol. Ecol. Japan 21: 35–52.

Kohda, M. 1991. Intra- and interspecific social organization among three herbivorous cichlid fishes in Lake Tanganyika. Japan. J. Ichthyol. 38: 147–163.

Kohda, M. 1995a. Does male-mating attack in the herbivorous cichlid, *Petrochromis polyodon*, facilitate the coexistence of congeners? Ecol. Fresh. Fish. 4: 152–159.

Kohda, M. 1995b. Territoriality of male cichlid fishes in Lake Tanganyika. Ecol. Fresh. Fish. 4: 180–184.

Kohda, M. & Y. Yanagisawa. 1992. Vertical distributions of two herbivorous cichlid fishes of the genus *Tropheus* in Lake Tanganyika. Ecol. Fresh. Fish. 1: 99–103.

Kuwamura, T. 1986. Parental care and mating systems of cichlid fishes in Lake Tanganyika: a preliminary field survey. J. Ethol. 4: 129–146.

Kuwamura, T. 1987. Male mating territory and sneaking in a maternal mouthbrooder, *Pseudosimochromis curvifrons* (Pisces : Cichlidae). J. Ethol. 5: 203–206.

Kuwamura, T. 1992. Overlapping territories of *Pseudosimochromis curvifrons* males and other herbivorous cichlid fishes in Lake Tanganyika. Ecol. Res. 7: 43–53.

McKaye, K.R. 1984. Behavioural aspects of cichlid reproductive strategies: patterns of territoriality and brood defence in Central American substratum spawners and African mouth brooders. pp. 245–273. *In:* R.J. Wooton & G.W. Potts (ed.) Fish Reproduction: Strategies and Tactics, Academic Press, London.

Myrberg, A.A. & R.E. Thresher. 1974. Interspecific aggression and its relevance to the concept of territoriality in reef fishes. Amer. Zool. 14: 81–96.

Ochi, H. 1993. Maintenance of separate territories for mating and feeding by males of a maternal mouthbrooding cichlid, *Gonathochromis pfefferi*, in Lake Tanganyika. Japan. J. Ichthyol. 40: 173–182.

Pianka, E.R. 1978. Evolutionary ecology. Harper and Row, New York. 397 pp.

Poll, M. 1986. Classification des Cichlidae du lac Tanganyika: tribus, generes et especes. Mem. Acad. Roy. Belgium (Sci.) 45: 1–163.

Robertson, D.R. 1984. Cohabitation of competiting territorial damselfishes on a Caribbean coral reef. Ecology 65: 1121–1135.

Robertson, D.R. & S.D. Gaines. 1986. Interference competition structures habitat use in a local assemblage of coral reef surgeonfishes. Ecology 67: 1372–1383.

Robertson, D.R. & B. Lassing. 1980. Spatial distribution patterns and coexistence of a group of territorial damselfishes from the Great Barrier reef. Bull. Mar. Sci. 30: 187–203.

Robertson, D.R., Y. Nicholas, C. Polunin & K. Leighton. 1979. The behavioral ecology of three Indian Ocean surgeonfishes (*Acanthurus lineatus, A. leucosternon* and *Zebrasoma scopsa*): their feeding strategies, and social and mating systems. Env. Biol. Fish. 4: 125–170.

Rossiter, A. 1994. Territory, mating success, and the individual male in a lekking cichlid fish. pp. 43–55. *In:* P.J. Jarman & A. Rossiter (ed.) Animal Societies, Kyoto University Press, Kyoto.

Sale, P.F. 1974. Mechanism of co-existing in a guild of territorial fishes at Heron Island. Proc. Inter. Coral Reef Symp. 1: 193–206.

Sale, P.F. 1977. Maintenance of high diversity in coral reef fish communities. Amer. Nat. 111: 337–359.

Sale, P.F. 1978. Coexistence of coral reef fishes – a lottery for living space. Env. Biol. Fish. 3: 85–102.

Sale, P.F. 1982. Stock-recruit relationships and regional coexistence in a lottery competitive system: a simulation study. Amer. Nat. 120: 139–159.

Sale, P.F. 1991. Reef fish communities: open nonequilibrial systems. pp. 564–598. *In:* P.F. Sale (ed.) The Ecology of Fishes on Coral Reefs, Academic Press, San Diego.

Takamura, K. 1984. Interspecific relationships of Aufwuchs-eating fishes in Lake Tanganyika. Env. Biol. Fish. 10: 225–241.

Thresher, R.E. 1976. Field analysis of the territoriality of the threespot damselfish, *Eupomacentrus planifrons* (Pomacentridae). Copeia 1976: 266–276.

Tilman, D. 1982. Resource competition and community structure. Princeton University Press, Princeton. 296 pp.

Yamaoka, K. 1982. Morphology and feeding behaviour of five species of genus *Petrochromis* (Teleostei : Cichlidae). Physiol. Ecol. Japan 19: 57–75.

Yamaoka, K. 1983. A revision of the cichlid fish genus *Petrochromis* from Lake Tanganyika, with description of a new species. Japan. J. Ichthyol. 30: 129–141.

Yamaoka, K. 1991. Feeding relationships. pp. 151–172. *In:* M.H.A. Keenleyside (ed.) Cichlid Fishes, Chapman & Hall, London.

Yanagisawa, Y. & M. Nishida. 1991. The social and mating system of the maternal mouthbrooder *Tropheus moorii* (Cichlidae) in Lake Tanganyika. Japan. J. Ichthyol. 38: 271–282.

Environmental Biology of Fishes **52**: 243–250, 1998.

Dyadic relationships in nesting males of the three-spined stickleback, *Gasterosteus aculeatus*

Seiichi Mori
Center for Ecological Research, Kyoto University, Shimo-sakamoto, Otsu 520-01, Japan
(e-mail: smori@gifu-keizai.ac.jp)

Received 30.9.1996 Accepted 26.5.1997

Key words: dyad, aggressive interaction, social relationships, nesting males, fish

Synopsis

Multidimensional scaling analysis of nest site location, nesting interval, and aggressive and mating interactions revealed that male three-spined sticklebacks, *Gasterosteus aculeatus*, formed nesting clusters comprising five patterns of dyadic relationships. These patterns were defined by intervening variables such as size difference between the two males, nest site locations, synchronization of nesting cycles (egg presence in the nest), distance between two nests, and nest raiding behaviours. Several patterns of dyadic relationships were integrated among nesting males and their constellations indicated an expression of the social structure. Through an understanding the social structure, I described effects of social context or situations on individual reproductive success.

Introduction

The study of social phenomena involves discriminating and analyzing three levels of complexity: interactions, relationships, and social structure (Hinde 1974, 1982). Figure 1 is a diagrammatic representation of the relation between the three levels. To describe an interaction, it must be specified what the individuals are doing together and how they are doing it. Hence, one has directly observed the behaviours of individuals, usually in interaction with each other (1 and 2 in Figure 1). Each interaction occupies a strictly limited span of time at each encounter between individuals. Relationships can be discussed in the context of a much longer time period covering all encounters. Since a relationship involves a series of interactions between the members of a dyadic that are known to each other, one can abstract a description of the relationship from observations of interactions over time (small-case letters in Figure 1).

Many studies on behaviour of the three-spined stickleback, *Gasterosteus aculeatus*, have been conducted to interpret the proximate features that influence behaviour within the ethological context of sign stimuli and innate releasing mechanisms (Tinbergen 1951). Although studies on the functional significance of behaviour have subsequently flourished (Bell & Foster 1994), the social structure that includes inter-individual interactions in an aggregation remains poorly understood. Here, I analyze social structure through the developing processes of inter-individual relationships and their generalized patterns. Male three-spined sticklebacks nest in colonies or in aggregations (FitzGerald 1983, FitzGerald & Whoriskey 1985, Mori 1985, 1993). Males that come onto a breeding ground in spring do not begin to build their nests simultaneously, nor do they

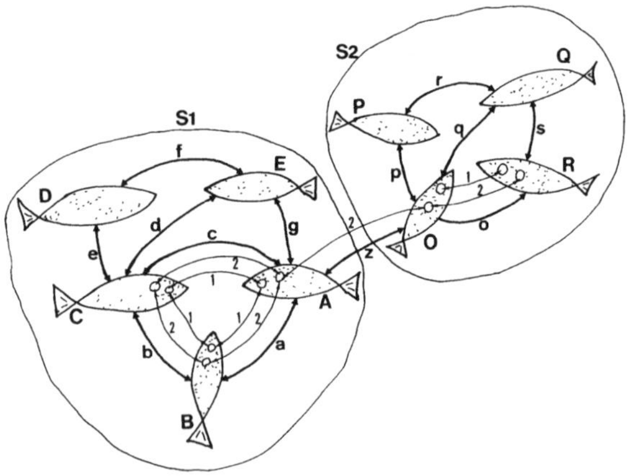

Figure 1. Diagrammatic representation of the relations between interactions, relationships and social structure (A, B, C, D, E, O, P, Q, and R = individuals; 1 and 2 = dyadic interactions; a, b, c, d, e, f, g, o, p, q, r, s, and z = dyadic relationships; S1 and S2 = clusters).

mate with females all at once. There is much temporal and spatial variation in the breeding activities among individuals in an aggregation. Several field studies have shown that this variation in activities has led to individual variation in reproductive success (e.g. Kynard 1978, Whoriskey & FitzGerald 1985, Mori 1993, 1995).

This study identifies (1) five pattern of dyadic relationships of nesting males, (2) the behavioural criteria that determine the five patterns of dyadic relationships, and (3) a cluster indicating a social structure, comprising five dyadic relationships limited in time and space. The social structure can be approached by the component dyads of the cluster. Furthermore, through an understanding of the social structure, I also described the effects of social relationships on individual reproductive success.

Study area and methods

The study area was part of an inlet fed by a spring in the Tsuya River, Gifu Prefecture, Central Japan. The detailed topography was described in Mori (1985). I set up an enclosure made of a net fence, 6 m in length and 1.3 m in width, along the shore of the inlet. Cover such as aquatic vegetation and gravel was cleaned away in the enclosure, so that any effect of nest cover on hatching success was removed in this experiment. The present study started in late April 1991 and ended in early June when all males finished breeding in this area. Additional observations showed that after mid June, the frequency of interactions in dyadic relationships were maintained at a low level and this level changed little over time (Mori unpublished data). Interactions ceased in late June, and all males had died by early July.

At the beginning of the study, 25 males of the three-spined stickleback (*leiurus* form) were introduced into the enclosure from the inlet, after being individually identified by sticking coloured silicone tubes on the dorsal spines. I also sketched the pattern of the black pigmented area on the left body side posterior to the lateral plates, so all individual males could be recognized without the tubes. Nuptial colouration was scored using a 7-point scale on the basis of the combination of colouration of the head and lateral body side (Mori 1993). I ensured that several gravid females were always present in the enclosure. After spawning, females were removed to the outside. Locations of nest sites were marked by numbered flags. Nests were located in open areas or in thick vegetation, and at various distances from the shore (Mori 1993).

Aggressive and mating behaviours of males were observed daily for 6 or 7 hours from the shore terrace. Each male was observed for 10 minutes to record its behaviour and 5 minutes to record its pattern of movement. I observed each male from one to four times daily. Data used here were the averaged behavioural frequency per 10 minute period. Synchronization represents the overlap in the number of nesting days in which both nests in a dyad were present relative to the total nesting days of both males. This includes days when the nests were empty.

The reproductive success (RS) index was calculated on the basis of the total number of young in and around the nest(s) during 4 days after hatching per male (Mori 1993) in 'cluster S' (see below). Fewer than 50 young, 50 to 200 young and more than 200 young were given a score of 1, 2 and 3, respectively. The four estimates were summed to yield an RS-index that ranged between 4 and 12.

Table 1. Variables to analyze the social structure which is formed by inter-individual interactions among sticklebacks.

A. Independent variables

A-1. Individual variables	A-2. Condition variables
Body size	Initiation of nest-building
Nuptial color	Location of nest sites
Mating frequency	Intruder's frequency
Aggressive frequency	Female visit

↘ ↙

B. Intervening variables for the two males
- Size differences between the two
- Nuptial color difference between the two
- Distance between the two nests
- Difference in time of initiating nest-building
- Difference in nest site types
- Synchrony rate of nesting cycles (egg presence in the nest)
- Presence of raiding behaviour

↓

C. Dependent variables
- Dyadic relationships through change of aggressive levels

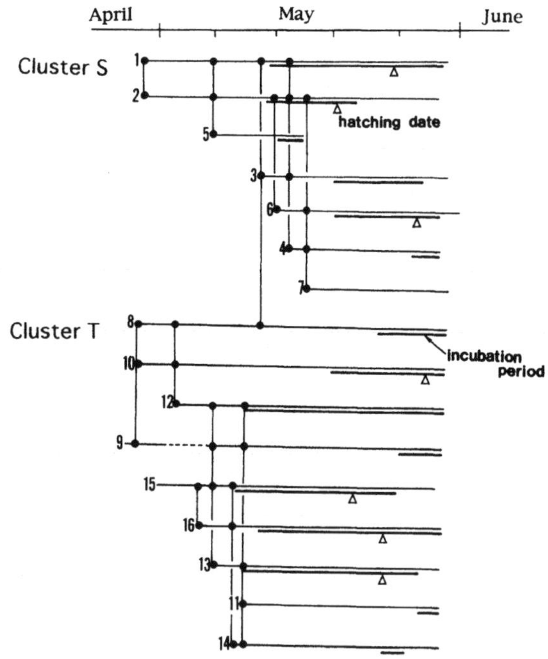

Figure 2. The nesting cycles of 16 male sticklebacks. Vertical lines between black circles show the start of face-to-face relationships among nesting males (horizontal lines = the nesting period; broken line = interval between nesting cycles; numbers = codes for individual males).

Data analysis

Independent, intervening (derived measurement), and dependent variables used to exhibit correlations between dyadic relationships are shown in Table 1. Intervening variables are defined by characteristics and behaviours of two individuals. Variables were transformed by numerical coding into three scores, creating trichotomous variables. To integrate many variables and standardize different units, these data were analyzed using the quantification method (Hayashi 1952, 1954, Tanaka & Wakimoto 1983), a form of multidimensional scaling (Shepard 1980). This method allows a transformation from qualitative data to quantitative data. Quantification method III is a patterning classification that does not use external criteria. After this statistical treatment, the discriminated patterns were analyzed as to determine the contribution (coefficient of determination) of intervening variables to dependent variables by using quantification method II.

Results

Nesting clusters

The nesting cycles of 16 males and the location of nest sites are shown in Figures 2 and 3. Daily observations revealed that fish interacting with the nest owner directly were almost always neighbouring nesting males. A nesting aggregation was regarded as a cluster, which was defined as a subset of members which are more closely identified with one another than they are with remaining members. The males constituted two clusters: 'cluster S' included seven nesting males forming 12 dyads, and nine males with 16 dyads in 'cluster T'.

Once males completed a nest, they tended to settle within that territory throughout their breeding cycles. For instance, males 1 and 2, which were the brightest males in the cluster but not the largest, first nested in neighbouring sites, developed 'relationship A', and initiated a form of 'cluster S' (Figure 2). About 10 days later, males 3, 4, and 5 took up

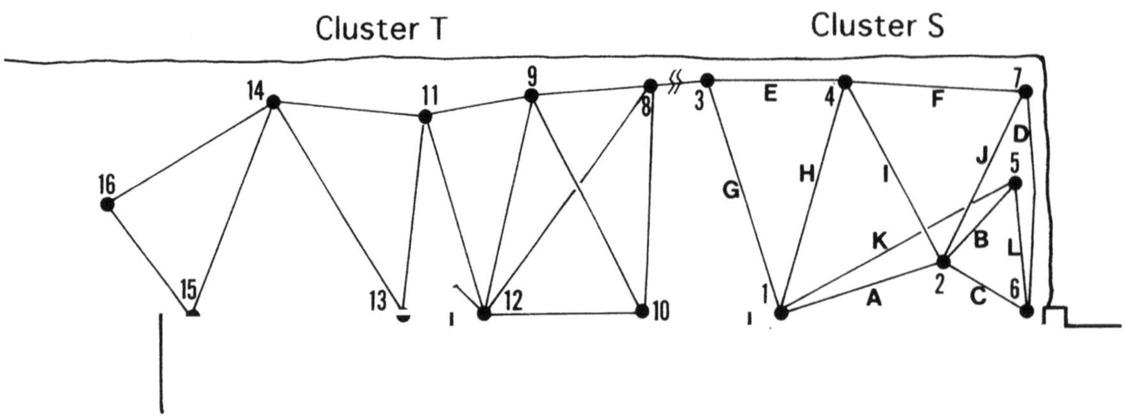

Figure 3. The location of nest sites (black circles). Individuals and dyads are identified by numerals and capital letters, respectively. Two nesting aggregations are labeled clusters S and T.

territories and built their nests near the nests of males 1 and 2, increasing the membership of cluster S. Before male 3 nested, all members in clusters S and T had been isolated. Male 6 had first settled at another site in the left part of Figure 3, and then came to nest on the right side as a member of 'cluster S'.

Daily change of aggressive level in dyads

In most cases, aggressive interactions were observed between two territorial and/or nesting males, while the frequency of aggression simultaneously undertaken by three or more males was very few in all daily observations (mean ± SD = 0.6 ± 1.3 times

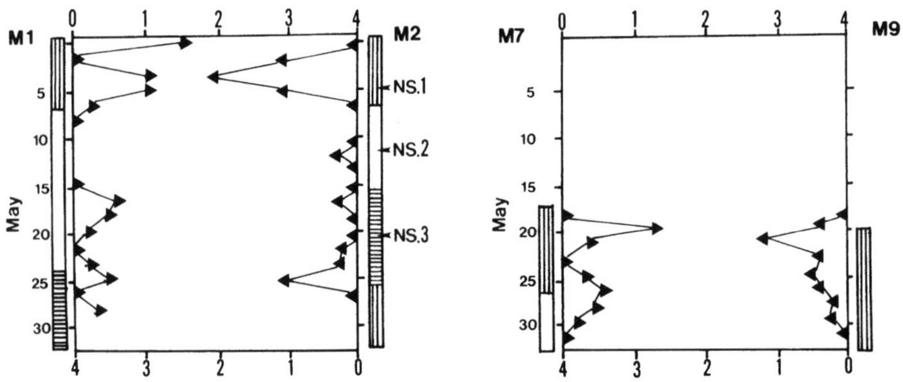

Figure 4. Four examples of change in the daily frequency of aggressive encounters for dyads, including both attack and threat. The number with M indicates the name of male. NS. 1 = male building a nest and holding a territory, but not yet mated. NS. 2 = male guarding eggs; NS. 3 = male guarding young.

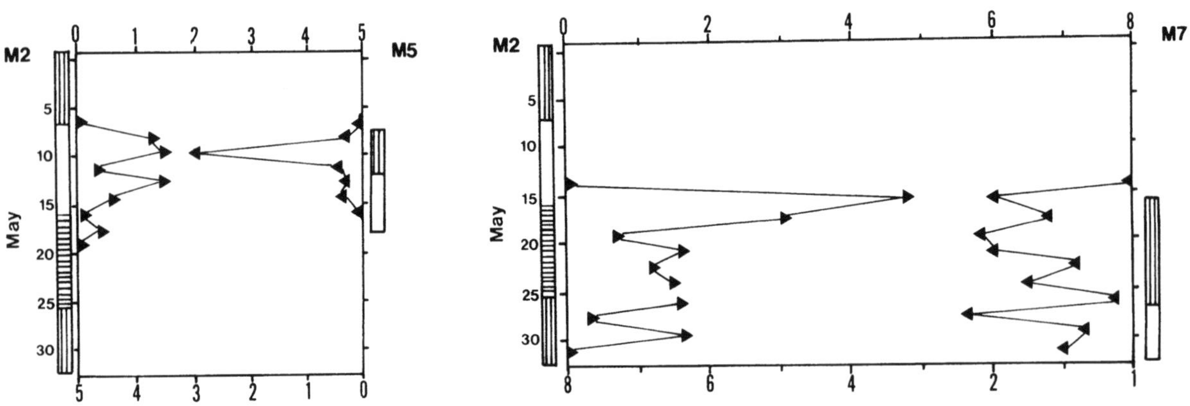

per day, N = 42). These results indicate that the relationship of breeding males fundamentally consisted of a dyad constituting two individuals.

Figure 4 shows the change in the daily frequency of aggressive encounters, including both attack and threat, for each male in four dyads. At the beginning of most dyadic relationships, the two males frequently attacked each other. There were various temporal patterns in the relationships, though aggression generally decreased with time. Dyadic relationships in this sense, are described by the temporal changes in aggression between two nesting males.

Patterns of dyadic relationship

Intervening variables derived from measurements on pairs of nesting males were transformed into trichotomous variables by a numerical coding then used to assign dyads (Table 2).

On the basis of the data in Table 2, the classification of dyadic relationships is shown in Figure 5. Abscissa indicates the intensity of the relationship, that is the frequency of aggressive encounters, and ordinate indicates the duration of relationship measured as the length of the synchronous period between two nesting cycles. For example, 'zone III' in the upper part of Figure 5 was an area of the lowest intensity of relationship. All 28 dyadic relationships were generally discriminated into five relation-patterns (I to V) for each cluster, but relationship P was omitted from the analysis because male 9 suddenly stopped nesting and some days later rebuilt his nest.

Males 1 and 2 forming 'dyad A' started to nest first during the breeding season, and synchronously maintained nests for longer time than in other dyads with a little intensity of aggressive interaction, so they belonged to relation-pattern I. It could be said that the two clusters mapped isomorphically onto the relational system.

The lower section of Figure 5 shows three males nesting along the shore that tended to form a relation-pattern I. Relation-pattern IV was observed for two males nesting along the shore that had started nesting in the middle period of breeding season. Two males which formed relation-pattern II nested at close distance. Males nesting on opposite shores

Table 2. Qualitative data transformed by coding into trichotomous variable (1, 2 and 3) used to assign dyads in cluster S by the quantitative method (see Data analysis section). Body size = size difference between the two males; Nuptial color = difference in score of nuptial color; Location = difference in nest site types; D.N.I. = difference in time of initiating nest-building; Distance = distance between the two nests; Raid = presence of raiding behaviour; Synchrony = synchrony rate of nest cycles.

Dyad	Body size	Nuptial color	Location	D.N.I.	Distance	Raid	Synchrony* (%)
A 1–2	1	1	1	1	2	3	1 (96.2)
B 2–5	1	2	2	2	1	3	3 (21.4)
C 2–6	1	2	1	2	1	3	2 (51.4)
L 5–6	2	1	2	2	1	3	3 (15.4)
G 1–3	1	1	3	2	2	3	2 (55.4)
H 1–4	2	2	3	2	2	3	2 (44.5)
I 2–4	1	2	3	2	2	3	2 (44.5)
J 2–7	1	3	3	2	2	3	2 (41.4)
E 3–4	1	1	1	3	2	1	1 (81.2)
F 4–7	1	3	1	3	2	2	1 (89.0)
D 6–7	3	1	1	3	2	3	1 (77.4)
K 1–5	1	2	2	2	3	2	3 (21.4)

Three categories of differences between the two males or nests: body size: 1 = same, 2 = < 3 mm, 3 = ⩾ 3 mm; nuptial color: 1 = same, 2 = score I, 3 = score II; nest location: 1 = both along the shore, 2 = one along the shore, 3 = each at opposite shore; D.N.I.: 1 = both nests early, 2 = one nest early, 3 = both nests late; distance: 1 = short (< 30 cm), 2 = medium (30–60 cm), 3 = long (60 cm ⩾); raid: 1 = raid in both, 2 = raid in one, 3 = none; * =Synchronous rate = $\frac{N(i,j) \times 2}{N(i) + N(j)}$,

N(i) = the number of nesting days for individual i, N(i, j) = the number of nesting days in which both individuals were present.

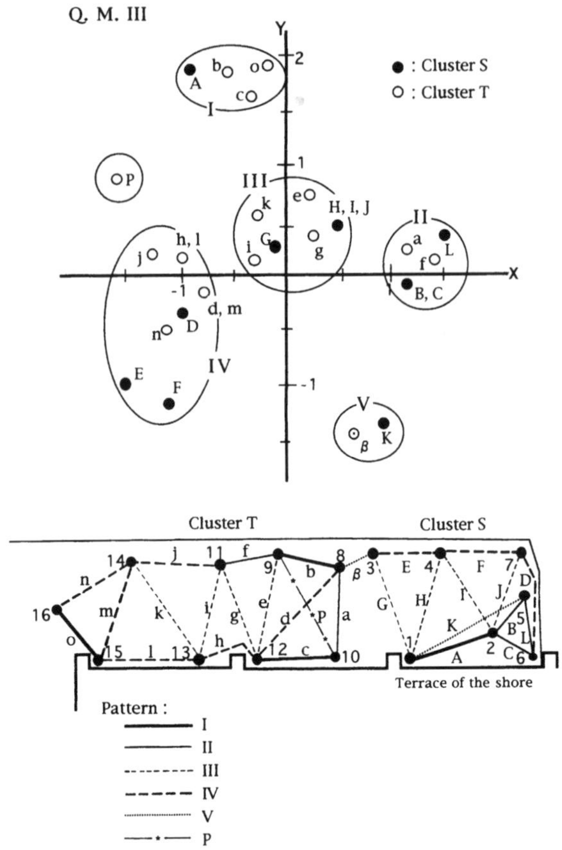

Figure 5. Similarity of dyadic relationships. Locations of circles define the constellations among dyadic relationships. Capital letters (black circles) are dyadic relationships of cluster S, lower case letters (white circles) are relationships of cluster T. The lower figure shows locations of nest sites. β = dyadic relationship connected between the two clusters. Pattern = relationship pattern of dyads.

(e.g., males 1 and 3) formed relation-pattern III. Five temporal patterns of aggressive relationship were specified in Figure 6. The lower section of this figure indicates characteristics of each relation-pattern.

Behavioural characteristics defining the dyadic relationship

Figure 7 indicates the contribution of factors to the five patterns of dyadic relationship defined by quantification method II using discriminating external criteria.

The most important factors determining the relationships are the synchronous period of nesting cycles, the location of nest sites, and the distance between nests. Another important factor was raiding behaviour. Just after raiding, the relationship between the owner and the raider became highly aggressive.

Five patterns of social relationships

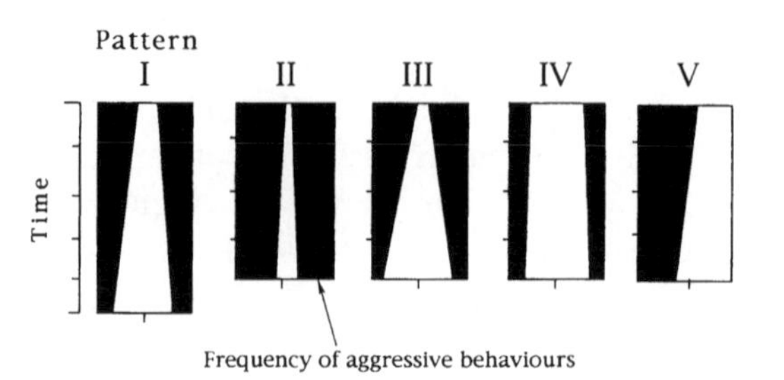

Pattern	Aggressive level	Synchronization	No. of dyads
I	constant low *	long	4
II	constant high	middle	5
III	sharp decline	middle	8
IV	constant low	middle	9
V	high in one side	short	2

* high at the starting

Figure 6. Five patterns in the aggressive relationship within dyads. Black areas = temporal change of aggressive levels between the two individuals. Pattern = relationship pattern of dyads.

Reproductive success

Most males of 'cluster S' (85.7%, 6/7 nests) and 'cluster T' (90%, 9/10 nests) obtained eggs. Fifty

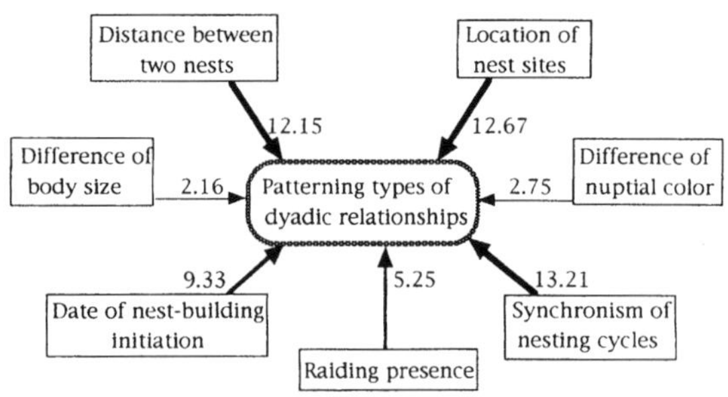

Figure 7. Contribution of intervening variables to patterning dyadic relationships on the base of quantification method II. The numerals and thickness of lines represent degrees of effective contribution.

percent of males with eggs in 'cluster S' and 44.4% in 'cluster T' succeeded in hatching embryos. The rate of hatching success was not significantly different between the two clusters (χ^2 test, $p > 0.05$).

The two males, males 1 and 2 (mean ± S.D. of RS index = 8.2 ± 0.4), in relation-pattern I had significantly higher reproductive success than males 3, 4, 5, and 6 (RS index = 6.1 ± 1.5), classified into relation-patterns II, III, IV and V (Mann-Whitney's U test, $p < 0.01$). While, male 7 which had started nesting last in the breeding season, forming relation-pattern III and IV, had no hatching success. Male 7 was a small dull-coloured fish but not the dullest.

Discussion

Functional definitions of sociality with a single underlying principle – i.e. individuals in aggregation act to minimize costs while maximizing benefits – have been extensively discussed since the late 1970's (Krebs & Davis 1978, Clutton-Brock et al. 1982, Rubenstein & Wrangham 1986, Dunber 1988). However, there are few studies on social structure in fish, including sticklebacks, that describe development of inter-individual interactions compared with the mammals (Kummer 1971, Eisenberg et al. 1972, Lee 1994). Here I studied one dimension of male stickleback behaviour, the dyadic inter-individual aggressive interaction, which is likely to be important for understanding social structure of male breeding aggregations.

Whereas each interaction occupies a strictly limited span of time at each encounter between individuals, relationships can be discussed in the context of a much longer time period covering all encounters. Dyads involve a series of interactions between two individuals which have known each other over a period of time (Hinde 1982).

Based on observations of many interactions in dyads over time, I found that males formed nesting clusters containing dyads with similar qualitative properties. Five patterns in the dyadic relationships among all members of a cluster were recognized. The patterns of dyadic relationships depicted in Figure 6 and their similarity as shown by coordinates in Figure 5 (Hayashi 1954) may indicate the expression of structure within social clusters. I identified some factors which determine the patterns: nest site locations, synchronization of nesting cycles (e.g., egg presence in the nest, Mori unpublished data), distance between two nests, and nest raiding behaviours. Therefore, the relationships are contingent on the position and temporal ordering of male settlement.

The study provided evidence that relatively isolated males were more aggressive toward another male than males with neighbours (Peeke & Dark 1990). When two males in a cluster nest and breed simultaneously (e.g., males 1 and 2, Figure 4), the frequency of aggressive interactions between them follows a cyclical pattern: high during nest building and courtship, low during egg guarding, and high during young guarding, as described for isolated males (Wootton 1971). This endogenous pattern of aggressive behaviour, however, is disturbed when neighbouring males settle out in an asynchronous manner. For example, male 2 was disrupted during his egg guarding period by the establishment of male 5 (early stages of egg guarding) and male 7 (late stages of egg guarding) in nearby territories (Figures 4 and 5). If hatching success is related, in part, to the amount of time a male spends fanning and caring for the eggs, then male 2 was at a definite disadvantage over males 1 and 5 because he had to spend more time away from the nest defending his territory during that critical period. As a result, in spite of belonging to the same relation-pattern, male 2 might have lower success in reproduction than male 1.

Males 1 and 2, which synchronously maintained nests along the shore for longer than in other relation-patterns, formed relation-pattern I and had the highest reproductive success. In contrast, males 3, 4, 6, and 7 which nested later and were asynchronous had low reproductive success. This suggested that stable social relationships increase individual reproductive success. An extremely important finding is that the degree of synchronization and various nest site variables (location, distance, etc.) have more influence on the patterns of dyadic relationships than do male-based characteristics such as body size and nuptial colour (Figure 7). These traits of external appearance may be more important earlier in the interaction. For example, the first males to settle out, males 1 and 2, were the brightest males, and male 7 was a dully coloured fish in the cluster at the beginning of breeding period. Furthermore, these traits might play a larger role in male-female interactions than they do in male-male interactions.

Acknowledgements

I am greatly indebted to R.J. Wootton of the University College of Wales, T.C.M. Bakker of Universität Bern and D.A. McLennan of the University of Toronto for critical comments and valuable corrections on this manuscript. Many thanks to H. Kawanabe of Kyoto University for generous support during my study, and to K. Ide, Y. Ikawa, and M. Nagoshi of Nara Women's University for considerable help in collecting field data.

References cited

Bell, M.A. & S.A. Foster (ed.). 1994. The evolutionary biology of the threespine sticklebacks. Oxford Science Publications, Oxford. 571 pp.

Clutton-Brock, T.H., F.E. Guinness & S.D. Albon. 1982. Red deer. University of Chicago Press, Chicago. 378 pp.

Dunber, R.I.M. 1988. Primate social systems. Cornell University Press, Ithaca. 373 pp.

Eisenberg, J.F., N.A. Muckenhirm & R. Rudran. 1972. The relation between ecology and social structure in primates. Science 176: 863–874.

FitzGerald, G.J. 1983. The reproductive ecology and behaviour of three sympatric sticklebacks (Gasterosteidae) in a saltmarsh. Biol. Behav. 8: 67–79.

FitzGerald, G.J. & R.J. Wootton. 1986. Behavioural ecology of sticklebacks. pp. 409–432. *In:* T.J. Pitcher (ed.) The Behaviour of Teleost Fishes, Croon Helm, London.

Hayashi, C. 1952. On the prediction of phenomena from qualitative data and the quantification and qualitative data from the mathematico-statistical point of view. Ann. Inst. Statist. Math. Japan 3: 69–98.

Hayashi, C. 1954. Multidimensional quantification with the applications to analysis of social phenomena. Ann. Inst. Statist. Math. Japan 5: 121–143.

Hinde, R.A. 1974. Biological bases of human social behaviour. McGraw-Hill, New York. 462 pp.

Hinde, R.A. 1982. Ethology. Fontana Paperbacks, William Collins Sons & Co. Ltd., Glasgow. 320 pp.

Kummer, H. 1971. Primate societies. Aldine Publishing Co. (in Japanese, translated by H. Mizuhara, Shakai shiso-sha, Tokyo. 218 pp).

Krebs, J.R. & N.B. Davis. 1978. Behavioural ecology. Sinauer Associates, Sunderland. 494 pp.

Kynard, B.E. 1978. Breeding behaviour of a lacustrine population of threespine sticklebacks (*Gasterosteus aculeatus* L.). Behaviour 67: 178–207.

Lee, P.C. 1994. Social structure and evolution. pp. 266–303. *In:* P.J.B. Slater & T.R. Halliday (ed.) Behaviour and Evolution, Cambridge University Press, Cambridge.

Mori, S. 1985. Reproductive behaviour of the landlocked three-spined stickleback, *Gasterosteus aculeatus microcephalus*, in Japan. Behaviour 93: 21–35.

Mori, S. 1993. The breeding system of the three-spined stickleback, *Gasterosteus aculeatus* (forma *leuira*), with reference to spatial and temporal patterns of nesting activity. Behaviour 126: 97–124.

Mori, S. 1994. Nest site choice by the three-spined stickleback, *Gasterosteus aculeatus*, in spring-fed waters. J. Fish. Biol. 45: 279–289.

Mori, S. 1995. Spatial and temporal variations in nesting success and the causes of nest losses of the freshwater three-spined stickleback, *Gasterosteus aculeatus*. Env. Biol. Fish. 43: 323–328.,

Peeke, H.V.S. & K.A. Dark. 1990. The effects of isolation on the sensitization and habituation of aggression in the threespine stickleback (*Gasterosteus aculeatus*). Ethology 85: 35–42.

Rubenstein, D.I. & R.W. Wrangham (ed.). 1986. Ecological aspects of social evolution. Princeton University Press, Princeton. 551 pp.

Shepard, R.N. 1980. Multidimensional scaling, tree-fitting, and clustering. Science 210: 390–398.

Tanaka, Y. & K. Wakimoto. 1983. Methods of multivariate statistical analysis. Gendai Sugaku-sha, Kyoto. 296 pp.

Tinbergen, N. 1951. The study of instinct. Clarendon Press, Oxford. 228 pp.

Whoriskey, F.G. & G.J. FitzGerald. 1985. Nest sites of the threespine stickleback: can site characteristics alone protect the nest against egg predators and are nest sites a limiting resource? Can. J. Zool. 63: 1991–1994.

Whoriskey, F.G. & G.J. FitzGerald. 1987. Intraspecific competition in stickleback (Gasterosteidae: Pisces): does mother nature concur? J. Anim. Ecol. 56: 939–947.

Wootton, R.J. 1971. Measures of the aggression of parental male three-spined sticklebacks. Behaviour 40: 228–262.

Wootton, R.J. 1976. The biology of sticklebacks. Academic Press, London. 387 pp.

Wootton, R.J. 1984. A functional biology of sticklebacks. Croon Helm, London. 265 pp.

Environmental Biology of Fishes **52**: 251–260, 1998.

Sex determination system of the rosy bitterling, *Rhodeus ocellatus ocellatus*

Kouichi Kawamura
National Research Institute of Aquaculture (NRIA), Nakatsuhamaura, Nansei, Watarai, Mie 516-01, Japan (e-mail: barabote@nria.affrc.go.jp)

Received 10.1.1996 Accepted 4.3.1997

Key words: backcross, crossover, gynogenesis, meiosis, triploid, fish

Synopsis

The rosy bitterling, *Rhodeus ocellatus ocellatus*, is the only fish species known in which artificial triploids are always male, regardless of the kind of polyploidization technique used. In order to elucidate the genetic sex determination system of the rosy bitterling, two kinds of gynogenesis were carried out: retention of the second polar body (GRSPB) and suppression of the first cleavage (GSFC). The sex ratio of progeny was nearly 7:1 (♂:♀) for both GRSPB and GSFC, while those of control and parental fish were almost 1:1. In backcrosses of female progeny by GRSPB and normal diploid males, male progeny were observed at low frequency (one or two individuals in each experiment), except in one experiment where the appearance rate of males was about 50%. From results of gynogenesis and backcrosses, the following conclusions can be made. The genetic sex determination system of the rosy bitterling is a heterogametic female system (ZW). Survival rate of super-females (WW), produced by gynogenesis, is much lower than that of males (ZW).There is a possibility that crossovers between sex determining genes and a centromere occur in the first meiosis. With repect to the mechanism of unisexuality (male) of artificial triploids of the rosy bitterling, only males (ZZZ and ZZW) are presumed viable, while females (ZWW) are probably inviable.

Introduction

The genetic sex determination system of teleost fishes is variable, thereby differing from those of mammals and birds (Yamazaki 1983). The rosy bitterling, *Rhodeus ocellatus ocellatus*, is a small cyprinid fish which inhabits small ponds and brooks, and is distributed in temperate regions of Eurasia (Nichols 1943, Okada 1960). It was accidentally introduced into Japan from China during the 1940s, mixed in with the grass carp, *Ctenopharyngodon idellus*, and the silver carp, *Hypophthalmichthys molitrix* (Nakamura 1955). Because bitterlings are beautiful ornamental fishes, many aquarists and researchers have attempted to make even more beautiful strains using hybridization. Contrary to their expectations, hybrid bitterlings tended to be male, and in many crossbreedings offspring became strictly male (Suzuki & Hibiya 1986, Suzuki & Jeon 1988a). These unisexual bitterlings (male) are probably sterile, not producing motile spermatozoa that can fertilize eggs (Suzuki 1961, Suzuki & Jeon 1988b). In addition to the unisexuality of hybrids, artificial triploids of the rosy bitterling, produced by cold shock technique, were also reported to be all male (Ueno & Arimoto 1982). Although unisexuality has been reported in many wild populations of fish species (Vrijenhoek et al. 1989), in artificially produced triploid fishes it has only been reported in the rosy bitterling. The unisexuality of hybrid or triploid bitterlings may provide valuable information regarding the mechanism and evolution of sex determination systems in teleost fishes.

Bitterlings are easy to feed and breed, and their reproductive cycle can be controlled (Asahina & Hanyu 1983, Shimizu & Hanyu 1982). For some bitterling species, maturation may occur within three months. The sex of mature bitterlings is easily identified by secondary sexual characters, such as the nuptial colorations of males and the development of a long ovipositor in females. Thus, bitterlings seem to be good experimental fishes, especially for developmental studies. This study aims to elucidate two points: the sex determination system of the rosy bitterling, including artificial triploids, and the mechanisms of unisexuality and sterility in hybrid bitterlings. We report on a series of experiments related to triploidy and gynogenesis, and attempt to explain the mechanism of unisexuality in artificial triploids of the rosy bitterling.

Materials and methods

Parental fish

Parental fish of the rosy bitterling used in this study were obtained in the suburbs of Ogaki City, located in the northwest region of central Honshu, Japan, on 15 April 1993. Fish were raised in aquaria at 21–24° C, with a photoperiod of L14:D10 hours. The timing of ovulation in females could be judged with considerable precision by the length of the ovipositor. Eggs and milt were collected in separate Petri dishes (90 mm in diameter × 15 mm in height) containing physiological saline by pressing the abdomens of parental fish, using the methods of Ueda et al. (1990). Insemination was done by mixing eggs and milt in Petri dishes, with rearing water that was sterilized by boiling. After insemination, eggs were rinsed twice with rearing water, and kept in small dishes in a dark, temperature-controlled incubator at 22° C. Fertilized eggs and embryos were kept in the incubator for approximately four weeks until free swimming. As they became juveniles, fish were transferred to fish tanks (45 cm in width × 30 cm in depth × 30 cm in height), and fed naupli of brine shrimp, *Artemia salina*, and pellets.

Production of artificial triploids

Artificial triploids of the rosy bitterling produced by a cold shock technique have been reported to be all male (Ueno & Arimoto 1982). In order to examine the influence of the actual techniques which lead to the retention of the second polar body on the sex of artificial triploids, artificial triploids of the rosy bitterling were produced from 20 April to 30 June 1993, using three different techniques: cold shock, heat shock, and hydrostatic pressure shock. The cold shock technique of Ueno & Arimoto (1982) was slightly modified such that eggs were placed in cold water of 0 to 4° C for 60 minutes, 5 minutes after insemination. In the heat shock technique, eggs were placed at 40° C for 3 minutes, 5 minutes after insemination. In the hydrostatic pressure shock technique, inseminated eggs were subjected to hydrostatic pressure of 650 kg cm^{-2} for 6 minutes (Hyon-Sob et al. 1991). These experiments were conducted five times, each using four batches of eggs including a control. To ascertain the ploidy of shock-treated fish, the DNA volume of erythrocytes was measured by microfluorometry (Komaru et al. 1988).

Extremely high and low water temperatures are known to strongly affect sex differentiation in some species of teleost fishes (Oshiro 1987, Yamamoto 1995). To examine the influence of water temperature on sex differentiation in artificial triploids of the rosy bitterling, controls and artificial triploids were produced by cold shock on 10 May 1993. Each experiment was conducted by using 30 batches of eggs. Progeny were reared at four different temperatures (16, 21, 25 and 32° C) for 6 months. The constant temperature, regulated by thermometers, was maintained after insemination until the 'sex check', except for the time required for the cold shock treatment.

Gynogenesis and backcross

Gynogenesis is one of the most powerful methods to examine the genetic sex determination system of bisexual species. In this study, two kinds of gynogenesis of the rosy bitterling were performed from 20

April to 30 June 1993, to examine the genetic sex determination of the rosy bitterling. One is gynogenesis by retention of the second polar body after haploid activation of eggs (GRSPB), resulting in diploidization of eggs. The other is the suppression of the first cleavage after karyogamy (GSFC). The Taiwan bitterling, *Tanakia himantegus* (Wu), was found to produce inviable hybrid embryos with the rosy bitterling, reciprocally (unpublished data). Consequently, spermatozoa of the Taiwan bitterling were used to induce both kinds of gynogenesis. The procedure of gynogenesis is as follows. Stripped sperm, diluted with physiological saline, were poured into small plastic Petri dishes to a depth of 1 mm and inactivated by ultraviolet (UV) irradiation; the intensity was fixed at 3 000erg mm^{-2} s^{-1}, based on previously obtained high survival rates (unpublished data). In GRSPB, eggs were exposed to a cold shock treatment for 60 minutes, 5 minutes after insemination with inactivated spermatozoa, while in GSFC, the cold shock treatment was also performed for 60 minutes, but at 30 to 40 minutes after insemination. Experiments were conducted by using 15 batches of eggs, consisting of 5 groups: GRSPB, GSFC, two groups inseminated with either untreated or genetically inactivated spermatozoa of the Taiwan bitterling, and controls which were inseminated with spermatozoa of the rosy bitterling. About 6 months later, when the gynogenetic progeny had matured, backcross experiments were carried out between female progeny of gynogenesis (GRSPB) and normal diploid males from 8 October 1993 to 16 December 1993. In GSFC, no mature female progeny were obtained.

Sex check

A 'sex check' was performed for all progeny produced in the experiments. Histological observation of gonads at about 6 months of age was performed in all but the GRSPB progeny, where secondary sexual characters and gametes were used for the sex check, since the progeny were used in backcross experiments. Mature males have nuptial coloration, while mature females develop an ovipositor without nuptial coloration.

Results

Sex ratio in artificial triploids

The survival to hatching of eggs treated by the three triploid production techniques was: 69.8% (37 of 53 eggs) from cold shock, 14.8% (13 of 88 eggs) from heat shock, and 77.7% (73 of 94 eggs) from hydrostatic pressure shock, while that of controls was 88.4% (38 of 43 eggs) (Table 1). Although the survival to hatching of pressure shock treated embryos was the highest of the three shock treatments, deformed embryos comprised about 70% (51 of 73 individuals) in this group, and they did not survive to six months after hatching. Meanwhile, in cold

Table 1. Sex of artificial triploids of the rosy bitterling produced by cold, heat, and hydrostatic pressure shocks. Data were pooled from five experiments, each containing a control and three shock groups. Fish were examined at 6 months after hatching.

Treatment	Eggs treated	Embryos hatched (%)*	Fish examined (%)**	Triploid in fish examined (%)***	Males in fish examined (%)***	Triploid males (%)****	Diploid males
Control	43	38 (88.4)	15 (39.5)	0 (0.0)	7 (46.7)	0 (0.0)	7
Cold shock	53	37 (69.8)	16 (43.2)	14 (87.5)	14 (87.5)	14 (100.0)	15
Heat shock	88	13 (14.8)	10 (76.9)	6 (60.0)	6 (60.0)	6 (100.0)	8
Hydrostatic pressure shock	94	73 (77.7)	10 (13.7)	7 (70.0)	7 (70.0)	7 (100.0)	8

* Percentage of eggs treated.
** Percentage of embryos hatched.
*** Percentage of fish examined.
**** Percentage of triploid fish examined.

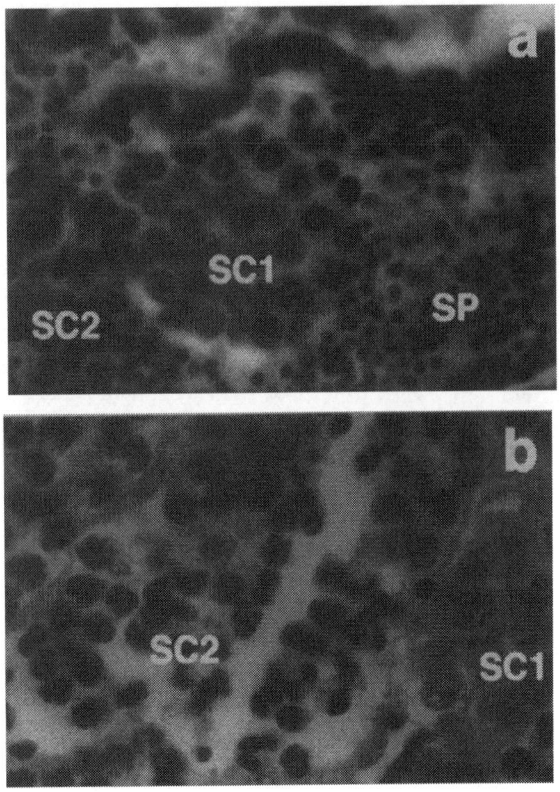

Figure 1. Histological sections of testes in the rosy bitterling, *Rhodeus ocellatus ocellatus*: a – diploid control; b – artificial triploid produced by cold shock (SC1 = primary spermatocyte; SC2 = secondary spermatocyte; SP = spermatozoon).

shock, heat shock and controls, deformed embryos were rarely seen. The low number of almost 6 month-old fish examined from cold shock treatment was due to a decrease caused by pathogens. From the heat shock treatment, the survival to hatching was very low (14.8%), compared with the other three treatments. Most unhatched embryos from the heat shock treatment appeared, from microscopic observation, to have died soon after the treatment.

The percentage of triploids, determined by microfluorometry, was: 87.5% (14 of 16 individuals) from cold shock, 60% (6 of 10 individuals) from heat shock, and 70% (7 of 10 individuals) from hydrostatic pressure shock. The frequency of triploid production in the rosy bitterling was the highest (87.5%) by cold shock treatment. Aneuploids were not seen in any experiments.

All triploids were male, regardless of the type of shock treatment. Histologically, all triploids had testes, and none had either hermaphroditic organs or ovaries. Spermatogonia and spermatocytes were observed, but spermatid-like and spermatozoon-like cells were not, although all were seen in diploid controls (Figure 1).

As shown in Table 1, cold shock is the most effective of the three treatments for producing triploids in the rosy bitterling. Therefore, artificial triploids produced by cold shock and controls were reared at

Table 2. Sex of diploids and artifical triploids of the rosy bitterling, produced by cold shock, reared at four different temperatures. Data were pooled from a total of 30 batches of eggs. Fish were examined at 6 months after hatching.

Treatment	Characteristic	Rearing temperature (°C)			
		16	21	25	32
Control (diploid)	Eggs treated	92	90	91	92
	Embryos hatched (%)	39 (42.4)	72 (80.0)	67 (73.6)	71 (77.2)
	Fish examined (%)	2 (5.1)	33 (45.8)	32 (47.8)	16 (22.5)
	Males in fish examined (%)	1 (50)	19 (57.6)	18 (56.2)	10 (62.5)
Cold shock	Eggs treated	62	62	62	62
	Embryos hatched (%)	21 (33.9)	36 (58.1)	38 (61.3)	7 (11.3)
	Fish examined (%)	0 (0.0)	16 (44.4)	25 (65.8)	4 (57.1)
	Triploid (%)	0 (0.0)	14 (87.5)	19 (76.0)	4 (100.0)
	Males in fish examined (%)	0 (0.0)	15 (93.8)	22 (88.0)	4 (100.0)
	Triploid males (%)	0 (0.0)	14 (100.0)	19 (100.0)	4 (100.0)
	Diploid males	0 (0.0)	1 (42.4)	3 (42.4)	0 (0.0)

four different temperatures (16, 21, 25 and 32° C). In preliminary experiments, no embryos of the rosy bitterling hatched at rearing temperatures over 35° C or below 15° C. The survival to hatching of control eggs was approximately 80%, except at 16° C, and about 60% for cold-shocked eggs, except at 16 and 32° C (Table 2). At 16° C, probably due to the low water temperature, both survival to hatching (33.9% in cold shock treated and 42.4% in controls) and survival rate at 6 months of age (0% in cold shock treated and 5.1% in controls) were very low. At 32° C, the survival to hatching was very low in cold shock treated eggs (11.3%). In controls, it was high (77.2%), but about one third of hatched embryos (45 individuals) were deformed and did not survive to six months after hatching. Thus, the survival rate of controls at 32° C was very low (22.5%). By six months after insemination, all but two individuals of the 16° C controls had matured. By histological observation, these were one male and one female. Although the ratio of males in the controls was 57.6% at 21° C, 56.2% at 25° C and 62.5% at 32° C, only triploid males appeared following cold shock treatment, regardless of water temperature.

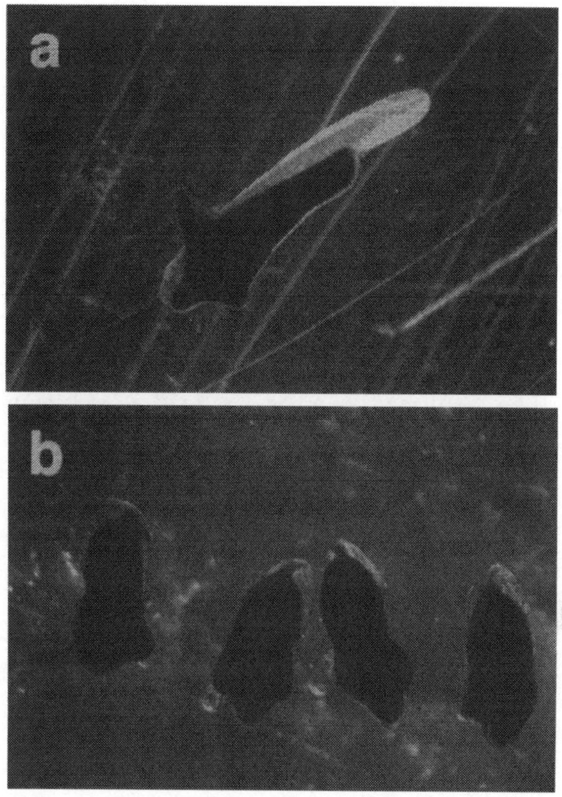

Figure 2. Hatched embryos of the rosy bitterling: a – diploid embryos (control); b – haploid embryo (from egg inseminated with UV-irradiated sperm).

Sex ratios in gynogenetic progeny

The fertilization rate of rosy bitterling eggs inseminated with Taiwan bitterling spermatozoa was about 90% (109 of 120 eggs), similar to that of controls (182 of 200 eggs). Under microscopic observation, successfully fertilized hybrid eggs began embryogenesis, but development was arrested during gastrulation or neurulation. None of the hybrid em-

Table 3. Sex of rosy bitterling gynogenetic offspring. Experimental eggs were inseminated with spermatozoa of the Taiwan bitterling, *Tanakia himantegus*. GRSPB = gynogenesis by retention of second polar body; GSFC = gynogenesis by suppression of first cleavage. Data were pooled from 15 experiments, using 15 batches of eggs. In some experiments, no GSFC offspring were obtained, due to high mortality. Fish were examined at 6 months after hatching.

	Eggs treated	Embryos hatched (%)	Deformed embryos hatched (%)	Fish on 30th day after hatching (%)	Fish examined (%)	Males in fish examined (%)
Control	200	158 (79.0)	2 (1.3)	106 (67.1)	38 (35.8)	20 (52.6)
Inseminated with untreated sperm	120	0 (0)	0 (0)	0 (0)	0 (0)	0 (0)
Inseminated with inactivated sperm	200	88 (44.0)	88 (100.0)	0 (0)	0 (0)	0 (0)
GRSPB	493	156 (31.6)	58 (37.2)	96 (61.5)	30 (31.3)	26 (86.7)
GSFC	1 309	233 (17.8)	215 (92.3)	13 (5.6)	8 (61.5)	7 (87.5)

bryos hatched, compared to 79% (158 of 200 eggs) of the controls (Table 3). In preliminary experiments, spermatozoa of the Taiwan bitterling were genetically inactivated, but their mobility was unimpaired by UV irradiation at a dose of 2 000–3 000 erg mm^{-2} (unpublished data). Eggs of the rosy bitterling, inseminated with UV-irradiated spermatozoa of the Taiwan bitterling, had a survival to hatching of 44% (88 in 200 eggs). All hatched embryos had short bodies, distorted notochords and extremely swollen abdomens (Figure 2), which are characteristic of the haploid syndrome. None of these deformed embryos reached free swimming, and all died within 20 days after hatching, as did the deformed triploid embryos produced by hydrostatic pressure shock. In contrast, 67.1% of control hatched embryos survived to the 30th day after hatching. The survival to hatching of treated eggs in GRSPB and GSFC was 31.6% and 17.8%, respectively. Abnormalities were found in 37.2% of GRSPB and 92.3% of GSFC hatched embryos. Although there was a low survival to hatching and a high rate of abnormalities in GRSPB embryos compared with controls, the situation was worse in GSFC embryos. The deformities seen in the GRSPB and GSFC embryos were very similar to those of the embryos produced using UV-irradiated spermatozoa from the Taiwan bitterling. The survival rate at six months after hatching was 35.8% for controls, 31.3% for GRSPB, and 61.5% for GSFC; the low rates in all experiments were due to diseases caused by pathogens.

Males, as identified by secondary sexual characteristics, comprised 86.7% of GRSPB, 87.5% of GSFC, and 52.6% of controls. Thus, sex ratios in both GRSPB and GSFC were extremely biased towards males (almost 7:1 [♂:♀]), whereas that of controls was nearly 1:1 (♂:♀). According to a contingency chi-square analysis, significant differences from the sex ratio of 1:1 (♂:♀) were seen in GRSPB ($\chi^2 = 16.1$, $p < 0.01$) and GSFC ($\chi^2 = 4.5$, $p < 0.05$), but not in controls ($\chi^2 = 0.1$, $p = 0.75$). The sex of all individuals in GRSPB and GSFC could clearly be identified not only by secondary sexual characters, but also by the type of gametes they produced. In GRSPB, spermiation was recognized in all males, and ovulation was observed in all females. In GSFC, spermiation was recognized in all males (7 individuals). The one female-like individual with an ovipositor died before maturity. When subjected to histological examination, this individual was found to have oogonia in the gonad, demonstrating that it was a complete female. No intersexes were ever seen in GRSPB, GSFC, or controls.

Sex ratios in backcrosses between gynogenetic females (GRSPB) and normal diploid males

All four female progeny from GRSPB matured and ovulated, and were used in five backcross experiments with normal diploid males derived from wild fishes. Survival to hatching ranged from 66.7% (F IV) to 93.6% (F II b), with a mean ± SD of 72.1 ± 19.1%. According to a contingency chi-square analysis, no significant differences in survival were seen between eggs of gynogenetic progeny and those of normal diploids ($\chi^2 = 0.23$ to 3.33, $p > 0.05$), except for F II b ($\chi^2 = 6.05$, $p < 0.05$) (Tables 3, 4). The survival rate at 30 days after hatching was very low in F III (33.3%) and F II b (27.3%), mainly due to mortality caused by pathogens. Six months after hatching, the surviving fish matured and displayed secondary sexual characters. When secondary sexual characters and gametes were examined, no intersexes were detected in any of these fish. The numbers of males produced was low in all experiments (6.3% in F II a, 0% in F II b, 22.2% in F III, 12.5% in F IV) except F I, which produced many males (54.5%).

Discussion

Factors related to sex determination in triploids

The unisex of artificial triploids of the rosy bitterling was first reported by Ueno & Arimoto (1982). Their triploids were produced only using a cold shock technique. In the present study, three different techniques (cold shock, heat shock and hydrostatic pressure shock) were employed to examine the effect of the treatment itself on the sex ratio. Artificial triploids of the rosy bitterling were exclusively male, regardless of the type of shock treatment applied.

Water temperature, particularly at the time of sex determination, was another factor which could conceivably affect the sex of rosy bitterling triploids. The goldfish, *Carassius auratus*, also a cyprinid species, has been reported to be easily sex-reversed to male by high water temperatures (Oshiro 1987). However, when diploid controls and artificial triploids were reared at various temperatures, the sex ratio of controls was nearly 1:1 at 16° C ($\chi^2 = 0$, $p = 1$), 21° C ($\chi^2 = 0.76$, $p = 0.35$), 25° C ($\chi^2 = 0.5$, $p = 0.52$) and 32° C ($\chi^2 = 1$, $p = 0.31$), whereas the triploids were exclusively male, regardless of temperature (Table 2). The sex ratio in the wild fish was also almost 1:1 (♂:♀ = 806:751) ($\chi^2 = 1.94$, $p = 0.16$).

The rearing water temperature for all subsequent experiments was maintained at 21–24° C, and all other environmental conditions were kept constant. Therefore, the unisex of triploid rosy bitterlings cannot be explained by environmental factors such as water temperature. Genetic factors, e.g. triploidy, are more likely to affect sex differentiation.

Sex determination system

Gynogenesis is known to be a useful technique for the elucidation of the genetic sex determining systems of bisexual species (Nakamura 1988). In fishes,

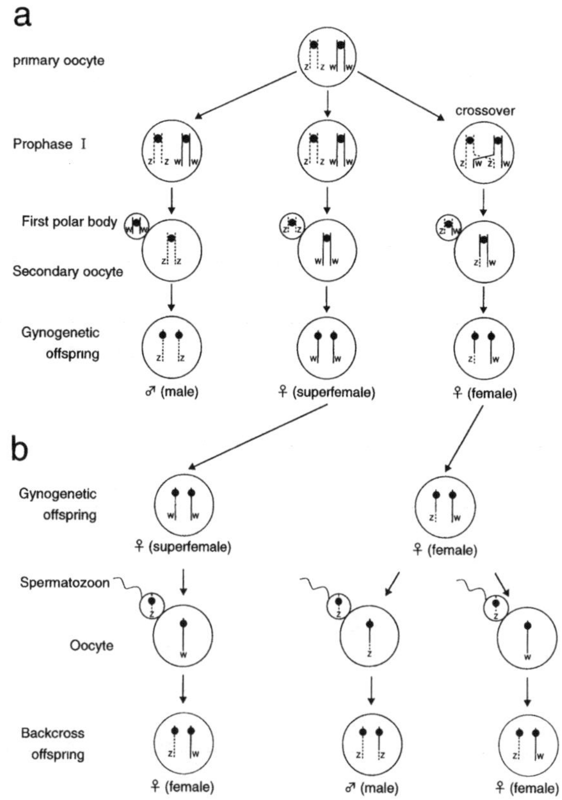

Figure 3. Schematic illustration of proposed models for gynogenesis (a) and backcrosses (b) in the rosy bitterling: a – three sex genotypes (ZZ, WW and ZW) would appear following gynogenesis induced by retention of the second polar body (GRSPB); b – one sex genotype (ZW) would appear in backcrosses of gynogenetic superfemales with normal diploid males, and two sex genotypes (ZZ and ZW) would appear in backcrosses of gynogenetic females, through a crossover, with normal diploid males.

Table 4. Sex of progeny from backcrosses between gynogenetic female offspring and normal diploid males of the rosy bitterling. Fish were examined at 6 months after hatching.

Gynogenetic female*	Eggs inseminated	Embryos hatched (%)**	Fish on 30th day after hatching (%)***	Fish examined (%)****	Males in fish examined (%)*****
F I	14	12 (85.7)	11 (91.7)	11 (100.0)	6 (54.5)
F II a	24	18 (75.0)	16 (88.9)	16 (100.0)	1 (6.3)
F II b	47	44 (93.6)	12 (27.3)	8 (66.7)	0 (0.0)
F III	31	27 (87.1)	9 (33.3)	9 (100.0)	2 (22.2)
F IV	36	24 (66.7)	16 (66.7)	8 (50.0)	1 (12.5)

* F I ~ F IV denote the individual females from which eggs were sampled for backcrosses with wild diploid males.
** Percentage of eggs fertilized.
*** Percentage of embryos hatched.
**** Percentage of fish on 30th day after hatching.
***** Percentage of fish examined.

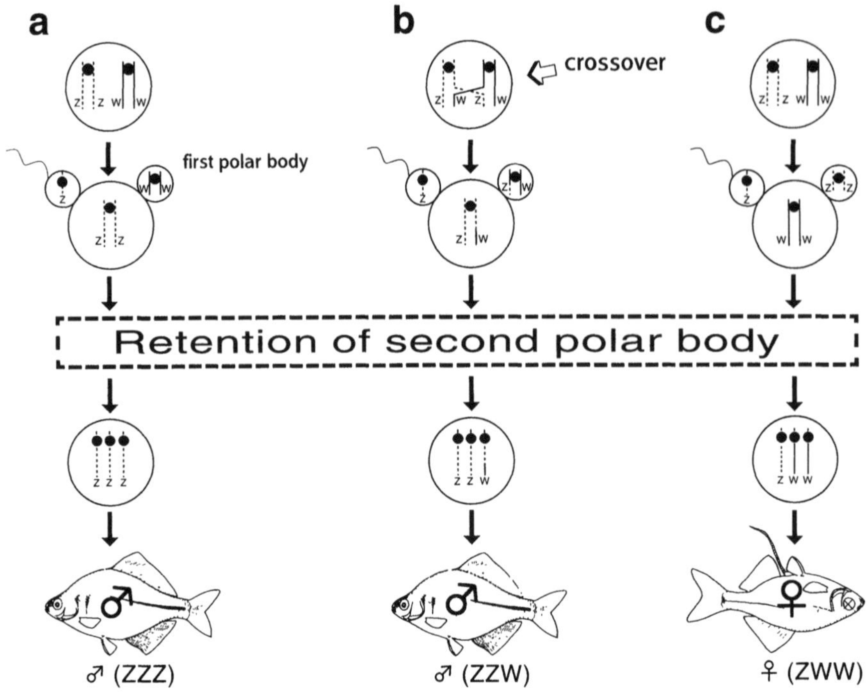

Figure 4. Schematic illustration of the proposed mechanism yielding only males in artificial triploid rosy bitterlings. The sex genotypes ZZZ and ZZW would be male and viable, whereas the sex genotype ZWW would be female and lethal, due to the genetic effects of homogamety of the W sex determining gene.

unlike mammals and birds, a variety of genetic sex determination systems (i.e. XY, ZW and XO) are known (Yamazaki 1983). In the rosy bitterling, the sex ratio of gynogenetic progeny was extremely biased towards males (♂:♀ = 7:1), whereas the sex ratios of control and parental fish from the wild were almost 1:1 (Table 3). From these results, the genetic sex determination system can be deduced to be a female heterogametic sex determination system (ZW). In teleost fishes, the XY sex determination system appears to be common; the ZW system is rare, and restricted to certain groups, i.e. poeciliid fishes (Poeciliidae), sticklebacks (Gasterosteidae), lizard fishes (Synodontidae) and tilapine cichlids (Cichlidae) (Chen & Ebeling 1968, Chen & Reisman 1970, Nishikawa & Sakamoto 1978, Trombka & Avtalion 1993). An XY system has been found only in the ostaryophysan fishes that have been studied to date, namely a Japanese loach, *Misgurunus anguilicaudatus* (Suzuki et al. 1985), and goldfish, *Carassius auratus* (Oshiro 1987); other sex determining systems have not been reported for this group. Thus, the rosy bitterling is the first ostaryophysan fish that has been found to have a ZW system. In the tilapine genus *Oreochromis*, both XY (*O. mossambicus* and *O. niloticus*) and ZW (*O. macrochir, O. urolepis hornorum* and *O. aureus*) systems have been found (Trombka & Avtalion 1993). Thus, it is possible that male (XY) and female (ZW) heterogametic sex determination systems could coexist in cyprinid fishes as well. To answer that question, it will be necessary to perform a detailed analysis of the genetic sex determination systems in the cyprinid fishes, including other bitterling species, as well as an analysis of the evolution and differentiation of plural genetic sex determination systems in the cyprinid fishes. The southern platyfish, *Xiphophorus maculatus*, has been reported to have three sex chromosomes: X, Y and W (Kallman 1984). The W chromosome induces female sex differentiation, and is superior to the Y chromosome; it is believed to have arisen as a mutation of the X chromosome. Kallman (1984) noted that such a mutation, related to sex genes or chromosomes, also has occurred in the lemming, *Myopus schisticolor*. The change in sex chromosome function does not appear to be restricted to specific groups, and could lead to a change in genetic sex determination systems. Thus, it is possible that, during the evolution of cyprinid fishes, including the bitterlings, a male heterogametic sex determination system changed to a female heterogametic system.

Crossover of sex determining genes during meiosis

In backcrosses of gynogenetic females with normal diploid males, male progeny were produced at rates ranging from 0 to 54.5% (Table 4). From the preceding discussion, the genetic sex determination system of the rosy bitterling has been deduced to be female heterogametic (ZW). Therefore, in theory, all progeny from backcrosses of gynogenetic females (superfemales WW) and normal diploid males (ZZ) must be female (ZW). However, males were produced in all but one (F II b) of the experiments (Table 4). One possibility is that an external factor, such as water temperature, induced a sex reversal. As previously discussed, however, environmental factors appear to have only minimal influence on the sex of the rosy bitterling. In experiment F I (Table 4), a high number and proportion of males were produced (6 males; 54.5%). While the occurrence of males in all other experiments could be explained by environmentally-induced secondary sex reversals, it seems unlikely that sex reversal was the sole cause of all males produced in experiment F I.

Crossovers of sex determining genes during meiosis have been reported to occur at high frequencies in other vertebrates, i.e. a tilapia, *Oreochromis aureus* (Avtalion & Don 1990), and the South African clawed frog, *Xenopus laevis* (Nakamura 1988). If such crossovers also occur in the rosy bitterling, the results of experiment F I could easily be explained (Figure 3). If one crossover in a pair of chromosomes involving sex genes occurs in the first meiosis, not only the W gene but also the Z gene still remains in the secondary oocyte. In gynogenesis, these type of eggs yield female progeny with the ZW sex genotype. Then, in backcrosses of ZW gynogenetic females with normal diploid males, two sex genotypes, ZZ (male) and ZW (female) will appear with a sex ratio of 1:1.

Mechanism of the unisexuality of artificial triploids

The sex ratios in both types of gynogenesis in the rosy bitterling were extremely biased towards males (Table 3). In a medaka, *Oryzias latipes*, a dominant sex determining gene (Y) lessens the survival rate of YY supermales (Yamamoto 1964). In the rosy bitterling, if a dominant sex determining gene (W) were to similarly lessen the survival rate of WW superfemales, the low ratio of females following gynogenesis could be explained. The absence of female artificial triploids also could be explained, as depicted in Figure 4. In the rosy bitterling, with a ZW sex determination system, the genotypes of artificial triploids should be ZZZ and ZWW. Following crossovers of sex determining genes in the first meiosis, ZZW also would appear. ZZZ would produce males, as would ZZW, since the Z genes are dominant over W in number. Conversely, in ZWW, the W genes are dominant over the Z in number. This genotype would produce females, but would be lethal due to the genetic effects of the WW genes. Only genotypes ZZZ and ZWW would survive, thereby producing only males. To verify this hypothesis, it will be necessary to examine the sex differentiation of triploid rosy bitterlings during early development.

Acknowledgements

K. Hosoya (NRIA), H. Onozato (Shinsyu University) and K.T. Wada (NRIA) encouraged me to carry out this study and gave valuable comments and helpful suggestions during various stages of this manuscript preparation. H. Maeda, Tokyo Sea Life Park of Tokyo Metropolitan, kindly provided the fish. Ann Graveson corrected the English. To all the above mentioned persons, I would like to express my sincere gratitude.

References cited

Asahina, K. & I. Hanyu. 1983. Role of temperature and photoperiod in annual reproductive cycle of the rose bitterling *Rhodeus ocellatus ocellatus*. Bull. Japan. Soc. Sci. Fish. 49: 61–67.

Avtalion, R.R. & J. Don. 1990. Sex-determining genes in tilapia: a model of genetic recombination emerging from sex ratio results of three generations of diploid gynogenetic *Oreochromis aureus*. J. Fish Biol. 37: 167–173.

Chen, T.R. & A.W. Ebeling. 1968. Karyological evidence of female heterogamety in the mosquitofish, *Gambusia affinis*. Copeia 1968: 70–75.

Chen, T.R. & H.M. Reisman. 1970. A comparative chromosome study of the North American species of sticklebacks (Teleostei: Gasterosteidae). Cytogenetics 9: 321–332.

Hyon-Sob, H., N. Taniguchi & A. Tsujimura. 1991. Production of clonal ayu by chromosome manipulation and confirmation by isozyme marker and tissue grafting. Nippon Suisan Gakkaishi 57: 825–832.

Kallman, K.D. 1984. A new look at sex determination in poeciliid fishes. pp. 95–172. *In:* B.J. Turner (ed.) Evolutionary Genetics of Fishes, Plenum Press, New York.

Komaru, A., Y. Uchimura, H. Ieyama & W.T. Wada. 1988. Detection of induced triploid scallop, *Chlamys nobilis*, by DNA microfluorometry with DAPI staining. Aquaculture 69: 245–253.

Nakamura, M. 1955. Immigrated fishes reproducing in Kanto Plain. Bull. Biogeogr. Soc. Japan 16–19: 333–337 (in Japanese).

Nakamura, T. 1988. Female heterogametic sex-determination in *Xenopus laevis* as reconfirmed by repeated diploid gynogenesis. Zool. Sci. 5: 187–189.

Nichols, J.T. 1943. The fresh-water fishes of China. Amer. Mus. Nat. Hist., New York. 322 pp.

Nishikawa, S. & K. Sakamoto. 1978. Comparative studies on the chromosomes in Japanese fishes – IV. Somatic chromosomes of two lizardfishes. J. Shimonoseki Univ. Fish. 27: 113–117 (in Japanese).

Okada, Y. 1960. Studies on the freshwater fishes of Japan. J. Fac. Fish. Pref. Univ. Mie 4: 400–405.

Oshiro, T. 1987. Sex ratio of diploid gynogenetic progeny derived from five different females of goldfish. Nippon Suisan Gakkaishi 53: 1899.

Shimizu, A. & I. Hanyu. 1982. Environmental regulation of annual reproductive cycle in a spring-spawning bitterling *Acheilognathus tabira*. Bull. Japan Soc. Sci. Fish. 48: 1563–1568.

Suzuki, N. & T. Hibiya. 1986. Hybridization experiments in Acheilognathine fishes (Cyprinidae), intergeneric hybrids between *Acheilognathus lanceolata* and *A. limbata*. Bull. Coll. Agr. & Vet. Med., Nihon Univ. 43: 151–164 (in Japanese).

Suzuki, N. & S.R. Jeon. 1988a. Hybridization experiments in Acheilognathine fishes (Cyprinidae), an intergeneric hybrid between *Rhodeus suigensis* and *R. uyekii*. Kor. J. Lim. 21: 57–78.

Suzuki, N. & S.R. Jeon. 1988b. Hybridization experiments in Acheilognathine fishes (Cyprinidae), and intrageneric hybrid between *Rhodeus uyekii* and *R. ocellatus smithi*. Kor. J. Lim. 21: 129–142.

Suzuki, R. 1961. External characters of artificial intergeneric hybrids among Japanese bitterlings (cyprinid fish). Bull. Japan Soc. Sci. Fish. 27: 418–424.

Suzuki, R., T. Oshiro & T. Nakanishi. 1985. Survival, growth and fertility of gynogenetic diploids induced in the cyprinid loach, *Misgurnus anguillicaudatus*. Aquaculture 48: 45–55.

Trombka, T. & R. Avtalion. 1993. Sex determination in tilapia – a review. Israel J. Aquacult. – Bamidgeh 45: 26–37.

Ueda, T., R. Kawahara & J. Kobayashi. 1990. Rose bitterling as a laboratory animal for study of cell engineering and for teaching materials. Bull. Takaharayama Inst. Nature Educ. 9: 8–17 (in Japanese).

Ueno, K. & B. Arimoto. 1982. Induction of triploids in *Rhodeus ocellatus ocellatus* by cold shock treatment of fertilized eggs. Experientia 38: 544–546.

Vrijenhoek, R.C., R.M. Dawley, C.J. Cole & J.P. Bogart. 1989. A list of the known unisexual vertebrates. pp. 19–23. *In:* R.M. Dawley & J.P. Bogart (ed.) Evolution and Ecology of Unisexual Vertebrates, New York State Museum, New York.

Yamazaki, F. 1983. Sex control and manipulation in fish. Aquaculture 33: 329–354.

Yamamoto, T. 1964. The problem of YY zygotes in the medaka, *Oryzias latipes*. Genetics 50: 45–58.

Yamamoto, E. 1995. Study about artificial sex control and clone production of a Japanese flounder, *Paralichthys olivaceus*. D.Sc. Thesis, University of Hokkaido, Hakodate. 200 pp. (in Japanese).

Environmental Biology of Fishes **52**: 261–269, 1998.

Reproductive success in female *Neolamprologus mondabu* (Cichlidae): influence of substrate types

Yasuhiro Takemon[1] & Katsuyuki Nakanishi[2]
[1] *Department of Life Sciences, University of Osaka Prefecture, Sakai, Osaka, 599 Japan (e-mail: takemon@el.cias.osakafu-u.ac.jp)*
[2] *Ise Agriculture, Forestry and Fisheries Office of Mie Prefecture, Ise, Mie, 516 Japan*

Received 4.6.1996 Accepted 5.3.1997

Key words: substrate brooder, body size, home range size, habitat preference, Lake Tanganyika, fish

Synopsis

Individual variation in body size, home range size, and reproductive success among *Neolamprologus mondabu* (Cichlidae) females were investigated with reference to substrate types in the littoral zone of northern Lake Tanganyika. Larger females occupied sandy substrates and smaller ones stony substrates. Female reproductive success, estimated as the number of offspring successfully released per month, was higher on sandy substrates than on stony substrates. This may be attributable to higher spawning intensity and better protection of broods against predators in open, sandy habitats.

Introduction

Habitat structure, particularly the substrate type, is important in the life cycle of fish since it affects the distribution and availability of food, spawning ground, and refuges from predation (Keenleyside 1979, Chundler et al. 1985, Kuwamura 1987, Hourigan 1989, Inoue & Nakano 1994, Tupper & Boutilier 1995). A number of studies have investigated the choice of spawning sites in relation to substrate types (Colgan & Ealey 1973, Balon 1975, 1981, Crisp & Ealey 1989, Gashagaza 1991, Leman 1993) which may include factors such as dissolved oxygen concentration (Peterson & Quinn 1996) and supply of ground water (Curry & Noakes 1995, Curry et al. 1995). Many of these studies have speculated that selection of certain spawning sites results in higher reproductive success, though few have substantiated this with evidence. Exceptions may be in salmonid fishes, where physical characteristics of spawning redds have been shown to affect the survival of offspring (Colgan & Ealey 1973, Crisp & Carling 1989, Leman 1993, Curry et al. 1995). However, these works have not considered the reproductive success of individual spawners, an important aspect from behavioural and evolutionary points of view.

Cichlid fishes inhabiting Lake Tanganyika are suitable for analysing individual reproductive success because of their residency and territorial social structures. Because their territorial mosaics are highly stable in spatial arrangement (Hori 1987, 1991, Kohda 1991, Gashagaza 1991, Kohda & Yanagisawa 1992), it is possible to monitor individual reproductive success. Many of them, in particular substrate nesters, breed within their home ranges repeatedly throughout the year with spawning periods synchronised with the lunar cycle (Nakai et al. 1990, Gashagaza 1991). In this paper, relations among substrate types, body size, home range size, and reproductive success in *Neolamprologus mondabu* (Cichlidae) females were examined based on

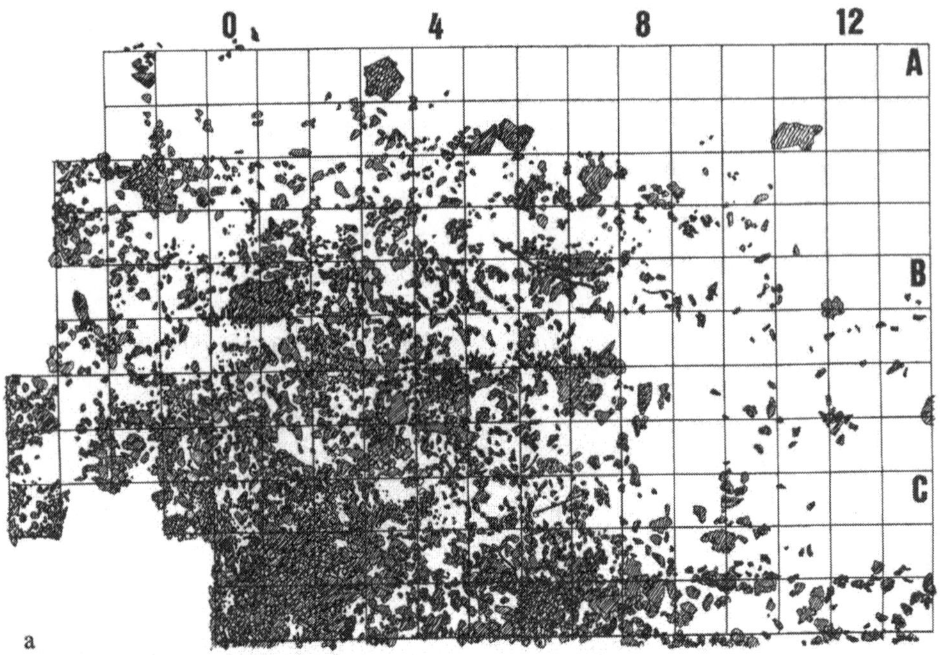

a

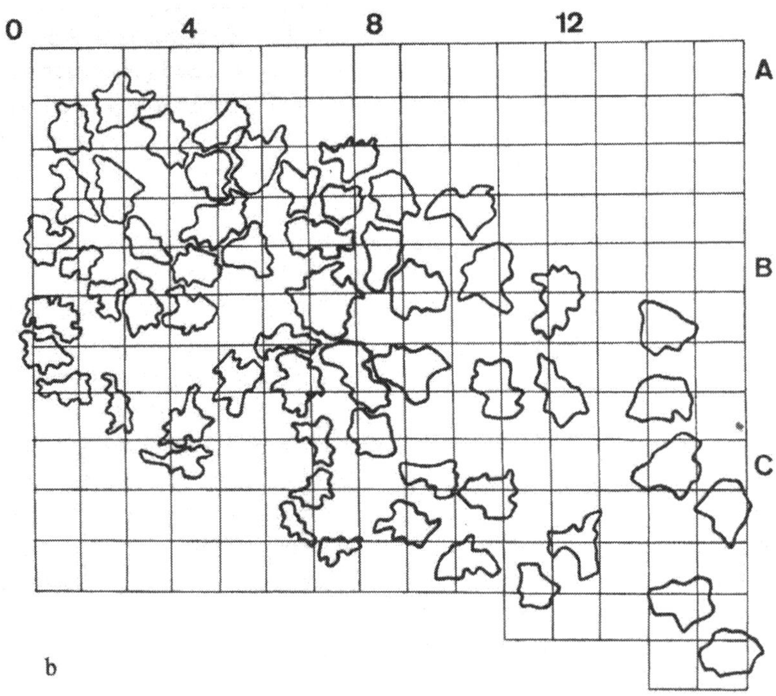

b

Figure 1. Substrate map (*a*) and home ranges of 58 female *Neolamprologus mondabu* (*b*) in the study area in the littoral zone of northern Lake Tanganyika at Pemba. Grid squares define 1 m × 1 m areas. The top and bottom lines of the substrate map correspond to ca. 3 m and 6 m water depth, respectively. Note in (*a*) that the substrate gradually changes from sandy flat (unshaded) at the upper right to stony substrate (shaded) at the lower left. Columns and rows with common symbols between (*a*) and (*b*) represent the same location in the map.

field observations using SCUBA in the littoral zone of the lake.

Habitat characteristics determined by the kinds and composition of substrate material are crucially important for substrate-nesting and mouthbrooding cichlid fishes, which require species-specific types of substrate for spawning (Brichard 1978, 1989, Nagoshi 1983, Kuwamura 1986, Yanagisawa 1986, 1987, Gashagaza 1991). Variation in reproductive success due to substrate structure is thought to influence intra- and interspecific social relations in the cichlid fish community. This is particularly true for *N. mondabu* which is considered a keystone species in the cichlid fish community on sandy substrates, due to their monopolisation of habitat space through strong territorial behaviour (Yuma 1993). Therefore, this study attempted to clarify the effects of substrate type on reproductive success of *N. mondabu* in order to understand the distribution pattern of individuals in natural habitat.

Materials and methods

The fish

Neolamprologus mondabu is a common cichlid fish in the northern part of Lake Tanganyika, which inhabits sandy and stony substrates in littoral areas (Brichard 1978, 1989, Hori 1983, Yuma 1993, 1994). The fish feeds on a wide range of benthos among which chironomid larvae tend to be most important (Hori 1983, Yuma 1993, 1994). Each individual has a stable home range maintained for at least three months and within which both feeding and breeding activities are carried out (Nakanishi unpublished). Adult females make a nest by digging a hole under or beside a stone to lay eggs on an exposed stone surface in the hole. Males of some chichlids, in particular monogamous species, assume an active role in caring for the brood (Kuwamura 1986, Yanagisawa 1987, Barlow 1991, Keenleyside 1991, Keenleyside & Machereth 1992). *N. mondabu*, however, has a polygynous mating system and the males do not demonstrate parental care (Nakanishi unpublished). Therefore, it is reasonable to focus on female-related factors to understand variation in reproductive success.

The Lamprologini species of the lake show periodicity in breeding activity which is synchronous with the lunar cycle (Nakai et al. 1990, Gashagaza 1991). *N. mondabu* starts its breeding activity during the first half moon (Takemon unpublished), so reproductive data can be collected over one month periods. When a female succeeds in breeding, it takes care of the brood in the home range for ca. two months until offspring grow to 2.5–3.0 cm total length, whereas, those which fail to breed attempt to breed in subsequent months (Nagoshi 1983, Takemon unpublished).

The study area

The study was conducted in the littoral zone of northern Lake Tanganyika at Pemba, 25 km south of Uvira (32° 4′ S, 29° 10′ E), Zaire. A study area of ca. 11 m by 20 m was established on the sandy to stony bottom of a wave-cut platform at a depth of 3–6 m. The area was divided into 1 m by 1 m grids using ropes and a substrate map was drawn on plastic graph paper at 1 : 20 scale (Figure 1a). This was used for recording fish behaviours and for analysis of substrate types.

The substrate in the study area was composed of sandy flat in the shallower part, stony bottom in the deeper part, and transitional zone between sandy and stony substrates where submerged stones occurred in the sandy bottom. Stony substrates with several layers of stones were found only in the deepest part of the area. Although a few pieces of woody debris were found on the bottom, most of the hard substrates were composed of stones and rocks.

Observations of home range and breeding behaviour

Field observations were conducted between 9 February and 27 April 1989. All 73 females showing a pair-bond display with a male (i.e. mutual 'greeting' and 'chasing') at intervals of ca. 30 min throughout a day, were selected for study. These females always had a fixed home range with one or a few nest sites and were assumed to be potential breeders. All females of this type inhabiting the study area were selected. Other small and presumably immature individuals that were chased away by males were excluded.

Home ranges of the 73 females were determined by observing each female continuously for 10 minutes and recording its position on the substrate map. The home range was demarcated by connecting the outer most points of fish locations (Figure 1b). Previous observations revealed that 10 minutes was sufficient to represent the whole home range (Takemon unpublished). Although individual identification based on scars was possible for only 5 of 73 females, females were assumed to occupy the same home ranges during the study due to their known territoriality and relatively stable social structures (Nakanishi unpublished, Yuma 1993).

Reproductive states of the 73 females were recorded every few days, at most. All females monitored were collected on 27 April 1989 using gill nets, preserved in 10% formalin, and their standard length measured.

Measurement of spawning intensity and reproductive success

Spawning intensity was calculated as the number of mating bouts per female per month. Of the 73 females observed, 43, 21, and 9 were monitored for three, two and one spawning periods, respectively. Therefore, the set of values possible for spawning intensity was 0, 0.33, 0.5 (i.e. for females that spawned once during two months of observation), 0.67, and 1.0. High spawning intensity does not always lead to high reproductive success because successful breeders skip spawning at least once during their brood care.

Reproductive success was represented by the number of offspring successfully released per female per month. Timing of offspring release was indicated by a female starting to chase young out of her home range. When we found this chasing behaviour, the number of young protected by the female at the last observation was assumed to be the number of offspring successfully released. The number of offspring that was clearly visible in or around the nest was counted during each observation. Unfinished matings at the end of the study period were not included in the calculation of reproductive success.

Data analyses

Two analyses were conducted. First, four variables (i.e., standard length, area of home range, spawning intensity, and number of offspring successfully released per month) were compared among female groups associated with different substrate types. Substrate types were classified by the proportion of area of sandy bottom within the home range, which was estimated from the map using a scanner and computer. The 73 females were divided into five categories (S1–S5), each with 14 or 15 females. The five categories represented the following ranges in

Table 1. Variation in standard length (mm) among five groups of female *Neolamprologus mondabu* (S1–S5) associated with different substrate types within the home-range. See text for definition of the groups.

	S1	S2	S3	S4	S5
n	15	15	15	14	14
Mean	79.9	79.2	77.1	70.2	62.8
SD	3.6	3.8	4.0	4.9	10.3
S5	**	**	**	*	
S4	**	**	**		
S3	*	NS			
S2	NS				

** = $p < 0.01$, * = $p < 0.05$, and NS = not significant, by Mann-Whitney U test.

Table 2. Variation in home range size (m^2) among five groups of female *Neolamprologus mondabu* (S1–S5) associated with different substrate types within the home range.

	S1	S2	S3	S4	S5
n	15	15	15	14	14
Mean	0.86	0.81	0.82	0.66	0.52
SD	0.08	0.16	0.19	0.08	0.16
S5	**	**	**	**	
S4	**	**	**		
S3	NS	NS			
S2	NS				

** = $p < 0.01$, and NS = not significant, by Mann-Whitney U test.

the proportion of area with sandy bottom: S1 = 0.92–0.85, S2 = 0.84–0.76, S3 = 0.75–0.67, S4 = 0.66–0.55, S5 = 0.54–0.19. Differences among the five groups were tested by the nonparametric Mann-Whitney U test, because the data were either discontinuous (e.g. spawning intensity) or assumed not to be normally distributed. Secondly, Spearman rank correlation analysis was conducted between pairs of five factors (substrate type, standard length, home range area, spawning intensity, and number of offspring released per month), to clarify the patterns of correlation among them.

Results

Distribution of female home ranges

N. mondabu females did not occur on the sandy flat in the upper right nor on the stony bottom in the lower left of the study area (Figure 1b). Consequently they extended from the upper left to the lower right part of the study area, which corresponded to a transitional zone between sandy and stony substrates (Figure 1a).

Body size

Standard length of the females ranged from 40.0 mm to 85.7 mm ($\bar{x}$ = 74.0 mm, SD = 8.6 mm, n = 73). Mean standard length differed significantly among the five groups but not between S1–S2 and S2–S3 (Table 1). Females of S4 and S5 were substantially smaller than the others. This showed that larger females inhabited predominantly sandy substrates. The higher variability in standard length in S5 than the others ($p < 0.01$, F-test) reflects the occurrence of small individuals on the stony substrates that made no spawning attempt throughout the study period.

Table 3. Variation in spawning intensity (the number of matings per female per month) among five groups of female *Neolamprologus mondabu* (S1–S5) associated with different substrate types within the home range.

	S1	S2	S3	S4	S5
n	15	15	15	14	14
Mean	0.67	0.72	0.64	0.32	0.14
SD	0.35	0.31	0.31	0.38	0.25
S5	**	**	**	NS	
S4	*	**	*		
S3	NS	NS			
S2	NS				

** = $p < 0.01$, * = $p < 0.05$, and NS = not significant, by Mann-Whitney U test.

Table 4. Variation in the number of offspring successfully released per female per month among five groups of female *Neolamprologus mondabu* (S1–S5) associated with different substrate types within the home range.

	S1	S2	S3	S4	S5
n	15	15	15	14	14
Mean	3.74	1.47	1.13	0.04	0.14
SD	5.05	2.91	2.47	0.13	0.25
S5	*	NS	NS	NS	
S4	*	NS	NS		
S3	NS	NS			
S2	NS				

* = $p < 0.05$, and NS = not significant, by Mann-Whitney U test.

Home range size

The home range size of females ranged from 0.268 m^2 to 1.350 m^2 with a mean of 0.737 m^2 and SD of 0.186 m^2 ($n = 73$). The fish inhabiting stony substrates (S4 and S5) had significantly smaller home ranges than those inhabiting predominantly sandy substrates (S1–S3) (Table 2). There was no significant difference in home range size among the latter groups.

Spawning intensity

A total of 21, 8, 10, 14, and 20 females had 0, 0.33, 0.5, 0.67, 1.0 mating bouts per month, respectively. Mean spawning intensity was 0.51 (SD = 0.39, n = 73). Spawning intensity varied widely, ranging from no spawning throughout three months in 11 females to three successive mating bouts in 10 females (n = 43). Spawning intensity of females on stony substrates (S4 and S5) was significantly lower than on sandy substrates (S1–S3) (Table 3). The fish on stony substrates spawned on average no more than once during the three monthly spawning periods. Seventeen fish on these substrates made no mating attempts. In contrast, many fish associated with sandy substrates (S1–S3) spawned two or three times. Only four individuals on sandy substrates made no spawning attempts during the study period and most fish spawned every month when the previous breeding was unsuccessful.

Number of offspring successfully released

Of 87 cases of spawning by 52 females, offspring survived until the juvenile period in 34 cases (39.1%) but were successfully released in only 19 cases (21.8%). Therefore, the total brood was lost during the embryo period in 53 of 87 cases (60.9%), and during the juvenile period in 15 of the remaining 34 cases (44.1%). However, the mortality of offspring in each brood was unknown from the present data. The highest reproductive success in one breeding (46 offspring released) was recorded for a S1 female, but most of females released less than 20. The average reproductive success was significantly higher in S1 than in S4 or S5 (Table 4). Females on sandy substrates released an average of over three offspring per month but those on stony substrate released less than 0.2.

Table 5. Matrix of Spearman's rank correlation coefficients among substrate type (ST), standard length (SL), home range size (HRS), spawning intensity (SI), and number of offspring successfully released (NOSR) of 73 *Neolamprologus mondabu* females.

	ST	SL	HRS	SI
NOSR	0.356*	0.216^{NS}	0.346*	0.324*
SI	0.544**	0.581**	0.546**	
HRS	0.706**	0.852**		
SL	0.757**			

** = $p < 0.005$, * = $p < 0.01$, and NS = not significant.

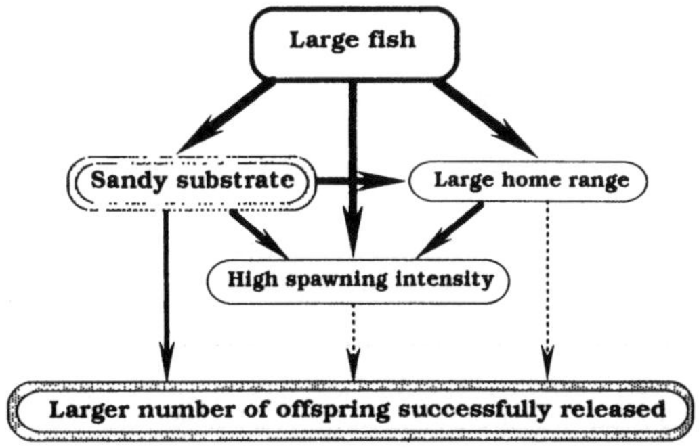

Figure 2. Causal links inferred from correlation analysis of possible factors relating to reproductive success in *Neolamprologus mondabu*. The width of arrows presents relative significance indicated by correlation coefficients (Table 5).

Correlation analysis

Home range size and spawning intensity correlated more strongly with standard length than with substrate type (Table 5). This suggests that larger females had larger home ranges and made more spawning effort. In contrast, reproductive success as measured by the number of offspring successfully released correlated most strongly with substrate type, but not with standard length. Thus, substrate type appears a more reliable predictor of reproductive success than body size.

Discussion

Effects of substrate type on reproductive success

Results of the present study suggest the following link between body size, substrate type, and reproductive success in *N. mondabu* (Figure 2): (1) larger fish had larger home ranges on more sandy substrates, (2) they also showed higher spawning intensity, but (3) substrate type explained more of the variation in reproductive success. Larger females of this species must have an advantage in breeding because fecundity increases with body size (Nagoshi 1983). Our study, however, indicates that substrate type is equally or perhaps more important for successful breeding.

Spawning intensity, which indicates the magnitude of reproductive investment, was significantly higher on sandy than on stony substrates. This was partly due to the presence of small females on stony substrates that showed no spawning activities. However, reproductive success was also higher on sandy substrates than on stony substrates. The advantage of sandy substrates for the egg and embryo stages was not clear, as these stages experienced a high mortality with ca. 61% of the total brood being lost. This mortality may be due to a respiration problem caused by low oxygen concentration in the nest hole, burying of the brood by sand inflow, and predation. The parent females were observed to alleviate the respiration problem by cleaning and fanning the egg-laden stone surface. Females also carried inflowing sand out of the nest hole by mouth. When inflow was very fast, e.g. during storms, they gave up breeding and left eggs and embryos in the sand. Predation by the crabs *Platythelphusa* spp. on eggs and/or embryos was observed on some occasions, but the data are insufficient to assess its overall impact.

Higher reproductive success on sandy substrates may be due primarily to earlier detection of predators than on stony substrates. Predation by piscivorous fishes may not be serious in the egg and embryo intervals when the brood is protected in the nest. However, young are vulnerable to several species of predatory cichlids known as 'young-eaters' in Lake Tanganyika, which have specialised techniques for catching them (Fryer & Iles 1972, Hori 1983, 1987, Kohda 1994). The risk of predation by fish is generally known to be higher on sandy substrates than on stony and rocky substrates, since the former lacks suitable refuges (Tupper & Boutilier 1995). This may also be true in the littoral zone of the lake if the young are not guarded by the parent fish. However, where parental guarding occurs as in *N. mondabu*, the efficiency of guarding does not depend on the availability of refuges but on the detectability of predators. Females of *N. mondabu* with broods furiously attack young-eaters such as *N. fasciatus*, *Lepidiolamprologus elongatus*, *L. attenuatus* and *L. profundicola* as far away as 1.5 m from the brood (Takemon unpublished). As any mode of young hunting requires an initial approach to a striking distance from the potential prey, such aggressive behaviour on the part of a guarding parent can keep the predator beyond striking range. Here, detection of predators is greatly facilitated by an unrestricted view on the open, sandy substrates. In contrast, parent fish may fail to detect predators on stony substrates because boulders can block their view.

Habitat preference by female N. mondabu

Many fishes change their habitats according to their needs for feeding and reproduction. Therefore, habitat selection during reproduction may not always relate to the social structure during non-reproductive periods. Females of *N. mondabu*, however, have resident home ranges in which both feeding and breeding occur. This type of territorial social system has also been reported for pomacentrid reef fishes (Hourigan 1986, 1989). In these cases, the relative importance of feeding vs. reproduction may be crucial for habitat selection. *N. mondabu* feeds on benthic animals, particularly chironomid larvae, by picking them from stone surfaces and also by filtering sand with their gill rakers (Hori 1983, Yuma 1994). Overall, its feeding sites are concentrated on stone surfaces covered by sand where tube-building chironomids are abundant (Takemon unpublished). Therefore, stony substrates may be a more suitable feeding site, but trade-offs may occur between energy gain and reproductive success. In this respect, preference by large females for predominantly sandy substrates seems to point to the supremacy of reproductive needs in their habitat selection.

Acknowledgements

We thank staffs of the C.R.S.N./Uvira, Republic of Zaire for providing facilities and assistance in the field work and M. Kohda and M. Yuma for their information on references. The manuscript was improved by comments from M. Tokeshi, School of Biological Sciences, Queen Mary and Westfield College (QMW) of London University. Data analysis and writing of the manuscript were conducted while the senior author was staying at QMW, funded by a Japan Society for the Promotion of Science Fellowship for Research at Centers of Excellence Abroad. This study was also supported by the Joint Study Project on Ecological, Limnological, and Fish-Biological Research in Lake Tanganyika and Its Environs for the Good and Stable Supply of Protein Resources (1986–1989) from the Japan International Cooperation Agency, and Grants-in-Aid for Overseas Scientific Survey (No. 61043032 and 04044088), for Scientific Research (No. 6148005), and for Scientific Research on Priority Areas (# 319) from the Japan Ministry of Education, Science, and Culture.

References cited

Balon, E.K. 1975. Reproductive guilds of fishes: a proposal and definition. J. Fish. Res. Board Can. 32: 821–864.

Balon, E.K. 1981. Addition and amendments to the classification of reproductive styles in fishes. Env. Biol. Fish. 6: 377–389.

Barlow, G.W. 1991. Mating system among cichlid fishes. pp. 173–190. *In:* M.H.A. Keenleyside (ed.) Cichlid Fishes: Behaviour, Ecology, and Evolution, Chapman and Hall, London.

Brichard, P. 1978. Fishes of Lake Tanganyika. T.F.H. Publications, Neptune City. 448 pp.

Brichard, P. 1989. Cichlids and other fishes of Lake Tanganyika. T.F.H. Publications, Neptune City. 544 pp.

Chundler, C.R., R.R. Sanders Jr. & A.M. Landry, Jr. 1985. Effects of three substrate variables on two artificial reef fish communities. Bull. Mar. Sci. 37: 129–142.

Colgan, P. & D. Ealey. 1973. Role of woody debris in nest site selection by pumpkinseed sunfish, *Lepomis gibbosus*. J. Fish. Res. Board Can. 30: 853–856.

Crisp, D.T. & P.A. Carling. 1989. Observations on siting, dimensions and structure of salmonid redds. J. Fish Biol. 34: 119–134.

Curry, R.A. & D.L.G. Noakes. 1995. Ground water and the selection of spawning sites by brook trout (*Salvelinus fontinalis*). Can. J. Fish. Aquat. Sci. 52: 1733–1740.

Curry, R.A., D.L.G. Noakes & G.E. Morgan. 1995. Ground water and the incubation and emergence of brook trout (*Salvelinus fontinalis*). Can. J. Fish. Aquat. Sci. 52: 1741–1749.

Fryer, G. & T.D. Iles. 1972. The cichlid fishes of the great lakes of Africa. T.F.H. Publications, Neptune City. 641 pp.

Gashagaza, M.M. 1991. Diversity of breeding habits in lamprologine cichlids in Lake Tanganyika. Physiol. Ecol. Japan 28: 29–65.

Hori, M. 1983. Feeding ecology of thirteen species of *Lamprologus* (Teleostei: Cichlidae) coexisting at a rocky shore of Lake Tanganyika. Physiol. Ecol. Japan 20: 129–149.

Hori, M. 1987. Mutualism and commensalism in the fish community of Lake Tanganyika. pp. 219–239. *In:* S. Kawano, J.H. Connel & T. Hidaka (ed.) Evolution and Cooperation in Biotic Communities, University of Tokyo Press, Tokyo.

Hori, M. 1991. Feeding relationships among cichlid fishes in Lake Tanganyika: effects of intra- and interspecific variations of feeding behavior on their coexistence. Ecol. Internat. Bull. 19: 89–101.

Hourigan, T.F. 1986. A comparison of haremic social systems in two reef fishes. pp. 23–28. *In:* L.C. Drickamer & I.E.C. Privat (ed.) Behavioral Ecology and Population Biology, Toulouse.

Hourigan, T.F. 1989. Environmental determinants of butterfly-fish social systems. Env. Biol. Fish. 25: 61–78.

Inoue, M. & S. Nakano. 1994. Physical environment structure of a small stream with special reference to fish microhabitat. Japan. J. Ecol. 44: 151–160 (in Japanese).

Keenleyside, M.H.A. 1979. Diversity and adaptation in fish behaviour. Springer-Verlag, Berlin. 208 pp.

Keenleyside, M.H.A. 1991. Parental care. pp. 191–208. *In:* M.H.A. Keenleyside (ed.) Cichlid Fishes: Behaviour, Ecology, and Evolution, Chapman and Hall, London.

Keenleyside, M.H.A. & R.W. Machereth. 1992. Effects of loss of male parent on brood survival in a biparental cichlid fish. Env. Biol. Fish. 34: 207–212.

Kohda, M. 1991. Intra- and interspecific social organization among three herbivorous cichlid fishes in Lake Tanganyika. Japan. J. Ichthyol. 38: 147–163.

Kohda, M. 1994. Individual specialized foraging repertories in the piscivorous cichlid fish *Lepidiolamprologus profundicola*. Anim. Behav. 48: 1123–1131.

Kohda, M. & Y. Yanagisawa. 1992. Vertical distribution of two herbivorous cichlid fishes in Lake Tanganyika. Ecol. Freshwat. Fish. 4: 180–184.

Kuwamura, T. 1986. Parental care and mating systems of cichlid fishes in Lake Tanganyika: a preliminary field survey. J. Ethol. 4: 129–146.

Kuwamura, T. 1987. Distribution of fishes in relation to the depth and substrate at Myako, east-middle coast of Lake Tanganyika. Afr. Stud. Monogr., Kyoto Univ. 7: 1–14.

Leman, V.N. 1993. Spawning sites of chum salmon, *Oncorhynchus keta*: microhydrological regime and viability of progeny in redds (Kamchatka River basin). J. Ichthyol. 33: 104–117.

Nagoshi, M. 1983. Distribution, abundance and parental care of the genus *Lamprologus* (Cichlidae) in Lake Tanganyika. Afr. Stud. Monogr., Kyoto Univ. 3: 39–47.

Nakai, K., Y. Yanagisawa, T. Sato, Y. Niimura & M.M. Gashagaza. 1990. Lunar synchronization of spawning in cichlid fishes of the tribe Lamprologini in Lake Tanganyika. J. Fish Biol. 37: 589–598.

Peterson, N.P. & T.P. Quinn. 1996. Spatial and temporal variation in dissolved oxygen in natural egg pockets of chum salmon, in Kennedy Creek, Washington. J. Fish Biol. 48: 131–143.

Tupper, M. & R.G. Boutilier. 1995. Effects of habitat on settlement, growth, and postsettlement survival of Atlantic cod (*Gadus morhua*). Can. J. Fish. Aquat. Sci. 52: 1834–1841.

Yanagisawa, Y. 1986. Parental care in a monogamous cichlid *Xenotilapia flavipinnis* in Lake Tanganyika. Japan. J. Ichthyol. 33: 249–261.

Yanagisawa, Y. 1987. Social organization of a polygynous cichlid *Lamprologus furcifer* in Lake Tanganyika. Japan. J. Ichthyol. 34: 82–90.

Yuma, M. 1993. Competitive and co-operative interactions in Tanganyikan fish communities. pp. 213–227. *In:* H. Kawanabe, J.E. Cohen & K.Iwasaki (ed.) Mutualism and Community Organization-Behavioural, Theoretical, and Food-web Approaches, Oxford University Press, Oxford.

Yuma, M. 1994. Food habits and foraging behaviour of benthivorous cichlid fishes in Lake Tanganyika. Env. Biol. Fish. 39: 173–182.

Environmental Biology of Fishes **52**: 271–280, 1998.

Parental care and social organization of the spiny eel, *Aethiomastacembelus platysoma*, in Lake Tanganyika

Naoya Abe
Intercultural Relations Department, Osaka International College for Women, 6-21-57, Tohda-cho, Moriguchi, Osaka 570, Japan (e-mail: abe@oiuw.oiu.ac.jp)

Received 30.9.1996 Accepted 15.12.1997

Key words: mastacembelid, reproduction, egg size, growth of young, foraging behavior, territorial behavior

Synopsis

This is the first report demonstrating the occurrence of parental care in mastacembelids. Social organization and parental care of a spiny eel *Aethiomastacembelus platysoma* were studied in Lake Tanganyika. Both males and females maintained individual territories though the frequency of aggressive interaction was low. The male guarded offspring in a rock hole within its territory. The egg size was large (2.5–2.7 mm in diameter) and the brood size in a nest was 5.7 on average in spite of more oocytes in the ovary (65 large oocytes on average). The duration of guarding was around 30 days after hatching and the young became independent just after they began to feed. Guarding males seldom attacked fishes that approached the nest, and often went out of the nest to forage though the stomach contents of guarding males were less than those of non-guarding males. Compared with Tanganyikan cichlid fishes that show prolonged parental care at open sites, the post-hatching guarding interval is short and the egg size is large, which seem to be traits common to fishes that utilize closed spaces as guarding sites in the lake.

Introduction

Lake Tanganyika has attracted many biologists because of the explosive speciation of cichlid fishes, and many works have been done on their ecology in the littoral zone (see Lowe-McConnell 1987). Less attention has been paid to non-cichlid fishes except for several pelagic species, although they comprise 45% of all fish species of the lake. More research is needed on non-cichlids to understand the structure and organization of the fish community.

Mastacembelids, or spiny eels, are one of the major groups of non-cichlid fishes in Lake Tanganyika (Coulter 1991) and are typical members of littoral fish communities (Kawabata & Mihigo 1982, Kuwamura 1987). Little is known about their ecology, especially their social organization and reproduction, not only in the lake but also in other localities (Travers 1984, Lowe-McConnell 1987).

Aethiomastacembelus platysoma, one of the endemic mastacembelids in Lake Tanganyika, inhabits rocky shores and feeds mainly on shrimp (Poll & Matthes 1962). In this paper I describe the social organization, parental care and several features of the reproductive ecology of *A. platysoma*, and compare some important traits with those of hole nesting cichlids of Lake Tanganyika. This is the first report to describe the occurrence of parental care in matacembelids.

Methods

Field studies were carried out at Mbemba (3° 36′ S,

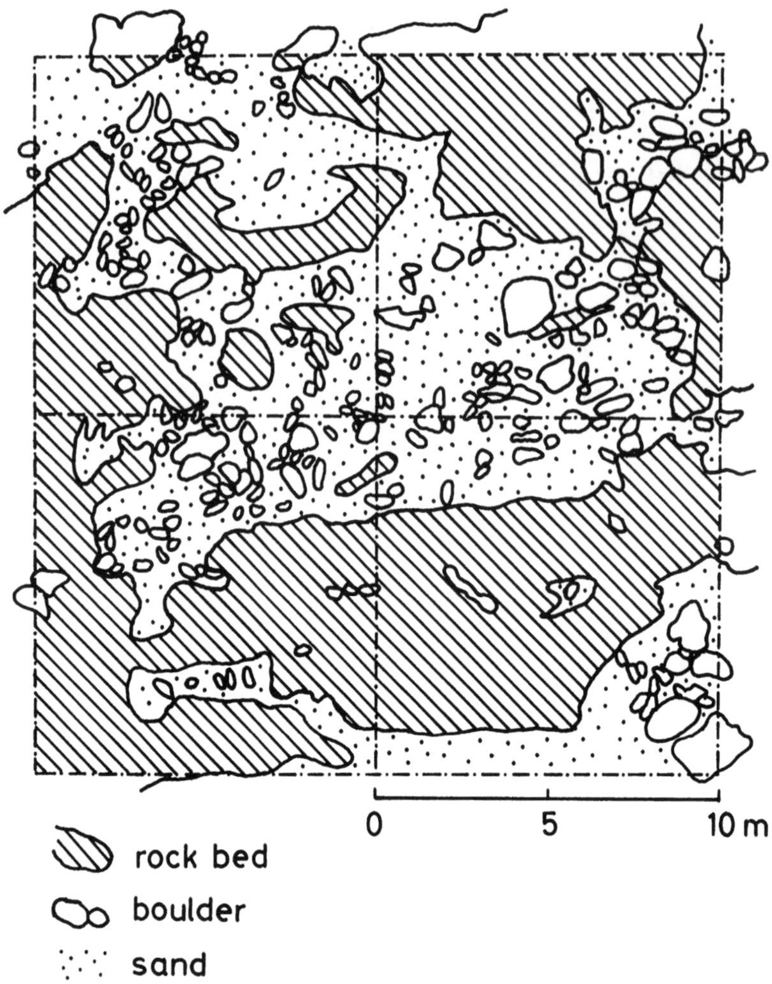

Figure 1. Study area at Mbemba, Zaire.

29° 10′ E), Zaire, on the northwestern part of Lake Tanganyika. A study area of 20 m × 20 m was established on a rocky shore at a depth of 5–13 m (Figure 1). A sand bed with boulders extended over the central part of the study area and was surrounded by rock beds.

Observations on the behavior of *A. platysoma* were done by SCUBA diving in the daytime (usually 10:00–12:00 h and 13:00–15:00 h) from March to September 1988. Most individuals found in the study area, and some large ones in the surrounding areas, were distinguished individually by body size and color pattern, especially the shapes of dorsal bands. Most small individuals, less than ca. 8 cm in standard length (SL), could not be distinguished individually because of their sporadic occurrence. In each survey, I followed a few individuals for 10 to 20 min each or until I lost sight of them; positions and behavior of the fish were recorded on a map (moving, staying in nest, under stone or in crevice, pecking prey, and fighting with conspecifics or other fish). Territories were defined by drawing a convex polygon connecting points where individuals were observed. Total observation time for each individual was 9–495 min (N = 32, mean = 142 min).

Offspring of *A. platysoma* in a nest hole were collected by tearing the hole open with an iron bar, or by blowing water into a nest hole using a pipette or a large (3400 cm^3) syringe afer pouring quinaldine anesthetic into the hole. Nest holes of four adult males were monitored by the latter method 9–10 times from 6 July to 25 August: a few young were

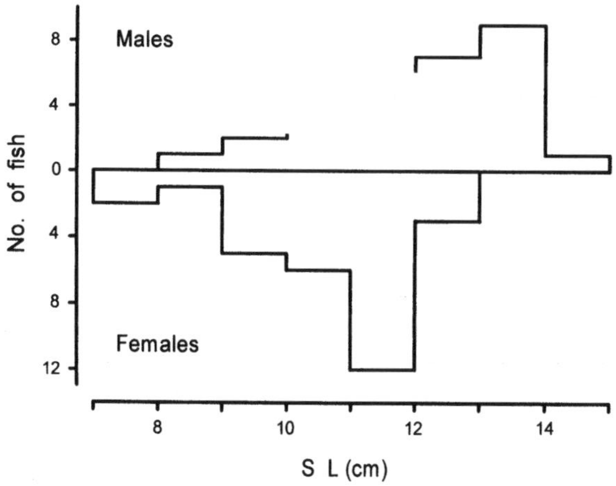

Figure 2. Standard lengths (SL) of both sexes of *A. platysoma*, captured at Mbemba in Lake Tanganyika during 1988. See text for sample sizes.

collected and the others were returned to the hole. A minimum amount of quinaldine anesthetic was used to lessen the effect on the subsequent development of the young. Total length (TL) of the young was measured; some were dissected to examine the stomach contents.

Twenty-seven individuals of *A. platysoma* were collected outside the study area from January to September 1988, and 29 in or around the study area in August at the end of the survey. Standard lengths and body and gonad weights were recorded. Stomach contents were measured as the number of food items because it was easy to count the black eyes of shrimp which constituted the majority of the stomach contents. For females, sizes of oocytes were measured and the large oocytes (≥ 1.5 mm in diameter) were counted. To compare the size and number of oocytes among species, other mastacembelids, *Caecomastacembelus albomaculatus, C. micropectus, C. moorii, C. ophidium, C. plagiostomus, C. tanganicae*, and *Aethiomastacembelus flavidus*, were collected using a screen net or by hand at Mbemba from January to September 1988 and the gonadal condition was examined for females.

Fifty-one eggs of *C. plagiostomus* guarded by a parent were collected from a rock-hole on 3 January 1988 and 49 eggs were put in an aquarium for 25 days. The water was aerated and changed every 3–4 days. The young were fed dry food (powdered Tetramin) after hatching. One or two individuals were

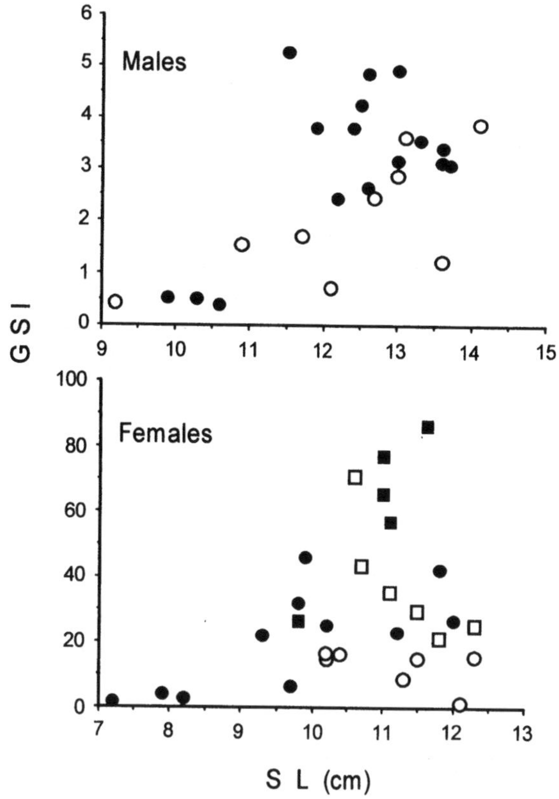

Figure 3. Gonado-somatic index (GSI) of *A. platysoma*, plotted against SL. The GSI was calculated as (wet gonad weight) × (wet body weight)$^{-1}$ × 1000. Open and filled points represent individuals collected in January – June 1988 and July – September 1988, respectively. Squares represent females with large oocytes (≥ 1.5 mm).

sampled every 1–7 days and they were measured (TL). Similarly, three of five young of *A. platysoma* found on 25 August were incubated in an aquarium after two were preserved; two of three young survived over a week and were sampled 8 and 21 days later.

Results

Size composition and size at maturity

The mean SL (± SD) of males was 12.2 ± 1.4 cm (range: 8.8–14.1 cm; N = 26) and that of females was 10.6 ± 1.3 cm (range: 7.2–12.3 cm; N = 30) (Figure 2). Males were longer than females (Mann-Whitney U-test, U = 151.5, $p < 0.001$). The color pattern of

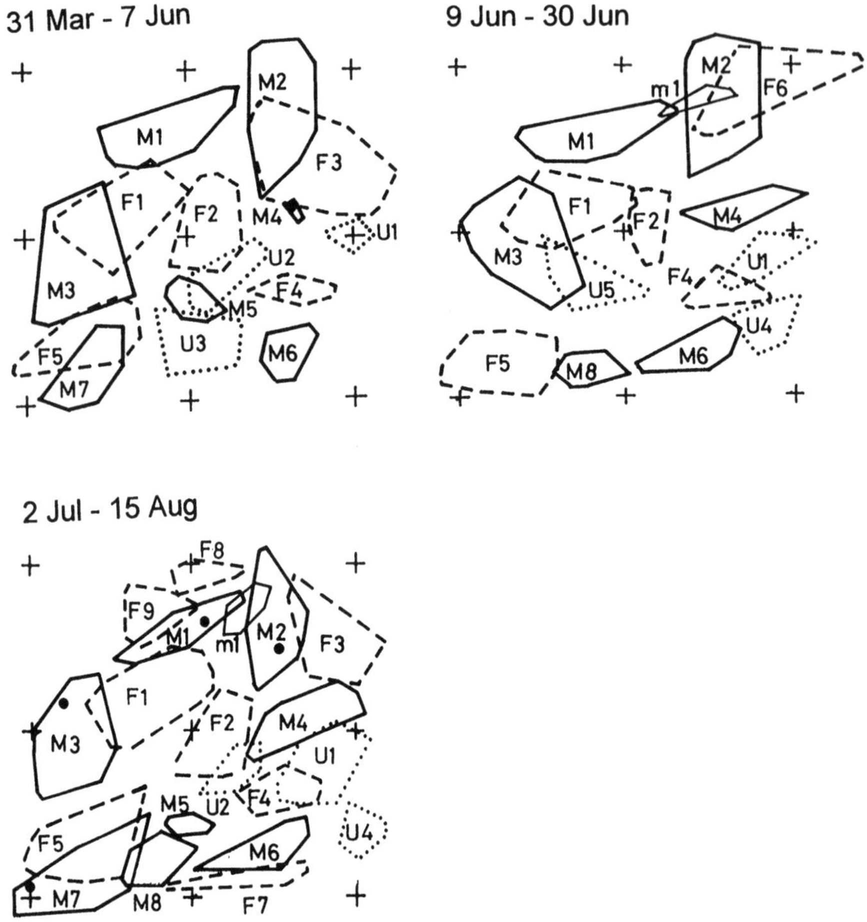

Figure 4. Spatial and temporal distribution of territories for adult males (M, bold solid lines), juvenile male (m, thin solid line), adult females (F, dashed lines) and individuals of unknown sex (U, dotted lines) in the study area between March and August 1988. Nest sites are marked with a solid circular symbol. Crosses show corners of the four 10 m × 10 m grids (see Figure 1).

large males differed from the others: large males (> ca. 12 cm SL) were wholly black, but females and small males had a red body with black dorsal bands.

The gonadosomatic indices (GSI's) of males ≥ 11 cm SL were much higher than those of males < 11 cm SL (Figure 3; Mann-Whitney U-test, $U = 2$, $N = 5, 20$, $p < 0.001$). The smallest male that guarded offspring was 11.9 cm SL. Thus, males seem to start reproduction when they attain 11.0–12.0 cm SL. The GSI's of females < 9 cm SL were nearly zero. The smallest female that had large oocytes (≥ 1.5 mm in diameter) was 9.8 cm SL, and females seem to mature around this size.

The GSI's of both large males and females caught during January through June were significantly lower than those caught during July to September (Figure 3; Mann-Whitney U-test, $U = 18.5$, $N = 7, 13$, $p < 0.05$ for males ≥ 11.5 cm SL; $U = 26$, $N = 13, 12$, $p < 0.005$ for females ≥ 9.5 cm SL). No brooding activities were observed until 20 May although the field study started in March.

Territory and nest hole

All of the individuals found in the field were benthic. Eight adult males, nine adult females, one juvenile male, and five small red individuals with unknown sex had territories in the study area (Figure 4). Territory size was not different between adult males and adult females (Table 1; Mann Whitney U-test, $U = 36$, $N = 8, 9$, $p > 0.05$). Territories partly overlapped each other. The area where two territories overlapped was 24% of the area of all the territories. Most overlapping was found between adult males and adult females. Little overlapping was found between territories of adult males or between adult females.

Most individuals maintained their territories through the study interval and the position of them changed only slightly (Figure 4). Six of 15 individuals that were found during the first period (31 March – 7 June) disappeared in the second period (9 June – 30 June), but five of them were found holding territories near their original territories in the third period (2 July – 15 August). After five individuals were removed on 26 August, no individuals invaded these territories, and no changes in the position of territories of the remaining individuals were observed during the following 17 days.

Territories were defended against conspecifics. Encounters with conspecifics were seen 24 times (0.3 times h^{-1} on average during 66.3 h of observations). No encounters were observed between adult males. When an adult male encountered an adult female, he chased the female (n = 4) or ignored her (n = 2). In the other cases the larger one chased the smaller one (n = 13), the smaller one escaped before being chased (n = 4) or they fought (n = 1). Escalated fighting that lasted for more than 3 min was observed between females F1 and F2, which were nearly the same size (11.1 and 11.0 cm SL, respectively), on the boundary of their territories. They raised their bodies, displaying the lateral side, and bit the mouth of each other. Finally, F1 bit the tail of F2 and carried her about 2 m.

Table 1. Mean (± SD) territory size (m^2) and degree of territory overlap for adult (M) and young males (m) of *A. platysoma*, adult females (F), and individuals of unknown sexes (U).

Sex	N	Territory size ± SD	Overlapping area (%)				
			M	F	m	U	Total
M	8	15.5 ± 10.0	1.0	19.9	1.3	2.5	24.4
F	9	15.6 ± 8.4	22.7	0.0	0.6	2.0	24.8
m	1	4.0	58.5	11.0	0.0	0.0	58.5
U	5	10.7 ± 3.8	12.4	6.5	0.0	0.0	18.4
All	23	14.0 ± 8.3					24.0

Many adult males were observed entering particular holes repeatedly after foraging excursions. I regarded these holes as nest sites because offspring were collected from most of them. Nest holes were 20.0 ± 4.1 cm deep (N = 7) and had large labyrinthine spaces inside. The entrance was narrow, 5.6 ± 1.4 cm long and 2.0 ± 1.0 cm wide. All of them were situated on the vertical surface of rocks. In the study area five of eight males were observed holding a nest hole (Figure 4).

No males changed positions of their holes. I destroyed the nest hole of M4 with an iron bar on 3 June to collect the young. This male extended his territory afterwards (Figure 4), but did not change the position of his territory nor find a new nest hole before the end of the study.

Paternal care and growth of offspring

Eggs and young of *A. platysoma* were found 29 times in 17 nest holes. Only two eggs were collected, from one nest hole. These yellowish eggs were demersal, not adhesive, and spherical with diameters of 2.6 and 2.7 mm. The age of the eggs when collected was unknown.

Size range of young under care was from 5.8 to 16.5 mm TL. Yolk had been completely absorbed in young larger than 15 mm TL. Stomachs of the large young (≥ 16 mm TL) contained food items such as copepods, insect larvae, and ostracods (N = 4), whereas those of smaller young that still had yolk-sac were empty (N = 7). Thus, young seem to begin to feed exogenously just before independence.

The size of young collected repeatedly from the nests of four males was plotted against the date of collection (Figure 5). The gradual increase in size provided an estimate of their growth rate. The mean growth rate was estimated as 0.40 mm day^{-1} in the young smaller than 12 mm TL and 0.20 mm day^{-1} in the larger ones (Figure 6). The growth rate was also calculated from four young collected from a nest of M1 and kept in an aquarium. The young grew at a rate of 0.39 mm day^{-1} during the first 8 days and 0.30 mm day^{-1} subsequently. These growth rates were similar to those estimated from the field survey. The guarding duration from hatching until

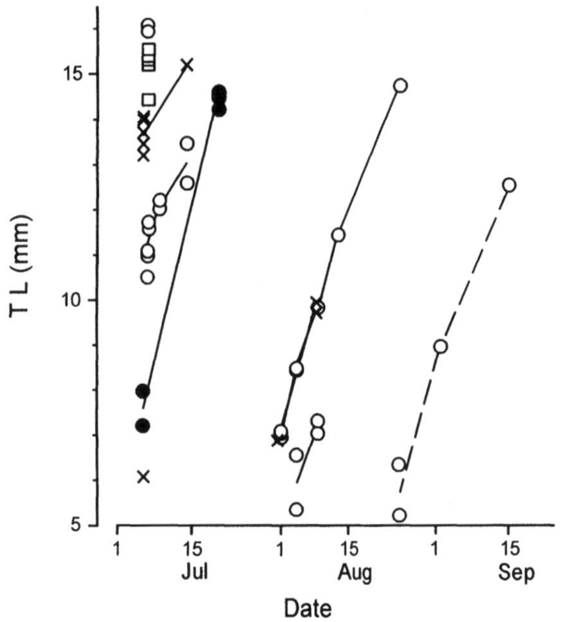

Figure 5. Size of young of *A. platysoma* under care, plotted against the date of collection. Solid and dashed lines represent the estimated growth of broods in the field and in an aquarium, respectively (see text). Different marks represent young guarded by different males (○ = M1, ● = M2, □ = M4, x = M9).

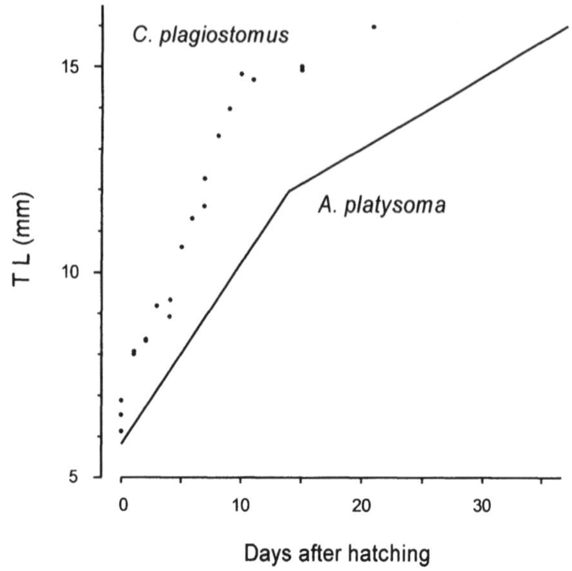

Figure 6. Growth curve of young of *A. platysoma* under care, and that of *C. plagiostomus* in the aquarium.

the young attained 16 mm TL was estimated to be about 30 days.

The number of brood cycles by a male during the monitoring period of 50 days varied from 0 to 5 (Figure 5). On three occasions fish M1 was observed to accept a new brood before the independence of the previous brood. Similarly, young of M9 once consisted of two size groups.

To determine the brood size, I tried to collect all eggs and young. The number of young in a brood at any time was 4.1 ± 4.8 (N = 7; range: 1–15) and was not correlated with size (Kendall's tau = – 0.038, p > 0.05). On the other hand, the mean number of large oocytes (≥ 1.5 mm in egg diameter) in the ovary was 62 ± 17 (N = 10; range: 32–88).

While caring for young, males usually stayed inside their nest holes, as described below. Behavior inside the dark holes could not be observed. One male, M9, often attacked other fishes that approached his nest hole during guarding (10 times h^{-1} on average). In contrast, the other guarding males rarely attacked other fishes (0.5 times h^{-1} on average). The nest of M9 was situated on a flat cliff and its opening was exposed to neighbors while those of the other males were more or less concealed by rugged rock. Therefore, position of nests may have affected the frequency of attacking behavior.

Telmatochromis bifrenatus and *Julidochromis marlieri* were attacked frequently when they approached within about 20 cm of a nest hole (Table 2). *Telmatochromis temporalis* were attacked when they came within 50 cm from a hole. *Caecomastacembelus albomaculatus* was attacked most severely, even when the distance of the fish from the nest exceeded 50 cm. I observed a *C. albomaculatus* enter a nest hole of *A. platysoma*, and the guarding male in the hole chased it away. *Neolamprologus*

Table 2. Frequency and mean distance at which other species were attacked by guarding males of *A. platysoma*.

Species	Frequency	Mean distance of attack from a nest hole ± SD (cm)
Julidochromis marlieri	24	24 ± 10
Telmatochromis bifrenatus	27	15 ± 9
T. temporalis	5	47 ± 6
Neolamprologus brichardi	7	6 ± 5
Cyprichromis sp.	1	2
Caecomastacembelus albomaculatus	3	50 ± 14

Table 3. Stomach contents (percentage by number) of 54 *A. platysoma* collected in Lake Tanganyika at Mbemba during January through September 1988.

Sex	No. examined	No. with food	Prey (%)			Total no. of food items ± SD
			Shrimp	Insect larva	Eggs	
Male						
Juvenile	6	6	96.2	3.8	0	27.0 ± 18.6
Adult without brood	13	13	92.0	6.1	1.8	31.6 ± 24.2
Adult with brood	6	4	41.5	15.1	43.4[1]	7.8 ± 8.6
Female						
Juvenile	3	3	92.7	7.3	0	21.3 ± 12.0
Adult	26	26	92.3	7.1	0.6	41.7 ± 34.7

[1] All were eggs of *A. platysoma*.

brichardi and *Cyprichromis* sp. were rarely attacked, though they were the most abundant fishes around the nest.

Food habits and foraging behavior

The main diet of *A. platysoma* was shrimp (Table 3). Feeding behavior was seen over rocks but not on the sandy bottom. The fish were observed to snap up each shrimp from a small pit after watching, while waving its body. Minor items in the diet were insect larvae and eggs. Most eggs were those of the cyprinodont *Lamprichthys tanganicanus*. I sometimes observed that *A. platysoma* dashed to where a pair of the cyprinodont descended to spawn, and then scrambled for eggs in small holes and crevices, often with some cichlid fishes.

One male, which was guarding three eggs in the nest, had 23 eggs of its own species in its stomach. Five eggs were found in the stomach of another male though it was not ascertained whether he was guarding eggs.

Adult males with broods had far fewer food items in the stomach than males without broods and adult females (Table 3; Mann-Whitney U-test, $U = 15.5$, $N = 6, 13$, $p < 0.05$ and $U = 8$, $N = 6, 26$, $p < 0.001$, respectively). Two of six males caught with broods had no food in the stomach. Number of food items was not different between adult males without broods and adult females (Mann-Whitney U-test, $U = 163$, $N = 13, 26$, $p = 0.24$).

The frequency of pecking at open sites was not different between non-guarding males and females (Table 4, Mann-Whitney U-test, $U = 220$, $N = 23, 24$, $p = 0.24$) and between guarding and non-guarding males (Mann-Whitney U-test, $z = 0.953$, $N = 23, 61$, $p = 0.34$). Adult females spent much more time at cryptic sites than non-guarding males (Mann-Whitney U-test, $U = 157.5$, $N = 24, 61$, $p < 0.001$), and guarding males stayed in the nest hole longer than non-guarding males (Mann-Whitney U-test, $z = 3.99$, $N = 25, 69$, $p < 0.001$).

Notes on reproduction of other spiny eels

In a hole on a large boulder I collected 51 eggs guarded by a *Caecomastacembelus plagiostomus* (sex unknown). Mean egg diameter was 3.0 ± 0.1 mm ($N = 4$), slightly larger than those of *A. platysoma* ($t = 7.75$, d.f. $= 4$, $p < 0.005$). The embryos hatched 2 days after collection. The growth rate of

Table 4. Frequency of pecking prey and time budget of adult *A. platysoma.*

	Time of observation (min)	Pecking rate in open site (times h^{-1})	% of time in a cryptic site[1]	% of time in nest
Non-guarding male	321	10.24	16.4	28.0
Guarding male	1211	9.96	11.7	60.0
Female	523	6.48	41.6	0

[1] Calculated from (time in cryptic site) × ((time of observation) − (time in nest))$^{-1}$ × 100.

the young in an aquarium was 0.8 mm day^{-1} during 10 days after hatching, but slowed to about 0.1 mm day^{-1} afterwards (Figure 6).

Several females of *Caecomastacembelus albomaculatus, C. micropectus, C. plagiostomus*, and *C. tanganicae* had large oocytes (≥ 1.5 mm in diameter) while *Aethiomastacembelus flavidus, C. moorii*, and *C. ophidium* had no large oocytes at all. The maximum oocyte sizes of the former four species and also *A. platysoma* were similar to each other (Table 5). *Caecomastacembelus micropectus*, the smallest mastacembelid in Lake Tanganyika, had only five large occytes on average and the GSI was also the smallest.

Discussion

The parental care of *A. platysoma* is characterized by male guarding in a deep hole on a rock until independence of the young. Such guarding in a hole has been also reported in some Tanganyikan cichlids of the tribe Lamprologini (Nakai et al. 1990, Gashagaza 1991). Females of *Lamprologus lemairii* and *Altolamprologus compressiceps* deposit eggs in holes and guard eggs and embryos there; females of *Lepidolamprologus profundicola* deposit eggs on a rock surface and transfer hatched embryos into a hole. These females guard the young just outside or within the holes until their independence. In contrast, other lamprologine cichlids guard the young at open sites for several months though some of them care for hatched embryos in holes until the absorption of the yolk (Nakai et al. 1990, Gashagaza 1991). I classify the fishes that guard their young in holes until independence as hole nesters and those that show prolonged parental care at open sites as rock and gravel nesters (Balon 1975, 1990).

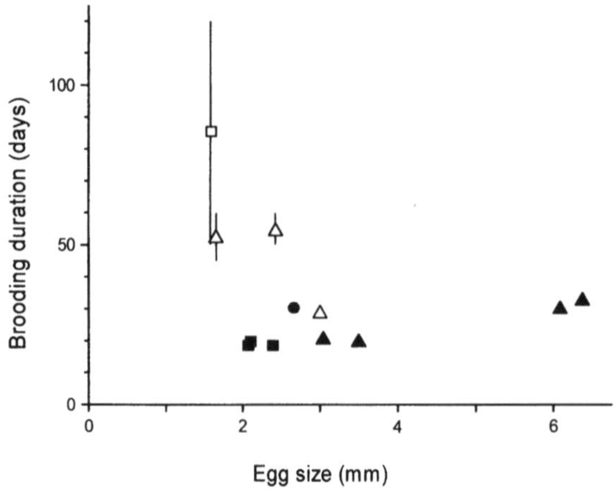

Figure 7. Comparison of egg size and guarding duration among Tanganyikan cichlids and *A. platysoma* . Vertical lines represent the range [● = *A. platysoma*, ■ = hole nesting cichlids (Gashagaza 1991), ▲ = mouthbrooding cichlids without guarding (Kuwamura et al. 1989, Yanagisawa & Sato 1990, Ochi 1993), □ = rock and gravel nesters (Gashagaza 1991), △ = mouthbrooding cichlids with buccal feeding (Yanagisawa & Nshombo 1983, Yanagisawa 1986, Kuwamura 1988)].

Some reproductive features of *A. platysoma* are comparable with those of the hole nesting cichlids, though the sex that guards is different. First, the free-swimming young under care are absent in hole nesting cichlids or occur for a very short duration in *A. platysoma*, and the guarding time is much shorter than that of rock and gravel nesting cichlids (Figure 7). Second, egg sizes of hole nesting cichlids and *A. platysoma* are larger than those of rock and gravel nesting cichlids. Lack of the free-swimming interval and large eggs are also seen in Tanganyikan mouth-

Table 5. Number and maximum size of eggs in the ovaries of five mastacembelid species collected in Lake Tanganyika at Mbemba in 1988. Species having large eggs (≥ 1.5 mm in diameter) were selected (GSI = gonadosomatic index).

Species	N	SL (mm) ± SD	Body weight ± SD (g)	Oocyte number ± SD	GSI ± SD	Maximum oocyte size (mm)
A. platysoma	10	11.2 ± 0.4	8.6 ± 1.2	62 ± 17	55.4 ± 21.9	2.2
C. albomaculatus	2	16.0 ± 1.9	13.2 ± 4.4	137 ± 57	66.4 ± 16.9	2.2
C. micropectus	5	7.5 ± 0.4	1.1 ± 0.2	5 ± 1	23.6 ± 8.2	2.0
C. plagiostomus	1	21.4	18.3	145	42.6	2.0
C. tanganicae	3	10.9 ± 0.6	2.8 ± 0.4	50 ± 5	60.3 ± 40	2.3

brooders. On the other hand, some mouthbrooders which show prolonged guarding at open sites have smaller eggs. These data suggest that holes and mouths have common characteristics as guarding sites.

Young of mouthbrooders are nearly free from predation as long as they stay in the mouth. In Lake Tanganyika only the mochokid *Synodontis multipunctatus* is known as a predator of young inside the mouth (Sato 1986). Holes also form an effective barrier against predators. The rate of disappearance of the brood before the free-swimming interval is lower in the hole nesting cichlids than in the rock and gravel nesting cichlids (Nakai et al. 1990). The rock hole used by *A. platysoma* forms a more closed space because the hole is narrower at the mouth and deeper inside that those of the hole nesting cichlids (Gashagaza 1991). Furthermore, in *A. platysoma* the guarder itself stays and guards inside the hole.

The lower young mortality of *A. platysoma* is also suggested by a model proposed by Sargent et al. (1987). The model predicts that the optimal egg size increases as parental care reduces mortality of young under care. Eggs of *A. platysoma* are larger than those of the hole nesting cichlids and smaller than those of the mouthbrooding cichlids (Figure 7).

One of the major assumptions of the model by Sargent et al. (1987) is that offspring from larger eggs take longer to become independent. This assumption is satisfied in the hole nesters and mouthbrooders because the duration of parental care increases with the egg size in these fishes (Figure 7). On the other hand, the rock and gravel nesters and the mouthbrooders that show prolonged guarding have smaller eggs in spite of the long duration of parental care. In these fishes egg size does not affect the size of young at independence directly because the young grow under parental care. They can produce a considerable number of large young by laying many small eggs and guarding them for a prolonged period (see also Crawford & Balon 1996).

Holes and mouths form closed spaces that isolate young not only from predators but also from prey. Some young inside mouths take food inhaled by the brooder, as seen in several species (Yanagisawa & Sato 1990, Yanagisawa & Ochi 1991). In hole nesters the density of prey inside the hole must quickly decrease after the young begin to feed because the immigration of prey through the small opening is limited. Therefore, the lack or decrease in time of guarding of the free-swimming young in hole nesters and mouthbrooders is probably related to limited food supply inside holes or mouths.

This study is the first report that mastacembelids show parental care. Spawning behavior of an Asian mastacembelid, *Macrognathus pancalus*, in an aquarium has been described by Job (1941) and Schoenebeck (1955). Parents did not guard offspring in this species.

Caecomastacebelus plagiostomus, as well as *A. platysoma*, show parental care for offspring because single individuals were found with eggs in each hole. No observations were made of breeding behavior in the other mastacembelids. However, eggs in ovaries of *C. micropectus, C. albomaculatus*, and *C. tanganicae* were as large as those of *A. platysoma* and *C. plagiostomus*. Most likely they also care for offspring by some means. High parental investment is expected in *C. micropectus* because of its small number of large oocytes. Although observing the behavior of these mastacembelids is difficult because of their cryptic habits (Abe 1997), further studies of their reproduction will be fruitful.

Acknowledgements

Thanks are due to the staff of the Centre de Recherche en Sciences Naturelles (CRSN), Uvira, Zaire for assisting in research and providing facilities. H. Kawanabe and other members of our research team kindly helped and advised me. Thanks are also due to Y. Yanagisawa and T. Kuwamura for critically reading the manuscript. This work was supported by the Japan International Cooperation Agency (No. 4860445) and forms a part of guest scientist activities, Center for Ecological Research, Kyoto University.

References cited

Abe, N. 1997. Ecology of non-cichlids in the littoral zone of Lake Tanganyika. pp. 241–256. *In:* H. Kawanabe, M. Hori & M. Nagoshi (ed.) Fish Communities in Lake Tanganyika, Kyoto University Press, Kyoto.

Balon, E.K. 1975. Reproductive guilds of fishes: a proposal and definition. J. Fish. Res. Board Can. 32: 821–864.

Balon, E.K. 1990. Epigenesis of an epigeneticist: the development of some alternative concepts on the early ontogeny and evolution of fishes. Guelph Ichthyol. Rev. 1: 1–42.

Coulter, G.W. 1991. Lake Tanganyika and its life. Oxford University Press, Oxford and British Museum (Natural History), London, 354 pp.

Crawford, S.S. & E.K. Balon. 1996. Cause and effect of parental care in fishes: an epigenetic perspective. pp. 53–107. *In:* J.S. Rosenblatt & C.T. Snowdon (ed.) Parental Care: Evolution, Mechanisms, and Adaptive Significance, Advances in the Study of Behavior 25, Academic Press, San Diego.

Gashagaza, M.K. 1991. Diversity of breeding habits in lamprologine cichlids in Lake Tanganyika. Physiol. Ecol. Japan 28: 29–65.

Job, T.J. 1941. Life-history and bionomics of spiny eel, *Mastacembelus pancalus* (Hamilton). Rec. Indian Mus. 43: 121–135.

Kawabata, M. & N.Y.K. Mihigo. 1982. Littoral fish fauna near Uvira, northwestern end of Lake Tanganyika. Afr. Stud. Monogr. 2: 133–143.

Kuwamura, T. 1987. Distribution of fishes in relation to the depth and substrate at Myako, east-middle coast of Lake Tanganyika. Afr. Stud. Monogr. 7: 1–14.

Kuwamura, T. 1988. Biparental mouthbrooding and guarding in a Tanganyikan cichlid *Haplotaxodon microlepis*. Japan. J. Ichthyol. 35: 62–68.

Kuwamura, T., M. Nagoshi & T. Sato. 1989. Female-to-male shift of mouthbrooding in a cichlid, *Tanganicodus irsacae*, with notes on breeding habits of two related species in Lake Tanganyika. Env. Biol. Fish. 24: 187–198.

Lowe-McConnell, R.H. 1987. Ecological studies in tropical fish communities. Cambridge University Press, London, 382 pp.

Nakai, K., Y. Yanagisawa, T. Sato, Y. Niimura & M.M. Gashagaza. 1990. Lunar synchronization of spawning in cichlid fishes of the tribe Lamprologini in Lake Tanganyika. J. Fish Biol. 37: 589–598.

Ochi, H. 1993. Maintenance of separate territories for mating and feeding by males of a maternal mouthbrooding cichlid, *Gnathochromis pfefferi*, in Lake Tanganyika. Japan. J. Ichthyol. 40: 173–182.

Poll, M. & H. Matthes. 1962. Trois poissons remarquables du Lac Tanganika. Ann. Mus. Roy. Afr. Centr., Zool. 111: 1–26.

Sato, T. 1986. A brood parasitic catfish of mouthbrooding cichlid fishes in Lake Tanganyika. Nature 323: 58–59.

Sargent, R.C., P.D. Taylor & M.R. Gross. 1987. Parental care and the evolution of egg size in fishes. Amer. Nat. 129: 32–46.

Travers, R.A. 1984. A review of the Mastacembeloidei, a suborder of synbranchiform teleost fishes. Part II: phylogenetic analysis. Bull. Br. Mus. Nat. Hist. (Zool.) 47: 83–150.

von Schoenebeck, K.J. 1955. Neues von den Stachelaalen. Aquar. Terrar. Z. 9: 315–317.

Yanagisawa, Y. 1986. Parental care in a monogamous mouthbrooding cichlid *Xenotilapia flavipinnis* in Lake Tanganyika. Japan. J. Ichthyol. 33: 249–261.

Yanagisawa, Y. & M. Nshombo. 1983. Reproduction and parental care of the scale-eating cichlid fish *Perissodus microlepis* in Lake Tanganyika. Physiol. Ecol. Japan 20: 23–31.

Yanagisawa, Y. & H. Ochi. 1991. Food intake by mouthbrooding females of *Cyphotilapia frontosa* (Cichlidae) to feed both themselves and their young. Env. Biol. Fish. 47: 191–201.

Yanagisawa, Y. & T. Sato. 1990. Active browsing by mouthbrooding females of *Tropheus duboisi* and *Tropheus moorii* (Cichlidae) to feed the young and/or themselves. Env. Biol. Fish. 27: 43–50.

Environmental Biology of Fishes **52**: 281–290, 1998.

Daily spawning and development of sensitivity to gonadotropin and maturation-inducing steroid in the oocytes of the bambooleaf wrasse, *Pseudolabrus japonicus*

Michiya Matsuyama[1], Sumito Morita[2,3], Takao Nasu[2] & Masaaki Kashiwagi[2]
[1] *Faculty of Agriculture, Kyushu University, Hakozaki, Fukuoka 812, Japan*
(e-mail: rinya_m@agr.kyushu-u.ac.jp)
[2] *Faculty of Bioresources, Mie University, Kamihama, Tsu 514, Japan*
[3] *Present address: Hyogo Prefectural Sea Farming Center, Akashi, Hyogo 678, Japan*

Received 21.12.1996 Accepted 1.12.1997

Key words: maturational competence, protogyny, teleost, fish

Synopsis

The cycle of oocyte development of the bambooleaf wrasse, *Pseudolabrus japonicus*, was studied to elucidate the endocrinological mechanism of oocyte maturation in a marine teleost. A single female reared with two males spawned every day for 17 days in captivity, indicating that this species is a daily spawner. Ovarian histology revealed that germinal vesicle migration of the largest oocytes progressed from 12:00 to 3:00 h, and germinal vesicle breakdown (GVBD) was completed at 6:00 h. Ovulation and spawning occurred between 6:00 and 9:00 h. The effectiveness of human chorionic gonadotropin (HCG) and 17α,20β-dihydroxy-4-pregnen-3-one (17,20β-P), which is one of the most potent steroidal inducers of GVBD in bambooleaf wrasse oocytes, in inducing final oocyte maturation was examined at eight different times of the day. The responsiveness of the oocyte to HCG and steroid differed at different times of the day. The GVBD could be induced by HCG but not 17,20β-P at 9:00 h. Between 12:00 and 18:00 h, not only HCG but also 17,20β-P induced GVBD. Both GVBD and ovulation spontaneously occurred between 0:00 and 6:00 h without any hormonal treatment. These results clearly showed that the oocyte of the bambooleaf wrasse possessed a diurnal maturation cycle. Responsiveness of oocytes to HCG appeared earlier than responsiveness to 17,20β-P. This suggests that sensitivity to 17,20 β-P is induced by gonadotropic hormone (GTH).

Introduction

Many endocrinological studies on final oocyte maturation have been carried out on teleosts, and it is accepted that oocytes undergo their final oocyte maturation under the influence of pituitary gonadotropic hormone (GTH) following the completion of vitellogenesis. The GTH stimulates production of maturation-inducing hormone (MIH) in the follicular cells, and MIH induces germinal vesicle breakdown (GVBD) of the oocytes (Nagahama 1987a). The compounds 17α,20β-dihydroxy-4-pregnen-3-one (17,20β-P) and 17α,20β,21-trihydroxy-4-pregnen-3-one (20β-S) are known to act as MIHs in teleosts (Nagahama & Adachi 1985, Trant et al. 1986). In addition to its steroidogenic actions, GTH is also involved in oocyte maturation. In the daily spawning marine teleosts Japanese whiting (kisu), *Sillago japonica* (Kobayashi et al. 1988), and a dragonet (tobinumeri), *Repomuceus beniteguri* (Zhu et al. 1989), it has been reported that the responsiveness of oocytes to GTH and 17,20β-P differs at different times of the day. In these species, *in vitro* responsiveness of oocytes to GTH appears earlier than responsiveness to 17,20β-P. Recent studies indicate that *in vitro* or *in vivo* treatment with GTH can induce the ability of oocytes to respond to MIH in Atlantic croaker, *Micropogonias undulatus* (Patino & Thomas 1990a), dragonet and kisu (Zhu et al. 1994), and red seabream, *Pagrus major* (Kagawa et al. 1994). This ability of oocytes to respond to MIH is termed 'maturational competence'.

The bambooleaf wrasse, *Pseudolabrus japonicus*, is among the most numerous labrids on coastal rocky reefs in southern Japan. Specimens are easily obtained by angling. The fish is resistant to handling and other stresses, and readily spawns in tanks without any environmental or hormonal treatment, unlike a number of other marine fish species which will not spawn in captivity. The bambooleaf wrasse exhibits diandric protogyny, with populations consisting of small initial-phase (IP) males (primary males), IP-females (primary females), and large terminal-phase (TP) males which may be derived either from females which have undergone a sex change to become males (secondary males) or from IP males (TP primary males; Nakazono 1979). In this species, the spawning behavior is mainly pair spawning performed by a TP male and a female; group spawning has not been observed. The TP males defend individual spawning territories around the perimeter of the rocky reef, and spawn exclusively with one female at a time between 8:00 and 10:00 h. It has been suggested that a single female visits and mates daily during the spawning period, based on long-term underwater observation (Nakazono 1979). Except for this work, however, very little information on the reproductive biology of bambooleaf wrasse is available, although we have described the annual reproductive cycle (Morita et al. 1997).

In this study, we first confirmed the occurrence of daily spawning by a single female bambooleaf wrasse, reared in captivity, with two TP males during the spawning season. Second, we examined the ovarian development of fish captured in the field or kept in captivity at 3 h intervals during the spawning season to clarify the daily reproductive cycle. Third, the relative effectiveness of nine steroids in GVBD was compared to identify the presumed MIH of bambooleaf wrasse. Finally, the responsiveness of oocytes to GTH and steroids at different times of the day was investigated.

Materials and methods

Spawning in captivity

Fish were caught by angling in coastal waters of the Fisheries Research Laboratory of Mie University in Ago Bay, Mie Prefecture, Japan on 19 October 1992, and were transferred to Mie University in Tsu City (ca. 100 km from the Fisheries Research Laboratory). A captive stock consisting of one female (54.3 g body weight, BW) and two TP males (109.1 and 116.4 g) was established in a round fiberglass tank of 500 l capacity, which was supplied with filtered recirculating sea water (14 l min^{-1}). A concrete block was placed on the bottom to provide sites for territories of the TP males. Fish were kept under natural daylength and water temperature, and were fed with shucked live shellfish twice a day. Spawning was measured each day by hourly counts of the number of spawned eggs collected in a net at the outflow. Spawned eggs were counted and the fertilization rate was determined. After the experiment, all fish were killed and their gonads were examined histologically in order to determine their sex; this revealed that the two TP males were secondary males. Detailed morphological features of the testes of the bambooleaf wrasse have been described (Matsuyama et al. 1997).

Ovarian reproductive cycle

For investigation of daily oocyte development, fish were collected in Ago Bay at 9:00, 12:00, 15:00 and 18:00 h on three days in early October 1992, and transferred to the laboratory within 40 minutes after sampling. Then, 5–10 IP fish captured at each sampling time were killed for study. Of the remaining fish, 54 IP fish and 7 TP males were held in an outdoor concrete tank (2.0 × 1.5 × 0.5 m) in which some concrete blocks had been placed. The tank was lit entirely by natural light, of which the direct rays were intercepted by the roof and three walls of a shed. The inflow rate of sea water at 21 °C varied between 30 and 60 l min^{-1}. Fish were fed with frozen shrimp or shucked live shellfish once a day. Females usually spawned daily between 6:00 and 9:00 h. Five to 10 IP fish in the tank were killed for study at each of four additional times, 21:00, 0:00, 3:00 and 6:00 h. After confirming individual sex by gonadal histology, total of 52 females weighing 23–85 g BW were used for investigation of the ovarian reproductive cycle.

After decapitation of these fish, the ovaries were dissected out and weighed, and the gonadosomatic index (GSI) was calculated as (gonad weight/body weight) × 100. Tissue samples were immersed in Bouin's solution dehydrated and embedded in Technovit resin (Kulzer). For light microscopy, 4 µm-thick sections were cut and stained with 1% toluidine blue solution. Developmental stages of oocytes and degenerative stages of postovulatory follicles (POF) were classified similarly to those of red seabream and kisu, both daily-spawning marine teleosts (Matsuyama et al. 1988, 1990).

Culture experiments

Hormones

Human chorionic gonadotropin (HCG) was purchased from Sankyo, Tokyo, Japan. The hormones 4-pregnen-3,20-dione (progesterone), 17α-hydroxy-4-pregnen-3,20-dione (17α-P), 20β-hydroxy-4-pregnen-3-one (20β-P), 17α,20β-dihydroxy-4-pregnen-3-one (17,20β-P), 17α,20α-dihydroxy-4-pregnen-3-one (17,20α-P), 17β-hydroxy-4-androsten-3-one (testosterone), 17β-hydroxy-4-androsten-3,11-dione (11-ketotestosterone), and 3,17β-dihydroxy-1,3,5[10]-estratrien (estradiol-17β) were purchased from Sigma (St. Louis, Missouri, U.S.A.), and 17α,20β,21-trihydroxy-4-pregnen-3-one (20β-S) from Steraloids (Witton, New Hampshire, U.S.A.).

Culture technique

Two experiments were performed. First to investigate the relative effectiveness of the nine steroids in inducing GVBD of bambooleaf wrasse oocytes *in vitro*, four females were sampled between 15:00 and 16:00 h, when GVBD could be induced by steroids in only the largest oocytes. Second, to investigate the sensitivity of oocytes to GTH and steroid, one or two females were sampled at 3 h intervals; 14 fish were used. The fish used in oocyte culture experiments had been reared in another outdoor concrete tank (2.0 × 1.5 × 0.5 m) with 5 TP males. These fish were sampled after confirming the daily spawning. Ovaries were removed and placed in ice-cold Leibovitz's L-15 culture medium (GIBCO, Rockville, Maryland, U.S.A.), buffered with 0.02 M HEPES with pH adjusted to 7.5 with 1 N NaOH. Gentamycin sulfate (200 mg l^{-1}, Sigma) was added at the time of setting up the assay. The ovaries were cut into small pieces weighing approximately 20 mg, and transferred separately to the wells of a 24-well plastic culture dish (Falcon, Franklin Lakes, New Jersey, U.S.A.), each containing 1 ml of incubation medium. Steroids were dissolved in ethanol, and HCG was dissolved and diluted with incubation medium. Ten microliters of hormone solution were added to the wells, and 10 µl of ethanol was added as a control for the steroid hormones. Cultures were maintained for 24 hours at 20 °C in a temperature controlled incubator in air.

Criteria of maturation

The response of each piece of ovary to various concentrations of hormones was judged as follows (Kobayashi et al. 1988); 0 = no response, 1 = some oocytes underwent GVBD but not ovulation, 2 = one to 10 of oocytes ovulated, 3 = more than 10 oocytes ovulated.

Results

Daily spawning of a single female

Figure 1 shows the eggs spawned by a single female bambooleaf wrasse, the spawning time, and the water temperature during the spawning season in cap-

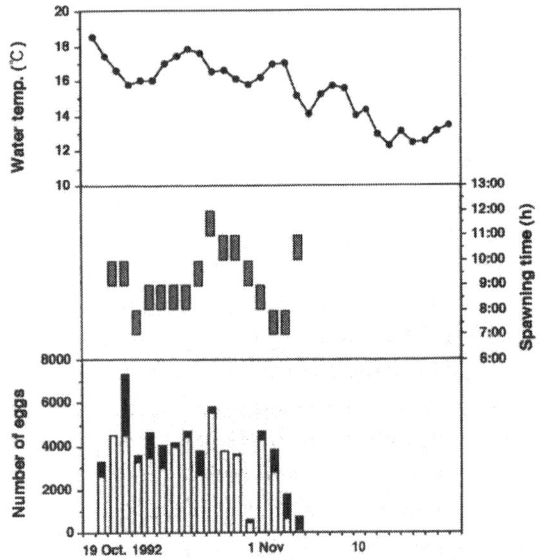

Figure 1. Number of eggs released by a single female bambooleaf wrasse reared with two males, spawning time, and changes in water temperature during the experimental period. Open and solid bars show fertilized and unfertilized eggs, respectively.

tivity. Spawning was first observed on the day following sampling, and daily spawning continued from 20 October to 5 November 1992. During this period, the water temperature gradually decreased from 17.8 to 15.1 °C. Daily spawning rates varied between 575 and 7320 eggs, amounting to a total of about 65 000 eggs over the experimental period. The mean number of eggs spawned per day was 3821. The percentage of fertilized eggs remained relatively high (mean = 78.8%) throughout the period. Spawning occurred between 7:00 and 12:00 h, and peaked between 8:00 and 9:00 h.

Diurnal cycle of oocyte development

The bambooleaf wrasse has an asynchronous type of ovary, i.e. oocytes at different stages of development are present simultaneously. Therefore, the ovarian stage was represented by the developmental stage of the largest oocytes in the ovary. The ovarian stages of fish sampled at any one of the eight different times of day showed a tendency to be similar (Table 1). Fish sampled at 9:00 h were in post-spawning condition with no ovulated eggs in the ovarian cavity and their POF were at stage I (newly-formed) and/or stage II (degeneration progressing). Between 9:00 and 15:00 h, in the largest oocytes, yolk accumulation was in progress (stage SY) or complete (stage TY). Migration of germinal vesicles (migratory nucleus, stage MN) occurred from 12:00 to 3:00 h, and GVBD (stage M) was observed at 6:00 h only. Degeneration of POF progressed as the time proceeded, and finally POF began to disappear in the mature ovary collected at 6:00 h. These results clearly indicate that GVBD occurred around 6:00 h and spawning was performed between 6:00 and 9:00 h daily.

Developmental stages of the second-largest oocytes in the ovaries varied with the developmental

Table 1. Changes in developmental stages of the largest oocytes and degenerative stages of postovulatory follicles (POF) in bambooleaf wrasse ovaries at various times of the day.

0:00 h		3:00 h		6:00 h		9:00 h	
Stage[1]	POF[1]	Stage	POF	Stage	POF	Stage	POF
MN	III	MN	III	MN	III	SY	I
MN	III	MN	III	M	III	SY	I
MN	III	MN	III	M	III	SY	I
MN	III	MN	III	M	III	SY	I
MN	III			M	III	SY	I
				M	III	TY	I, II
				M	–[2]	TY	I, II
				M	–	TY	I, II
				M	–	TY	II
						TY	II
12:00 h		**15:00 h**		**18:00 h**		**21:00 h**	
Stage	POF	Stage	POF	Stage	POF	Stage	POF
TY	I, II	TY	II	MN	II	MN	II, III
TY	II	TY	II	MN	II	MN	II, III
TY	II	TY	II	MN	II, III	MN	III
TY	II	TY	II	MN	II, III	MN	III
TY	II	MN	II, III	MN	II, III		
MN	II						
MN	II						
MN	II						
MN	II						
MN	II						

[1] See Matsuyama et al. (1988, 1990) for oocyte developmental stages and degenerative stages of postovulatory follicles (SY = secondary yolk stages; TY = tertiary yolk stage; MN = migratory nucleus stage; and M = mature stage). [2] Postovulatory follicles were absent from the ovary.

stage of the largest oocytes (Table 2). In the ovaries in which the largest oocytes were at stage of SY or TY, the second-largest oocytes were at stage PY. In ovaries with the largest oocytes undergoing germinal vesicle migration (stage MN), the second-largest oocytes were in the process of active vitellogenesis (stage SY). When GVBD was completed in the largest oocytes (stage M), the second largest oocytes were at stage SY.

The GSI of bambooleaf wrasse increased rapidly and reached a peak at 6:00 h (4.92 ±0.64, $p < 0.05$ vs. 15:00 h) when GVBD occurred (Figure 2). At 9:00 h, when spawning was completed GSI had decreased drastically to its lowest value (2.66 ± 0.11, $p < 0.01$ vs. 6:00 h), due to the release of the ovulated oocytes in the ovarian cavity. Thereafter, GSI showed a tendency to increase with time, but the changes in GSI between 9:00 and 3:00 h were not statistically significant ($p > 0.05$). Thus, GSI values reflected the ovarian condition, particularly GVBD and ovulation, in the diurnal reproductive cycle of bambooleaf wrasse.

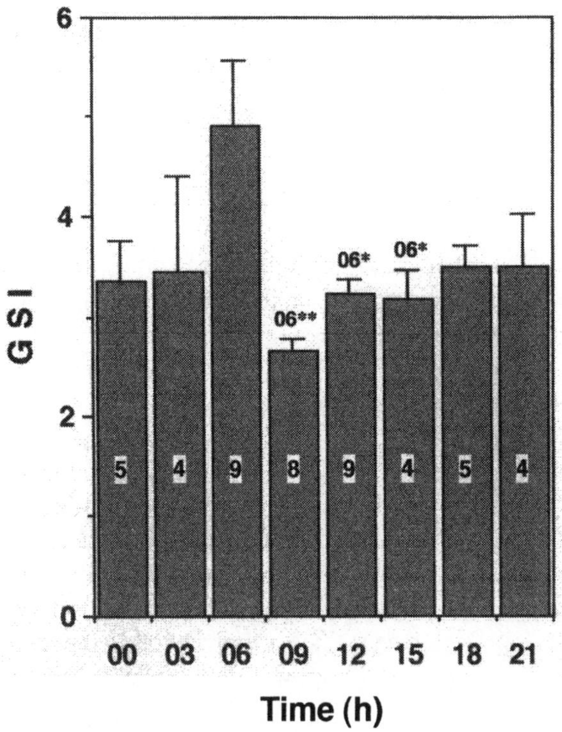

Figure 2. Diurnal changes in gonadosomatic index (GSI). Asterisk(s) indicates the significance (*$p < 0.05$, **$p < 0.01$) of differences in GSI between two times using Duncan's multiple range test. Numeral in each bar indicates sample size.

In vitro effects of various steroids on GVBD and ovulation

The relative effectiveness of the nine steroids administered is summarized in Table 3. Progesterone, 17α-P, 20β-P, 17,20β-P, and 20β-S were effective in inducing oocyte maturation *in vitro*. Progesterone at 1.0 ng ml^{-1} was less effective than 17α-P, 20β-P, 17,20β-P, and 20β-S. At concentrations of 10 ng ml^{-1}, 17,20β-P, and 20β-S were the most potent in inducing GVBD and ovulation. In contrast, 17,20α-P, es-

Table 2. Developmental stages of the second-largest oocytes in ovaries with largest oocytes at different developmental stages, in bambooleaf wrasse.

Time (h)	Developmental stages of oocytes	
	Largest	Second-largest
9:00	SY	PY
9:00–15:00	TY	PY
12:00–6:00	MN	SY
6:00	M	SY

See Matsuyama et al. (1988, 1990) for oocyte developmental stages (PY = primary yolk stage; SY = secondary yolk stage; TY = tertiary yolk stage; MN = migratory nucleus stage; and M = mature stage).

Table 3. In vitro effectiveness of various steroid hormones on GVBD and ovulation in bambooleaf wrasse oocytes. Numbers are mean response of oocytes for three tests on each of four fish. See text for criteria for oocyte maturation.

Hormone	Dose (ng ml^{-1})				
	10	1	0.1	0.01	0[1]
Progesterone	1.3	0.3	0	0	
17α-P	1.8	0.7	0	0	
20β-P	1.3	0.7	0	0	
17,20β-P	2.1	0.8	0	0	
20β-S	2.2	0.7	0	0	
17,20α-P	0	0	0	0	
Testosterone	0	0	0	0	
11-ketotestosterone	0	0	0	0	
Estradiol-17β	0	0	0	0	
Control					0

[1] Control medium contains 1% ethanol.

tradiol-17β, testosterone, and 11-ketotestosterone were totally ineffective at concentrations as high as 10 ng ml^{-1}.

Development of sensitivity to GTH and steroid

As described above, 17,20β-P and 20β-S showed almost the same potency in inducing GVBD in bambooleaf wrasse oocytes *in vitro*, so only the results obtained using 17,20β-P are presented in this study. The changes in sensitivity of oocytes to HCG and 17,20β-P are shown in Table 4. Fish sampled at 9:00 h responded weakly to HCG at high concentrations, but 17,20β-P was not effective in inducing GVBD and ovulation at that time. At 12:00 h, HCG induced GVBD and ovulation, while 17,20β-P only showed an effect on GVBD and not ovulation. At 15:00 h and 18:00 h, HCG and 17,20β-P induced both GVBD and ovulation. Fish sampled at 21:00 h underwent spontaneous GVBD in the absence of HCG or 17,20β-P, but did not ovulate; however, in the presence of HCG or 17,20β-P, the pieces of ovary released large numbers of ovulated oocytes. All the fish sampled at 0:00, 3:00, and 6:00 h underwent GVBD and ovulation spontaneously without any hormonal treatment. The ovarian oocytes of bambooleaf wrasse can be classified into three stages based on their responsiveness (GVBD and ovulation) to HCG and 17,20β-P as follows. Oocytes sampled at 9:00 h can be induced to undergo final oocyte maturation by GTH but not by steroid; these oocytes were at the tertiary yolk stage (TY) just after the completion of vitellogenesis. Oocytes collected at 12:00 h, 15:00 h, and 18:00 h can be induced to undergo final oocyte maturation by both GTH and steroid; these oocytes were at an early stage of germinal vesicle migration (MN). Oocytes collected at 21:00, 0:00, 3:00, and 6:00 h spontaneously undergo final oocyte maturation without and hormonal treatment; these oocytes were between late MN and mature (M) stages.

Discussion

In the present study, a single female reared with two TP males in captivity continued to spawn every day for 17 days from 20 October 1992, and daily spawning took place between 7:00 and 12:00 h, peaking between 8:00 and 9:00 h. Fertilization rate was relatively high. Although only spawning data from one captive stock was presented in this study, three other experiments on spawning by a single female in captivity showed similar results to that of this study, i.e. the single female spawned almost every

Table 4. Development of sensitivity to HCG and 17,20β-P in bambooleaf wrasse oocytes. Numbers are average response of oocytes duplicate tests. See text for criteria for oocyte maturation.

Time (h)	Individual no.	HCG (IU ml^{-1})					17,20β-P (ng ml^{-1})				
		100	10	1	0.1	0	100	10	1	0.1	0
0:00	1	2.5	3.0	3.0	3.0	3.0	3.0	3.0	3.0	3.0	2.0
	2	2.5	3.0	3.0	2.5	3.0	3.0	3.0	2.0	1.5	1.5
3:00	3	2.0	2.5	2.0	2.0	2.0	2.5	2.5	2.0	2.0	2.0
	4	3.0	3.0	3.0	2.5	3.0	3.0	3.0	3.0	3.0	2.5
6:00	5	3.0	3.0	3.0	3.0	3.0	3.0	3.0	3.0	3.0	3.0
	6	3.0	3.0	3.0	3.0	3.0	3.0	3.0	3.0	3.0	3.0
9:00	7	1.0	0	0	0	0	0	0	0	0	0
	8	1.5	2.0	0	0	0	0	0	0	0	0
12:00	9	2.0	2.0	2.0	0	0	1.0	1.0	1.0	0	0
15:00	10	2.5	2.0	0	0	0	2.5	0	0	0	0
	11	2.0	2.5	0.5	0	0	1.5	1.0	0	0	0
18:00	12	2.5	1.5	0	0	0	1.5	1.0	0.5	0	0
	13	2.5	2.5	2.0	1.0	0	2.5	1.0	0.5	1.0	0
21:00	14	3.0	3.0	2.5	2.0	1.0	3.0	3.0	2.5	1.0	1.0

day between 8:00 and 9:00 h during October (unpublished data). According to underwater observations, the spawning season of bambooleaf wrasse is relatively short, about two weeks from mid to late October, and spawning behavior was observed between 8:00 and 10:00 h (Nakazono 1979). This indicates that spawning in our recirculating tank system occurred normally. Thus, the spawning experiment revealed that a single female of bambooleaf wrasse spawns daily and spawning regularly occurred in the morning, mainly between 8:00 and 9:00 h, during the spawning season.

Photoperiod and water temperature, or the combination of both factors, are accepted as the principal environmental cues which mediate gonadal condition in fishes (Lam 1983). Daily spawning was discontinued when water temperatures fell below 15 °C. In our other spawning experiment (Matsuyama et al. 1997), in which two females were reared with one TP male and the water temperature was fixed at 20 °C, spawning continued for about three months, from 1 October to 28 December. These results suggest that low water temperature may play a role in terminating the spawning season of the bambooleaf wrasse. Further investigations on external cues regulating gonadal cycles, including maturation, are necessary to understand fully the reproductive cycle of this species.

Two C19 steroids (testosterone and 11-ketotestosterone) and a C18 steroid (estradiol-17β) were totally ineffective in inducing oocyte maturation in bambooleaf wrasse. In agreement with structure-activity studies in other teleosts (Scott & Canario 1987), only C21 steroids hydroxylated at position 20β or in combination with a 17α hydroxyl group had high potencies. The results obtained show that 17,20β-P and 20β-S are equally potent and highly effective steroids at inducing GVBD in bambooleaf wrasse oocytes *in vitro*. Nagahama & Adachi (1985) definitively showed that 17,20β-P is the MIH in salmonids, and this steroid has been identified as one of the most potent maturation-inducing steroids in many other fish species (Scott & Canario 1987). Similarly, 20β-S has been positively identified as the MIH in the perciform family Sciaenidae (Trant et al. 1986, Thomas & Trant 1989, Trant & Thomas 1989, Patino & Thomas 1990b), and it is present at substantial levels in blood plasma of some other teleosts (Scott et al. 1990, Zairin et al. 1992, King et al. 1994). Female bambooleaf wrasse also produce both 17,20β-P and 20β-S; serum 20β-S levels increased only during the spawning season, but 17,20β-P was present at high and constant levels throughout a year, suggesting the possibility of a role of 20β-S in oocyte maturation in bambooleaf wrasse (Morita et al. 1997). Although we used 17,20β-P as a tentative MIH of the bambooleaf wrasse in the present study, the true MIH of this species has not been conclusively identified. The equipotence of these two steroids could be due to conversion of 17,20β-P to 20β-S, as in Atlantic croaker (Patino & Thomas 1990c), or it could result from relative nonspecificity of 17,20β-P and 20β-S for the MIH receptor, as in rainbow trout, *Oncorhynchus mykiss* (Yoshikuni et al. 1993). Further studies are necessary to determine the true MIH of bambooleaf wrasse.

The present study demonstrates that bambooleaf wrasse have a diurnal cycle of oocyte development, and further clearly shows differences in responsiveness to HCG and 17,20β-P which were related to the time of day when oocytes were isolated. Similar results on differential sensitivity to hormones have been reported in kisu (Kobayashi et al. 1988), dragonet (Zhu et al. 1989), and red seabream (Kagawa et al. 1994). These species spawn every day in the evening during the spawning season. In kisu and dragonet, the ability of the oocytes to undergo GVBD in response to 17,20β-P (maturational competence) *in vitro* is acquired around 12–18 hours before spawning (Kobayashi et al. 1988, Zhu et al. 1989). However, in red seabream, maturational competence appeared around 28 h before spawning (Kagawa et al. 1994). In the bambooleaf wrasse, oocytes collected at 12:00, 15:00, and 18:00 h underwent GVBD in response to 17,20β-P, suggesting maturational competence appeared around 12–21 hours before spawning, similar to kisu and dragonet. In kisu (Matsuyama et al. 1990) and dragonet (Zhu et al. 1989), oocytes undergo GVBD about 4–5 hours before spawning, and bambooleaf wrasse oocytes showed GVBD within 4 hours before spawning. In contrast, oocytes of red seabream undergo GVBD 9–13 hours before spawning (Mat-

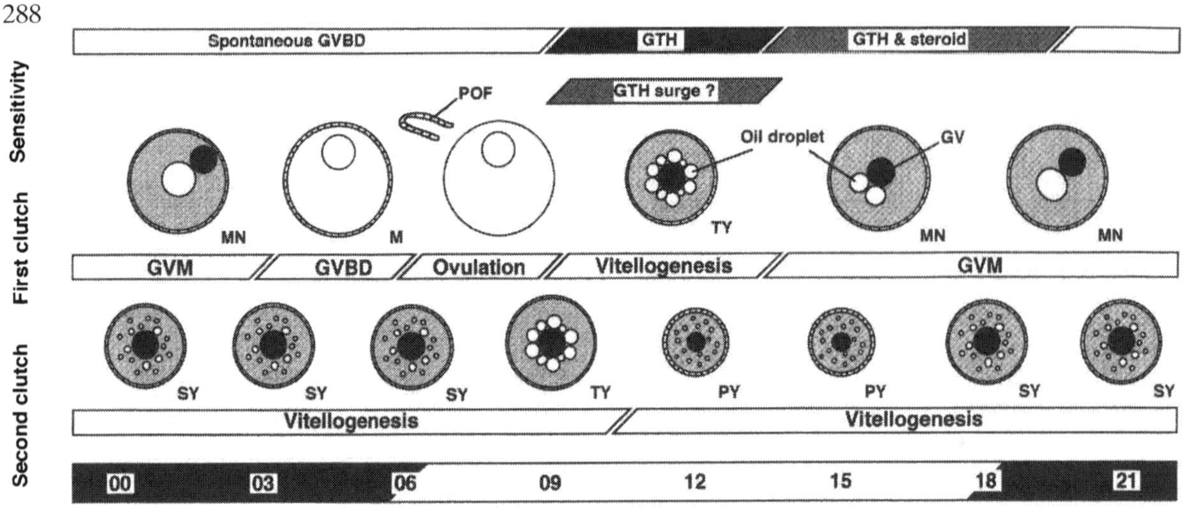

Figure 3. Diagrammatic representation of the diurnal cycle of oocyte development, sensitivity of the first clutch of oocytes to pituitary gonadotropic hormone (GTH) and steroid, and gonadosomatic index (GSI) of the bambooleaf wrasse. See Table 2 for explanation of oocyte stages. GV = germinal vesicle; GVBD = germinal vesicle breakdown; GVM = germinal vesicle migration; and POF = post-ovulatory follicle.

suyama et al. 1988). Thus, the differences in time of appearance of maturational competence among species may be caused by the different patterns of oocyte development.

It is generally accepted that, in teleosts, GTH stimulates follicular production of MIH, which directly acts on the oocytes and induces final oocyte maturation (Nagahama 1987b). Recent studies indicated that GTH develops sensitivity of oocytes to MIH in addition to inducing the production of MIH in dragonet and kisu (Zhu et al. 1994) and red seabream (Kagawa et al. 1994). Thomas & Patino (1991) showed that the concentration of MIH receptors in the ovary of spotted seatrout, *Cynoscion nebulosus,* a species in which 20β-S is the MIH, increased in response to *in vitro* GTH treatment and, concomitantly, oocytes acquired maturational competence. This fact strongly suggests that an increase in MIH receptor is one of the essential components of the development of GTH-induced maturational competence.

Sensitivity of bambooleaf wrasse oocytes to GTH and 17,20β-P changed with time of day, indicating that they have a daily developmental cycle of sensitivity to hormones, as in kisu and dragonet. Oocytes collected at 9:00 h underwent GVBD and ovulation after HCG treatment, but not in the presence of 17,20β-P. This suggests that follicles have already acquired GTH receptors, and that GTH stimulates production of the MIH in the follicles while inducing MIH receptors in the oocytes. Oocytes collected at 12:00, 15:00, and 18:00 h can be induced to undergo final oocyte maturation by both GTH and steroid. This suggests that these oocytes had already acquired receptors not only for GTH but also for steroid. Oocytes collected at 21:00, 0:00, 3:00, and 6:00 h spontaneously undergo final oocyte maturation without any hormonal treatment, which indicates that the oocytes have already been exposed to a surge of both endogenous GTH and endogenous MIH. Considering these results taken together, a GTH surge is estimated to occur at some point between 9:00 and 12:00 h, 18–21 hours prior to spawning, since the oocytes of bambooleaf wrasse began to acquire sensitivity to MIH during this period. Daily changes in two clutches of oocytes, the largest and the second-largest oocytes, GSI, and sensitivity of the largest oocytes to hormones are summarized in Figure 3. In kisu (Kobayashi et al. 1988) and dragonet (Zhu et al. 1989), a GTH surge is estimated to occur at some point between 10:00 and 13:00 h (7–10 hours before spawning), and 1:00 and 7:00 h (11–17 hours before spawning), respectively. The time of the daily GTH surge, which is fixed in different species, probably controls the diurnal maturation cycle of oocytes.

It is still unknown how GTH increases the number of MIH receptors on the oocyte membrane. In Atlantic croaker (Patino & Thomas 1990a) and red seabream (Kagawa et al. 1994), it has been shown that inhibition of the synthesis RNA (by actinomycin D) and protein (by cycloheximide) blocked GTH-induction of maturational competence, suggesting that synthesis of new protein(s) through a mechanism dependent on RNA is essential for the development of maturational competence in oocytes.

In summary, the present study demonstrated that single female bambooleaf wrasse spawn daily during the spawning season and possess a diurnal cycle of oocyte development, and also showed that bambooleaf wrasse oocytes have a daily developmental cycle of sensitivity to GTH and MIH. The daily cycle of ovarian development in this species constitutes an advantage for experimenters in that physiologically different oocytes can be obtained on the same day as demand dictates. These features make bambooleaf wrasse a good model for studies of the endocrine control of oocyte maturation, including maturational competence, which will provide interesting information on hormonal control of oocyte maturation of teleosts in the future.

Acknowledgements

We are grateful to A. Nakazono for useful information on spawning behavior of wild wrasse and some critical comments and advice on our study. This study was supported in part by grants-in-aid for scientific research from the Ministry of Education, Science, and Culture of Japan.

References cited

Kagawa, H., H. Tanaka, K. Okuzawa & K. Hirose. 1994. Development of maturational competence of oocytes of red seabream, *Pagrus major,* after human chorionic gonadotropin treatment *in vitro* requires RNA and protein synthesis. Gen. Comp. Endocrinol. 94: 199–260.

King, W.V., P. Thomas, R.M. Harrell, R.G. Hodson & C.V. Sullivan. 1994. Plasma levels of gonadal steroids during final oocyte maturation of striped bass, *Morone saxatilis* L. Gen. Comp. Endocrinol. 95: 178–191.

Kobayashi, M., K. Aida, K. Furukawa, Y.K. Law, T. Moriwaki & I. Hanyu. 1988. Development of sensitivity to maturation-inducing steroids in the oocytes of the daily spawning teleost, the kisu *Sillago japonica*. Gen. Comp. Endocrinol. 72: 264–271.

Lam, T.J. 1983. Environmental influences on gonadal activity in fish. pp. 65–116. *In:* W.S. Hoar, D.J. Randall & E.M. Donaldson (ed.) Fish Physiology, Vol. 9B, Academic Press, New York.

Matsuyama, M., S. Adachi, Y. Nagahama, K. Maruyama & S. Matsuura. 1990. Diurnal rhythm of serum steroid hormone levels in the Japanese whiting, *Sillago japonica*, a daily-spawning teleost. Fish Physiol. Biochem. 8: 329–338.

Matsuyama, M., S. Adachi, Y. Nagahama & S. Matsuura. 1988. Diurnal rhythm of oocyte development and plasma steroid hormone levels in the red sea bream, *Pagrus major,* during the spawning season. Aquaculture 73: 357–372.

Matsuyama, M., S. Morita, N. Hamaji, M. Kashiwagi, K. Ohta & Y. Nagahama. 1997. Diurnal spermatogenesis and spawning in the secondary male of a protogynous wrasse, *Pseudolabrus japonicus* (Teleostei, Labridae). Zool. Sci. 14: 1001–1008.

Morita, S., M. Matsuyama, T. Nasu & M. Kashiwagi. 1997. Seasonal changes of gonadal histology and serum steroid hormone levels in the bambooleaf wrasse *Pseudolabrus japonicus.* Nippon Suisan Gakkaishi 63: 694–700 (in Japanese).

Nagahama, Y. 1987a. Gonadotropin action on gametogenesis and steroidogenesis in teleost gonads. Zool. Sci. 4: 209–222.

Nagahama, Y. 1987b. 17α,20β-dihydroxy-4-pregnen-3-one: a teleost maturation-inducing hormone. Dev. Growth Differ. 29: 1–12.

Nagahama, Y. & S. Adachi. 1985. Identification of a maturation-inducing steroid in a teleost, the amago salmon (*Oncorhynchus rhodurus*). Dev. Biol. 109: 428–435.

Nakazono, A. 1979. Studies on the sexual reversal and spawning behavior of five species of Japanese labrid fishes. Report of the Fishery Research Laboratory, Kyushu University 4: 1–64 (in Japanese).

Patino, R. & P. Thomas. 1990a. Effects of gonadotropin on ovarian intrafollicular processes during the development of oocyte maturational competence in teleost, the Atlantic croaker: evidence for two distinct stages of gonadotropic control of final oocyte maturation. Biol. Reprod. 43: 818–827.

Patino, R. & P. Thomas. 1990b. Gonadotropin stimulates 17α,20β,21-trihydroxy-4-pregnen-3-one production from endogenous substrates in Atlantic croaker ovarian follicles undergoing final maturation *in vitro.* Gen. Comp. Endicrinol. 78: 474–478.

Patino, R. & P. Thomas. 1990c. Induction of maturation of Atlantic croaker oocyte by 17α,20β,21-trihydroxy-4-pregnen-3-one *in vitro*: consideration of some biological and experimental variables. Exp. Zool. 255: 97–109.

Scott, A.P. & A.V. Canario. 1987. Status of oocyte maturation-inducing steroids in teleosts. pp. 224–234. *In:* D.R. Idler, L.W. Crim & J.M. Walsh (ed.) Proceeding of the Third International Symposium on the Reproductive Physiology of Fish, Memorial University of Newfoundland, St. John's.

Scott, A.P., A.V.M. Canario & F. Prat. 1990. Radioimmunoassay of ovarian steroids in plasmas of ovulating sea bass (*Discentrarchus labrax*). Gen. Comp. Endocrinol. 46: 444–451.

Thomas, P. & R. Patino. 1991. Changes in 17α,20β, 21-trihydroxy-4-pregnen-3-one membrane receptor concentrations in ovaries of spotted seatrout during final oocyte maturation. pp. 80–82. *In:* A.P. Scott, J.P. Sumpter, D.E. Kime & M.S. Rolfe (ed.) Proceedings of the 4th International Symposium on Reproductive Physiology of Fish, University of East Anglia, Norwich.

Thomas, P. & J.M. Trant. 1989. Evidence that 17α,20β, 21-trihydroxy-4-pregnen-3-one is a maturation-inducing steroid in spotted sea trout. Fish Physiol. Biochem 7: 185–191.

Trant, J.M. & P. Thomas. 1989. Isolation of a novel maturation-inducing steroid produced *in vitro* by ovaries of Atlantic croaker. Gen. Comp. Endocrinol. 75: 397–404.

Trant, J.M., P. Thomas & C.H.L. Shackleton. 1986. Identification of 17α,20β,21-trihydroxy-4-pregnen-3-one as the major ovarian steroid produced by the teleost *Micropogonias undulatus* during final oocyte maturation. Steroids 47: 89–99.

Yoshikuni, M., N. Shibata & Y. Nagahama. 1993. Specific binding of [^{3}H] 17α,20β-dihydroxy-4-pregnen-3-one to oocyte cortices of rainbow trout (*Oncorhynchus mykiss*). Fish Physiol. Biochem. 11: 15–24.

Zairin, M., Jr., K. Asahina, K. Furukawa & K. Aida. 1992. Plasma steroid hormone profiles during HCG induces ovulation in female walking catfish *Clarias batrachus* Zool. Sci. 9: 607–617.

Zhu, Y., K. Aida, K. Furukawa & I. Hanyu. 1989. Development of sensitivity to maturation-inducing steroids and gonadotropins in the oocytes of the tobinumeri-dragonet. *Repomucenus beniteguri*, Callionymidae (Teleostei). Gen. Comp. Endocrinol. 76: 250–260.

Zhu, Y., M. Kobayashi, K. Furukawa & K. Aida. 1994. Gonadotropin develops sensitivity to maturation-inducing steroid in the oocytes of daily spawning teleosts, tobinumeri-dragonet *Repomucenus beniteguri* and kisu *Sillago japonica.* Fisheries Science 60: 541–545.

Environmental Biology of Fishes **52**: 291–297, 1998.

Effects of sandy substrate and light on hypermelanosis of the blind side in cultured Japanese flounder *Paralichthys olivaceus*

Nakahiro Iwata & Kotaro Kikuchi
Central Research Institute of Electric Power Industry, Abiko, Chiba 270–11, Japan
(e-mail: m-iwata@criepi.denken.or.jp)

Received 31.1.1997 Accepted 19.7.1997

Key words: ambicoloration, color anomaly, flatfish, rearing condition

Synopsis

Rearing experiments were carried out to clarify the effects of sandy substrate and light irradiation on hypermelanosis of the blind side (the staining type of ambicoloration) in cultured Japanese flounder. Fish were reared in three experimental conditions: (1) no sandy substrate into which fish could bury themselves and with upward light irradiating their blind sides, (2) no sandy substrate and no upward light, and (3) sandy substrate (transparent glass sand) with upward light irradiation. Pigmented areas on the blind side were measured by an image analyzing system. Flounder from the third condition (sandy substrate with light irradiation) showed the lowest pigmentation on the blind side. In contrast, fish from the second condition (no sandy substrate and no light irradiation), showed the highest pigmentation coverage. These results indicate that sandy substrate on the bottom in culture tanks is more important than light irradiation as a factor affecting hypermelanosis of the blind side in cultured Japanese flounder.

Introduction

Hypermelanosis of the blind side (ambicoloration) is known in Japanese flounder *Paralichthys olivaceus* (Yamaguchi 1955, Maruyama 1976, Yamamoto & Oda 1991, Seikai 1991, Takahashi 1994, Suzuki 1994) as well as other flatfish species (Cunningham 1891, 1893, 1895, Osborn 1940, 1941, Houde 1971, Stickney & White 1975, Masaki & Ito 1984, Seikai 1985, Gartner 1986, Burton 1988). This color anomaly of Japanese flounder occurs in nearly all hatchery-reared flounder (Yamamoto & Oda 1991, Seikai 1991, Takahashi 1994), although it is rarely found in wild fish (Yamaguchi 1955, Maruyama 1976). Such unusual coloration is one of the most important problems in production of juveniles for land based culture and stock enhancement in this species, because it sometimes reduces the market price of the fish.

Occurrence of hypermelanosis of the blind side in Japanese flounder has been suggested to be affected by several factors such as light irradiation to the blind side, source of fish, kind of food and stocking density (Seikai 1991, Takahashi 1994). Among them, light irradiation to the blind side has been regarded as the most important causative factor by several experimental efforts not only with Japanese flounder (Seikai 1991) but also with other flatfish species (Cunningham 1891, 1893, 1895, Osborn 1940, Stickney & White 1975). However, most of these results were produced by experiments using fish tanks in which the bottom was covered with sandy substrate to prevent upward light irradiation (Cunningham 1891, 1893, Stickney & White 1975, Seikai 1991). Therefore, the question still remains as to whether it is not light but the sandy substrate which affects the blind side pigmentation. This study was

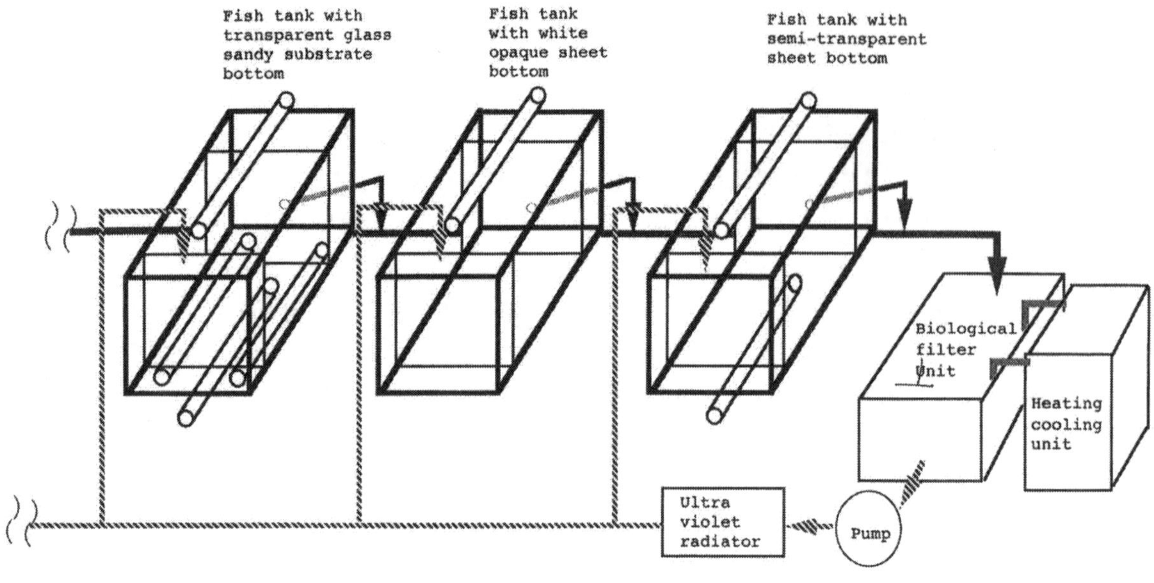

Figure 1. Diagram of the test apparatus. Three of six fish tanks are shown in the figure.

designed to clarify which is the main factor causing hypermelanosis of the blind side in cultured Japanese flounder.

Materials and methods

Fish

The flounder used for experiments were obtained from two commercial hatcheries in Mie (Anagawa Syokusan Co., Ltd.) and Aich (Nissin Marine Tech Co., Ltd.) prefectures, and were transported to our laboratory in Chiba Prefecture at a size of about 50 mm total length and 1 g body weight. Fish were reared for about a month (fish from Mie) or a week (fish from Aich) in a recirculating seawater aquarium of 1000 l volume, and fed a commercial pellet diet until the start of experiments. The aquarium was placed in an experimental room under conditions of 12 h light with a 40 W fluorescent lamp 1.5 m above the culture tank. Water temperature was kept at approximately 23 °C.

Experimental apparatus

Experiments were carried out in a closed recirculating aquarium (Figure 1). It consisted of 6 fish tanks ($60 \times 40 \times 30$ cm), a biological filter unit (800 l volume), an ultra violet radiator (30 W), a circulation pump and a heating-cooling unit. The acrylic fish tanks (Figure 2) with transparent bottoms (5 mm in thickness) were placed on transparent acrylic plates (5 mm in thickness). The insides of tank walls were made of matted black vinyl chloride plates, and the top of the tank was covered with a transparent acrylic plate of 3 mm thickness. Two baffle plates of transparent vinyl chloride (3 mm thick) were installed at the ends of each tank. An 18 W fluorescent lamp (Matsushita Electric Industrial Co. Ltd., Osaka, FL20SSEX-N/18) was set 10 cm above water level of each fish tank.

Experimental condition

Effects of light irradiation and sandy substrate on the hypermelanosis of the blind side were examined under three experimental conditions as follows. In the first condition, a semi-transparent polypropylene sheet of 0.75 mm thickness (Acrysunday Co., Ltd., Tokyo, No. 17001) was set on the bottom of tank, and an 18 W fluorescent lamp was installed 40 cm below the tank. Therefore, fish could not bury themselves and their blind sides were irradiated

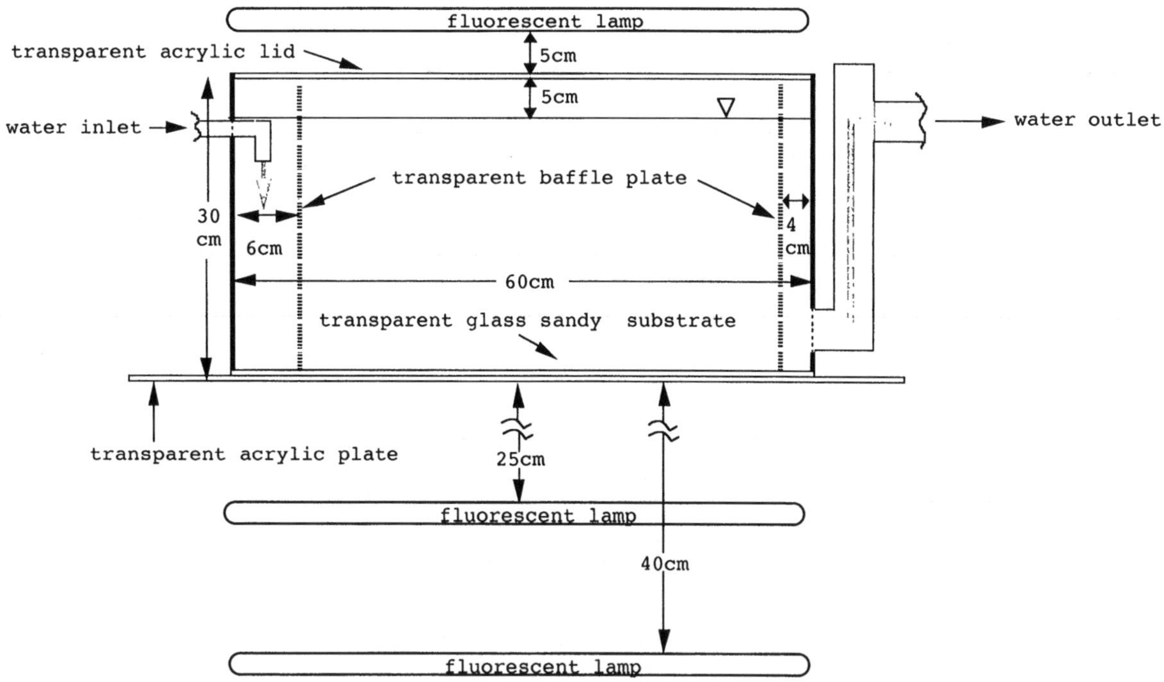

Figure 2. Cross section view of a fish tank with transparent glass sandy substrate bottom.

by a light under the tank. In the second condition, a white opaque polypropylene sheet of 0.75 mm thickness (Acrysunday Co., Ltd, No. 17002) was set on the bottom, and no light source was installed under the tank, hence fish could not bury themselves as in the first condition but the blind sides were lighted only by reflection from the tank bottom. In the third condition, the tank bottom was covered with transparent glass sandy substrate (particle size: 0.5–0.7 mm, Toshiba Ballotini Co., Ltd., Tokyo, GP-500B, hereafter referred to as 'glass sand') 1 cm thick, and three 18 W fluorescent lamps were installed under the tank (one was 40 cm and the others 25 cm below the tank bottom). In this condition, fish could bury themselves into a sandy substrate, and their blind sides were irradiated by lights under the tank. The fluorescent lights above and under the fish tank were turned on from 8:00 to 20:00 (12 h). Each experimental unit was separated by silver blackout films. Rearing temperature was controlled at 23 °C ± 1 °C through the experiments.

The rearing experiment was conducted twice as experiment 1 and experiment 2. Experiment 1 was carried out for 12 weeks from 23 May 1994 with fish from Mie Prefecture, and experiment 2 was for 16 weeks from 14 November 1994 with fish from Aichi Prefecture. In both experiments, fish which had no visible pigmentation on the blind side were chosen and placed in three fish tanks. The number, total length (TL) and body weight (BW) of fish used are shown in Tables 1 and 2. Fish were fed a commercial pellet diet for the flounder (Higashimaru Co., Ltd.) twice a day at about 3% of their body weight from Tuesday to Friday. On each Monday, fish were transferred to the other three tanks with cleaned bottom sheets or new glass sand, so as to keep bottom conditions constant. All fish were weighed every second Monday to calculate the amount of feed.

Measurement of pigmented area coverage on the blind side

At the end of each experiment, all fish were anesthetized with 250 ppm ethyl 3-aminobenzoate methanesulfonic acid salt (Aldrich Chemical Inc.), and photographs of their blind sides were taken individually. With these photographs, total areas of the

Table 1. Experimental conditions and results of experiment 1.

Bottom material of fish tank	1 semi-transparent sheet	2 white opaque sheet	3 transparent glass sandy substrate
Upward photon flux density at the bottom (μmol m^{-2} sec^{-1})[1]			
start of experiment	15	0	23
end of experiment	11	0	18
Illumination at the bottom of fish tank (lux)[2]			
downward	1800	1800	1900
upward	1000	10	1400
Number of fish[3]			
start of experiment	25	25	25
end of experiment	17	18	22
Body weight of fish (g)[3]			
start of experiment	3.9 ± 0.7	3.8 ± 1.1	4.0 ± 0.9
end of experiment	43.8 ± 9.4	43.1 ± 18.5	47.9 ± 15.1
Total length of fish (mm)[3]			
start of experiment	79 ± 5	78 ± 7	79 ± 6
end of experiment	171 ± 13	166 ± 20	175 ± 19
Percentage of pigmented area (%)[3,4]	7.3 ± 7.1	23.0 ± 13.1	1.6 ± 0.8

[1] The average of upward photon flux density in water measured at nine points on the fish tank bottom.
[2] The average of illumination measured at nine points on the fish tank bottom without water, before the experiment.
[3] Average ± standard deviation.
[4] Coverage of pigmented area on the blind side except dorsal, anal, and caudal fin.

Table 2. Experimental conditions and results of experiment 2.

Bottom material of fish tank	1 semi-transparent sheet	2 white opaque sheet	3 transparent glass sandy substrate
Upward photon flux density at the bottom (μmol m^{-2} sec^{-1})[1]			
start of experiment	13	0	22
end of experiment	8	0	10
Number of fish			
start of experiment	20	20	20
end of experiment	17	12	17
Body weight of fish (g)[2]			
start of experiment	1.0 ± 0.3	1.1 ± 0.3	1.0 ± 0.3
end of experiment	42.4 ± 12.5	46.2 ± 18.7	44.6 ± 15.7
Total length of fish (mm)[2]			
start of experiment	53 ± 5	53 ± 4	52 ± 4
end of experiment	171 ± 15	169 ± 26	169 ± 20
Percentage of pigmented area (%)[2,3]	10.2 ± 8.2	16.0 ± 9.1	1.6 ± 0.8

[1] The average of upward photon flux density in water measured at nine points on the fish tank bottom.
[2] Average ± standard deviation.
[3] Coverage of pigmented area on the blind side except dorsal, anal, and caudal fin.

blind side except caudal, dorsal and anal fins were measured as well as pigmented areas, and coverages of pigmented areas were calculated. Measurements were carried out with an image analyzing system consisting of a personal computer (Power Macintosh 8100, Apple Inc.), an image scanner (GT9000ART, Seiko Epson Co., Ltd.), and image analyzing software (IPLab Spectrum, Signal Analytics Co., Ltd., Vienna, VA).

Measurement of light intensity and spectrum

Photon flux density in the water was measured at nine points on each tank bottom using a quantum meter (LI-COR Co., LI-192SA, Lincoln, NE) before and after the experiments, with a flat sensor placed on the bottom.

Upward and downward illumination at the bottom without rearing water were measured at the start of experiment 1 with an illumination meter with a hemisphere sensor (Minolta Co., Ltd., T-1H).

Photon flux distribution of upward light was measured in tanks with the semi-transparent sheet bottom and the glass sand bottom using a multi-spectrometer (Opto-research Co. Ltd., Tokyo). Measurements were carried out about 1 cm above the bottom of tanks without rearing water before the start of experiment 1.

Results

Results of experiment 1 and 2 are shown in Tables 1 and 2, respectively. There were no significant differences in the total length and body weight at the end of each experiment between fish from different tanks. Therefore, differences in rearing conditions such as light and sandy substrate were not considered to affect the growth of fish. The number of individuals at the end differed among experimental groups. Almost all the fish lost were those that jumped out of tanks, because there was a small space between the tank and lid of every tank.

Upward light intensity in the tank measured as photon flux density and illumination were the highest in the tank with glass sand, and next highest in the tank with the semi-transparent sheet in both experiments. Upward light in the tank with the white opaque sheet (10 lux) in experiment 1 originated from reflection of downward light. Downward light intensity in experiment 1 was similar among the 3 experimental groups.

Coverages of pigmented areas on the blind sides at the end of experiment 1 were 7.3% in the semi-transparent sheet tank, 23.0% in the white opaque sheet tank, and 1.6% in the glass sand tank. The coverage in experiment 2 was highest in the white opaque sheet tank (16.0%), next highest in the semi-transparent sheet tank (10.2%), and lowest in the glass sand tank (1.6%). The coverage in the glass sand tank was significantly lower than the other two in both experiments (Mann-Whitney U-test, $p < 0.05$). The pigmentation area of fish was the largest in the tank with the least upward light. On the contrary, very small values were produced in the tank with the strongest upward light. The transparent glass sandy substrate used in this study did not affect the spectrum distribution of upward light. Glass sand may work as a submersible substrate for fish.

Discussion

Norman (1934) showed that ambicoloration of flatfish could be divided into 3 types by its characteristics; staining (the development of chromatophores in certain regions of the skin of the blind side produces irregular patches of dull pigment which give the lower surface a stained appearance), spotting (the blind side exhibits a varying number of black or brown spots), and true-ambicoloration (the coloration of the blind side closely resembles that of the ocular surface, even to the characteristic spots and markings). Similar to other observations (Seikai 1991), pigmentation occurred at the caudal peduncle, the anterior trunk adjacent to dorsal and anal fins, and the base of the pectoral fin of the blind side in this study (Figure 3), and was considered to be the staining type. Therefore, it is considered that sandy substrate on the bottom of culture tank is more important than light irradiation as a factor for the staining type of ambicoloration in Japanese flounder.

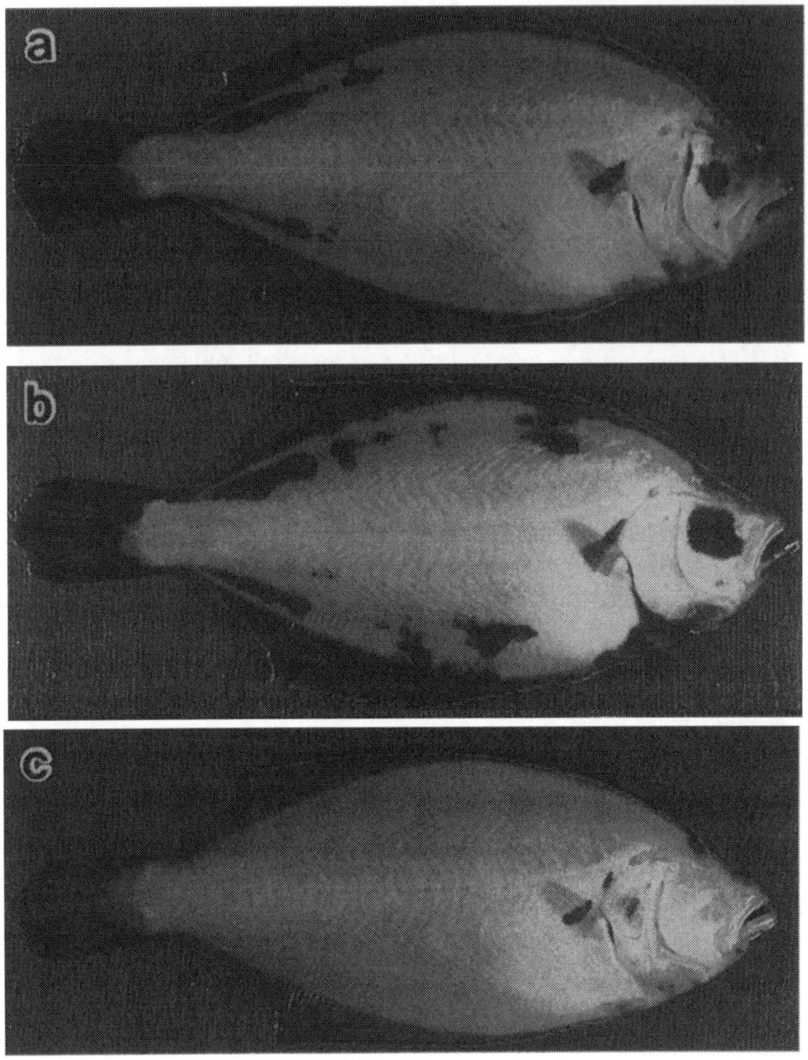

Figure 3. Blind sides of Japanese flounders at the end of experiment 2: a – Fish reared in the tank with semi-transparent sheet bottom. Coverage of pigmented area on the blind side = 8.6%, TL = 196 mm, BW = 66.7 g. b – Fish reared in the tank with white opaque sheet bottom. Coverage of pigmented area on the blind side = 19.4%, TL = 178 mm, BW = 50.5 g. c – Fish reared in the tank with transparent glass sandy substrate bottom. Coverage of pigmented area on the blind side = 1.7%, TL = 172 mm, BW = 44.8 g.

In previous reports indicating that upward light was the main reason for the pigmentation of other flatfish species such as *Pleuronectes flesus* (Cunningham 1891, 1893) and *Paralichthys dentatus* (Stickney & White 1975), the inhibition of upward light was done by covering the bottom of fish tank with coarse sand or silt. Furthermore, from photographs and description of the characteristics of pigmentation, the coloration in these studies was considered to be of the staining type. This suggests that it is not light but sandy substrate which affects the staining type ambicoloration of other flatfish species.

On the other hand, Osborn (1940, 1941) reported that blind side pigmentation of *Paralichthys dentatus* could be inhibited by preventing upward light even if there was not a sandy substrate (the bottom of the fish tank was frosted glass). Though it is not clear which type of ambicoloration occurred in Osborn's study, the differences in light effect between this and Osborn's study suggest that there are different mechanisms governing ambicoloration on the blind side of flatfishes.

From this study, we conclude that sandy substrate significantly affects hypermelanosis of the blind side in cultured Japanese flounder. However, visible pigmentation still remained in some small areas in the presence of sandy substrate, even though there was less pigmentation coverage on the blind side compared to the other 2 groups. Therefore, more study is required to clarify the reason and develop techniques to prevent hypermelanosis on the blind side in the cultured flounder.

Acknowledgements

We wish to thank David Noakes and Tadahisa Seikai for their valuable advice and critical reading of the manuscript. We also acknowledge Yutaka Hagiwara, Hiromichi Hanyu, Masaki Honda, and Haruo Honda for their technical help.

References cited

Burton, D. 1988. Melanophore comparisons in different forms of ambicoloration in the flatfish *Pseudopleuronectes americanus* and *Reinhardtius hippoglossoides*. J. Zool., Lond. 214: 353–360.

Cunningham, J.T. 1891. An experiment concerning the absence of color from the lower sides of flat-fishes. Zool. Anzeiger 14: 27–32.

Cunningham, J.T. 1893. Researches on the coloration of the skins of flat-fishes. J. Mar. Biol. Assoc. U.K. 3 (N.S.): 111–118.

Cunningham, J.T. 1895. Additional evidence on the influence of light in producing pigments on the lower sides of flat fishes. J. Mar. Biol. Assoc. U.K. 4 (N.S.): 53–59.

Gartner, J.V., Jr. 1986. Observations on anomalous conditions in some flatfishes (Pisces: Pleuronectiformes), with a new record of partial albinism. Env. Biol. Fish. 17: 141–152.

Houde, E.D. 1971. Developmental abnormalities of the flatfish *Achirus lineatus* reared in the laboratory. U.S. Fish. Bull. 69: 537–544.

Maruyama, K. 1976. Abnormal specimens of flatfishes collected in Iwate Prefecture. Japan J. Ichthyol. 23: 118–120 (in Japanese).

Masaki, Y. & H. Ito. 1984. Stock identification of finespotted flounder, *Pleuronichthys cornutus* in the western Seto inland sea. Bull. Nansei Reg. Fish. Res. Lab. 16: 11–51 (in Japanese).

Norman, J.R. 1934. A systematic monograph of the flatfishes (Heterosomata). Volume 1. Brit. Mus., London. 459 pp.

Osborn, C.M. 1940. The experimental production of melanin pigment on the lower surface of summer flounders (*Paralichthys dentatus*). Proc. Nat. Acad. Sci. U.S.A. 26: 155–161.

Osborn, C.M. 1941. Studies on the growth of integumentary pigment in the lower vertebrates I. The origin of artificially developed melanophores on the normally unpigmented ventral surface of the summer flounder (*Paralichthys dentatus*). Biol. Bull. Mar. Biol. Lab., Woods Hole 81: 341–351.

Seikai, T. 1985. Effect of population density and tank color on the abnormal coloration of hatchery-reared mud dab, *Limanda yokohamae.* Suisanzoshoku 33: 119–128 (in Japanese).

Seikai, T. 1991. Influences of fluorescent light irradiation, ocular side pigmentation, and source of fishes on the blind side pigmentation in the young Japanese flounder, *Paralichthys olivaceus.* Suizanzoshoku 39: 173–180 (in Japanese).

Stickney, R.R. & D.B. White. 1975. Ambicoloration in tank cultured flounder, *Paralichthys dentatus.* Trans. Amer. Fish. Soc. 104: 158–160.

Suzuki, N. 1994. Ultrastructure of the skin on reverse side of hatchery-reared Japanese flounder, *Paralichthys olivaceus* with reference to the pigmentation. Bull. Nansei Natl. Fish. Res. Inst. 27: 113–128 (in Japanese).

Takahashi, Y. 1994. Influence of stocking density and food at the late phase of the larval period on hypermelanosis on the blind body side in juvenile Japanese flounder. Nippon Suisan Gakkaishi 60: 593–598 (in Japanese).

Yamaguchi, M. 1955. Anomalies observed in flatfishes *Paralichthys olivaceus* and *Paralichthys stellatus.* Saishuu to Shiiku 17: 244 (in Japanese).

Yamamoto, S. & T. Oda. 1991. The effects of several diets on survival, growth and color anomalies in juvenile flounders *Paralichthys olivaceus.* Bull. Fish. Exp. Sta. Okayama Pref. 6: 100–108 (in Japanese).

Environmental Biology of Fishes **52**: 305–312, 1998.

Growth of dark chub, *Zacco temmincki* (Cyprinidae), with a discussion of sexual size differences

Osamu Katano
National Research Institute of Fisheries Science, 1088, Komaki, Ueda, Nagano 386, Japan
(e-mail: katano@nrifs-u.affrc.go.jp)

Received 4.8.1995 Accepted 3.2.1997

Key words: age, body size, growing season, opercular ring, polygynous mating system

Synopsis

I examined the annual and seasonal growth of dark chub, *Zacco temmincki,* in a Japanese river. Investigation of opercular rings showed that the fish reached a maximum age of 8 years. There was no significant sexual size difference at younger ages (3–5 years), but males were larger than females at older ages (6–7 years). Annual increments of length and weight for males that were recaptured were also larger than those for females. The fish spawn from June to August. Females grew for a short period from April to May, but males on average grew for a longer period from April to August. There was no sexual difference in growth rate except during the spawning period. Annual growth rate was negatively correlated with fish length in each sex. The sexual size differences at older ages of the fish might be due to the polygynous mating system in which most mature males could not obtain females and invested for somatic growth in the spawning period, and a short growing season that was overlapped considerably with the spawning period.

Introduction

Sexual size dimorphism is found in many fishes. Females are larger than males of the same age in a variety of species (e.g., anadromous brown trout, *Salmo trutta*: Berg & Jonsson 1990, yellow perch, *Perca flavescens*: Diana & Salz 1990, flatfish, *Pleuronectes platessa*: Deniel 1990, lingcod, *Ophiodon elongatus:* Smith & McFarlane 1990, greenback mullet, *Liza subviridis*: Al-Daham & Wahab 1991, Australian sharpnose shark, *Rhizoprionodon taylori*: Simpfendorfer 1993). Generally, large females can develop more eggs and thereby achieve higher reproductive success compared with small females, but increase in body size is not necessarily correlated with higher male reproductive success (Parker 1992, Andersson 1994). Therefore, females in many species might have developed larger body size.

On the other hand, males are larger than females in several fishes although the reason for the larger body size is not clear (e.g. Japanese minnow, *Pseudorasbora parva:* Nakamura 1969, gudgeon, *Gobio gobio*: Mann 1980, pumpkinseed, *Lepomis gibbosus*: Deacon & Keast 1987, saddleback darter, *Percina vigil:* Heins & Baker 1989, filefish, *Brachaluteres ulvarum:* Akagawa et al. 1995). Several authors have proposed that polygynous mating systems are associated with the evolutionary development of larger male size (Trivers 1985, Andersson 1994). In polygynous mating systems, there is severe competition among males for access to females, and only a limited number of large, dominant and/or territorial males can monopolize females by repelling small males and/or by being chosen as mates by females. Thus, natural selection should favor large male size (Parker 1992, Andersson 1994).

then gain a new level of information regarding eel migration.

The purpose of this study was to determine the influence of environmental factors on otolith Sr/Ca ratios in Japanese eel elvers under various water temperature and salinity conditions.

Materials and methods

A total of 340 glass-eels were used: 100 collected from the first dam of the Saigo River in Fukuoka Prefecture in May 1994 and 240 collected from the mouth of the Saigo River in March 1995. To mark the commencement of this period, the elvers were immersed in a 50 ppm solution of alizarincomplexion (ALC) in seawater to produce a fluorescent ALC mark on the otoliths. The freshwater used in this study was well water from the Kyushu University campus, while the seawater was obtained from the Fishery Research Laboratory of Kyushu University.

To examine the relationship between Sr and Ca concentrations in otoliths and various ambient salinity conditions, four groups of 25 elvers were each reared for 15 days, starting from 28 May 1994, in either freshwater, one-third seawater, two-thirds seawater, or seawater (34‰). To examine the relationship between Sr and Ca concentrations in otoliths and different ambient temperatures, 240 elvers were reared for 58 days starting from 29 March 1995. Eight groups of 30 elvers each were reared at temperatures of 12, 17, 22 or 27 °C in either seawater or freshwater tanks, respectively. The elvers were reared under light conditions of 12L/12D within an enclosed tank and fed to satiation with tubificid oligochaetes, *Limnodrilus* sp., once a day.

Sagittal otoliths were extracted under a dissecting microscope and rinsed twice in deionized water. After being dried overnight, they were embedded in a cyanoacrylate resin on a glass slide, ground and then polished successively by three-, one-, and one quarter-micrometer diamond paste to just outside the nucleus. The location of the fluorescent ALC mark was determined under a UV light microscope fitted with a fluorescence attachment. The samples were coated with carbon.

We first determined whether Sr/Ca ratios in otoliths of individual eels changed as the fish grew. Two eels, reared at 22 °C in seawater and freshwater, respectively, were selected for study. Their otoliths were sampled at eight points arranged along a radius from the nucleus to the otolith edge, at 10 μm intervals from the ALC mark outwards. These samples were analyzed for Sr and Ca concentrations by spot analysis using a wavelength dispersive X-ray electron microprobe (JEOL, JXL-8900, Tokyo). This analysis used an accelerating voltage of 15 kv, a beam current of 0.05 μA and a diameter of 1 μm focused with a counting time of 10 seconds. $CaSiO_3$ and $SrTiO_3$ were used as the standards. The mean Sr and Ca concentrations within the samples were calculated from these seven points.

Using the same methods, we then analyzed how Sr/Ca ratios varied with salinity and temperature in the two experiments. For this, we randomly chose a location on each otolith midway between the ALC and the outer edge, and conducted spot analysis at seven points at that location.

We also analyzed the Sr and Ca concentrations of the tubificid oligochaetes, seawater, and freshwater used in the present experiment. The Ca concentrations of the samples were determined using an Atomic Absorption Photometer (Hitachi, 180–80) at the Laboratory for Waste Water Treatment, Ky-

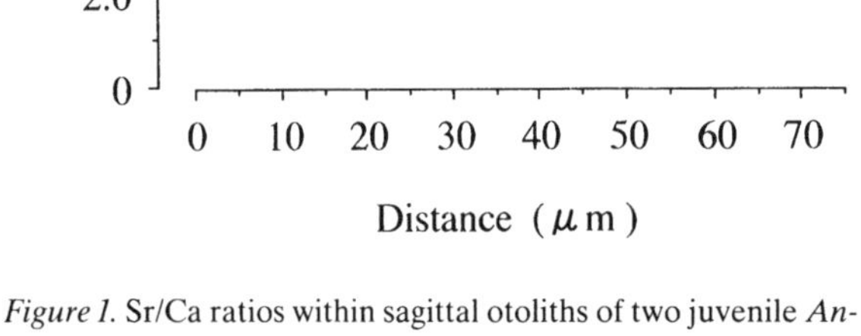

Figure 1. Sr/Ca ratios within sagittal otoliths of two juvenile *Anguilla japonica*, each kept in different rearing water for 58 days. Plotted points were located along a radius from the fluorescent ALC mark (commencement of experiment) to the edge of the otolith, and represent the mean of three measurements (○ = 22 °C seawater, ● = 22 °C freshwater).

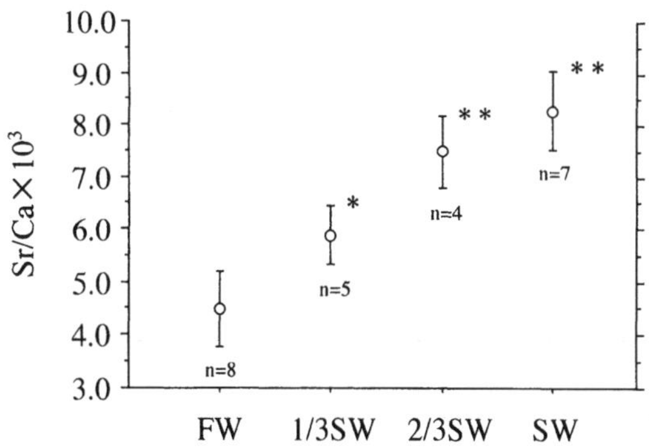

Figure 2. Sr/Ca ratios within sagittal otoliths of juvenile *Anguilla japonica* kept in different dilutions of seawater and in freshwater for 15 days at 22 °C. Plotted values are the mean ± SD for each group (sample sizes are below bars). Data for each individual were the mean of seven points (see text; FW = freshwater, SW = seawater, * = < 0.005, and ** = $p < 0.0001$ for comparisons to FW by Fisher's Least Significant Difference test).

ushu University. The Sr concentrations of the samples were determined using an Inductively Coupled Plasma-Mass Spectrometer (Yokogawa, PM-2000) at the Center of Advanced Instrumental Analysis, Kyushu University.

Results

The Sr/Ca ratio analyses for each otolith midway from a precise fluorescent ALC mark to the edge of the otolith are shown in Figure 1. There was a clear difference in mean Sr/Ca concentration ratios in the

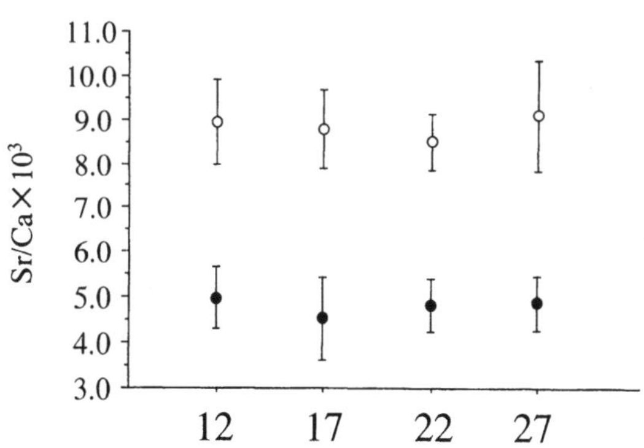

Figure 3. Sr/Ca ratios in sagittal otoliths of juvenile *Anguilla japonica* kept at various temperatures in seawater or freshwater for 58 days. Plotted values are the mean ± SD of five fish. Data for each fish were the mean of seven points (○ = seawater, ● = freshwater).

freshwater and seawater groups which remained constant throughout the rearing periods.

The otolith Sr/Ca ratios in the freshwater, one-third seawater, two-thirds seawater and pure seawater groups increased significantly with salinity (one-way ANOVA $p < 0.0001$, Figure 2). The Sr/Ca ratios in groups reared at 12, 17, 22 and 27 °C in pure seawater were 9.0 ± 0.96, 8.8 ± 0.89, 8.5 ± 0.64 and 9.1 ± 1.26, respectively (Figure 3), and these were not significantly different ($p = 0.78$ by one-way ANOVA). The Sr/Ca ratios at 12, 17, 22 and 27 °C in freshwater were 5.0 ± 0.68, 4.5 ± 0.91, 4.8 ± 0.59 and 4.9 ± 0.60, respectively (Figure 3), and similarly, no significant difference was seen ($p = 0.76$ by one-way ANOVA).

The results of analysis of Sr and Ca concentrations within food, seawater and freshwater are presented in Table 1 and the Sr/Ca ratios of each sample were 18.6, 19.5 and 4.7, respectively. Because well water was used in this study, the Ca concentration was about 3.7 times higher than the mean value of freshwater from rivers in Kyushu. The mean value Ca concentration of freshwater from these rivers were 9.8 ppm and 3.37–35.90 ppm, respectively (Kobayashi 1952).

Discussion

This study revealed a significant increase in the otolith Sr/Ca ratios with increased salinity (Figure 2). It is therefore possible that otolith Sr/Ca ratios could be used to reconstruct the environmental salinity history of the Japanese eel. Radtke et al. (1988) indicated a dramatic drop in otolith Sr/Ca ratios, reflecting movement from a seawater to a freshwater

Table 1. Ca and Sr concentrations (mean ± SD) in the food (tubificid oligochaetes), seawater, and freshwater used in this study. Sample sizes are shown in parentheses.

Sample	Concentration (ppm)		Sr/Ca × 10^3
	Ca	Sr	
Food	563.82 ± 26.71 (5)	10.54 ± 0.54 (5)	18.6
Seawater	433.10 ± 4.43 (5)	8.46 ± 0.27 (5)	19.5
Freshwater	36.42 ± 0.28 (5)	0.17 ± 0.00 (5)	4.7

environment in the goby, *Stenogobius genivittatus*. Secor (1992) suggested that a change in ambient environmental salinity had affected the otolith Sr/Ca ratios in the striped bass, *Morone saxatilis*. Our results regarding the otolith Sr/Ca ratios in *A. japonica* elvers were similar to these earlier findings.

The Sr/Ca ratios of seawater, of food used in this experiment, and of the otoliths of elvers reared in seawater were 19.5, 18.6 and 8.3, respectively (Table 1, Figure 2), with the Sr/Ca ratio of seawater and food being about twice that in the otoliths. This shows the well-known effect of biopurification (Boroughs et al. 1957) in which an organism discriminates against trace elements (e.g., Sr) which are chemically similar to a host element (e.g., Ca). On the other hand, the Sr/Ca ratio of freshwater and of the otolith of elvers reared in freshwater were 4.7 and 4.5, respectively (Table 1, Figure 2). These results suggest that Sr incorporation into the otolith of the elver is different in seawater vs. freshwater. Sr incorporation into the otolith of goldfish, *Carassius auratus*, was at first a linear function of ambient Sr (range, 0.02–7.58 ppm), gradually reaching an upper limit with the increased concentration; Sr/Ca ratios showed a similar relationship (Mugiya & Tanaka 1995). Considering their results, it appears that the Sr/Ca ratio of the goldfish otolith would be influenced by the concentration of Sr (0.17 ppm) in the freshwater used in this experiment. It is likely that the Sr incorporation mechanism for the Japanese eel is different from that of the goldfish, but we do not know enough about Sr and Ca incorporation mechanisms to say whether their concentrations are caused by ambient water or food. These topics remain a subject for future research.

Investigators have reported an inverse relationship between temperature and otolith Sr/Ca ratio in the marine fish, *Fundulus heteroclitus* (Radtke 1989), the Atlantic herring, *Clupea harengus* (Townsend et al. 1989, Radtke et al. 1990), and the Atlantic cod, *Gadus morhua* (Townsend et al. 1995). On the other hand, in the goldfish otolith Sr/Ca ratios increased linearly with ambient temperature (Mugiya & Tanaka 1995). Kalish (1989, 1991) found no temperature relationship for otolith Sr/Ca ratio and suggested that seasonal changes in fish physiology can cause an incidental correlation between them. We also found no temperature relationship for eel otolith Sr/Ca ratios in either freshwater or seawater groups (Figure 3). Tzeng (1994) suggested that Sr/Ca ratios in *A. japonica* otoliths were inversely correlated with water temperature, in both laboratory-reared and field-caught eels. He suggested that changes in Sr/Ca ratios towards the edge of the otolith (high-low-high) correspond with a yearly cyclic low-high-low temperature history. However, it is difficult to compare our results with those of Tzeng (1994), because of differences in the rearing conditions. He reared fish under two temperature regimes; natural conditions and 5 °C above natural. Because his eels were not reared under uniform conditions, it is likely that the Sr/Ca ratios in his study fluctuated depending on the sampling time.

Mugiya & Tanaka (1995) reared goldfish at various temperatures ranging from 12 to 28 °C and suggested that the peak Sr/Ca ratio occurred at 24 °C. They also proposed that the optimal temperature for this species may be around 25 °C. Moreover, they further suggested that a difference in environmental salinity may result in the opposite results for freshwater vs. marine fish species. In this study we used four different temperatures ranging from 12 to 27 °C, and found no difference in eel otolith Sr/Ca ratios among them. The reason for this is that the Japanese eel is able to adapt to a wide variety of habitats from the tropics to the subarctic zone (Tesch 1977). Therefore, it is likely that the physiological mechanism of Sr and Ca incorporation into the Japanese eel otolith changed little across the range of temperatures used in this experiment.

In contrast, when reconstructing the environmental history of the Japanese eel using the Sr/Ca ratios, we must take into consideration salinity as an environmental parameter. Additional biochemical studies on the metabolism of Sr and Ca within the eel otolith are necessary.

Acknowledgements

We are indebted to M. Watanabe, the Center of Advanced Instrumental Analysis, Kyushu University for measurement of strontium concentrations on an Inductively Coupled Plasma-Mass Spectrometer and T. Ariga, the Laboratory for Waste Water Treatment, Kyushu University for measurements of calcium concentrations on an Atomic Absorption Photometer.

References cited

Boroughs, H., S.J. Townsley & R.W. Hiatt. 1957. The metabolism of radionuclides by marine organisms. III. The uptake of calcium-45 in solution by marine fish. Limnol. Oceanogr. 2: 28–32.

Deelder, C.L. 1958. On the behaviour of elvers (*Anguilla vulgaris* Turt.) migrating from the sea into fresh water. J. Cons. Int. Explor. Mer. 24: 135–146.

Kalish, J.M. 1989. Otolith microchemistry: validation of the effects of physiology, age and environment on otolith composition. J. Exp. Mar. Biol. Ecol. 132: 151–178.

Kalish, J.M. 1991. Determination of otolith chemistry: seasonal variation in the composition of blood plasma, endolymph and otoliths of bearded rock cod *Pseudophycis barbatus*. Mar. Ecol. Prog. Ser. 74: 137–159.

Kobayashi, J. 1952. Chemical investigation on the water of rivers in Japan. II. On the quality of water in the Kyushu district. Nogaku Kenkyu 41: 27–49 (in Japanese).

Mugiya, Y. & S. Tanaka. 1995. Incorporation of water-borne strontium into otoliths and its turnover in the goldfish *Carassius auratus*: effects of strontium concentrations, temperature, and 17β-Estradiol. Fish. Sci. 61: 29–35.

Radtke, R.L. & T.E. Targett. 1984. Structural and chemical rhythmic patterns in otoliths of the Antarctic fish *Notothenia larseni* and their application to age determination. Polar Biol. 3: 203–210.

Radtke, R.L., R.A. Kinzie. III. & S.D. Folsom. 1988. Age at recruitment of Hawaiian freshwater gobies. Env. Biol. Fish. 23: 205–213.

Radtke, R.L. 1989. Strontium-calcium concentration ratios in fish otoliths as environmental indicators. Comp. Biochem. Physiol. 92A: 189–193.

Radtke, R.L., D.W. Townsend, S.D. Folsom & M.A. Morrison. 1990. Strontium:calcium ratios in otoliths of herring larvae as indicators of environmental histories. Env. Biol. Fish. 27: 51–61.

Radtke, R.L. & D.J. Shafer. 1992. Environmental sensitivity of fish otolith microchemistry. Aust. J. Mar. Freshwater Res. 43: 935–951.

Sadovy, Y. & K. Severin. 1992. Trace elements in biogenic aragonite: correlation of body growth rate and strontium levels in the otoliths of the white grunt, *Haemulon plumieri* (Pisces: Haemulidae). Bull. Mar. Sci. 50: 237–257.

Secor, D.H. 1992. Application of otolith microchemistry analysis to investigate anadromy in Chesapeake Bay striped bass *Morone saxatilis*. U.S. Fish. Bull. 90: 798–806.

Tesch, F.-W. 1977. The eel. Biology and management of anguillid eels. Chapman & Hall, London. 434 pp.

Townsend, D.W., R.L. Radtke, S. Corwin & D.A. Libby. 1992. Strontium:calcium ratios in juvenile Atlantic herring *Clupea harengus* L. otoliths as a function of water temperature. J. Exp. Mar. Biol. Ecol. 160: 131–140.

Townsend, D.W., R.L. Radtke, D.P. Malone & J.P. Wallinga. 1995. Use of otolith strontium:calcium ratios for hindcasting larval cod *Gadus morhua* distributions relative to water masses on Georges Bank. Mar. Ecol. Prog. Ser. 119: 37–44.

Tzeng, W.N. 1994. Temperature effects on the incorporation of strontium in otoliths of Japanese eel *Anguilla japonica*. J. Fish Biol. 45: 1055–1066.

Tzeng, W.N. & Y.C. Tsai. 1994. Change in otolith microchemistry of the Japanese eel, *Anguilla japonica*, during its migration from the ocean to the rivers of Taiwan. J. Fish Biol. 45: 671–683.

Environmental Biology of Fishes **52**: 305–312, 1998.

Growth of dark chub, *Zacco temmincki* (Cyprinidae), with a discussion of sexual size differences

Osamu Katano
National Research Institute of Fisheries Science, 1088, Komaki, Ueda, Nagano 386, Japan
(e-mail: katano@nrifs-u.affrc.go.jp)

Received 4.8.1995 Accepted 3.2.1997

Key words: age, body size, growing season, opercular ring, polygynous mating system

Synopsis

I examined the annual and seasonal growth of dark chub, *Zacco temmincki,* in a Japanese river. Investigation of opercular rings showed that the fish reached a maximum age of 8 years. There was no significant sexual size difference at younger ages (3–5 years), but males were larger than females at older ages (6–7 years). Annual increments of length and weight for males that were recaptured were also larger than those for females. The fish spawn from June to August. Females grew for a short period from April to May, but males on average grew for a longer period from April to August. There was no sexual difference in growth rate except during the spawning period. Annual growth rate was negatively correlated with fish length in each sex. The sexual size differences at older ages of the fish might be due to the polygynous mating system in which most mature males could not obtain females and invested for somatic growth in the spawning period, and a short growing season that was overlapped considerably with the spawning period.

Introduction

Sexual size dimorphism is found in many fishes. Females are larger than males of the same age in a variety of species (e.g., anadromous brown trout, *Salmo trutta*: Berg & Jonsson 1990, yellow perch, *Perca flavescens*: Diana & Salz 1990, flatfish, *Pleuronectes platessa*: Deniel 1990, lingcod, *Ophiodon elongatus:* Smith & McFarlane 1990, greenback mullet, *Liza subviridis*: Al-Daham & Wahab 1991, Australian sharpnose shark, *Rhizoprionodon taylori*: Simpfendorfer 1993). Generally, large females can develop more eggs and thereby achieve higher reproductive success compared with small females, but increase in body size is not necessarily correlated with higher male reproductive success (Parker 1992, Andersson 1994). Therefore, females in many species might have developed larger body size.

On the other hand, males are larger than females in several fishes although the reason for the larger body size is not clear (e.g. Japanese minnow, *Pseudorasbora parva:* Nakamura 1969, gudgeon, *Gobio gobio*: Mann 1980, pumpkinseed, *Lepomis gibbosus*: Deacon & Keast 1987, saddleback darter, *Percina vigil:* Heins & Baker 1989, filefish, *Brachaluteres ulvarum:* Akagawa et al. 1995). Several authors have proposed that polygynous mating systems are associated with the evolutionary development of larger male size (Trivers 1985, Andersson 1994). In polygynous mating systems, there is severe competition among males for access to females, and only a limited number of large, dominant and/or territorial males can monopolize females by repelling small males and/or by being chosen as mates by females. Thus, natural selection should favor large male size (Parker 1992, Andersson 1994).

However, even if large body size is advantageous in intermale competition females can also increase fitness by growing large. Since fish show indeterminate growth, where individuals continue to grow throughout life, fish body sizes are strongly affected by food conditions and environmental factors (Wootton 1990). It is important to know how both females and males invest energy into reproduction because there is a trade-off between present reproductive success and somatic growth in the spawning period (Williams 1966, Schaffer 1974). Thus, it is necessary to investigate sexual differences in reproductive investment and growth patterns in life history to explain sexual size dimorphism in fish.

Dark chub, *Zacco temmincki,* are widely distributed in the rivers and lakes of eastern Asia (Miyadi et al. 1976). In Japan, individuals aggregate in the shallow parts of rivers during the spawning period, which lasts from June to August. During the mating act, a male presses laterally against a female and stirs up the river bed by vibrating his anal fin (Katano 1990b). A single female conducts several matings in a day, repeating one to four mating sequences throughout the entire spawning period (Katano 1992).

The age and growth of dark chub has been reported in a few papers (Maki 1976, Takemoto 1979), but in these studies, sex was not determined and age was determined only by using a size-frequency distribution, presumably resulting in much error in estimating age of older fish. In the present study, I measured the seasonal and yearly growth of dark chub and observed their mating behavior by marking individuals in a river. Here I analyse sexual differences in growth rate and body size, and then estimate energy allocation to reproduction and growth for both sexes.

Methods

The study was conducted along a 114.7 m length of the Kiyotaki River near Kyoto, Japan. In this area, water depth was commonly less than 40 cm, reaching 80 cm in pools. The river bed was a mix of sand, gravel, pebble, and rock.

I captured fish in the study area using baited traps and a hand-net on six occasions: 15–20 May, 9–16 August, 30 October–4 November 1982, 4–9 April, 15–21 May and 9–14 August 1983. I measured standard length (SL) and wet weight of captured fish to the nearest 0.1 cm and 0.1 g, respectively, and recorded individual characteristics such as spots on the body for future identification of individuals. Some small parts of fins except the pectorals were cut in an individually distinct manner. Although cut fins regenerated, their slight deformities enabled individual identification throughout the study period (until August 1983). The number of fish (larger than 4.0 cm in SL) captured and released into the river after marking in each period was 189 in May, 216 in August, 195 in October-November 1982, 119 in April, 281 in May, and 246 in August 1983.

Sex was determined from observations of mating behavior, a ripeness check by gentle abdominal pressure (release of sperm or eggs), or by the development of secondary sexual characters such as tubercles and breeding color (Katano 1990a). I made daily observations through the mating period from 25 May to 16 August 1982 except when the water was turbid (Katano 1990b). Mating activities occurred from 1 June to 13 August in 1982.

Supplementary sampling was conducted on 19 May ($n = 58$) and 1 July 1984 ($n = 69$), and number of opercular rings was counted to estimate age following the methods of Casselman (1983) and Treasurer (1994). Examination of seasonal changes in size distributions of fish smaller than 8.0 cm long has shown that individuals of this species reach 1.5 to 2.5 cm SL at age 1 year, 3.0 to 5.0 cm SL at age 2 years, and that the age of individuals which have a single opercular ring (SL = 6.2–7.1 cm in July) is 3 years (Katano 1993).

I calculated annual increment in standard length and wet weight as L_2-L_1, where L_1 = initial length and weight, L_2 = length and weight after one year, and the specific growth rate (%) as $100 (\log_e L_2-\log_e L_1)/(t_2-t_1)$, where L_1 = length or weight at time t_1, and L_2 = length or weight at time t_2.

Annual growth was calculated for individuals captured in May or August in 1982 and recaptured after one year (in May or August) in 1983. Almost all of these individuals might be mature in the spawning period because the minimum standard length of the fish that reproduced was 7.5 cm in females and 7.0 cm in males (Katano unpublished data).

Results

Age and size

As judged from an examination of opercular rings, some individuals of both sexes reached a maximum age of 8 years (Table 1). The standard length and wet weight of the largest 8-year-old female and male was, respectively, 12.3 cm and 41.0 g, and 15.7 cm and 66.9 g. There was no significant sexual difference in length or weight at ages 3–5 years, but males were significantly larger in SL than females at ages 6 and 7. Males were also heavier than females at age 6.

Seasonal growth pattern

The specific growth rates in both standard length and weight of dark chub were largest from April to May in the pre-spawning period in both sexes (Figure 1). Weight decreased from May to November in females and from August to November in males. From May to August males grew more rapidly than females (Mann-Whitney U-test, length: $z = 5.65$, $p < 0.001$; weight: $z = 5.21$, $p < 0.001$). However, the specific growth rates in both length and weight did not differ between the sexes in other times (Mann-Whitney U-test, $p > 0.05$). The short growing season (April–August) of dark chub was overlapped considerably with the spawning period (May–August).

The specific growth rate in weight per day for males from May to August ranged widely from – 0.23 to 0.62% while that for females ranged from only – 0.47 to 0.09%, indicating extreme individual variation in male growth during the spawning period.

Females exhibited multiple matings; individual females conducted 1 to 4 mating sequences in the course of a spawning period in 1982 (Katano 1992). The specific growth rate was calculated for 10 of 41 females that were captured in both May and August during the spawning period. The specific growth rate was not significantly correlated with the number of mating sequences observed (Kendall's rank correlation test, $n = 10$, length: $tau = -0.111$, $p > 0.05$; weight: $tau = -0.333$, $p > 0.05$).

Annual growth

Annual growth was investigated for 26 females and 22 males. The initial standard length of these individuals ranged from 7.5 to 12.1 cm for females and from 7.0 to 13.3 cm for males (no significant sexual difference, Mann-Whitney U-test, $z = 1.30$, $p > 0.1$), although their ages were not determined. The mean

Table 1. Comparison of standard length (SL) and body weight (BW) between female and male *Z. temmincki*. Samples were collected on 19 May and 1 July 1984. Since mating activities occurred from 1 June to 13 August in 1982 (Katano 1990b), an age of 3 years indicates that the fish lived 3 years ± 86 days. Data for males are from Katano (1993).

Age	Female			Male			Sexual difference (Mann Whitney U-test)	
	n	SL (cm) ($\bar{x} \pm SD$)	BW (g) ($\bar{x} \pm SD$)	n	SL (cm) ($\bar{x} \pm SD$)	BW (g) ($\bar{x} \pm SD$)	SL	BW
3	13	6.5 ± 0.6	4.8 ± 1.5	10	6.6 ± 0.7	5.6 ± 1.7	NS	NS
4	11	8.1 ± 0.5	10.5 ± 2.4	14	8.2 ± 0.5	10.0 ± 2.0	NS	NS
5	16	9.5 ± 0.5	17.0 ± 2.7	21	9.4 ± 0.8	16.3 ± 3.7	NS	NS
6	7	10.5 ± 0.8	23.8 ± 3.6	15	11.4 ± 0.9	28.6 ± 6.0	$U_{cal} = 21.0$*	$U_{cal} = 23.0$*
7	5	10.9 ± 0.7	29.3 ± 5.0	10	12.4 ± 1.1	38.1 ± 10.4	$U_{cal} = 7.5$*	NS
8	4	11.6 ± 0.5	33.0 ± 5.8	1	15.7	66.9	–	–

NS = not significant at the 5% level, * = $p < 0.05$, – = sample sizes were too small to test statistically.

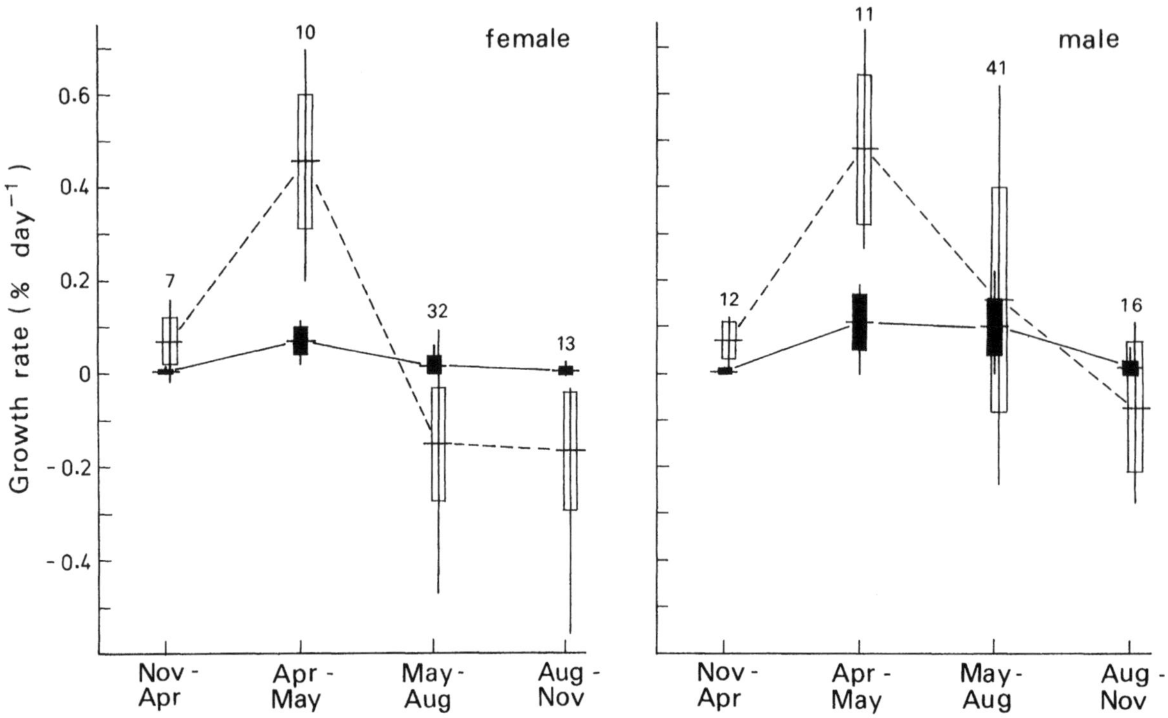

Figure 1. Seasonal changes in specific growth rate in standard length and body weight in each sex of *Z. temmincki.* Numerals indicate sample sizes, vertical lines ranges, horizontal bars means and rectangles standard deviations. Solid rectangles and lines show growth rate in length while open rectangles and broken lines growth rate in weight. Data in 1982 and 1983 are combined for calculation of growth rate from May to August. Data for males are from Katano (1993).

annual increment in length and weight for females was significantly less than those for males (Table 2, Mann-Whitney U-test, length: $z = 4.84$, $p < 0.001$; weight: $z = 4.25$, $p < 0.001$).

The annual increment in female length was negatively correlated with initial female length (Pearson's correlation coefficient, $n = 26$, $r = -0.622$, $p < 0.001$, regression equation is $Y = -0.29X + 3.66$, where Y is annual increment and X is initial length in cm). The annual increment in male length was not correlated with initial length ($p > 0.05$). There were also no significant correlations between annual weight increment and initial length in either sex ($p > 0.05$).

The annual specific growth rate in both length and weight was negatively correlated with initial standard length in each sex (Figure 2). The slopes of the regression equations did not differ significantly between females and males (length: $F_{1,44} = 1.34$, $p > 0.2$; weight: $F_{1,44} = 1.05$, $p > 0.3$ by analysis of covariance). However, the position of the regression line for males was significantly higher than that for females (length: $F_{1,45} = 37.2$, $p < 0.001$; weight: $F_{1,45} =$

Table 2. Mean annual increases in standard length and body weight of *Z. temmincki.*

	n	Initial length (cm)	Annual increments					
			Length (cm)			Weight (g)		
			Mean	SD	Range	Mean	SD	Range
Female	26	7.5–12.1	0.8	0.5	0.2–2.1	4.0	3.7	– 2.2–13.2
Male	22	7.0–13.3	1.8	0.6	0.9–3.1	10.3	5.1	3.4–22.9

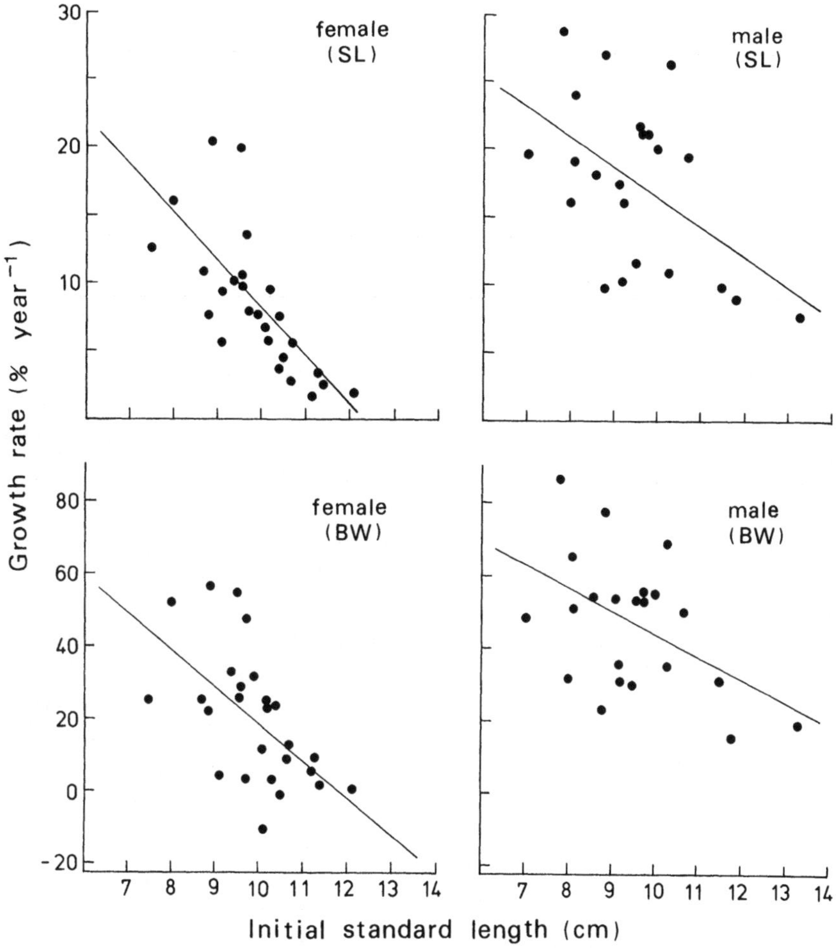

Figure 2. Relationships between initial standard length and annual specific growth rate in standard length (SL) or body weight (BW) in each sex of *Z. temmincki*. Female (SL): r = – 0.719, p < 0.001, Y = – 3.44X + 42.32; male (SL): r = – 0.508, p < 0.02, Y = – 2.17X + 38.09; female (BW): r = – 0.583, p < 0.002, Y = – 10.06X + 119.23, male (BW): r = – 0.475, p < 0.05, Y = – 6.11X + 105.20 (X is initial length and Y is growth rate in regression equations).

27.3, p< 0.001), indicating the smaller annual growth rate in females.

Discussion

Data presented here demonstrate that males were larger than females at older ages in *Z. temmincki*. Annual growth rate in males was also greater than that for females. Fish somatic growth is generally affected by food abundance, social interactions for foraging, reproductive investment and environmental conditions such as water temperature, oxygen and current velocity (Elliot 1975, Brett 1979, Rubenstein 1981, Allen & Wootton 1982, East & Magnan 1987, Kramer 1987, Wootton 1990). Because both females and males inhabited the same reaches of a river with overlapping home ranges (Katano 1987), food abundance and environmental factors might not be important for sexual differences in growth rate.

Thus, sexual size differences might be due to two factors in the life history of the fish, i.e., sexual differences in social status (dominance rank) and reproductive investment. Both females and males exhibited aggressive behaviors against each other in foraging activities (Katano 1985, 1987). In the dominance hierarchy of the fish, (1) larger individuals tended to be more dominant than smaller ones and (2) males were more dominant than females of the same size. However, because foraging success was also affected by individual variations in feeding menu selection (algae, aquatic insects or terrestrial invertebrates), active time and home range use, the growth rate of subordinates was not necessarily less than that of dominants. Furthermore, there were no significant sexual differences in growth rate in non-mating seasons, indicating that dominance order was not directly associated with individual foraging success and sexual size differences.

Males on average grew more rapidly than females in the spawning period from May to August although individual variation in growth rate was extreme in males. In *Z. temmincki,* only a small number of large dominant males can obtain spawning females and participate in pair mating while other smaller males behave as sneakers or do not participate in mating activities. The growth rate of large paired males was low during the mating period, while most other males grew rapidly. Some sneakers or non-participating males participated in mating as paired males in the following year (Katano 1993), indicating that these categories of male mating status were not determined genetically.

On the other hand, almost all females mated one to four times during the spawning period in the study area (Katano 1992). The number of mating sequences in females was not correlated with either their size or body condition in the prespawning period (Katano & Maekawa 1995). It also was not correlated with their growth rate in the spawning period, indicating that females which succeeded in feeding during the spawning period invested more energy in oocyte production than in somatic growth. Thus, individual variation in female growth rate in the spawning period might be small. These sexual differences in reproductive investment might be responsible for sexual size differences of *Z. temmincki* at older ages.

Larger females developed more oocytes than smaller females (Katano 1992). The gonadosomatic index was strongly correlated with female length in the prespawning period (April–May), but this correlation was negative or not significant in males (Katano 1990a). Here, annual growth rate was negatively correlated with fish length for both sexes. Annual increment in female length was also negatively correlated with initial female length. Larger females evidently invested more energy into reproduction rather than somatic growth compared with smaller females. Thus the explanation for smaller growth rates of very large individuals differed between females and males; large males invested more energy in mating behavior and male-male interactions without allocating adequate time for feeding while large females invested more in production and development of oocytes than small females.

In two Spanish cyprinids, *Rutilus lemmingii* and *Chondrostoma polylepis willkommi*, spawning occurred from March to May and growth occurred for several months after the spawning period (Velasco et al. 1990, Herrera & Fernández-Delgado 1994). The annual growth did not differ between the sexes. Although the mating system of these species has not been described, the comparatively long growing season might mask the influence of mating activities on sexual size differences. On the other hand, the growing season of *Z. temmincki* was very short (April to August) and overlapped considerably with the spawning period. Consequently the influence of individual mating activities on annual growth and the development of sexual size differences might be large in *Z. temmincki.*

Maki (1976) reported that in a small tributary of the River Kishigawa *Z. temmincki* increased their length from April to September, but did not grow from October to the following March, although sexual growth differences were not examined. These results coincide with my findings in this study, but it is not evident why the species stops growing from autumn to winter. Dark chub are widely distributed in Japan, Taiwan, Korea and China (Miyadi et al. 1976), and my study area (Kyoto Prefecture: 35° N, 136° E) was near the boundary of its northern distribution in Japan (Toyama Prefecture: 36–37° N, 137–138° E, Tanaka 1993). During winter the fish hide in crevices under rocks and boulders without exhibiting foraging activities. Thus a short growing season in my study area may result from environmental factors such as low water temperature and a lack of suitable food items except in spring and summer.

In pumpkinseed, *Lepomis gibbosus*, males are larger than females only at older ages (Deacon & Keast 1987) as in *Z. temmincki*. Large males show territorial behavior and achieve a high reproductive success, whereas small males behave as sneakers (Gross 1982). In fishes whose mating system is polygynous, the majority of males that cannot obtain females may invest for future reproductive success by increasing somatic growth. Sexual size differences in relation to sexual and intermale differences in growth rate, as reported in this study, should be common in fishes whose mating system is polygynous and when the growing season is short.

Acknowledgements

I thank S. Inoue and A. Horisaka for their help during the field work. Financial aid was provided by a Grant-in-Aid for Scientific Research (No. 01790298) from the Ministry of Education, Science and Culture of Japan.

References cited

Akagawa, I., Y. Tsukamoto & M. Okiyama. 1995. Sexual dimorphism and pair spawning into a sponge by the filefish, *Brachaluteres ulvarum*, with a description of the eggs and larvae. Japan. J. Ichthyol. 41: 397–407.

Al-Daham, N.K. & N.K. Wahab. 1991. Age, growth and reproduction of the greenback mullet, *Liza subviridis* (Valenciennes), in an estuary in Southern Iraq. J. Fish Biol. 38: 81–88.

Allen, J.R.M. & R.J. Wootton. 1982. The effect of ration and temperature on the growth of the three-spined stickleback, *Gasterosteus aculeatus* L. J. Fish Biol. 20: 409–422.

Andersson, M. 1994. Sexual selection. Princeton University Press, Princeton. 599 pp.

Berg, O.K. & B. Jonsson. 1990. Growth and survival rates of the anadromous trout, *Salmo trutta*, from the Vardnes River, northern Norway. Env. Biol. Fish. 29: 145–154.

Brett, J.R. 1979. Environmental factors and growth. pp. 599–675. *In:* W.S. Hoar, D.J. Randall & J.R. Brett (ed.) Fish Physiology, Vol. 8, Academic Press, London.

Casselman, J.M. 1983. Age and growth assessment of fish from their calcified structures – techniques and tools. National Oceanic and Atmospheric Administration, Nat. Mar. Fish. Serv. Technical Report 8: 1–17.

Deacon, L.I. & J.A. Keast. 1987. Patterns of reproduction in two populations of pumpkinseed sunfish, *Lepomis gibbosus*, with differing food resources. Env. Biol. Fish. 19: 281–296.

Deniel, C. 1990. Comparative study of growth of flatfishes on the west coast of Brittany. J. Fish Biol. 37: 149–166.

Diana, J.S. & R. Salz. 1990. Energy storage, growth, and maturation of yellow perch from different locations in Saginaw Bay, Michigan. Trans. Amer. Fish. Soc. 119: 976–984.

East, P. & P. Magnan. 1987. The effect of locomotor activity on the growth of the brook charr, *Salvelinus fontinalis* Mitchill. Can. J. Zool. 65: 843–846.

Elliott, J.M. 1975. The growth rate of brown trout (*Salmo trutta* L.) fed on reduced rations. J. Anim. Ecol. 44: 823–842.

Gross, M.R. 1982. Sneakers, satellites and parentals: polymorphic mating strategies in North American sunfishes. Z. Tierpsychol. 60: 1–26.

Heins, D.C. & J.A. Baker. 1989. Growth, population structure, and reproduction of the percid fish *Percina vigil*. Copeia 1989: 727–736.

Herrera, M. & C. Fernández-Delgado. 1994. The age, growth and reproduction of *Chondrostoma polylepis willkommi* in a seasonal stream in the Guadalquivir River basin (southern Spain). J. Fish Biol. 44: 11–22.

Katano, O. 1985. Aggressive behaviour and dominance relationships of the dark chub, *Zacco temmincki*, with special reference to their individual recognition. Japan. J. Ichthyol. 32: 225–238.

Katano, O. 1987. Social structure of the dark chub, *Zacco temmincki*, in a small pond in relation to individual differences. Physiol. Ecol. Japan 24: 1–132.

Katano, O. 1990a. Seasonal, sexual and individual variations in gonad weight and secondary sexual characters of the dark chub, *Zacco temmincki*. Japan. J. Ichthyol. 37: 246–255.

Katano, O. 1990b. Dynamic relationships between the dominance of male dark chub, *Zacco temmincki*, and their acquisition of females. Anim. Behav. 40: 1018–1034.

Katano, O. 1992. Cannibalism on eggs by dark chub, *Zacco temmincki* (Temminck & Schlegel) (Cyprinidae). J. Fish Biol. 41: 655–661.

Katano, O. 1993. Factors affecting mating status of the male dark chub, *Zacco temmincki*. Physiol. Ecol. Japan 29: 118–131.

Katano, O. & K. Maekawa. 1995. Individual differences in egg cannibalism in female dark chub (Pisces: Cyprinidae). Behaviour 132: 237–253.

Kramer, D.L. 1987. Dissolved oxygen and fish behaviour. Env. Biol. Fish. 18: 81–92.

Maki, I. 1976. Seasonal changes of population numbers, growth and productivity of Kawamutu (*Zacco temmincki*, Pisces, Cyprinidae) in a small tributary of the River Kishigawa, Wakayama. Mem. Fac. Edu., Wakayama Univ., Nat. Sci. 25: 5–14 (in Japanese).

Mann, R.H.K. 1980. The growth and reproductive strategy of the gudgeon, *Gobio gobio* (L.), in two hard-water rivers in southern England. J. Fish Biol. 17: 163–176.

Miyadi, D., H. Kawanabe & N. Mizuno. 1976. Coloured illustrations of freshwater fishes of Japan (2nd ed.). Hoikusha, Osaka. 462 pp. (in Japanese).

Nakamura, M. 1969. Cyprinid fishes of Japan. Research Institute for Natural Resources, Tokyo. 455 pp. (in Japanese).

Parker, G.A. 1992. The evolution of sexual size dimorphism in fish. J. Fish Biol. 41 (Supplement B): 1–20.

Rubenstein, D.I. 1981. Individual variation and competition in the Everglades pygmy sunfish. J. Anim. Ecol. 50: 337–350.

Schaffer, W.M. 1974. Optimal reproductive effort in fluctuating environments. Amer. Nat. 108: 783–790.

Simpfendorfer, C.A. 1993. Age and growth of the Australian sharpnose shark, *Rhizoprionodon taylori*, from north Queensland, Australia. Env. Biol. Fish. 36: 233–241.

Smith, B.D. & G.A. McFarlane. 1990. Growth analysis of Strait of Georgia lingcod by use of length-frequency and length-increment data in combination. Trans. Amer. Fish. Soc. 119: 802–812.

Takemoto, K. 1979. On the population density and the growth of Kawamutu (*Zacco temmincki*, Cyprinidae) in a rivulet, Asegawa, Wakayama. Nanki Seibutsu 21: 72–77 (in Japanese).

Tanaka, S. (ed.) 1993. Fishes in rivers and lakes of Toyama. C.A.P., Toyama. 289 pp. (in Japanese).

Treasurer, J.W. 1994. The distribution, age and growth of wrasse (Labridae) in inshore waters of west Scotland. J. Fish Biol. 44: 905–918.

Trivers, R. 1985. Social evolution. Benjamin Cummings, Menlo Park. 462 pp.

Velasco, J.C., P.A. Rincón & J. Lobón-Cerviá. 1990. Age, growth and reproduction of the cyprinid *Rutilus lemmingii* (Steindachner, 1866) in the River Huebra, Central Spain. J. Fish Biol. 36: 469–480.

Williams, G.C. 1966. Natural selection, the costs of reproduction, and a refinement of Lack's principle. Amer. Nat. 100: 687–690.

Wootton, R.J. 1990. Ecology of teleost fishes. Chapman & Hall, London. 404 pp.

Environmental Biology of Fishes **52**: 313–319, 1998.

Analyses of growth processes of pond smelt, *Hypomesus nipponensis*, in Lake Ogawara, Japan, through the use of daily otolith increments

Satoshi Katayama[1], Michio Omori[1] & Richard L. Radtke[2]
[1] *Department of Applied Bio-science, Faculty of Agriculture, Tohoku University, 1-1 Tsutsumidori Amamiya-machi, Aoba-ku, Sendai 981, Japan*
[2] *University of Hawaii at Manoa, Hawaii Institute of Geophysics, and Planetology School of Ocean and Earth Science and Technology, 2525 Correa Road, Honolulu, Hi 96822, U.S.A. (e-mail: radtke@hawaii.edu)*

Received 19.8.1997 Accepted 24.11.1997

Key words: anadromous, resident, life history

Synopsis

It has been reported that pond smelt, *Hypomesus nipponensis*, in Lake Ogawara, Japan, appear in small and large size groups during spawning despite being an annual fish. It is hypothesized that pond smelt have a bimodal life history, anadromous for large size groups and resident in the lake for small size groups. We calculated the body length and growth rate of the small and large size groups through the use of daily otolith increments and compared growth parameters between groups. In addition, the growth processes of resident fish in Lake Ogawara and anadromous fish in the adjacent sea were examined and compared with small and large size groups, respectively. We found that the two size groups diverged after 40–50 days from hatching with significant size groups present after 50 days. Through otolith increment analysis the growth processes of the small and large size groups were correlated with resident and anadromous forms respectively. These results revealed the utility of otolith increment analyses to clarify population structure of this species.

Introduction

Pond smelt, *Hypomesus nipponensis* McAllister (revised by Saruwatari et al. 1997), are distributed in freshwater, brackish water and coastal seas. Pond smelt are important commercial fish throughout the lakes of Japan. Harvest of pond smelt in Lake Ogawara between the years 1985–1995 reached 500–700 tons a year. Moreover, unlike many other lakes, pond smelt have not been artificially introduced and stocked into Lake Ogawara. Recent ecological research in Lake Ogawara has revealed the coexistence of two size groups (Katayama & Okata 1995, Katayama et al. 1996a,b). Here pond smelt show a bimodal body length distribution during the spawning season from March to May, in spite of being an annual (Bagon et al. 1996) fish. In Lake Ogawara the large size group of pond smelt are present only during the spawning season while the small size group is resident year round (Katayama et al. 1996b). Both size groups breed in the lake and the inflowing rivers (Katayama & Okata 1995).

Seasonal changes in the body length of pond smelt have been examined in many lakes. Body lengths of 0+ fish during spawning season, range widely from 40 mm for Lake Ogawara (Katayama et al. 1996b) to 120 mm for Lake Abashiri (Uto & Sakazaki 1987). Moreover, the variation of body length among year classes is known for several other lakes (Shiraishi 1961). In Lake Ogawara, however, the small and large size groups occur every year within a year class (Katayama et al. 1996b). Large pond smelt are also distributed in the coastal sea adjacent to Hachinohe (Katayama & Kawasaki 1994). Despite the lack of direct evidence, it is hypothesized that the pond smelt have a bimodal life history; anadromy for the larger size group and residency for the smaller group.

It is known for several species of fish that anadromous and resident individuals can coexist in a single population. Salmonids are well known for this type of population. In addition, residency-anadromy has been reported for stickleback, *Gasterosteus aculeatus* (Hagen 1967, Ziuganov et al. 1987, Higuchi et al. 1996) and river-sculpin, *Cottus amblystomopsis* and *C. nozawae* (Goto & Andoh 1990). Heterogeneity in life history and/or morphology provide the maximum fitness of a population facing temporal and geographical changes in their environment (Levins 1962, 1963, Iwasa 1982).

Analyses of otolith annuli are used to clarify the variation of growth rates in a fish population, including stocks exhibiting different growth rates (Hanabuchi 1989, Svenning et al. 1992). However, for annual fishes, such as pond smelt, it is necessary to observe daily increments and examine growth processes that occur sub-annually.

Via the application of daily otolith increments (Radtke & Dean 1982) and comparative growth between size groups we observed the back-calculated body length and growth rate of small and large size groups present during the spawning season in Lake Ogawara. Furthermore, the growth process of sampled resident fish in lake Ogawara and sampled anadromous fish in the adjacent sea were compared between small and large size groups, respectively. We further attempted to clarify whether the groups were resident and anadromous based upon their growth processes as an index.

Methods

Lake Ogawara is a brackish water lake located on the Pacific side of northernmost Honshu, Japan that covers an area of 63.3 km^2 (Figure 1). The northern section of the lake is connected to the Pacific Ocean by the 6-km-long River Takase through which sea water is brought into the lake by tidal flow, making the lake somewhat salty, 0.5—0.6 (Cl^- g l^{-1}) (Sato 1953, Kawasaki & Ito[1]). The southern part of Lake Ogawara has inflow from the Shichinohe and Sadoro rivers.

Samples of pond smelt, n=30 for the small size group and n=29 for the large size group, were collected during the spawning season from the set net fisheries in Lake Ogawara on 10 April 1991. Small and large size groups of the fish in the spawning season have different body length distributions, while their distributions overlap each other. The following classification of these groups are assumed: the fish with a body length of less than 70 mm were considered the small size group and fish of 70 mm and greater were considered the large size group (Katayama & Okata 1995, Katayama et al. 1996b). For

Kawasaki, T. & K. Ito. 1995. Conservation of the environment and intervention by man – a case study on Lake Ogawara. Proceedings of 6th International Conference on the Conservation and Management of Lake Kasumigaura: 196–198.

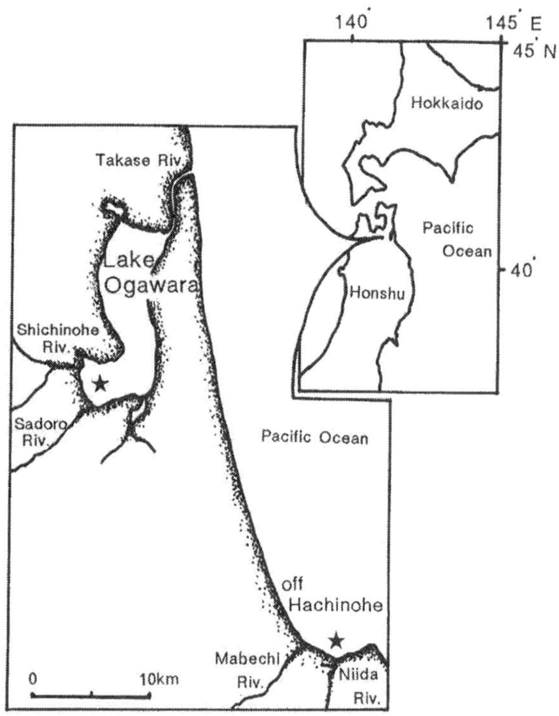

Figure 1. Location of sampling sites (asterisks) on Lake Ogawara and adjacent areas.

comparison, samples of resident pond smelt (n=14) were collected from the set net fisheries outside the spawning season on 2 October 1990 and anadromous fish (n=8) were captured by angling on 23 November 1991 off Hachinohe. The pond smelt which inhabit the waters off Hachinohe are an anadromous stock. They are reported to be present in the sea almost all year round except during the spawning period (Katayama & Kawasaki 1994) and they appear to have come from a separate stock in Lake Ogawara. The samples were frozen for preservation. Standard length (SL) was measured and age was determined by using the otolith phase method (Katayama & Kawasaki 1994). Measurement and analyses were carried out for 0+ aged fish only from Lake Ogawara (year class 1990) and from the sea (year class 1991).

Daily otolith increments of pond smelt have been validated in sagitta by Ohama (1990) and for the lapillus by Okada.[2] We utilized the sagitta due to ease of handling and larger increments. Sagittae were dissected and then cleaned with a 30% sodium hypochlorite solution. After rinsing with freshwater, otoliths were dried and mounted in polyethylene plastic. Otolith width, which corresponds to the short axis of an ellipse, was measured microscopically with reflected light and used for back-calculations. These samples were ground from both sides of the anterior and posterior edges of the otolith using 600–2000 grit grinding paper. The resulting section was 30–40 μm of thickness and revealed a sagittal section through the core region. Daily increment enumerations and radius measurements were made on the transect from the core to the dorsal edge due to enhanced increment readability. Otolith increment measurements were accomplished using an otolith increment analysis system, which consisted of a light microscope, video camera, monitor, and an image analysis system (Ratok System Engineering Inc.) used by the Tohoku National Fisheries Research Institute of Japan. Back-calculated body lengths were estimated by multiplying each radius by the ratio, otolith width/distance from the core to the dorsal edge of the section. The calculated values were applied to regression formulae of otolith width to body length as derived from various sized fish.

Back-calculated body lengths were estimated for 0—150 days with the consideration that the daily periodicity of otolith increments have been validated for pond smelt from 0 to approximately 145 days of age (Ohama 1990). All analysis in this study combined data for males and females, because growth is known not to differ significantly between sexes (Shiraishi 1961).

[2] Okada, K. 1992. Daily increment of the lapillus of pond smelt. Abstract of 1992 Spring Meeting of Japanese Society of Fisheries Science: 365 (in Japanese).

Results

The relationship of body length to otolith width of pond smelt indicated an inflection point at approximately 0.98 mm of otolith width (Figure 2). Data conformed well to linear regression for both ranges, <0.98 mm and ≥0.98 mm, with r^2 values of 0.978 and 0.968, respectively. Using these formulae the body length was back-calculated for each day of age. The distributions of the calculated body lengths on 10 April including small and large size groups, are

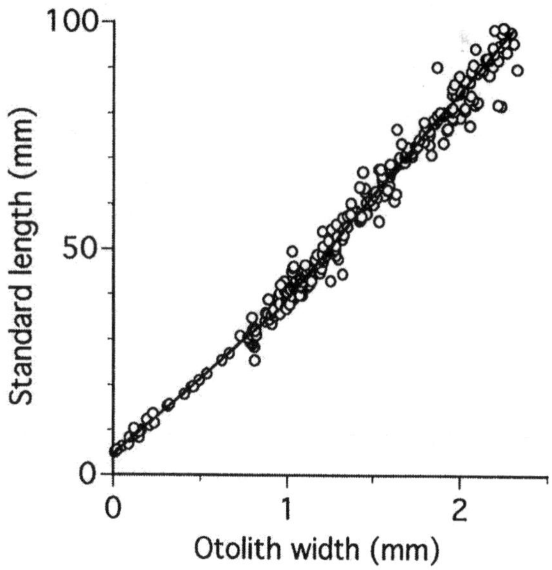

Figure 2. Scaling between standard length (mm) and otolith width (mm) of pond smelt, *Hypomesus nipponensis*. Linear regressions were fit to the range from 0 to 0.98 (linear regression, Y=34.7X+4.33, r^2=0.978) and from 0.98 to 2.45 (Y=43.5X–3.25, r^2=0.968).

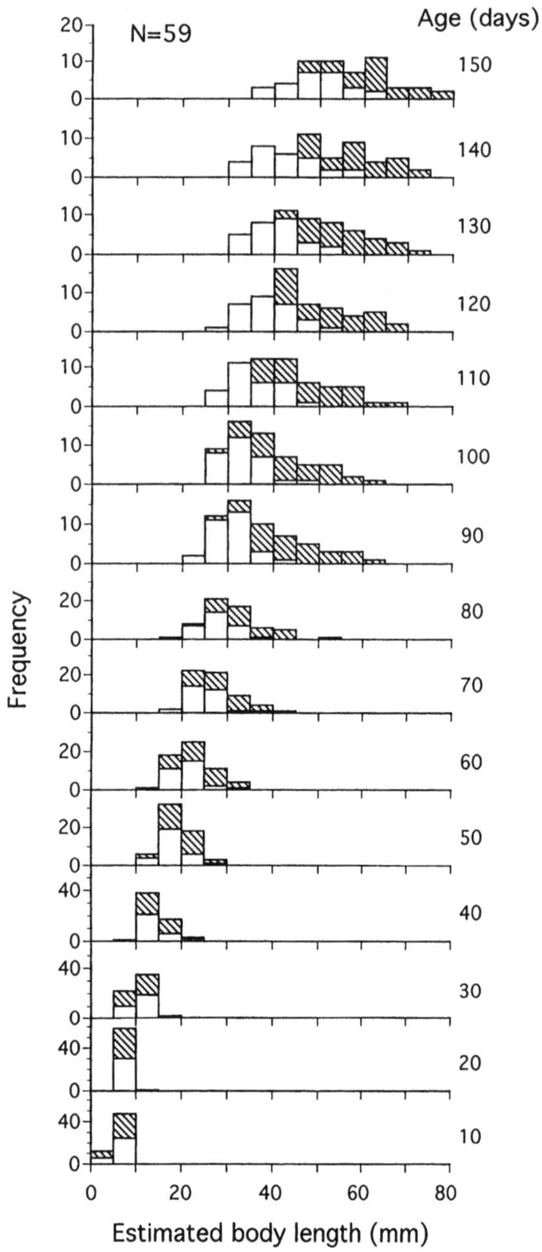

Figure 3. Back-calculated size frequency distributions for every 10 days after hatching for fish <70 mm (open bars) and ≥70 mm (shaded bars).

shown in Figure 3. Body lengths exemplify a normal distribution from 0 to 60 days of age. At 90 days and greater, however, the body length of large size group became larger than that of small size group and the distribution was skewed to a larger size class. The mean body length and mean growth rate for every 10 days was used to compare the growth processes after hatching between the size groups <70 mm and ≥70 mm (Figure 4). Figure 4 also illustrates the results of the t-test on the difference between averages.

Body lengths of the small size group increased gradually. However, body length of the large size group became larger than the small size groups at 40 days of age. The difference between these size groups was significant ($p<0.05$) at 50 days age. The growth rate at 10 days of age for the small size group increased up to 50–60 days of age and then decreased gradually. Conversely, the growth rate of the large size group continued to increase until 60–70 days of age. The difference between average growth rate was significant ($p<0.01$) at 40–50 days.

Growth was compared between the small and large size groups from Lake Ogawara to resident fish and anadromous fish from the adjacent coastal sea off Hachinohe. Growth rates were calculated for body length every 10 days after hatching and were displayed for each comparison in Figure 5. The growth rates of the small size group differed slightly. The growth rates of the large size group were very similar to one another with few differences. It follows from these results that the growth of the small and large size groups were approximately the same as that of the resident and the anadromous fish, respectively.

Discussion

We can recognize from otolith increment analysis that the small and large size groups of pond smelt that appear during the spawning season in Lake Ogawara were resident and anadromous fish, respectively. There was a significant difference in growth in body length at 50 days which was distinct from the growth rate at 40–50 days of age. However, this does not directly imply that anadromous fish migrate downstream to the sea at 40–50 days after hatching. It is unknown whether the greater growth rates associated with migration occur before or after migration to the sea. As a result of living in the sea, anadromous fish grow to a larger maximum size than resident fish. Greater growth of anadro-

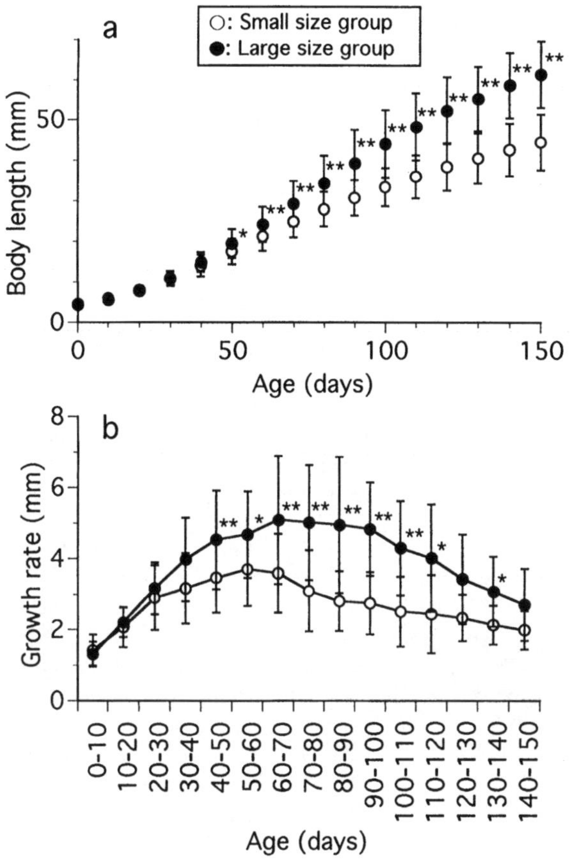

Figure 4. Comparison of size at age (a) and growth rates (b) for every 10 days between the small and large size groups. Bars indicate one standard deviation. * = p<0.05, ** = p<0.01.

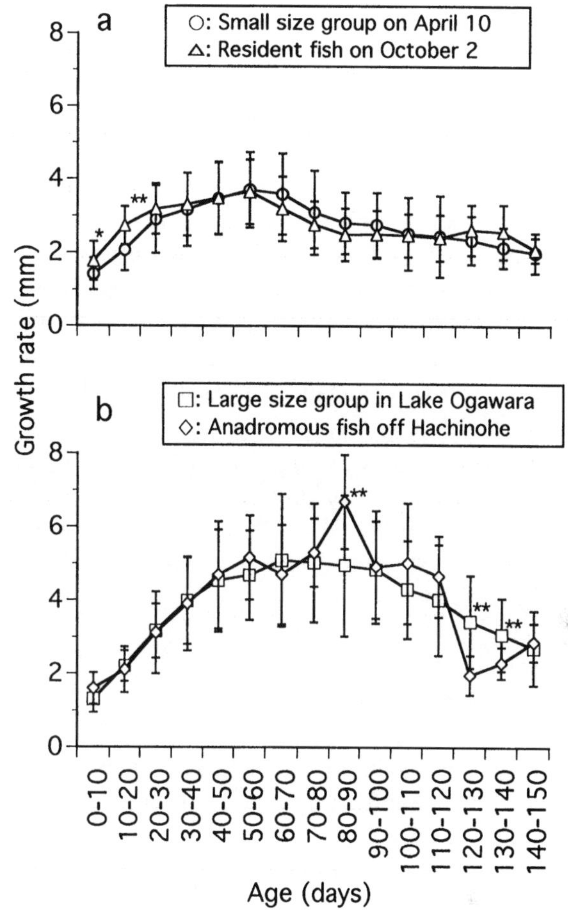

Figure 5. Comparison of growth rates for 10 days: a – between the small size group on 10 April 1991, and resident fish in the lake on 2 October 1990, b – between large size group on 10 April and anadromous fish off Hachinohe on 23 November 1991. Bars indicate one standard deviation (* = p<0.05, ** = p<0.01).

mous fish is known for many fishes: three-spined stickleback, *Gasterosteus aculeatus* (Narver 1969, Mori 1990, Snyder 1991), herring, *Clupea pallasii* (Kanno 1980), 4 species of the genus *Salmo*, 3 species of *Salvelinus* and 1 species of *Oncorhynchus* (reviewed in Hutchings & Morris 1985). This has been interpreted, from an aspect of evolutionary biology, to exceed migration cost and lead to increased fitness (Baker 1978, Gross 1987). However, it would be untrue to say that the greater growth of anadromous pond smelt was caused only by living in seawater. In other fresh water lakes resident pond smelt can be seen to grow as large as the anadromous pond smelt in Lake Ogawara (Shiraishi 1961). Likewise, a negative relationship between fish size and abundance in each year class has been found in several lakes (Shiraishi 1961). Maybe both biotic and abiotic environmental factors caused the growth differences. The growth advantage of anadromous pond smelt can be considered to be caused by such environmental factors as prey abundance and prey types and sizes. Still, it remains unresolved whether the faster growth rates associated with migration occurred before or after immigrating to the sea. This problem can be seen as a mechanism of divergence between the resident and anadromous groups. Divergence mechanisms have been suggested in several fishes. Only the faster growing males become precocious and resident in the Atlantic salmon, *Salmo salar* (Dalley et al. 1983, Myers 1984) and masu salmon, *Oncorhynchus masou* (Kubo 1974, Utoh 1976). In contrast, larger individuals migrate downstream in Arctic charr, *Salvelinus alpinus* (Strand & Heggberget 1994).

Our research revealed the usefulness of otolith increment analyses to clarify population structure in fishes where size groups exhibit differences in growth. Recent progress in analytical techniques of trace elements of fish otolith may make it possible for the nutritional and environmental history of individual fish to be disclosed. An excellent example is the utilization Sr/Ca concentration ratios of otoliths for discriminating anadromous fish from residential fish (Casselman 1982, Radtke et al. 1988, 1996, Kalish 1990, Secor 1992, Rieman et al. 1994). Microchemistry could also be useful in clarifying the timing and size of downstream migrations of anadromous fish and examining the divergence mechanism of resident and anadromous groups in pond smelts.

Acknowledgements

Our special thanks are due to A. Okata and T. Kawasaki for helpful suggestions through our research. The authors wish to express our gratitude to Lake Ogawara Fisheries Cooperative Associations for helping us in sampling of the pond smelt. Gratitude is also express to Y. Ozeki, Central National Fisheries Research Institute of Japan, and D. Kitagawa, Tohoku National Fisheries Research Institute of Japan, for valuable advice on the observation of otoliths and providing the convenience of using the otolith increment analysis system. Thanks are extended to D. Shafer, S. Nunnery and two anonymous reviewers whose editorial comments greatly increased the clarity of this manuscript. Their help was instrumental to the completion of this study. This study was funded in part by National Science Foundation grants DPP-9530081 and DEB-93-22618. This is Hawaii Institute of Geophysics and Planetology contribution No. 980 and SOEST contribution No. 4586.

References cited

Bagon, M., J.H. Harper & C.R. Townsend. 1996. Ecology: individuals, populations and communities, 3rd ed. Blackwell Science, London. 1068 pp.

Baker, R.R. 1978. The evolutionary ecology of animal migration. Holmes and Meier, New York. 1012 pp.

Casselman, J.M. 1982. Chemical analysis of the optically different zones in eel otoliths. pp. 74–82. *In*: K.H Loftus (ed.) Proceedings of the 1980 North American Eel Conference, Ontario Fish. Tech. Rep. 4.

Dalley, E.L., C.W. Andrews & J.M. Green. 1983. Precocious male Atlantic salmon parr (*Salmo salar*) in Insular Newfoundland. Can. J. Fish. Aquat. Sci. 40: 647–652.

Goto, A. & T. Andoh. 1990. Genetic divergence between the sibling species of river-sculpin, *Cottus amblystomopsis* and *C. nozawae*, with special reference to spesiation. Env. Biol. Fish. 2: 257–266.

Gross, M.R. 1987 Evolution of diadromy in fishes. Amer. Fish. Soc. Symposium 1: 14–25.

Hagen, D.W. 1967. Isolating mechanisms in threespine stickleback (Gasterosteus). J. Fish. Res. Board Can. 24: 1637–1692.

Hanabuchi, S. 1989. Age and growth of cutlassfish, *Trichiurus lepturus*, in the Tsushima Waters. Bull. Seikai Reg. Fish. Res. Lab. 67: 37–57 (in Japanese).

Higuchi, M., A. Goto & F. Yamazaki. 1996. Genetic structure of threespine stickleback, *Gasterosteus aculeatus*, Lake Harutori, Japan, with reference to coexisiting anadromous and freshwater forms. Ichthyol. Res. 43: 349—358.

Hutchings, J.A. & D.W. Morris. 1985. The influence of phylogeny, size and behaviour on patterns of covariation in salmonid life histories. Oikos 45: 118–124.

Iwasa, Y. 1982. Why does the polymorphism occur? pp. 87–105. *In*: Y. Ito (ed.) Introduction to Social Ecology, Tokyo University Press, Tokyo (in Japanese).

Kalish, J.M. 1990. Use of otolith microchemistry to distinguish the progeny of sympatric anadromous and non-anadromous salmonids. U.S. Fish. Bull. 88: 657–666.

Kanno, Y. 1980. On the life history of herring, *Clupea pallasii* Cuvier et Valenciennes, in the brackish Lake Notoro, Hokkaido. 2. The relationship between land-locked and migratory herring based on scale analysis. Bull. Fac. Fish. Hokkaido Univ. 31: 175–183 (in Japanese).

Katayama, S. & T. Kawasaki. 1994. Age determination of pond smelt using otolith phase. Tohoku Journal of Agricultural Research 44: 91–106.

Katayama, S. & A. Okata. 1995. Pond smelt spawning in the inflowing river into Lake Ogawara. Tohoku Journal of Agricultural Research 45: 87–102.

Katayama, S., M. Omori & A. Okata. 1996a. Morphometric and reproductive characteristics of the pond smelt population in Lake Ogawara. Tohoku Journal of Agricultural Research 46: 125–140.

Katayama, S., M. Omori & A. Okata. 1996b. Selection on the spawning ground of pond smelt spawners in Lake Ogawara. Kaiyo Monthly 28: 308–314 (in Japanese).

Kubo, T. 1974. Notes on the phase differentiation and smolt transformation of juvenile masu salmon (*Oncorhynchus masou*). Scientific Reports of the Hokkaido Salmon Hatchery 28: 9–26 (in Japanese).

Levins, R. 1962. Theory of fitness in a heterogeneous environment. The fitness set and adaptive function. Amer. Nat. 891: 361–373.

Levins, R. 1963. Theory of fitness in a heterogeneous environment. Developmental flexibility and niche selection. Amer. Nat. 893: 75–90.

Myers, R.A. 1984. Demographic consequences of precocious maturation of Atlantic salmon, *Salmo salar*. Can. J. Fish. Aquat. Sci. 41: 1349–1353.

Mori, S. 1990. Two morphological types in the reproductive stock of three-spined stickleback, *Gasterosteus aculeatus*, in Lake Harutori, Hokkaido Island. Env. Biol. Fish. 27: 21–31.

Narver, D.W. 1969. Phenotypic variation in threespine sticklebacks (*Gasterosteus aculeatus*) of the Chignik River system, Alaska. J. Fish. Res. Board Can. 26: 405–412

Ohama, H. 1990. Age determination of Japanse smelt with daily otolith increment. Nippon Suisan Gakkaishi 56: 1053–1057 (in Japanese).

Radtke, R.L. & J.M. Dean. 1982. Increment formation in the otoliths of embryo, larvae, and juveniles of the mummichog, *Fundulus heteroclitus*. U.S. Fish. Bull. 80: 201–215.

Radtke, R.L., R.A. Kinzie & S.A. Folsom. 1988. Age at recruitment of Hawaiian freshwater gobies. Env. Biol. Fish. 23: 205–213.

Radtke, R.L., M. Svenning, D. Malone, A. Klementsen, J. Ruzicka & D. Fey. 1996. Migration in an extreme northern population of Arctic charr, *Salvelinus alpinus*: insights from otolith microchemistry. Mar. Ecol. Prog. Ser. 136: 13–23.

Rieman, B.E., D.L. Myers & R.L. Nielsen. 1994. Use of otolith microchemistry to discriminate *Onchorhynchus nerka* of resident and anadromous origin. Can. J. Fish. Aquat. Sci. 55: 68–77.

Saruwatari, T., J.A. L—pez & T.W. Pietsch. 1997. A revision of osmerid genus *Hypomesus* Gill (Teleostei: Salmoniformes), with the description of a new species from the southern Kurill Islands. Spesies Diversity 2: 59–82.

Strand, R. & T.G. Heggberget. 1994. Growth and sex distribution in an anadromous population of Arctic char in Northern Norway. Trans. Amer. Fish. Soc. 123: 377–384.

Sato, R. 1953. Biological observation on the Pond smelt, *Hypomesus olidus* (PALLAS), in Lake Ogawara, Aomori Pref., Japan. 3. Annual cycle of ecological elements in relation to production of food organisms of the fish. Tohoku Journal of Agricultural Research 4: 71–90.

Secor, D.H. 1992. Application of otolith microchemistry analysis to investigate anadromy in Chesapeake Bay striped bass *Morone saxatilis*. U.S. Fish. Bull. 90: 798–806.

Shiraishi, Y. 1961. The fisheries biology and population dynamics of pond smelt, *Hypomesus olidus* (Pallas). Bulletin of Freshwater Fisheries Research Laboratory 10: 1–263 (in Japanese).

Snyder, R.J. 1991. Migration and life histories of the threespine stickleback: evidence for adaptive variation in growth rate between populations. Env. Biol. Fish. 31: 381–388.

Svenning, M., A. Smith-Nilsen & M. Jobling. 1992. Sea water migration of arctic charr (*Salvelinus alpinus* L.) – correlation between freshwater growth and seawater migration, based on back-calculation from otoliths. Nordic J. Freshw. Res. 67: 18–26.

Utoh, H. 1976. Study of the mechanism of differentiation between the stream resident form and the seaward migratory form in masu salmon, *Oncorhynchus masou* Brevoot. 1. Growth and sexual maturity of precocious masu salmon parr. Bull. Fac. Fish. Hokkaido Univ. 26: 321–326.

Utoh, H. & S. Sakazaki. 1987. Life history of the smelt, *Hypomesus transpacificus nipponensis* McAllister, in Abashiri. 3. Seaward and upstream migration. Bulletin of Fisheries Experimental Station of Hokkaido 29: 1–16 (in Japanese).

Ziuganov, V.V., G.J. Golovatuk, K.A. Savvaitova & V.F. Bygaev. 1987. Genetically isolated sympatric forms of threespine stickleback, *Gasterosteus aculeatus*, in Lake Azabachije (Kamchatka-peninsula, USSR). Env. Biol. Fish. 18: 241–247.

Environmental Biology of Fishes **52**: 321–329, 1998.

Starvation of drifting goby larvae due to retention of free embryos in upstream reaches

Akihisa Moriyama[1], Yasunobu Yanagisawa, Nobuhiko Mizuno & Koji Omori
Department of Biology, Ehime University, Matsuyama 790-77, Japan (e-mail: yyana@sci.ehime-u.ac.jp)
[1]*Present address: Ocean Research Institute, University of Tokyo, Minamidai, Nakano-ku, Tokyo 164, Japan*

Received 28.9.1995 Accepted 21.1.1997

Key words: amphidromy, diel drift pattern, otolith, *Rhinogobius*, Gobiidae

Synopsis

The embryonic drift pattern of amphidromous *Rhinogobius* species varies along river courses. It peaks soon after sunset in reaches with gentle gradient on the plains, whereas it occurs throughout the day in steep reaches. Examination of the size of the sagittal otolith and notochord length indicated that the age of embryos drifting in steep reaches varied widely, ranging from 0 to 7 days. Many had exhausted their yolk. We attributed the all-day drift in steep reaches to retention of drifting embryos in slack waters or eddies near the banks. In contrast, most embryos drifting in reaches with gentle gradients were recently-hatched individuals. We estimated that in normal or low river flow most embryos from the upstream reaches perish through starvation before they reach the sea.

Introduction

Many freshwater gobies in Japan descend from streams to the sea (or lake) soon after hatching, and ascend streams in the early juvenile period (Dotu 1955, 1961, Dotu & Mito 1955, Mizuno 1960). The embryonic drift of several species is known to peak after sunset at localities near the river mouth (Iguchi & Mizuno 1990). However, Iguchi & Mizuno (1990) found that the diel drift pattern of the common goby *Rhinogobious brunneus* varied along a river course. The drift peaked soon after sunset in reaches with gentle gradient on the plains, whereas it occurred throughout the day in steep reaches upstream. Iguchi & Mizuno (1991) explained this diel variation as follows. Free embryos (see Balon 1990, Balon & Bruton 1994) of *R. brunneus* have poor swimming ability, but show positive phototaxis to weak light of about 500 lux and negative phototaxis to strong light. In steep reaches where valleys, trees or boulders interrupt sunlight, the embryos experience illumination in the daytime that is low enough to trigger movement from the nest. Then strong currents carry them downstream once they leave the nest so embryonic drift occurs all day. On the other hand, along low gradient reaches on the plains, strong ambient light inhibits newly-hatched embryos from leaving the nest in the daytime, and their migration starts at twilight. Similarly, when they reach the well-illuminated lower reaches, embryos drifting from upper courses halt in pools to avoid light in the daytime and resume migration at twilight. Since their nighttime drifting gradually ceases owing to their becoming inactive and gravitating to the bottom, drift peaks soon after sunset.

This hypothesis attaches great importance to insolation as an environmental factor. However, topographical features of water-courses per se may also

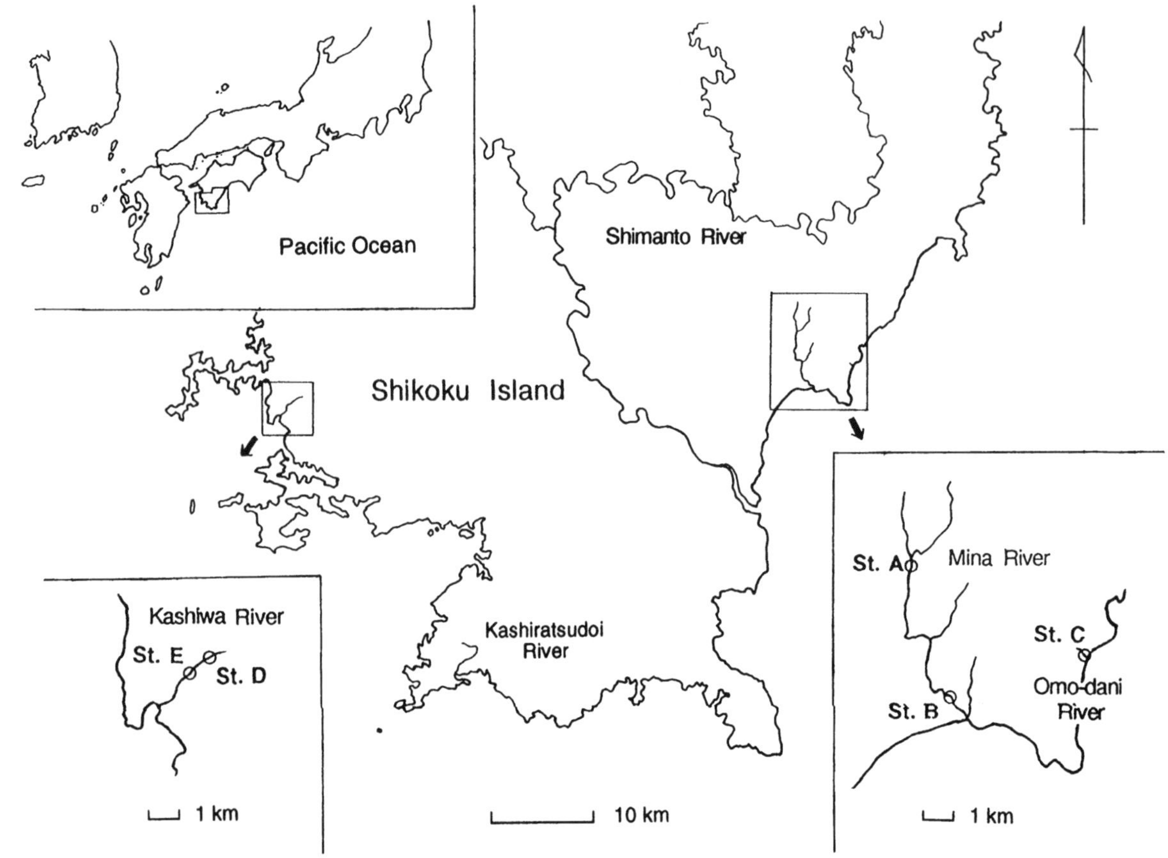

Figure 1. Map of rivers surveyed on Shikoku Island in southwestern Japan.

cause variation in the diel drift pattern. Stream flow, especially in upstream reaches, is complex. A single stretch will include parcels of water each moving at very different speeds (Moss 1980). In steep courses in mountains, which are often classified as Aa-type reaches (Kani 1944, Mizuno & Kawanabe 1981), small falls form vertical eddies in the pool downstream as well as horizontal eddies and slack water near the pool banks. Therefore, embryos detained by these eddies take longer to descend the stream. This is one possible reason why drift of embryos occurs throughout the day in steep reaches.

Table 1. Topographical features and sampling dates for each location.

River	Station	Reach type	Distance from from the river mouth (km)	Altitude (m)	Sampling date
Mina	A	Aa	7.3	91.4	25–26 May 1993
					5– 6 Aug 1993
	B	Bb	0.5	2.6	12 May 1993
					28–29 Jun 1993
Omo-dani	C	Aa	0.1	8.3	16–17 Jun 1993
Kashiwa	D	Aa	2.3	85.0	6– 7 Jul 1994
	E	Aa	1.5	40.0	6– 7 Jul 1994

In this study, we estimated the ages of drifting free embryos of *R. brunneus* from the size of the sagittal otolith, notochord length, and yolk size to examine whether drift is retarded in steep courses. If so, embryos may starve before reaching the sea. To examine this possibility, we also collected drifting embryos near the river mouth and estimated their ages.

Materials and methods

Fish

Rhinogobius brunneus is a species aggregate whose members are sometimes sympatric and genetically independent (Masuda et al. 1989). Four types are distributed in southwestern Shikoku, Japan (Mizuno 1976), where our study was conducted. They are tentatively designated as *Rhinogobius* sp. CB (cross-band type), DA (dark type), LD (large-dark type) and CO (cobalt type) (Kawanabe & Mizuno 1989). We mainly studied *Rhinogobius* sp. DA, which is abundant in steep reaches of small rivers (Mizuno et al. 1979).

Study sites and sampling methods

Drifting free embryos were collected at five sampling stations on three small rivers, the Mina River (12 km long), the Omo-dani River (0.5 km long) and the Kashiwa River (3.2 km long) (Figure 1, Table 1). *Rhinogobius* species in these rivers are distributed as far as 11, 0.2 and 2.7 km upstream of the river mouth, respectively. We determined the reach type of each station following Kani (1944), who defined reach types by the number of riffle/pool sequences in stream segments between meander bends (Type A have more than one riffle/pool sequence between bends, Type B have only one), and the gradient and turbulence of riffles or cascades connecting the pools (Type a are small waterfalls, b are turbulent high-gradient riffles, and c are smooth low-gradient riffles). Thus, a steep reach having two or more plunge pools separated by small falls in each segment between meander bends is defined as Type Aa, whereas a reach of moderate gradient with only one pool and turbulent riffle between bends is defined as Type Bb (see also Mizuno & Kawanabe 1981).

A modified drift net with 0.5 × 0.3 m mouth, 1.0 m length, and 64 μm mesh was set at the central part of a riffle for 10 min periods for collecting embryos. Sampling dates are given in Table 1. Netting was conducted at 19:00, 20:00, 21:00, 1:00, 2:00, 3:00, 7:00, 8:00, 9:00, 13:00, 14:00 and 15:00 h at stations A and C where drifting was observed throughout the day and 19:00, 20:00 and 21:00 hours at station B where drifting occurred exclusively soon after sunset. To confirm all-day drifting, netting was conducted at

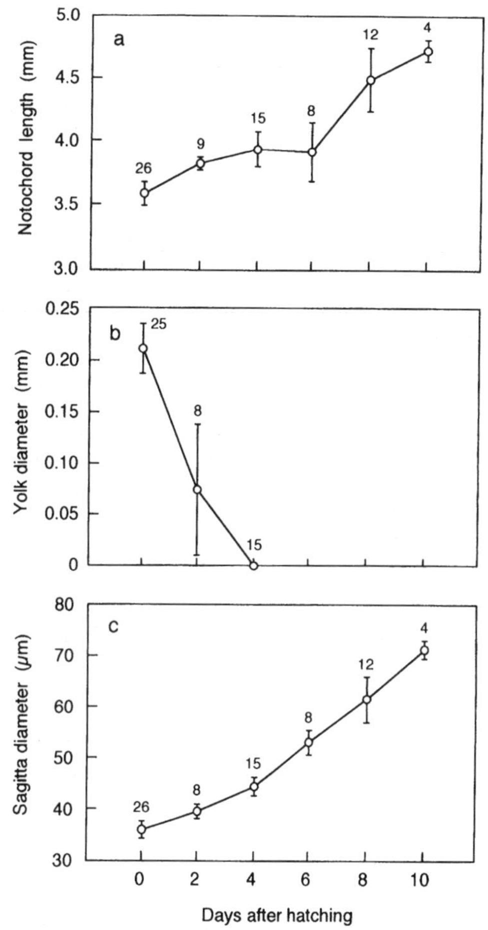

Figure 2. Changes in the notochord length (a), yolk diameter (b), and sagitta diameter (c) of free embryos of *Rhinogobius* sp. DA after hatching. Vertical bars indicate one standard deviation of the mean and numerals show sample size.

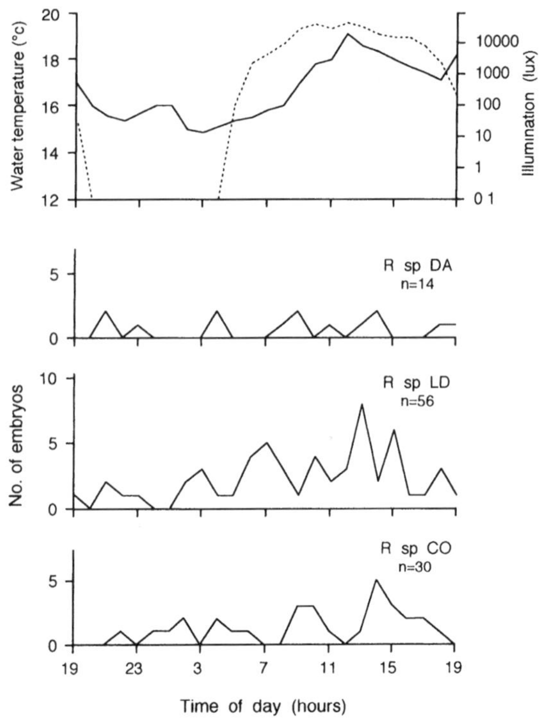

Figure 3. Diel changes in the number of drifting embryos of *Rhinogobius* sp. DA, LD and Co at station A on 25 and 26 May 1993. The average volume of water filtered by the net was 30.2 m^3 per 10 min.

hourly intervals over a 24-hour period at station A. At stations D and E in the Kashiwa River, netting was done simultaneously at 16:00, 20:00, 0:00, 4:00 and 13:00 hours. Additional embryos were collected at other times. At each netting, current velocity, water temperature and intensity of illumination adjacent to the net were measured. Samples were fixed in 95% ethanol immediately after catching for later sorting and measurement in the laboratory.

Age estimation

To estimate the age of drifting embryos, we compared sagitta size, notochord length, and yolk size between embryos caught in the field and those reared in an aquarium. *Rhinogobius* sp. DA from the Kashiratsudoi River (Figure 1) spawned in an aquarium. The hatched embryos were kept in diluted seawater (1 freshwater: 2 seawater) under a 14-hour light (500 lux) and 10-hour dark photoperiod and larvae were fed the rotifer *Brachionus plitilis* ad libitum. Water temperature ranged between 20 and 22° C. Every two days after hatching, 4 to 29 specimens were sampled and fixed in 95% ethanol.

The longitudinal diameter of the yolk was used as an index of yolk size. Specimens whose yolk was not detectable externally were dissected for measurement. The elliptical shape of the sagitta varied among individuals, so the geometric mean of the maximum and minimum diameters was used to estimate size. Daily rings on the sagitta were not used for age determination, because rings in the central part were indistinct.

Results

Embryos reared in the aquarium

Newly-hatched embryos of *Rhinogobius* sp. DA

Table 2. Drifting embryos of *Rhinogobius* species at each station and the proportion of free embryos whose yolk was used up.

Station	Species	Embryos examined	Embryos without yolk (%)
A	DA	15	8 (53.3)
	LD	45	39 (86.7)
	CO	23	15 (65.2)
B	DA	45	2 (4.4)
	CB	53	4 (7.5)
	LD	1	1 (100.0)
	giurinus	14	2 (14.3)
C	DA	18	2 (11.1)
D	DA	29	26 (89.7)
E	DA	27	22 (81.5)

were on average 3.77 mm (SD = 0.080, n = 24) in total length and 3.59 mm (SD = 0.087) in notochord length. The yolk and sagitta diameters were 0.21 mm (SD = 0.024) and 35.4 μm (SD = 1.65), respectively. Notochord length was negatively correlated with yolk diameter (r = – 0.583, p < 0.01), but not significantly correlated with sagitta diameter (r = 0.163, p = 0.46).

Growth of the notochord was retarded between day 4 and day 6 after hatching (Figure 2a); the difference in notochord length was not significant between these two days (t-test, t = 0.324, p = 0.75). Yolk size decreased with time and by day 4 all embryos had used up their yolk (Figure 2b). Sagitta size increased progressively with time (Figure 2c) and sagitta growth followed the Equation

$$\ln S = 3.572 + 0.0707D,$$

where S is sagitta diameter and D is days after hatching. We used this equation to estimate the age of drifting embryos.

Drifting free embryos

In two series of netting at station A, a total of 104 embryos were collected, composed of *Rhinogobius* sp. DA (16.3%), LD (51.9%) and CO (31.7%). Embryos of each species drifted throughout the day without any significant peaks (Figure 3), as was also found by Iguchi & Mizuno (1990). In netting at station B after sunset, 104 embryos collected consisted almost exclusively of *Rhinogobius* sp. CB (60.8%) and *Rhinogobius giurinus* (36.6%). At stations C, D and E, only embryos of *Rhinogobius* sp. DA were collected (Table 2), numbering 28, 29 and 31 individuals, respectively. No diel peaks of drifting were observed at these stations.

Embryos of *Rhinogobius* sp. DA collected at station A were on average 3.67 mm (SD = 0.247, n = 17) in notochord length. About half had no yolk (Table 2), suggesting that they were older than 2 days. Their saggitae ranged from 25.1 to 53.1 μm in diameter (Figure 4), averaging 39.4 μm (SD = 7.55, n = 16). The four smallest sagittae were smaller than the average sagitta (35.6 μm) of newly-hatched embryos in the aquarium, whereas the largest one yielded an age estimate of 6 days.

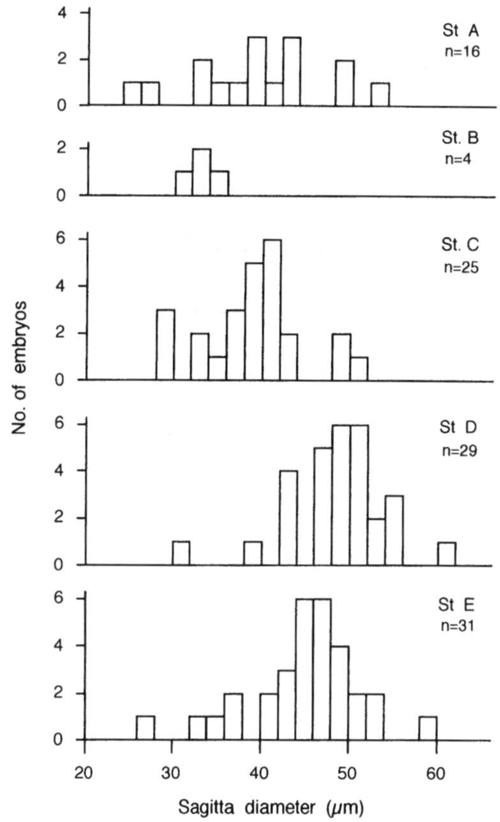

Figure 4. The frequency distribution of sagitta diameter for embryos of *Rhinogobius* sp. DA at each sampling station.

At station B, embryos of *Rhinogobius* sp. DA were 3.87 mm (SD = 0.105, n = 4) in notochord length. Almost all retained the yolk (Table 2). Their saggitae were all smaller (33.0 μm ± 1.29 SD, n = 4) than those of the average newly-hatched embryos in the aquarium and were of similar size (Figure 4). These data indicate that most of these embryos were captured soon after hatching.

At station C, embryos were 3.70 mm (SD = 0.136, n = 28) in notochord length. A few had used up the yolk (Table 2). Their sagittae ranged from 28.2–50.1 μm in diameter (Figure 4), averaging 39.1 μm (SD = 5.68, n = 25). They were estimated to be 0 to 5 days old, with a mode at 2 days.

The development of free embryos collected at stations D and E was very similar. Their mean notochord length was 3.73 mm (SD = 0.147, n = 28)

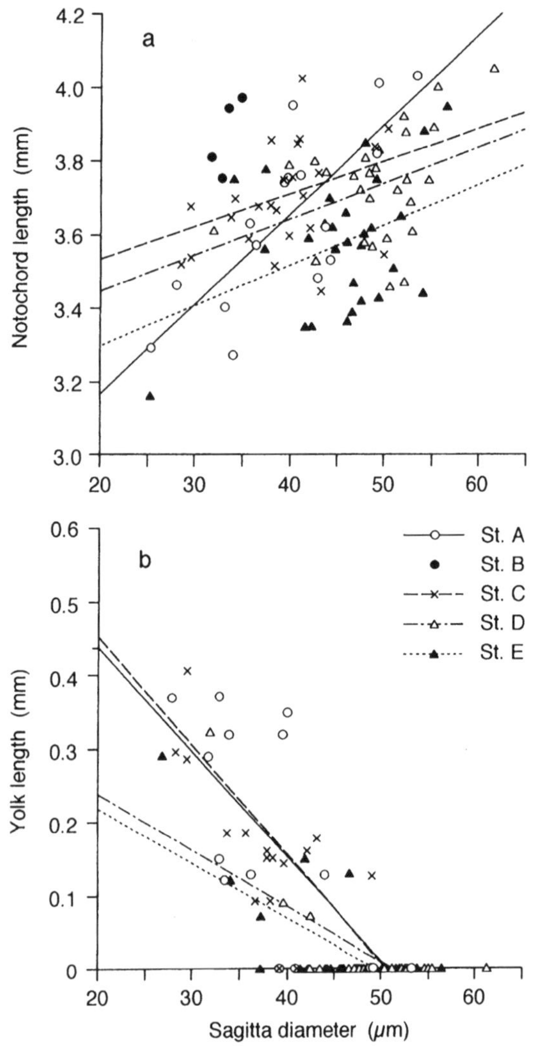

Figure 5. Notochord length (a) and yolk diameter (b) against sagitta diameter in embryos of *Rhinogobius* sp. DA at each sampling station. Notochord length-st. A: r = 0.773, n = 16, p < 0.001; st. C: r = 0.360, n = 24, p = 0.084; st. D: r = 0.384, n = 28, p = 0.044; st. E: r = 0.377, n = 27, p < 0.053. Yolk diameter-st. A: r = – 0.682, n = 12, p < 0.02; st. C: r = – 0.668, n = 16, p < 0.01; st. D: r = – 0.692, p < 0.001; st. E: r = – 0.688, n = 26, p < 0.001.

and 3.58 mm (SD = 0.184, n = 31), respectively, and more than 80% had no yolk (Table 2). The sagitta ranged from 31.8 to 61.2 μm in diameter with an average of 48.7 μm (SD = 5.57, n = 28) at station D, and ranged from 26.7 to 56.3 μm with an average of 45.0 μm (SD = 6.52, n = 31) at station E (Figure 4). Embryos at station D were estimated to be 0 to 7 days old with a mode at 5 days, and those at station E were 0 to 6 days old with a mode at 4 days.

At each sampling station, notochord length of embryos was positively correlated with sagitta size (Figure 5a). The relationship between these parameters, however, varied considerably among sampling stations (ANCOVA, p < 0.001). The yolk size was negatively correlated with sagitta size (Figure 5b). No embryos estimated as 4 days old or more from sagitta size had yolk.

Embryos of other species were generally at similar developmental stages to those of *Rhinogobius* sp. DA collected at the same station (Table 2). Most embryos of *Rhinogobius* sp. LD and CO at station A had no yolk. In contrast, most embryos of *Rhinogobius* sp. CB and *R. giurinus* at station B had not consumed their yolk.

Discussion

Age estimate

For the age estimate of drifting embryos, we used a growth curve based on embryos of known age reared in the aquarium. Growth of the notochord was retarded between day 4 and day 6 after hatching. This retardation may be due to the shift from endogenous to exogenous feeding that occurred slightly earlier than day 4. In contrast, sagitta size increased nearly linearly with time. It is known that growth of the otoliths is little influenced by starvation (Campana & Neilson 1985). Starved fishes continue to increase otolith size even after cessation of linear body growth (Brother 1981, Marshall & Parker 1982, Campana 1983a). Otolith growth is more or less independent of somatic growth, because otoliths can enlarge by means of calcium uptake from the aquatic environment (Campana 1983b). Therefore, otolith size may be a more accurate index of embryonic age than notochord length.

The development of embryos is influenced by environmental factors. Egg size within the same species varies geographically and seasonally and the time from fertilization to hatching can vary greatly in relation to ambient water temperature (e.g., Blaxter & Hempel 1963, Ware 1975, Iguchi & Yamaguchi 1994). Accordingly, otolith size at hatching may vary under different environmental conditions. It is known in the ayu, *Plecoglossus altivelis*, that high water temperature shortens the time from fertilization to hatching, resulting in embryos at hatching having smaller otoliths (Tsukamoto & Kajihara 1987). In the present study, some drifting free embryos of *Rhinogobius* sp. DA had smaller sagittae than did newly-hatched embryos reared in the aquarium. This difference probably cannot be attributed only to water temperature, because no consistent relations were found between sagittal size of newly-hatched embryos and water temperature at the sampling stations and in the laboratory. To determine embryonic age more accurately, a detailed investigation of variation in otolith size and the influence of water temperature and other environmental factors on otolith growth is needed.

Retention of drifting embryos

Our results confirmed the finding of Iguchi & Mizuno (1990) that the diel drift pattern of goby free embryos is related to the reach type: drift peaked soon after sunset in the Bb-type reach (station B), whereas it occurred throughout the day in the Aa-type reach (stations A, C, D and E). Most embryos collected in the Bb-type reach had recently hatched, whereas in the Aa-type reach, the age of embryos ranged widely and many had exhausted their yolk. Therefore, variation in the diel drift pattern may be explained by a difference the time embryos take to descend a river. The wide age variation of drifting embryos in the Aa-type reach would not result only from their behavioural responses to ambient light (Iguchi & Mizuno 1991).

Retention of drift in the Aa-type reach may be attributed to its topographical features, although the empirical data have not been obtained. As noted earlier, in the Aa-type reach, small falls form vertical eddies in pools and slack or eddying water at a number of places near the banks. Since free embryos of *R. brunneus* have poor swimming ability (Iguchi & Mizuno 1991), their movements may be passive. Once embryos stray into slack or eddying water, they can leave there only by chance. Therefore, we suspect this to be the primary reason for retention.

The modal age of drifting embryos differed among sampling stations in the Aa-type reach (Figure 4). Embryos at station C were younger than those of other stations. This is probably due to the narrow distribution range of *Rhinogobius* sp. DA and the short length of the Omo-dani River, resulting in low retention of embryos.

Starvation of drifting embryos

For free embryos of amphidromous fishes, the time taken to descend to their feeding site (the sea or lake) is critical (Balon & Bruton 1994). The present study showed that many drifting embryos of *Rhinogobius* sp. DA, LD and CO in the Aa-type reach had exhausted their yolk. Despite this, in a station near the river mouth (station B), embryos of these species mostly retained their yolk. In the Shimanto River (196 km long), the longest river in Shikoku, *Rhinogobius* species are distributed as far as 120 km upstream of the river mouth. A body shape analysis of *Rhinogobius* sp. CB embryos collected in the Bb-type reach of this river indicates that many of them are starved (K. Iguchi personal communication). These data indicate that most goby free embryos which descended from the upstream reaches died on the way. High mortality of drifting embryos is also reported in another Japanese amphidromous fish *Plecoglossus altivelis* and their death has been attributed to starvation and predation (Tsukamoto 1991).

In some amphidromous fishes such as *P. altivelis* and *Cottus hangiogensis*, mature adults migrate downstream in the breeding season (Shiraishi & Suzuki 1962, Goto 1986, 1988). This behaviour may serve to decrease the mortality of drifting embryos (Goto 1986). In *Rhinogobius* species, however, no such downward migration has been observed. On the contrary, *Rhinogobius* sp. CB and LD move upstream as they grow (Tamada 1986, personal communication).

This leads us to ask why *Rhinogobius* species spawn in the steep courses at high risk of starvation of their offspring. The answer probably lies in the fluctuation of flow level. Netting in this study was conducted on low or normal flow days. Iguchi & Mizuno (1990) showed that, at a site below a weir in the middle course, free embryos of *Rhinogobius* sp. CB rarely drift in the daytime in low flow but do so abundantly in high flow. Even in Bb-type reaches on the plains, embryos of *Rhinogobius* species drift abundantly in the daytime on high flow days after heavy rain (H. Yuge unpublished data). These observations suggest that retardation of drift does not occur in high flow. The breeding season of *Rhinogobius* species (May–July) includes the rainy season when the river level often rises. If hatching coincides with high flow, the loss of drifting embryos through retention will be minimal. However, it is unlikely that gobies can adjust their spawning activity such that their embryos hatch during high flow, since the fluctuation of flow level is sudden and unpredictable.

It is known that individuals of *Rhinogobius* sp. CB inhabiting upper courses are larger than those of the same age inhabiting lower courses (Tamada 1986). It remains to be investigated if growth and/or survival advantages accruing to gobies inhabiting upper courses is sufficient to compensate for the high mortality of their offspring.

Acknowledgements

We are indebted to K. Iguchi and A. Rossiter for constructive comments on the manuscript. We are also grateful to K. Tsukamoto for advice on the otolith analysis, to T. Shimizu for help in rearing embryos in the aquarium, and to T. Osugi and Y. Tamura for helping to collect fish.

References cited

Balon, E.K. 1990. Epigenesis of an epigeneticist: the development of some alternative concepts on the early ontogeny and evolution of fishes. Guelph Ichthyol. Rev. 1: 1–42.

Balon, E.K. & M.N. Bruton. 1994. Fishes of the Tatinga River, Comoros, with comments on freshwater amphidromy in the goby *Sicyopterus lagocephalus*. Ichthyol. Explor. Freshwaters 5: 25–40.

Blaxter, J.H.S. & G. Hempel. 1963. The influence of egg size on herring larvae (*Clupea harengus* L.). J. Cons. Perm. Int. Explor. Mer. 28: 211–240.

Brothers, E.B. 1981. What can otolith microstructure tell us about daily and subdaily events in the early life history of fish? Rapp. P-V. Reun. Cons. Int. Explor. Mer. 178: 393–394.

Campana, S.E. 1983a. Feeding periodicity and the production of daily growth increments in the otoliths of steelhead trout (*Salmo gairdneri*) and starry flounder (*Platichthys stellatus*). Can. J. Zool. 61: 1591–1597.

Campana, S.E. 1983b. Calcium deposition and otolith check formation during periods of stress in coho salmon, *Oncorhynchus kisutch*. Comp. Biochem. Physiol. 75A: 215–300.

Campana, S.E. & J.D. Neilson. 1985. Microstructure of fish otoliths. Can. J. Fish. Aquat. Sci. 42: 1014–1032.

Dotu, Y. 1955. The life history of goby, *Chaenogobius urotaenia* (Hilgendorf). Sci. Bull. Agr. Kyushu Univ. 15: 367–374. (in Japanese).

Dotu, Y. 1961. The binomics and life history of the gobioid fish, *Rhinogobius giurinus* (Rutter). Bull. Fac. Fish. Nagasaki Univ. 10: 120–126. (in Japanese).

Dotu, Y. & S. Mito. 1955. Life history of a gobioid fish, *Sicydium japonicus* Tanaka. Sci. Bull. Agr. Kyushu Univ. 15: 213–221. (in Japanese).

Goto, A. 1986. Movement and population size of the river sculpin, *Cottus hangiongensis* in the Daitobetsu River of southern Hokkaido. Japan. J. Ichthyol. 32: 421–430.

Goto, A. 1988. Reproductive behavior and homing after downstream spawning migration in the river sculpin, *Cottus hangiongensis*. Japan. J. Ichthyol. 34: 488–496.

Iguchi, K. & N. Mizuno. 1990. Diel changes of larval drift among amphidromous gobies in Japan, especially *Rhinogobius brunneus*. J. Fish Biol. 37: 255–264.

Iguchi, K. & N. Mizuno. 1991. Mechanisms of embryonic drift in the amphidromous goby, *Rhinogobius brunneus*. Env. Biol. Fish. 31: 295–300.

Iguchi, K. & M. Yamaguchi. 1994. Adaptive significance of inter and intra-populational egg size variation in ayu *Plecoglossus altivelis* (Osmeridae). Copeia 1994: 184–190.

Kani, T. 1944. Ecology of torrent-inhabiting insects. pp. 171–317. *In:* H. Furukawa (ed.) Insects 1, Kenkyu-sha Press, Tokyo. (in Japanese).

Kawanabe, H. & N. Mizuno (ed.). 1989. Freshwater fishes of Japan. Yamatokeikoku-sha Press, Tokyo. 719 pp. (in Japanese).

Marshall, S.L. & S.S. Parker. 1982. Pattern identification in the microstructure of sockeye salmon (*Oncorhynchus nerka*) otoliths. Can. J. Fish. Aquat.Sci. 39: 542–547.

Masuda, Y., T. Ozawa & S. Enami. 1989. Genetic differentiation among eight color types of the freshwater goby, *Rhinogobius brunneus*, from western Japan. Japan. J. Ichthyol. 36: 30–41.

Mizuno, N. 1960. Study on a freshwater goby, *Rhinogobius similus* Gill, with a proposition on the relationships between landlocking and speciation of some freshwater gobies in Japan. Mem. Coll. Sci. Univ. Kyoto Ser. B 27: 97–115.

Mizuno, N. 1976. Study on a freshwater goby, *Rhinogobius brunneus* (Pisces : Gobiidae), III. Distribution of four colour types in Shikoku and Kyushu Islands. Physiol. Ecol. Japan 17: 373–381. (in Japanese).

Mizuno, N. & H. Kawanabe. 1981. A topographical classification of streams, with an introduction of the system widely used in Japan. I. Reach type, stream zone and stream type. Verh. internat. Verein Limnol. 21: 913.

Mizuno, N., S. Uehara & M. Maki. 1979. Studies on the freshwater fish, *Rhinogobius brunneus* (Gobiidae), IV. Habitat segregation among sympatric populations of 4 colour types. Japan. J. Ecol. 29: 137–147. (in Japanese).

Moss, B. 1980. Ecology of fresh waters. Blackwell, Oxford. 332 pp.

Shiraishi, Y. & N. Suzuki. 1962. The spawning activity of ayu-fish, *Plecoglossus altivelis*. Bull. Freshw. Fish. Res. Lab. 12: 84–107. (in Japanese).

Tamada, K. 1986. Size variations of a freshwater goby, *Rhinogobius brunneus*, crossband type, along the river length in terms of age and growth. The Nanki Seibutu 28: 23-34. (in Japanese).

Tsukamoto, K. 1991. Age and growth of ayu larvae *Plecoglossus altivelis* collected in the Nagara, Kiso and Tone River during the downstream migration. Nippon Suisan Gakkaishi 57: 2013–2022. (in Japanese).

Tsukamoto, K. & T. Kajihara. 1987. Age determination of ayu with otolith. Nippon Suisan Gakkaishi 53: 1985–1997.

Ware, D.M. 1975. Relation between egg size, growth, and natural mortality of larval fish. J. Fish. Res. Board Can. 32: 2503–2512.

Environmental Biology of Fishes **52**: 331–343, 1998.

Feeding of a benthic goby in a river where nektonic fishes are absent

Tomonori Osugi, Yasunobu Yanagisawa & Nobuhiko Mizuno
Department of Biology, Ehime University, Matsuyama 790-77, Japan (e-mail: yyana@sci.ehime-u.ac.jp)

Received 18.10.1995 Accepted 17.1.1997

Key words: feeding behaviour, ecological release, social structure, river flow, invertebrate drift, *Rhinogobius*, Gobiidae, Pisces

Synopsis

The social structure and feeding behaviour of the benthic goby *Rhinogobius* sp. DA (dark type) were investigated in a small river in Shikoku, Japan, where nektonic fishes are absent. The gobies had overlapped home ranges but were often aggressive against conspecifics, especially of the same sex and similar body size. Their diet consisted mainly of aquatic invertebrates, but terrestrial invertebrates were also an important item. This contrasted with few terrestrial invertebrates in the diet of the conspecifics in an adjacent river harbouring a nektonic cyprinid fish. An increase in flow level greatly changed their feeding behaviour; on high flow days, a greater portion of the population appeared from the interstices of the bottom and more often rose into the water column to catch drift than on normal flow days. Plasticity of their feeding behaviour is discussed in terms of ecological release.

Introduction

Most fishes living in Japanese rivers are euryphagous (Kawanabe 1969). They generally exploit a variety of food items and can shift their food habits according to ecological and environmental changes. Their euryphagy is attributed primarily to instability of fluvial environments (Kawanabe 1959). Rivers in Japan are generally short and in steep gradients compared with continental rivers. There, rainfall may suddenly elevate the water level, resulting in the change of food conditions for fishes. Specialization in food items may be maladaptive under such conditions.

Rhinogobius species are benthic gobies that inhabit almost all courses of Japanese rivers. They are omnivorous, eating algae and aquatic insect larvae on the bottom (Kodama 1961, Sawara 1978). Kawanabe (1973) found plasticity of their feeding behaviour at Okinawa Island; in the upper stream above a big falls where nektonic fishes are absent, they fed on fallen and drifting terrestrial insects, by rising into the water column. After his finding, however, no detailed studies have been conducted on the feeding ecology of *Rhinogobius* species in relation to the presence or absence of nektonic fishes.

In the present study, we investigated the feeding ecology of *Rhinogobius* sp. DA (dark type) (Kawanabe & Mizuno 1989) in a small river where nektonic fishes are absent. Food habits of this population were compared with those of another population in a river where nektonic fishes are present. Particular attention was paid to how the fluctuation of river flow influences the feeding behaviour of the goby.

Material and methods

Study sites and sampling

Observations and sampling were conducted at the Kashiwa River (3.2 km long), southwestern Shikoku, Japan. The upper courses of this river are in steep gradients and 2–4 m across. In the middle courses, there are many weirs for agricultural use and dense patches of a ditch reed. The river dries up in the lower courses from 0.3 to 1.1 km above the river mouth except during and for a brief period following heavy rain. Fish fauna in the middle courses, where the observation site was set, consisted of only three benthic fishes, two gobies *Rhinogobius* sp. DA, *Sicyopterus japonicus* and an eel *Anguilla japonica*. Specimens of *Rhinogobius* sp. DA were collected near the observation site both on normal flow days (8, 13 and 14 September 1994) and on a high flow day (30 September 1994).

Sampling was also conducted at the Kiku River (6.0 km long), 5 km south of the Kashiwa River. This river is dammed by a debris barrier 3.3 km above the river mouth. Above the barrier, *Rhinogobius* sp. DA was landlocked and a nektonic cyprinid fish *Zacco temmincki* was distributed abundantly together with two other benthic gobies *Rhinogobius* sp. CO (cobalt type) and *S. japonicus*. Specimens of *Rhinogobius* sp. DA and *Z. temmincki* were collected at a site 0.8 km above the barrier on normal flow days (24–27 September 1994).

Collected specimens were immediately fixed in 20% formalin and then preserved in 10% formalin. After measurements of total length (TL) and wet weight, the stomach (the anterior part of the gut just before the first bend) was dissected out and contents sorted and identified to order. The number of each item was recorded and its volume percentage of the contents was estimated visually under a binocular microscope. Then the total contents and the other body parts were dried (60° C, > 24 h) and weighed separately, and the stomach fullness [(dry weight of contents)/(total dry body weight) × 1000] was calculated. The weight of each item was estimated as the dry weight of the total contents multiplied by its volume percentage.

Drift was quantitatively measured both on normal days (8 times at the Kashiwa and once at the Kiku) and on high flow days (3 times at the Kashiwa) by setting a drift net with 0.5 × 0.3 m mouth and 1.0 m length at the head of the pool for 10 min. The volume of water that passed through the net was calculated from the submerged area of the net mouth and the current velocity at the centre of the net mouth submerged. Animal drift collected was sorted and identified to order. The number and weight of each item were measured in the same way as the stomach contents.

Census and observation

The behaviour of *Rhinogobius* sp. DA was observed at a 3 × 4 m quadrat in a pool of the Kashiwa River (Figure 1) during the period from 12 July to 17 September 1994, when its breeding activity had finished. The quadrat was divided into 50 × 50 cm grids whose corners were marked by stones painted white. The water level fluctuated greatly during the observation period (Figure 2); it was rapidly elevated by rain and lowered soon after rain. Based on the water level, the observation period was divided into normal flow days and high flow days. The water current velocity and direction at the water surface were measured at each 25 × 25 cm subgrid on a normal flow day (16 July) and on a high flow day (21 August) (Figure 3). The head of the pool was defined as an area within 1 m of the inlet from the upper rapid where the current velocity was > 5 cm sec^{-1} on the normal flow day.

Before observations and daily censuses, 67 gobies caught in the quadrat were measured in total length, sexed by the elongate first dorsal fin in males, and marked individually by injecting acrylic paint under the skin. In 31 daily censuses, the positions of marked gobies were plotted on a plastic sheet of the quadrat map and classified either as in the open or in the shade. Staying on the upper surfaces of substrates was regarded as being in the open and staying on the lateral side or underside of a stone or rock as being in the shade. Many members of the marked gobies were missed during each census; they were judged to have hidden themselves

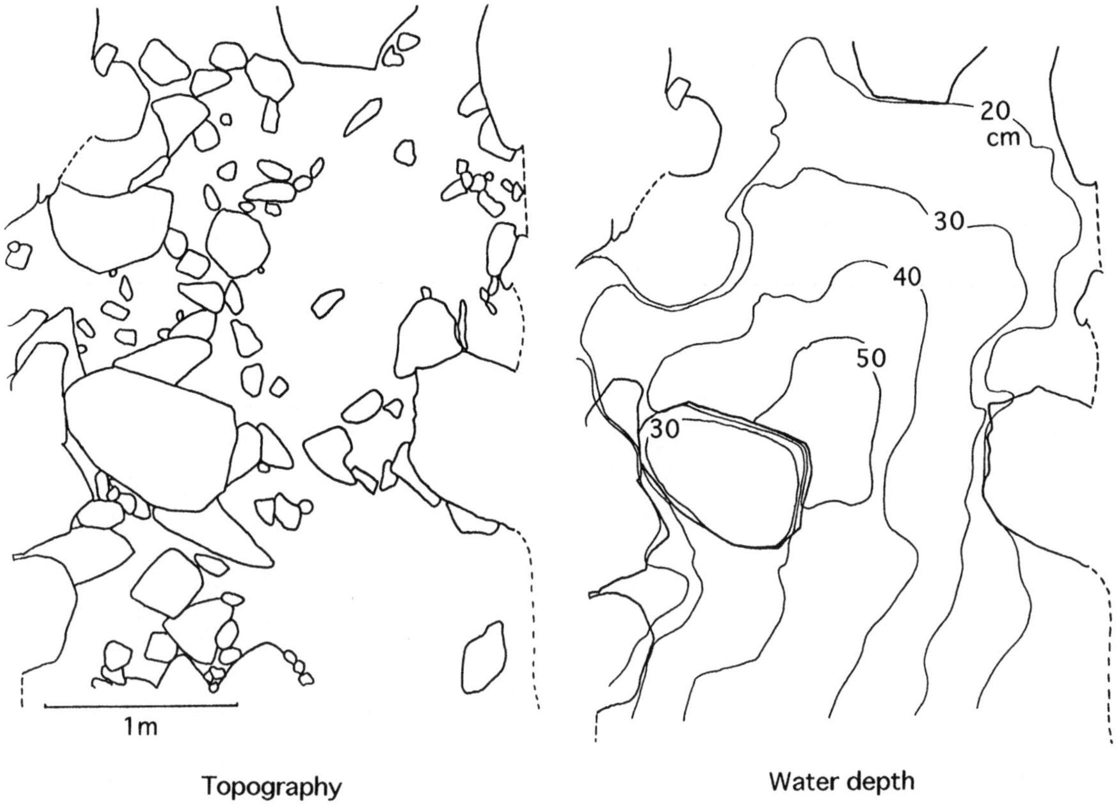

Figure 1. Topography of the quadrat and water depth on a normal flow day (16 July).

in the interstices of the bottom, unless they had emigrated elsewhere.

In four censuses, the number of unmarked gobies in the quadrat was also recorded. From the ratio between marked and unmarked gobies, 62% of gobies were estimated to be marked.

After daily censuses, the behaviour of each randomly selected marked goby was observed for 10 min. During the observation, their foraging tracks were drawn on the map and the positions of the following three feeding actions were recorded each time they occurred:

Looking = looking at the bottom to find food. This action was distinguishable from other actions by a downward or sideward head shaking involving an ocular movement.

Picking = picking at benthic organisms. This action was usually preceded by looking.

Rising = rising up to 40 cm, usually 5–20 cm, into the water column to take drifting particles.

At the same time, their interactions with conspecifics were also recorded. Interactive behaviour was categorized into 12 actions (Table 1) and the interac-

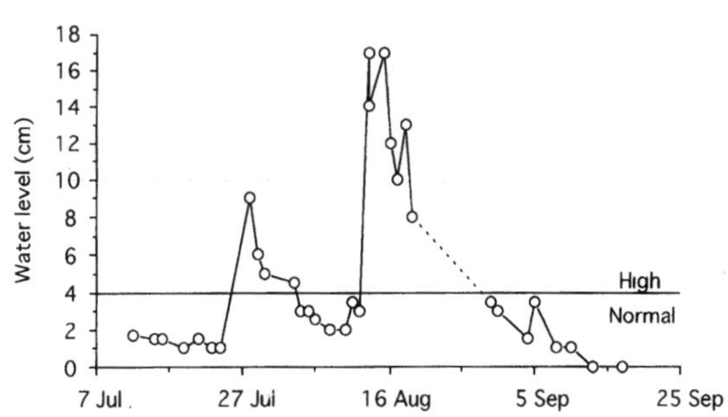

Figure 2. Changes of water level during the study period. The lowest water level (17 September) was set at 0 cm and days > 4 cm high were classed as high flow days.

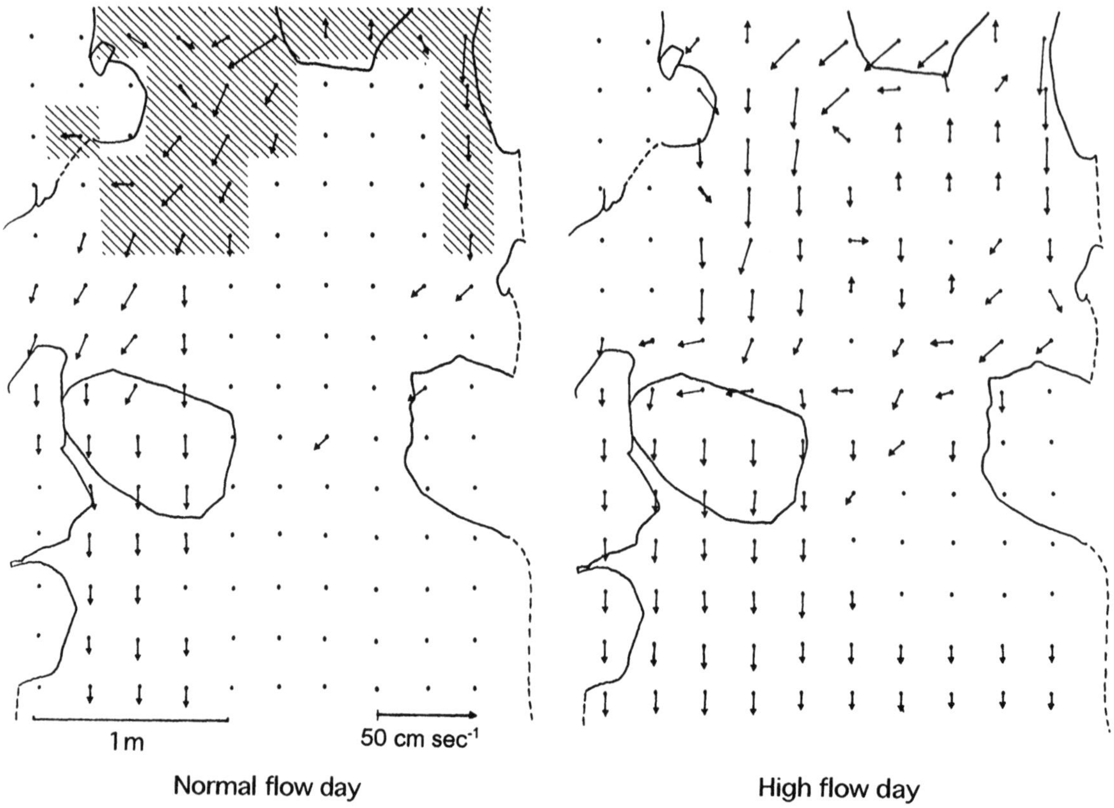

Figure 3. Water current velocity and direction in the quadrat on a normal and high flow day (16 July and 21 August). The shaded area shows the head of the pool.

tions were classified into six ranks according to the strength of aggressiveness (Table 2).

These 10-min observations were conducted 101 times in total. The home range was expressed as the minimum convex polygon covering all positions in daily censuses and traces in 10-min observations. In-

Table 1. List of interactive behaviours of *Rhinogobius* sp. DA.

Category	Action	Movements
Aggressive behaviour	Glare (Gl)	Directs its eyes to an approaching opponent
	Approach (Ap)	Comes up to an opponent
	Fin up (Fu)	Erects its first dorsal fin
	Mouth open (Mo)	Opens its mouth widely
	Lateral display (Ld)	Shows its body to the opponent with the unpaired fins expanded and the mouth opened
	Drive off (Do)	Rushes to the opponent
	Chase (Ch)	Follows the opponent rapidly
Defeated behaviour	Motionless (Ml)	Remains motionless with the dorsal fins folded
	Leave (Le)	Leaves the opponent voluntarily
	Flee (Fl)	Flees from the opponent
	Colour change (Cc)	Stays still while a dark lateral line appears on the body
	Head shake (Hs)	Shakes its head laterally

Table 2. Interactions of *Rhinogobius* sp. DA ranked according to their strength. See Table 1 for abbrevation of each action.

Rank	Action of the winner	Action of the loser
1	Ld	Ld
2	Do and Ch	Fl
3	Do	Fl
4	Ap, Fu, Mo, Ld*	Cc and/or Hs
5	Ap, Fu*	Le
6	Gl	Ml and/or Le

* These actions occurred in sequence and the loser often reacted in the middle of the sequence.

dividuals > 50% of whose home ranges lay in the head of the pool were called pool-head dwellers.

Results

Social structure and interaction

Of 67 marked gobies, 40 were males and 27 females. Size of males (71.4 mm TL ± 12.6 SD) was larger than that of females (58.7 mm TL ± 5.5 SD; t-test, $t = 4.90$, $p < 0.001$).

Thirty-six marked gobies were found four or more times in censuses on normal flow days (Figure 4). Their home range size varied among individuals

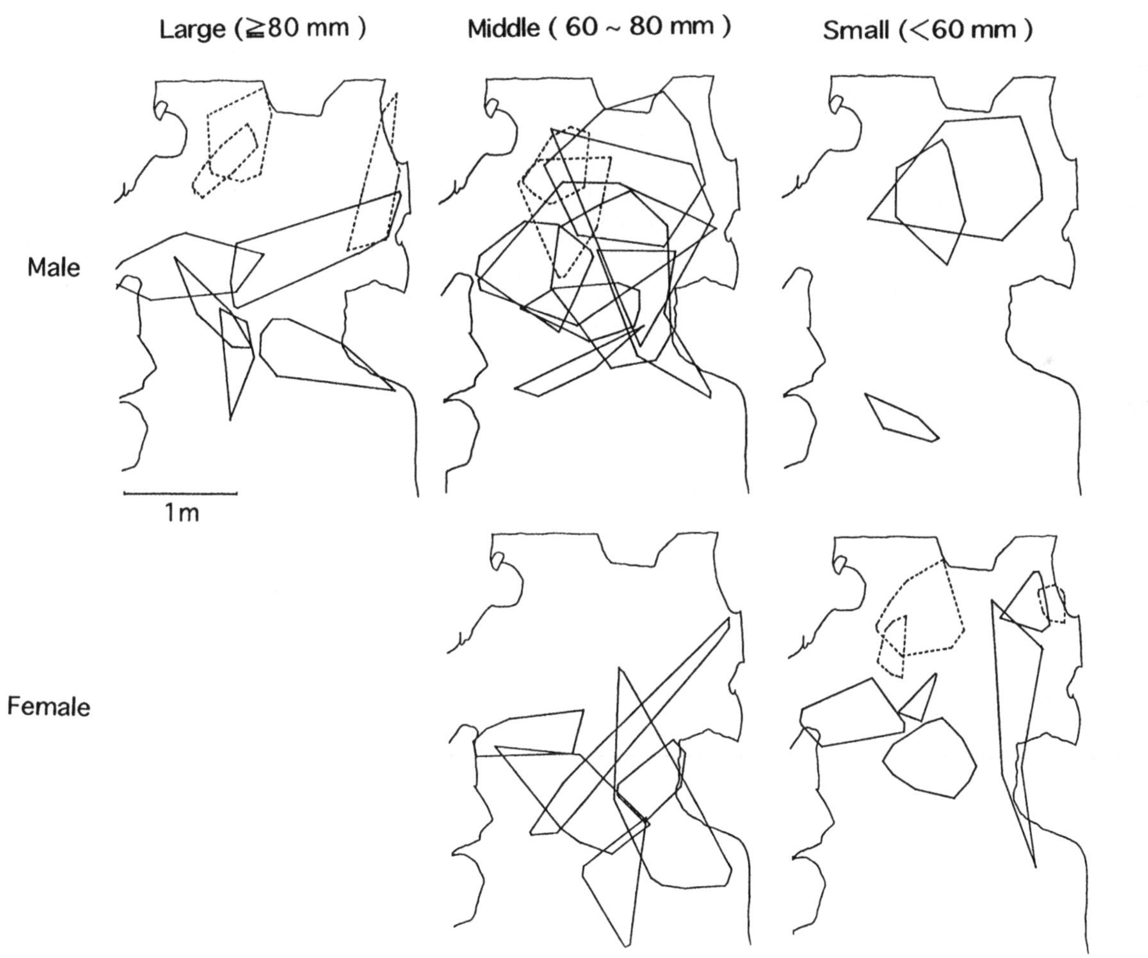

Figure 4. Home ranges of marked males (upper) and females (lower) on normal flow days. Broken lines show home ranges of pool-head dwellers.

Table 3. Home range sizes of males and females on normal and high flow days. Only individuals which were found at least four times in daily censuses of each period were included. The size difference was significant between males and females on high flow days (Mann-Whitney U test, $z = 2.53$, $p < 0.05$).

Period	Individual	n	Home range (cm^2) ($x \pm SD$)
Normal flow	Male	22	5236 ± 4329
	Female	14	3416 ± 2416
High flow	Male	25	7786 ± 6075
	Female	12	2840 ± 2628

and size differences between the sexes were not significant (Table 3). Each home range overlapped with 1–26 ($x = 12.6$) other home ranges. Five males and three females were pool-head dwellers (Figure 4). Their home ranges (2503 cm^2 ± 1550 SD, n = 8) tended to be smaller than those of the others (5107 cm^2 ± 4031 SD, n = 28), though the difference was not significant (Wilcoxon signed rank test, $z = 1.75$, $p = 0.08$). The pool-head dwellers (84.2 mm TL ± 11.6 SD, n = 5) were larger than the others (71.9 mm TL ± 10.0 SD, n = 20, $t = 2.34$, $p < 0.05$), if individuals of small size class (< 60 mm TL) were excluded.

On high flow days, gobies became active and often moved beyond the home ranges of normal flow days. Male home ranges were significantly larger than those on normal flow days, but female home ranges were not (Table 3). Home range overlap of each individual between normal and high flow days,

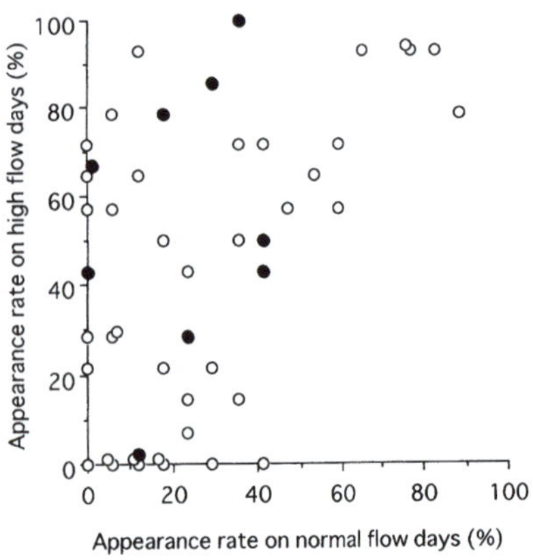

Figure 5. Appearance rate per census on normal and high flow days for each marked goby. Solid circles show individuals ≥ 80 mm TL.

represented as the overlapped area in the home range of normal flow days, was on average 52.0% (± 35.5 SD, n = 22). Pool-head dwellers significantly enlarged their home ranges on high flow days (8494 cm^2 ± 5324 SD, n = 4, $z = 2.55$, $p < 0.05$).

The number of marked gobies found per census was greater on high flow days (22.9 ± 2.5 SD, n =14) than on normal flow days (17.2 ± 5.0 SD, n = 17) ($t = 3.83$, $p < 0.001$). Most individual gobies appeared more frequently on high flow days, and those which appeared frequently on normal flow days also did so on high flow days (Figure 5). Although all large-sized individuals (≥ 80 mm TL) concealed themselves on more than half of normal flow days, some of them appeared very often on high flow days. The proportion of gobies which appeared in the open to the total gobies found was greater on high flow days than on normal flow days (97.8% ± 2.3 SD vs. 82.7% ± 8.6 SD, $z = 4.31$, $p < 0.001$).

Social interactions occurred 1.8 times per 10-min observation, on average. Interactions of moderate strength (i.e. ranks 3, 4 and 5; Table 2) accounted for 69% of the total (Table 4). The frequency of intra-sexual interactions was greater than that expected from random occurrences proportional to the sex ratio ($\chi^2 = 12.38$, $p < 0.01$). The strongest interaction, i.e., mutual display, occurred only between consexuals. The winner in an interaction was larger than the loser in most cases (166/189). Strong aggressive interactions tended to occur between individuals of similar body size (Figure 6).

Feeding behaviour

The occurrence pattern of the three feeding actions

Table 4. The occurrence of intra- and intersexual interactions classified into six ranks (see Table 2).

Rank	Sexes of winner and loser				Total
	Male-Male	Male-Female	Female-Male	Female-Female	
1	3	0	0	3	6
2	13	3	5	1	22
3	18	17	7	7	49
4	20	3	6	4	33
5	19	7	11	5	42
6	18	6	5	0	29
Total	91	36	34	20	181
(expected)*	(68.26)	(46.07)	(39.80)	(26.87)	

* Expected value, presuming that an observed goby interacts with males and females in proportion to the sex ratio.

differed between normal and high flow days (Figure 7). Feeding actions related to benthic feeding, looking and picking, were common on normal flow days, whereas rising was the commonest feeding action on high flow days. On normal flow days, there were no significant differences in the frequencies of these three actions among size classes, except for rising being more frequent than picking in large individuals ($z = 2.02$, $p < 0.05$). On high flow days, rising was more frequent than looking ($z = 3.30$, $p < 0.01$) and picking ($z = 3.3$, $p < 0.001$), when all size classes were combined.

Rising mainly occurred at sites where the current velocity was between 5 and 15 cm sec^{-1}. On normal flow days, about half of the rising (54/112) occurred in the head of the pool (Figure 8). On high flow days, rising rarely occurred in the head of the pool, where the current velocity usually exceeded 15 cm sec^{-1} (Figure 3), but it frequently occurred above some particular protruding stones below the head of the pool (Figure 8). When not rising gobies alternately waited on such stones for drifting matter. Of gobies observed for 10 min, 81.5% (22/27) showed rising at least once on high flow days, but

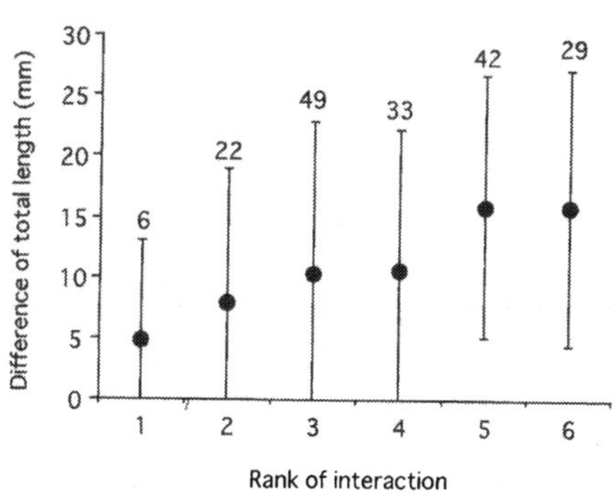

Figure 6. Body size differences of gobies which interacted in relation to the ranks of interactions. For the rank of interaction see Table 2. Vertical lines show SD and numerals sample size. The difference of rank 5 with ranks 1, 2, 3 and 4 and that of rank 6 with ranks 1 and 2 were significant at $p < 0.05$ (t-test).

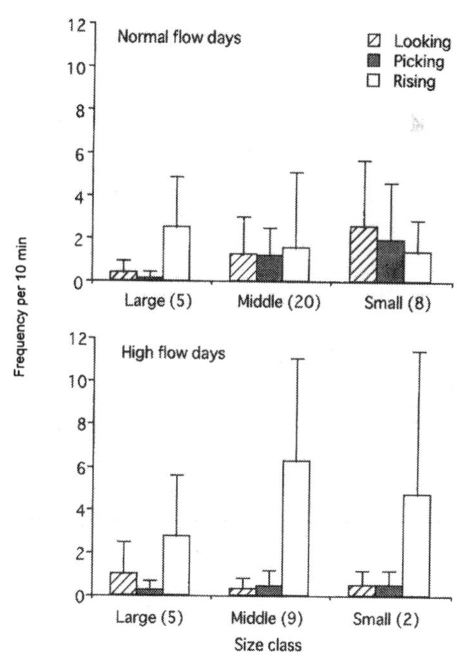

Figure 7. Frequencies of looking, picking and rising in gobies of large (≥ 80 mm TL), middle (60–80) and small (< 60) size classes on normal and high flow days. Looking followed by picking was excluded from looking. Vertical lines show SD and sample size is in parentheses.

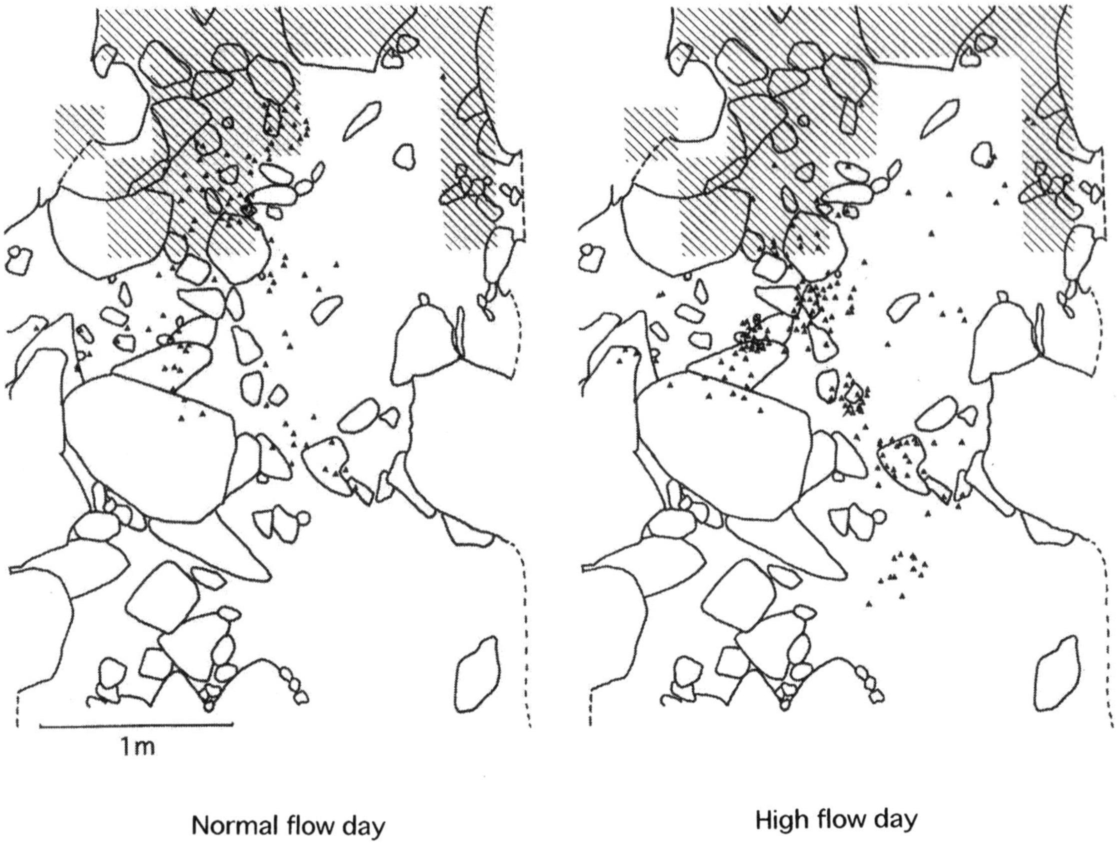

Figure 8. Rising points (triangles) observed on normal and high flow days. The shaded area shows the head of the pool.

only 47.3% (35/74) did so on normal flow days (χ^2 = 34.47, p < 0.001).

Food habits

Stomach fullness of *Rhinogobius* sp. DA in the Kashiwa River was greater on high flow days than on normal flow days (Figure 9). On normal flow days, stomach fullness of individuals collected in the open (2.75 ± 1.87 SD, n = 20) was greater than that in the shade (1.33 ± 0.83 SD, n = 20, t = 3.11, p < 0.01). On both normal and high flow days, the stomach contents mainly consisted of aquatic invertebrates, but terrestrial invertebrates were also an important dietary component (12.5 and 21.3% of the total weight on normal and high flow days, respectively) (Table 5). Items eaten on normal and high flow days were similar (Table 6), showing similar diversity. Invertebrate drift collected by the net was more abundant on high flow days than on normal

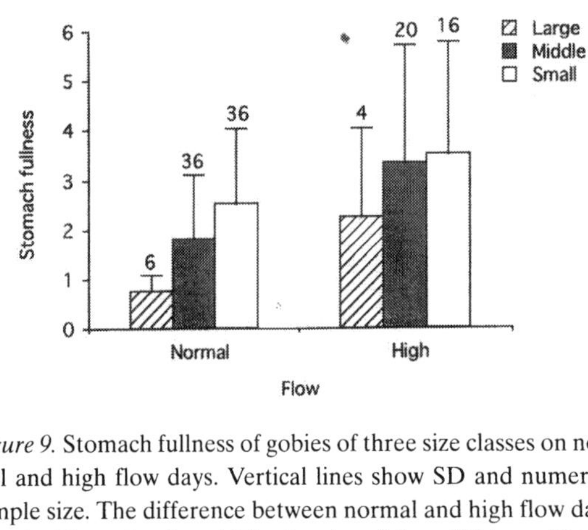

Figure 9. Stomach fullness of gobies of three size classes on normal and high flow days. Vertical lines show SD and numerals sample size. The difference between normal and high flow days was significant for the middle size class (Mann-Whitney U test, z = 2.94, p < 0.01) and for all size classes combined (z = 3.54, p < 0.001).

Table 5. Stomach contents of *Rhinogobius* sp. DA on normal and high flow days in the Kashiwa River and *Rhinogobius* sp. DA and *Zacco temmincki* on normal flow days in the Kiku River. Food items other than insects were combined at the class level. Dry weight (x10^{-2} mg) per individual is followed by the mean number in parentheses. For diversity index, Shannon-Weaver function (H') was used.

Species	*Rhinogobius* sp. DA			*Zacco temmincki*
River	Kashiwa	Kashiwa	Kiku	Kiku
Flow	Normal	High	Normal	Normal
Sample size	78	40	30	28
Mean TL (mm)	71.2	64.8	61.6	105.1
Mean stomach fullness	2.03	3.28	2.95	3.76
Diversity index (H')	2.38	2.39	2.20	1.24
Aquatic invertebrates				
Ephemeroptera	18.70 (0.848)	25.36 (0.775)	27.11 (1.200)	33.60 (0.500)
Odonata	0	0	9.88 (0.075)	0
Plecoptera	0	0	2.65 (0.050)	28.30 (0.071)
Megaloptera	0.26 (0.015)	0	0	0
Trichoptera	8.86 (0.346)	3.32 (0.175)	7.38 (0.275)	0.43 (0.036)
Lepidoptera	0.29 (0.012)	11.56 (0.050)	0	0
Diptera	9.37 (1.583)	17.83 (2.325)	15.70 (3.025)	0.26 (0.107)
Ploima	0.30 (0.108)	0.51 (0.250)	0.67 (0.325)	0
Annelida	0.02 (0.013)	0	0	0
Gastropoda	4.43 (0.259)	14.17 (0.325)	2.61 (0.200)	0
Conchostraca	7.42 (1.682)	10.94 (2.550)	3.53 (0.450)	0
Cyclopoida	0.00 (0.013)	0	0.20 (0.150)	0
Isopoda	7.38 (0.225)	34.85 (0.750)	0	0
Subtotal	57.07 (5.103)	118.54 (7.200)	69.72 (5.750)	62.60 (0.714)
Terrestrial invertebrates				
Ephemeroptera	0	0	0	7.43 (0.107)
Orthoptera	0	21.98 (0.050)	0	0
Thysanoptera	0.10 (0.012)	0.11 (0.025)	0	0.77 (0.036)
Hemiptera	0.43 (0.038)	0.75 (0.025)	0	0.45 (0.036)
Trichoptera	0.38 (0.013)	1.39 (0.050)	1.68 (0.100)	12.05 (0.214)
Lepidoptera	3.79 (0.089)	6.60 (0.025)	0	0
Coleoptera	2.36 (0.013)	0	0	0
Hymenoptera	0.82 (0.038)	3.12 (0.050)	0	37.85 (0.714)
Diptera	0.03 (0.013)	0	0	4.37 (0.464)
Araneae	3.41 (0.100)	0.28 (0.025)	1.50 (0.025)	3.62 (0.107)
Acarina	0.25 (0.049)	0.88 (0.350)	0	0
Geophilomorpha	0	1.31 (0.025)	0	0
Subtotal	11.56 (0.363)	36.42 (0.625)	3.18 (0.125)	66.54 (1.679)
Others				
Fish scales	0	0	15.73	2.06
Algae	12.99	7.92	16.57	42.14
Land plant fragments	10.01	6.38	9.56	366.11
Unknown	0.89	1.76	0.26	4.04
Sand	0.32	0.35	0	0.77
Total	92.84 (5.466)	171.36 (7.825)	115.01 (5.875)	544.25 (2.393)

Table 6. Similarity of food items between *Rhinogobius* sp. DA in the Kashiwa River and *Rhinogobius* sp. DA and *Zacco temmincki* in the Kiku River. Kendall's rank correlation coefficient ($-1 \leq \tau \leq 1$) was used as a similarity index.

Site, species and river flow	a	b	c	d
Kashiwa River				
Rhinogobius sp. DA (a) (normal flow)		0.551	0.322	0.205
Rhinogobius sp. DA (b) (high flow)	< 0.001		0.195	0.085
Kiku River				
Rhinogobius sp. DA (c) (normal flow)	< 0.01	n.s.		0.263
Zacco temmincki (d) (normal flow)	n.s.	n.s.	< 0.05	

flow days, although taxonomic diversity of both groups was similar (Table 7). Similarity of invertebrate drift and animal items eaten by the goby was higher on normal flow days than on high flow days.

In the Kiku River, *Rhinogobius* sp. DA collected on normal flow days also had fed mainly on aquatic invertebrates (Table 5). They scarcely ate terrestrial animals (2.8% of the total weight). Similarity of animal food items to invertebrate drift was relatively low (Table 7). Stomach contents of the cyprinid *Zacco temmincki* in the Kiku River mainly consisted of fragments of land plants (Table 5). This cyprinid ate terrestrial invertebrates as much as aquatic ones.

Discussion

Social structure

The social structure of fishes in steep streams has been best studied in salmonid fishes. Individuals in pools exhibit a social hierarchy or territoriality (Jenkins 1969, Bachman 1984, Furukawa-Tanaka 1985, Tanida et al. 1989, Hughes 1992). Large, dominant ones occupy the head of the pool where they have high opportunities for drift feeding, while subordinates locate themselves behind them (Furukawa-Tanaka 1992). In Japanese streams, nektonic cyprinids such as *Leusiscus* (*Tribolodon*) *hakonensis* also show a similar social structure if salmonid fishes are absent (T. Furukawa-Tanaka personal com-

Table 7. The quantity of drift on normal and high flow days in the Kashiwa River and on normal flow days in the Kiku River, and similarity of invertebrate drift to animal items eaten by *Rhinogobius* sp. DA and *Zacco temmincki*. The drift net was set at sites of similar water current velocity. For diversity index Shannon-Weaver function (H') was used and for similarity index Kendall's rank correlation coefficient ($-1 \leq \tau \leq 1$) was used.

River River flow Current velocity (cm s^{-1})	Kashiwa Normal 22.3	Kashiwa High 22.9	Kiku Normal 20.7
Dry weight of invertebrate drift (aquatic + terrestrial) (mg m^{-3})	1.685 (0.660 + 1.025)	3.697 (2.107 + 1.590)	1.163 (0.524 + 0.640)
Dry weight of total drift (mg m^{-3})	99.097	827.961	152.772
Diversity index of invertebrate drift (H')	2.381	2.379	1.622
Similarity to diet (τ)			
Rhinogobius sp. DA	0.368 ($p < 0.01$)	0.023 (n.s.)	0.147 (n.s.)
Zacco temmincki	–	–	0.253 ($p < 0.05$)

munication). In the Kashiwa River, no nektonic fishes were present and a benthic goby *Rhinogobius* sp. DA predominated in the pool area. Unlike territorial salmonid fishes, home ranges of this goby greatly overlapped with each other and the head of the pool was not always occupied by large individuals. Nevertheless, those which occupied the head of the pool were on average larger than others and drift feeding by rising frequently occurred there on normal flow days, suggesting that the head of the pool is a good feeding site even for this benthic goby.

Space-related behaviour has not been investigated in detail for stream gobies. The present study showed that *Rhinogobius* sp. DA are often aggressive against conspecifics and the aggressiveness is greater when their body sizes are similar. The encounters were disproportionally frequent between consexuals. The fact that large males were more or less spaced out within a pool may be a consequence of their aggressiveness. On the other hand, different-sized individuals had more overlap in home ranges and their aggressiveness was weaker. A large individual's tolerance of small ones in its territory or home range is known in many marine and lacustrine fishes (e.g., Robertson 1972, Kuwamura 1984, 1992, Yanagisawa 1987, Ochi 1989). Its tolerance is usually due to the difference in resource demand or social status with small ones, or to the ease it should have in repulsing small ones from its site. In the case of *Rhinogobius* sp. DA, the latter may be the primary reason, because their habitat abounds in shelters on the bottom.

Influence of river flow on feeding behaviour

For stream fishes, elevation of flow after heavy rain must be the most important agent that influences their lives. However, little attention has been paid to how fishes respond to such an environmental change. The sudden rising of a river usually has a destructive effect on fish populations (Meffe 1984, Chapman & Kramer 1991). However, the present study showed that it promotes the feeding activities of *Rhinogobius* sp. DA. The amount of invertebrate drift per unit volume of water doubled on high flow days. An increase of invertebrate drift after rain must be a common phenonemon, although quantitative data are still not available (Miyasaka 1950).

Most individuals of the goby appeared in the open more often on high flow days than on normal flow days, and some virtually foraged only on high flow days (Figure 5). This may mean that food conditions on normal flow days are so poor that the goby depends on invertebrate drift on high flow days for food.

Plasticity of feeding behaviour

From his field studies on food habits of Japanese stream fishes, Kawanabe (1959) emphasized the importance of interspecific interactive food segregation for their coexistence; competitive interactions would narrow the diet of a species which potentially has a wide dietary range. A well-known example is the interaction of the cyprinid *Zacco platypus* with the annual algal-feeder *Plecoglossus altivelis* in the middle reaches of water-courses (Miyadi 1960). The former move from the central part of a riffle to its periphery near the banks, shifting their main food from algae to benthic animals, when the latter have established feeding territories in the riffle. Kawanabe (1973) believed that plasticity of feeding behaviour found in freshwater fishes is based on their potentially wide feeding repertoire that may or may not be partially suppressed by interaction with competitors.

Plasticity of feeding behaviour in *Rhinogobius* species was reported first from Okinawa Island (Kawanabe 1973). In the upper stream above a large falls where nektonic fishes were absent, they fed on fallen and drifting terrestrial insects, which exceeded 30% of their stomach contents in volume. They sometimes rose to the water surface to eat fallen insects and on occasions even broke the water surface to catch insects. On the other hand, the gobies in the lower stream fed exclusively on benthic organisms, as reported from many rivers (Kawanabe et al. 1956, Kodama 1961, Sawara 1978). In the present study, in the Kashiwa River where nektonic fishes are absent, the gobies often rose into the water column, though rarely up to the water surface (we observed this behaviour only three times), and terrestrial insects constituted their most important diet item. This contrasts with a very small proportion of terrestrial insects in the diet of the gobies in the Kiku River harbouring a nektonic fish *Z. temmincki*. This difference cannot be attributed only to the difference in the availability of terrestrial insects between the two rivers, because *Z. temmincki* in the Kiku River ate a substantial amount of terrestrial insects. In the Kiku River, rising of the goby also occurred but at a very low rate (personal observation). It is likely that the presence of *Z. temmincki* influences the feeding behaviour of the goby, though the nature of their interaction has not been studied.

We propose that drift feeding of *Rhinogobius* sp. DA in this study is obviously a case of ecological release, i.e., the enlargement of the niche of a species due to the absence of a competitor or other restricting species (Lincoln et al. 1982). In aquatic environments, which always possess a set of pelagic and benthic habitats and resources, benthic species most often advance into the pelagic zone to enlarge their niches (Schulter & McPhail 1993, Robinson & Wilson 1994). Morphologies of *Rhinogobius* gobies are well adapted to benthic life (e.g., sucker-life pelvic fins). It is of interest that in spite of their anatomical constraint they still have a potential for using pelagic resources. This may be due to instability of fluvial environments where fluctuation of water flow changes food conditions.

Acknowledgements

We are indebted to K. Omori for constructive suggestions and to D. Takahashi for his help in the field.

References cited

Bachman, R.A. 1984. Foraging behavior of free-ranging wild and hatchery brown trout in a stream. Trans. Amer. Fish. Soc. 113: 1–32.

Chapman, L.J. & D.L. Kramer. 1991. The consequences of flooding for the dispersal and fate of poeciliid fish in an intermittent tropical stream. Oecologia 87: 299–306.

Furukawa-Tanaka, T. 1985. The ecology of salmonid fishes in Japanese mountain streams. I. Food conditions and feeding habit of Japanese charr, *Salvelinus leucomaenis* (Pallas). Japan. J. Ecol. 35: 481–504.

Furukawa-Tanaka, T. 1992. Optimal feeding position for stream fishes in relation to invertebrate drift. Human and Nature 1: 63–81.

Hughes, N.F. 1992. Ranking of feeding positions by drift-feeding Arctic grayling (*Thymallus arcticus*) in dominance hierarchies. Can. J. Fish. Aquat. Sci. 49: 1994–1998.

Jenkins, T.M. Jr. 1969. Social structure, position choice and microdistribution of two trout species (*Salmo trutta* and *Salmo gairdneri*) resident in mountain streams. Anim. Behav. Monogr. 2: 57–123.

Kawanabe, H. 1959. Food competition among fishes in some rivers of Kyoto Prefecture, Japan. Mem. Coll. Sci. Univ. Kyoto 26: 253–268.

Kawanabe, H. 1969. Fishes in rivers and lakes. Chuokoron-sha, Tokyo. 196 pp. (in Japanese).

Kawanabe, H. 1973. Food and feeding habits of a freshwater goby, *Rhinogobius brunneus* in Okinawa Island, the Ryukyus, with some discussion on interactive food segregation. Biol. Mag. Okinawa 10: 1–6. (in Japanese).

Kawanabe, H., D. Miyadi, S. Mori, E. Harada, Y. Mizuhara & R. Ogushi. 1956. Ecology of natural stock of ayu, *Plecoglossus altivelis* I. Contr. Physiol. Ecol. Kyoto Univ. 79: 1–37. (in Japanese).

Kawanabe, H. & N. Mizuno. 1989. Freshwater fishes of Japan. Yama-Kei, Tokyo. 720 pp. (in Japanese).

Kodama, H.N. 1961. Food habits of a fresh-water goby, *Rhinogobius similis*. Gill. Japan. J. Ecol. 11: 226–231. (in Japanese).

Kuwamura, T. 1984. Social structure of the protogynous fish *Labroides dimidiatus*. Publ. Seto Mar. Biol. Lab. 29: 117–177.

Kuwamura, T. 1992. Overlapping territories of *Pseudosimochromis curvifrons* males and other herbivorous cichlid fishes in Lake Tanganyika. Ecol. Res. 7: 43–53.

Lincoln, R.J., G.A. Boxshall & P.F. Clark. 1982. A dictionary of ecology, evolution and systematics. Cambridge University Press, Cambridge. 298 pp.

Meffe, M.K. 1984. Effects of abiotic disturbance on coexistence of predator-prey fish species. Ecology 65: 1525–1534.

Miyadi, D. 1960. Ayu no hanashi (On the ecology of ayu *Plecoglossus altivelis*). Iwanami, Tokyo. 226 pp. (in Japanese).

Miyasaka, H. 1950. Seasonal changes in the amount of fish and invertebrate drift in the Tikuma River. Nippon Suisan Gakkaishi 16: 14–21. (in Japanese).

Ochi, H. 1989. Acquisition of breeding space by nonbreeders in the anemonefish *Amphiprion clarkii* in temperate waters of southern Japan. Ethology 83: 279–294.

Robertson, D.R. 1972. Social control of sex reversal in a coral-reef fish. Science 177: 1007–1009.

Robinson, B.W. & D.S. Wilson. 1994. Character release and displacement in fishes: a neglected literature. Amer. Nat. 144: 596–627.

Sawara, Y. 1978. Ecological studies on the common freshwater goby, *Rhinogobius brunneus*, especially on the growth, food

habits and feeding activity. J. Fac. Sci. Univ. Tokyo, Sec IV 14: 201–236.
Schluter, D. & J.D. McPhail. 1993. Character displacement and replicate adaptive radiation. Trends Ecol. Evol. 8: 197–200.
Tanida, K., T. Maruyama & Y. Saito. 1989. Feeding ecology of Japanese charr (*Salvelinus leucomaenis*) in a high moor and adjacent stream in central Japan. Physiol. Ecol. Japan Spec. Vol. 1: 279–294.
Yanagisawa, Y. 1987. Social organization of a polygynous cichlid *Lamprologus furcifer* in Lake Tanganyika. Japan. J. Ichthyol. 34: 82–90.

Environmental Biology of Fishes **52**: 345–355, 1998.

Competitive interactions for foraging microhabitat among introduced brook charr, *Salvelinus fontinalis*, and native bull charr, *S. confluentus*, and westslope cutthroat trout, *Oncorhynchus clarki lewisi*, in a Montana stream

Shigeru Nakano[1], Satoshi Kitano[2], Katsuki Nakai[3] & Kurt D. Fausch[4]
[1] *Tomakomai Research Station, Hokkaido University Forests, Takaoka, Tomakomai, Hokkaido 053, Japan (e-mail: shige@exfor.agr.hokudai.ac.jp)*
[2] *Research Institute of North Pacific Fisheries, Faculty of Fisheries, Hokkaido University, Hakodate, Hokkaido 041, Japan*
[3] *Lake Biwa Museum Project Office, Shiga Prefecture Board of Education, Otsu, Shiga 520, Japan*
[4] *Department of Fishery and Wildlife Biology, Colorado State University, Fort Collins, CO 80523, U.S.A.*

Received 31.8.1995 Accepted 2.3.1997

Keywords: interspecific dominance hierarchy, removal experiment, foraging rate, foraging distance, cover use, salmonid fishes

Synopsis

Competitive interactions for foraging microhabitat among introduced brook charr, *Salvelinus fontinalis*, and native bull charr, *S. confluentus*, and westslope cutthroat trout, *Oncorhynchus clarki lewisi*, were studied by species removal experiments in a tributary of the Flathead Lake and River system, northwestern Montana, focusing on brook charr influences on bull charr. When the three species were in sympatry, they interacted with each other, forming a size-structured, mixed-species dominance hierarchy in two stream pools. The influences of interference interactions were examined by measuring changes in five characteristics of foraging microhabitat and behavior, focal point height and velocity, cover use, and foraging rate and distance, after the successive removal of two species. Cutthroat trout removal resulted in increased foraging rates and distances, and decreased cover use for brook charr, but no changes for bull charr. After removal of brook charr from the two-species system, bull charr also increased foraging rates and distances and occupied more exposed positions. Moreover, total fish densities, which had initially decreased owing to the removal experiments, were partly compensated for by subsequent bull charr immigration, implying that competitive interactions with brook charr are an important factor in the mechanisms responsible for the regulation of bull charr densities, at least on a local scale.

Introduction

Many species of stream-dwelling salmonids have been introduced to lotic habitats to which they were not indigenous, sometimes with detrimental consequences for native species (see Fausch 1988, Krueger & May 1991, for reviews). Studies that have explored competitive relationships between such pairs of salmonid species that did not evolve in sympatry have generally found that they use similar niches and compete strongly for limited resources (e.g. Fausch & White 1981, 1986, Griffith 1988, Lohr & West 1992). In many regions, populations of native species have declined in abundance and been displaced from much of their range by introduced salmonids (Larson & Moore 1985, Griffith 1988, Fausch 1988, 1989).

Bull charr, *Salvelinus confluentus*, which are widely distributed in northwestern North America, are a comparatively recently recognized species distinct from dolly varden, *S. malma* (Cavender 1978, 1980, Haas & McPhail 1991). Bull charr are now thought to be extinct in California and are considered a 'species of special concern' being listed as 'endangered' throughout most of their remaining range (Williams et al. 1989). Cumulative effects resulting from human activities, such as logging and hydropower development are thought to be responsible for the dramatic declines in their populations (Fraley & Shepard 1989, Ratliff & Howell 1992, Rieman & McIntyre[1]).

Interactions with introduced, non-native salmonids, including brook charr, *S. fontinalis*, brown trout, *Salmo trutta*, rainbow trout, *Oncorhynchus mykiss*, and lake charr, *S. namaycush*, are also believed to have displaced bull charr populations (Markle 1992, Ratliff & Howell 1992, Ziller 1992, Donald & Alger 1993, Rieman & McIntyre[1]). Congeneric brook charr, in particular, may strongly influence many bull charr populations in areas where the two species occur in sympatry, both through intensive hybridization (Leary et al. 1983, Markle 1992, Kitano et al. 1994) and interspecific competition (Buckman et al. 1992, Dambacher et al. 1992, Ratliff & Howell 1992). However, the mechanisms and effects of competition are still unclear.

There is a growing trend for conducting removal experiments of potential competitors, which provide the most direct evidence of resource competition (Schoener 1983). Although many workers have tested interspecific competition by removal experiments under laboratory conditions (Fausch & White 1986, Dolloff & Reeves 1990, Grossman & Boule 1991, Wang & White 1994), such approaches have been attempted for only a few stream salmonid systems in natural habitats (Everest & Chapman 1972, Fausch & White 1981, Lohr & West 1992). In this study, we examined competitive interactions among introduced brook charr, and native bull charr and westslope cutthroat trout, *O. clarki lewisi*, in a tributary of the Swan River, a major tributary of the Flathead Lake and River system in northwestern Montana. The focus was primarily on exploring competitive influences of brook charr on foraging behavior and microhabitat of bull charr.

Background

Brook charr were introduced from eastern North America to the Flathead Lake and River system in the late 1800s (MacCrimmon & Campbell 1969). Only two native salmonids, bull charr and westslope cutthroat trout, originally occurred in sympatry in the basin. The lifecycle of bull charr in this drainage is largely fluvial-lacustrine. Most fish migrate downstream after 1–4 years of riverine life, grow to maturity in lakes, and later migrate upstream to spawn (Fraley & Shepard 1989). In contrast, two life-history forms are known to occur in cutthroat trout; fluvial-lacustrine, migrating downstream primarily at ages 2 and 3, and fluvial, residing in streams throughout their life. Populations of brook charr in the system are wholly fluvial (Mike Enk personal communication). Because the three salmonids inhabit stream habitats for significantly long periods of their life-history, there is a strong potential for resource-limiting competitive interactions to occur among them.

Nakano et al. (1992) observed that resource use patterns differed significantly between bull charr and cutthroat trout, which occurred in natural sympatry in a tributary of the Flathead River that lacked introduced brook charr. They speculated that such differentiation in resource use allowed coexistence of these two native salmonids. On the other hand, cutthroat trout were displaced from their native range throughout the Rocky Mountains following introductions of brook charr, strong competitive interactions of the two species being thought to have played an important role (Griffith 1972, 1988, Fausch 1989; see Likness & Graham 1988, for the Flathead River drainage). In contrast,

1 [1] Rieman, B.E. & J.D. McIntyre. 1993. Demographic and habitat requirements for conservation of bull trout. U.S. Dept. Agric., Forest Service, Gen. Tech. Rep. INT-302, Intermountain Research Station, Ogden. 37 pp.

no reports have dealt directly with competitive interactions between sympatrically-occurring bull charr and brook charr. Such background provided a focus for the present research and design of the field experiments.

Materials and methods

Study site

The study was conducted in Elk Creek (47°32′ N, 113°46′ W), a third-order tributary of the Swan River. The study reach, located approximately 2.6 km upstream from the confluence with the Swan River, was a moderate gradient (1.4%), 5–10 m wide mountain stream with pools separated by high-gradient riffles or rapids. The streambed comprised mainly gravel, with small areas of sand. The daily fluctuations of both maximum (range = 14.3 – 15.2 °C, n = 14) and minimum water temperatures (9.6 – 10.1 °C, n = 14) were small during the study period in early to mid August 1992. Stream discharge, measured in the study reach every four days using the U.S. Geological Survey midsection method (Orth 1983), was also stable throughout the study period (range = 1.31 – 1.38 $m^3 s^{-1}$, n = 4).

Underwater observations of fish behavior and removal experiments were performed in two pools (A and B; hereafter termed observation pools) from 8 to 22 August 1992 (Table 1). Although the pools were separated from adjacent pools by riffles of about 10–30 m long, fish could move freely between them. Prior to the observations, a 1 m^2 grid was set up on the streambed of each observation pool by placing colored markers, and detailed topographic maps were drawn for each pool (scale 1:40). Water depth and current velocity, measured with a portable current meter (Tanida et al. 1985) at the surface, middle and bottom, were measured at each corner of the grid cells (Table 1). The distribution and total area of instream and overhead cover (sensu Dolloff & Reeves 1990) were also recorded, and the proportion by area was subsequently calculated for each cell. The mean current velocity in pool B was significantly greater than that in pool A, when combining the surface, middle and bottom (t-test on $\log_{10}$ transformed data, $p < 0.001$). The proportion of cover area in pool B were significantly larger than that in pool A (t-test on arcsine-square-root transformed data, $p < 0.001$), there being no difference in mean depth (t-test on $\log_{10}$ transformed data, $p = 0.054$).

Observations

Focal animal observations (Altmann 1974) were made for each fish over 70 mm in fork length (FL) at various times between 7:00–19:00 h. In the present study, all observed fish over 80 mm were individually identified by their characteristically shaped parr marks, spot patterns and scars (e.g., see Bachman 1984). Fork lengths (± 1 mm) were measured for fish captured during or after removal experiments (see below). Fork lengths of fish not captured were estimated to the nearest 5 mm by comparison with fish of known length. The error in length estimations was less than ± 5% (Shigeru Nakano unpublished data).

The observers lay motionless at the downstream end or stream margin of the pools and allowed several minutes to elapse before beginning observations, thus enabling fish to resume their normal behavior patterns. The observations lasted for 5 min

Table 1. General characteristics, represented as mean ± SE for the three habitat variables, of each observation pool in Elk Creek (see text). Sample sizes are shown in parentheses.

Pool	Length (m)	Width (m)	Area (m^2)	Habitat variables		
				Depth (cm)	Velocity (cm sec^{-1})	Cover (%)
A	19	6 – 9	163	49.5 ± 1.8 (163)	35.1 ± 1.3 (489)	10.0 ± 0.02 (163)
B	21	5 – 9	102	54.4 ± 2.3 (102)	42.8 ± 1.0 (306)	32.7 ± 0.03 (102)

for each fish, during which time all foraging and aggressive acts were recorded. Most individuals of the three species were observed to be drift foragers, typically holding focal points in the stream flow and intercepting drifting food items, with occasional feeding attempts on the bottom. If a fish subject to aggression simply fled without any display, it was regarded as a one-sided attack. All other contacts were recorded as mutually aggressive acts. A dominance hierarchy was constructed only for individuals which were continuously identified, from the results of one-sided attacks observed in each pool, following Katano (1990).

The locations (horizontal positions and vertical distance from the streambed) at which foraging and aggressive events occurred, and focal points were held, were also recorded on the maps. Graduated steel rods set in the streambed were used for estimating vertical distance. During the study, 408 focal animal observations were carried out on the three species in the two pools.

Three characteristics of foraging microhabitat were calculated for each species in each pool (during each period): (1) mean vertical distance between the focal point and the streambed (termed focal point height), (2) mean focal point current velocity, measured using a current meter, and (3) mean percentage of cover in the grid cells used by the fish (cover use). Two characteristics of foraging behavior were also calculated: (1) mean distance between focal points and the points at which foraging actions were observed (foraging distance), and (2) mean frequency of foraging attempts per 5 min period (foraging rate).

Removal experiments

Competitive influences of brook charr were tested by measuring changes in the five characteristics of foraging behavior and microhabitat of bull charr following the experimental removal of brook charr. If interspecific competition with brook charr is not influential, removal of brook charr should not result in any changes in these characteristics of bull charr even after the reduction in total fish densities.

Two separate removal experiments were conducted in the observation pools using underwater angling. Individual fish were selectively removed from pools by drifting live benthic invertebrates impaled on hooks past their positions while simultaneously observing them underwater. Initially, all cutthroat trout were removed on 14 August to provide simpler, two-species (bull charr and brook charr) competitive systems. Subsequently, to examine the significance of the competitive influences of brook charr on resource use by bull charr, all brook charr were removed on 18 August. Accordingly, the observation period was divided into three parts based on changes in species composition in the pools as follows: period I, the three species in sympatry (initial condition); period II, bull charr and brook charr in sympatry; period III, bull charr in allopatry. Any individuals of the species already removed that subsequently migrated into the pools were removed immediately after being sighted. Following each removal experiment, observations were repeated after 48 h of acclimation.

Data analyses

The five independent variables of foraging microhabitat and behavior were potentially affected by three factors (species, pools, and periods). To compare differences in the foraging microhabitats and behavior between species, separate analyses were performed on periods I and II using a two-way ANOVA model (factors = species, pools). In contrast, to analyze the effects of the species removals on the five parameters for each species (bull charr and brook charr), a two-way ANOVA model (factors = periods, pools) was used for each species. Tukey HSD multiple comparisons were carried out when significant differences were encountered. Arcsine-square-root transformations were applied to percentages. Logarithmic transformations ($\log_{10}$) were used for other variables to stabilize variances. Nonparametric tests (Kruskal-Wallis, Mann-Whitney U, and Spearman's rank correlation tests) were used to compare size and dominance rank of fish. All statistical tests were two-tailed. In all cases, statistical significance was evaluated at $p < 0.05$. Reduced alpha levels ($p = 0.05/n$, where n = number of

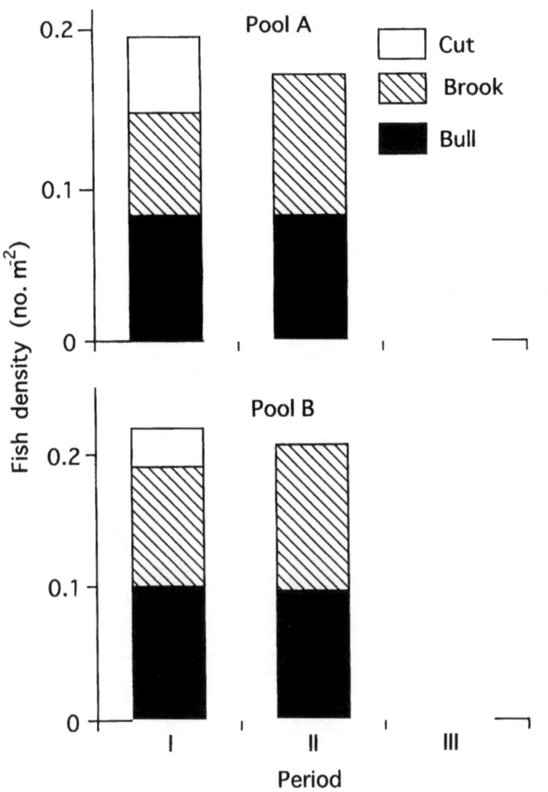

Figure 1. Fish densities (no m^{-2}) in observation pools A and B, during each observation period in Elk Creek. Cut = cutthroat trout, Brook = brook charr, Bull = bull charr.

pairwise comparisons) were used to correct for experiment-wise error.

Results

Fish sizes and densities

Fork lengths of fish from both pools ranged from 72 to 150 mm for bull charr, 78 to 234 mm for brook charr, and 76 to 228 mm for cutthroat trout, being significantly different among the three species in each pool (Kruskal-Walis test: A, $p = 0.013$; B, $p = 0.010$). Bull charr were significantly smaller than cutthroat trout in pool A (Mann-Whitney U-test, $p = 0.002$) and brook charr in pool B ($p = 0.015$). There were no significant differences between other species pairs in either of the pools ($p > 0.017$ [0.05/3] for those pairs).

Differences in species composition between the pools were not significant in either period I (χ^2 test, $p = 0.592$) or II ($p = 0.804$, Figure 1). Total fish densities were nearly equal between pools A and B in both periods I and II, whereas the bull charr density was higher in pool B in period III . Total fish densities, which had decreased as a result of the experimental removals, were partly compensated for in each pool, by immigrating brook charr during period II and its initial acclimation period, and immigrating bull charr during period III and its acclimation period.

Interspecific dominance hierarchies

Interspecific, size-structured dominance hierarchies were recognized among individuals over 80 mm, continuously identified in the same pools during period I (Figure 2). Two dominance matrices were constructed from the results of 118 one-sided interactions in pool A, and 84 such interactions in pool B. Individuals were classified into 11 (pool A)

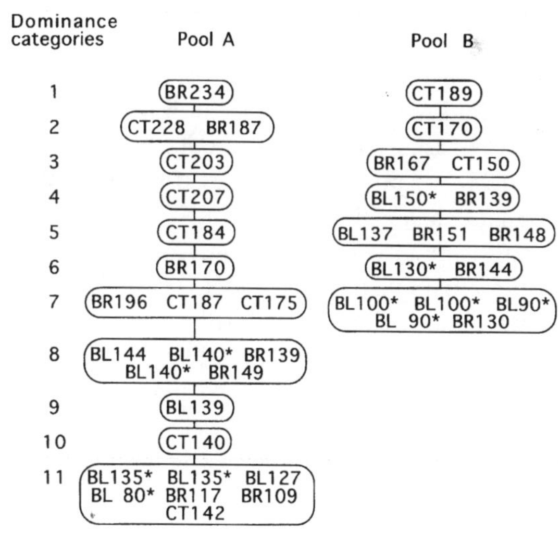

Figure 2. Hierarchical relationships among individuals inhabiting pools A and B, respectively. The species and fork length (mm) of each individual is shown. CT = cutthroat trout, BR = brook charr, BL = bull charr. Individuals within the same frame belong to the same dominance category. Asterisks indicate the fork lengths of fish estimated by underwater observations (see text).

Table 2. Results of two-way ANOVA tests (factors = species, pools) for differences in foraging microhabitat and behavior in periods I and II. Asterisks indicate significance at p = 0.05.

Variables	Effects								
	Species			Pools			Species-by-pools		
	F	df	p	F	df	p	F	df	p
					Period I				
Foraging microhabitat									
Focal point height	51.70	2,208	< 0.001*	0.02	1,208	0.885	1.76	2,208	0.174
Focal point velocity	15.74	2,75	< 0.001*	0.20	1,75	0.656	0.12	2,75	0.894
Cover use	5.79	2,216	0.017*	69.91	1,216	< 0.001*	0.13	2,216	0.717
Foraging behavior									
Foraging rate	39.71	2,215	< 0.001*	0.27	1,215	0.606	0.67	2,215	0.515
Foraging distance	147.12	2,871	< 0.001*	4.65	1,871	0.031*	1.52	2,871	0.220
					Period II				
Foraging microhabitat									
Focal point height	18.83	1,146	< 0.001*	5.73	1,146	0.018*	0.07	1,146	0.786
Focal point velocity	6.41	1,46	0.015*	0.76	1,46	0.388	0.11	1,46	0.742
Cover use	16.05	1,190	< 0.001*	53.65	1,190	< 0.001*	0.08	1,190	0.783
Foraging behavior									
Foraging rate	75.19	1,145	< 0.001*	< 0.01	1,145	0.982	0.75	1,145	0.388
Foraging distance	87.20	1,843	< 0.001*	11.67	1,843	0.001*	1.64	1,843	0.201

and 7 (B) dominance categories, their fork lengths being significantly correlated with their dominance category in both pools (A, $r_s = 0.92$, $p < 0.001$; B, $r_s = 0.93$, $p < 0.001$). A significant difference in dominance rank occurred among the three species in pool B (Kruskal-Walis test, p = 0.015), but not in A (p = 0.05). In pool B, cutthroat trout were more dominant than bull charr (Mann-Whitney U-test, p = 0.014), but no significant difference was found for the remaining pairs of species (p > 0.017 [0.05/3] for those pairs). Although statistical tests of aggressive encounters were not possible owing to the small sample size in both pools during periods II (A, n = 34; B, n = 23) and III (A, n = 13; B, n = 7), dominant-subordinate relationships remained fairly stable for individuals of the species present.

Foraging microhabitats and behavior

During period I, all five characteristics of foraging microhabitat and behavior were significantly different among the species (Table 2). In contrast, the effect of pools was significant only for cover use and foraging distance. None of the interaction effects were significant. Cutthroat trout had the greatest focal point height, followed by brook charr and bull charr, respectively (Tukey HSD test, p < 0.001 between each pair of species; Figure 3). The focal point velocity of cutthroat trout was significantly greater than those of brook charr (p = 0.002) and bull charr (p < 0.001), with that of brook charr being greater than that of bull charr (p = 0.031). Cover use of cutthroat trout was significantly lower than for bull charr (p = 0.001) and brook charr (p = 0.002), with no significant difference being evident between values for bull charr and brook charr (p = 0.696). Cover use of fish in pool B was significantly higher than that in pool A. Cutthroat trout had a higher foraging rate than brook charr (p = 0.003) and bull charr (p < 0.001) , with brook charr also having a higher rate than bull charr (p = 0.003; Figure 4). The foraging distance of cutthroat trout was the greatest, followed by brook charr and bull charr, respectively (p < 0.001 between each pair of spe-

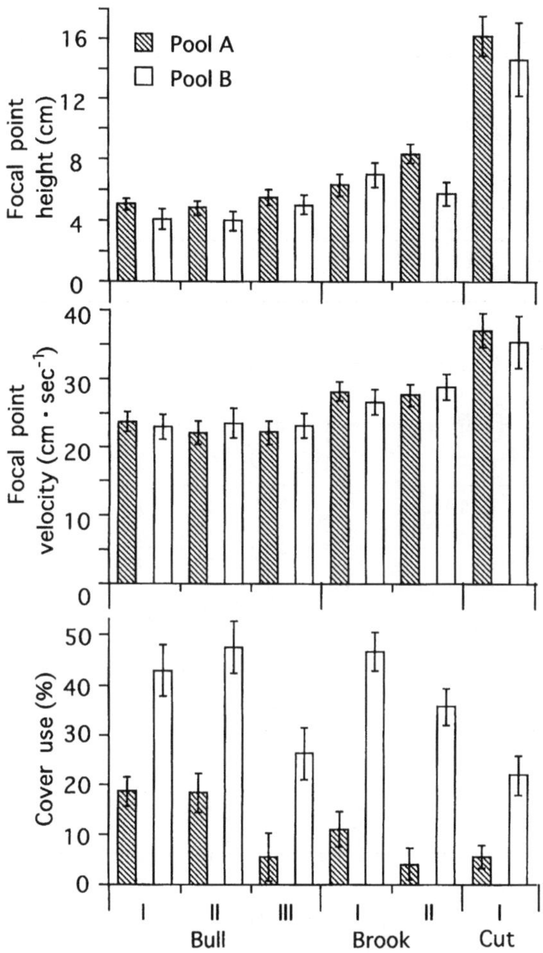

Figure 3. Focal point height, focal point velocity, and cover use by each species in pools A and B during each observation period. Vertical bars indicate ± 1 SE (see Figure 1 for species abbreviations).

cies). Foraging distance of fish in pool A was significantly greater than that in pool B.

During period II, all five characteristics of foraging microhabitat and behavior were significantly different between brook charr and bull charr (Table 2). Focal point height, cover use and foraging distance were significantly different between the pools. None of the interaction effects were significant. Brook charr had a greater focal point height and velocity, lower cover use, higher foraging rate, and longer foraging distance than bull charr (Table 2, Figures 3, 4). Cover use in pool B was significantly

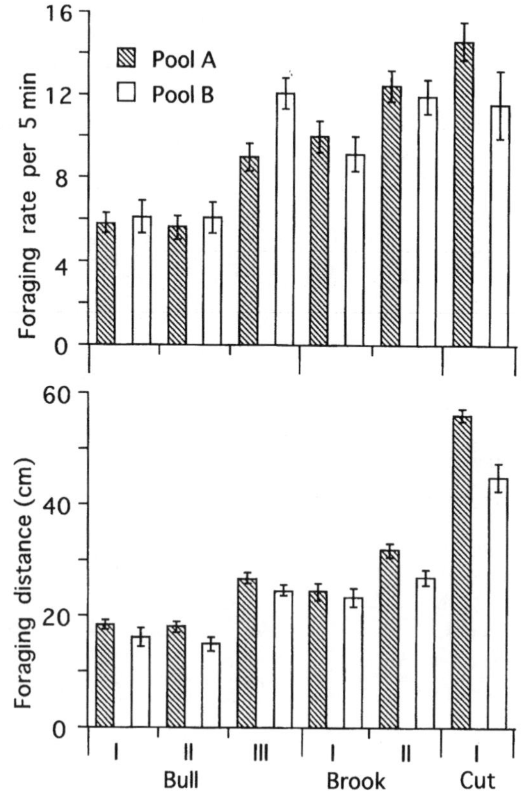

Figure 4. Foraging rate and foraging distance of each species in pool A and B during each observation period. Vertical bars indicate ± 1 SE (see Figure 1 for species abbreviations).

higher than that in pool A, whereas the focal point height and foraging distance were shorter in pool B.

After removal of cutthroat trout, the cover use of brook charr decreased, but foraging rates and distances increased (Table 3, Figures 3, 4), as indicated by the significant effect of periods on these three characteristics. In contrast, cover use and foraging distance were significantly different between pools. Interaction effect was significant only for focal point height.

For bull charr, cover use and foraging rate and distance were significantly different among periods, and cover use and foraging distance were significantly different between pools (Table 3). No interaction effects were significant. Cutthroat trout removal did not result in significant changes in cover use (Tukey HSD test, $p = 0.700$), foraging rate ($p = 1.000$) or foraging distance ($p = 0.892$) of bull charr (Figures 3, 4). However, brook charr removal

Table 3. Results of two-way ANOVA tests (factors = periods, pools) for differences in foraging microhabitat and behavior of bull charr and brook charr. Asterisks indicate significance at p = 0.05.

Variables	Effects								
	Periods			Pools			Periods-by-pools		
	F	df	p	F	df	p	F	df	p
					Brook charr				
Foraging microhabitat									
Focal point height	0.02	1,163	0.894	0.10	1,163	0.754	6.45	1,163	0.012*
Focal point velocity	0.07	1,59	0.789	0.02	1,59	0.885	0.39	1,59	0.391
Cover use	5.79	1,216	0.017*	60.91	1,216	< 0.001*	0.13	1,216	0.717
Foraging behavior									
Foraging rate	9.03	1,168	0.003*	0.51	1,168	0.477	0.13	1,168	0.716
Foraging distance	29.69	1,900	< 0.001*	13.11	1,900	< 0.001*	3.70	1,900	0.055
					Bull charr				
Foraging microhabitat									
Focal point height	2.79	2,173	0.064	2.21	1,173	0.139	0.89	2,173	0.413
Focal point velocity	0.06	2,65	0.943	0.49	1,65	0.485	0.21	2,65	0.812
Cover use	9.91	2,244	< 0.001*	36.11	1,244	< 0.001*	0.59	2,244	0.556
Foraging behavior									
Foraging rate	20.19	2,173	< 0.001*	3.85	1,173	0.051	1.32	2,173	0.270
Foraging distance	66.73	2,889	< 0.001*	15.72	1,889	< 0.001*	1.97	2,889	0.140

resulted in significant increases in both foraging rate ($p < 0.001$) and distance ($p < 0.001$) of bull charr, and a decrease in bull charr cover use ($p < 0.001$). For both species, cover use was significantly higher in pool B, but foraging distance was significantly greater in pool A.

Discussion

When the three species were in sympatry, bull charr, brook charr and cutthroat trout interacted directly with each other forming size-structured, mixed-species dominance hierarchies, as has been reported for other combinations of stream salmonid species (e.g. Fausch & White 1986, Glova 1986, Nakano 1995a). Such mixed, interspecific interference can change resource use by individuals in a local group (a pool) and, eventually, a whole population (Morse 1974, Nakano & Furukawa-Tanaka 1994). Under potential influence of interspecific interference, foraging microhabitats and behavior differed considerably between the three species in the present study, with bull charr having the lowest foraging rate and shortest foraging distance, while maintaining focal points closest to the streambed where current velocity was lowest. Some resource use characteristics, however, changed dramatically in response to the experimental removal of potential competitors.

Cutthroat trout removal resulted in changes in cover use, foraging rate and distance of brook charr, but not of bull charr. On the other hand, drastic changes were evident in these three parameters by bull charr following brook charr removal. However, the strength of the competitive influences on bull charr could not be compared between the two potential competitors, because the effect of species difference was not distinguishable from that of the experimentally lowered total fish densities (see Fausch 1988). Therefore, one can only speculate that cutthroat trout were more influential on brook charr than on bull charr under natural conditions. Brook charr probably have a fundamental niche that is intermediate to that of cutthroat trout and bull charr, so cutthroat trout may have a larger potential overlap in resource use with brook charr than with bull charr (see Griffith 1972, 1974, Dambacher et al. 1992, Nakano et al. 1992).

When brook charr were removed from the remaining two-species systems, bull charr made more feeding forays, and to greater distances from their focal points. The increase in both foraging rate and distance in the experimentally allopatric bull charr appeared to result from the release from interference competition formerly exerted by brook charr in the interspecific dominance hierarchy. Moreover, such increases in foraging should result in greater gross energy gain and therefore increase in growth and survival, of bull charr. The significant pools effect on foraging distance was probably due to the higher velocity in pool B, which might well have reduced the foraging distance.

Two foraging microhabitat parameters, focal point height and velocity, of bull charr did not differ during experimental allopatry and sympatry with brook charr. In reality, these variables are regarded to be closely associated with each other, since current velocity decreases downward in the water column of stream pools. It is well known that current velocity at focal points is the most important microhabitat component for many water-column, drift-foraging fish species, being closely linked with species-specific morphologies and physiological constraints (Grossman & Freeman 1987, Hill & Grossman 1993). These characteristics can be rather fundamental niche components and therefore little influenced by competition.

Bull charr occupied positions with less cover following the removal of brook charr. This result, however, was contrary to several observations of competitive influences on microhabitat use in other stream salmonid trophic guilds (Fausch & White 1981, Dolloff & Reeves 1990, but see Lohr & West 1992). For example, Fausch & White (1981) found that brook charr chose foraging positions closer to cover during experimental allopatry than when sympatric with brown trout. The reaches of Elk Creek studied here had abundant instream and overhead cover which covered 10–30% of the area in observation pools. Thus, cover was apparently not in short supply, at least for use by resident fish. Instead, decrease in use of cover by experimentally allopatric bull charr appeared to result from the removal of interference competition with brook charr. In Elk Creek, bull charr appeared to avoid interference by brook charr by occupying positions beneath cover. Visual isolation between fish provided by instream cover, and shaded areas provided by overhead cover, may reduce aggressive encounters (Dolloff 1986, Burgert et al. 1991, Fausch 1993). The significant differences in cover use between pools A and B appeared to be due to the large difference in the abundance of cover between pools.

Two elements of our experimental design prevented us from fully testing interspecific competition among the three salmonids, and comparing it with the strength of intraspecific competition. First, we had no unmanipulated control pools. However, because there were no apparent changes in stream discharge or temperature regimes during the 14 day study, such external factors are unlikely to explain the niche shifts we observed in both pools. Second, changes in fish density between periods prevented us from measuring the strength of interspecific competition relative to that of intraspecific competition (see Fausch 1988). However, reductions in fish densities caused by brook charr removal were partly compensated for by bull charr immigration from other local habitats. This response of bull charr to lowered fish densities also implies that competitive interactions with brook charr play an important role in regulating bull charr densities, at least on a local scale. Several authors have suggested that intraspecific aggressive interactions strongly influence population regulation of stream salmonids (Elliott 1990, Nakano 1995b), and similar mechanisms may operate in sympatric populations of bull charr and brook charr (cf. Fausch & White 1986).

Acknowledgements

We are most grateful to Tom Weaver (Montana Department of Fish, Wildlife, and Parks), Mike Enk and Doug Russell (U.S. Forest Service, Flathead National Forest), Makoto Nagoshi and Keiko Ide (Nara Women's University), and Chuck DeLaney and Eric Ruhlen for their help in the field. Jack

Stanford (Flathead Lake Biological Station, University of Montana) and Leo Marnell (Glacier National Park) provided useful field information. We are also grateful to Hiroya Kawanabe (Kyoto University) for administration of research funding. This research was jointly funded by the Japan Society for the Promotion of Science (grant to Hiroya Kawanabe, Makoto Nagoshi, Koji Maekawa, Shigeru Nakano, and Katsuki Nakai) and the National Science Foundation (grant INT-9016512 from the International and Ecology Programs to Kurt Fausch) under the U.S.-Japan Cooperative Science Program, and supported partly by a Grant-in-aid (No. 07306022, 07308042, 07740592 and 08454246) from the Ministry of Education, Science and Culture of Japan.

References cited

Altmann, J. 1974. Observational study of behavior: sampling methods. Behaviour 49: 226–267.

Bachman, R.A. 1984. Foraging behavior of free-living wild and hatchery brown trout in a stream. Trans. Amer. Fish. Soc. 113: 1–32.

Buckman, R.C., W.E. Hosford & P.A. Dupee. 1992. Malheur River bull trout investigations. pp. 45–57. *In:* P.J. Howell & D.V. Buchanan (ed.) Proceedings of the Gearhart Mountain Bull Trout Workshop, Oregon Chapter of the American Fisheries Society, Corvallis.

Burgert, R.M., T.C. Bjornn & W.R. Meehan.1991. Summer habitat use by young salmonids and their responses to cover and predators in a small southeast Alaska stream. Trans. Amer. Fish. Soc. 120: 474–485.

Cavender, T.M. 1978. Taxonomy and distribution of the bull trout, *Salvelinus confluentus* (Suckley), from the American Northwest. Calif. Fish Game 64: 139–174.

Cavender, T.M. 1980. Systematics of *Salvelinus* from the north Pacific basin. pp. 295–322. *In:* E.K. Balon (ed.) Charr: Salmonid Fishes of the Genus *Salvelinus*, Dr W. Junk Publishers, The Hague.

Dambacher, J.M., M.W. Buktenica & G.L. Larson. 1992. Distribution, abundance, and habitat utilization of bull trout and brook trout in Sun Creek, Crater Lake National Park, Oregon. pp. 30–36. *In:* P.J. Howell & D.V. Buchanan (ed.) Proceedings of the Gearhart Mountain Bull Trout Workshop, Oregon Chapter of the American Fisheries Society, Corvallis.

Dolloff, C.A. 1986. Effects of stream cleaning on juvenile coho salmon and dolly varden in southeast Alaska. Trans. Amer. Fish. Soc. 115: 743–755.

Dolloff, C.A. & G.H. Reeves. 1990. Microhabitat partitioning among stream-dwelling juvenile coho salmon, *Oncorhynchus kisutch*, and dolly varden, *Salvelinus malma*. Can. J. Fish. Aquat. Sci. 47: 2297–2306.

Donald, D.B. & D.J. Alger. 1993. Geographic distribution, species displacement, and niche overlap for lake trout and bull trout in mountain lakes. Can. J. Zool. 71: 238–247.

Elliott, J.M. 1990. Mechanisms responsible for population regulation in young migratory trout, *Salmo trutta*. III. The role of territorial behaviour. J. Anim. Ecol. 59: 803–818.

Everest, F.H. & D.W. Chapman. 1972. Habitat selection and spatial interaction by juvenile chinook salmon and steelhead trout in two Idaho streams. J. Fish. Res. Board Can. 29: 91–100.

Fausch, K.D. 1988. Tests of competition between native and introduced salmonids in streams: what have we learned? Can. J. Fish. Aquat. Sci. 45: 2238–2246.

Fausch, K.D. 1989. Do gradient and temperature affect distributions of, and interactions between, brook charr (*Salvelinus fontinalis*) and other resident salmonids in streams? Physiol. Ecol. Japan. Spec. Vol. 1: 303–322.

Fausch, K.D. 1993. Experimental analysis of microhabitat selection by juvenile steelhead (*Oncorhynchus mykiss*) and coho salmon (*O. kisutch*) in a British Columbia stream. Can. J. Fish. Aquat. Sci. 50: 1198–1207.

Fausch, K.D. & R.J. White. 1981. Competition between brook trout (*Salvelinus fontinalis*) and brown trout (*Salmo trutta*) for positions in a Michigan stream. Can. J. Fish. Aquat. Sci. 38: 1220–1227.

Fausch, K.D. & R.J. White. 1986. Competition among juveniles of coho salmon, brook trout, and brown trout in a laboratory stream, and implications for Great Lakes tributaries. Trans. Amer. Fish. Soc. 115: 363–381.

Fraley, J.J. & B.B. Shepard. 1989. Life history, ecology and population status of migratory bull trout (*Salvelinus confluentus*) in the Flathead Lake and River system, Montana. Northwest Sci. 63: 133–143.

Glova, G.J. 1986. Interaction for food and space between experimental populations of juvenile coho salmon (*Oncorhynchus kisutch*) and coastal cutthroat trout (*Salmo clarki*) in a laboratory stream. Hydrobiol. 131: 155–168.

Griffith, J.S. 1972. Comparative behavior and habitat utilization of brook trout (*Salvelinus fontinalis*) and cutthroat trout (*Salmo clarki*) in small streams in northern Idaho. J. Fish. Res. Board Can. 29: 265–273.

Griffith, J.S. 1974. Utilization of invertebrate drift by brook trout (*Salvelinus fontinalis*) and cutthroat trout (*Salmo clarki*) in small streams in Idaho. Trans. Amer. Fish. Soc. 103: 440–447.

Griffith, J.S. 1988. Review of competition between cutthroat trout and other salmonids. Am. Fish. Soc. Symp. 4: 134–140.

Grossman, G.D. & V. Boule. 1991. An experimental study of competition for space between rainbow trout (*Oncorhynchus mykiss*) and rosyside dace (*Clinostomus funduloides*). Can. J. Fish. Aquat. Sci. 48: 1235–1243.

Grossman, G.D. & M.C. Freeman. 1987. Microhabitat use in a stream fish assemblage. J. Zool. London 212: 151–176.

Haas, G.R. & J.D. McPhail. 1991. Systematics and distribution of dolly varden (*Salvelinus malma*) and bull trout (*Salvelinus *confluentus*) in North America. Can. J. Fish. Aquat. Sci. 48: 2191–2211.

Hill, J. & G.D. Grossman. 1993. An energetic model of microhabitat use for rainbow trout and rosyside dace. Ecology 74: 685–698.

Katano, O. 1990. Dynamic relationships between the dominance of male dark chub, *Zacco temmincki*, and their acquisition of females. Anim. Behav. 40: 1018–1034.

Kitano, S., K. Maekawa, S. Nakano & K.D. Fausch. 1994. Spawning behavior of bull trout in the Upper Flathead Drainage, Montana, with special reference to hybridization with brook charr. Trans. Amer. Fish. Soc. 123: 988–992.

Krueger, C.C. & B. May. 1991. Ecological and genetic effects of salmonid introduction in North America. Can. J. Fish. Aquat. Sci. 48 (Suppl. 1): 66–77.

Larson, G.L. & S.E. Moore. 1985. Encroachment of exotic rainbow trout into stream populations of native brook trout in the southern Appalachian Mountains. Trans. Amer. Fish. Soc. 114: 195–203.

Leary, R.F., F.W. Allendorf & K.L. Knudsen. 1983. Consistently high meristic counts in natural hybrids between brook trout and bull trout. Syst. Zool. 32: 369–376.

Likness, G.A. & P.J. Graham. 1988. Westslope cutthroat trout in Montana: life history, status, and management. Amer. Fish. Soc. Symp. 4: 53–60.

Lohr, S.C. & J.L. West. 1992. Microhabitat selection by brook and rainbow trout in a southern Appalachian stream. Trans. Amer. Fish. Soc. 121: 729–736.

MacCrimmon, H.R. & J.S. Campbell. 1969. World distribution of brook trout, *Salvelinus fontinalis*. J. Fish. Res. Board Can. 26: 1699–1725.

Markle, D.F. 1992. Evidence of bull charr X brook charr hybrids in Oregon. pp. 58–67. *In:* P.J. Howell & D.V. Buchanan (ed.) Proceedings of the Gearhart Mountain Bull Trout Workshop, Oregon Chapter of the American Fisheries Society, Corvallis.

Morse, D.H. 1974. Niche breadth as a function of social dominance. Amer. Nat. 108: 818–830.

Nakano, S. 1995a. Competitive interactions for foraging microhabitats in a size-structured interspecific dominance hierarchy of two sympatric stream salmonids in a natural habitat. Can. J. Zool. 73: 1845–1854.

Nakano, S. 1995b. Individual differences in resource use, growth and emigration under the influence of a dominance hierarchy in fluvial red-spotted masu salmon in a natural habitat. J. Anim. Ecol. 64: 75–84.

Nakano, S., K.D. Fausch, T. Furukawa-Tanaka, K. Maekawa & H. Kawanabe. 1992. Resource utilization by bull char and cutthroat trout in a mountain stream in Montana, U.S.A. Japan. J. Ichthyol. 39: 211–217.

Nakano, S. & T. Furukawa-Tanaka. 1994. Intra- and interspecific dominance hierarchies and variation in foraging tactics of two species of stream-dwelling chars. Ecol. Res. 9: 9–20.

Orth, D.J. 1983. Aquatic habitat measurements. pp. 61–84. *In:* L.A. Nielsen & D.L. Johnson (ed.) Fisheries Techniques, American Fisheries Society, Bethesda.

Ratliff, D.E. & P.J. Howell. 1992. The status of bull charr populations in Oregon. pp. 10–17. *In:* P.J. Howell & D.V. Buchanan (ed.) Proceedings of the Gearhart Mountain Bull Trout Workshop, Oregon Chapter of the American Fisheries Society, Corvallis.

Schoener, T.W. 1983. Field experiments on interspecific competition. Amer. Nat. 122: 240–285.

Tanida, K., K. Yamashita & A. Rossiter. 1985. A portable current meter for field use. Japan. J. Limnol. 46: 219–221.

Wang, L. & R.J. White. 1994. Competition between wild brown trout and hatchery greenback cutthroat trout of largely wild parentage. N. Amer. J. Fish. Manage. 14: 475–487.

Williams, J.E., J.E. Johnson, D.A. Hendrickson, S. Contreras-Balderas, J.D. Williams, M. Navarro-Mendoza, D.E. McAllister & J.E. Deacon. 1989. Fish of North America: endangered, threatened, or of special concern. Fisheries 14 (6): 2–20.

Ziller, J.S. 1992. Distribution and relative abundance of bull charr in the Sprague River subbasin, Oregon. pp. 18–29. *In:* P.J. Howell & D.V. Buchanan (ed.) Proceedings of the Gearhart Mountain Bull Trout Workshop, Oregon Chapter of the American Fisheries Society, Corvallis.

Environmental Biology of Fishes **52**: 357–370, 1998.

Temporal variation in foraging group structure of a size-structured stream fish community

Usio Nisikawa[1] & Shigeru Nakano[2]
Biodiversity Group, Institute of Low Temperature Science, Hokkaido University, Sapporo 060, Japan
[1] *Present address: Department of Zoology, University of Otago, P.O. Box 56, Dunedin, New Zealand (e-mail: nisikawa.usio@stonebow.otago.ac.nz)*
[2] *Tomakomai Research Station, Hokkaido University Forests, Takaoka, Tomakomai 053, Japan*

Received 31.1.1996 Accepted 4.3.1997

Key words: stream fish assemblage, foraging behaviour, omnivore, niche shifts, resource dynamics

Synopsis

Temporal variation in foraging group structure of a fish assemblage was examined in a flood-prone stream in southern Hokkaido, Japan. Foraging behaviour was observed underwater for four species which inhabit the water column: ayu, *Plecoglossus altivelis*, white-spotted charr, *Salvelinus leucomaenis*, masu salmon, *Oncorhynchus masou*, and Japanese dace, *Tribolodon hakonensis*, with each species being categorized into five size classes (species-size group; SSG). Based on foraging behaviour, each SSG of the fish assemblage was classified into one of four foraging groups: algae grazers, drift foragers, benthos-drift foragers, and omnivores, defined as SSG exhibiting similar foraging behaviour. All size classes of ayu, and of charr and salmon were categorized as algae grazers and drift foragers, respectively, throughout the study period. In contrast, size classes of dace were categorized as drift foragers, benthos-drift foragers, or omnivores with the same size classes often assigned to different foraging groups from month to month. Digestive tract contents of the fishes in the four foraging groups reflected their observed foraging behaviour, and foraging groups were therefore regarded as representing trophic groups. Abundance and membership of each foraging group varied in accordance with changes in abundance of SSG due to their growth, immigration, emigration, and/or mortality. Moreover, due to numerical dominance within the assemblage, plasticity in foraging behaviour of small- and medium-sized dace also played a key role in determining variability in the foraging group structure. Relative frequencies of two types of foraging behaviour, algae nipping and benthos foraging, of the small-sized dace were significantly correlated with the level of each resource, whereas no significant relationship was detected between foraging frequencies of the medium-sized dace and either resource. Fluctuations in foraging group structure within this assemblage occurred through niche shifts of some component members and by changes in SSG composition.

Introduction

The persistence or variability of trophic structure in stream fish assemblages has been examined at various spatial and temporal scales (e.g. Horwitz 1978, Schlosser 1982a, b, Grossman et al. 1982, 1985, Bowlby & Roff 1986, Chipps et al. 1994). In these studies, both biotic forces, such as predation or competition, and abiotic forces, such as unpredictable environmental changes, have been considered to contribute to shaping the trophic group composition (Schlosser 1987a). Traditionally, such studies have allocated species to known trophic groups a priori, and have assumed all individuals of the same species or age class to have a fixed niche under all conditions (e.g. Grossman et al. 1982, Schlosser 1982a, b). Consequently, whenever trophic structure varied, it was considered to have been due to changes in species or age class composition of the community. In reality most stream fishes are potentially able to take many different sorts of food, so they can respond to changes in food availability (Wooton 1992). However, many studies on tropic structure have ignored such plasticity in feeding habits of fishes.

Fishes often vary their food habits or habitat use according to the season (Whitaker 1976, Winemiller 1989, Magalhães 1993), habitat structural heterogeneity (Diehl 1992, 1993), resource abundance or availability (Brabrand 1985, Garman & Moring 1993, Lobón-Cerviá & Rincón 1994), position in the shoal (Krause 1993), and presence or absence of competitors (Langeland et al. 1991, Bergman & Greenberg 1994, Nakano & Furukawa-Tanaka 1994) or predators (Power 1987, Schlosser 1987b, 1988, Werner & Hall 1988). Moreover, fishes undergo drastic niche shifts throughout their ontogeny so that different-sized individuals of the same species may function as different species (Werner & Gilliam 1984, Werner 1986), the so-called size-structured population (cf. Ebenman & Persson 1988). Such ontogenetic niche shifts greatly increase the complexity of species interactions and have major consequences for community dynamics (Werner & Gilliam 1984).

Therefore, trophic structure of a stream fish assemblage is expected to change over the short term, even if the species or age class composition remains stable. In particular, in temperate streams, temporal changes in resource abundance associated with seasonality, unpredictble disturbance and/or local interactions within and among trophic levels may promote niche shifts of component members as well as their relative abundances within the community.

Our objective in this study was to clarify the short-term variability of foraging group structure resulting from change in component members of a stream fish assemblage, and to examine the niche shift associated with body size and resource dynamics.

Materials and methods

Study area

The study was conducted from June to September 1994, in an 80 m stretch of the Kokamotu Stream (41°32′ N, 140°00′ E), which discharges into the Sea of Japan on the north coast of the Oshima Peninsula, southern Hokkaido, Japan.

The study reach, located approximately 300 m upstream from the river mouth, was 5–15 m wide, and included three pools and two low-gradient riffles. The streambed was composed mainly of cobbles and pebbles, with small areas of exposed bedrock. The riparian zone vegetation consisted of grass (*Miscanthus sinensis* and *Zizania latifolia*). We measured stream discharge in the study reach three to five times monthly using the U.S. Geological Survey midsection method (Orth 1983). The

Table 1. Environmental conditions in the study reach from June to September 1994. Data on water discharge and water temperature indicate mean ± 1 SE.

Month	Water discharge (m^3 sec^{-1})	Water temperature (° C)		Water surface area (m^2)	Frequency of heavy precipitation*
		Minimum	Maximum		
June	3.8 ± 0.4	12.0 ± 0.4	18.3 ± 0.5	760	0
July	2.9 ± 0.5	16.0 ± 0.0	23.7 ± 0.3	620	2
August	3.7 ± 0.4	18.8 ± 0.2	24.0 ± 0.5	640	4
September	10.6 ± 1.6	17.8 ± 0.4	20.7 ± 1.1	830	5

* Days that received more than 20 mm day^{-1} rainfall.

flow regime fluctuated throughout the study period, and water surface area changed accordingly (Table 1). Water temperature, recorded four or five times monthly with a maximum-minimum mercury thermometer, also varied.

Environmental conditions in the Kokamotu Stream fluctuated greatly due to flooding following heavy precipitation. Stream discharge often exceeded 45 m^3 sec^{-1}, which was 4–15 times as much as summer base flow, during heavy precipitation (> 20 mm day^{-1}; see Table 1).

Life history of fishes

Four fishes which inhabit the water column were studied: ayu, *Plecoglossus altivelis*, white-spotted charr, *Salvelinus leucomaenis*, masu salmon, *Oncorhynchus masou*, and Japanese dace, *Tribolodon hakonensis*. The ayu is amphidromous, and all individuals descend the stream and enter the sea immediately after hatching in late fall and ascend streams the next spring. Ayu inhabit streams from early spring to mid or late fall during their one year of life (Kawanabe & Mizuno 1989). In contrast, the life cycles of the other three species are typically anadromous, with both stream-resident and sea-run type individuals. Stream-residents remain in streams throughout their lives, whereas sea-run fish descend streams after one to several years of riverine life (Kawanabe & Mizuno 1989). During the overall study period, both stream-resident and/or sea-run charr, salmon, and dace were observed in the study site. However, adult sea-run salmon and charr were omitted from the study, since they were rarely seen in the study reach. The stream was also inhabited by floating goby, *Chaenogobius urotaenia*, striped-floating goby, *Chaenogobius* sp., common freshwater goby, *Rhinogobius* sp., flathead goby, *Luciogobius guttatus*, and river sculpin, *Cottus hangiongensis*. We also excluded these benthic fishes from the present study because of the difficulty of direct observations.

Observation of fish behaviour

We made direct underwater observations of fish behaviour from 7:00–19:00 h for a four- or five-day period every month from June to September. The standard length of each fish was estimated to the nearest 10 mm underwater by reference to a nearby ruler. Each species was subsequently classified into five size classes: (1) = standard length ≤ 40 mm (SS); (2) = 50–90 mm (S); (3) = 100–140 mm (M); (4) = 150–190 mm (L); (5) = ≥ 200 mm (LL) (hereafter termed species-size group; SSG). Each SSG was treated separately for each month when analyzed. During the course of the study, a total of 36 SSG, 13 in June, 7 in July, 6 in August, and 10 in September, was recognized in the study reach.

The observers lay motionless in an upstream direction at the downstream end of each pool or riffle, and waited for at least five minutes until fish resumed normal activities. We adopted the focal animal observation method (Altmann 1974); a focal individual was randomly chosen and observed for 3 minutes, during which all foraging acts were recorded. Individuals that were observed for at least 1 minute were also included in the analysis. In each month we observed 150 to 200 individuals, and a total of 703 observations was made during the study. Foraging behaviour was classified into four categories: 1 – nipping (filamentous) algae; 2 – grazing (unicellular) algae; 3 – intercepting drifting invertebrates the water column; and 4 – picking or sucking benthic invertebrates. Since fish were only rarely observed to jump out of the water to catch flying insects, this behaviour was included in drift foraging. After several observations had been made at one location, the observers moved slowly upstream.

The SSG were grouped into several foraging groups based on characteristics of their foraging behaviour. Here we define 'foraging groups' as SSG exhibiting similar foraging behaviour. The SSG for which any of the behaviours comprised > 60% of their feeding bouts were regarded as specialists on that particular resource. Other SSG which failed to meet the above criteria were regarded as generalists using two or more types of resources combined. The generalists were further grouped based on the two most frequent foraging behaviours.

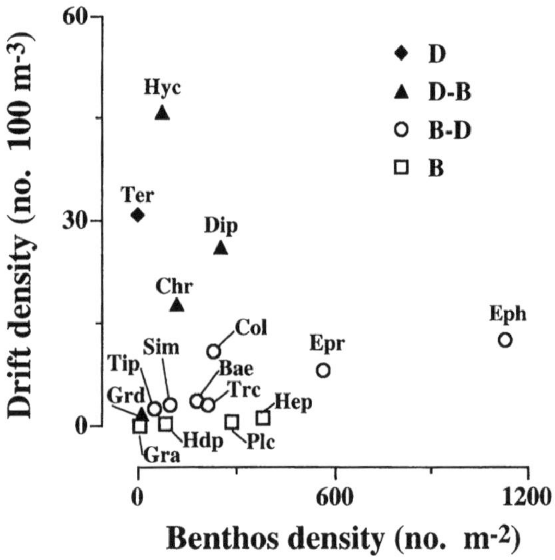

Figure 1. Relationship between drift and benthos density in each invertebrate taxon during June to September 1994. Drift ratio was calculated by mean drift density (n = 24) divided by mean benthos density (n = 20). Aquatic invertebrates were arbitrarily classified into three groups based on the drift ratio: benthos (abbreviated B: drift ratio < 0.01); benthos-drift (B-D: drift ratio 0.01–0.09); and drift-benthos (D-B: drift ratio ≥ 0.10). Terrestrial invertebrates and fish larvae were categorized as drift items (D). Hep = Heptageniidae, Gra = *Goera* (Trichoptera), Hdp = Hydropsychidae, Plc = total Plecoptera, Col = total Coleoptera, Epr = Ephemerellidae, Bae = Baetidae, Eph = total Ephemeroptera, Sim = Simuliidae, Tip = Tipulidae, Grd = *Goerodes* (Trichoptera), Trc = total Trichoptera, Chr = Chironomidae, Dip = total Diptera, Hdc = Hydrachnellae, and Ter = Terrestrial invertebrates.

The number of individuals in each SSG was surveyed from the downstream to the upstream end of the study reach by underwater observation once or twice daily, usually before and/or after daily behavioural observations. We conducted a total of 23 surveys during the study period. The SSG for which mean densities were estimated to be less than 0.005 per m^2 in each month were omitted from the present analysis.

Diet and resource

Representative specimens of each species were sampled twice, in late June and early September during the daytime within an approximately 1 km reach upstream of the study area, with an electrofishing unit (Model 12, Smith-Root Inc.) and a cast net. Captured fish were measured in standard length to the nearest 1 mm after being anesthetized with FA100 (Tanabe Pharmaceutical Co. Ltd). Their digestive tract contents were collected using a stomach pump, and the fish were then released after recovering from the anesthesia. We were able to sample the entire stomach contents for salmon and charr, and the entire alimentary tract contents for dace (Nisikawa & Nakano unpublished data). However, owing to difficulty in sampling stomach contents of ayu with the pump, collected ayu were immediately preserved in 10% formalin solution. Gut contents, preserved in 5% formalin, were examined under a binocular microscope. A volumetric proportion of each prey item of the total prey consumed by each individual was visually estimated on a Petri dish with 5 mm grids. Food items were classified as unicellular or filamentous algae, or terrestrial or aquatic invertebrates, and the aquatic invertebrates were identified to genus or family. Some of these aquatic invertebrate genus/family designations were subsequently converted to a higher taxonomic level for analysis.

Potential food resources for the fishes (drifting and benthic invertebrates, and unicellular and filamentous algae) were assessed in the study reach prior to the observation period each month. Stream drift was sampled three times a day, at early morning, noon, and dusk. In each sampling period, one drift net (mouth opening = 0.04 m^2, length = 1 m, mesh = 100 μm) was set for 20 min in each of the two riffles where water depth was < 20 cm. Care was taken to sample the entire water column, from streambed to water surface. Each drift sample was preserved in 5% formalin and later sorted in the laboratory. Drifting invertebrates were expressed as density (per 100 m^3), which was calculated by dividing the total number of animals in the net by the volume of water sieved (Allan & Russek 1985). The latter factor was estimated from the area of submerged net mouth, current velocity at the center of the net mouth, and the duration of the sampling period. Current velocity was measured with a portable current meter (Tanida et al. 1985) at the end of each sampling period.

Benthic invertebrates were collected with a Sur-

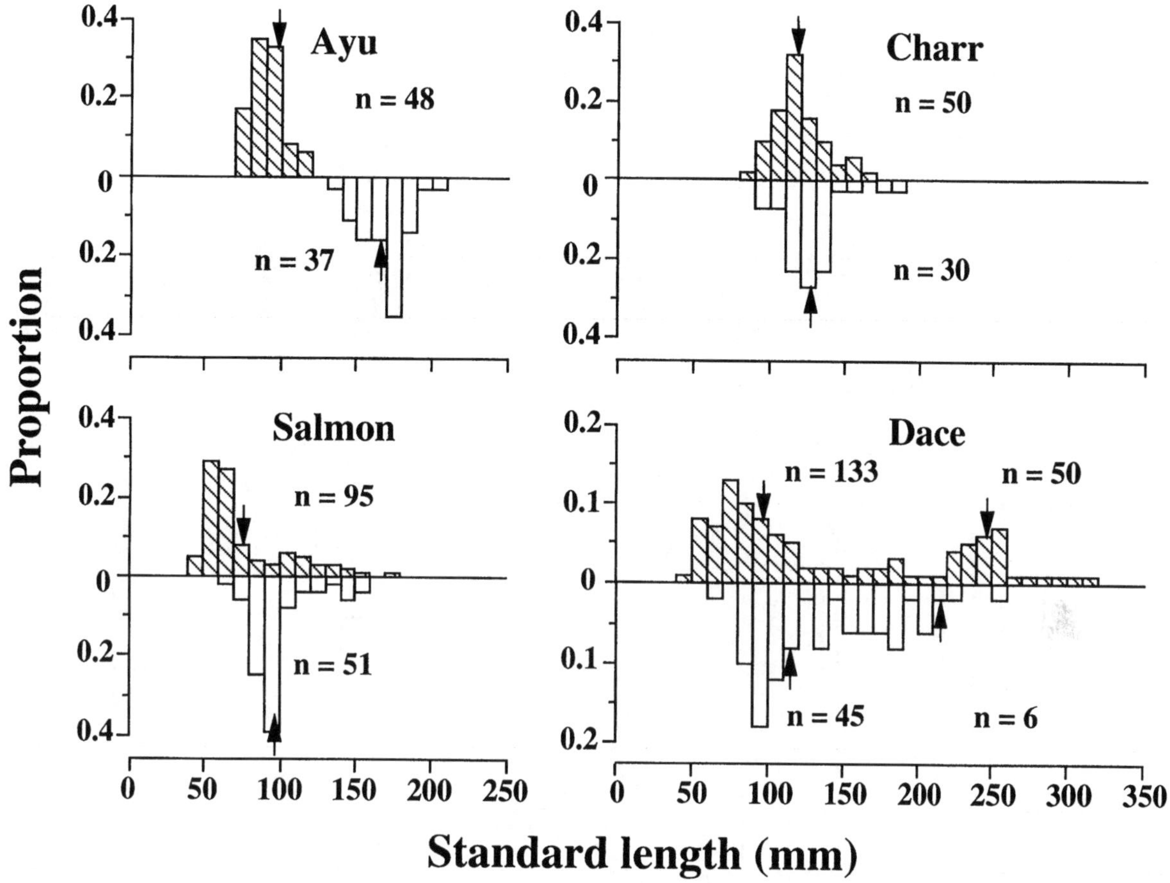

Figure 2. Frequency distributions of standard lengths of four fish species captured in late June and early September 1994. Shaded and open bars indicate June samples and September samples, respectively, arrows show mean standard lengths, n denotes sample size. Since the length frequency distribution for dace was bimodal, they were classified into small (standard length < 200 mm) or large (≥ 200 mm) size classes, and sample size and mean standard length are presented separately for each.

ber net sampler (quadrat area = 0.0625 m^2, length = 1 m, mesh = 100 μm). Five samples were collected every month, one from a random location in each of the three pools and two low-gradient riffles in the study reach. Samples were preserved in 5% formalin for later examination. Benthos was expressed as density (no. m^{-2}), which was calculated by dividing the total number of invertebrates in the net by the quadrat area.

Both drifting and benthic invertebrates were identified following the methods used in dietary analysis. We estimated drift ratio for benthic invertebrates by dividing mean drift density (no. 100 m^{-3}) by mean benthos density (no. m^{-2}) with all the samples combined. Invertebrates were then categorized into benthos (drift ratio < 0.01), benthos-drift (drift ratio 0.01–0.09), and drift-benthos (drift ratio ≥ 0.10) items according to their drift ratio (Figure 1). Terrestrial invertebrates and fish embryos and larvae, which did not occur in benthos samples or appeared only in the digestive tract contents, respectively, were categorized as drift items.

We established ten equally spaced (each 5 m apart) 1 m^2 grids along the center of the stream channel by placing painted cobbles on the streambed every month. Densities of unicellular and filamentous algae were evaluated by assigning points from 1 to 10 (1 = very sparse, 10 = very dense) for each of the ten grids.

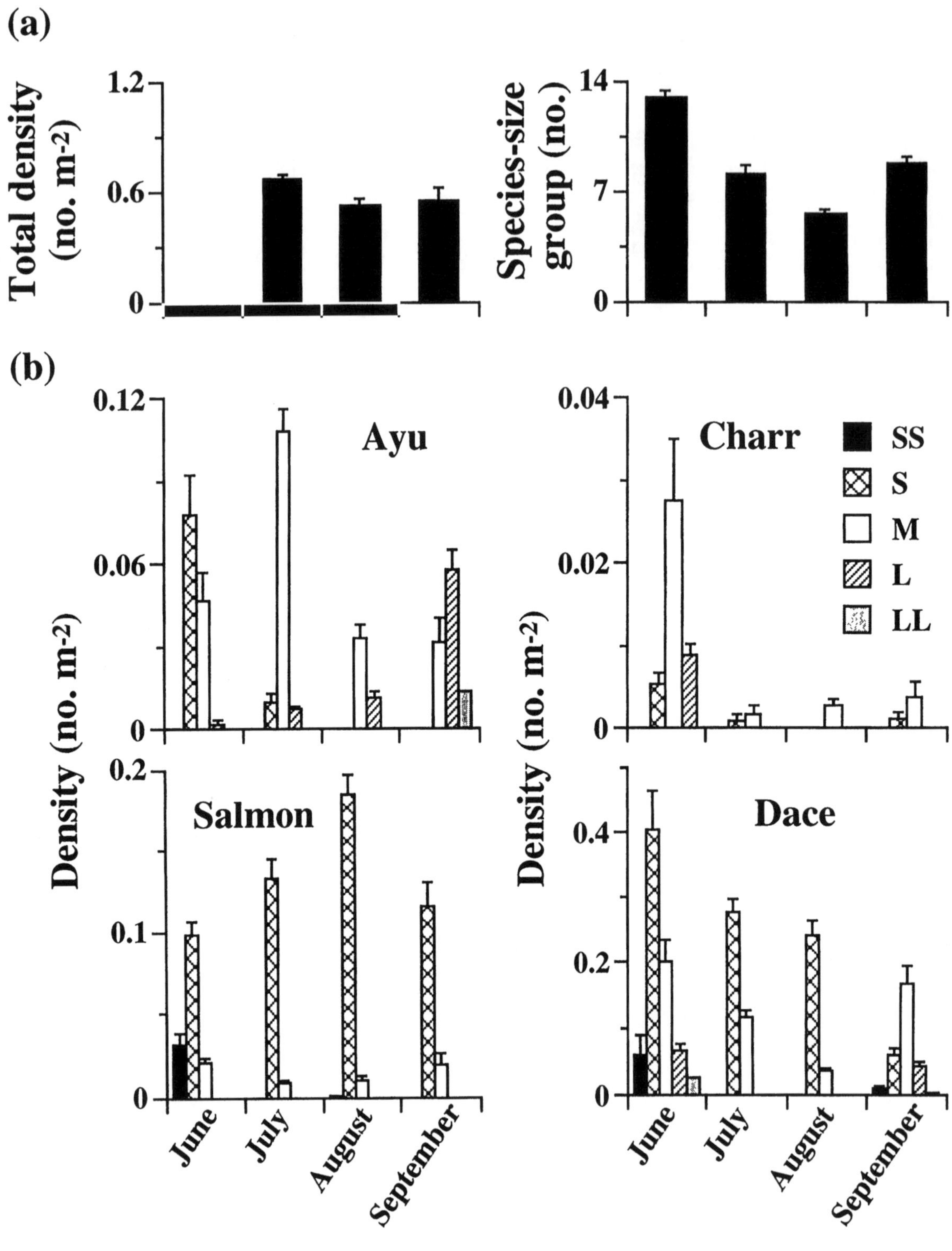

Figure 3. Monthly changes in mean total density and number of species-size groups (a) and mean density and the size-group composition for each species (b) in the study reach from June to September 1994. Data are mean + 1 SE.

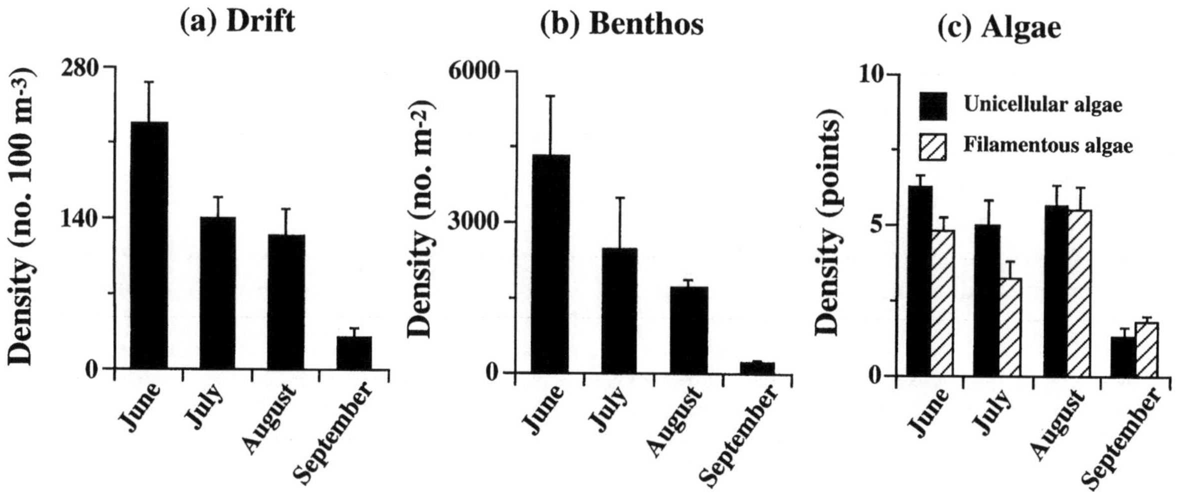

Figure 4. Monthly mean abundance of (a) drift, (b) benthos, and (c) unicellular and filamentous algae in the study reach. Data are mean + 1 SE.

Statistical analyses

Absolute numbers were square-root transformed, and relative frequencies were arcsin-square-root transformed to stabilize variances for statistical analyses (Sokal & Rohlf 1981). All statistical tests were two-tailed. In all cases, statistical significance was evaluated at $p < 0.05$.

Results

Body size and population density

In ayu, charr, and salmon, the mean standard length in September was significantly larger than that in June (t-test; ayu: $t = 90.82$, $p < 0.0001$; charr: $t = 9.23$, $p < 0.0001$; salmon: $t = 22.13$, $p < 0.0001$) with their relative frequency distributions being unimodal in each month (Figure 2). Dace showed bimodality in the length frequency distribution and was classified into small (< 200 mm standard length) and large (≥ 200 mm) size classes, which were regarded as 'stream-resident' and 'sea-run' types, respectively (Nisikawa unpublished data). As in the other species, the mean length of the stream-resident type dace was larger in September than that in June ($t = 19.67$, $p < 0.0001$) but was not in the sea-run type ($t = 0.17$, $p = 0.87$). Thus, growth was evident during June through September in all fishes except the sea-run type of dace.

We detected four size-classes in ayu (S to LL), three in charr (S to L) and salmon (SS to M), and five in dace (SS to LL), during the overall study period. The total number of SSG varied markedly within the four study months (one-way ANOVA: $F_{3,19} = 31.27$, $p < 0.0001$; Figure 3). They decreased from June to August, but increased in September.

The total fish density decreased significantly from June to September when all the SSG were combined (one-way ANOVA: $F_{3,19} = 9.79$, $p = 0.0004$; Figure 3). The densities of all species except salmon decreased as the season proceeded, whereas salmon density increased from June to August, but then decreased slightly in September.

The composition of SSG varied considerably from month to month in all species (MANOVA; ayu: Hotelling-Lawley Trace = 20.20, $F_{12,44} = 24.69$, $p < 0.0001$; charr: Hotelling-Lawley Trace = 41.51, $F_{9,47} = 72.25$, $p < 0.0001$; salmon: Hotelling-Lawley Trace = 10.23, $F_{9,47} = 17.80$, $p < 0.0001$; dace: Hotelling-Lawley Trace = 188.55, $F_{15,41} = 171.79$, $p < 0.0001$). Thus, size-structures and densities of the assemblage members varied within the four-month period.

Resources

Drift, benthos, and algae density varied markedly throughout the study period (one-way ANOVA; drift: $F_{3,20} = 16.76$, $p < 0.0001$; benthos: $F_{3,16} = 11.07$, $p = 0.0004$; unicellular algae: $F_{3,36} = 30.09$, $p < 0.0001$; filamentous algae: $F_{3,36} = 13.26$, $p < 0.0001$; Figure 4). Mean densities of drift and benthos peaked in June and decreased as the season proceeded. Mean densities of unicellular and filamentous algae were relatively high in June through August, and declined in September.

Foraging group structure

Among SSG, which were observed throughout the four months, foraging habits of S- and M-size dace

Table 2. Mean proportions (± 1 SE) of four foraging behaviour: algae nipping (AN), algae grazing (AG), drift foraging (DF), and benthos foraging (BF) exhibited by each species-size group (SSG) from June to September 1994. The SSG were defined as combinations of each of five body size classes (SS to LL) and of four species: ayu, charr, salmon, and dace. n = sample size.

Month	SSG	n	Foraging behaviour			
			AN	AG	DF	BF
June	S-ayu	37	–	0.944 ± 0.037	0.056 ± 0.037	–
	M-ayu	11	–	0.964 ± 0.034	0.036 ± 0.034	–
	S-charr	12	–	–	0.749 ± 0.077	0.251 ± 0.077
	M-charr	33	–	–	0.969 ± 0.015	0.031 ± 0.015
	L-charr	7	–	–	0.917 ± 0.030	0.083 ± 0.030
	SS-salmon	10	–	–	0.992 ± 0.007	0.007 ± 0.007
	S-salmon	34	–	–	0.995 ± 0.004	0.004 ± 0.004
	M-salmon	12	–	–	1.000 ± 0.000	–
	SS-dace	9	0.157 ± 0.140	–	0.450 ± 0.202	0.393 ± 0.208
	S-dace	46	0.088 ± 0.035	0.033 ± 0.021	0.505 ± 0.061	0.375 ± 0.061
	M-dace	19	0.093 ± 0.057	–	0.558 ± 0.103	0.349 ± 0.093
	L-dace	6	–	0.167 ± 0.144	0.583 ± 0.217	0.250 ± 0.217
	LL-dace	4	–	–	1.000 ± 0.000	–
July	S-ayu	20	–	0.909 ± 0.051	0.091 ± 0.051	–
	M-ayu	41	–	0.982 ± 0.009	0.018 ± 0.009	–
	L-ayu	6	–	0.995 ± 0.004	0.005 ± 0.004	–
	S-salmon	35	–	–	0.985 ± 0.008	0.015 ± 0.008
	M-salmon	5	–	–	0.933 ± 0.060	0.067 ± 0.060
	S-dace	40	0.039 ± 0.026	0.014 ± 0.007	0.674 ± 0.055	0.272 ± 0.051
	M-dace	18	0.056 ± 0.053	0.083 ± 0.038	0.610 ± 0.078	0.251 ± 0.073
August	M-ayu	27	–	1.000 ± 0.000	–	–
	L-ayu	12	–	1.000 ± 0.000	–	–
	S-salmon	42	–	–	0.994 ± 0.006	0.006 ± 0.006
	M-salmon	12	–	–	0.992 ± 0.007	0.008 ± 0.007
	S-dace	65	0.314 ± 0.053	0.002 ± 0.001	0.526 ± 0.050	0.158 ± 0.034
	M-dace	6	0.297 ± 0.173	–	0.542 ± 0.175	0.161 ± 0.128
September	M-ayu	15	–	0.900 ± 0.070	0.100 ± 0.070	–
	L-ayu	21	–	0.903 ± 0.064	0.097 ± 0.064	–
	LL-ayu	4	–	0.657 ± 0.200	0.343 ± 0.200	–
	S-salmon	21	–	–	1.000 ± 0.000	–
	M-salmon	16	–	–	1.000 ± 0.000	–
	SS-dace	4	–	–	0.463 ± 0.137	0.537 ± 0.137
	S-dace	11	0.028 ± 0.019	0.048 ± 0.037	0.749 ± 0.078	0.175 ± 0.068
	M-dace	30	0.142 ± 0.054	0.023 ± 0.012	0.579 ± 0.064	0.256 ± 0.051
	L-dace	8	0.067 ± 0.060	–	0.833 ± 0.094	0.100 ± 0.089
	LL-dace	4	–	–	1.000 ± 0.000	–

varied month by month, whereas foraging behaviour of M-size ayu, and S- and M-size salmon was nearly uniform throughout the study period (Table 2). Excepting these five SSG, the others appeared to exhibit either specialized or generalized foraging habits only temporarily. In the present study, we identified two specialist foraging groups, algae grazers and drift foragers, and two generalist groups, benthos-drift foragers and algae-drift foragers (Table 3). The algae-drift foragers were here termed omnivores since they foraged both on animals and plants. The composition of SSG in each foraging group varied monthly (Table 3). The algae grazers group consisted solely of ayu throughout the study period, whereas the drift foragers group was composed primarily of charr and salmon. However, the latter group also included various size groups of dace, particularly in July and September. The benthos-drift foragers and omnivorous groups consisted solely of dace in June and September, and in August, respectively.

Fishes in the algae grazers group fed exclusively on unicellular algae (95% of the diet; Figure 5), whereas in the drift foragers group over 40% of the diet comprised drift items with a considerable amount of benthos-drift items. The diet of the benthos-drift foragers group consisted mainly of benthos-drift, drift-benthos, and drift items, with small amounts of benthos items and filamentous algae. The omnivores group consumed approximately equal proportions of invertebrate items (benthos, benthos-drift, and drift-benthos) and filamentous algae. The diet of fish in each group reflected their foraging behaviour.

The variation in foraging group structure was examined at the assemblage level by considering monthly mean abundance of each SSG (Figure 6, Table 3). The two most abundant species-size groups in the assemblage remained nearly constant during the study period (usually S-size dace and M-size dace). Nevertheless, the most abundant foraging group varied month by month due to the assignment of S-size and/or M-size dace to different foraging groups. As a result of variation in the component members of both abundant and scarce SSG, the foraging group composition varied significantly

Table 3. Members of the four foraging groups and their proportional abundance in the assemblage from June to September 1994. See Table 2 for definition of species size groups (SSG).

Foraging groups	June		July		August		September	
	SSG	Proportion	SSG	Proportion	SSG	Proportion	SSG	Proportion
Algae grazers	S-ayu	0.08	S-ayu	0.01	M-ayu	0.06	M-ayu	0.06
	M-ayu	0.05	M-ayu	0.16	L-ayu	0.02	L-ayu	0.11
			L-ayu	0.01			LL-ayu	0.03
Drift foragers	SS-salmon	0.03	S-salmon	0.20	S-salmon	0.36	S-salmon	0.22
	S-salmon	0.10	M-salmon	0.01	M-salmon	0.02	M-salmon	0.04
	M-salmon	0.02	S-dace	0.42			S-dace	0.11
	S-charr	0.01	M-dace	0.18			L-dace	0.08
	M-charr	0.03					LL-dace	0.01
	L-charr	0.01						
	LL-dace	0.03						
Benthos-drift foragers	SS-dace	0.02					SS-dace	0.02
	S-dace	0.39					M-dace	0.32
	M-dace	0.22						
	L-dace	0.05						
Omnivores					S-dace	0.47		
					M-dace	0.07		

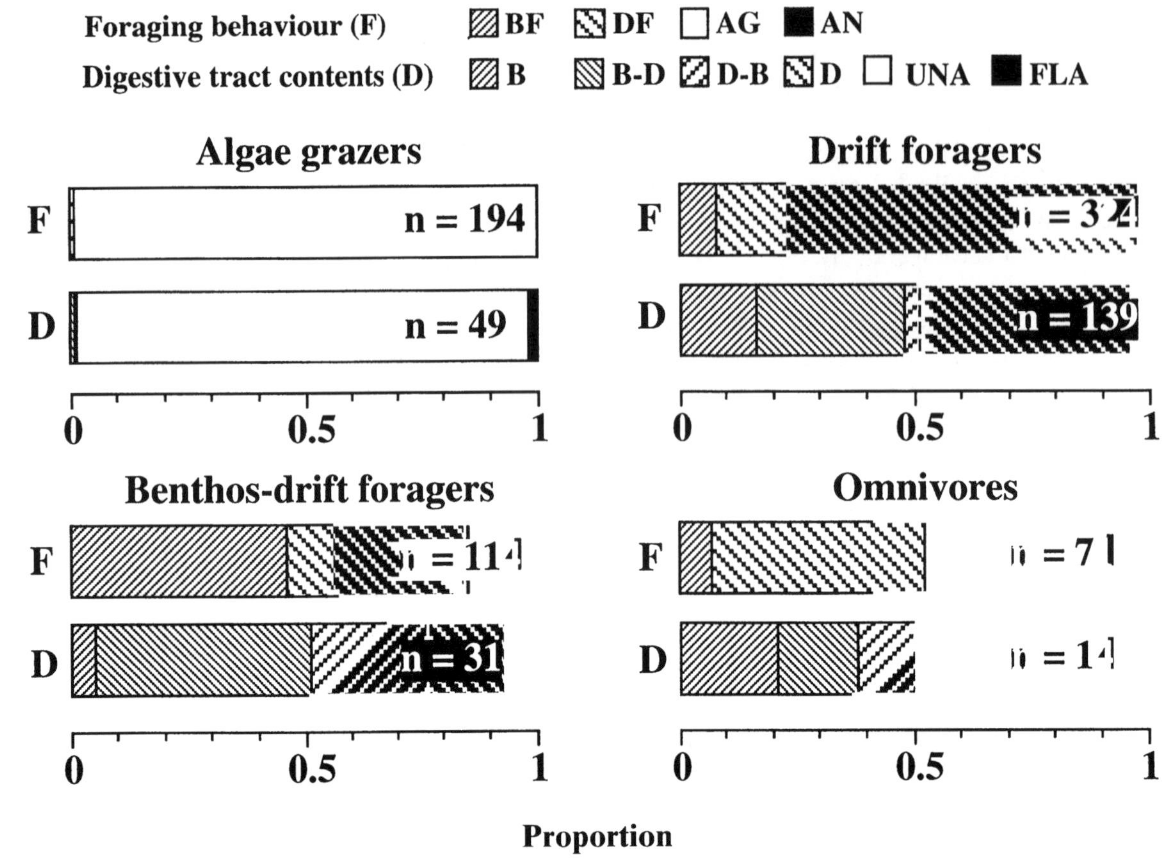

Figure 5. Foraging behaviour (F) and diets (D) of four foraging groups (n = sample size, AN = nipping (filamentous) algae, AG = grazing (unicellular) algae, DF = drift foraging, BF = benthos foraging, B = benthos items, B-D = benthos-drift items, D-B = drift-benthos items, D = drift items, UNA = unicellular algae, and FLA = filamentous algae). See Figure 1 for the definitions of B, B-D, D-B, and D items.

among the four months (MANOVA: Hotelling-Lawley Trace = 37.44, $F_{12,44} = 45.76$, $p < 0.0001$).

To examine variation in the foraging group structure, relationships between resource densities and proportion of each foraging behaviour adopted by five SSG: M-size ayu, S-size salmon, M-size salmon, S-size dace, and M-size dace which were observed throughout the study period, were considered (Table 4). The result of simple linear regression revealed that relative frequencies of algae grazing in M-size ayu were significantly positively correlated with unicellular algae densities. In contrast, relative frequencies of drift foraging in S- and M-size salmon were not significantly correlated with drift densities. Relative frequencies of benthos foraging and algae nipping in S-size dace were significantly positively correlated with benthos and filamentous algae densities, respectively. However, in M-size dace, none of their foraging acts were correlated with any of the resource densities.

Discussion

The SSG composition of the fish assemblage in our study reach showed temporal fluctuations during the four-month study period. The size structures of fishes changed seasonally, partly due to growth (Figure 2). Immigration, emigration, and/or mortality were other causes of the temporal variation in size structure. For example, smaller-sized (SS or S) groups of charr, salmon, and dace reappeared after a temporary disappearance from the study reach, suggesting immigration and emigration of these SSG (Figure 3). Moreover, the increased densities of ayu from August to September, and salmon from

July to August, revealed their immigration, whereas the decreased densities of the other two species from June to August indicated their emigration and/or mortality at the species-level. Such growth, immigration, and emigration and/or mortality also resulted in seasonal fluctuations in the SSG composition of the fishes studied.

Within stream movements are known for various fishes (cf. Schlosser 1995) including salmonids (Nakano et al. 1990, Gowan et al. 1994), cyprinids (Fang et al. 1993), and osmerid ayu (Kawanabe 1970). Although many causes (e.g. feeding, spawning, and wintering) are known (see Northcote 1978, for a review), fish movement may be more common in variable or harsh systems than in stable and benign systems (Gowan et al. 1994). In general, unpredictable environmental disturbances such as floods and droughts strongly affect stream fish assemblage structure (cf. Harrell 1978, Grossman et al. 1982, Schlosser & Toth 1984, Meffe & Sheldon 1990). These harsh physical forces often destabilize the previous environmental conditions such as flow regime, streambed morphology, and resource availability, and cause fishes to emigrate or to immigrate (Schlosser 1982a). The study reach suffered severe floods due to heavy precipitation during the summer (Table 1), which may have caused emigration, immigration, and/or mortality of fishes.

The fish community in the study stream was composed of four foraging groups: algae grazers, drift foragers, benthos-drift foragers, and omnivores. Diets expected from observed foraging behaviour were nearly consistent with their digestive tract contents. Thus, these foraging groups could be regarded as trophic groups. Fish communities in North American streams include trophic groups similar to those detected in the present study. Schlosser (1982a, b) classified North American stream fishes into six trophic groups: herbivore-detritivores, omnivores, general insectivores, surface and water column insectivores, benthic insectivores, and insectivore-piscivores. Although insectivore-piscivores were absent in the Kokamotu Stream, other groups were common. The herbivore-detritivores group corresponds to the algae grazers group of the present study, omnivores to omnivores, general insectivores to benthos-drift

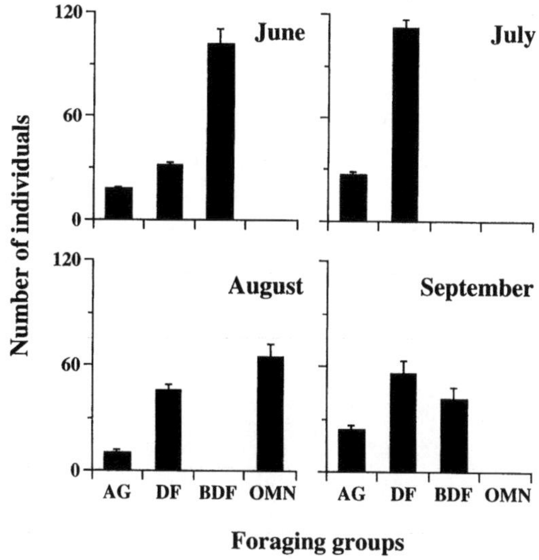

Figure 6. Monthly mean abundance (+ 1 SE) of four foraging groups in the study reach from June to September 1994. DF = drift foragers, AG = algae grazers, BDF = benthos-drift foragers, and OMN = omnivores.

Table 4. Correlation coefficients between relative frequency of each foraging mode exhibited by species-size groups (SSG) and each resource level. Correlation analysis was performed only for SSG which were observed throughout the study period from June to September 1994. U-algae = unicellular algae, F-algae = filamentous algae.

SSG	Resource			
	Drift	Benthos	U-algae	F-algae
M-ayu	-0.193^{NS}	–	0.232*	–
S-salmon	-0.090^{NS}	0.102^{NS}	–	–
M-salmon	-0.081^{NS}	0.095^{NS}	–	–
S-dace	-0.140^{NS}	0.240***	-0.086^{NS}	0.271***
M-dace	0.000^{NS}	0.056^{NS}	-0.025^{NS}	-0.041^{NS}

* $p < 0.05$, *** $p < 0.001$, NS = not significant.

foragers, and surface and water column insectivores to drift foragers. Although the five benthic insectivore fish species present were excluded from observations, this trophic group and the four other studied may be standard components of stream fish communities in the temperate zone.

Water column fishes observed in the study reach included both ecologically conservative fishes such as ayu, charr, and salmon, and ecologically plastic fish such as dace; the former were always classified in a single foraging group, whereas the latter were often assigned to different groups. Ayu and the two salmonid species were very conservative in foraging habits, with the former always belonging to algae grazers and the latter the drift foragers group. Ayu graze algae as soon as they ascend streams, and the formation of their comb-like teeth is consistent with their commencement of riverine life (Miyadi 1960). The observation that the two salmonid fishes primarily intercepted drifting terrestrial or aquatic invertebrates agrees with previous findings (Furukawa-Tanaka 1985, Tanida et al. 1989, Nakano 1995a). However, these salmonid fishes shift to benthos foraging depending on the social status (Nakano & Furukawa-Tanaka 1994, Nakano 1995b, Fausch et al. 1997). The two salmonid fishes and ayu may primarily have morphological and/or physiological constraints that inhibit plasticity in their foraging habits, but under particular environmental circumstances (e.g. low resource levels), they can potentially shift their foraging behaviour (Miyadi 1960, Fausch et al. 1997, Nakano et al. unpublished data). At least under the range of the resource levels observed in this study, they were regarded as conservative in foraging habits.

Dace, however, is known as a typical omnivorous fish, consuming a wide range of animal and plant material (Mizuno et al. 1958, Nakamura 1969, Katano 1987). In contrast to the other three fishes, marked shifts in foraging niche were observed in dace of various size groups in the present study. Plant material is reported to dominate in the diet of cyprinid fishes when the invertebrate food supply declines (Brabrand 1985, Lobón-Cerviá & Rincón 1994). In the present study, there was a positive relationship between the proportion of (filamentous) algae nipping behaviour exhibited by S-size dace and filamentous algae density, and benthos foraging behaviour and benthos density. Moriyama (unpublished data) reported that the Japanese dace shows much plasticity in resource utilization and concluded that it likely changes its diet selection according to changes in resource availability. However, foraging niche shifts of M-size dace were not explained by resource availability. Other factor(s) (e.g. presence of competitors) may have governed their foraging choices.

The composition of each foraging group varied considerably from month to month due to variations in the SSG composition of the assemblage. Studies to date support the idea that the trophic structure of a stream fish assemblage may vary with species or age-class composition (e.g. Schlosser 1982a, b). Furthermore, since dace were abundant in the assemblage (Table 3), their plastic foraging habits directly lead to fluctuations in the total number of the drift foragers group, and to the temporary appearance of the benthos-drift foragers group in June and September and the omnivorous group in August (Table 3, Figure 6). Therefore, foraging niche shifts of S- and M-size dace also played a major role in determining the foraging group composition of the fish assemblage. The appearance or disappearance of both the benthos-drift foragers and omnivorous groups were considered to be partly governed by the resource availability because of the influence of resource levels on the foraging niche of S-size dace. The foraging behaviour of the other fishes had no major effects on the variability of the foraging group composition; rather, variation in their SSG composition and abundance contributed greatly.

In conclusion, this study revealed that the foraging group structure of a stream fish assemblage susceptible to physical disturbance or change varied considerably over the short term due to growth, immigration, emigration, and/or mortality, and niche shifts of smaller-sized (S- and M-size) dace.

Acknowledgements

We are grateful to J. Masanori Toda, A. Rossiter, Y. Taniguchi, and the members of the Biodiversity

Group at the Institute of Low Temperature Science, Hokkaido University for critical comments on an earlier draft of the manuscript. Sincere thanks go to A. Katagiri and T. Segawa for construction of research equipment, and to N. Tuno for assistance in data input. In the field and laboratory we were assisted by Y. Hirota, S. Imai, N. Kuhara, T. Masuyama, K. Satô, K. Simoda, R. Tamura, T. Tamura, Y. Takahasi, N. Tuno, K. Utikawa, S. Wada, and S. Yamamoto. We wish to thank all of them. We must offer special thanks to S. Natume, H. Sugawara, and M. Syôji of Hiyama Experimental Forest, Hokkaido University, for their hospitality during our work. This study was supported partly by a Grant-in-aid (No. 07306022, 07308042, 07740592, 08454246, and 09NP1501) from the Ministry of Education, Science, and Culture of Japan.

References cited

Allan, J.D. & E. Russek. 1985. The quantification of stream drift. Can. J. Fish. Aquat. Sci. 42: 210–215.

Altmann, J. 1974. Observational study of behavior: sampling methods. Behavior 49: 226–267.

Bergman, E. & L.A. Greenberg. 1994. Competition between a planktivore, a benthivore, and a species with ontogenetic diet shifts. Ecology 75: 1233–1245.

Bowlby, J.N. & J.C. Roff. 1986. Trophic structure in Southern Ontario streams. Ecology 67: 1670–1679.

Brabrand, Å. 1985. Food of roach (*Rutilus rutilus*) and ide (*Leusiscus idus*): significance of diet shift for interspecific competition in omnivorous fishes. Oecologia 66: 461–467.

Chipps, S.R., W.B. Perry & S.A. Perry. 1994. Fish assemblages of the central Appalachian Mountains: an examination of trophic group abundance in nine West Virginian streams. Env. Biol. Fish. 40: 91–98.

Diehl, S. 1992. Fish predation and benthic community structure: the role of omnivory and habitat complexity. Ecology 73: 1646–1661.

Diehl, S. 1993. Effects of habitat structure on resource availability, diet, and growth of benthivorous perch, *Perca fluviatilis*. Oikos 67: 403–414.

Ebenman, B. & L. Persson (ed.) 1988. Size structured populations: ecology and evolution. Springer-Verlag, Berlin. 283 pp.

Fang, L.-s., I.-s. Chen, C.-h. Yang, J.-j. Li, J.-t. Wang & M.-c. Liu. 1993. The fish community of a high mountain stream in Taiwan and its relation to dam design. Env. Biol. Fish. 38: 321–330.

Fausch, K.D., S. Nakano & S. Kitano. 1997. Experimentally induced foraging mode shift by sympatric charrs in a Japanese mountain stream. Behav. Ecol. 8: 414–420.

Furukawa-Tanaka, T. 1985. The ecology of salmonid fishes in Japanese mountain streams, I. Food condition and feeding habit of Japanese charr, *Salvelinus leucomaenis* (Pallas). Japan. J. Ecol. 35: 481–504.

Garman, G.C. & J.R. Moring. 1993. Diel and annual production of two boreal river fishes following clearcut logging. Env. Biol. Fish. 36: 301–311.

Gowan, C., M.K. Young, K.D. Fausch & S.C. Riley. 1994. Restricted movement in resident stream salmonids: a paradigm lost? Can. J. Fish. Aquat. Sci. 51: 2626–2637.

Grossman, G.D., P.B. Moyle & J.O. Whitaker, Jr. 1982. Stochasticity in structural and functional characteristics of an Indiana stream fish assemblage. Amer. Nat. 120: 423–454.

Grossman, G.D., M.C. Freeman, P.B. Moyle & J.O. Whitaker, Jr. 1985. Stochasticity and assemblage organization in an Indiana stream fish assemblage. Amer. Nat. 126: 275–285.

Harrell, H.L. 1978. Response of the Devil's River (Texas) fish community to flooding. Copeia 1978: 60–68.

Horwitz, R.J. 1978. Temporal variability patterns and the distributional patterns of stream fishes. Ecol. Monogr. 48: 307–321.

Katano, O. 1987. 'Individual mosaic' hypothesis on ecology of omnivorous cyprinids. Freshwater Fish Final Volume: 55–62 (in Japanese).

Kawanabe, H. 1970. Social behaviour and production of ayufish in the River Ukawa between 1955 and 1969, with reference to the stability of its territoriality. Japan. J. Ecol. 20: 144–151.

Kawanabe, H. & N. Mizuno (ed.) 1989. Freshwater fishes of Japan. Yama-Kei Publishers, Tokyo. 719 pp. (in Japanese).

Krause, J. 1993. The relationship between foraging and shoal position in a mixed shoal of roach (*Rutilus rutilus*) and chub (*Leuciscus cephalus*): a field study. Oecologia 93: 356–359.

Langeland, A., J.H. L'Abee-Lund, B. Jonsson & N. Jonsson. 1991. Resource partitioning and niche shift in arctic charr *Salvelinus alpinus* and brown trout *Salmo trutta*. J. Anim. Ecol. 60: 895–912.

Lobón-Cerviá, J. & P.A. Rincón. 1994. Trophic ecology of red roach (*Rutilus arcasii*) in a seasonal stream; an example of detritivory as a feeding tactic. Fresh. Biol. 32: 123–132.

Magalhães, M.F. 1993. Feeding of an Iberian stream cyprinid assemblage: seasonality of resource use in a highly variable environment. Oecologia 96: 253–260.

Meffe, G.K. & A.L. Sheldon. 1990. Post-defaunation recovery of fish assemblages in southeastern blackwater streams. Ecology 71: 657–667.

Miyadi, D. 1960. A story of ayu. Iwanami-shoten, Tokyo. 226 pp. (in Japanese).

Mizuno, N., H. Kawanabe, D. Miyadi, S. Mori, H. Kodama, R. Ohgushi, A. Kusakabe & Y. Furuya. 1958. Life history of some stream fishes with special reference to four cyprinid species. Bulletin of Physiology and Ecology, College of Science, University of Kyoto 81: 1–48 (in Japanese).

Nakamura, M. 1969. Cyprinid fishes of Japan – studies on the life history of cyprinid fishes of Japan. Research Institute for Natural Resources Press, Tokyo. 455 pp. (in Japanese).

Nakano, S. 1995a. Individual differences in resource use, growth, and emigration under the influence of a dominance hierarchy

in fluvial red-spotted masu salmon in a natural habitat. J. Anim. Ecol. 64: 75–84.
Nakano, S. 1995b. Competitive interactions for foraging microhabitats in a size-structured interspecific dominance hierarchy of two sympatric stream salmonids in a natural habitat. Can. J. Zool. 73: 1845–1854.
Nakano, S., T. Kachi & M. Nagoshi. 1990. Restricted movement of the fluvial form of red-spotted masu salmon, *Oncorhynchus masou rhodurus*, in a mountain stream, central Japan. Japan. J. Ichthyol. 38: 1–5.
Nakano, S. & T. Furukawa-Tanaka. 1994. Intra- and interspecific dominance hierarchies and variation in foraging tactics of two species of stream-dwelling chars. Ecol. Res. 9: 9–20.
Northcote, T.G. 1978. Migratory strategies and production in freshwater fishes. pp. 326–359. *In:* S.D. Gerking (ed.) Ecology of Freshwater Fish Production, Halsted Press, New York.
Orth, D.J. 1983. Aquatic habitat measurements. pp. 61–84. *In:* L.A. Neilsen & D.L. Johnson (ed.) Fisheries Techniques, Amer. Fish. Soc., Bethesda.
Power, M.E. 1987. Predator avoidance by grazing fishes in temperate and tropical streams: importance of stream depth and prey size. pp. 333–351. *In:* W.C. Kerfoot & A. Sih (ed.) Predation: Direct and Indirect Impacts on Aquatic Communities, University Press of New England, Hanover.
Schlosser, I.J. 1982a. Fish community structure and function along two habitat gradients in a headwater stream. Ecol. Monogr. 52: 395–414.
Schlosser, I.J. 1982b. Trophic structure, reproductive success, and growth rate of fishes in a natural and modified headwater stream. Can. J. Fish. Aquat. Sci. 39: 968–978.
Schlosser, I.J. 1987a. A conceptual framework for fish communities in small warmwater streams. pp. 17–24. *In:* W.J. Matthews & D.C. Heins (ed.) Community and Evolutionary Ecology of North American Stream Fishes, University of Oklahoma Press, Norman.
Schlosser, I.J. 1987b. The role of predation in age- and size-related habitat use by stream fishes. Ecology 68: 651–659.
Schlosser, I.J. 1988. Predation risk and habitat selection by two size classes of a stream cyprinid: experimental test of a hypothesis. Oikos 52: 36–40.
Schlosser, I.J. 1995. Dispersal, boundary processes, and trophic-level interactions in streams adjacent to beaver ponds. Ecology 76: 908–925.
Schlosser, I.J. & L.A. Toth. 1984. Niche relationships and population ecology of rainbow (*Etheostoma caeruleum*) and fantail (*E. flabellare*) darters in a temporally variable environment. Oikos 42: 229–238.
Sokal, R.R. & F.J. Rohlf. 1981. Biometry. W.H. Freeman and Company, San Francisco. 859 pp.
Tanida, K., K. Yamashita & A. Rossiter. 1985. A portable current meter for field use. Japan. J. Limn. 46: 219–221.
Tanida, K., T. Maruyama & Y. Saito. 1989. Feeding ecology of Japanese charr (*Salvelinus leucomaenis*) in a high moor and adjacent streams in central Japan. Physiol. Ecol. Japan. Spec. Vol. 1: 279–294.
Werner, E.E. 1986. Species interactions in freshwater communities. pp. 344–358. *In:* J.M. Diamond & T.J. Case (ed.) Community Ecology, Harper & Row, New York.
Werner, E.E. & J.F. Gilliam. 1984. The ontogenetic niche and species interactions in size-structured populations. Ann. Rev. Ecol. Syst. 15: 393–425.
Werner, E.E. & D.J. Hall. 1988. Ontogenetic habitat shifts in bluegill: the foraging rate-predation risk trade-off. Ecology 69: 1352–1366.
Whitaker, Jr. J.O. 1976. Seasonal changes in food habits of some cyprinid fishes from the White River at Petersburg, Indiana. Amer. Midl. Nat. 97: 411–418.
Winemiller, K.O. 1989. Ontogenetic diet shifts and resource partitioning among piscivorous fishes in the Venezuelan llanos. Env. Biol. Fish. 26: 177–199.
Wooton, R.J. 1992. Fish ecology. Chapman and Hall, New York. 212 pp.

Environmental Biology of Fishes **52**: 371–378, 1998.

Food resources of shrimp-eating cichlid fishes in Lake Tanganyika

Masahide Yuma[1], Tetsuya Narita[1], Michio Hori[2] & Takaki Kondo[3]
[1] *Center for Ecological Research, Kyoto University, Otsu 520–01, Japan (e-mail: yuhma@ecology.kyoto-u.ac.jp)*
[2] *Department of Zoology, Faculty of Science, Kyoto University, Kyoto 606-01, Japan*
[3] *Division of Natural Science, Osaka Kyoiku University, Kashiwara 582, Japan*

Received 14.4.1996 Accepted 25.3.1997

Key words: prey species, prey size, diurnal variation of prey abundance, Atyidae

Synopsis

Species, size composition, and diurnal variation in abundance of shrimp (Atyidae) were examined and compared with food habits of 14 species of shrimp-eating cichlids in Lake Tanganyika. In the daytime when the cichlids foraged actively, only one species of shrimp, *Limnocaridina latipes*, was present on rock surfaces. Eleven of the 14 species of shrimp-eaters examined preyed mainly on *L. latipes*; the shrimp-eaters foraging only on rock surfaces preyed exclusively on *L. latipes*, whereas shrimp-eaters also taking foods from the interstices between or under stones and debris preyed on various species of shrimp. Although three kinds of shrimp-eaters, each with a unique prey catching technique and preferred shrimp size, were recognized, all shrimp-eaters preyed on shrimp of greatly overlapping sizes. This fact confirms that partitioning of food resources takes place via different capturing techniques. Moreover, the utilization of the same food resource in a different manner may promote group foraging in some shrimp-eaters, because the shrimp is a suitable animal to hunt in groups and is a quickly renewable food of potential abundance.

Introduction

The cichlid fishes of the African Great Lakes exploit a wide variety of prey by specializing in the utilization of their respective foods (Fryer & Iles 1972). There are, however, many examples showing that groups of cichlids use similar food resources but capture the food in different manners and forage at different feeding sites. Such segregation in feeding behavior has been considered to be one of the factors which supports coexistence of species that depend on similar food resources (Ribbink et al. 1981, Hori 1983, Witte 1984, Yamaoka 1991, Rossiter 1995). Furthermore, in Tanganyikan cichlids exploiting similar food resources, there are several examples of facultative commensalism in feeding behavior that might be a factor which enhances their coexistence (Hori 1987, 1991, Hori et al. 1993, Yuma 1993, 1994). An example is reported in shrimp-eaters; several species of shrimp-eaters forage in a group, and a fish in a mixed-species group gets more foods than when foraging singly. Their various searching and capturing habits might serve to develop group foraging by utilizing a prey which another fish might have overlooked (Hori 1987, Yuma 1994). A critical assumption for understanding these fish behaviors is that shrimp are superabundant. This assumption has been based on observations made during the night. Despite all this, the species, size composition, and diurnal variation in abundance of shrimp exploited by cichlids has not been examined.

In the lake, 13 species of Atyidae shrimp have been recorded (Calman 1899, 1906, Bouvier 1925,

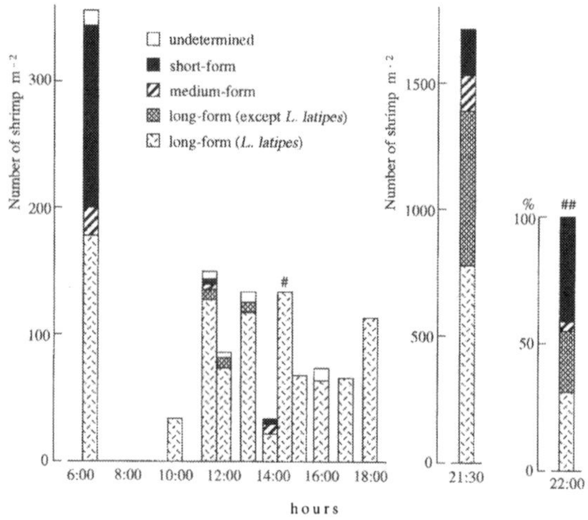

Figure 1. Diurnal changes in the abundance of shrimp on rock surfaces (# = the abundance of shrimp on the undersurface of rocks, ## = the relative abundance of shrimp swimming in the water column).

Coulter 1991). Bouvier (1925) described that these Tanganyikan shrimp were categorized into three types by the proportional size of the sixth abdominal somite; i.e. the long-form shrimp (length of the sixth abdominal somite is more than 2.5 times its width), the middle-form ones (length is about 2 times its width) and the short-form ones (length is less than 1.5 times its width). Preliminary observations suggested that short-form shrimp of *Caridella* were abundant at night on rock surfaces but were rare in the daytime, and that long-form shrimp consisting of *Limnocaridina* species appeared on rock surfaces in the daytime (Narita unpublished). Because the cichlids feed during the daytime, *Limnocaridina* shrimp may be the principal food resource for shrimp-eaters.

This paper describes the actual abundance of shrimp in the daytime as a food resource of shrimp-eating cichlids of the rocky shore of Lake Tanganyika and compares the potential food resource and amount of shrimp consumed by fishes in relation to the species and size composition of the shrimp.

Materials and methods

Collection of shrimp

Underwater sampling of shrimp was carried out with the aid of SCUBA along the rocky shore at Luhanga and Mbemba on the northwestern end of Lake Tanganyika. A 30 × 30 cm quadrat was set on a smooth rock surface that lacked large holes or crevices, at a depth of 3–5 m. Shrimp in two or three quadrats were collected hourly from 10:00 to 18:00 h at Luhanga on 6, 9, and 13 December 1985, by scooping the benthic animals on the rock surface with a handnet of 0.5 mm mesh. Shrimp on the undersurfaces of rocks were collected from two areas of 20 × 50 cm at 14:30 h on 20 February 1980. An early morning sample at 6:30 h of shrimp on a rock surface was collected by scooping in a 30 × 30 cm quadrat at Mbemba on 24 October 1985. A night sample of shrimp at 21:30 h was collected by scooping across the rock surface of an area of 20 × 500 cm. Free-swimming shrimp above the rock surface at 22:00 h were randomly scooped with a handnet at Luhanga on 9 February 1980. Specimens in each sample were preserved in 10% formalin.

Shrimp from the stomachs of cichlids

Specimens of cichlids for stomach contents analyses were collected with a gillnet near the sampling sites for shrimp in 1985. A small amount of concentrated formalin was injected into the abdomen of each fish specimen immediately after collection. Additional fish specimens collected in 1979–1993 at Luhanga were examined. The stomach contents were analyzed by the points method (Hynes 1950).

Fourteen species of cichlids were analyzed in this study: *Altolamprologus compressiceps* (n = 18; standard length (SL) 27–99 mm), *Gnathochromis pfefferi* (n = 19; SL 57–91 mm), *Lamprologus callipterus* (n = 12; SL 61–90 mm), small *Lamprologus lemairii* (n = 10; SL 27–66 mm), small *Lepidiolamprologus elongatus* (n = 16; SL 66–69 mm), small *Lepidiolamprologus profundicola* (n = 2; SL 66–69 mm), *Lobochilotes labiatus* (n = 19; SL 54–262 mm), and small *Neolamprologus fasciatus* (n = 3; SL 57–

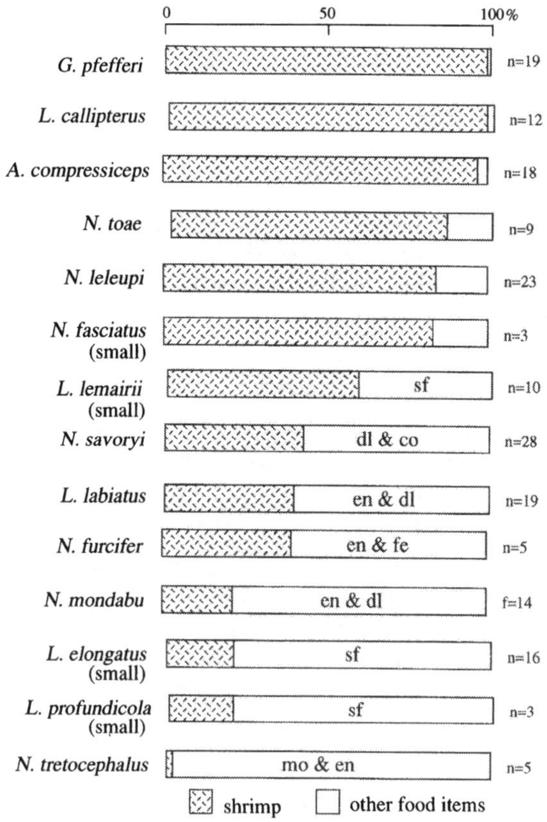

Figure 2. Relative abundance of shrimp in the stomach contents of fourteen benthivorous cichlid fishes. Shaded areas show the proportion of shrimp by volume. Letters in open boxes show the main items found in stomach contents (dl = Diptera larvae, en = Ephemeroptera nymphs, mo = molluscs, fe = fish eggs, sf = small fishes, and n = the number of fishes examined).

85 mm), *Neolamprologus furcifer* (n = 5; SL 53–99 mm), *Neolamprologus leleupi* (n = 23; SL 44–81 mm), *Neolamprologus mondabu* (n = 14; SL 37–76 mm), *Neolamprologus savoryi* (n = 28; SL 31–64 mm), *Neolamprologus toae* (n = 9; SL 56–65 mm) and *Neolamprologus tretocephalus* (n = 5; SL 50–82 mm). These fishes swallowed each prey item whole, so we were able to identify and measure accurately most of the shrimp from the stomachs of cichlids.

Identification and measurement of shrimp

Shrimp collected in the field and from the stomachs of cichlids were identified to the species level ac-

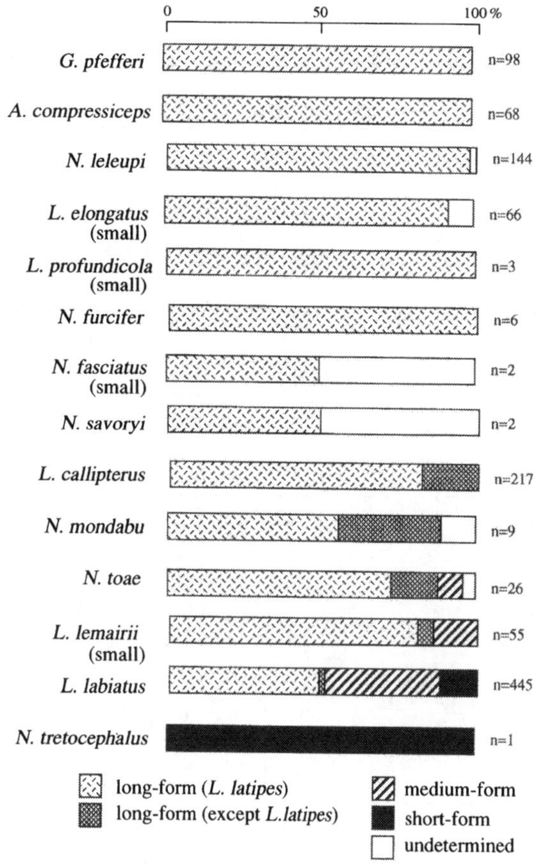

Figure 3. Shrimp composition in the stomach contents of fourteen species of benthivorous cichlid fishes (n = the number of shrimp examined).

cording to Bouvier (1925). One species of *Caridella* and six species of *Limnocaridina* were found in this study. Each species was categorized into one of three types by the proportion of length / width of the sixth abdominal somite, following Bouvier (1925): (1) *Limnocaridina latipes, Limnocaridina parvula, Limnocaridina similis,* and *Limnocaridina tanganyikae* as long-form species, (2) *Limnocaridina socius* and *Limnocaridina spinipes* as medium-form species, and (3) *Caridella minuta* as a short-form species.

The length of the sixth abdominal somite was measured to the nearest 0.025 mm. In this paper, the length of the sixth abdominal somite was adopted as a surrogate of shrimp size because the carapace was frequently damaged in specimens taken from fish stomach samples.

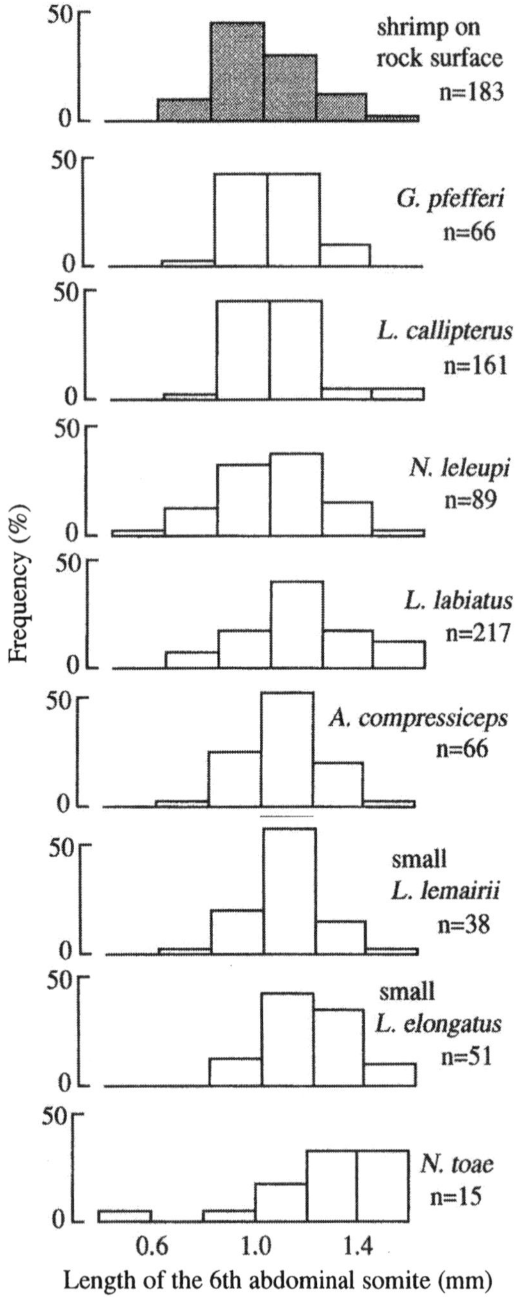

Figure 4. Size frequency distribution of *Limnocaridina latipes* from rock surfaces in the daytime and from the stomach contents of eight species of shrimp-eating cichlid fishes (n is the number of shrimp measured).

Results

Shrimp resources for cichlids

Diurnal changes in the abundance of shrimp are shown in Figure 1. The long-form shrimp, which mainly consisted of *L. latipes*, were present on rock surfaces throughout the day (mean density = 82.7 indiv. m^{-2}, SD = 59.9), whereas the short-form ones were found primarily in the early morning. Species composition and abundance of shrimp on the undersurface of rocks were similar to those on rock surfaces. The long-form shrimp other than *L. latipes* and the medium-form shrimp were rare throughout the day.

Shrimp were much more abundant on the rock surface at night than during the day (Figure 1). The density of *L. latipes* at night was about five times that during the day. The long-form shrimp other than *L. latipes*, and medium- and short-form ones were also abundant at night. Also, many shrimp were found swimming in the water column, where short-form shrimp were relatively more predominant than the long-form ones. These results reveal that the resource available for shrimp-eating cichlids foraging on rock surfaces in the daytime was primarily long-form shrimp, which consisted almost exclusively of *L. latipes*.

Feeding habits of cichlids

Stomach contents of fourteen cichlids, given as volumetric proportions of total stomach contents, are compared in Figure 2. Five species (*G. pfefferi, L. callipterus, A. compressiceps, N. toae, N. leleupi,* and small *N. fasciatus*) preyed almost exclusively on shrimp (more than 80% of total stomach contents). Shrimp were also the main food item for small *L. lemairii* (more than 50% of total stomach contents).

The other cichlids were less dependent on shrimp as their food. Shrimp were, however, a more popular food item than Ephemeroptera nymphs and Diptera larvae in *L. labiatus*, and also shrimp was a numerically more abundant food item than small fish in small *L. elongatus,* and small *L. profundicola.*

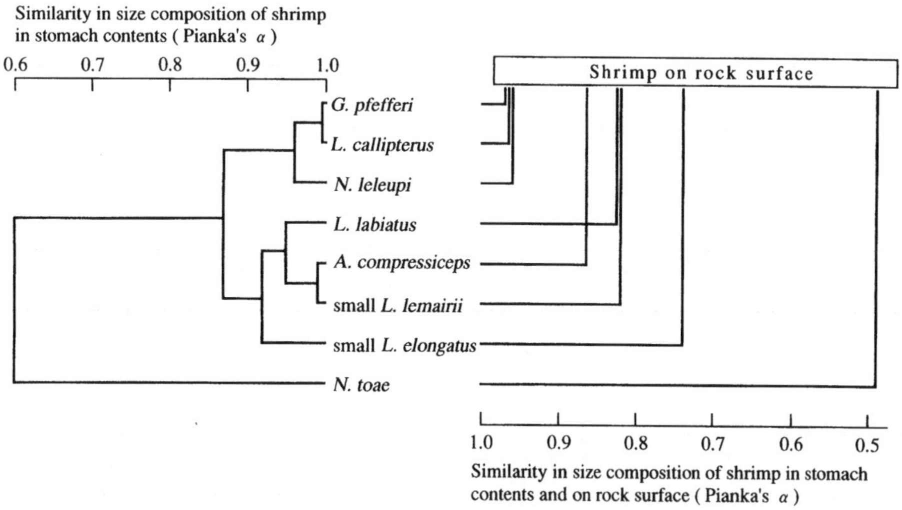

Figure 5. Similarity in the size composition of *Limnocaridina latipes* eaten by 8 species of shrimp-eating cichlid fishes (left side), and similarity in size composition of shrimp in the diet of each cichlid species and those sampled from rock surfaces in the daytime (right side). Similarity was calculated by the α index of Pianka (1973) and the clusters of similarities were analyzed by the method of Mountford (1962).

N. tretocephalus was not a shrimp-eater, and only one fish was recorded as having eaten a shrimp.

Shrimp taken by cichlids

The species composition of shrimp eaten by cichlids is shown in Figure 3. *L. latipes*, being the main potential resource on rock surfaces in daytime, was the main shrimp species consumed by all cichlids except *N. tretocephalus. Lamprologus callipterus*, small *L. lemairii, N. mondabu, N. toae* and *Lobochilotes labiatus* also preyed on a small number of long-form shrimp such as *L. tanganyikae, L. similis,* and *L. parvula*.

The medium-form shrimp, *L. spinipes* and *L. socius*, were eaten by *N. toae*, small *L. lemairii,* and *L. labiatus*. Nearly 40% of shrimp eaten by *L. labiatus* consisted of the medium-form ones. The short-form shrimp were rarely eaten by the cichlids, though they were abundant at night (Figure 1). Only two species of cichlids, *L. labiatus* and *N. tretocephalus*, preyed on short-form shrimp (*C. minuta*).

Size of shrimp as potential and actual food resources

Size-frequency distributions of *L. latipes*, which was the main prey species of cichlids, were compared between those sampled from rock surfaces and those from the stomachs of cichlids (Figure 4). Small *L. profundicola*, small *N. fasciatus, N. furcifer, N. savoryi,* and *N. mondabu* were excluded from this analysis because of their small sample sizes.

Shrimp taken by *G. pfefferi, L. callipterus,* and *N. leleupi* were quite similar in size to each other (Pianka's $\alpha = 0.97$–0.99, Figure 5). Their sizes were similar also to the sizes of shrimp collected from rock surfaces (Pianka's $\alpha = 0.96$–0.99, Figure 5; Kolmogorov-Smirnov test, $p > 0.05$, Figure 6). *L. labiatus, A. compressiceps*, small *L. lemairii,* and small *L. elongatus* ate shrimp of the same size (Pianka's $\alpha = 0.92$–0.99, Figure 5), but they took significantly larger shrimp compared to those on the rock surfaces ($p < 0.001$), and those taken by *G. pfefferi, L. callipterus,* and *N. leleupi* ($p < 0.001$). *N. toae* ate larger shrimp than the other cichlids.

The size-composition of shrimp from rock surfaces and those eaten by the cichlids are very similar except for those taken by *N. toae* (Figure 5). Among the cichlids analyzed, 3 groups of shrimp-eaters were recognized based on the size of shrimp eaten.

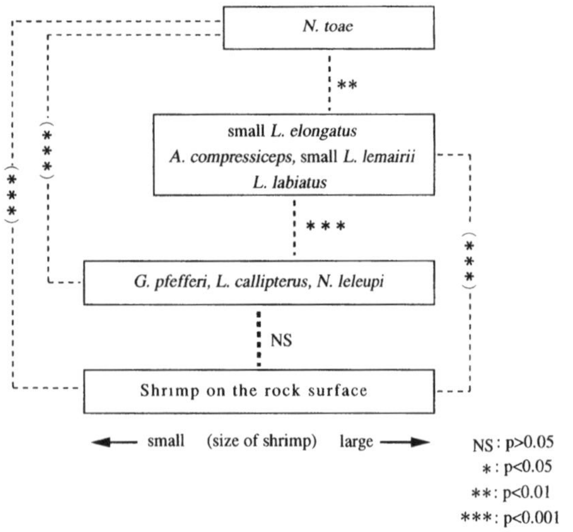

Figure 6. Relationship in the size of shrimp as food resource and utilized by 8 species of shrimp-eating cichlid fishes. Significance was analyzed by the Kolmogorov-Smirnov test.

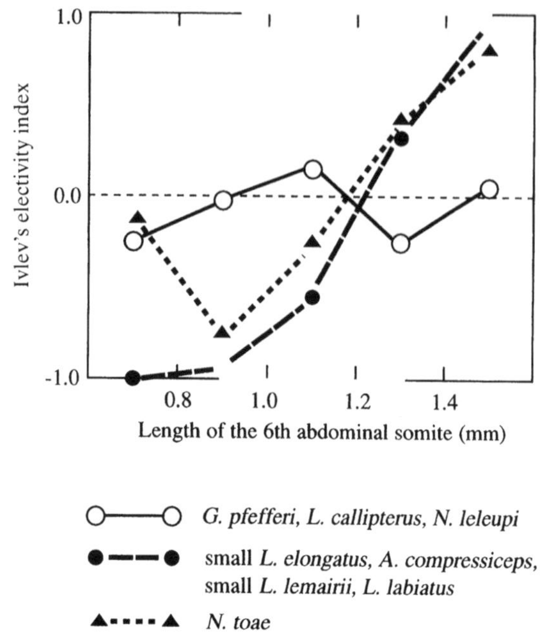

Figure 7. Prey size selectivity of shrimp-eating cichlids.

Species were grouped that had at least 90% similarity in size composition of shrimp eaten based on Pianka's α. Fishes of the first group of shrimp-eaters, *G. pfefferi, L. callipterus,* and *N. leleupi*, seem to take shrimp randomly from rock surfaces based on shrimp size, whereas the fishes of the other two groups selected larger shrimp (Figure 7).

Discussion

Potential food resources for shrimp-eating cichlids

Many shrimp-eating cichlids coexist on the rocky shores of Lake Tanganyika. One of the factors which might enable their coexistence may be the superabundance of shrimp, especially at night (Hori 1983, Yuma 1994). This study revealed that less than one fifth of the density of *L. latipes* captured at night occurred on the rock surface during the day (Figure 1), and that other shrimp present at night were nearly absent during the day. It is believed that they hide under the rocks in daytime. One consequence of the diurnal variation in abundance and availability of shrimp was that most shrimp-eating cichlids preyed exclusively on *L. latipes*, which appeared on the rock surface in the daytime (Figure 3).

Some shrimp-eaters preyed on shrimp other than *L. latipes*. *L. callipterus* and *N. mondabu* frequently foraged around the boundaries between stone and sand (Hori 1983, Yuma 1993, Takemon & Nakanishi 1998), and it is probable that they hunt long-form shrimp other than *L. latipes* in such areas. Although the feeding behaviors of *N. toae* and small *L. lemairii* are unclear, it is possible that they feed during evening or hunt in dark areas such as in holes (Hori 1983), and thus capture a few long-form shrimp other than *L. latipes* and medium-form ones. *L. labiatus* obtains food by sucking debris from various sites (Yuma 1994, Kohda & Tanida 1996), and thus may eat many medium- and short-form shrimp hidden in the debris. Therefore, the shrimp species eaten by shrimp-eaters depends on their mode of foraging; i.e., shrimp-eaters foraging only on rock surfaces preyed exclusively on *L. latipes*, while the shrimp-eaters taking foods also from the interstices of, or under, stones and debris preyed also on other shrimp species.

N. tretocephalus is a sluggish picker of benthic animals and never shows any quick darting behavior towards food items (Yuma 1994). This species is therefore supposed to experience difficulty in capturing agile shrimp, but can capture some less agile short-form shrimp. This means, conversely, that the short-form shrimp on rock surfaces are susceptible to predation. The short-form shrimp are, however, rarely eaten by any shrimp-eaters despite their considerable abundance at night (Figure 1). They hide in deeper areas under rocks in the daytime. On the other hand, the long-form shrimp might have a higher ability to escape from foraging fish by longer fleeing distance made possible with the proportionally longer abdominal somite, and therefore they may be more active on and around rock surfaces in the daytime.

A single species of shrimp, *L. latipes*, was eaten nearly exclusively by cichlids while other shrimp species seemed to be rather free from predation by fish. This might reflect some differences in life histories among shrimp species. To clarify this view, the habitat use and behavior of each shrimp species should be studied in detail.

Partitioning of shrimp resources among cichlids

Figure 6 summarizes comparisons of the size composition of potential and actual food resources exemplified by the shrimp species, *L. latipes*. The small-shrimp eaters (*G. pfefferi, L. callipterus,* and *N. leleupi*) captured shrimp randomly from rock surfaces (Figure 7). The middle-size-shrimp eaters (*L. labiatus, A. compressiceps*, small *L. lemairii,* and small *L. elongatus*) and the large-shrimp eater (*N. toae*) selected larger shrimp or preyed on larger ones as a result of the interaction between shrimp and foraging fish (see below).

G. pfefferi, L. callipterus, and *N. leleupi* capture shrimp by picking them up with no pursuit or only a short dart (Yuma, 1994). The former two species frequently forage with other shrimp-eaters, but the latter does not. Irrespective of their foraging behaviors, these fishes are supposed to capture shrimp that remain stationary and thus eat prey of the same size as those on the rock surface.

A. compressiceps frequently hover near other foraging fish and aim at a shrimp which have escaped from other foragers. Small *L. elongatus* frequently situate themselves behind foraging fish and capture shrimp escaping from the fish in front by a long dart ahead (Hori 1993, Yuma 1994). It is probable that *A. compressiceps* and small *L. elongatus* capture larger shrimp because they flee for a longer distance and are thus easier to find.

L. labiatus is not a picker but a sorter, which takes animals hidden in debris by sorting the edible matter through gill rakers. This species frequently follows other shrimp-eaters and selects a foraging site nearby where some shrimp might take refuge from other foraging fish. It is probable that *L. labiatus* thereby takes larger shrimp which flee for a longer distance from other foraging fish. Therefore, the difference in size composition of shrimp available and eaten is partly caused by size-selection of the cichlids themselves but possibly also due to the relationship between the escaping ability of shrimp and the searching and capturing modes of the cichlids.

Despite the significant differences in shrimp sizes eaten by different fish species, the size ranges overlapped greatly; Pianka's α index was ≥ 0.87 among all species except for *N. toae* (Figure 5). This means that fish utilize greatly overlapping food resources and strongly supports the idea that partitioning of the food resource takes place via different capturing techniques (Hori 1987, Rossiter 1995). Moreover, utilization of the same food resource in a different manner may promote group foraging in some shrimp-eaters (Hori 1987, Yuma 1994), because the shrimp is a suitable animal to hunt in groups and is a quickly renewable food resource of potential abundance. This study substantiates in part facultative commensalism (Hori 1987) and also exploitative mutualism (Matsuda et al. 1993) among the shrimp-eating cichlids in Lake Tanganyika.

Acknowledgements

We thank M.K. Kwetuenda, M.M. Gashagaza, M. Nshombo and other members of the Center for Hydrological Research at Uvira, Zaire, for providing field and laboratory facilities. Thanks go also to M. Nagoshi, H. Kawanabe, and many other Japanese colleagues for their encouragement and discussion during the field work. K. Yamaoka and T. Sato kindly provided some of the fish samples. This study was supported in part by Grants-in-Aid for Overseas Scientific Research (No. 148006, 58041043, 60041036, 02041058, 07044194) and for Cooperative Scientific Research (No. 05304004) from the Japan Ministry of Education, Science, and Culture.

References cited

Bouvier, E.L. 1925. Recherches sur la morphologie, les variations et la distribution systematique des crevettes d'eau douce de la famille des Atyides. Encyclopedie entomologique 4: 1–370.

Calman, W.T. 1899. On two species of macrurous crustacea from Lake Tanganyika. Proc. Zool. Soc. London 1899: 704–712.

Calman, W.T. 1906. Report on the macrurous crustaceas from Lake Tanganyika. Proc. Zool. Soc. Lond. 1906: 704–712.

Coulter, G.W. 1991. Lake Tanganyika and its life. Oxford University Press, London. 354 pp.

Fryer, G. & T.D. Iles. 1972. The cichlid fishes of the great lakes of Africa. Oliver & Boyd, Edinburgh. 461 pp.

Hori, M. 1983. Feeding ecology of thirteen species of *Lamprologus* (Teleostei; Cichlidae) coexisting at a rocky shore of Lake Tanganyika. Physiol. Ecol. Japan 20: 129–149.

Hori, M. 1987. Mutualism and commensalism in a fish community in Lake Tanganyika. pp. 219–239. *In:* S. Kawano, J.H. Connell & T. Hidaka (ed.) Evolution and Coadaptation in Biotic Communities, University of Tokyo Press, Tokyo.

Hori, M. 1991. Feeding relationship among cichlid fishes in Lake Tanganyika: effects of intra- and interspecific variations of feeding behavior on their coexistence. Ecol. Inter. Bull. 19: 89–101.

Hori, M., M.M. Gashagaza, M. Nshombo & H. Kawanabe. 1993. Littoral fish communities in Lake Tanganyika: irreplaceable diversity supported by intricate interactions among species. Conserv. Biol. 7: 657–666.

Hori, M., K. Yamaoka & K. Takamura. 1983. Abundance and micro-distribution of cichlid fishes on a rocky shore of Lake Tanganyika. Afr. Stud. Monogr. 3: 25–38.

Hynes, H.B.N. 1950. The food of fresh-water stickleback (*Gasterosteus aculeatus* and *Pygosteus pungitus*), with a review of methods used in studies of the food of fishes. J. Anim. Ecol. 19: 36–58.

Kohda, M. & K. Tanida. 1996. Overlapping territory of the benthophagous cichlid fish, *Lobochilotes labiatus*, in Lake Tanganyika. Env. Biol. Fish. 45: 13–20.

Matsuda, H., P.A. Abrams & M. Hori. 1993. The effect of adaptive anti-predator behavior on exploitative competition and mutualism between predators. Oikos 68: 549–559.

Mountford, M.D. 1962. An index of similarity and its application to classificatory problems. pp. 43–50. *In:* P.W. Marphy (ed.) Progress in Soil Zoology, Butterworths, London.

Pianka, E.R. 1973. The structure of lizard communities. Ann. Rev. Ecol. Syst. 4: 53–74.

Ribbink, A.J., B.A. Marsh, A.C. Marsh, A.C. Ribbink & B.J. Sharp. 1983. A preliminary survey of the cichlid fishes of rocky habitats in Lake Malawi. S. Afr. J. Zool. 18: 149–310.

Rossiter, A. 1995. The cichlid fish assemblages of Lake Tanganyika: ecology, behaviour and evolution of its species flock. Advances in Ecological Research 26: 187–252.

Takemon, Y & K. Nakanishi. 1998. Reproductive success in female *Neolamprologus mondabu* (Cichlidae): influence of substrate types. Env. Biol. Fish. 52: 261–269 (this volume).

Witte, F. 1984. Ecological differentiation in Lake Victoria haplochromines: comparison of cichlid species flocks in African lakes. pp. 155–167. *In:* A.A. Echelle & I. Kornfield (ed.) Evolution of Fish Species Flocks, University of Maine at Orono Press, Orono.

Yamaoka, K. 1991. Feeding relationship. pp. 151–172. *In:* M.H.A. Keenleyside (ed.) Cichlid Fishes, Chapman & Hall, London.

Yuma, M. 1993. Competitive and cooperative interactions in Tanganyikan fish communities. pp. 213–227. *In:* H. Kawanabe, J.E. Cohen & K. Iwasaki (ed.) Mutualism and Community Organization: Behavioural, Theoretical and Food Web Approaches, Oxford University Press, London.

Yuma, M. 1994. Food habits and foraging behaviour of benthivorous cichlid fishes in Lake Tanganyika. Env. Biol. Fish. 39: 173–182.

Environmental Biology of Fishes **52**: 379–389, 1998.

Feeding habits of largemouth bass in a non-native environment: the case of a small lake with bluegill in Japan

Mikio Azuma[1] & Yasunari Motomura[2]
[1] *Biological Laboratory, Faculty of Education, Nagasaki University, 1-14 Bunkyo-Machi, Nagasaki 852-8521, Japan (e-mail: azumai@net.nagasaki-u.ac.jp)*
[2] *Kyushu Bunka Gakuen Senior High School, 1-8 Yatake-Machi, Sasebo 857-0047, Japan*

Received 6.10.1995 Accepted 2.12.1997

Key words: regurgitation, gobiid fishes, early appearance of piscivory, food selectivity, food availability, ecological release

Synopsis

Little is known about the ecology of largemouth bass, *Micropterus salmoides* in non-native environments. The stomach contents of a total of 381 bass ranging from 21 to 283 mm TL in a small lake in Japan were examined monthly from April to November 1992 by the 'points' method. The proportion of bass with empty stomachs increased in larger fish. The principal foods of the bass regardless of size consisted of two numerically dominant gobiids, the trident goby, *Tridentiger obscurus*, and paradise goby, *Rhinogobius giurinus*. The young-of-the-year (YOY) bass fed exclusively and selectively on the abundant young of paradise goby. As they grew, the bass also consumed fewer but larger trident goby. Bass of all sizes preferred these gobies from spring to autumn when several successive gobiid cohorts recruited from offshore and settled inshore. In contrast, YOY bluegill, *Lepomis macrochirus*, of 10-20 mm TL were selected only during summer. Bass switched from zooplanktivory to piscivory at 30-40 mm TL, much smaller than previously reported in their native range. Presumably because of the high availability and vulnerability of the gobies, the bass became piscivorous earlier and grew quickly. The gobies also allowed the bluegill to largely escape predation risk from the bass.

Introduction

The largemouth bass, *Micropterus salmoides* (Centrarchidae), was introduced into Japan from the United States in 1925. Eighty adult bass around 250 mm in total length were released into Lake Kawahara-oike from Suwanoike Pond near the volcanic Mount Unzen in 1977. The aim of releasing bass into the lake was to control the growing population of bluegill, *Lepomis macrochirus,* and enhance the declining population of deep-bodied crucian carp, *Carassius cuvieri*, for game fishing. The bluegill were stocked in the lake by an unknown person around 1972 and rapidly became dominant. They were thought by anglers to strongly affect the population of deep-bodied crucian carp. However, this attempt at population control of bluegill by introducing bass seemed doomed to failure. At present, populations of both alien fishes seem to have reached stable coexistence, with many bluegill and fewer bass.

Of 11 indigenous fish species in the lake, two landlocked species of gobies, the trident goby, *Tridentiger obscurus*, and paradise goby, *Rhinogobius giurinus*, were abundant enough to make up significant portions of the diets of bass, and age 1+ and older bluegill (Azuma 1992). The population densities of the two gobies have been decreasing annually. Although many studies have been made on feeding ecology of bass in their native range in North America (e.g. Debets 1954, Lewis et al. 1961, 1974, Wright 1970, Paragamian 1978, Adams et al. 1982, Smagula & Adelman 1982, Howick & O'Brien 1983, Storck 1986, Hoyle & Keast 1987, Hambright 1991), little is known about the ecology of this popular game fish in non-native environments (Hickley et al. 1994). In addition, studies of the fish in Japan have been published almost exclusively in Japanese.

The purposes of this paper are first to report on feeding habits of bass in the lake, and second to consider effects of their exclusive foraging on two gobiid fishes and on the bluegill population under these unique conditions. The study is part of a series of investigations in Lake Kawahara-oike on problems of introducing alien fishes to new environments.

Materials and methods

Study sites

Kawahara-oike, a vegetated eutrophic lake of 13 ha and maximum depth of 9 m, is located in Nagasaki Peninsula, the western part of Japan. Koike, a small swamp connected with the northern part of the lake through a narrow inlet, was reclaimed in 1977. Koike was suitable spawning and nursery ground for deep-bodied crucian carp stocked from Osaka Prefecture for game fishing in 1957. The geographical location and a bathymetric map of the lake with survey stations are shown in Azuma (1992) and Figure 1.

The lake has 14 fish species, including three introduced species, the two species of centrarchids, and the deep-bodied crucian carp. Native fishes include four cyprinids (*Carassius gibelio langsdorfi, Cyprinus carpio, Zacco temmincki, Pseudorasbora parva*), four gobiids (*T. obscurus, R. giurinus, R. brunneus, Chaenogobius urotaenia*), medaka, *Oryzias latipes*, Japanese eel, *Anguilla japonica*, and striped mullet, *Mugil cephalus cephalus*. The last two fish species enter the lake from the sea through a sluicegate. There are also four species of shrimp in the lake: *Neocardina denticulata denticulata, Caridina leucosticta, Palaemon paucidens*, and *Macrobrachium nipponense. N. denticulata* is by far the dominant atyid shrimp in the littoral zone. Eight species of submerged macrophytes flourish along the littoral zone almost all year, except for the rocky shores in the southern part of the lake.

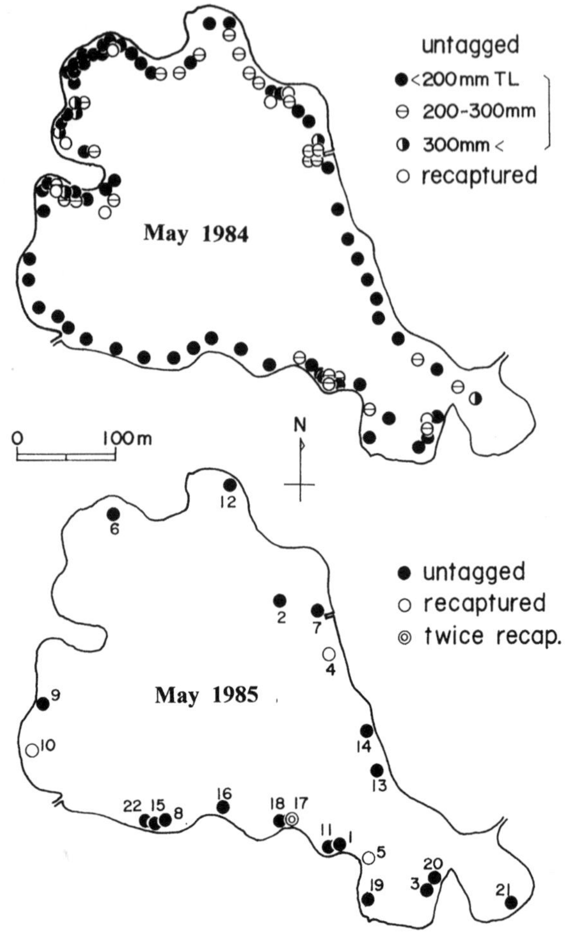

Figure 1. Maps showing the sites of tagged recaptures and untagged bass caught with lures, and the positions of observed untagged bass of each size on 19 May 1984 (top), and the sites of tagged recaptures and untagged bass caught with lures on 18 May (numbers by circles: 1–11) and 19 May (12–22), 1985 (bottom).

In our nighttime survey (unpublished data), phantom midge, *Chaoborus crystallinus*, larvae are distributed all over the limnetic zone, whereas larvae of various chironomids occupy the littoral zone. The same distribution pattern is also found in the dominant *Daphnia longispina* and *Neutrodiaptomus formosus* and subdominant species *Sida crystallina* and *Macrocyclops albidus* of Cladocera and

Copepoda; the dominant species is limnetic, and the subdominant one littoral.

Estimation of population size and seasonal growth

In May 1984 and 1985, bass were caught with lures around the littoral zone of the lake. They were measured (total length = TL, mm), weighed (wet weight, g), marked with numbered anchor tags, and released. Lures were used to recapture bass about two weeks after marking. Recapture sites and positions of observed bass were recorded on a 10 m × 10 m mesh grid map.

From April 1992 to February 1993, bass were collected with lures occasionally, and with a seine 10 m long and 2 m high with 3 mm mesh size (seine II) at 12 stations along the littoral zone once or twice a month. Fish longer than 100 mm TL were measured, contents of their stomachs regurgitated using Miranda's (1986) method, marked with numbered anchor tags, and immediately released where captured. Fish less than 100 mm TL were preserved in 10% neutralized formalin for measurement and examination of stomach contents in the laboratory. Marked fish that were recaptured were released after recording tag number, total length, and collecting the stomach contents by regurgitation.

Data for 1984 and 1985 were used to estimate population sizes of age 1+ and older bass, using the Chapman's modification of the Petersen estimator (Ricker 1975). Population size in 1992 was estimated from the multiple censuses, using Schumacher & Eschmeyer's method (Ricker 1975).

To estimate seasonal growth of bass, the fish caught were allocated to several year-classes based on age structure estimated from size-frequency distributions by Tsutsumi & Tanaka's (1987) method. All samples were pooled from 1984 to 1992, including additional samples collected in 1988 with seine-I (5 mm mesh size). Ages were not determined by scales, otoliths, or other bony structures.

Stomach analysis

The stomach contents regurgitated from each bass larger than 100 mm (145 fish captured by seine, 90 by lures) were collected in a tub of fresh water and preserved in 10% formalin for later examination. Stomach contents of the fish smaller than 100 mm collected by seining were examined by dissection. Modification of Hynes' (1950) 'points' method was employed for enumerating stomach contents of fish of both size groups. Average body weight corresponding to total length of a prey type was used as the point for each food item.

Food selectivity

The selective feeding of bass was quantified using Ivlev's (1961) method. The percentage composition of each prey item in the environment, p_i in Ivlev's electivity index (E_i), was calculated based on a series of collections of zooplankters with a Kitahara quantitative net (Nakai 1962), of epiphytic and terrestrial invertebrates and small fishes with a CM-larval net, and of fishes and macroinvertebrates including shrimps with seine-II. For each food item, the number of individuals per m^3 based on each sampling gear was multiplied by the point. The relative content of any ingredient in the ration, r_i in Ivlev's index, was calculated using percentage points for each food animal in the ration.

Results

Distribution, population size and seasonal growth

Almost all bass were distributed close to the lake shore in 1984 and 1985 (Figure 1). More precisely, their main habitat was found along the outer margin of and above the submerged vegetation that covered about 19% of the lake area and grew thickly throughout the year. Additional angling and gill netting in offshore habitats confirmed that most bass inhabited the edges of littoral aquatic vegetation (unpublished data).

Based on the capture-mark-and-recapture data in May 1984 and 1985, the population size of age 1+ and older bass was estimated at 310 and 266 individuals, respectively (Table 1). Differences between es-

Table 1. Estimates of population size of largemouth bass in Lake Kawahara-oike.

Fishing gears and years	Effective tags recaptured	No. of untagged fish caught	Effective tags at large	Estimates in number	95% range
Lure					
1984*	7	52	46	310	140– 480
1985*	4	22	57	266	85– 446
Seine II					
1992*	5	125	113	1,002	570– 4,110
1992**	8	369	203	3,199	1,671–37,594***

* age 1+ and older.
** whole population including YOY.
*** 90% range.

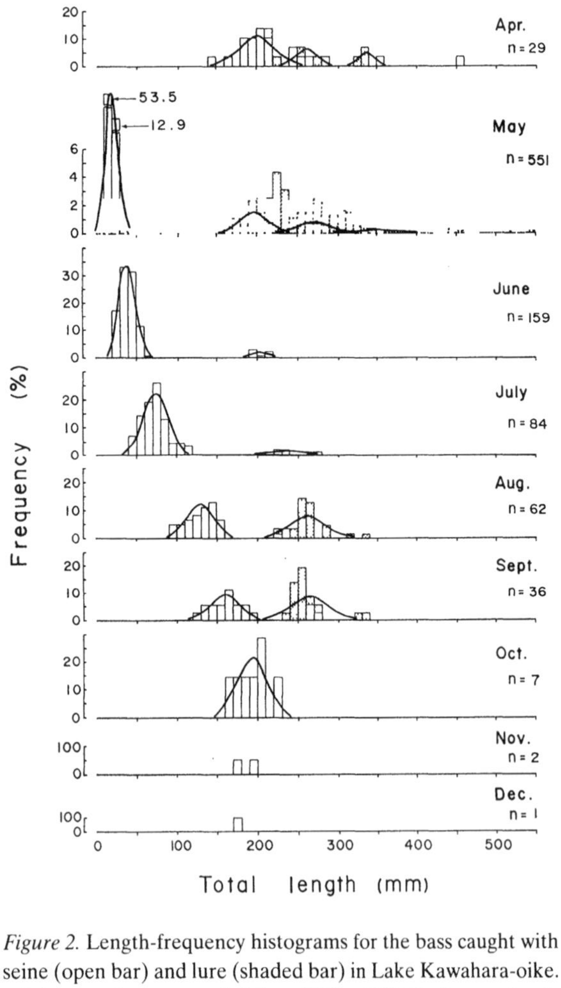

Figure 2. Length-frequency histograms for the bass caught with seine (open bar) and lure (shaded bar) in Lake Kawahara-oike. All the samples were pooled from 1984 to 1992. Each year-class is shown by normal distribution curves fitted to the observed length-frequency distributions of 1992.

timates for these two years were not significant ($p > 0.05$). In 1992, the population size of bass including young-of-the-year (YOY) was estimated at 3199 fish based on the data from 13 successive censuses between April and September, whereas that of 1+ and older bass was 1002 individuals based on the data from intensive censuses in April and May. Thus the proportion of YOY to the total bass population was 69%. This value is approximately coincident with that obtained from the size composition for the individuals collected in May (72%). The abundance of age 1+ and older bass in 1992 was about three times that in 1984–1985.

Seasonal change in length-frequency histograms show that newly recruited young bass began to appear in the littoral zone in May and grew quickly, attaining about 200 mm TL by October (Figure 2). Normal distribution curves were fitted to the histograms, and approximate growth curves developed from the modes (Figure 3). The growth curves show that little growth occured for any year-class from October to April, and that the approximate total length of each year-class in May was estimated at 200 mm for age 1, 270 mm for age 2, and 350 mm for age 3, respectively. For fish of age 3+ and older, annual growth increments were about 50 mm. The largest fish caught during the study was 542 mm TL and was assumed to be age 7+. The approximate growth rate of bass in this study was nearly equal to that in Lake Biwa, the largest lake in Japan (Yamanaka 1989).

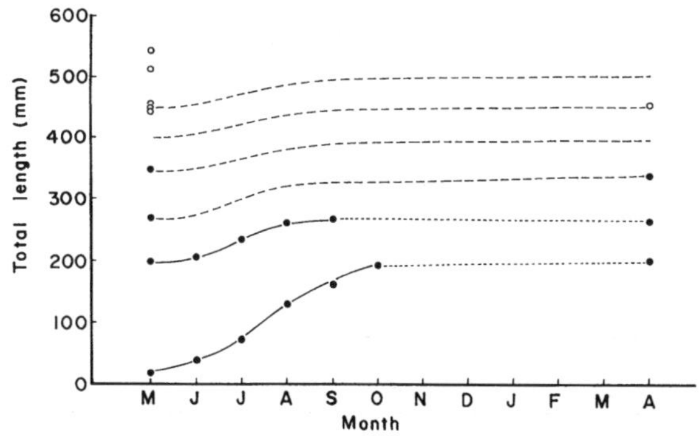

Figure 3. Growth curves for each year-class of largemouth bass in Lake Kawahara-oike obtained from the monthly increase of length. Solid circle: average from the normal distribution curve (Figure 2); open circle= observed value.

Incidence of individuals with empty stomachs

The specimens collected with seine and lure in 1992 were grouped into 20 mm TL size classes. Based on the stomach contents analysis, the proportion of individuals with empty stomachs was investigated in relation to season and size (Figure 4).

Age 1+ and older bass had a high proportion of empty stomachs in April and May, which corresponds to the spawning season when bass were engaged in nest-defense, and seldom fed. In contrast, the proportion of age 1+ and older bass with empty stomachs declined steeply in June and July when bass fed more actively, perhaps recovering from starvation during reproduction. Thereafter, the proportion became higher again in August and September. The YOY bass that appeared in the littoral zone in May apparently foraged actively until October, after which the proportion with empty stomachs increased during winter when density of bass in the littoral zone decreased with falling water temperature. In April and May, most bass above 300 mm TL had completely empty stomachs. This may reflect the high reproductive activity of larger bass in this period. After the reproductive period, the proportion of bass with empty stomachs tended to increase with total length.

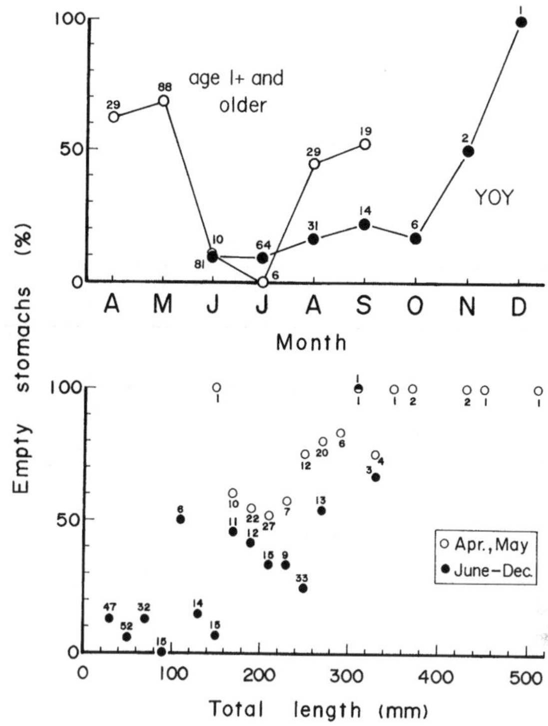

Figure 4. The seasonal incidence of empty stomachs (top) of YOY (solid circle) and age 1+ and older bass (open circle), and the relationship between incidence of empty stomachs and total length (bottom) of bass caught with seine and lure in 1992. Number by circle indicates the number of bass examined.

Dietary composition with size of bass

Diet composition of the bass (Figure 5) showed that their principal food was two gobiid fishes, the trident and paradise gobies, regardless of size of the bass. Surprisingly, even for YOY bass less than 30 mm TL 30% of the diet was young trident goby newly recruited from offshore, whereas about 60% of the diet was cladocerans (mainly *S. crystallina* and *Diaphanosoma brachyurum*). Bass 30 to 100 mm TL ate mainly trident goby as well as some paradise goby, the young of bluegill, the floating goby, *C. urotaenia*, juveniles of cyprinid fishes such as the topmouth gudgeon, *P. parva,* and the silver crucian carp, *C. gibelio langsdorfi*, zygopteran nymphs, and the dominant atyid shrimp, *N. denticulata*. The larger bass 100 mm TL and longer fed principally on the trident and paradise gobies, but also ate bluegill, zygopteran-anisopteran adults, palaemonid shrimp, *P. paucidens*, stink bugs, and young of striped mullet, *M. cephalus*. The only bluegill eaten by bass were YOY individuals, and constituted only

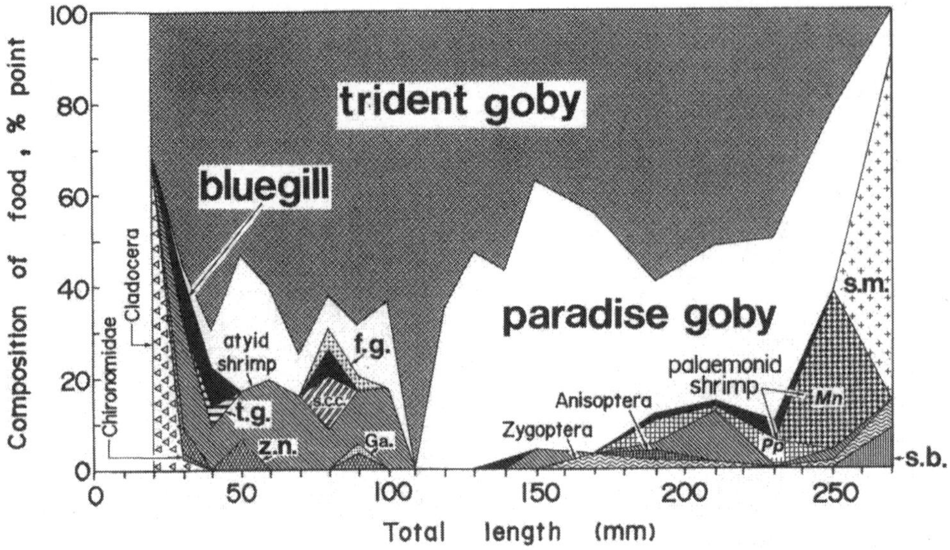

Figure 5. Changes in dietary composition ('points' method) with size of bass (f.g. = floating goby, s.c.c. = silver crucian carp, t.g. = topmouth gudgeon, s.m. = striped mullet, *Mn* = *Macrobrachium nipponense, Pp* = *Palaemon paucidens*, Ga. = Gammaridea, z.n. = zygopteran nymph, s.b. = stink bugs).

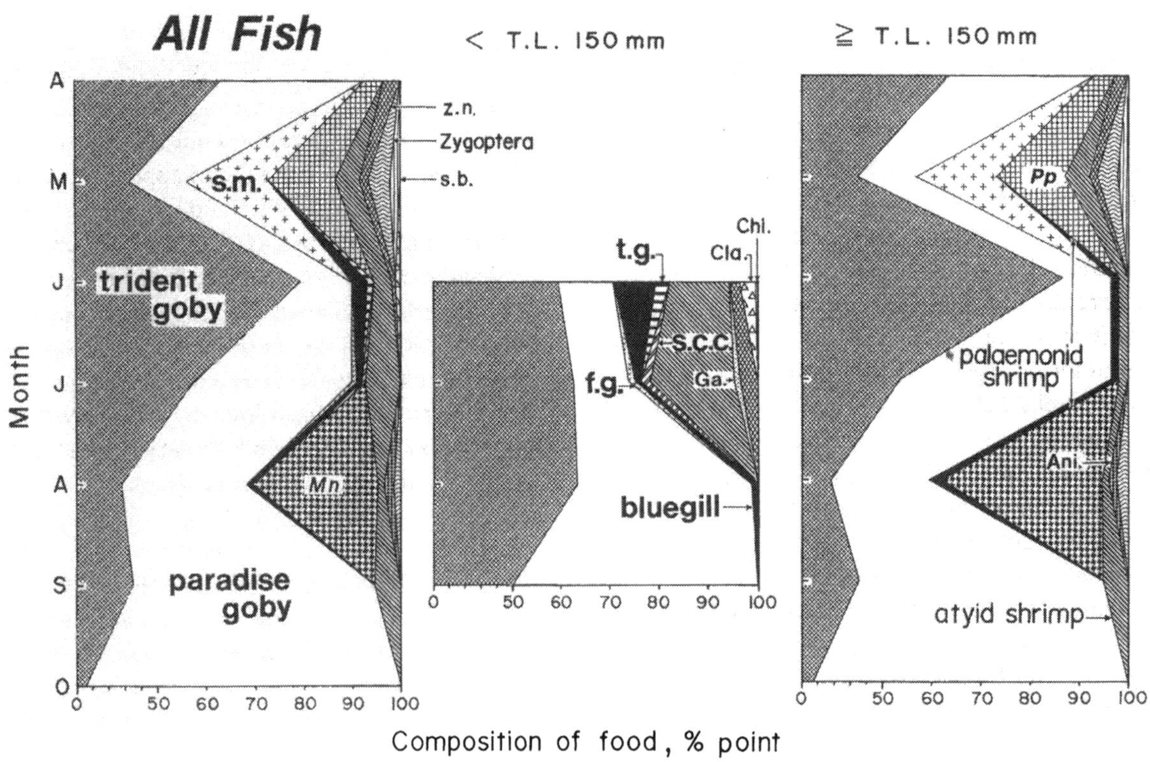

Figure 6. Monthly changes in diet of bass, based on the 'points' method. Left: all fish, middle: smaller fish, right: larger fish (f.g. = floating goby, s.c.c. = silver crucian carp, t.g. = topmouth gudgeon, s.m. = striped mullet, *Mn* = *Macrobrachium nipponense, Pp* = *Palaemon paucidens*, Ga. = Gammaridea, Cla. = Cladocera, Chi. = Chironomidae, z.n. = zygopteran nymph, Ani. = Anisoptera, s.b. = stink bugs).

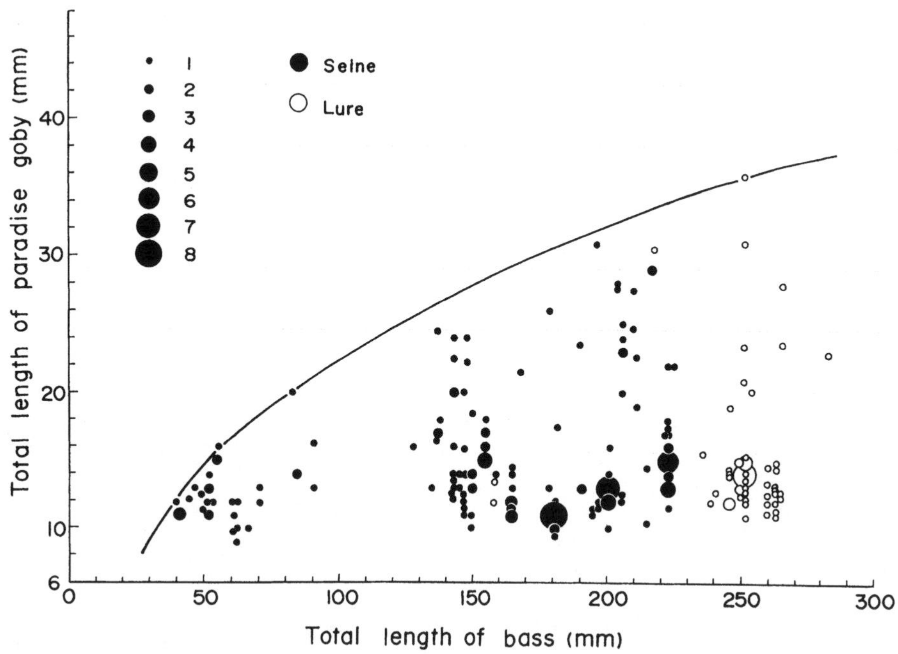

Figure 7. Size distribution of paradise goby consumed by bass of various sizes caught with seine and lure. The numbers of the goby are represented by the size of circles.

a small percentage of the stomach contents (5% or less).

Seasonal change in diet

To examine the seasonal change in diet of the bass, fish were divided into two size groups. Those less than 150 mm TL consumed a variety of prey items, whereas those larger than 150 mm TL ate primarily the two gobiid fishes (Figure 6). The trident goby which made up nearly 80% of the stomach contents of bass in June was replaced thereafter with paradise goby, which accounted for over 90% of the contents in October. Alternation of the principal dietary items seemed to correspond with the seasonal abundance of these two gobiid fishes; i.e. recruitment of the trident goby to the littoral zone occurred in June, about one month before the paradise goby. The YOY bluegill were also eaten by bass less than 150 mm TL from June to August when the bluegill appeared (Azuma 1992). The bass larger than 150 mm TL also fed on YOY bluegill in summer, but the bluegill made up a low percentage of their diet. Juveniles of fishes such as the topmouth gudgeon, silver crucian carp and floating goby were eaten by the smaller size bass only in June and July when these prey fishes began to appear (Azuma et al. unpublished). Young striped mullet were eaten by the larger bass in May. The dominant atyid shrimp was taken by smaller bass during June and July, while larger bass consumed these prey from May to September in relatively low percentage. One palaemonid shrimp, *P. paucidens*, was eaten mainly by smaller bass in June, whereas another *M. nipponense*, was eaten only by larger bass in August.

Relation between size of bass and their prey

Total lengths of prey fish in bass stomachs were estimated by substituting the length of the prey's head remaining in their stomachs into the allometric relationship between total length and head length of intact prey fish. Unidentifiable gobies were distributed into either of two species, according to the monthly ratio of identifiable individuals.

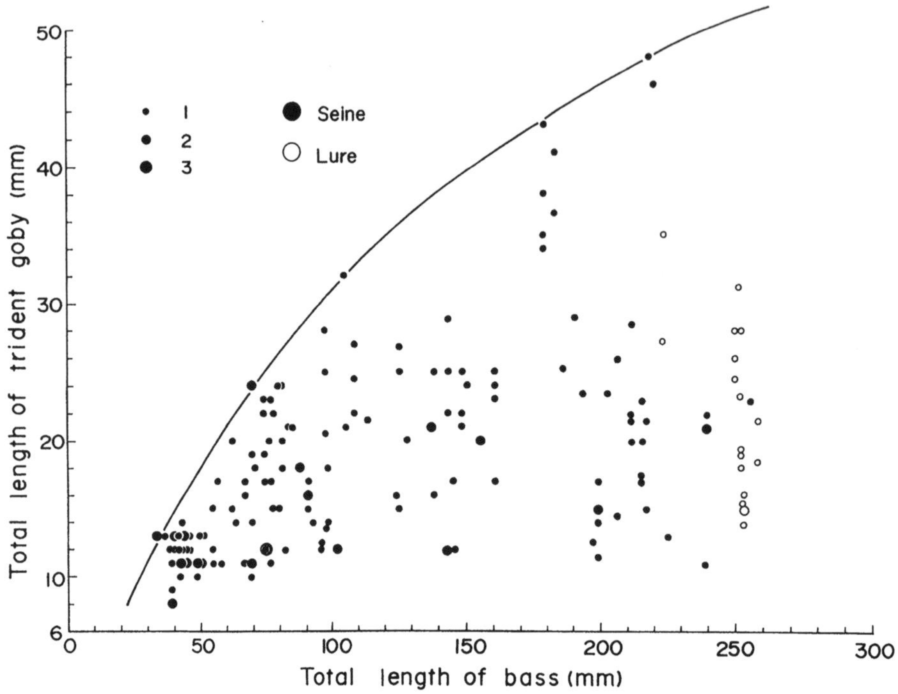

Figure 8. Size distribution of trident goby consumed by bass of various sizes caught with seine and lure. The numbers of the goby are represented by the size of circles.

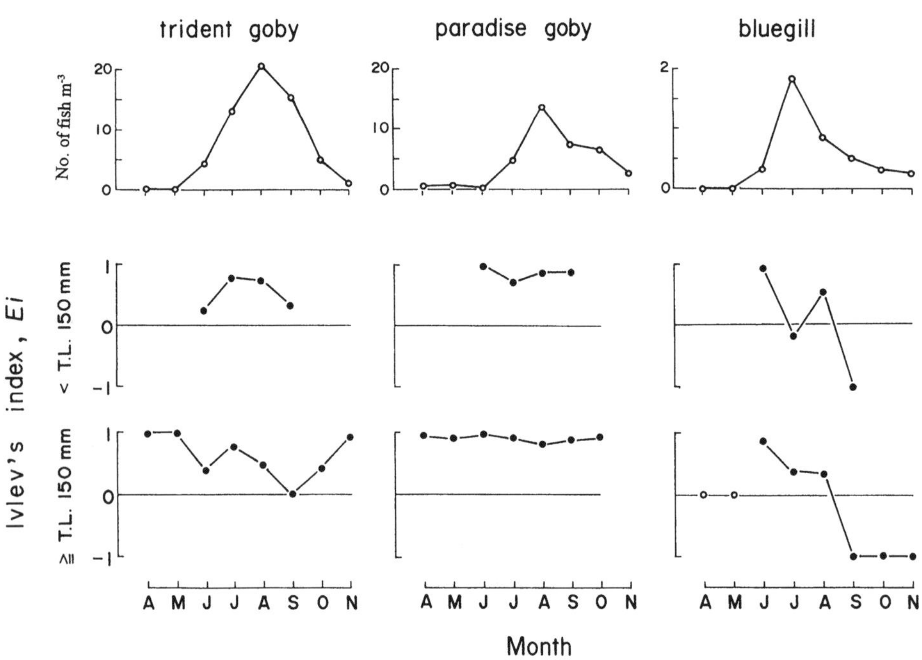

Figure 9. Monthly abundance (top), and monthly changes in food selectivity (E_i) by bass of two size groups (middle and bottom) for two gobiids and YOY bluegill. Two open circles on a zero line of bluegill by larger-sized bass indicate E_i =0, because of p_i =0, r_i =0.

Maximum size of paradise goby eaten by bass increased asymptotically with increasing size of bass (Figure 7) but less steeply than trident goby (Figure 8). Minimum size of the prey remained similar in both species. All sizes of bass ate primarily small paradise goby (10–15 mm TL). In contrast, bass ate all sizes of trident gobies (8–48 mm TL). Maximum and mean number of gobies per bass stomach were 41 and 3.3 fish for paradise goby, whereas these values were 22 and 2.1 fish for trident goby. Bass of a wide range of sizes preferred relatively small paradise gobies and ate many of them, whereas bass had no apparent size preference for the trident goby and ate relatively fewer. Sizes of all other prey fishes, except for one topmouth gudgeon, were distributed below the upper limit of sizes of trident goby eaten.

Food selectivity

Largemouth bass selected trident goby and paradise goby in all months when they were present (Figure 9), except that large bass showed neutral selection of trident goby in September when they had declined from peak abundance. These gobiids were not only highly available to bass of all sizes throughout the growing season, but were selected even when they were not abundant. In contrast, bass of both size classes selected YOY bluegill only during summer when they appeared in the littoral zone, and showed negative values in autumn when YOY bluegill had grown too large to be eaten.

Discussion

Eighty individuals of adult bass introduced into Lake Kawahara-oike in 1977 increased their abundance by three or four times in seven years. Thereafter, the population estimate of age 1+ and older bass in 1992 was about three times more than that of 1984–1985. The latter increase might be partly a result of thinning the bluegill population by angling and seining during our investigations from 1986 to 1989, because bluegill are known to be an effective predator on eggs and smaller bass juveniles (Werner 1977). Although there is little direct evidence for nest predation by bluegill, Azuma (1992, and unpublished data) suggested that age 4+ bluegill fed exclusively on fish eggs, probably those of bass. The approximate population sizes of bluegill were estimated at 80 000 in 1989, 73 000 in 1992, and 45 000 in 1993. Nevertheless, as subsequent surveys in the lake (unpublished data) showed, abundance of bass did not increase after 1992 despite the decrease in number of bluegill by seining from 1992 to 1993. Abundance of bass may have nearly attained carrying capacity. Standing crop of bass was less than 8% of that of bluegill in 1992, which is smaller than the 20% obtained in natural or nearly natural bodies of water in the Great Lakes region of North America (Werner 1977). Therefore, it is highly unlikely that bass interrupted the bluegill's shift from inshore vegetated habitat to the offshore pelagic zone (Mittelbach 1981, 1986, Werner & Hall 1988, Osenberg et al. 1994) in this lake, presumably because the bass is much less abundant than in native environments. Azuma (1992) reported that a sufficient supply of available prey and the absence of any congeneric species or other competitor or predator was the primary reason why the bluegill remained in vegetated habitats throughout their life.

One of the most notable ontogenetic diet shifts of bass in this lake is the early piscivory; e.g., bass only 26 mm TL fed on trident goby. Bass between 30 mm and 40 mm TL had diets of >80% fish prey. This is considerably smaller than the previously known size at which bass became piscivorous. For example, Paragamian (1978) reported that bass switched to piscivory at about 125 mm TL in Big Creek Lake, Iowa. Shelton et al. (1979) found that bass of 40–50 mm TL fed increasingly on fish, and became primarily piscivorous at 75–100 mm. Keast & Eadie (1985) also reported that fish comprised 10–34% of the diet of age 0 bass in Lake Opinicon, Ontario, and that they shifted almost entirely to piscivory by the beginning of their second summer (80–142 mm TL). In Lake Naivasha, Kenya, bass over 260 mm FL fed principally on crayfish, *Tilapia zilli,* and bass juveniles (Hickley et al. 1994). The size at which bass became piscivorous might be regulated primarily by availability of prey fish, as was the case in Lake Naivasha (Hickley et al. 1994).

Another important factor is whether or not the

predators besides bass inhabit the same water body. Generally, there are no genuine piscivorous fish in Japanese freshwaters, except for the piscivorous chub, *Opsariichthys uncirostris,* of Lake Biwa origin. The early appearance of piscivory in bass, not only in Lake Kawahara-oike but also in the other bodies of water in Japan, may reflect the high availability of suitable, small prey fishes such as gobies; the trident and paradise gobies in the present study and the common freshwater goby, *Rhinogobius brunneus,* in Lake Haruna (Yoshizawa et al. 1980), Lake Biwa (Maehata 1992) and other lakes and ponds. The extension of piscivory to smaller sizes implies a type of ecological release in feeding behavior that might have occurred in the bass, as has been reported in bluegills under congener-free conditions (Azuma 1992). Aggus & Elliott (1975) suggested that age 0+ bass that began eating fish early in their growing season attained a much greater length than those that continued feeding exclusively on invertebrates, so it is possible that the early transition to piscivory accounts for the high growth rate of bass in Lake Kawahara-oike. Age-0 bass grew to about 200 mm TL by October of their first year.

Several cohorts of gobiids recruited from offshore and settled along the shore in a seasonal sequence. The trident goby began to settle in the littoral zone in June, about one month earlier than the paradise goby. Age-0 bass selected the young of paradise goby throughout the summer. As they grew, bass consumed fewer but larger trident gobies, which sometimes hovered above submerged macrophytes. Consequently, these fish were more easily attacked by bass than the larger paradise goby, which did not hover but rather settled on bare substrate among the vegetation (Azuma & Mizutani unpublished data).

Gobies were selected by bass from spring to autumn which contrasted markedly to YOY bluegill, which were selected only during summer. This may be due to the bluegill's deeper body and habit of hiding among aquatic plants. Gobies also comprised a substantial proportion of the diets of age 1+ and older bluegill (Azuma 1992). Young bluegill showed a remarkable preference for copepods in the littoral zone, contrasting with bass less than 30 mm TL, which fed more on cladocerans (Figure 5). This phenomenon might reflect interactions between young bass and bluegill for zooplankton. Overall, the high availability and vulnerability of the gobies apparently alows bluegill to escape from heavy predation. In addition, the submerged vegetation provided a complex habitat refuge for the bluegill, which also probably reduced predation risk, as Anderson (1984) and Savino & Stein (1982, 1989) argued. This represents a unique case where indigenous fishes reduce the effects of a nonnative predator on its normal, but also nonnative, prey.

Acknowledgements

We are grateful to many undergraduates who assisted us in the field and laboratory; in particular we thank Toshihide Sato, Masahide Sasa, and Tatsuro Ito for their help, and Ayako Ota for her assistance in preparing the diagrams. Yasuo Iwamoto of Nagasaki Shikimi Senior High School kindly provided us with the unpublished data in Figure 1. We also thank Haruo Tetsumura and Tetsushi Senta of Nagasaki University and the anonymous reviewers for their helpful comments on and improvements of the manuscript.

References cited

Adams, S.M., R.B. McLean & M.M. Huffman. 1982. Structuring of a predator population through temperature-mediated effects on prey availability. Can. J. Fish. Aquat. Sci. 39: 1175–1184.

Aggus, L.R. & G.V. Elliott. 1975. Effects of cover and food on year-class strength of largemouth bass. pp. 317–322. *In*: R.H. Stroud & H. Clepper (ed.) Black Bass Biology and Management, Sport Fishing Institute, Washington, D.C.

Anderson, O. 1984. Optimal foraging by largemouth bass in structured environments. Ecology 65: 851–861.

Azuma, M. 1992. Ecological release in feeding behaviour: the case of bluegills in Japan. Hydrobiologia 243/244: 269–276.

Debets, H. 1954. Feeding habits of the largemouth bass as revealed by a gastroscope. Prog. Fish-Cult. 16: 134–136.

Hambright, K.D. 1991. Experimental analysis of prey selection by largemouth bass: role of predator mouth width and prey body depth. Trans. Amer. Fish. Soc. 120: 500–508.

Hickley, P., R. North, S.M. Muchiri & D.M. Harper. 1994. The diet of largemouth bass, *Micropterus salmoides*, in Lake Naivasha, Kenya. J. Fish Biol. 44: 607–619.

Howick, G.L. & W.J. O'Brien. 1983. Piscivorous feeding behavior of largemouth bass: an experimental analysis. Trans. Amer. Fish. Soc. 112: 508–516.

Hoyle, J.A. & A. Keast. 1987. The effect of prey morphology and size on handling time in a piscivore, the largemouth bass (*Micropterus salmoides*). Can. J. Zool. 65: 1972–1977.

Hynes, H.B.N. 1950. The food of fresh-water sticklebacks (*Gasterosteus aculeatus* and *Pygosteus pungitius*), with a review of methods used in studies of the food of fishes. J. Anim. Ecol. 19: 36–58.

Ivlev, V.S. 1961. Experimental ecology of the feeding of fishes. Yale University Press, New Haven. 302 pp.

Keast, A. & J.M. Eadie. 1985. Growth depensation in year-0 largemouth bass: the influence of diet. Trans. Amer. Fish. Soc. 114: 204–213.

Lewis, W.M., G.E. Gunning, E. Lyles & W.L. Bridges. 1961. Food choice of largemouth bass as a function of availability and vulnerability of food items. Trans. Amer. Fish. Soc. 90: 227–280.

Lewis, W.M., R. Heidinger, W. Kirk, W. Chapman & D. Johnson. 1974. Food intake of largemouth bass. Trans. Amer. Fish. Soc. 103: 277–280.

Maehata, M. 1992. Food habits of largemouth bass in the southern basin of Lake Biwa. Ann. Rep. Biwako Bunkakan 10: 1–8 (in Japanese).

Miranda, L.E. 1986. Removal of stomach contents from live largemouth bass using hydrogen peroxide. North Amer. J. Fish. Manag. 6: 285–286.

Mittelbach, G.G. 1981. Foraging efficiency and body size: a study of optimal diet and habitat use by bluegills. Ecology 62: 1370–1386.

Mittelbach, G.G. 1986. Predator-mediated habitat use: some consequences for species interactions. Env. Biol. Fish. 16: 159–169.

Nakai, Z. 1962. Apparatus for collecting macroplankton in the spawning surveys of Iwashi (sardine, anchovy, and round herring and others). Bull. Tokai Reg. Fish. Res. Lab. 9: 221–237.

Osenberg, C.W., M.H. Olson & G.G. Mittelbach. 1994. Stage structure in fish: resource productivity and competition gradients. pp.151–170. *In*: D.J. Stouder, K.L. Fresh & R.J. Feller (ed.) Theory and Application in Fish Feeding Ecology, University of South Carolina Press, Columbia.

Paragamian, V.L. 1978. Food habits of largemouth bass (*Micropterus salmoides*) at Big Creek Lake. Proc. Iowa Acad. Sci. 85: 31–34.

Ricker, W.E. 1975. Computation and interpretation of biological statistics of fish populations. Bull. Fish. Res. Board Can. 191:1–382.

Savino, J.F. & R.A. Stein. 1982. Predator-prey interaction between largemouth bass and bluegills as influenced by simulated, submersed vegetation. Trans. Amer. Fish. Soc. 111: 255–266.

Savino, J.F. & R.A. Stein. 1989. Behavior of fish predators and their prey: habitat choice between open water and dense vegetation. Env. Biol. Fish. 24: 287–293.

Shelton, W.L., W.D. Davies, T.A. King & T.J. Timons. 1979. Variation in the growth of the initial year class of largemouth bass in West Point Reservoir, Alabama and Georgia. Trans. Amer. Fish. Soc. 108: 142–149.

Smagula, C.M. & I.R. Adelman. 1982. Day-to-day variation in food consumption by largemouth bass. Trans. Amer. Fish. Soc. 111: 543–548.

Storck, T.W. 1986. Importance of gizzard shad in the diet of largemouth bass in Lake Shelbyville, Illinois. Trans. Amer. Fish. Soc. 115: 21–27.

Tsutsumi, H. & M. Tanaka. 1987. A new method for measurement of body size and analysis of generation structure of population, using a micro-computer. Benthos Research 31: 18–28 (in Japanese).

Werner, E.E. 1977. Species packing and niche complementarity in three sunfishes. Amer. Nat. 111: 553–578.

Werner, E.E. & D.J. Hall. 1988. Ontogenetic habitat shifts in bluegill: the foraging rate-predation risk trade-off. Ecology 69: 1352–1366.

Wright, L.D. 1970. Forage size preference of the largemouth bass. Prog. Fish-Cult. 32: 39–42.

Yamanaka, O. 1989. Report of collective surveys for measures against largemouth bass in 1985–1987. Sci. Rep. Shiga Pref. Fish. Exp. Sta. 40: 63–92 (in Japanese).

Yoshizawa, K., Y. Takayanagi, M. Mogi, S. Kobayashi, K. Nobosawa, A. Sato, T. Ikeda & M. Murata. 1980. Resource ecological studies on a warm water piscivore-IV. Feeding ecology of largemouth bass in Lake Haruna. Bull. Gunma Pref. Fish. Exp. Sta. 29: 32–45 (in Japanese).

Environmental Biology of Fishes **52**: 391–394, 1998.

A new perspective on lakes: Kawanabe's latest achievements

David L.G. Noakes
Department of Zoology and Axelrod Institute of Ichthyology, University of Guelph, Guelph, Ontario, N1G 2W1, Canada (e-mail: dnoakes@uoguelph.ca)

In October 1996 the Lake Biwa Museum opened in Oroshimo, Japan after 10 years of planing and construction. It is a most ambitious concept and a very impressive facility. By June of 1997 it had attracted almost 800 000 visitors – almost twice the attendance predicted for the entire first year of operation.

The founding Director General of the Lake Biwa Museum is Hiroya Kawanabe. After a distinguished career as professor of ecology at Kyoto University, Hiroya Kawanabe founded the Centre for Ecological Research at Otsu. After retirement from the university he moved to his new position at the Lake Biwa Museum. Kawanabe-san is internationally recognized for his pioneering and collaborative research on the ayu, *Pleoglossus altivelis*, Lake Biwa and the ecology of Lake Tanganyika fishes (Kawanabe 1996, Kawanabe et al. 1997). The Lake Biwa Museum is not only the pinnacle of his personal and professional achievements, but also the physical embodiment of his philosophy of science and life.

The Lake Biwa Museum is situated on the shore of the southern basin of Lake Biwa, the largest and oldest lake in Japan. It is dramatic in design and innovative in concept (Figure 1 and figures on p. 34 of this volume). The Museum is a unique combination of public displays, education, research and public service. The Lake Biwa Museum is remarkable because of its mandate from Shiga Prefecture, the source of its financial support. It combines human cultural, ecological and biological aspects. With a permanent staff of more than 20 scientists, and technical staff the Museum is housed in more than 23 000 m^2 of floor space. This includes the large public display area, a facility of captive breeding of fishes and other aquatic animals, and the main administration and research offices.

The International Conference on Ancient Lakes (ICAL '97) was hosted by the Lake Biwa Museum from 20–29 June 1997, the first major conference held there. The Conference was organized and hosted by Hiroya Kawanabe. It could not have been held anywhere else in the world without the venue of the remarkable new Lake Biwa Museum. As one would expect in Japan, there is a very strong awareness and emphasis on human culture, human values and human history. Because Lake Biwa is about 2 000 000 years old it encompasses examples of much of the history of Japan, as well as serving as a microcosm for modern Japan.

The Conference began and ended with a typhoon – not unexpected events during the rainy season in the heart of Japan. The typhoon on the day before the conference opened literally served to clear the air, so that the meeting was held under near ideal conditions of pleasant, sunny weather. The typhoon that moved across Lake Biwa during the closing session of the conference suggested both the gathering clouds over ancient lakes, and the strong winds of change sweeping towards those lakes.

ICAL '97 brought together about 200 delegates from more than 20 countries, dealing with subjects as diverse as fossil footprints, fireflies, endemic fishes and traditional fishing methods. ICAL '97 was also uniquely suited to the venue at the Lake Biwa Museum since it is Japanese scientists and researchers who have carried out extensive collaborative research programs over many years on many of the ancient lakes of the world, particularly on Lake

Figure 1. Aerial view of the Lake Biwa Museum. Photograph courtesy of Hiroya Kawanabe.

Tanganyika in the east African Rift Valley and Lake Baikal. There is a perception in some circles that Japan is a closed society, looking mostly inward and dedicated to increasing industrialization. ICAL '97 clearly demonstrated that this is not the case. ICAL '97 was organized and hosted by Japanese scientists at a new, world-class Museum and stressed both scientific and cultural aspects of ancient lakes. In fact, the full title of the conference emphasized this point strongly, 'Ancient Lakes: their biological and cultural diversities'. The conference highlighted the extensive international scientific and cultural collaborations of a large number of Japanese researchers, demonstrating that Japan is far ahead of other countries in these activities.

As the full title of the Conference would suggest, a good deal of time and attention was devoted to cultural aspects of ancient lakes. Only in Japan would the first commemorative paper be devoted to questions of endemism and speciation and the second paper to considerations of poetry, folklore and myths about lakes. A mid-conference excursion on Lake Biwa, with special visits to religious temples and shrines, and a tour of a traditional fishing village on Okishima Island was the clearest demonstration of the interrelationships between biological and cultural diversity.

For readers of this volume the main interest in ICAL '97 would be the numerous sessions on various aspects of fish biology and limnology. From the first words of the first speaker in the first plenary session, it was clear that views and viewpoints on ancient lakes are changing. Christian Léveque wasted no time in making the point that intralacustrine speciation was an inescapable conclusion from studies of the fauna of ancient lakes.

The endemic fish faunas of Lake Baikal and the east African Rift Valley lakes (Tanganyika, Victoria and Malawi/Nyassa) were continuing focal points for the Conference. The news on the fish faunas of these lakes is both good and bad (Bruton 1990, Lowe-McConnell 1996, Ogutu-Ohwayo et al. 1997). Even more fish species are being discovered in the east African lakes. Witte et al. (1997) have reported on at least 200 more species of cichlids in Lake Victoria than were known only a decade ago. The bad news in this case is that many of those species may have been very recently driven to extinction as a result of human activities. In general, the pressures on the endemic faunas of ancient lakes are manifold and increasing (Fryer 1996). Those pressures are exacerbated by human population growth, poverty and increasing exploitation of fish stocks for food.

Perhaps the strongest message to emerge from ICAL '97 was the need to take human cultural values and concerns into account as part of any program to study, or manage, ancient lakes. In particular, the traditional knowledge, customs and values of indigenous riparian people must be understood and appreciated before they are lost in a rush to import and enforce external (often western European or North American) assumptions. The simplest, but perhaps most dramatic, example of this contrasts typical North American values for recreational aspects of lakes with the values of indigenous peoples around Lake Titicaca in South America or Lake

Figure 2. Five fisher brothers on the Kenyan shore of Lake Victoria with their beach seinse catch of two Nile perch in 1983. Photograph by Nicholas Farrar Hughes.

Victoria in east Africa (Balon & Bruton 1986). North American values led to the stocking of exotic rainbow trout, *Oncorhynchus mykiss*, in Lake Titicaca – to the detriment of both the endemic fishes and the indigenous people. While trout stocks lasted, they supplied commercial canning operations that exported fish outside the country, with concurrent loss of dietary protein for the native people. When that operation finally came to an end it was replaced by comparable efforts to produce rainbow trout in cage culture in the lake – again for the purpose of foreign export, with concomitant loss of dietary protein for the local people. In Lake Victoria the most dramatic recent change has been the tremendous increase in exotic Nile perch, *Lates niloticus*, – overall fish harvest from the Lake has increased by 5–10 times! A number of major fishing ventures have been launced to capitalize on this new fishery (Figure 2), along with major processing and export of the final product (Reidmiller 1994). Processed fish is now second only to coffee as a foreign export for Uganda. Of course a few people prosper as a result, but the majority of people around the Lake are probably in a worse situation as regards dietary intake of fish protein and independence of livelihood as fishers (e.g. Witte et al. 1997).

The edited proceedings of ICAL '97 are to be published in the near future and will surely take a rightful place in the forefront of a growing literature on lakes of the world. They will also bring Hiroya Kawanabe to his rightful place as a major international scientist responsible for new perspectives on lakes and humans.

References cited

Balon, E.K. & M.N. Bruton. 1986. Introduction of alien species or why scientific advice is not heeded? Env. Biol. Fish. 16: 225–230.

Bruton, M.N. 1990. The conservation of the fishes of Lake Victoria, Africa: an ecological perspective. Env. Biol. Fish. 27: 161–175.

Fryer, G. 1996. Endemism and adaptive radiation in great lakes. Env. Biol. Fish. 45: 109–131.

Kawanabe, H. 1996. Asian great lakes – especially Lake Biwa. Env. Biol. Fish. 47: 219–234.

Kawanabe, H., M. Hori & M. Nagoshi (ed.) 1997. Fishes of Lake Tanganyika. University of Kyoto Press, Kyoto. 327 pp.

Lowe-McConnell, R. 1996. Fish communities in the African great lakes. Env. Biol. Fish. 45: 219–235.

Ogutu-Ohwayo, R., R.E. Hecky, A.S. Cohen & L. Kaufman.

1997. Human impacts on the African Great Lakes. Env. Biol. Fish. 50: 117–131.
Riedmiller, S. 1994. Lake Victoria fisheries: the Kenyan reality and environmental implications. Env. Biol. Fish. 39: 329–338.
Witte, F., T. Goldschmidt, J. Wanink, M. van Oijen, K. Goudswaard, E. Witte-Maas & N. Bouton. 1997. The destruction of an endemic species flock: quantitative data on the decline of the haplochromine cichlids of Lake Victoria. Env. Biol. Fish. 34: 1–28.

Environmental Biology of Fishes **52**: 395–405, 1998.

Species and subject index*

* Prepared by Samantha J. Rayroux and Kurt D. Fausch, Department of Fishery and Wildlife Biology, Colorado State University, Fort Collins, CO 80523, U.S.A.

GPSR Compliance

The European Union's (EU) General Product Safety Regulation (GPSR) is a set of rules that requires consumer products to be safe and our obligations to ensure this.

If you have any concerns about our products, you can contact us on ProductSafety@springernature.com

In case Publisher is established outside the EU, the EU authorized representative is:

Springer Nature Customer Service Center GmbH
Europaplatz 3
69115 Heidelberg, Germany

FSC
www.fsc.org
MIX
Papier | Fördert gute Waldnutzung
FSC® C083411

Kolibri 360 GmbH
Ferdinand-Jühlke-Straße 7
99095 Erfurt, Deutschland
produktsicherheit@kolibri360.de